Formula Weights

AgBr	187.78	K_2CrO_7	294.19
AgCl	143.32	KHC_2O_4	128.13
Ag_2CrO_4	331.73	$KHC_2O_4 \cdot H_2C_2O_4$	218.16
AgI	243.77	$KHC_8H_2O_4(KHP)$	204.23
$AgNO_3$	169.87	$KH(IO_3)_2$	389.92
AgSCN	165.95	K_2HPO_4	174.18
$Al(C_9H_6ON)_3(AlOx_3)$	459.46	KH_2PO_4	136.09
Al_2O_3	101.96	$KHSO_4$	136.17
$Al_2(SO_4)_3$	342.14	KI	166.01
As_2O_3	197.85	KIO_3	214.00
$BaCO_3$	197.35	KIO_4	230.00
$BaCl_2$	208.25	$KMnO_4$	158.04
$BaCl_2 \cdot 2H_2O$	244.27	KNO_3	101.11
$BaCrO_4$	253.33	KOH	56.11
BaO	153.34	KSCN	97.18
$BaSO_4$	233.40	K_2SO_4	174.27
Bi_2O_3	466.0	$MgCl_2$	95.22
$C_6H_{12}O_6$ (glucose)	180.16	$Mg(C_9H_6ON)_2(MgOx_2)$	312.60
CO_2	44.01	$MgNH_4PO_4$	137.35
$CaCl_2$	110.99	MgO	40.31
$CaCO_3$	100.09	$Mg_2P_2O_7$	222.57
CaC_2O_4	128.10	$MgSO_4$	120.37
CaF_2	78.08	MnO_2	86.94
CaO	56.08	Mn_2O_3	157.88
$CaSO_4$	136.14	Mn_3O_4	228.81
CeO_2	172.12	$Na_2B_4O_7 \cdot 10H_2O$	381.37
$Ce(SO_4)_2$	332.25	NaBr	102.90
$(NH_4)_2Ce(NO_3)_6$	548.23	$Na(C_6H_5)_4B$	342.20
$(NH_4)_4Ce(SO_4)_4 \cdot 2H_2O$	632.6	$NaC_2H_3O_2$	82.03
Cr_2O_3	151.99	$Na_2C_2O_4$	134.00
CuO	79.54	NaCl	58.44
Cu_2O	143.08	NaClO	74.44
$CuSO_4$	159.60	NaCN	49.01
$Fe(NH_4)_2(SO_4)_2 \cdot 6H_2O$	392.14	Na_2CO_3	105.99
FeO	71.85	$NaHCO_3$	84.01
Fe_2O_3	159.69	$Na_2H_2EDTA \cdot 2H_2O$	372.23
Fe_3O_4	231.54	NaOH	40.00
HBr	80.92	NaSCN	81.07
$HC_2H_3O_2$ (acetic acid)	60.05	Na_2SO_4	142.04
HCl	36.46	$Na_2S_2O_3 \cdot 5H_2O$	248.18
$HClO_4$	100.46	$Ni(C_4H_7O_2N_2)_2(Ni\text{-}DMG_2)$	288.94
$H_2C_2O_4$	90.04	NH_3	17.03
$H_2C_2O_4 \cdot 2H_2O$	126.07	NH_4Cl	53.49
H_5IO_6	227.94	$(HOCH_2)_3CNH_2(THAM; Tris)$	121.14
HNO_3	63.01	$NH_2CO_2NH_2$ (Urea)	60.05
H_2O	18.015	$(NH_4)_2C_2O_4 \cdot H_2O$	142.11
H_2O_2	34.01	NH_4NO_3	80.04
H_3PO_4	98.00	$(NH_4)_2SO_4$	132.14
H_2S	34.08	$(NH_4)_2S_2O_8$	228.18
H_2SO_3	82.08	NH_2SO_3H	97.09
H_2SO_4	98.08	$PbCrO_4$	323.18
HSO_3NH_2 (sulfamic acid)	97.09	$PbSO_4$	303.25
HgO	216.59	P_2O_5	141.94
Hg_2Cl_2	472.09	Sb_2O_3	291.50
$HgCl_2$	271.50	SiO_2	60.08
$Hg(NO_3)_2$	324.61	$SnCl_2$	189.60
KBr	119.01	SnO_2	150.69
$KBrO_3$	167.01	$SrSO_4$	183.68
$K(C_6H_5)_4B$	358.31	SO_2	64.06
KCl	74.56	SO_3	80.06
$KClO_3$	122.55	TiO_2	79.90
KCN	65.12	V_2O_5	181.88
K_2CrO_4	194.20	$Zn_2P_2O_7$	304.68

analytical chemistry

TO THE STUDENT: A Solutions Manual for this textbook is available through your college bookstore under the title Solutions Manual to accompany Analytical Chemistry, Fourth Edition by Gary D. Christian. The Solutions Manual can help you with course material by acting as a tutorial, review, and study aid. If the Solutions Manual is not in stock, ask the bookstore manager to order a copy for you.

analytical chemistry

4th EDITION

GARY D. CHRISTIAN
UNIVERSITY OF WASHINGTON

JOHN WILEY & SONS

New York · Chichester · Brisbane · Toronto · Singapore

Library of Congress Cataloging in Publication Data:

Christian, Gary D.
  Analytical chemistry.

  Includes index.
  1. Chemistry, Analytic—Quantitative.
I. Title.

QD101.2.C57 1986   543   85-12133
ISBN 0-471-8874-6

Printed in the United States of America

10 9 8 7 6 5 4 3 2

preface

This text is designed for college students majoring in chemistry and fields related to chemistry. It covers most major areas of modern quantitative analysis, beginning with an introductory description of the analytical process and ending with chapters on applications to specific types of samples (clinical, forensic, and environmental analyses). A practical but fundamental approach has been taken, with examples drawn from the life sciences, clinical chemistry, air and water pollution, occupational health and safety, and industrial analyses. Experience has shown that analytical chemistry becomes more meaningful when the student realizes that an incorrect blood analysis may endanger a patient's life, or that an error in quality-control analysis may result in serious financial loss for a manufacturer.

While this fourth edition is extensively revised from the third, many of the features developed in previous editions remain. **Boldface type** is used for key terms and important equations and concepts are boxed to aid in review, as are example problems. Each chapter is introduced with a summary paragraph that lists the topics to be covered, giving the student a broad overview of each subject. The use of dimensional analysis is emphasized throughout to give the student a better feel for the proper setting up of problems. The use of SI units or symbols (e.g., L, mL, mol, and s) is now emphasized. Analytical concentrations are represented by C_A, rather than [A']. The concept of normality and equivalents is introduced, but the emphasis remains on the use of molarity and moles.

A number of chapters have been consolidated and rearranged. The material on expression of analytical results is now included in the introductory chapter. The techniques for obtaining and preparing a solution of the sample for analysis are described in Chapter 2, along with the tools required for the preparation and analysis. Discussion of acid–base equilibria in Chapter 5 now precedes the coverage of all analysis techniques so that this important material may be available for later discussion or application. The discussion of acid–base reactions in nonaqueous solvents is now combined with the description of nonaqueous titrations in Chapter 8. The kinetics chapter has been moved to just before the chapter on clinical chemistry, since the two are related.

All problems and recommended references have been grouped by topic, for ease in assignment. Many new ones have been added.

Experiments are now placed at the end of the text, grouped by topic, as suggested by several reviewers and users of the text. Each one continues to contain a description of the principles and chemical reactions involved, so the student gains an overview of what is being analyzed and how. Solutions and reagents to prepare in advance of the experiment are listed. Experiments are designed where possible to avoid the use of asbestos, chloroform, carbon tetrachloride, or benzene, consistent with occupational health and safety. Laboratory safety is still of prime importance, and the appendix contains a section on safety in the laboratory.

The logarithm tables have been moved from the back cover of the text to an appendix to make room for useful tables on tolerances of glassware, chemical grades, and concentrations of commercial acids and bases.

A number of new topics and changes have been added. For example, the electronic balance is described in Chapter 2. Propagation of error is treated in detail in Chapter 3, including logarithm and significant figures in propagation of error. Equilibria are introduced in terms of Gibbs free energy in Chapter 4, and general rules are listed for solving chemical equilibrium problems. Buffers and calculations are presented in terms of proton acceptor/donor (Chapter 5).

In Chapter 11, the description of the cell voltage and use of the Nernst equation for electrode potentials is presented to reflect the fact that the reference electrode is connected to the common (ground) of the potential measuring device and the sign on the voltage scale denotes the relative polarity of the indicator electrode with respect to the reference electrode; and the voltage increases with increased cation concentration or decreases with increased anion concentration.

The use of internal standards and standard addition for atomic absorption spectrophotometry is given in Chapter 14. The description of capacity factor, resolution and the Kovats retention index in gas chromatography is added in Chapter 16, and the important technique of ion chromatography is included. Chapter 17 includes the rapidly growing technique of flow injection analysis. The description of immunoassay techniques in Chapter 19 has been expanded to include newer techniques.

Some topics or sections have necessarily been deleted or consolidated to avoid substantial expansion. Thus, electrodes of the third kind are no longer described in Chapter 11, and description of the Van Slyke manometric apparatus has been deleted in Chapter 19.

Special thanks go to users of the text who have contributed comments and suggestions for changes or corrections. These are always welcome. A number of colleagues have aided immeasurably in providing specific suggestions for revision of this text. Special mention goes to Professors Agnes A. Alich, College of St. Scholastica; James D. Carr, University of Nebraska; Geoffrey N. Coleman, University of Alabama; Lee C. Coombs, California Polytechnic State University, San Luis Obispo; Lawrence H. Daly, State University of New York at Albany; John C. Gutweiler, University of Missouri; Raymond T. O'Donnell, State University of New York at Oswego; and William L. Switzer, North Carolina State University at Raleigh.

My wife, Sue, endured the arduous and lengthy task of translating my handwritten notes into typed text, and her encouragement, support, and understanding will be forever remembered.

Solutions Manual. A comprehensive solutions manual is available for use by students, in which all problems are completely worked out and all questions are answered.

Gary D. Christian

contents

CHAPTER 1 **Introduction** 1

1.1 What Is Analytical Chemistry? 1
1.2 The Analytical Process 2
1.3 Range 7
1.4 Review of Fundamental Concepts 7
1.5 Concentrations of Solutions 10
1.6 Expressions of Analytical Results 15

CHAPTER 2 **Basic Tools of Analytical Chemistry** 23

2.1 The Laboratory Notebook 23
2.2 Laboratory Materials and Reagents 25
2.3 The Analytical Balance 25
2.4 Volumetric Glassware 33
2.5 Other Apparatus 42
2.6 Obtaining the Sample 49
2.7 Operations of Drying and Preparing a Solution of the Analyte 51

CHAPTER 3 **Data Handling** 59

3.1 Significant Figures 59
3.2 Rounding Off 64
3.3 Accuracy and Precision 64

3.4   Determinate Errors   64
3.5   Indeterminate Errors   65
3.6   Ways of Expressing Accuracy   66
3.7   Standard Deviation   67
3.8   Propagation of Error   70
3.9   Significant Figures and Propagation of Error   75
3.10   Control Charts   76
3.11   The Confidence Limit   77
3.12   Tests of Significance   79
3.13   Rejection of a Result   84
3.14   Linear Least Squares   87
3.15   The Correlation Coefficient   88

CHAPTER 4   **General Concepts of Chemical Equilibrium   97**

4.1   Chemical Reactions: The Rate Concept   97
4.2   Gibbs Free Energy and the Equilibrium Constant   99
4.3   Le Châtelier's Principle   100
4.4   Temperature Effects on Equilibrium Constants   100
4.5   Pressure Effects on Equilibria   100
4.6   Effect of Concentrations on Equilibria   101
4.7   Catalysts   101
4.8   Completeness of Reactions   101
4.9   Equilibrium Constants for Dissociating Species   101
4.10   Calculations Using Equilibrium Constants   102
4.11   The Common Ion Effect   106
4.12   Activity and Activity Coefficients   107
4.13   The Diverse Ion Effect: The Thermodynamic Equilibrium Constant   110

CHAPTER 5   **Acid–Base Equilibria   115**

5.1   Aqueous Acid–Base Theories   115
5.2   Acid–Base Equilibria in Water   116
5.3   The pH Scale   118
5.4   pH at Elevated Temperatures: Blood pH   122
5.5   Weak Acids and Bases   122
5.6   Salts of Weak Acids and Bases   124
5.7   Buffers   128
5.8   Polyprotic Acids and Their Salts   133
5.9   Physiological Buffers   141
5.10   Buffers for Biological and Clinical Measurements   143
5.11   The Diverse Ion Effect on Acids and Bases: K_a^0 and K_b^0 145

CHAPTER 6   **Gravimetric Analysis   150**

6.1   Precipitation Equilibria: the Solubility Product   150
6.2   The Diverse Ion Effect on Solubility: K_{sp}^0 154

6.3   Unit Operations in Gravimetric Analysis   **155**
6.4   Calculations   **164**
6.5   Examples of Gravimetric Analysis   **168**
6.6   Organic Precipitates   **169**

CHAPTER 7   **Principles of Volumetric Analysis   174**

7.1   General Principles   **174**
7.2   Molarity Calculations   **177**
7.3   Normality Calculations   **186**
7.4   Titer   **192**

CHAPTER 8   **Acid–Base Titrations   196**

8.1   Strong Acid Versus Strong Base   **196**
8.2   Detection of the End Point: Indicators   **198**
8.3   Weak Acid Versus Strong Base   **201**
8.4   Weak Base Versus Strong Acid   **205**
8.5   Titration of Sodium Carbonate   **206**
8.6   Titration of Polyprotic Acids   **209**
8.7   Mixtures of Acids or Bases   **210**
8.8   Titration of Amino Acids   **212**
8.9   Kjeldahl Analysis   **213**
8.10   Preparation of Standard Base Solutions   **215**
8.11   Preparation of Standard Acid Solutions   **216**
8.12   Nonaqueous Solvents   **217**
8.13   Titrations in Nonaqueous Solvents   **219**

CHAPTER 9   **Precipitation and Complexometric Titrations   223**

9.1   Complexes: Formation Constants   **223**
9.2   Chelates: EDTA   **225**
9.3   EDTA Titration Curves   **230**
9.4   Detection of the End Point: Indicators   **231**
9.5   Fraction of Dissociating Species in Polyligand Complexes: β-Values   **235**
9.6   Effect of Acidity on Solubility of Precipitates   **237**
9.7   Effect of Complexation on Solubility   **238**
9.8   Precipitation Titrations   **239**

CHAPTER 10   **Reduction–Oxidation Titrations   248**

10.1   Principles   **248**
10.2   Electrochemical Cells   **249**
10.3   The Nernst Equation   **254**
10.4   Calculation of the Equilibrium Constant of a Reaction   **257**
10.5   Titration Curves   **260**

10.6   Detection of the End Point   **263**
10.7   Formal Potential   **265**
10.8   Limitations of Electrode Potentials   **267**
10.9   Titrations Involving Iodine: Iodimetry and Iodometry   **267**
10.10   Titrations with Oxidizing Agents   **272**
10.11   Titrations with Reducing Agents   **274**
10.12   Preparation of the Analyte Solution   **274**

CHAPTER 11   **Potentiometry and Potentiometric Titrations   280**

11.1   Electrodes of the First Kind   **280**
11.2   Electrodes of the Second Kind   **282**
11.3   Redox Electrodes   **283**
11.4   Cells Without Liquid Junction   **285**
11.5   Cells with Liquid Junction   **286**
11.6   Reference Electrodes: The Saturated Calomel Electrode   **288**
11.7   Measurement of Potential   **291**
11.8   Determination of Concentrations from Potential Measurements   **293**
11.9   The Residual Liquid-Junction Potential   **294**
11.10   Accuracy of Direct Potentiometric Measurements   **294**
11.11   The Glass pH Electrode   **295**
11.12   Standard Buffers   **299**
11.13   Accuracy of pH Measurements   **301**
11.14   Measurements with the pH Meter   **301**
11.15   pH Measurement of Blood   **303**
11.16   pH Measurements in Nonaqueous Solvents   **303**
11.17   Ion-Selective Electrodes   **304**
11.18   Potentiometric Titrations (Indirect Potentiometry)   **313**

CHAPTER 12   **Electrolytic Methods   325**

12.1   Electrogravimetry   **326**
12.2   Controlled-Potential Coulometry   **330**
12.3   Controlled-Current Coulometry   **332**
12.4   Voltammetry   **335**
12.5   Polarography   **340**
12.6   Anodic-Stripping Voltammetry   **347**
12.7   Pulse Polarography   **348**
12.8   The Static Mercury Drop Electrode (SMDE)   **349**
12.9   Amperometry   **349**

CHAPTER 13   **Spectrometry   357**

13.1   Interaction of Electromagnetic Radiation with Matter   **357**
13.2   Electronic Spectra and Molecular Structure   **364**
13.3   Infrared Absorption and Molecular Structure   **369**
13.4   Solvents for Spectrometry   **369**

13.5   Quantitative Calculations   **371**
13.6   Principles of Instrumentation   **377**
13.7   Types of Instruments   **385**
13.8   Spectrometric Error   **388**
13.9   Deviations from Beer's Law   **389**
13.10   Fluorometry   **392**

CHAPTER 14   **Atomic Spectrometric Methods   404**

14.1   Emission Spectroscopy   **404**
14.2   Flame Emission Spectrometry   **406**
14.3   Plasma Emission Spectrometry   **407**
14.4   Distribution Between Ground and Excited States   **408**
14.5   Atomic Absorption Spectrophotometry   **409**
14.6   Internal Standard and Standard Addition Calibration   **420**

CHAPTER 15   **Solvent Extraction   425**

15.1   The Distribution Coefficient   **425**
15.2   The Distribution Ratio   **426**
15.3   The Percent Extracted   **427**
15.4   Solvent Extraction of Metals   **429**
15.5   Analytical Separations   **433**
15.6   Multiple Batch Extractions   **434**
15.7   Countercurrent Distribution   **436**

CHAPTER 16   **Chromatographic Methods   441**

16.1   Principles   **442**
16.2   Classification of Chromatographic Techniques   **445**
16.3   Techniques of Column Chromatography   **446**
16.4   Size Exclusion Chromatography   **449**
16.5   Ion Exchange Chromatography   **451**
16.6   Gas Chromatography   **456**
16.7   High-Performance Liquid Chromatography   **469**
16.8   Paper Chromatography   **477**
16.9   Thin-Layer Chromatography   **479**
16.10   Electrophoresis   **481**

CHAPTER 17   **Automation in the Laboratory   485**

17.1   Principles of Automation   **485**
17.2   Automated Instruments: Process Control   **486**
17.3   Automatic Instruments   **488**
17.4   Semiautomatic Instruments   **489**
17.5   The AutoAnalyzer   **490**

17.6 The Centrifugal Analyzer **493**
17.7 Flow Injection Analysis **495**
17.8 Microprocessor-Controlled Instruments: "Smart" Instruments **496**
17.9 Computers in Analytical Chemistry **498**

CHAPTER 18 Kinetic Methods of Analysis 502

18.1 Kinetics **502**
18.2 Catalysis **505**
18.3 Enzyme Catalysis **508**

CHAPTER 19 Clinical Chemistry 520

19.1 The Composition of Blood **520**
19.2 Collection and Preservation of Samples **522**
19.3 Clinical Analysis **523**
19.4 Immunoassay **526**
19.5 The Blood Gas Analyzer **533**
19.6 Trace Elements in the Body **533**

CHAPTER 20 Drug Analysis: Narcotics and Dangerous Drugs 537

20.1 Classification of Drugs **537**
20.2 Gas Chromatography Screening **540**
20.3 Thin-Layer Chromatography Screening **543**
20.4 Spectrophotometric Methods of Measurement **545**

CHAPTER 21 Environmental Analysis 548

21.1 Air Analysis **548**
21.2 Water Analysis **554**

EXPERIMENTS 556

experiments

Use of Apparatus 556

1 Use of the analytical balance **556**
2 Use of the pipet and buret **557**

Gravimetry 559

3 Gravimetric determination of chloride **559**
4 Gravimetric determination of SO_3 in a soluble sulfate **561**
5 Gravimetric determination of nickel in a nichrome alloy **563**

Acid–Base Titrations 564

6 Determination of replaceable hydrogen in an acid by titration with sodium hydroxide **564**
7 Determination of total alkalinity of soda ash **566**
8 Determination of bicarbonate in blood using back-titration **568**

Precipitation Titrations 570

9 Determination of silver in an alloy: Volhard's method **570**
10 Determination of chloride in a soluble chloride: Fajans' method **572**

Complexometric Titrations 573

11 Mercurimetric determination of blood or urine chloride **573**
12 Determination of water hardness with EDTA **576**

Reduction–Oxidation Titrations 577

13 Analysis of an iron alloy or ore by titration with potassium dichromate **577**
14 Analysis of commercial hypochlorite or peroxide solution by iodometric titration **580**
15 Iodometric determination of copper **583**
16 Determination of antimony by titration with iodine **584**

Potentiometric Measurements 586

17 pH Titration of unknown soda ash **586**
18 Potentiometric titration of a mixture of chloride and iodide **587**
19 Determination of the pH of hair shampoos **589**
20 Potentiometric determination of fluoride in drinking water, using a fluoride ion-selective electrode **590**

Electrochemical Measurements 592

21 Electrolytic determination of copper in an ore **592**
22 Polarographic determination of a trace metal **593**
23 Amperometric titration of lead with dichromate **595**
24 Coulometric titration of arsenic with iodine **596**

Spectrometric Measurements 598

25 Spectrophotometric determination of iron **598**
26 Determination of nitrate nitrogen in water **599**
27 Spectrophotometric determination of lead on leaves using solvent extraction **601**
28 Spectrophotometric determination of inorganic phosphorous in human serum or urine **602**
29 Spectrophotometric determination of manganese and chromium in mixture **604**
30 Ultraviolet spectrophotometric determination of aspirin, phenacetin, and caffeine in APC tablets using solvent extraction **606**
31 Infrared determination of a mixture of xylenes **607**
32 Fluorometric determination of riboflavin (vitamin B_2) **608**
33 Determination of calcium by atomic absorption spectrophotometry **609**
34 Determination of mercury in laboratory air or water using flameless atomic absorption spectrophotometry **611**
35 Flame emission spectrometric determination of sodium **613**

Chromatography 615

36 Anion ion exchange chromatography separation of cobalt and nickel, followed by EDTA titration of the metals using back-titration **615**
37 Determination of sulfate in urine by precipitation titration following ion exchange separation **617**
38 Paper chromatography separation and identification of metal ions **618**
39 Thin-layer chromatography separation of amino acids **620**
40 Gas chromatography analysis of a tertiary mixture **621**
41 Qualitative and quantitative analysis of fruit juices for vitamin C, using high-performance liquid chromatography **622**
42 Analysis of analgesics using high-performance liquid chromatography **623**

Kinetic Measurements 624

43 Catalytic determination of traces of selenium **624**
44 Enzymatic determination of glucose in blood **625**

APPENDIX A **Literature of Analytical Chemistry** **628**

APPENDIX B **Review of Mathematical Operations:**
Exponents, Logarithms, The Quadratic Formula, and Calculators **632**

APPENDIX C **Balancing Reduction–Oxidation Reactions** **639**

APPENDIX D **Tables of Constants** **644**

Table D.1 Dissociation Constants for Acids **644**
Table D.2 Dissociation Constants for Bases **645**
Table D.3 Solubility Product Constants **646**
Table D.4 Formation Constants for Some EDTA Metal Chelates **647**
Table D.5 Some Standard and Formal Reduction Electrode
Potentials **648**

APPENDIX E **Safety in the Laboratory** **651**

APPENDIX F **Answers to Even Numbered Problems** **657**

Index **661**

analytical chemistry

chapter 1

INTRODUCTION

1.1 WHAT IS ANALYTICAL CHEMISTRY?

Analytical chemistry can be thought of as comprising two branches, qualitative and quantitative. **Qualitative analysis** deals with finding *what* constituent or constituents are in an analytical sample, and **quantitative analysis** deals with the determination of *how much* of a given substance is in the sample. In the latter case, a history of the sample composition will generally be known; or else the analyst will have performed a qualitative test. With today's instrumentation and with the large variety of chemical measurements available, specificity or sufficient selectivity[1] can often be achieved so that the quantitative measurement serves as a qualitative measurement. However, simple qualitative tests are usually more rapid than quantitative procedures. Qualitative analysis is composed of two fields: inorganic and organic. The former is usually covered in introductory chemistry courses, whereas the latter is best left until after the student has had a course in organic chemistry.

This text deals principally with quantitative analysis. In the consideration of applications of different techniques, examples are drawn from the life sciences, clinical chemistry, environmental chemistry, occupational health and safety applications, and industrial analysis.

---

[1]A clear distinction should be made between the terms **specific** and **selective.** A specific reaction or test is one that occurs only with the substance of interest, while a selective reaction or test is one that can occur with other substances but exhibits a degree of preference for the substance of interest. Few reactions are specific but many exhibit selectivity.

1.2   THE ANALYTICAL PROCESS

The analytical process involves a logical sequence: (1) defining the problem, (2) obtaining and pretreating the sample, (3) performing the required separations, (4) making the appropriate measurements, and (5) data presentation. The unit operations of analytical chemistry that are common to most types of analyses are considered in more detail below.

Defining the Problem

In defining the analytical problem, the analyst asks such questions as: What information is required? How sensitive must the method be? How accurate and precise[2] must it be? What interferences are likely and what separations are required? How soon must the results be available? How many samples must be run? What equipment is available? What is the cost?

Examples of the manner in which the analysis of particular types of samples are made are given in Chapters 19 through 21. These chapters describe commonly performed clinical, drug, and environmental analyses. The various techniques described in this text are utilized for the specific analyses. Hence, it will be useful for you to read through these applications chapters both now and after completing the majority of this course to gain an appreciation of what goes into analyzing real samples and why the analyses are made.

For a particular analysis, one or more procedures may already have been developed. For example, blood chloride is routinely determined by titration with silver ion, produced by electro-oxidation of silver metal at controlled current. The only decision, then, is which available method to use. The method is often dictated by the availability of chemicals and equipment in the laboratory and by the experience of the analyst. The literature references in Appendix A describe various procedures for the analysis of different substances.

Once the problem has been defined, the following steps can be started.

Sampling

The type of sample that must be taken for analysis will depend on the information desired. Suppose a core sample from the earth is taken for geological study. A scientist may wish to know how a particular substance varies along the core, that is, at different depths in the earth, as a measure of the change with time. Therefore, analysis of slices of the core would be performed. On the other hand, a grain company may have to know the average protein content of a grain shipment. The sample analyzed, which will be only a small fraction of the shipment, must be representative of the whole. The method of obtaining the sample may involve taking a random sample from each sack of grain, or from every tenth sack, if the shipment is large, and so forth. The sampled portions would be mixed together and reduced further in size until small enough to analyze. Methods of sampling solids, liquids, and gases are discussed in Chapter 2.

---

[2]**Accuracy** is the degree of agreement between a measured value and a true value. **Precision** is the degree of agreement between replicate measurements of the same quantity and does not necessarily imply accuracy. These terms are discussed in more detail in Chapter 3.

In the case of biological fluids, the conditions under which the sample is collected can be important, for example, whether a patient has just eaten. The composition of blood varies considerably before and after meals, and for many analyses a sample is collected after the patient has fasted for a number of hours. Preservatives such as sodium fluoride for glucose preservation and anticoagulants may be added to blood samples when they are collected; these may affect a particular analysis.

Blood samples may be analyzed as whole blood, or they may be separated to yield plasma or serum according to the requirements of the particular analysis. Most commonly, the concentration of the substance external to the red cells (the extracellular concentration) will be a significant indication of physiological condition, and so serum or plasma is taken for analysis.

If whole blood is collected and allowed to stand for several minutes, the soluble protein **fibrinogen** will be converted by a complex series of chemical reactions (involving calcium ion) into the insoluble protein **fibrin,** which forms the basis of a gel, or **clot.** The red and white cells of the blood become caught in the meshes of the fibrin network and contribute to the solidarity of the clot, although they are not necessary for the clotting process. After the clot forms, it shrinks and squeezes out a straw-colored fluid, **serum,** which does not clot but remains fluid indefinitely. The clotting process can be prevented by adding a small amount of an **anticoagulant,** such as heparin or a citrate salt (i.e., a calcium complexor). An aliquot of the unclotted whole blood can be taken for analysis, or the red cells can be centrifuged to the bottom, and the light pinkish-colored **plasma** remaining can be analyzed. Plasma and serum are essentially identical in chemical composition, the chief difference being that fibrinogen has been removed from the latter.

Details of sampling other materials are available in reference books on specific areas of analysis. See the references at the end of the chapter for a few citations.

Handling the Sample

Certain precautions should be taken in handling and storing samples to prevent or minimize contamination, loss, decomposition, or matrix change. In general, one must prevent contamination or alteration of the sample by (1) the container, (2) the atmosphere, or (3) light.

The sample may have to be protected from the atmosphere or from light. It may be an alkaline substance, for example, which will react with carbon dioxide in the air. Blood samples to be analyzed for CO_2 should be protected from the atmosphere.

The stability of the sample must be considered. Glucose, for example, is unstable, and a preservative such as sodium fluoride is added to blood samples. The preservation must not, of course, interfere in the analysis. Proteins and enzymes tend to denature on standing and should be analyzed without delay. Trace constituents may be lost during storage by adsorption onto the container walls.

Urine samples are unstable, and calcium phosphate precipitates out, entrapping metal ions or other substances of interest. Precipitation can be prevented by keeping the urine acidic (pH 4.5), usually by adding 1 or 2 mL glacial acetic acid per 100 mL sample. Store under refrigeration. Urine, as well as whole blood, serum, plasma, and tissue samples, can also be frozen for prolonged storage. Deproteinized blood samples are more stable than untreated samples.

Corrosive gas samples will often react with the container. Sulfur dioxide, for example, is troublesome. In the case of automobile exhaust, it is also lost by dissolving in condensed water vapor from the exhaust. In such cases, it is best to analyze the gas by a stream process.

Measuring the Sample

The amount of sample taken for analysis usually must be known. Either the sample will be *weighed* or the *volume* of sample will be determined, depending on whether the analyst wants to know the amount of **analyte** (substance being determined) per unit weight or per unit volume in the bulk material from which the sample was taken.

Solid samples will usually contain a variable amount of adsorbed water that must be removed or measured before weighing the sample. Removal is generally achieved by drying the sample in an oven, provided the sample is not decomposed at a temperature slightly above the boiling point of water. Some samples require higher temperatures because of the great affinity of moisture for their sample surface.

Preparation of the Sample for Measurement: Dissolving and Separation

If the sample is a solid, it usually must be put into solution before the analysis can be completed, and this preparation of a solution of the sample is discussed in Chapter 2. Generally, aqueous dissolution of solids is accomplished via their acid-base, redox, or complexation properties. Biological fluids may sometimes be analyzed directly. Often, however, proteins interfere and must be removed. Various destruction techniques such as acid digestion or dry ashing accomplish such removal. Or proteins may be precipitated to give a **protein-free filtrate** (PFF), in a separation procedure to eliminate interference.

The dissolved sample of analyte will usually have to undergo some form of preliminary treatment to separate the analyte from interfering constituents and/or to render it in a form suitable for measurement. For example, the gravimetric analysis of iron as Fe_2O_3 requires that all the iron be present as iron(III), its usual form. A volumetric determination by reaction with dichromate ion, on the other hand, requires that all the iron be converted to iron(II) before reaction, and the reduction step will have to be included in the sample preparation.

There are a number of separation procedures for removing interferences and achieving the desired selectivity. Almost all chemical or physical changes are candidates for separation mechanisms, that is, precipitation reactions, solubility, volatility, freezing, and so forth.

Measurement of the Analyte

The method employed for the actual quantitative measurement of the analyte will depend on a number of factors, not the least important being the amount of analyte present and the accuracy required. Many available techniques possess varying degrees of selectivity, sensitivity, accuracy and precision, and rapidity. **Gravimetric analysis** usually involves the selective separation of the analyte by precipitation, followed by the very nonselective measurement of mass (of the precipitate). In **volumetric,** or **titrimetric, analysis,** the analyte reacts with a measured volume of reagent of known concentration, in a process called **titration.** A change in some physical or chemical property signals the completion of the reaction. There are large numbers of **instrumental analysis** measurement methods, based on the measurement of a physical property of the sample, for example, an electrical property or the absorption of electromagnetic radiation.

Modern instrumental methods generally involve sophisticated instrumentation and are more sensitive than gravimetric and volumetric methods. They often possess a high degree of selectivity and hence may require fewer separation steps in the analysis. Like gravimetric analysis, however, an instrumental technique may include a separation. Gas chromatography, for example, involves a separation based on the interaction of solute gases with liquid or solid stationary phases, followed by measurement of the separated gases by a nonselective detector.

Table 1.1 compares the various analytical methods to be described in this text, with respect to sensitivity, precision, selectivity, speed, and cost. The numbers given may be exceeded in specific applications, and the methods may be applied to other uses, but these are representative of typical applications. The lower concentrations determined by titrimetry require the use of an instrumental technique for measuring the completion of the titration. The selection of a technique, when more than one is applicable, will depend, of course, on the availability of equipment and personal experience and preference of the analyst.

The various methods of determining an analyte can be classified as either **absolute** or **relative.** Absolute methods rely upon accurately known fundamental constants for calculating the amount of analyte, for example, atomic weights. In gravimetric analysis, for example, an insoluble derivative of the analyte of known chemical composition is prepared and weighed, as in the formation of AgCl for chloride determination. The precipitate contains a known fraction of the analyte, in this case, fraction of Cl = at. wt. Cl/f. wt. AgCl = 35.453/143.32 = 0.24737. Hence, it is a simple matter to obtain the amount of Cl contained in the weighed precipitate. Most methods, however, are relative in that they require comparison against some solution of known concentration. In titrimetric analysis, for example, the analyte is reacted with the solution of a reagent in a known stoichiometric ratio. Hydrochloric acid, for example, reacts with sodium hydroxide in a 1:1 ratio:

$$HCl + NaOH \rightarrow NaCl + H_2O$$

The volume of sodium hydroxide solution required to just completely react with the hydrochloric acid sample is measured. If we know the concentration of the sodium hydroxide solution in moles per liter, then the number of moles of NaOH added can be calculated (volume × molarity), and so we know the number of moles of HCl in the sample. Therefore, in this relative method, it is necessary to prepare a reacting solution (sodium hydroxide) of accurately known concentration.

Most instrumental methods of analysis are relative. Instruments register a signal due to some physical property of the solution. Spectrophotometers, for example, measure the fraction of electromagnetic radiation from a light source that is absorbed by the sample. This fraction must be related to the analyte concentration by comparison against the fraction absorbed by a known concentration of the analyte. In other words, the instrumentation must be **calibrated.** A standard procedure for calibrating instruments is to prepare a **working curve** known as an **analytical calibration curve** by plotting the measured signal as a function of analyte concentration.

TABLE 1.1
Comparison of Different Analytical Methods

Method	Approx. Range, mol/L	Approx. Precision, %	Selectivity	Speed	Cost	Principal Uses
Gravimetry	10^{-1}–10^{-2}	0.1	Poor–moderate	Slow	Low	Inorg.
Titrimetry	10^{-1}–10^{-4}	0.1–1	Poor–moderate	Moderate	Low	Inorg., org.
Potentiometry	10^{-1}–10^{-6}	2	Good	Fast	Low	Inorg.
Electrogravimetry, coulometry	10^{-1}–10^{-4}	0.01–2	Moderate	Slow–moderate	Moderate	Inorg., org.
Voltammetry	10^{-3}–10^{-10}	2–5	Good	Moderate	Moderate	Inorg., org.
Spectrophotometry	10^{-3}–10^{-6}	2	Good–moderate	Fast–moderate	Low–moderate	Inorg., org.
Fluorometry	10^{-6}–10^{-9}	2–5	Moderate	Moderate	Moderate	Org.
Atomic spectroscopy	10^{-3}–10^{-9}	2–10	Good	Fast	Moderate–high	Inorg., multielement
Chromatography	10^{-3}–10^{-9}	2–5	Good	Fast–moderate	Moderate–high	Org., multicomponent
Kinetic methods	10^{-2}–10^{-10}	2–10	Good–moderate	Fast–moderate	Moderate	Inorg., org., enzymes

Calculation of the Amount of Analyte

There are various ways of expressing the amount or the concentration of analyte in the sample that represents the bulk material from which it was taken. The information required, the method employed, or convention will dictate the methods of expressing analytical results, as described in Section 1.6. Good laboratory records are essential in analytical chemistry. This assures that the correct data are used for calculating the reported answer and presenting the precision of the data, an important way of communicating the validity of an analysis to others. The presentation of data is discussed in Chapter 3.

1.3 RANGE

Analytical methods are often classed according to size of sample. Such classification is arbitrary and there is no sharp dividing line. The analysis may be classed as **meso, semimicro, micro,** or **ultramicro.** Table 1.2 gives approximate classifications according to sample weight or volume. The volume classifications are those employed in clinical laboratories. Special handling techniques and balances for weighing are required for micro and ultramicro operations.

Analyze Versus Determine

The terms analyze and determine represent two different meanings. We say a sample is **analyzed** for part or all of its constituents. The substances measured are called the **analytes.** The process of measuring the analyte is called a **determination.** Hence, in analyzing blood for its chloride content, we determine the chloride concentration.

The constituents in the sample may be classified as **major** (>1%), **minor** (0.1–1%), or **trace** (<0.1%). A few parts per million of a constituent might be classed as **ultratrace.**

An analysis may be **complete** or **partial;** that is, either all constituents or only selected constituents may be determined. Most often, the analyst is requested to report on a specified chemical or chemicals.

1.4 REVIEW OF FUNDAMENTAL CONCEPTS

Quantitative analysis is based on a few fundamental atomic and molecular concepts, which are reviewed below. You have undoubtedly been introduced to these in your general chemistry course, but we briefly review them here, since they are so fundamental to quantitative calculations.

Atomic, Molecular, and Formula Weights

The gram-atomic weight for any element is the weight of a specified number of atoms of that element, and that number is the same from one element to another. A gram-atomic weight of any element contains exactly the same number of atoms

TABLE 1.2
Classification of Analytical Methods According to Size of Sample

Method	Sample Weight, mg	Sample Volume, μL^a
Meso	>100	>100
Semimicro	10–100	50–100
Micro	1–10	<50
Ultramicro	<1	

$^a\mu L$ = Microliter. Sometimes the symbol λ (lambda) is used in place of μL.

of that element as there are carbon atoms in exactly 12 g of carbon 12. This number is Avogadro's number, 6.022×10^{23} atoms/g-at wt.

Since naturally occurring elements consist of mixtures of isotopes, the chemical atomic weights will be an average of the isotope weights of each element, taking into account their relative naturally occurring abundances. Thus, none of the elements has an integral weight. Gram-molecular weight (g.m.w.) is the sum of the atomic weights of the atoms that make up a compound. The term "gram-formula weight" (g.f.w.) is a more accurate description for substances that don't exist as molecules but exist as ionic compounds (strong electrolytes—acids, bases, salts).

Daltons

Biologists and biochemists sometimes use the unit **dalton** to report masses of large proteinaceous substances such as chromosomes, ribosomes, viruses, and mitochondria, where the term molecular weight would be inappropriate. The mass of a single carbon-12 atom is equivalent to 12 daltons, and one dalton is therefore 1.661×10^{-24} g, the reciprocal of Avogadro's number. The number of daltons in a single molecule is numerically equivalent to the molecular weight (g/mol). Strictly speaking, it is not correct to use the dalton as a unit of molecular weight, and it should be reserved for the types of substances mentioned above. For example, the mass of an *Escherichia coli* bacterium cell is about 1×10^{-12} g, or 6×10^{11} daltons.

Moles

The chemist knows that atoms and molecules react in definite proportions. Unfortunately, one cannot conveniently count the number of atoms or molecules. But since the chemist has determined their relative masses, their reactions can be described on the basis of the relative masses of atoms and molecules reacting, instead of the number of atoms and molecules reacting. For example, in the reaction

$$\mathrm{Ag^+ + Cl^- \rightarrow \underline{AgCl}}$$

we know that one silver ion will combine with one chloride ion. We know further, since the atomic weight of silver is 107.870 compared with the atomic weight of chlorine at 35.453, that 107.870 mass units of silver will combine with 35.453 mass units of chlorine. To simplify calculations, the chemist has operationally defined the **mole** as the atomic, molecular, or formula weight of a substance expressed in **grams.**[3]

Now, since a mole of any substance contains the same number of atoms or molecules as a mole of any other substance, atoms will react in the same mole ratio as their atom ratio in the reaction. In the above example, one silver ion reacts with one chloride ion, and so each mole of silver ion will react with one mole of chloride ion. (Each 107.87 g will react with 35.453 g.)

---

[3]Actually, the term *gram-atomic weight* is more correct for atoms, *gram-formula weight* for ionic substances, and *gram-molecular weight* for molecules, but we will use *moles* in a broad sense to include all substances. In place of gram-formula weight we will simply use formula weight (f.wt.).

EXAMPLE 1.1 Calculate the number of grams in one mole of $CaSO_4 \cdot 7H_2O$.

SOLUTION
One mole is the formula weight expressed in grams. The formula weight is

Ca	40.08
S	32.06
11 O	176.00
14 H	14.11

262.25 g/mol

The number of moles of a substance is calculated from

$$\text{moles} = \frac{\text{grams}}{\text{formula weight (g/mol)}} \tag{1.1}$$

where formula weight represents the atomic or molecular weight of the substance. Thus,

$$\text{moles } Na_2SO_4 = \frac{g}{\text{f. wt.}} = \frac{g}{142.04 \text{ g/mol}}$$

$$\text{moles } Ag^+ = \frac{g}{\text{f. wt.}} = \frac{g}{107.870 \text{ g/mol}}$$

Since we usually work with millimole quantities, a more convenient form of Equation 1.1 is

$$\text{millimoles} = \frac{\text{milligrams}}{\text{formula weight (mg/mmol)}} \tag{1.2}$$

Conversely, we can calculate the grams of material from the number of moles:

$$g\ Na_2SO_4 = \text{moles} \times \text{f. wt.} = \text{moles} \times 142.04 \text{ g/mol}$$

$$g\ Ag = \text{moles} \times \text{f. wt.} = \text{moles} \times 107.870 \text{ g/mol}$$

Again, we usually work with millimole quantities, and so

$$\text{milligrams} = \text{millimoles} \times \text{formula weight (mg/mmol)} \tag{1.3}$$

EXAMPLE 1.2 Calculate the number of moles in 500 mg Na_2WO_4 (sodium tungstate).

SOLUTION
$$\frac{500 \text{ mg}}{293.8 \text{ mg/mmol}} \times 0.001 \text{ mol/mmol} = 0.00170 \text{ mol}$$

EXAMPLE 1.3 How many milligrams are in 0.250 mmol Fe_2O_3 (ferric oxide)?

SOLUTION
$$0.250 \text{ mmol} \times 159.7 \text{ mg/mmol} = 39.9 \text{ mg}$$

1.5 CONCENTRATIONS OF SOLUTIONS

There are a number of ways in which solution concentrations may be expressed. Some are more useful than others in quantitative calculations. We will review here the common concentration units used. Their use in quantitative volumetric calculations will be treated in more detail in Chapter 7.

Molarity

The mole concept is useful in expressing concentrations of solutions, especially in analytical chemistry, where we need to know the volume ratios in which solutions of different materials will react. A one-**molar** solution is defined as one that contains one mole of substance in each liter of a **solution.** It is prepared by dissolving one mole of the substance in the solvent and diluting to a final volume of 1 liter in a volumetric flask; or a fraction or multiple of the mole may be dissolved and diluted to the corresponding fraction or multiple of a liter (for example, 0.01 mol in 10 mL). More generally, the **molarity** of a solution is expressed as moles/liter or as millimoles/milliliter. **Molar** is abbreviated as *M,* and we talk of the *molarity* of a solution when we speak of its concentration. A one-molar solution of silver nitrate and a one-molar solution of sodium chloride will react on an equal-volume basis. We can be more general and calculate the moles of substance in any volume of the solution:

$$
\boxed{
\begin{aligned}
\text{moles} &= (\text{moles/liter}) \times \text{liters} \\
&= \text{molarity} \times \text{liters}
\end{aligned}
}
\qquad (1.4)
$$

The liter is an impractical unit for the relatively small quantities encountered in titrations, and we normally work with milliliters. So,

$$
\boxed{
\begin{aligned}
\text{millimoles} &= \text{molarity} \times \text{milliliters} \\
&(\text{or mmol} = M \times \text{mL})
\end{aligned}
}
\qquad (1.5)
$$

EXAMPLE 1.4   A solution is prepared by dissolving 1.26 g $AgNO_3$ in a 250-mL volumetric flask and diluting to volume. Calculate the molarity of the silver nitrate solution. How many millimoles $AgNO_3$ were dissolved?

SOLUTION

$$
M = \frac{1.26 \text{ g}/169.9 \text{ g/mol}}{0.250 \text{ L}} = 0.0297 \text{ mol/L (or 0.0297 mmol/mL)}
$$

Then,

$$
\text{millimoles} = (0.0297 \text{ mmol/mL})(250 \text{ mL}) = 7.42 \text{ mmol}
$$

It should be emphasized that the units in a calculation must combine to give the proper units in the answer. Thus, in this example, grams cancel to leave the proper unit, moles/liter, or molarity. The use of units in the calculation to check if the final units are proper is called **dimensional analysis.** Such dimensional analysis is very useful in properly setting up computations.

EXAMPLE 1.5 How many grams per milliliter of NaCl are contained in a 0.250 M solution?

SOLUTION

$$0.250 \text{ mol/L} = 0.250 \text{ mmol/mL}$$
$$0.250 \text{ mmol/mL} \times 58.4 \text{ mg/mmol} \times 0.001 \text{ g/mg} = 0.0146 \text{ g/mL}$$

EXAMPLE 1.6 How many grams Na_2SO_4 should be weighed out to prepare 500 mL of a 0.100 M solution?

SOLUTION

$$500 \text{ mL} \times 0.100 \text{ mmol/mL} = 50.0 \text{ mmol}$$
$$50.0 \text{ mmol} \times 142 \text{ mg/mmol} \times 0.001 \text{ g/mg} = 7.10 \text{ g}$$

EXAMPLE 1.7 Calculate the concentration of potassium ion in grams/liter after mixing 100 mL of 0.250 M KCl and 200 mL of 0.100 M K_2SO_4.

SOLUTION

$$\text{mmol K}^+ = \text{mmol KCl} + 2 \times \text{mmol K}_2SO_4$$
$$= 100 \text{ mL} \times 0.250 \text{ mmol/mL}$$
$$+ 2 \times 200 \text{ mL} \times 0.100 \text{ mmol/mL}$$
$$= 65.0 \text{ mmol in 300 mL}$$
$$\frac{65.0 \text{ mmol} \times 39.1 \text{ mg/mmol} \times 0.001 \text{ g/mg} \times 1000 \text{ mL/L}}{300 \text{ mL}} = 8.47 \text{ g/L}$$

Normality

A sometimes useful unit of concentration in quantitative analysis is **normality** (N). A one-**normal** solution contains one equivalent/liter. An **equivalent** represents the mass of material providing Avogadro's *number of reacting units,* and so the number of equivalents is given by the number of moles multiplied by the number of reacting units per molecule or atom; the **equivalent weight** is the formula weight divided by the number of reacting units. Table 1.3 lists the reacting units used for different types of reactions. For acids and bases, the number of reacting units is based on the number of protons (i.e., hydrogen ions) an acid will furnish or a base will react with, while for oxidation–reduction reactions it is based on the number of electrons an oxidizing or reducing agent will take on or supply. Thus, for example, sulfuric acid, H_2SO_4, has two reacting units of protons, that is, there are two equivalents of protons in each mole. Therefore,

$$\text{equivalent weight} = \frac{98.08 \text{ g/mol}}{2 \text{ eq/mol}} = 49.04 \text{ g/eq}$$

The normality of a sulfuric acid solution is twice its molarity, that is, $N = $ (g/eq. wt.)/L. The number of equivalents is given by

$$\boxed{\text{number of equivalents (eq)} = \frac{\text{wt. (g)}}{\text{eq. wt. (g/eq)}} = \text{normality (eq/L)} \times \text{volume (L)}}$$

$$(1.6)$$

TABLE 1.3
Reacting Units in Different Reactions

Reaction Type	Reacting Unit
Acid–base	H^+
Oxidation–reduction	Electron
Precipitation or complexometric	Charge

Again, we usually work with milliequivalent quantities, and

$$\text{meq} = \frac{\text{mg}}{\text{eq. wt. (mg/meq)}} = \text{normality (meq/mL)} \times \text{mL} \qquad (1.7)$$

In addition to acid–base and oxidation–reduction reactions, we may have titrations involving formation of a precipitate (precipitation titrations) or of a soluble complex (complexometric titrations). In these cases, there is no exchange of a reacting unit, but merely a combination of the analyte (usually a metal ion) with the reactant. The charge on the metal ion is taken as the reacting unit, and the equivalent weight is equal to the atomic weight divided by the charge. The equivalent weight of the precipitating or complexing agent is its formula weight divided by the number of equivalents of metal ion with which each mole of the anion reacts. For the titration of lead ion with chloride ion,

$$Pb^{2+} + 2Cl^- \rightarrow \underline{PbCl_2}$$

$$\text{equivalent weight Pb} = \frac{\text{formula weight } Pb^{2+}}{2}$$

$$= \frac{207.2 \text{ g/mol}}{2 \text{ eq/mol}}$$

$$= 103.6 \text{ g/eq}$$

Since each chloride ion reacts with one equivalent of lead ion,

$$\text{equivalent weight Cl} = \frac{\text{formula weight } Cl^-}{1}$$

$$= \frac{35.45 \text{ g/mol}}{1 \text{ eq/mol}}$$

$$= 35.45 \text{ g/eq}$$

EXAMPLE 1.8 Calculate the equivalent weights of the following substances: (a) NH_3, (b) $H_2C_2O_4$ (in reaction with NaOH), (c) $KMnO_4$ (Mn is reduced to Mn^{2+}), (d) Na_3PO_4 (in precipitating Ag_3PO_4).

SOLUTION
(a)

$$\text{eq. wt.} = \frac{NH_3 \text{ g/mol}}{1 \text{ eq/mol}} = \frac{17.03}{1} = 17.03 \text{ g/eq}$$

(b)

$$\text{eq. wt.} = \frac{H_2C_2O_4 \text{ g/mol}}{2 \text{ eq/mol}} = \frac{90.04}{2} = 45.02 \text{ g/eq}$$

(c) The Mn undergoes a five-electron change, from valence +7 to +2:

$$\text{eq. wt.} = \frac{KMnO_4 \text{ g/mol}}{6 \text{ eq/mol}} = \frac{158.04}{5} = 31.608 \text{ g/eq}$$

d) Each mole of phosphate ion reacts with three equivalents of Ag^+ ($3Ag^+ + Na_3PO_4 \rightarrow \underline{Ag_3PO_4} + 3Na^+$):

$$\text{eq. wt.} = \frac{Na_3PO_4 \text{ g/mol}}{3 \text{ eq/mol}} = \frac{163.94}{3} = 54.647 \text{ g/eq}$$

EXAMPLE 1.9 Calculate the normality of the solutions containing the following: (a) 5.300 g/L Na_2CO_3 (when the CO_3^{2-} reacts with two protons), (b) 5.267 g/L $K_2Cr_2O_7$ (the Cr is reduced to Cr^{3+}), (c) 2.68 g/L H_2SO_4 used as a precipitating agent for Ba^{2+}.

SOLUTION

(a) CO_3^{2-} reacts with $2H^+$ to H_2CO_3:

$$N = \frac{5.300 \text{ g/L}}{(105.99/2) \text{ g/eq}} = 0.1000 \text{ eq/L}$$

(b) Each Cr^{6+} is reduced to Cr^{3+}, a total change of 6 e^-/molecule of $K_2Cr_2O_7$:

$$N = \frac{5.267 \text{ g/L}}{(294.19/6) \text{ g/eq}} = 0.1074 \text{ eq/L}$$

(c) Each SO_4^{2-} reacts with one Ba^{2+} (which contains 2 eq/mol):

$$N = \frac{2.68 \text{ g/L}}{(98.1/2) \text{ g/eq}} = 0.0546 \text{ eq/L}$$

The advantage of working with equivalents and normality is that species react on a 1:1 equivalent ratio but frequently not on a 1:1 mole ratio. It must be emphasized that the number of equivalents, or the normality, depends upon the specific reaction. Na_2CO_3, for example, may react with either one H^+ ($CO_3^{2-} + H^+ \rightarrow HCO_3^-$) or two H^+ ($CO_3^{2-} + 2H^+ \rightarrow H_2CO_3$), and in the first case there is one reacting unit, compared to two in the second case. Thus, there is a danger in the use of the units normality, equivalents, or equivalent weight, that is, they are entirely dependent on the particular reaction, and this must be specified. If you encounter a 0.1000 N solution of $KMnO_4$, you may not know if this refers to reduction of permanganate to Mn^{2+} (the usual reaction in acid solution) or to MnO_2 (occurs in neutral solution). The first corresponds to a five-electron change (number of reacting units) and the second to a three-electron change. If the molarity is given, there is no question what the concentration is.

In clinical chemistry, equivalents are frequently defined in terms of the number of **charges** on an ion rather than on the number of reacting units. Thus, for example, the equivalent weight of Ca^{2+} is one-half its atomic weight and the number of equivalents is twice the number of moles. This use is convenient for electroneutrality calculations. More detail is given in Section 1.6 below.

Formality

One sometimes sees the term **formality** used for solutions of ionic salts that do not exist as molecules in the solid or in solution. The concentration is given as **formal** (*F*). Operationally, formality is identical to molarity; the former is sometimes reserved for describing makeup concentrations of solutions (i.e., total analytical concentration), and the latter for equilibrium concentrations. For convenience, we shall use molarity exclusively, a common practice.

Molality

Another useful concentration unit is **molality,** *m*. A one-**molal** solution contains one mole per 1000 grams of **solvent.** The molal concentration is convenient in physicochemical measurements of the colligative properties of substances, such as freezing point depression, vapor pressure lowering, and osmotic pressure; colligative properties depend solely on the number of solute particles present in solution per mole of solvent. Molal concentrations are not temperature dependent as molar and normal concentrations are (since the solvent volume is temperature dependent).

Density Calculations

The concentrations of commercial acids and bases are usually given in terms of percent by weight. It is frequently necessary to prepare solutions of a given approximate molarity from these. In order to do so, the density is required for a calculation of the molarity. **Density** is the weight per unit volume at the specified temperature, usually g/mL or g/cm^3 at 20°C. (One milliliter is the volume occupied by 1 g or water at its maximum density, 3.98°C, and is equal to 1.000028 cm^3. For all practical purposes, these two units are interchangeable.) The density of water is 1.00000 g/mL at 4°C and 0.99823 g/mL at 20°C.

Sometimes **specific gravity** is listed, rather than density. This is defined as the ratio of the mass of a body (e.g., a solution), usually at 20°C, to the mass of an equal volume of water at 4°C (or sometimes 20°C). That is, specific gravity is the ratio of the densities of the two substances; it is a dimensionless quantity. Since the density of water at 4° is 1.00000 g/mL, density and specific gravity are equal when referred to water at 4°. When specific gravity is referred to water at 20°, density is equal to specific gravity × 0.99823.

EXAMPLE 1.10 How many milliliters of concentrated sulfuric acid, 94.0% (g/100 g solution), density 1.831 g/cm^3, are required to prepare 1 liter of a 0.100 *M* solution?

SOLUTION
Consider 1 cm^3 = 1 mL. The concentrated acid contains 0.940 g H_2SO_4 per gram of solution, and the solution weighs 1.831 g/mL. The product of these two numbers, then, gives the grams H_2SO_4 per milliliter of solution:

$$M = \frac{0.940 \text{ g/g} \times 1.831 \text{ g/mL}}{98.1 \text{ g/mol}} \times 1000 \text{ mL/L} = 17.5 \text{ mol/L}$$

We must dilute this solution to prepare 1 liter of a 0.100 *M* solution. The same number of millimoles of H_2SO_4 must be taken as will be contained in the final solution. Since mmol = *M* × mL and mmol dilute acid = mmol concentrated acid,

$$0.100\ M \times 1000\ \text{mL} = 17.5\ M \times x\ \text{mL}$$

$$x = 5.71\ \text{mL concentrated acid to be diluted to 1000 mL}$$

Molarity and normality are the most useful concentrations in quantitative analysis, and calculations using these are discussed in more detail in Chapter 7.

1.6 EXPRESSIONS OF ANALYTICAL RESULTS

Results of an analysis may be reported in many ways, and the beginning analytical chemist should be familiar with some of the common expressions and units of measure employed. Results will nearly always be reported as concentration, on either a weight or a volume basis: the quantity of analyte per unit weight or per volume of sample. The units used for the analyte will vary.

We shall first review the common units of weight and volume in the metric system and then describe methods of expressing results. The gram (g) is the basic unit of mass and is the unit employed most often in macro analyses. The kilogram (kg) is 10^3 g. For small samples or trace constituents, smaller units are employed. The milligram (mg) is 10^{-3} g, the microgram (μg) is 10^{-6} g, and the nanogram (ng) is 10^{-9} g. The basic unit of volume is the liter (L). The milliliter (mL) is 10^{-3} L, and is used commonly in volumetric analysis. The microliter (μL) is 10^{-6} L (10^{-3} mL), and the nanoliter (nL) is 10^{-9} L (10^{-6} mL). (Prefixes for even smaller quantities include pico for 10^{-12} and femto for 10^{-15}.)

Solid Samples

Calculations for solid samples are based on weight.[4] The most common way of expressing the results of macro determinations is to give the weight of analyte as a **percent** of the weight of sample (weight/weight basis). The weight units of analyte and sample are the same. For example, a limestone sample weighing 1.267 g and containing 0.3684 g iron (solute) would contain

$$\frac{0.3684\ \text{g}}{1.267\ \text{g}} \times 100\% = 29.08\%\ \text{Fe}$$

The general formula for calculating percent on a weight/weight basis, which is the same as parts per hundred, then is

$$\% \ (\text{wt/wt}) = \left[\frac{\text{wt solute (g)}}{\text{wt sample (g)}} \right] \times 10^2 \quad (\%/\text{g solute/g sample}) \qquad (1.8)$$

Note that in such calculations, grams of solute do not cancel with grams of solution; the fraction represents grams of solute per gram of sample. Multiplication by 10^2 (grams of sample per gram of solute) converts to gram of solute per 100 grams of sample. Since the conversion factors for converting weight of solute and weight of sample (weights expressed in any units) to grams of solute and grams of sample

---

[4]They are really based on mass, but the term "weight" is commonly used. See Section 2.3 for a description and determination of mass and weight.

are always the same, the conversion factors will always cancel. Thus, any weight may be used in the definition.

Trace concentrations are usually given in smaller units, such as **parts per thousand** (ppt, ‰), **parts per million** (ppm), or **parts per billion** (ppb). These are calculated in a manner similar to parts per hundred (%):

$$\text{ppt (wt/wt)} = \left[\frac{\text{wt solute (g)}}{\text{wt sample (g)}}\right] \times 10^3 \quad \text{(ppt/g solute/g sample)} \tag{1.9}$$

$$\text{ppm (wt/wt)} = \left[\frac{\text{wt solute (g)}}{\text{wt sample (g)}}\right] \times 10^6 \quad \text{(ppm/g solute/g sample)} \tag{1.10}$$

$$\text{ppb (wt/wt)} = \left[\frac{\text{wt solute (g)}}{\text{wt sample (g)}}\right] \times 10^9 \quad \text{(ppb/g solute/g sample)} \tag{1.11}$$

The weight units can be any other units so long as both analyte and sample weights are in the same units. **Parts per trillion** (parts per 10^{12} parts) is also abbreviated ppt, so be careful to define which you mean. In the above example, we have 29.08 parts per hundred of iron in the sample, or 290.8 parts per thousand; 2908 parts per ten thousand; 29,080 parts per hundred thousand; and 290,800 per million (290,800 g of iron per 1 million grams of sample, 290,800 pounds of iron per 1 million pounds of sample, etc.) Working backward, 1 ppm corresponds to 0.0001 part per hundred, or 10^{-4} percent. Table 1.3 summarizes the concentration relationships for ppm and ppb. Note that ppm is simply mg/kg or μg/g and that ppb is μg/kg, or ng/g.

EXAMPLE 1.11 A 2.6-g sample of plant tissue was analyzed and found to contain 3.6 μg zinc. What is the concentration of zinc in the plant in ppm? In ppb?

SOLUTION

$$\frac{3.6\ \mu\text{g}}{2.6\ \text{g}} = 1.4\ \mu\text{g/g} \equiv 1.4\ \text{ppm}$$

$$\frac{3.6 \times 10^3\ \text{ng}}{2.6\ \text{g}} = 1.4 \times 10^3\ \text{ng/g} \equiv 1400\ \text{ppb}$$

One ppm is equal to 1000 ppb. One ppb is equal to 10^{-7}%.

Clinical chemists sometimes prefer to use the unit **milligram percent** (mg%) rather than ppm for small concentrations. This is defined as mg of analyte per 100 g of sample. The sample in Example 1.11 would then contain $(2.6 \times 10^{-3}\ \text{mg}/2.6\ \text{g}) \times 100\ \text{mg\%} = 0.14\ \text{mg\%}$ zinc.

TABLE 1.3
Common Units For Expressing Trace Concentrations

Unit	Abbreviation	Wt/Wt	Wt/Vol	Vol/Vol
Parts per million	ppm	mg/kg	mg/L	μL/L
(1 ppm = 10^{-4}%)		μg/g	μg/mL	nL/mL
Parts per billion	ppb	μg/kg	μg/L	nL/L
(1 ppb = 10^{-7}% = 10^{-3} ppm)		ng/g	ng/mL	pL/mL[a]
Milligram percent	mg%	mg/100 g	mg/100 mL	

[a]pL = Picoliter = 10^{-12} L.

Liquid Samples

Results for liquid samples may be reported on a weight/weight basis, as above, or they may be reported on a **weight/volume basis.** The latter is probably more common, at least in the clinical laboratory. The calculations are similar to those above. Percent on a weight/volume basis is equal to g of analyte per 100 mL of sample, while mg% is equal to mg of analyte per 100 mL of sample. This latter unit is often used by clinical chemists for biological fluids, and their accepted terminology is milligrams per deciliter (mg/dL) to distinguish from mg% on a weight/weight basis. Whenever a concentration is expressed as a percentage, it should, if not clear, be specified whether this is wt/vol or wt/wt. Parts per million and parts per billion can also be expressed on a weight/volume basis; ppm is calculated from mg/L or μg/mL; and ppb is calculated from μg/L or ng/mL. Or, the following fundamental calculations are used:

$$\% \text{ (wt/vol)} = \left[\frac{\text{wt. solute (g)}}{\text{vol. sample (mL)}} \right] \times 10^2 \quad \text{(\%/g solute/mL sample)} \tag{1.12}$$

$$\text{ppt (wt/vol)} = \left[\frac{\text{wt. solute (g)}}{\text{vol. sample (mL)}} \right] \times 10^3 \quad \text{(ppt/g solute/mL sample)} \tag{1.13}$$

$$\text{ppm (wt/vol)} = \left[\frac{\text{wt. solute (g)}}{\text{vol. sample (mL)}} \right] \times 10^6 \quad \text{(ppm/g solute/mL sample)} \tag{1.14}$$

$$\text{ppb (wt/vol)} = \left[\frac{\text{wt. solute (g)}}{\text{vol. sample (mL)}} \right] \times 10^9 \quad \text{(ppb/g solute/mL sample)} \tag{1.15}$$

Note that % (wt/vol) is not pounds/100 gallons of solution; the units must be expressed in grams of solute and milliliters of solution.

EXAMPLE 1.12   A 25.0-μL serum sample was analyzed for glucose content and found to contain 26.7 μg. Calculate the concentration of glucose in ppm and in mg/dL.

SOLUTION

$$25.0 \; \mu\text{L} \times \frac{1 \text{ mL}}{1000 \; \mu\text{L}} = 2.50 \times 10^{-2} \text{ mL}$$

$$26.7 \; \mu\text{g} \times \frac{1 \text{ g}}{10^6 \; \mu\text{g}} = 2.67 \times 10^{-5} \text{ g}$$

$$\text{ppm} = \frac{2.67 \times 10^{-5} \text{ g}}{2.50 \times 10^{-2} \text{ mL}} \times 10^6 \; \mu\text{g/g} = 1.07 \times 10^3 \; \mu\text{g/mL}$$

or

$$\frac{26.7 \; \mu\text{g}}{0.0250 \text{ mL}} = 1.07 \times 10^3 \; \mu\text{g/mL} = \text{ppm}$$

$$\text{mg/dL} = \frac{26.7 \; \mu\text{g} \times 10^{-3} \text{ mg/}\mu\text{g}}{0.025 \text{ mL}} \times 100 \text{ ml/dL} = 107$$

(Note the relationship between ppm (wt/vol) and mg/dL.)

The **approximate** relationship between concentrations in parts per million (or parts per billion) and in moles/liter can be seen by assuming a formula weight of

100 for an analyte. Then, since 1 ppm $= 10^{-3}$ g/L, it is equal to $(10^{-3}$ g/L$)$ $(10^2$ g/mol$) = 10^{-5}$ mol/L. Similarly, 1 ppb $= 10^{-8}$ mol/L. Note that this latter concentration is smaller than the hydrogen ion concentration in pure water $(10^{-7}$ mol/L$)$! Of course, this relationship is approximate and will vary with the formula weight. One part per million solutions of zinc and copper, for example, will not be the same molarity. Conversely, equal molar solutions of different species will not be equal in terms of ppm unless the formula weights are equal. The former concentration is based on the number of molecules per unit volume, while the latter is based on the weight of the species per unit volume.

It should be stressed that solutions of equal concentrations on a weight/weight or weight/volume basis do not have the same number of molecules or reaction species, but solutions of the same molarity do.

EXAMPLE 1.13 Calculate the molar concentrations of 1.00 ppm solutions each of Li^+ and Pb^{2+}.

SOLUTION

Li concentration $= 1.00$ ppm $= 1.00$ mg/L   Pb concentration $= 1.00$ ppm $= 1.00$ mg/L

$$M_{Li} = \frac{1.00 \text{ mg/L} \times 10^{-3} \text{ g/mg}}{6.94 \text{ g/mol}} = 1.44 \times 10^{-4} \text{ mol/L}$$

$$M_{Pb} = \frac{1.00 \text{ mg/L} \times 10^{-3} \text{ g/mg}}{207 \text{ g/mol}} = 4.83 \times 10^{-6} \text{ mol/L}$$

Because lead is much heavier than lithium, a given weight contains a fewer number of moles and its molar concentration is less.

If the analyte is a liquid dissolved in another liquid, the results may be expressed on a **volume/volume** basis, but this is rare in situations you are likely to encounter. The calculations would be handled in the same manner as those above, with the same volume units used for solute and sample. Gas analyses may be reported on a weight/weight, weight/volume, or volume/volume basis.

Clinical chemists frequently prefer to use a unit other than weight for expressing the amount of major electrolytes in biological fluids (Na^+, K^+, Ca^{2+}, Mg^{2+}, Cl^-, $H_2PO_4^-$, etc.). This is the unit **milliequivalent** (meq). In this context, it is defined as the number of millimoles of analyte multiplied by the charge on the analyte ion. Results are generally reported as meq/L. This concept is employed to give an overall view of the electrolyte balance. The physician can tell at a glance if total electrolyte concentration has increased or decreased markedly. Obviously, the milliequivalents of cations will be equal to the milliequivalents of anions. One mole of a monovalent cation (1 eq) and half a mole of a divalent anion (1 eq) have the same number of positive and negative charges (one mole each). As an example of electrolyte or charge balance, Table 1.4 summarizes the averages of major electrolyte compositions normally present in human blood plasma and urine. Chapter 19 discusses the ranges and physiological significance of some chemical constituents of the human body.

The milliequivalents of a substance can be calculated from its weight in milligrams simply as follows (similar to how we calculate millimoles):

TABLE 1.4
Major Electrolyte Composition of Normal Human Plasma[a]

Cations	Meq/L	Anions	Meq/L
Na^+	143	Cl^-	104
K^+	4.5	HCO^{3-}	29
Ca^{2+}	5	Protein	16
Mg^{2+}	2.5	HPO_4^-	2
Total	155	SO_4^{2-}	1
		Organic acids	3
		Total	155

[a]Reproduced from *Clinical Chemistry*, 3rd edition, by Joseph S. Annino, Little, Brown and Company (Boston, 1964).

$$\text{meq} = \frac{\text{mg}}{\text{eq. wt. (mg/meq)}} = \frac{\text{mg}}{\text{f. wt. (mg/mmol)}/n(\text{meq/mmol})}$$

$$n = \text{charge on ion}$$

(1.16)

The equivalent weight of Na^+ is 23.0 (mg/mmol/1 (meq/mmol) = 23.0 mg/meq.
The equivalent weight of Ca^{2+} is 40.1 (mg/mmol)/2(meq/mmol) = 20.0 mg/meq.

EXAMPLE 1.14 The concentration of zinc ion in blood serum is about 1 ppm. Express this as meq/L.

SOLUTION

$$1 \text{ ppm} = 1 \text{ μg/mL} = 1 \text{ mg/L}$$

The equivalent weight of Zn^{2+} is 65.4 (mg/mmol/2(meq/mmol) = 32.7 mg/meq. Therefore,

$$\frac{1 \text{ mg/L}}{32.7 \text{ mg/meq}} = 3.06 \times 10^{-2} \text{ meq/L}$$

This unit is usually used for the major electrolyte constituents as in Table 1.4, rather than the trace constituents, as in the example here.

Expressing Concentrations as Equivalents of Substances

We have more or less implied thus far that the analyte is determined in the form it exists or for which we want to report the results. This is often not true. In the analysis of the iron content of an ore, we may determine the iron in the form of Fe_2O_3 and then report it as % Fe. Or we may determine the iron in the form of Fe^{2+} (for example, by titration) and report it as % Fe_2O_3. This is perfectly proper so long as we know the relationship of what is measured to what we wish. Hence, if we analyze for the calcium content of water, we may wish to report it as parts per million (mg/L) of $CaCO_3$ (this is the typical way of expressing water hardness). We know that each gram of Ca^{2+} is equivalent to (or could be converted to) f. wt. $CaCO_3$/f. wt. Ca^{2+} grams of $CaCO_3$. That is, multiplication of the milligrams of Ca^{2+} determined by 100.09/40.08 will give the equivalent number of milligrams of $CaCO_3$. The calcium does not have to exist in this form (we may not even know in what form it actually exists); we simply have calculated the weight that could exist and will report the result as if it did. Specific operations necessary for calculating the weight of the desired constituent will be described in Chapters 6 and 7.

At this point we should mention some of the different weight criteria used for expressing results with biological tissues and solids. The sample may be weighed in one of three physical forms: wet, dry, or ashed. This can apply also to fluids, although fluid volume is usually employed. The wet weight is taken on the fresh, untreated sample. The dry weight is taken after the sample has been dried by heating, desiccation, freeze-drying, and so on. If the test substance is unstable to heat, the sample should not be dried by heating. The weight of the ash residue after the organic matter has been burned off is the third basis of weight. This can obviously be used only for mineral (inorganic) analysis.

QUESTIONS

1. Outline the steps commonly employed in an analytical procedure. Briefly describe each step.
2. List some of the common measuring techniques employed in analytical chemistry.
3. List some separation procedures employed in analytical chemistry.
4. Define instrumental analysis.
5. Distinguish between the expression of concentration on weight/weight, weight/volume, and volume/volume bases.
6. Express ppm and ppb on weight/weight, weight/volume, and volume/volume bases.
7. Define the term "equivalent weight," used for electrolytes in clinical chemistry. Why is this used?

PROBLEMS

Weight/Mole Calculations
8. Calculate the grams of substance required to prepare the following solutions: (a) 250 mL of 5.00% (wt/vol) $NaNO_3$; (b) 500 mL of 1.00% (wt/vol) NH_4NO_3, (c) 1,000 mL of 10.0% (wt/vol) $AgNO_3$.
9. What is the wt/vol % of the solute in each of the following solutions? (a) 52.3 g Na_2SO_4/L, (b) 275 g KBr in 500 mL, (c) 3.65 g SO_2 in 200 mL.
10. Calculate the formula weights of the following substances: (a) $BaCl_2 \cdot 2H_2O$, (b) $KHC_2O_4 \cdot H_2C_2O_4$, (c) $Ag_2Cr_2O_7$, (d) $Ca_3(PO_4)_2$.
11. Calculate the number of millimoles contained in 500 mg of each of the following substances: (a) $BaCrO_4$, (b) $CHCl_3$, (c) $KIO_3 \cdot HIO_3$, (d) $MgNH_4PO_4$, (e) $Mg_2P_2O_7$, (f) $FeSO_4 \cdot C_2H_4(NH_3)_2SO_4 \cdot 4H_2O$.
12. Calculate the number of grams of each of the substances in Problem 11 that would have to be dissolved and diluted to 100 mL to prepare a 0.200 M solution.
13. Calculate the number of milligrams of each of the following substances you would have to weigh out in order to prepare the listed solutions: (a) 1.00 L of 1.00 M NaCl, (b) 0.500 L of 0.200 M sucrose ($C_{12}H_{22}O_{11}$), (c) 10.0 mL of 0.500 M sucrose, (d) 0.0100 L of 0.200 M Na_2SO_4, (e) 250 mL of 0.500 M KOH, (f) 250 mL of 0.900% NaCl (g/100 mL solution).

14. The chemical stockroom is supplied with the following stock solutions: 0.100 M HCl, 0.0200 M NaOH, 0.0500 M KOH, 10.0% HBr (wt/vol), 5.00% Na_2CO_3 (wt/vol). What volume of stock solution would be needed to obtain the following amounts of solutes? (a) 0.0500 mol HCl, (b) 0.0100 mol NaOH, (c) 0.100 mol KOH, (d) 5.00 g HBr, (e) 4.00 g Na_2CO_3, (f) 1.00 mol HBr, (g) 0.500 mol Na_2CO_3.

Molarity Calculations

15. Calculate the molar concentrations of all the cations and anions in a solution prepared by mixing 10.0 mL each of the following solutions: 0.100 M $Mn(NO_3)_2$, 0.100 M KNO_3, 0.100 M K_2SO_4.

16. A solution containing 10.0 mmol $CaCl_2$ is diluted to 1 liter. Calculate the number of grams of $CaCl_2 \cdot 2H_2O$ per milliliter of the final solution.

17. Calculate the molarity of each of the following solutions: (a) 10.0 g H_2SO_4 in 250 mL of solution, (b) 6.00 g NaOH in 500 mL of solution, (c) 25.0 g $AgNO_3$ in 1.00 L of solution.

18. Calculate the number of grams in 500 mL of each of the following solutions: (a) 0.100 M Na_2SO_4, (b) 0.250 M $Fe(NH_4)_2(SO_4)_2 \cdot 6H_2O$, (c) 0.667 M $Ca(C_9H_6ON)_2$.

19. Calculate the grams of each substance required to prepare the following solutions: (a) 250 mL of 0.100 M KOH, (b) 1.00 L of 0.0275 M $K_2Cr_2O_7$, (c) 500 mL of 0.0500 M $CuSO_4$.

20. How many milliliters of concentrated hydrochloric acid, 38.0% (wt/wt), specific gravity 1.19, are required to prepare 1 liter of a 0.100 M solution?

21. Calculate the molarity of each of the following commercial acid or base solutions: (a) 70.0% $HClO_4$, specific gravity 1.668, (b) 69.0% HNO_3, specific gravity 1.409, (c) 85.0% H_3PO_4, specific gravity 1.689, (d) 99.5% $HC_2H_3O_2$ (acetic acid), specific gravity 1.051, (e) 28.0% NH_3, specific gravity 0.898.

PPM Calculations

22. A solution contains 6.0 micromoles Na_2SO_4 in 250 mL. How many ppm sodium does it contain? Of sulfate?

23. A solution (100 mL) containing 325 ppm K^+ is analyzed by precipitating it as the tetraphenyl borate, $K(C_6H_5)_4B$, dissolving the precipitate in acetone solution, and measuring the concentration of tetraphenyl borate ion, $(C_6H_5)_4B^-$, in the solution. If the acetone solution volume is 250 mL, what is the concentration of the tetraphenyl borate in ppm?

24. Calculate the molar concentrations of 1.00 ppm solutions of each of the following. (a) $AgNO_3$, (b) $Al_2(SO_4)_3$, (c) CO_2, (d) $(NH_4)_4Ce(SO_4)_4 \cdot 2H_2O$, (e) HCl, (f) $HClO_4$.

25. Calculate the ppm concentrations of 2.50×10^{-4} M solutions of each of the following. (a) Ca^{2+}, (b) $CaCl_2$, (c) HNO_3, (d) KCN, (e) Mn^{2+}, (f) MnO_4^-.

26. You want to prepare 1 liter of a solution containing 1.00 ppm Fe^{2+}. How many grams ferrous ammonium sulfate, $FeSO_4 \cdot (NH_4)_2SO_4 \cdot 6H_2O$, must be dissolved and diluted in 1 liter? What would be the molarity of this solution?

27. A 0.456-g sample of an ore is analyzed for chromium and found to contain 0.560 mg Cr_2O_3. Express the concentration of Cr_2O_3 in the samples as (a) percent, (b) parts per thousand, (c) parts per million.

28. How many grams NaCl should be weighed out to prepare 1 liter of a 100 ppm solution of (a) Na^+, (b) Cl^-?

29. You have a 250 ppm solution of K^+ as KCl. You wish to prepare from this a 0.00100 M solution of Cl^-. How many milliliters must be diluted to 1 L?

30. One liter of a 500 ppm solution of $KClO_3$ contains how many grams K^+?

Equivalent Weight Calculations

31. Calculate the equivalent weights of the following substances as acids or bases: (a) HCl, (b) $Ba(OH)_2$, (c) $KH(IO_3)_2$, (d) H_2SO_3, (e) $HC_2H_3O_2$ (acetic acid).

32. Calculate the molarity of a 0.250 N solution of each of the acids or bases in Problem 31.

Charge Equivalent Calculations

33. A chloride concentration is reported as 300 mg/dL. What is the concentration in meq/L?

34. A calcium concentration is reported as 5.00 meq/L. What is the concentration in mg/dL?

35. A urine specimen has a chloride concentration of 150 meq/L. If we assume that the chloride is present in urine as sodium chloride, what is the concentration of NaCl in g/L?

RECOMMENDED REFERENCES

Standard Reference Materials

1. R. Alvarez, S. D. Rasberry, and G. A. Uriano, "NBS Standard Reference Materials: Update 1982," *Anal. Chem.,* **54** (1982) 1239A.

2. R. S. Barratt, "The Preparation of Standard Gas Mixtures: A Review," *Analyst,* **106** (1981) 817.

3. R. W. Seward and R. Mavrodineanu, "Standard Reference Materials: Summary of the Clinical Laboratory Standards Issued by the National Bureau of Standards," *NBS Special Publications* 260–271, Washington, D.C., 1981, 168 pp.

4. *Standard Coal Samples.* Available from U.S. Department of Energy, Pittsburgh Mining Technology Center, P.O. Box 10940, Pittsburgh, PA 15236. Samples have been characterized for 14 characteristics, including major and trace elements.

Analytical Calculations

5. Q. Fernando and M. D. Ryan, *Calculations in Analytical Chemistry.* New York: Harcourt Brace Jovanovich, 1982.

chapter 2

BASIC TOOLS AND OPERATIONS OF ANALYTICAL CHEMISTRY

Throughout the text, specific analytical equipment and instrumentation available to the analyst will be discussed as they pertain to specific measuring techniques. Several standard items, however, are common to most analyses. These are described in this chapter. They include the analytical balance and volumetric glassware, and items such as drying ovens, filters, and so forth. Detailed explanation of the physical manipulation and use of this equipment is best done by your laboratory instructor, when you can see and practice with the actual equipment, particularly since the type and operation of equipment will vary from one laboratory to another. Some of the general procedures of good laboratory technique will be mentioned as we go along.

2.1 THE LABORATORY NOTEBOOK

You should first realize that in the analytical laboratory, more than anywhere else, cleanliness and neatness are of the utmost importance. This also applies to the keeping of an orderly notebook. All data should be recorded permanently in ink *when it is collected.* When you go into the analytical laboratory, you will find that this orderliness is to your advantage. First of all, there is a saving of time in not having to reorganize and rewrite the data. There is probably an additional saving of time, since you will be more organized in carrying out the operations of the analysis if you have trained yourself to put the data down in an orderly fashion. Chances for a mistake are reduced. Second, if you make an immediate record, you will be able to detect possible errors in measurements or calculations. And data will not be lost or transferred incorrectly if they are recorded directly in a notebook instead of collected on scraps of paper.

An example of a well-kept notebook with properly recorded data is illustrated below for the volumetric analysis of a soda ash unknown in your laboratory:

Date: 5 February, 1985
Analysis of soda ash unknown
Principle: The soda ash is dissolved in water and titrated to a bromcresol green end point with standard hydrochloric acid. The hydrochloric acid is standardized against primary standard sodium carbonate. Weigh sodium carbonate and soda ash unknown by difference.
Reference: Experiment 7

Titration Reaction $CO_3^{2-} + 2H^+ = H_2CO_3$

Standardization

$$M\ (HCl) = \frac{mg\ Na_2CO_3}{f.\ wt.\ Na_2CO_3 \times \frac{1}{2}\ (mmol\ Na_2CO_3/mmol\ HCl) \times mL\ HCl}$$

$$= \frac{mg\ Na_2CO_3}{105.99\ (mg/mmol) \times \frac{1}{2} \times mL\ HCl}$$

	#1	#2	#3
Bottle + sample	24.2689 g	24.0522 g	23.8597 g
Less sample	24.0522 g	23.8597 g	23.6269 g
g Na$_2$CO$_3$	0.2167 g	0.1925 g	0.2328 g
mg Na$_2$CO$_3$	216.7 mg	192.5 mg	232.8 mg
Buret reading	40.26 mL	35.68 mL	43.29 mL
Initial reading	0.03 mL	0.00 mL	0.02 mL
Net volume	40.23 mL	35.68 mL	43.27 mL
Molarity:	0.1016$_4$ M	0.1018$_0$ M	0.1015$_2$ M

Mean: 0.1016$_5$
Std. devn: 1.6 ppt
Range: 2.8 ppt

Soda Ash

$$\%\ Na_2CO_3 = \frac{M \times mL \times f.\ wt.\ Na_2CO_3 \times \frac{1}{2}\ (mmol\ Na_2CO_3/mmol\ HCl)}{mg\ sample} \times 100$$

$$= \frac{0.1016_5\ (mmol/mL) \times mL \times 105.99\ (mg/mmol) \times \frac{1}{2}}{mg\ sample} \times 100$$

Bottle + sample	25.6728 g	25.4673 g	25.2371 g
Less sample	25.4673 g	25.2371 g	25.0027 g
g sample	0.2055 g	0.2302 g	0.2344 g
mg sample	205.5 mg	230.2 mg	234.4 mg
Buret reading	35.67 mL	40.00 mL	40.70 mL
Initial reading	0.00 mL	0.01 mL	0.05 mL
Net volume	35.67 mL	39.99 mL	40.65 mL
% Na$_2$CO$_3$	93.50%	93.58%	93.42%

Mean: 93.50
Std. devn: 0.9 ppt
Range: 1.7 ppt

The above example is an abbreviated version in which actual calculation or numerical setups are omitted. For complete record keeping, you should include the computational setups in your notebook so an error can be tracked down later, if necessary.

Rather than fill all the space in the laboratory notebook, it is recommended you leave alternate pages for scratch pages (e.g., the left page, leaving the right page for summarizing data). It is also important that you record your data to the proper number of significant figures. Significant figures are discussed in the next chapter, and you should review this material before beginning in the laboratory.

2.2   LABORATORY MATERIALS AND REAGENTS

Table 2.1 lists the properties of materials used in the manufacture of common laboratory apparatus. Borosilicate glass (brand names: Pyrex; KIMAX) is the most commonly used material for laboratory apparatus such as beakers, flasks, pipets, and so forth. It is stable to hot solutions and to rapid changes in temperature. For more specific applications, there are several other materials employed that may possess advantage with respect to chemical resistance, thermal stability, and so forth.

The different grades of chemicals are listed in the back cover of the text. In general, only A.C.S. reagent-grade or primary standard chemicals should be used in the analytical laboratory. The reagent-grade chemicals, besides meeting minimum requirements of purity, may be supplied with a report of analysis of the impurities (printed on the label). Primary standard chemicals are generally at least 99.95% pure. They are analyzed and the results are printed on the label. They are more expensive than reagent-grade chemicals and are used only for the preparation of standard solutions or for the standardization of a solution by reaction (titration) with it. Not all chemicals are available in primary standard grade. There are special grades of solvents for special purposes, for example, spectral grades or chromatographic grades. These are specifically purified to remove impurities that might interfere in the particular application.

In addition to commercial producers, the National Bureau of Standards in Washington, D.C., supplies primary standard chemicals. *NBS Special Publication 260* catalogues standard reference materials. Reference standards are complex materials, such as alloys, that have been carefully analyzed for the ingredients and are used to check or calibrate an analytical procedure.

The concentrations of commercially available acids and bases are listed on the inside of the back cover.

2.3   THE ANALYTICAL BALANCE

Weighing is a required part of almost any analysis, both for measuring the sample and for preparing standard solutions. In analytical chemistry we deal with rather small weights, on the order of a few grams to a few milligrams or less. Weighings must be made to three or four significant figures, and so the weighing device must be both accurate and sensitive. There are various sophisticated ways of achieving this, but the most useful and versatile device used is the **analytical balance.**

TABLE 2.1
Properties of Laboratory Materials

Material	Max. Working Temperature, °C	Sensitivity to Thermal Shock	Chemical Inertness	Notes
Borosilicate glass	200	150°C change OK	Attacked somewhat by alkali solutions on heating	Trademarks: Pyrex (Corning Glass Works) KIMAX (Owens-Illinois)
Soft glass		Poor	Attacked by alkali solutions	
Alkali-resistant glass		More sensitive than borosilicate		Boron-free. Trademark: Corning
Fused quartz	1050	Excellent	Resistant to most acids, halogens	Quartz crucibles used for fusions
High-silica glass	1000	Excellent	More resistant to alkalis than borosilicate	Similar to fused quartz. Trademark: Vycor (Corning)
Porcelain	1100 (glazed) 1400 (unglazed)	Good	Excellent	
Platinum	ca. 1500		Resistant to most acids, molten salts. Attacked by aqua regia, fused nitrates, cyanides, chlorides at >1000°C. Alloys with gold, silver, and other metals	Usually alloyed with iridium or rhodium to increase hardness. Platinum crucibles for fusions and treatment with HF
Nickel and iron			Fused samples contaminated with the metal	Ni and Fe crucibles used for peroxide fusions
Stainless steel	400–500	Excellent	Not attacked by alkalis and acids except conc. HCl, dil. H_2SO_4, and boiling conc. HNO_3	
Polyethylene	115		Not attacked by alkali solutions or HF. Attacked by many organic solvents (acetone, ethanol OK)	Flexible plastic
Polystyrene	70		Not attacked by HF. Attacked by many organic solvents	Somewhat brittle
Teflon	250		Inert to most chemicals	Useful for storage of solutions and reagents for trace metal analysis

Principle

We really deal with masses rather than weights. The **weight** of an object is the force exerted on it by the gravitational attraction. This force will differ at different locations on the earth. **Mass,** on the other hand, is the quantity of matter of which the object is composed, and is invariant.

The analytical balance is a first-class level that compares two masses. Figure 2.1 illustrates such a balance. The fulcrum A lies between the points of application of forces B and C. M_1 represents the unknown mass and M_2 represents a known mass.

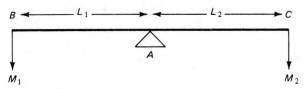

FIGURE 2.1
Principle of analytical balance.

The principle of operation is based on the fact that at balance, $M_1L_1 = M_2L_2$. If L_1 and L_2 are constructed to be as nearly equal as possible, then, at balance, $M_1 = M_2$. A pointer is placed on the *beam* of the balance to indicate on a scale at the end of the pointer when a state of balance is achieved. The operator adjusts the value of M_2 until the pointer returns to its original position on the scale when the balance is unloaded. Although mass is determined, the ratio of masses is the same as the ratio of weights when an equal-arm balance is used. It is customary, then, to use the term *weight* instead of *mass* and to speak of the operator as *weighing*. The known masses are called **standard weights.**

Single-Pan Balances

Most analytical weighings in the modern laboratory are made using a single-pan balance because of the rapidity with which measurements can be made compared to the double-pan balance as described above. A schematic diagram of a typical single-pan balance is shown in Figure 2.2. As usual, a first-class (unsymmetrical) lever is pivoted on a knife edge, and a pan is at one end in which the object is placed. However, there is no pan at the other end for placing weights. When the balance is not in use, a series of weights totaling 160 to 200 g are on the pan end of the beam. These are counterbalanced by a single weight on the other end of the beam, which also acts as part of a damping piston. When an object is placed on the pan, individual weights are *removed* from this end of the beam to restore it to equilibrium. This is accomplished by means of knobs on the front of the balance that lift weights or combinations of weights from the beam. Thus, the weights are never handled. These weights will be equal to the weight of the object on the pan.

Actually, the beam is not brought completely to balance, but weights are removed only to the nearest whole gram or 0.1 g, depending on the balance. The imbalance of the beam is registered optically and automatically on an illuminated vernier scale by a light ray reflected from an engraved optical scale on the beam. The last digits (nearest 0.1 mg) are read from this scale. Alternatively, the imbalance may be read on a digital counter.

The original no-load reading or position is called the **zero point,** and the position under load is called the **rest point.** In operation, the rest point is made to coincide with the zero point. The zero point is generally adjusted to read zero by adjusting the vernier by means of a knob.

Single-pan balances are under constant load of 160 or 200 g, a required feature since they are not brought back to a state of balance. The sensitivity of a balance varies with the load because it is governed by the center of gravity of the beam; the beam bends slightly under load causing a change in the center of gravity and the sensitivity. Calibration of the vernier or digital readout of a single-pan balance to read the amount of imbalance is done at a given sensitivity, that is, at a given load. Therefore, the load must remain constant.

FIGURE 2.2
Schematic diagram of a typical single-pan balance.

All weights of a single-pan balance are concealed and are removed by means of control knobs on the front of the balance: one for tens (10–90 g, for example), one for units (1–9 g), and, if applicable, one for 0.1 units (0.1–0.9 g). The weights removed are registered on a counter on the front of the balance. The beam is brought to rest rapidly by means of an air piston damper.

Care must be taken not to damage the knife edges while the balance is not in operation and while objects are being placed or removed from the pan. A three-position beam-arrest knob is used to protect the knife edges and beam. The center position arrests the pan and beam; a second position partially releases the pan for use while finding the approximate weight of the object on the pan; and a third position completely releases the pan to allow the balance to come to rest.

A typical single-pan balance is shown in Figure 2.3. Weighings can be made in less than a minute with these balances. Although they involve a reduced number of operations and consequently less time and less change for error, single-pan balances are not necessarily more accurate or precise than double-pan balances.

Semimicro and Micro Balances

The discussion thus far has been limited to conventional "macro" or "analytical" balances. These perform weighings to the nearest 0.1 mg, and loads of up to 100 or 200 g can be handled. These are satisfactory for most routine analytical weighings. All of the above classes of balances can be made more sensitive by changing the parameters affecting the sensitivity, such as decreasing the mass of the beam and pans, increasing the length of the beam, and changing the center of gravity of

FIGURE 2.3
Typical single-pan balance. (Courtesy of Arthur H. Thomas Company.)

the beam. Lighter material can be used for the beam since it need not be as sturdy as the beam of a conventional balance.

The *semimicro* balance is sensitive to about 0.01 mg, and the *micro* balance is sensitive to about 0.001 mg (1 µg). The load limits of these balances are correspondingly less than the conventional balance, and greater care must be taken in their use.

Zero Point Drift

The zero point of a balance is not a constant that can be determined or set and forgotten. It will drift for a number of reasons, including temperature changes, humidity, and static electricity. The zero point should therefore be checked at least once every half-hour during the period of using the balance.

Electronic Balances

Modern electronic balances offer added convenience in weighing and are subject to fewer errors or mechanical failures. The operation of dialing weights, turning and reading micrometers, and beam and pan arrest are eliminated. A digital-display electronic balance is shown in Figure 2.4, and the operating principle of an electronic balance is illustrated in Figure 2.5. There are no weights or knife edges. The pan sits on the arm of a movable hanger (2), and this movable system is compensated by a constant electromagnetic force. The position of the hanger is monitored by an electrical position scanner (1), which brings the weighing system back to the zero position. The compensation current is proportional to the weight placed on the pan. This is sent in digital form to a microprocessor that converts it into the corresponding weight value, which appears as a digital display. The weight of the container can be automatically subtracted.

A single control bar is used to switch the balance on and off, to set the display to zero, and to tare a container automatically on the pan. Since results are avail-

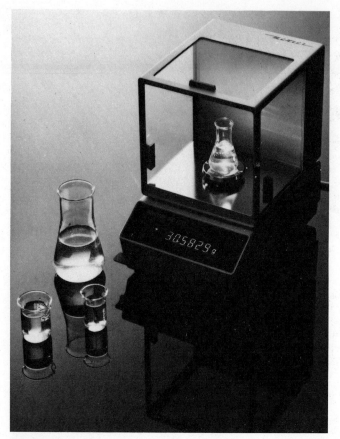

FIGURE 2.4
Electronic analytical balance. (Courtesy of Arthur H. Thomas Company.)

able as an electrical signal, they can be readily processed by a personal computer and stored. Weighing statistics can be automatically calculated.

Electronic analytical balances can be purchased with different weighing ranges and readability. A macrobalance will have a range on the order of 160 g, readable to 0.1 mg, and a semimicrobalance will have a range of about 30 g, readable to 0.01 mg. Ultramicrobalances are available that are sensitive to 0.1 µg or less.

FIGURE 2.5
Operating principle of electronic balance: 1, position scanner; 2, hanger; 3, coil; 4, temperature sensor. (From K. M. Lang, *American Laboratory,* March, 1983, p. 72. Reproduced by permission of American Laboratory, Inc.)

Weight in a Vacuum

The weighings that are made on a balance will, of course, give the weight in air. When an object displaces its volume in air, it will be buoyed up by the weight of air displaced (**Archimedes' principle**). If the density of the weights and the density of the object being weighed are the same, then they will be buoyed up by the same amount, and the recorded weight will be equal to the weight in a vacuum, where there is no buoyancy. If the densities are markedly different, the differences in the buoyancies will lead to a small error in the weighing: one will be buoyed up more than the other, and an unbalance will result. Such a situation arises in the weighing of very dense objects [for example, platinum vessels (density = 21.4) or mercury (density = 13.6)] or light, bulky objects [for example, water (density ≈ 1)]; and in very careful work, a correction should be made for this error. The density of brass weights is 8.4 and that of stainless steel weights is 7.8. You should consult more advanced texts for details of correction (e.g., References 8 and 9). (Reference 4 describes the calibration of the weights in a single-pan balance.)

Actually, in most cases, a correction is not necessary because the error resulting from the buoyancy will cancel out in percent composition calculations. The same error will occur in the numerator (as the concentration of a standard solution or weight of a gravimetric precipitate) and in the denominator (as the weight of the sample). Of course, all weighings must be made with the materials in the same type of container (same density) to keep the error constant.

An example where correction in vacuum is used is in the calibration of glassware. The mass of water or mercury delivered or contained by the glassware is measured. From a knowledge of the density of the liquid at the specified temperature, its volume can be calculated from the mass. Even in these cases, the buoyancy correction is only about one part per thousand. For most objects weighed, buoyancy errors can be neglected. Handbooks contain tables for converting weight of water or mercury in air to volume at different temperatures, using brass weights.

Sources of Error in Weighing

Several possible sources of error have been mentioned, including zero point drift, the weights, and buoyancy. Change in ambient temperature or temperature of the object being weighed is probably the biggest source of error, causing a drift in the zero or rest point due to air current convections. Hot or cold objects must be brought to ambient temperature before being weighed.

General Rules for Weighing

We have seen that there are several types of balances, and the operation of these will differ according to the manufacturer. The specific operation of your particular balance will be explained by your instructor. The main objectives are to (1) protect the knife edges, (2) protect all parts from dust and corrosion, (3) avoid contamination or change in load (of sample or container), and (4) avoid draft (air convection) errors. Some general rules you should familiarize yourself with before weighing with any type of analytical balance are:

1. Never handle objects to be weighed with the fingers. A piece of clean paper or tongs should be used.
2. Weigh at room temperature, and thereby avoid air convection currents.
3. Never place chemicals directly on the pan, but weigh them in a vessel

(weighing bottle, weighing dish) or on powder paper. Always brush spilled chemicals off immediately with a soft brush.

4. Always close the balance case door before making the weighing. Air currents will cause the balance to be unsteady.

5. Never place objects or weights on the pans or remove them without securing the beam arrest and the pan arrest.

Weighing of Solids

Solid materials are usually weighed and dried in a **weighing bottle.** Some of these are shown in Figure 2.6. They have standard tapered ground-glass joints, and hygroscopic samples (which take on water from the air) can be weighed with the bottle kept tightly capped. Replicate weighings are most conveniently carried out by **difference.** The sample in the weighing bottle is weighed, and then a portion is removed (e.g., by tapping) and quantitatively transferred to a vessel appropriate for dissolving of the sample. The weighing bottle and sample are reweighed and from the difference in weight, the weight of sample is calculated. The next sample is removed and the weight is repeated to get its weight by difference, and so on. This is illustrated under The Laboratory Notebook for the soda ash experiment.

It is apparent that by this technique an average of only one weighing for each sample, plus one additional weighing for the first sample, is required. However, each weight represents the difference between two weighings, so that the total experimental error is given by the combined error of both weighings. Weighing by difference *with the bottle capped* must be employed if the sample is hygroscopic or cannot otherwise be exposed to the atmosphere before weighing. If there is no danger from atmospheric exposure, the bottles need not be capped.

For **direct weighing,** a **weighing dish** is used (Figure 2.7). The dish is weighed empty and then with the added sample. This requires two weighings for each sample. The weighed sample is transferred with the aid of camel's-hair brush after tapping. Direct weighing is satisfactory only if the sample is nonhygroscopic.

When making very careful weighings (e.g., to a few tenths of a milligram), you must take care not to contaminate the weighing vessel with extraneous material that may affect its weight. Special care should be taken not to get perspiration from the hands on the vessel, because this can be quite significant. It is best to handle the vessel with a piece of paper. Solid samples must frequently be dried to a constant weight. Samples must always be allowed to attain the temperature of the balance room before weighing.

Weighing of Liquids

Weighing of liquids is usually done by direct weighing. The liquid is transferred to a weighed vessel (e.g., a weighing bottle), which is capped to prevent evaporation during weighing, and is then weighed. If a liquid sample is weighed by difference

FIGURE 2.6
Weighing bottles.

FIGURE 2.7
Weighing dish.

by pipetting out an aliquot from the weighing bottle, the inside of the pipet must be rinsed several times after transferring. Care should be taken not to lose any sample from the tip of the pipet during transfer.

Types of Weighing

There are two types of weighing done in analytical chemistry, **rough** and **accurate.** Rough weighings to two or three significant figures are normally used when the amount of substance to be weighed need only be known to within a few percent. Examples are reagents to be dissolved and standardized later against a known standard, or the apportioning of reagents that are to be dried and then later weighed accurately, or simply added as is, as for adjusting solution conditions. That is, only rough weighings are needed when the weight is not involved in the computation of the analytical result. Rough weighings should not be done on analytical balances but on triple-beam, top-loading, or torsion balances.

Accurate weighings are reserved for obtaining the weight of a sample to be analyzed, the weight of the dried product in gravimetric procedures, or the weight of a dried reagent being used as a standard in a determination, all of which must generally be known to four significant figures to be used in calculating the analytical result. **These are performed on an analytical balance, usually to the nearest 0.1 mg.** An exact predetermined amount of reagent is rarely weighed (e.g., 0.5000 g), but rather an approximate amount (about 0.5 g) is weighed accurately (e.g., to give 0.5129 g). Some chemicals are never weighed on an analytical balance. Sodium hydroxide pellets, for example, are so hygroscopic that they continually absorb moisture. The weight of a given amount of sodium hydroxide is not reproducible (and its purity is not known). To obtain a solution of known sodium hydroxide concentration, the sodium hydroxide is weighed on a rough balance and dissolved, and the solution is standardized against a standard acid solution.

2.4   VOLUMETRIC GLASSWARE

Although accurate volume measurements of solutions can be avoided in gravimetric methods of analysis, they are required for almost any other type of analysis involving solutions.

Volumetric Flasks

Volumetric flasks are used in the diluting of a sample or solution to a certain volume. They come in a variety of sizes, from 1 L or more to 1 mL. A typical flask is shown in Figure 2.8. These flasks are designed **to contain** an accurate volume at the specified temperature (20 or 25°C) when the bottom of the meniscus (the concave curvature of the upper surface of water in a column caused by capillary action—see Figure 2.15) just touches the etched "fill" line across the neck of the glass. The coefficient of expansion of glass is small, and for ambient temperature fluctuations the volume can be considered constant. These flasks are marked with **"TC"** to indicate "to contain." Other, less accurate containers, such as graduated cylinders, are also marked "TC."

When using a volumetric flask, a solution should be diluted stepwise. The solution to be diluted is added to the flask, and then diluent (usually distilled water) is added to fill the flask about two-thirds (taking care to rinse down any reagent on the ground glass lip). It helps to swirl the solution before diluent is added to the

FIGURE 2.8
Volumetric flask.

neck of the flask to obtain most of the mixing (or dissolving in the case of a solid). Finally, diluent is added so that the bottom of the meniscus is even with the middle of the calibration mark (at eye level). If there are any droplets of water on the neck of the flask above the meniscus, take a piece of tissue and blot these out. Also, dry the ground-glass stopper joint.

The solution is finally thoroughly mixed as follows. Keeping the stopper on securely by using the thumb or palm of the hand, invert the flask and swirl or shake it *vigorously* for 5 to 10 seconds. Turn right side up and allow the solution to drain from the neck of the flask. Repeat at least 10 times.

Note. Should the volume of liquid go over the calibration mark, it is still possible to save the solution as follows. Paste against the neck of the flask a thin strip of paper and mark on it with a sharp pencil the position of the meniscus, avoiding parallax error. After removing the thoroughly mixed solution from the flask, fill the flask with water to the calibration mark. Then by means of a buret or small volume graduated pipet, add water to the flask until the meniscus is raised to the mark on the strip of paper. Note and record the volume so added.

Pipets

The pipet is used to transfer a particular volume of solution. As such, it is often used to deliver a certain fraction **(aliquot)** of a solution. To ascertain the fraction, the original volume of solution from which the aliquot is taken must be known, but it need not all be present, so long as it has not evaporated or been diluted. There are two common types of pipets, the **volumetric,** or **transfer, pipet** and the **measuring pipet** (see Figures 2.9 and 2.10). Variations of the latter are also called **clinical,** or **serological, pipets.**

Pipets are designed **to deliver** a specified volume at a given temperature, and they are marked **"TD."** Again, the volume can be considered to be constant with small changes in temperature. Pipets are calibrated to account for the drainage film remaining on the glass walls. This drainage film will vary somewhat with the time taken to deliver, and usually the solution is allowed to drain under the force of gravity and the pipet is removed shortly after the solution is delivered. A uniform drainage time should be adopted.

FIGURE 2.9
Transfer pipets.

FIGURE 2.10
Measuring pipets.

The volumetric pipet is used for accurate measurements, since it is designed to deliver only one volume and is calibrated at that volume. Accuracy to four significant figures is generally achieved, although with proper calibration, five figures may be obtained if necessary. See the table on the back cover for tolerances of class A transfer pipets. Measuring pipets are straight-bore pipets that are marked at different volume intervals. These are not as accurate, because nonuniformity of the internal diameter of the device will have a relatively larger effect on total volume than is the case for pipets with a bulb shape. Also, the drainage film will vary with the volume delivered. At best, accuracy to three significant figures can be expected from these pipets, unless you make the effort to calibrate the pipet for a given volume delivered.

Most volumetric pipets are calibrated to deliver with a certain small volume remaining in the tip. This should not be shaken or blown out. In delivering, the pipet is held vertically and the tip is touched on the side of the vessel to allow smooth delivery without splashing and so that the proper volume will be left in the tip. The forces of attraction of the liquid on the wall of the vessel will draw out a part of this.

Some pipets are **"blowout"** types (including measuring pipets calibrated to the entire tip volume. The final volume of solution must be blown out from the tip to deliver the calibrated amount. These pipets are easy to identify, as they will always have one or two **ground bands or rings** around the top. (These are not to be confused with a colored ring which is used only as a color coding for the volume of the pipet.) The solution is not blown out until the pipet has completely drained by gravity. Blowing to increase the rate of delivery will change the volume of the drainage film.

Volumetric pipets are available in sizes of 100 to 0.5 mL or less. Measuring and serological pipets range from a total capacity of 25 to 0.1 mL. Measuring pipets can be used for accurate measurements, especially for small volumes, if they are calibrated at the particular volume wanted. The larger measuring pipets usually deliver too quickly to allow drainage as fast as the delivery, and they have too large a bore for accurate reading.

In using a pipet, one should always wipe the tip dry after filling. If a solvent other than water is used, or if the solution is viscous, pipets must be recalibrated for the new solvent or solution to account for difference in drainage rate.

Pipets are filled by suction, using a rubber pipet bulb or other such pipetting device. Before using a pipet, practice filling it with water. Corrosive or toxic solutions must *never* be pipetted by mouth.

Syringe Pipets

These can be used for both macro and micro volume measurements. The calibration marks on the syringes may not be very accurate, but the reproducibility can be excellent if an automatic deliverer is used, such as a spring load device that draws the plunger up to the same preset level each time. The volume delivered in this manner is free from drainage errors, because the solution is forced out by the plunger. The volume delivered can be accurately calibrated. Microliter syringe pipets are used for introduction of samples into gas chromatographs. A typical syringe is illustrated in Figure 2.11. They are fitted with a needle tip, and the tolerances are as good as those found for other micropipets. In addition, any desired volume throughout the range of the syringe can be delivered.

The above syringe pipets are useful for accurate delivery of viscous solutions or volatile solvents; with these materials the drainage film would be a problem in conventional pipets. Syringe pipets are well suited to rapid delivery, and also for thorough mixing of the delivered solution with another solution as a result of the rapid delivery.

A second type of syringe pipet is that shown in Figure 2.12. This type is convenient for rapid, one-hand dispensing of fixed volumes in routine procedures and is widely used in the clinical chemistry laboratory. It contains a disposable nonwetting plastic tip (e.g., polypropylene) to reduce both film error and contamination. A thumb button operates a spring-loaded plunger, which stops at an intake or a discharge stop; the latter stop is beyond the former to ensure complete delivery. The sample never contacts the plunger and is contained entirely in the plastic tip. These pipets are available in volumes of 1 to 1000 μL and are reproducible to 1 to 2% or better.

The actual volume delivered by these and other micropipets frequently does not need to be known because they are used in relative measurements. For example,

FIGURE 2.11
Hamilton microliter syringe.

FIGURE 2.12
Syringe pipet with disposable tip. (Courtesy of Eppendorf Gerätebau, Netheler and Hinz Gmbh, Hamburg.)

the same pipet may be used to deliver a sample and an equal volume of a standard solution for calibrating the instrument used for the measurement. Precision in delivery is usually more important than the absolute volume delivered.

Burets

A buret is used for the accurate delivery of a variable amount of solution. Its principal use is in **titrations,** where a standard solution is added to the sample solution until the **end point** (the detection of the completion of the reaction) is reached. The conventional buret for macro titrations is marked in 0.1-mL increments from 0 to 50 mL; one is illustrated in Figure 2.13. The volume delivered can be read to the nearest 0.01 mL by interpolation (good to about ±0.02 or ±0.03 mL). Burets are also obtainable in 10-, 25-, and 100-mL capacities, and microburets are available in capacities of down to 2mL, where the volume is marked in 0.01-mL increments and can be estimated to the nearest 0.001 mL. Ultramicroburets of 0.1-mL capacity graduated in 0.001-mL (1-μL) intervals are used for microliter titrations. Micro and ultramicro *micrometer* burets, in which a Teflon plunger displaces the titrant from the buret, are also available. See Figure 2.14. The plunger is moved by turning a micrometer or a digital readout dial, which is calibrated in microliter intervals as small as 0.002 μL per division or less.

Drainage film is a factor with conventional burets, as with pipets, and this can be a variable if the delivery rate is not constant. The usual practice is to deliver at a fairly slow rate, about 15 to 20 mL per minute, and then to wait several seconds after delivery to allow the drainage to "catch up." In actual practice, the rate of delivery is only a few drops per minute near the end point, and there will be no time lag between the flow rate and the drainage rate. As the end point is ap-

FIGURE 2.13
Typical buret.

FIGURE 2.14
Ultramicro burets.

proached, fractions of a drop are delivered by just opening, or "cracking," the stopcock and then touching the tip of the buret to the wall of the titration vessel. The fraction of the drop is then washed down into the solution with distilled water.

Care and Use of Volumetric Glassware

We have mentioned a few precautions in the use of volumetric flasks, pipets, and burets. Your laboratory instructor will supply you with detailed instructions in the use of each of these tools. A discussion of some general precautions and good laboratory technique follows.

Cleanliness of glassware is of the utmost importance. If films of dirt or grease are present, liquids will not drain uniformly and will leave water breaks or droplets on the walls. Under such conditions the calibration will be in error. Initial cleaning should be with hot, dilute detergent, about 2%. Use of a buret or test tube brush aids the cleaning of burets and necks of volumetric flasks—but be careful of scratching the interior walls. Pipets should be rotated to coat the entire surface with detergent. If detergent does not do the job, then treatment with hot dichromate—sulfuric acid (ca. 60°C) cleaning solution may be effective. Prepare such solution by carefully adding 400 mL concentrated sulfuric acid to a solution of 20 g sodium dichromate in 15 mL water. Avoid getting water in the solution after it is prepared, because the useful lifetime of the solution would thereby be shortened. The mixture is very corrosive, and care must be taken to prevent skin and clothing burns. Rinse affected parts thoroughly with tap water whenever you get it on yourself, and treat with sodium bicarbonate.

A note of caution in using dichromate cleaning solution: When contacted with chlorides, volatile chromyl chloride, CrO_2Cl_2, can be formed. This is rated as a serious toxic hazard (high hazard as an acute local irritant, and both an acute and a chronic hazard in inhalation) and is a recognized carcinogen. Other inorganic ions such as bromide can also generate volatile toxic substances. Hence, glassware should always be well rinsed before cleaning with dichromate cleaning solution, and it is advisable to do the cleaning in a fume hood.

Small volumetric flasks can be filled with the cleaning solution, but larger ones may be partially filled and tilted and rotated to coat the entire surface. Pipets and burets should be inverted and filled by suction (not by mouth!); a clamped rubber tube on the end will hold the solution in. Warm cleaning solution should remain in contact with the glass for about 15 minutes. Cleaning solution may be used at room temperature but is then slower acting and may have to be left in contact overnight. Used cleaning solution should be returned to the original container. After any cleaning, the volumetric apparatus must be rinsed thoroughly with tap water and finally with several small portions of distilled or deionized water. Of course, chromic acid cleaning solution should never be used if you are to perform a chromium determination, especially at the trace level, because chromium will be leached from the glass for long periods of time.

Pipets and burets should be rinsed at least twice with the solution with which they are to be filled. If they are wet, they should be rinsed first with water, if they have not been already, and then a minimum of *three* times with the solution to be used; about one-fifth the volume of the pipet or buret is adequate for each rinsing. A volumetric flask, if it is wet from a previously contained solution, is rinsed with three portions of only water, since later it will be filled to the mark with water. It need not be dry.

Care in reading the volume will avoid parallax error, that is, error due to incorrect alignment of the observer's eye, the meniscus, and the scale. This applies in the reading of any scale, such as the pointer scale of an analytical balance. Correct position is with your eye at the same level as the meniscus. If the eye level is above the meniscus, the volume read will be smaller than that taken; the opposite will be true if the eye level is too low.

After glassware is used, it can usually be cleaned sufficiently by immediate rinsing with water. If the glassware has been allowed to dry, it should be cleaned with detergent. Volumetric flasks should be stored with the stopper inserted, and preferably filled with distilled water. Burets should be filled with distilled water and stoppered with a rubber stopper when not in use.

General Tips for Accurate and Precise Titrating

Before using a buret, you may have to grease the stopcock. A thin layer of stopcock grease (not silicone lubricant) is applied uniformly to the stopcock, using very little near the hole and taking care not to get any grease in the hole. The stopcock is

FIGURE 2.15
Meniscus illuminator for buret.

inserted and rotated. There should be a uniform and transparent layer of grease, and the stopcock should not leak. If there is too much lubricant, it will be forced into the barrel or may work into the buret tip and clog it. Grease can be removed from the buret tip and the hole of the stopcock by using a fine wire. If the buret contains a Teflon stopcock, it does not require lubrication. The buret is filled above the zero mark and the stopcock is opened to fill the tip. Check the tip for air bubbles. If any are present, they may work out of the tip during the titration, causing an error in reading. Work air bubbles out by rapid squirting of titrant through the tip or tapping the tip while solution is flowing. No bubbles should be in the barrel of the buret. If there are, the buret is probably dirty.

The initial reading of the buret is taken by allowing it to drain slowly to the zero mark. Wait a few seconds to make certain the drainage film has caught up to the meniscus. Read the buret to the nearest 0.02 mL (for a 50-mL buret). The initial reading may be 0.00 mL or greater. The reading is best taken by placing your finger just in back of the meniscus or by using a meniscus illuminator (Figure 2.15). The meniscus illuminator has a white and a black field, and the black field is positioned just below the meniscus. Avoid parallax error by making the reading at eye level.

The titration is performed with the sample solution in an Erlenmeyer flask. The flask is placed on a white background, and the buret tip is positioned in the neck of the flask. The flask is swirled with the right hand while the stopcock is manipulated with the left hand (Figure 2.16). This grip on the buret maintains a slight inward pressure on the stopcock to ensure that leakage will not occur. The solution can be more efficiently stirred by means of a magnetic stirrer and stirring bar.

As the titration proceeds, the indicator changes color in the vicinity where the titrant is added, owing to local excesses; but it rapidly reverts to the original color as the titrant is dispersed through the solution to react with the sample. As the end point is approached, the return to the original color occurs more slowly, since the dilute solution must be mixed more thoroughly to consume all the titrant. At this point, the titration should be stopped and the sides of the flask washed down with distilled water from the wash bottle. A drop from the buret is about 0.05 mL, and the volume is read to the nearest 0.02 mL. It is therefore necessary to split drops near the end point. This can be done by slowly turning the stopcock until a fraction of a drop emerges from the buret tip and then closing it. The fraction of drop is touched off onto the wall of the flask and is washed into the flask with the wash bottle, or it is transferred with a glass stirring rod. There will be a sudden and

FIGURE 2.16
Proper technique for titration.

"permanent" (lasting at least 30 seconds) change in the color at the end point when a fraction of a drop is added.

The titration is usually performed in triplicate. After performing the first titration, you can calculate the approximate volume for the replicate titrations from the weights of the samples and the molarity of the titrant. This will save time in the titrations. The volume should not be calculated to nearer than 0.1 mL in order to avoid bias in the reading.

After a titration is complete, unused titrant should never be returned to the original bottle but should be discarded.

If a physical property of the solution, such as potential, is measured to detect the end point, the titration is performed in a beaker with magnetic stirring so electrodes can be placed in the solution.

Tolerances and Precision

The National Bureau of Standards (NBS) has prescribed certain *tolerances,* or absolute errors, for different volumetric glassware, and some of these are listed on the back cover of the text. For volumes of greater than about 25 mL, the tolerance is within 1 part per thousand (ppt) relative, but it is larger for smaller volumes. The letter "A" stamped on the side of a volumetric flask, buret, and so forth, indicates that it complies with class A tolerances. This says nothing about the precision of delivery. Volumetric glassware that meets NBS specifications or that is certified by NBS can be purchased, but at a significantly higher price than uncertified glassware. Less expensive glassware may have tolerances double those specified by NBS. It is a simple matter, however, to calibrate this glassware to an accuracy as good as or exceeding the NBS specifications (see Experiment 2). You are referred to more advanced texts for a description of glassware calibration (e.g., see Reference 8).

The precision of reading a 50-mL buret is about ±0.02 mL. Since a buret is always read twice, the total absolute uncertainty may be as much as ±0.04 mL. The relative uncertainty will vary inversely with the total volume delivered. It becomes apparent that a titration with a 50-mL buret should involve about 40 mL titrant to achieve a precision of 1 ppt. Smaller burets can be used for increased precision at smaller volumes. Pipets will also have a certain precision of reading, but only one reading is required for volumetric pipets.

Selection of Glassware

As in weighing operations, there will be situations where you need to accurately know volumes of reagents or samples measured or transferred (accurate measurements), and others in which only approximate measurements are required (rough measurements). If you wish to prepare a standard solution of 0.1 M hydrochloric acid, it can't be done by measuring an accurate volume of concentrated acid and diluting to a known volume, because the concentration of the commercial acid is not known adequately. Hence, an approximate solution is prepared which is then standardized. We see in the table on the inside back cover that the commercial acid is about 12.4 M. To prepare 1 liter of a 0.1 M solution, about 8.1 mL needs to be taken and diluted. It would be a waste of time to measure this (or the water used for dilution) accurately. A 10-mL graduated cylinder or 10-mL measuring pipet will suffice, and the acid can be diluted in an ungraduated 1-L bottle. If, on the other hand, you wish to dilute a stock standard solution accurately, then a transfer pipet must be used and the dilution must be done in a volumetric flask. Any volumetric measurement that is a part of the actual analytical measurement must be done with the accuracy required of the analytical measurement. This generally means four-significant-figure accuracy, and transfer pipets and volumetric flasks are required. This includes taking an accurate portion of a sample, preparation of a standard solution from an accurately weighed reagent, and so forth. Burets are used for accurate measurement of variable volumes, as in a titration. Preparation of reagents that are to be used in an analysis just to provide proper solution conditions (e.g., buffers for pH control) need not be prepared highly accurately, and less accurate glassware can be used, for example, graduated cylinders.

2.5   OTHER APPARATUS

Besides apparatus for measuring mass and volume, there are a number of other items of equipment commonly used in analytical procedures.

Blood Samplers

Syringes are used to collect blood samples.[1] Stainless steel or aluminum needles are generally used with glass or plastic syringes. These generally present no problem of contamination, although special precautions may be required for certain trace element analyses. **Vacutainers** or similar devices are often used in place of syringes. These are evacuated test tubes with a rubber cap. The needle is pushed

---

[1]You should *not* attempt to collect a blood sample unless you have been specifically trained to do so. A trained technician will generally be assigned to this job.

FIGURE 2.17
Desiccator and desiccator plate.

through the cap after the other end has been inserted into the vein and the blood is drawn into the evacuated tube. The tube may contain an anticoagulating agent to prevent clotting of the blood if plasma or whole blood samples are to be analyzed.

A finger puncture, instead of a venipuncture, is used when small quantities of blood are to be collected for microprocedures. Up to 0.5 mL or more blood can be squeezed from the finger into a small collection tube by puncturing the finger with a sterilized sharp-pointed knifelike object.

Desiccators

A **desiccator** is used to keep samples dry while they are cooling and before they are weighed and, in some cases, to dry the wet sample. Dried or ignited samples and vessels are cooled in the desiccator. A typical glass desiccator is shown in Figure 2.17. A desiccator is an airtight container that maintains an atmosphere of low humidity. A desiccant such as calcium chloride is placed in the bottom to absorb the moisture. This desiccant will have to be changed periodically as it becomes "spent." It will usually become wet in appearance or caked from the moisture when it is time to be changed. A porcelain plate is usually placed in the desiccator to support weighing bottles, crucibles, and other vessels. An airtight seal is made by application of stopcock grease to the ground-glass rim on the top of the desiccator. A **vacuum desiccator** has a side arm on the top for evacuation so that the contents can be kept in a vacuum rather than just an atmosphere of dry air.

The top of a desiccator should not be removed any more than necessary, since the removal of moisture from the air introduced is rather slow, and continued exposure will limit the lifetime of the desiccant. A red-hot crucible or other vessel should be allowed to cool in the air about 60 seconds before it is placed in the desiccator. Otherwise, the air in the desiccator will be heated appreciably before the desiccator is closed, and as the air cools, a partial vacuum will be created. This will result in a rapid inrush of air when the desiccator is opened and possible spilling or loss of sample as a consequence. A hot weighing bottle should not be stoppered when placed in a desiccator because on cooling, a partial vacuum is created and the stopper may seize. The stopper should be placed in the desiccator with the weighing bottle.

Table 2.2 lists some commonly used desiccants and their properties. Aluminum oxide, magnesium perchlorate, calcium oxide, calcium chloride, and silica gel can be regenerated by heating at 150, 240, 500, 275, and 150°C, respectively.

TABLE 2.2
Some Common Drying Agents

Agent	Capacity	Deliquescent	Trade Name
$CaCl_2$ (anhydrous)	High	Yes	
$CaSO_4$	Moderate	No	Drierite (W. A. Hammond Drierite Co.)
CaO	Moderate	No	
$MgClO_4$ (anhydrous)	High	Yes	Anhydrone (J. T. Baker Chemical Co.); Dehydrite (Arthur H. Thomas Co.)
Silica gel	Low	No	
Al_2O_3	Low	No	
P_2O_5	Low	Yes	

Furnaces and Ovens

A **muffle furnace** (Figure 2.18) is used to ignite samples to high temperatures, either to convert precipitates to a weighable form or to burn organic materials prior to inorganic analysis. There should be some means of regulating the temperature, since losses of some metals may occur at temperatures in excess of 500°C (see Chapter 3). Temperatures up to about 1200°C can be reached with muffle furnaces.

A **drying oven** is used to dry samples prior to weighing. A typical drying oven is shown in Figure 2.19. These ovens are well ventilated for uniform heating. The usual temperature employed is about 110°C, but temperatures of 200 to 300°C may be obtained.

FIGURE 2.18
Muffle furnace. (Courtesy of Arthur H. Thomas Company.)

FIGURE 2.19
Drying oven. (Courtesy of Arthur H. Thomas Company.)

Hoods

A **fume hood** is used when chemicals or solutions are to be evaporated. When perchloric acid or acid solutions of perchlorates are to be evaporated, the fumes should be collected, or the evaporation should be carried out in fume hoods specially designed for perchloric acid work (i.e., constructed from components resistant to attack by perchloric acid).

When performing trace analysis, as in trace metal analysis, care must be taken to prevent contamination. The conventional fume hood is one of the "dirtiest" areas of the laboratory since laboratory air is drawn into the hood and over the sample. **Laminar flow hoods** or work stations are available for providing very clean work areas. Rather than drawing unfiltered laboratory air into the work area, the air is prefiltered and then flows over the work area and out into the room to create a positive pressure and prevent unfiltered air from flowing in. A typical laminar flow work station is shown in Figure 2.20. The high-efficiency particulate air filter

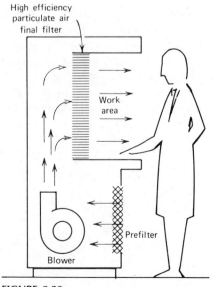

FIGURE 2.20
Laminar-flow work station. (Courtesy of Dexion, Inc., 344 Beltine Boulevard, Minneapolis, Minn.)

(a)   *(b)*

FIGURE 2.21
Wash bottles: *(a)* polyethylene, squeeze type; *(b)* glass, blow type.

removes all particles from the air larger than 0.3 μm. Vertical laminar flow stations are preferred when fumes are generated that should not be blown over the operator. Facilities are available to exhaust noxious fumes.

Wash Bottles

A **wash bottle** of some sort should be handy in any analytical laboratory, to be used for quantitative transfer of precipitates and solutions and to wash precipitates. These come in a variety of shapes and sizes, as seen in Figure 2.21. They may be constructed from a Florence flask and glass tubing, as in Figure 2.21B.

Centrifuges and Filters

A **centrifuge** has many useful applications, particularly in the clinical laboratory, where blood may have to be separated into fractions such as serum or plasma, and proteins may have to be separated by precipitation followed by centrifuging.

Filters for filtering precipitates are of various types. The Gooch crucible, sintered-glass crucible, and porcelain filter crucible are illustrated in Figure 2.22. The **Gooch crucible** is porcelain and has holes in the bottom; a glass filter disc is supported on top of it. In the past, an asbestos mat was usually prepared, but this is inconvenient and potentially hazardous since asbestos fibers in the air are carcinogenic, and the filter disc will handle fine precipitates. The **sintered-glass crucible** contains a sintered-glass bottom, which is available in fine (F), medium (M), or coarse (C) porosity. The **porcelain filter crucible** contains a porous unglazed

(a)   *(b)*   *(c)*

FIGURE 2.22
Filtering crucibles: *(a)* Gooch crucible; *(b)* sintered-glass crucible; *(c)* porcelain filter crucible.

FIGURE 2.23
Crucible holders.

bottom. Glass filters are not recommended for concentrated alkali solutions because of the possibility of attack by these solutions. See Table 2.1 for maximum working temperatures for different types of crucible materials.

Gelatinous precipitates such as hydrous iron oxide should not be filtered in filter crucibles because they clog the pores. Even with filter paper, the filtration of the precipitates can be slow.

Filter crucibles are used with a **crucible holder** mounted on a filtering flask (Figure 2.23). A safety bottle is connected between the flask and the aspirator.

Ashless filter paper is generally used for quantitative work in which the paper is ignited away and leaves a precipitate suitable for weighing (see Chapter 6). There are various grades of filter papers for different types of precipitates. These are listed in Table 2.3 for Whatman and for Schliecher and Schuell papers.

Techniques of Filtration

By proper fitting of the filter paper, the rate of filtration can be increased. A properly folded filter paper is illustrated in Figure 2.24. The filter paper is folded in the shape of a cone, with the overlapped edges of the two quarters not quite meeting

TABLE 2.3
Types of Filter Papers

Precipitate	Whatman	Schliecher and Schuell
Very fine (e.g., $BaSO_4$)	No. 42	No. 589, Blue or Red Ribbon
Small or medium (e.g., AgCl)	No. 40	No. 549, White Ribbon
Gelatinous or large crystals (e.g., $Fe_2O_3 \cdot xH_2O$)	No. 41	No. 589, Black Ribbon

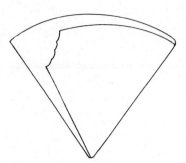

FIGURE 2.24
Properly folded filter paper.

($\frac{1}{8}$ inch apart). About $\frac{1}{4}$ inch is torn away from the corner of the inside edge. This will allow a good seal against the funnel to prevent air bubbles from being drawn in. After the folded paper is placed in the funnel, it is wetted with distilled water. The stem is filled with water and the top of the wet paper is pressed against the funnel to make a seal. With a proper fit, no air bubbles will be sucked into the funnel, and the suction supplied by the weight of the water in the stem will increase the rate of filtration. The filtration should be started immediately. The precipitate should occupy not more than one-third to one-half of the filter paper in the funnel, because many precipitates tend to "creep." Do not allow the water level to go over the top of the paper.

The precipitate should be allowed to settle in the beaker before filtration is begun. The bulk of the clear liquid can then be decanted and filtered at a rapid rate before the precipitate fills the pores of the filter paper.

Care must be taken in the decanting and the transferring of the precipitate to avoid losses. This is properly done by use of a stirring rod and a wash bottle, as illustrated in Figure 2.25. The solution is decanted by pouring it down the glass rod, which guides it into the filter without splashing. The precipitate is most readily washed while still in the beaker. After the mother liquor has been decanted off, wash the sides of the beaker down with several milliliters of the wash liquid, and then allow the precipitate to settle as before. Decant the wash liquid into the filter and repeat the washing operation two or three times. The precipitate is then transferred to the filter by holding the glass rod and beaker in one hand, as illustrated, and washing it out of the beaker with wash liquid from the wash bottle.

If the precipitate must be quantitatively collected, as in gravimetric analysis, the last portions of precipitate are removed by scrubbing the walls with a moistened **rubber policeman** (Figure 2.26). Wash the remainder of loosened precipitate from the beaker and from the policeman. If the precipitate is being collected in a filter paper, then instead of a rubber policeman, a small piece of the ashless filter paper can be rubbed on the beaker walls to remove the last bits of precipitate and added to the filter. This should be held with a pair of forceps.

After the precipitate is transferred to the filter, it is washed with five or six small portions of wash liquid. This is more effective than adding one large volume. Divert the liquid around the top edge of the filter to wash the precipitate down into the cone. Each portion should be allowed to drain before adding the next one. Check for completeness of washing by testing for the precipitating agent in the last few drops of the washings.

FIGURE 2.25
Proper technique for transfer of a precipitate.

FIGURE 2.26
Rubber policeman.

2.6 OBTAINING THE SAMPLE

Collecting a representative sample is an aspect of analytical chemistry that the beginning analytical student is often not concerned with, because the samples handed to him are assumed to be homogeneous and representative. Yet this process can be the most critical aspect of an analysis. The significance and accuracy of measurements can be limited by the sampling process. Unless sampling is done properly, it becomes the weak link in the chain of the analysis. A life could sometime depend on the proper handling of a blood sample during and after sampling.

Many professional societies have specified definite instructions for sampling given materials (for example, American Society for Testing Materials, Association of Official Analytical Chemists, American Public Health Association, and others). By appropriate application of experience and statistics, these materials can be sampled as accurately as the analysis can be performed. Often, however, the matter is left up to the analyst. The ease or complexity of sampling will, of course, depend on the nature of the sample.

The problem involves obtaining a sample that is representative of the whole. This sample is called the *gross sample*. Its size may vary from a few grams or less to several pounds, depending on the type of bulk material. Once a representative gross sample is obtained, it may have to be reduced to a sufficiently small size to be handled. This is called the *sample*. Once the sample is obtained, an aliquot, or portion, of it will be analyzed. This aliquot is called the *analysis sample*. Several replicate analyses on the same sample may be performed by taking separate aliquots.

In the clinical laboratory, the gross sample is usually satisfactory for use as the sample, because it is not large and it is homogeneous (blood and urine samples, for example). The analysis sample will usually be from a few milliliters to a fraction of a drop (a few microliters) in quantity.

Some of the problems associated with obtaining gross samples of solids, liquids, and gases are considered below.

1. SOLIDS   Inhomogeneity of the material, variation in particle size, and variation within the particle make sampling of solids more difficult than other materials. The easiest but usually most unreliable way to sample a material is the *grab sample,* which is one sample taken at random and assumed to be representative. The grab sample will be satisfactory only if the material from which it is taken is homogeneous. For most reliable results, it is best to take $\frac{1}{50}$ to $\frac{1}{100}$ of the total bulk for the gross sample, unless the sample is fairly homogeneous. The larger the particle size, the larger the gross sample should be.

The easiest and most reliable time to sample large bodies of solid materials is while they are being moved. In this way any portion of the bulk material can usually be exposed for sampling. Thus, a systematic sampling can be performed to obtain aliquots representing all portions of the bulk. Some samples follow.

In the loading or unloading of bags of cement, a representative sample can be obtained by taking every fiftieth or so bag, or by taking a sample from each bag. In the moving of grain by wheelbarrow, representative wheelbarrow loads or a shovelful from each wheelbarrow can be taken. All of these aliquots are combined to form the gross sample.

In the case of tissues, several small portions of an organ can be taken and combined. Often, the entire organ is taken as the gross sample, and this is ground up to use as the sample.

2. LIQUIDS.   These materials tend to be homogeneous and are much easier to sample. Liquids mix by diffusion only very slowly and must be shaken to obtain a homogeneous mixture. If the material is indeed homogeneous, a simple grab (single random) sample will suffice. For all practical purposes, this method is satisfactory for taking blood samples. Twenty-four-hour urine sample collections are generally more reliable than single specimens.

The timing of sampling of biological fluids is, however, very important. The composition of blood varies considerably before and after meals, and for many analyses a sample is collected after the patient has fasted for a number of hours. Preservatives such as sodium fluoride for glucose preservation and anticoagulants may be added to blood samples when they are collected.

Blood samples may be analyzed as whole blood, or they may be separated to yield plasma or serum according to the requirements of the particular analysis. Most commonly, the concentration of the substance external to the red cells (the extracellular concentration) will be a significant indication of physiological condition, and so serum or plasma is taken for analysis.

If liquid samples are not homogeneous, and if they are small enough, they can be shaken and sampled immediately. For example, there may be particles in the liquid that have tended to settle. Large bodies of liquids are best sampled after a transfer or, if in a pipe, after passing through a pump when it has undergone the most thorough mixing. Large stationary liquids can be sampled with a "thief" sampler, which is a device for obtaining aliquots at different levels. It is best to take the sample at different depths at a diagonal, rather than straight down. The separate aliquots of liquids can be analyzed individually and the results combined, or the aliquots can be combined into one gross sample and replicate analyses performed. This latter procedure is probably preferred because the analyst will then have some idea of the precision of the analysis.

3. GASES. The usual method of sampling gases involves displacement of a liquid. The liquid must be one in which the sample has little solubility and with which it does not react. Mercury is the liquid employed most commonly. The mercury is allowed to trickle from the bottom of the container, whereupon the gas is pulled in at the top. Such a procedure allows collection of an average sample over a relatively long period of time. A grab-type sample is satisfactory in some cases. In the collecting of a breath sample, for example, the subject could blow into an evacuated bag. Auto exhaust could be collected in a large evacuated plastic bag.

The volume of gross gas sample collected may or may not need to be known. Often, the *concentration* of a certain analyte in the gas sample is measured, rather than the *amount*. The temperature and pressure of the sample will, of course, be important in determining the volume and hence the concentration. Since the life scientist is rarely confronted with gas analysis, we will not go into detail concerning the problems associated with this rather complicated field. You are referred to more comprehensive books on the subject.

The gas sampling mentioned here does not concern gas constituents dissolved in liquids, such CO_2 or O_2 in blood. These are treated as liquid samples and are then handled accordingly to measure the gas in the liquid or to release it from the liquid for measuring.

2.7 OPERATIONS OF DRYING AND PREPARING A SOLUTION OF THE ANALYTE

After a sample has been collected, a solution of the analyte must usually be prepared before the analysis can be continued. Drying of the sample may be required and it must be weighed or the volume measured. If the sample is already a solution (e.g., serum, urine, or water), then extraction, precipitation, or concentration of the analyte may be in order, and this may also be true with other samples.

In this section we describe common means for preparing solutions of inorganic and organic materials. Included is the dissolution of metals and inorganic compounds in various acids or in basic fluxes (fusion), the destruction of organic and biological materials for determination of inorganic constituents (using wet digestion or dry ashing), and the removal of proteins from biological materials so they do not interfere in the analysis of organic or inorganic constituents.

Drying the Sample

Solid samples will usually contain variable amounts of adsorbed water. With inorganic materials, the sample will generally be dried before weighing. This is accomplished by placing it in a drying oven at 105 to 110°C for 1 or 2 hours. Other nonessential water, such as that entrapped within the crystals, may require higher temperatures for removal.

Decomposition or side reactions of the sample must be considered during drying. Material unstable under conditions of heat can be dried by setting it in a desiccator; using a vacuum desiccator will hasten the drying process. If the sample is weighed without drying, the results will be on an "as is" basis and should be reported as such.

Plant and tissue samples can usually be dried by heating. See the end of Chapter 1 for a discussion of the various weight bases (wet, dry, ash) used in connection with reporting analytical results for these samples.

Sample Dissolution

Before the analyte can be measured, some sort of sample alteration is generally necessary to get the analyte into solution or, for biological samples, to rid it of interfering organic substances, such as proteins. There are two types of sample preparation: those that totally destroy the sample matrix and those that are nondestructive or only partially destructive. The former type can generally be used only when the analyte is inorganic or can be converted to an inorganic derivative for measurement (e.g., Kjeldahl analysis, in which organic nitrogen is converted to ammonium ion—see below). The latter type must be used if the analyte to be measured is an organic substance.

Dissolving Inorganic Solids

Strong mineral acids are good solvents for many inorganics. Hydrochloric acid is a good general solvent for dissolving metals that are above hydrogen in the electromotive series. Nitric acid is a strong oxidizing acid and will dissolve most of the common metals, nonferrous alloys, and the "acid-insoluble" sulfides.

Perchloric acid, when heated to drive off water, becomes a very strong and efficient oxidizing acid in the dehydrated state. It dissolves most of the common metals and destroys traces of organic matter. It must be used with extreme caution, because it will react explosively with many easily oxidizable substances, especially organic matter.

Some inorganic materials will not dissolve in acids, and **fusion** with an acidic or basic **flux** in the molten state must be employed to render them soluble. The sample is mixed with the flux in a sample-to-flux ratio of about 1 to 10 or 20, and the combination is heated in an appropriate crucible until the flux becomes molten. When the melt becomes clear, usually in about 30 minutes, the reaction is complete. The cooled solid is then dissolved in dilute acid or in water. During the fusion process, insoluble materials react with the flux to form a soluble product. Sodium carbonate is one of the most useful basic fluxes, and acid-soluble carbonates are produced.

Destruction of Organic Materials

Animal and plant tissue, biological fluids, and organic compounds are usually decomposed by **wet digestion** with a boiling oxidizing acid or mixture of acids, or by **dry ashing** at a high temperature (400–700°C) in a muffle furnace. In wet digestion, the acids oxidize organic matter to carbon dioxide, water, and other volatile products, which are driven off, leaving behind salts or acids of the inorganic constituents. In dry ashing, atmospheric oxygen serves as the oxidant; that is, the organic matter is burned off, leaving an inorganic residue. Oxidizing aids may be employed in dry ashing.

1. DRY ASHING. Although various types of dry ashing and wet digestion combinations are used with about equal frequency by analysts for organic and biological materials, simple dry ashing with no chemical aids is probably the most commonly employed technique. Lead, zinc, cobalt, antimony, chromium, molybdenum, strontium, and iron traces can be recovered with little loss by retention or volatilization. Usually a porcelain crucible can be used. Lead is volatilized at temperatures in excess of about 500°C, especially if chloride is present, as in blood or urine. Platinum crucibles are preferred for lead for minimal retention losses.

If an oxidizing material is added to the sample, the ashing efficiency is enhanced. Magnesium nitrate is one of the most useful aids, and with this it is possible to recover arsenic, copper, and silver, in addition to the above listed elements.

Liquids and wet tissues are dried on a steam bath or by gentle heat before they are placed in a muffle furnace. The heat from the furnace should be applied gradually up to full temperature to prevent rapid combustion and foaming.

After dry ashing is complete, the residue is usually leached from the vessel with 1 or 2 mL hot concentrated or 6 M hydrochloric acid and transferred to a flask or beaker for further treatment.

Another dry technique is that of **low-temperature ashing.** A radio-frequency discharge is used to produce activated oxygen radicals, which are very reactive and will attack organic matter at low temperatures. Temperatures of less than 100°C can be maintained, and volatility losses are minimized. Introduction of elements from the container and the atmosphere is reduced, and so are retention losses. Radiotracer studies have demonstrated that 17 representative elements are quantitatively recovered after complete oxidation of organic substrate.

Elemental analysis in the case of organic compounds (e.g., for carbon or hydrogen) is usually performed by **oxygen combustion** in a tube, followed by an absorption train. Oxygen is passed over the sample in a platinum boat, which is heated to quantitatively convert carbon to CO_2, hydrogen to H_2O, and so forth. These combustion gases pass into the absorption train, where they are absorbed in preweighed tubes containing a suitable absorbent. For example, **Ascarite** (sodium hydroxide on asbestos) is used to absorb the CO_2, and **Dehydrite** (magnesium perchlorate) is used to absorb the H_2O. The gain in weight of the absorption tubes is a measure of the CO_2 and H_2O liberated from the sample. Details of this technique are important, and, should you have occasion to use it, you are referred to more comprehensive texts on elemental analysis.

2. WET DIGESTION. Wet digestion with a mixture of nitric and sulfuric acids is the second most often used oxidation procedure. Usually a small amount (e.g., 5 mL) of sulfuric acid is used with larger volumes of nitric acid (20–30 mL). Wet digestions are usually performed in a Kjeldahl flask. The nitric acid destroys the bulk of the organic matter, but it does not get hot enough to destroy the last traces. It is boiled off during the digestion process until only sulfuric acid remains and dense, white SO_3 fumes are evolved and began to reflux in the flask. At this point, the solution gets very hot, and the sulfuric acid acts on the remaining organic material. Charring may occur at this point if there is considerable or very resistant organic matter left. If the organic matter persists, more nitric acid may be added. Digestion is continued until the solution clears. All digestion procedures must be performed in a fume hood.

A much more efficient digestion mixture employs a mixture of nitric, perchloric, and sulfuric acids in a volume ratio of about 3:1:1. Ten milliliters of this mixture will usually suffice for 10 g fresh tissue or blood. The perchloric acid is an extremely efficient oxidizing agent when it is dehydrated and hot and will destroy the last traces of organic matter with relative ease. Samples are heated until nitric acid is boiled off and perchloric acid fumes, which are less dense than SO_3 but which fill the flask more readily, appear. The hot perchloric acid is boiled, usually until fumes of SO_3 appear, signaling the evaporation of all the perchloric acid. Sufficient nitric acid must be added at the beginning to dissolve and destroy the bulk of organic matter, and there must be sulfuric acid present to prevent the sample from going to dryness, or else there is danger of explosion from the perchloric acid. A hood specially designed for perchloric acid work should be used for all digestions incorporating perchloric acid.

This mixture is even more efficient if a small amount of molybdenum(VI) catalyst is added. As soon as water and nitric acid are evaporated, oxidation proceeds vigorously with foaming, and the digestion is complete in a few seconds. The digestion time is reduced considerably.

A mixture of nitric and perchloric acids is also commonly used. The nitric acid boils off first, and care must be taken to prevent evaporation of the perchloric acid to near dryness, or a violent explosion may result; this procedure *is not recommended* unless you have considerable experience in digestion procedures. **Perchloric acid should never be added directly to organic or biological material.** Always add an excess of nitric acid first. Explosions with perchloric acid are generally associated with formation of peroxides, and the acid turns dark in color (e.g., yellowish brown) prior to explosion. Certain organic compounds such as ethanol, cellulose, and polyhydric alcohols can cause hot concentrated perchloric acid to explode violently; this is presumably due to formation of ethyl perchlorate.

A mixture of nitric, perchloric, and sulfuric acids allows zinc, selenium, arsenic, copper, cobalt, silver, cadmium, antimony, chromium, molybdenum, strontium, and iron to be quantitatively recovered. Lead is often lost if sulfuric acid is used. The mixture of nitric and perchloric acids can be used for lead and all the above elements. Perchloric acid must be present to prevent losses of selenium. It maintains strong oxidizing conditions and prevents charring that would result in formation of volatile compounds of lower oxidation states of selenium. Samples containing mercury cannot be dry ashed. Wet digestion with heat applied must be done using a reflux apparatus because of the volatile nature of mercury and its compounds. Cold or room temperature procedures are often preferred to obtain partial destruction of organic matter. For example, in urine samples, which contain a relatively small amount of organic matter compared with blood, mercury can be reduced to the element with copper(I) and hydroxylamine hydrochloride and the organic matter destroyed by potassium permanganate at room temperature. The mercury can then be dissolved and the analysis continued.

Many nitrogen-containing compounds can be determined by **Kjeldahl digestion** to convert the nitrogen to ammonium sulfate. The digestion mixture consists of sulfuric acid plus potassium sulfate to increase the boiling point of the acid and thus increase its efficiency. A catalyst is also added (mercury, copper, or selenium). After destruction of the organic matter, sodium hydroxide is added to make the solution alkaline, and the ammonia is distilled into an excess of standard hydrochloric acid. The excess acid is back-titrated with standard alkali to determine the amount of ammonia collected. With a knowledge of the percent nitrogen composition in the compound of interest, the amount of the compound can be calculated from the amount of ammonia determined. This is the most accurate method for determining protein. Protein contains a definite percentage of nitrogen, which is converted to ammonium sulfate during the digestion. See chapter 8 for further details.

The relative merits of the oxidation methods have been studied extensively. However, there is still no agreement as to which is to be preferred. Dry ashing is recommended for its simplicity and relative freedom from positive errors (contamination), since few or no reagents are added. The potential errors of dry oxidation are volatilization of elements and losses by retention on the walls of the vessel. Adsorbed metals on the vessel may in turn contaminate future samples. Wet digestion is considered superior in terms of rapidity (although it does require more operator attention), low level of temperature maintained, and freedom from loss by retention. The chief error attributed to wet digestion is the introduction of impurities from the reagents necessary for the reaction. This problem has been minimized as commercial reagent-grade acids have become available in greater purity, and specially prepared high-purity acids can now be obtained commercially. The time required for ashing or digestion will vary with the sample and the technique employed. Two to 4 hours are common for dry ashing and $\frac{1}{2}$ to 1 hour is common for wet digestion.

Partial Destruction or Nondestruction of Sample Matrix

Obviously, when the substance to be determined is organic in nature, nondestructive means of preparing the sample must be employed. For the determination of metallic elements, it is also sometimes unnecessary to destroy the sample, particularly with biological fluids. For example, several metals in serum or urine can be determined by atomic absorption spectroscopy by direct aspiration of the sample or a diluted sample into a flame. Constituents of solid materials such as soils can sometimes be extracted by an appropriate reagent. Thorough grinding, mixing, and refluxing are necessary to extract the analyte. Many trace metals can be extracted from soils with 1 M ammonium chloride or acetic acid solution. Some, such as selenium, can be distilled as the volatile chloride or bromide.

Protein-Free Filtrates

Proteins in biological fluids interfere with many analyses and must be removed nondestructively. Several reagents will precipitate (coagulate) proteins. Trichloroacetic acid (TCA), tungstic acid (sodium tungstate plus sulfuric acid), and barium hydroxide plus zinc sulfate (a neutral mixture) are some of the common ones. A measured volume of sample (serum, for example) is usually treated with a measured volume of reagent. Following precipitation of the protein (approximately 10 minutes), the sample is filtered through dry filter paper without washing, or else it is centrifuged. An aliquot of the **protein-free filtrate** (PFF) is then taken for analysis. Details for preparing specific types of protein-free filtrates are given in Chapter 19 (under Collection and Preservation of Samples) as well as in experiments requiring them.

Dialysis

Small molecules and ions can be separated from macromolecules, such as proteins, by **dialysis.** The sample is placed in a bag of semipermeable material (collodion, cellophane, etc.) and the bag is placed in water or an appropriate reagent solution. The small molecules (e.g., glucose, urea) will diffuse through the membrane into the surrounding medium until equilibrium is reached, that is, until the concentration on each side of the membrane is equal. The Technicon Auto-Analyzer utilizes a continuous-flow dialysis to separate the analytes from the bulk sample; these are mixed with appropriate reagents in the flow system after being separated. This system is discussed in more detail in Chapter 17.

Laboratory Techniques

When a solid sample is to be dried in a weighing bottle, the cap is removed from the bottle and, to avoid spilling, both are placed in a beaker and covered with a ribbed watch glass. Some form of identification should be placed on the beaker.

The weighed sample may be dissolved in a beaker or Erlenmeyer flask. If there is any fizzing action, cover the vessel with a watch glass. After dissolution is complete, wash the walls of the vessel down with distilled water. Also wash the watch glass so the washings fall into the vessel. You may have to evaporate the solution to decrease the volume. This is best done by covering the beaker with a ribbed watch glass to allow space for evaporation. Low heat should be applied to prevent bumping; a steam bath or variable-temperature hot plate is satisfactory.

Use of a **Kjeldahl flask** for dissolution will avoid some of the difficulties of splattering or bumping. Kjeldahl flasks are also useful for performing digestions. They derive their name from their original use in digesting samples for Kjeldahl nitrogen analysis. They are well suited to all types of wet digestions of organic samples and acid dissolution of metals. Kjeldahl flasks come in assorted sizes from 10 to 800

FIGURE 2.27
Kjeldahl flasks.

mL. Some of these are shown in Figure 2.27. The sample and appropriate acids are placed in the round bottom of the flask and the flask is tilted while it is heated. In this way the acid can be boiled or refluxed with little danger of loss by "bumping." The flask may be heated with a flame or in special electrically heated Kjeldahl digestion racks, which heat several samples simultaneously.

QUESTIONS

1. Describe the basic pieces of apparatus used for volumetric measurements. List whether each is designed to contain or to deliver the specified volume.
2. Describe the principle and operation of the analytical balance.
3. Is a single-pan balance more accurate than a double-pan balance? Why?
3. Why is a microbalance more sensitive than an analytical balance?
5. What does "TD" on glassware mean? "TC"?
6. Explain weighing by difference.
7. Distinguish between the zero point and rest point of a balance.
8. Explain the operation of a single-pan balance.
9. List the general rules for the use of the balance.
10. Describe the principles of dry ashing and wet digestion of organic and biological materials. List the advantages of each.
11. What are the two principal means of dissolving inorganic materials?
12. What is a PFF? How would you prepare it?
13. What set of conditions must be carefully avoided to use perchloric acid safely for digesting organic materials?
14. What is a gross sample? Sample? Analysis sample? Grab sample?

RECOMMENDED REFERENCES

Reagents
1. J. Rosin, *Reagent Chemicals and Standards,* 5th ed. Princeton: D. Van Nostrand Company, 1967.

Analytical Balances
2. D. F. Rohrbach and M. Pickering, "A Comparison of Mechanical and Electronic Balances," *J. Chem. Educ.,* **59** (1982) 418.

Calibration of Weights
3. W. D. Abele, "Laboratory Note: Time-Saving Applications of Electronic Balances," *Am. Lab.,* **13** (1981) 154. Calibration of weights using mass standards calibrated by the National Bureau of Standards is discussed.
4. D. F. Swinehart, "Calibration of Weights in a One-Pan Balance," *Anal. Lett.,* **10** (1977) 1123.

Calibration of Volumetric Ware
5. G. D. Christian, "Coulometric Calibration of Micropipets," *Microchem. J.,* **9** (1965), 16.
6. M. R. Masson, "Calibration of Plastic Laboratory Ware," *Talanta,* **28** (1981) 781. Tables for use in calibration of polypropylene vessels are presented.
7. W. Ryan, "Titrimetric and Gravimetric Calibration of Pipettors: A Survey," *Am. J. Med. Technol.,* **48** (1982) 763. Calibration of pipettors of 1 to 500 μL is described.
8. D. A. Skoog and D. M. West, *Fundamentals of Analytical Chemistry,* 4th ed. New York: Saunders, 1982. Describes calibration of volumetric ware.

9. M. R. Winward, E. M. Woolley, and E. A. Butler, "Air Buoyancy Corrections for Single-Pan Balances," *Anal. Chem.,* **49** (1977) 2126.

Clean Laboratories

10. J. R. Moody, "The NBS Clean Laboratories for Trace Element Analysis," *Anal. Chem.,* **54** (1982) 1358A.

Sampling

11. J. A. Bishop, "An Experiment in Sampling," *J. Chem. Educ.,* **35** (1958) 31.

12. G. Ingram, "Treatment and Weighing of Samples," in C. L. Wilson and D. W. Wilson, eds., *Comprehensive Analytical Chemistry,* Vol. **1B,** 1960. Chapter VIII2.

13. I. M. Kolthoff and E. B. Sandell, *Textbook of Quantitative Analysis,* 3rd ed. New York: The Macmillan Company, 1952, Pp. 241–243, sampling.

14. H. A. Laitinen and W. E. Harris, *Chemical Analysis,* 2nd ed. New York: McGraw-Hill, 1975. Chapter 27, sampling. Discusses statistics of sampling.

15. J. R. Moody, "The Sampling, Handling and Storage of Materials for Trace Analysis," *Phil. Trans. Roy. Soc. London, Ser. A,* **305** (1982) 669.

16. R. C. Tomlinson, "Sampling," in C. L. Wilson and D. W. Wilson, eds., *Comprehensive Analytical Chemistry,* Vol. **IA.** New York: Elsevier Publishing, 1959. Chapter II3.

17. F. J. Welcher, ed., *Standard Methods of Chemical Analysis,* 6th ed., Vol. **2.** New York: D. Van Nostrand Co., 1963. Contains chapters on sampling of solids, liquids, and gases.

Sample Preparation and Dissolution

18. G. D. Christian, "Medicine, Trace Elements, and Atomic Absorption Spectroscopy," *Anal. Chem.,* **41** (1) (1969) 24A. Describes the preparation of solutions of biological fluids and tissues.

19. G. D. Christian, E. C. Knoblock, and W. C. Purdy, "Polarographic Determination of Selenium in Biological Materials," *J. Assoc. Offic. Agric. Chemists,* **48** (1965) 877; R. K. Simon, G. D. Christian, and W. C. Purdy. "Coulometric Determination of Arsenic in Urine," *Am. J. Clin. Pathol.,* **49** (1968) 207. Describes use of Mo(VI) catalyst in digestions.

20. C. E. Gleit and W. D. Holland, "The Use of Electrically Excited Oxygen for the Low Temperature Decomposition of Organic Substances," *Anal. Chem.,* **34** (1962) 1454.

21. T. T. Gorsuch, "Radiochemical Investigations on the Recovery for Analysis of Trace Elements in Organic and Biological Materials," *Analyst,* **84** (1959) 135.

22. S. Nobel and D. Nobel, "Determination of Mercury in Urine," *Clin. Chem.,* **4** (1958) 150. Describes room temperature digestion.

23. A. Steyermarch, *Quantitative Organic Microanalysis,* 2nd ed. New York: Academic Press, 1961. Describes procedures for determination of the elemental content (C, H, N, O, Cl, etc.) of organic compounds.

chapter 3

DATA HANDLING

Although data handling normally follows the collection of data in an analysis, it is treated early in the text because a knowledge of statistical analysis will be required as you perform experiments in the laboratory. Also, statistics are necessary to understand the significance of the data that are collected and therefore to set limitations on each step of the analysis. The design of experiments (including size of sample required, accuracy of measurements required, number of analyses needed, and so forth) is determined from a proper understanding of what the data will represent.

3.1 SIGNIFICANT FIGURES

The weak link in the chain of any analysis is governed by that measurement that can be made with the least accuracy. It is useless to extend an effort to make the other measurements of the analysis more accurately than this limiting measurement. The number of significant figures can be defined as **the number of digits necessary to express the results of a measurement consistent with the measured precision.** Each digit denotes the actual quantity it specifies. For example, in the figure 237, we have 2 hundreds, 3 tens, and 7 units.

The digit 0 can be a significant part of a measurement, or it can be used merely to place the decimal point. The number of significant figures in a measurement is independent of the placement of the decimal point. Take the number 92,067. This number has five significant figures, regardless of where the decimal point is placed. For example, 92,067 micrometers, 9.2067 centimeters, 0.92067 decimeter, and 0.092067 meter all have the same number of significant figures. They merely represent different ways (units) of expressing one measurement. The zero between the decimal point and the 9 in the last number is used only to place the decimal point. There is no doubt whether any zero that *follows* a decimal point is significant or is used to place the decimal point. In the number 727.0, the zero is not used to locate the decimal point, but is a significant part of the figure. Ambiguity can arise if a zero *precedes* a decimal point. If it falls between two other nonzero integers, then it will be significant. Such was the case with 92,067. In the number 936,600, it is impossible to determine whether one or both or neither of the zeros is used merely to place the decimal point or whether they are a part of the measurement. It is best in cases like this to write only the significant figures you are sure about and then to locate the decimal point by an expression of 10 raised to the appropriate power. Thus, 9.3660×10^5 has five significant figures, but 936,600 contains six digits, one to place the decimal.

EXAMPLE 3.1 List the proper number of significant figures in the following numbers and indicate which zeros are significant.

0.216; 90.7; 800.0; 0.0670

SOLUTION

0.216	three significant figures
90.7	three significant figures; zero is significant
800.0	four significant figures; all zeros are significant
0.0670	three significant figures; only the last zero is significant

---

The significance of the last digit of a measurement can be illustrated as follows. Assume that each member of a class measures the length of a rod, using the same meter stick. Assume further that the meter stick is graduated in 1-mm increments. The measurements can be estimated to the nearest 0.1 division (0.1 mm) by interpolation, but the last digit is uncertain, since it is only an estimation. A series of readings, for example, might be

36.4 mm
36.8 mm
36.0 mm
37.1 mm
—————
36.6 mm (average)

Multiplication and Division

In many measurements, one estimated digit that is uncertain will be included. This is the last significant figure in the measurement; any digits beyond it are meaningless. In multiplication and division, the uncertainty of this digit is carried through the mathematical operations, thereby limiting the number of certain digits in the answer. There is at least the degree of relative uncertainty in the answer of a multiplication or division as there is in the operator with the least degree of certainty, that is, the one with the least number of significant figures. We shall designate this limiting number as the **key number.** If there is more than one operator with the same lowest number of significant figures, then the one with the smallest absolute magnitude without regard to the decimal point is the key number (since its uncertainty is the greatest).

EXAMPLE 3.2   In the following pairs of numbers, pick the one that would represent the key number in a multiplication or division. (a) 42.67 or 0.0967; (b) 100.0 or 0.4570; (c) 0.0067 or 0.10.

SOLUTION
(a) 0.0967 (has three significant figures)
(b) 100.0 (both have four significant figures, but the uncertainty here is 1 part per thousand versus about 1 part in 4600)
(c) 0.10 (both have two significant figures, but the uncertainty here is 10% versus about 1 part in 70)

EXAMPLE 3.3   Give the answer of the following operation to the maximum number of significant figures and indicate the key number.

$$\frac{35.63 \times 0.5481 \times 0.05300}{1.1689} \times 100\% = 88.5470578\%$$

SOLUTION
The key number is 35.63. The answer is therefore 88.55%, and it is meaningless to carry the operation out to more than five figures (the fifth figure is used to round off the fourth). The 100% in this calculation is an absolute number, since it is used only to move the decimal point, and it has an infinite number of significant figures. Note that the key number has a relative uncertainty at best of 1 part in 3600, and so the answer has a relative uncertainty at best of 1 part in 3600; thus, the answer has a relative uncertainty at least of this magnitude (i.e., about 2.5 parts in 8900). The objective in a calculation is to express the answer to at least the precision of the least certain number, but to recognize the magnitude of its uncertainty. (Similarly, in making a series of measurements, one should strive to make each to about the same degree of relative uncertainty.)

If the magnitude of the answer without regard to decimal or sign is **smaller** than that of the key number, *one additional figure may be carried in the answer in order to express the minimum degree of uncertainty,* but it is written as a subscript to indicate that it is more doubtful.

EXAMPLE 3.4   Give the answer of the following operation to the maximum number of significant figures and indicate the key number.

$$\frac{42.68 \times 891}{132.6 \times 0.5247} = 546.57$$

SOLUTION
The key number is 891. Since the absolute magnitude of the answer is less than the key number, it becomes $546._6$. The last 6 is written as a subscript to indicate it is more doubtful. Again, the key number has a relative uncertainty of about 1 part in 900, so the answer has an uncertainty of at least 6 parts in 5500 (0.6 parts in 550).

In multiplication and division, the answer from each step of a series of operations can statistically be rounded to the number of significant figures to be retained in the final answer. But for consistency in the final answer, it is preferable to carry one additional figure until the end and then round off.

Addition and Subtraction

Additions and subtractions are handled in a somewhat different manner. Here, we do not have a key number, and the placing of the decimal point is important in determining how many figures will be significant. Suppose you wish to calculate the formula weight of Ag_2MoO_4 from the individual atomic weights:

Ag	107.87	0
Ag	107.87	0
Mo	95.94	
O	15.99	94
O	15.99	94
O	15.99	94
O	15.99	94
	———	——
	375.67	76

The atomic weight of molybdenum is known only to the nearest 0.01 atomic unit. Since this unit has an element of uncertainty in it, we cannot justifiably say that we know the formula weight of a compound containing molybdenum to any closer than 0.01 atomic unit. Therefore, the most accurately known value for the atomic weight of Ag_2MoO_4 is 375.68. All numbers being added or subtracted can be rounded to the least significant unit before adding or subtracting. But again, for consistency in the answer, one additional figure can be carried out and then the answer rounded to one less figure.

Summarizing the importance of significant figures, there are two questions to ask. First, how accurately do you have to *know* a particular result? If you only want to learn whether there is 12% or 13% of a substance in the sample, then you need only make all required measurements to two significant figures. If the sample weighs about 2 g, there is no need to weight it to more than 0.1 g. The second question is, how accurately can you *make* each required measurement? Obviously, if you can read the absorbance of light by a colored solution to only three figures (e.g., $A = 0.447$), it would be useless to weigh the sample to more than three figures (e.g., 6.67 g).

When a number in a measurement is small (without regard to the decimal point) compared with those of the other measurements, there is some justification in making the measurement to one additional figure. This can be visualized as follows. Suppose you wish to weigh two objects of essentially the same mass, and you wish to weigh them with the same precision, for example, to the nearest 0.1 mg, or 1 part per thousand. The first object weighs 99.8 mg, but the second weighs 100.1 mg. You have weighed both objects with equal accuracy, but you have retained an additional significant figure in one of them. This analogy can also be related to the justification for adding an additional significant figure when the answer of a mathematical operation is less than the key number.

When the key number in a series of measurements is known, then, the overall accuracy can be improved if desired either by making the key number larger (e.g., by increasing the sample size), or by making the measurement to an additional figure if possible (e.g., by weighing to one additional figure). This would be desirable when the number is small in magnitude compared to those of the other measurements (without regard to the decimal) in order to bring its uncertainty closer to that of the others.

In carrying out analytical operations, then, you should measure quantities to the same *absolute* uncertainty when adding or subtracting, and to the same *relative* uncertainty when multiplying or dividing.

If a computation involves both multiplication/division and addition/subtraction, then the individual steps must be treated separately. As good practice, retain one extra figure in the intermediate calculations until the final result (unless it drops out in a subsequent step). When a calculator is used, all digits can be kept in the calculator until the end.

EXAMPLE 3.5   Give the answer of the following computation to the maximum number of significant figures:

$$\frac{\left(\dfrac{97.7}{32.42} \times 100.0\right) + 36.04}{687}$$

SOLUTION

$$\frac{301._{36} + 36.08}{687} = \frac{337._4}{687} = 0.491_1$$

In the first operation, the key number is 97.7 and the result is $301._{36}$. We carried an additional figure until the addition step and then rounded, since the division has only three significant figures. In the division step, the key number is 687; but since the absolute magnitude of the answer is less, we carry one more figure. Note that if in the first step we had rounded to $301._4$, the numerator would have become $337._5$ and the final answer would be 0.491_3 (still within the experimental uncertainty).

Logarithms

In changing from logarithms to antilogarithms, and vice versa, the number being operated on and the logarithm mantissa have the same number of significant figures. (See Appendix B for a review of the use of logarithms.) All zeros in the mantissa are significant. Suppose, for example, we wish to calculate the pH of a 2.0×10^{-3} M solution of HCl from pH $= -\log [H^+]$. Then,

$$pH = -\log 2.0 \times 10^{-3} = -(-3 + 0.30) = 2.70$$

The -3 is the characteristic (from 10^{-3}), a pure number determined by the position of the decimal. The 0.30 is the mantissa for the logarithm of 2.0 and therefore has only two digits. So, even though we know the concentration to two figures, the pH (the logarithm) has three figures. If we wish to take the antilogarithm of a mantissa, the corresponding number will likewise have the same number of digits as the mantissa. The antilogarithm of 0.072 (contains three figures in mantissa .072) is 1.18, and the logarithm of 12.1 is 1.083 (1 is the characteristic, and the mantissa has three digits, .083).

3.2   ROUNDING OFF

If the digit following the last significant figure is greater than 5, the number is rounded up to the next higher digit. If it is less than 5, the number is rounded to the present value of the last significant figure:

$$9.47 = 9.5$$
$$9.43 = 9.4$$

If the last digit is a 5, the number is rounded off to the nearest even digit:

$$8.65 = 8.6$$
$$8.75 = 8.8$$
$$8.55 = 8.6$$

This is based on the statistical prediction that there is an equal chance that the last significant figure before the 5 will be even or odd. That is, in a suitably large sampling, there will be an equal number of even and odd digits preceding a 5.

3.3   ACCURACY AND PRECISION

Accuracy is the degree of agreement between the measured value and the true value. An absolute true value is seldom known. A more realistic definition of accuracy, then, would assume it to be the agreement between a measured value and the *accepted* true value.

We can, by good analytical technique, such as making comparisons against a known standard sample of similar composition, arrive at a reasonable assumption about the accuracy of a method, within the limitations of the knowledge of the "known" sample (and of the measurements). The accuracy to which we know the value of the standard sample is ultimately dependent on some measurement that will have a given limit of certainty in it.

Precision is defined as the degree of agreement between replicate measurements of the same quantity. That is, it is the repeatability of a result. Good precision does not assure good accuracy. This would be the case, for example, if there were a systematic error in the analysis. A weight used to measure each of the samples may be in error. This error does not affect the precision, but it does affect the accuracy. On the other hand, the precision can be relatively poor and the accuracy, more or less by chance, might be good. Since all real analyses are unknown, the higher the degree of precision, the greater the chance of obtaining the true value. It would be fruitless to hope that a value is accurate if the precision is poor, and the analytical chemist strives for repeatable results to assure the highest possible accuracy.

As we shall see later, the more measurements that are made, the more reliable will be the measure of precision. The number of measurements required will depend on the accuracy required and on the known reproducibility of the method.

3.4   DETERMINATE ERRORS

Two main classes of errors can affect the accuracy or precision of a measured quantity. **Determinate errors** are those that, as the name implies, are determinable and that presumably can be either avoided or corrected. They may be constant, as in the case of an uncalibrated weight that is used in all weighings. Or, they may be variable but of such a nature that they can be accounted for and corrected, such as a buret whose volume readings are in error by different amounts at different volumes.

The error can be proportional to sample size, or may change in a more complex manner. More often than not, the variation is unidirectional, as in the case of solubility, loss of a precipitate (negative error). It can, however, be random in sign. Such an example is the change in solution volume and concentration occurring with changes in temperature. This can be corrected for by measuring the solution temperature. Such measurable determinate errors are classed as *systematic errors*.

Some common determinate errors are:

1. Instrumental errors. These include faulty equipment, uncalibrated weights, and uncalibrated glassware.
2. Operative errors. These include personal errors and can be reduced by experience and care of the analyst in the physical manipulations involved. Operations in which these errors may occur include transfer of solutions, effervescence and "bumping" during sample dissolution, incomplete drying of samples, and so on. These are difficult to correct for. Other personal errors include mathematical errors in calculations and prejudice in estimating measurements.
3. Errors of the method. These are the most serious errors of an analysis. Most of the above errors can be minimized or corrected for, but errors that are inherent in the method cannot be changed unless the conditions of the determination are altered. Some sources of methodic errors include coprecipitation of impurities, slight solubility of a precipitate, side reactions, incomplete reactions, impurities in reagents, and so forth. Sometimes correction can be relatively simple, for example, by running a **reagent blank.** A blank determination is an analysis on the added reagents only. It is standard practice to run such blanks and to subtract the results from those for the sample. When errors become intolerable, another approach to the analysis must be made. Sometimes, however, we are forced to accept a given method in the absence of a better one.

Determinate errors may be additive or multiplicative, depending on the nature of the error or how it enters into the calculation.

3.5 INDETERMINATE ERRORS

The second class of errors includes the **indeterminate errors,** often called accidental or random errors, which represent the experimental uncertainty that occurs in any measurement. These errors are revealed by small differences in successive measurements made by the same analyst under virtually identical conditions, and they cannot be predicted or estimated. These accidental errors will follow a random distribution; therefore, mathematical laws of probability can be applied to arrive at some conclusion regarding the most probable result of a series of measurements.

It is beyond the scope of this text to go into mathematical probability, but we can say that indeterminate errors should follow a **normal distribution,** or **Gaussian curve.** Such a curve is shown in Figure 3.1. The symbol *s* represents the *standard*

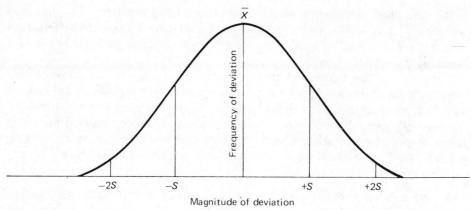

FIGURE 3.1
Normal error curve.

deviation of a population of measurements, and this measure of precision defines the spread of the normal population distribution as shown in Figure 3.1. It is apparent that there should be few very large errors and that there should be an equal number of positive and negative errors.

Indeterminate errors really originate in the limited ability of the analyst to control or make corrections for external conditions, or in his inability to recognize the appearance of factors that will result in errors. Some random errors stem from the more statistical nature of things, for example, nuclear counting errors. Sometimes, by changing conditions, some unknown error will disappear. Of course, it will be impossible to eliminate all possible random errors in an experiment, and the analyst must be content to minimize them to a tolerable or insignificant level.

3.6 WAYS OF EXPRESSING ACCURACY

There are various ways and units in which the accuracy of a measurement can be expressed, an accepted true value for comparison being assumed.

Absolute Errors

The difference between the true value and the measured value, with regard to the sign, is the **absolute error,** and it is reported in the same units as the measurement. If a 2.62-g sample of material is analyzed to be 2.52 g, the absolute error is -0.10 g. If the measured value is the average of several measurements, the error is called the **mean error.** The mean error can also be calculated by taking the average difference, with regard to sign, of the *individual* test results from the true value.

Relative Error

The absolute or mean error expressed as a percentage of the true value is the **relative error.** The above analysis has a relative error of $(-0.10/2.62) \times 100\% = -3.8\%$. The **relative accuracy** is the measured value or mean expressed as a percentage of the true value. The above analysis has a relative accuracy of $(2.52/2.62)$

× 100% = 96.2%. We should emphasize that neither number is known to be "true" and the relative error or accuracy is based on the mean of two sets of measurements.

The relative error can be expressed in units other than percentages. In very accurate work, we are usually dealing with relative errors of less than 1%, and it is convenient to use a smaller unit. A 1% error is equivalent to 1 part in 100. It is also equivalent to 10 parts in 1000. This latter unit is commonly used for expressing small uncertainties. That is, the uncertainty is expressed in **parts per thousand,** written as ppt. The number 23 expressed as parts per thousand of the number 6725 would be 23 parts per 6725, or 3.4 ppt. Parts per thousand is often used in expressing precision of measurement (see below).

EXAMPLE 3.6   The results of an analysis are 36.97 g, compared with the accepted value of 37.06 g. What is the relative error in parts per thousand?

SOLUTION

$$\text{absolute error} = 36.97 \text{ g} - 37.06 \text{ g} = -0.09 \text{ g}$$

$$\text{relative error} = \frac{-0.09}{37.06} \times 1000\text{‰} = -2.4 \text{ ppt}$$

‰ indicates parts per thousand, just as % indicates parts per hundred.

3.7 STANDARD DEVIATION

Each set of analytical results should be accompanied by an indication of the **precision** of the analysis. Various ways of indicating precision are acceptable.

The standard deviation σ of an infinite set of experimental data is theoretically given by

$$\sigma = \sqrt{\frac{\Sigma(x_i - \mu)^2}{N}} \tag{3.1}$$

where x_i represents the individual measurements and μ the mean of the infinite number of measurements (which should represent the "true" value). This equation holds strictly only as $N \rightarrow \infty$, where N is the number of measurements. In practice, we must calculate the individual deviations from the mean of a limited number of measurements, $\bar{x}$, in which it is anticipated that $\bar{x} \rightarrow \mu$, although we have no assurance this will be so; $\bar{x}$ is given by $\Sigma(x_i/N)$.

For a set of N measurements, it is possible to calculate N independently variable deviations from some reference number. But if the reference number chosen is the estimated mean, $\bar{x}$, the sum of the individual deviations (retaining signs) must necessarily add up to zero, and so values of $N - 1$ deviations are adequate to define the Nth value. That is, there are only $N - 1$ independent deviations from the mean; when $N - 1$ values have been selected, the last is predetermined. We have, in effect, used one degree of freedom of the data in calculating the mean, leaving $N - 1$ **degrees of freedom** for calculating the precision.

As a result, the **estimated standard deviation $\underline{s}$ of a finite set of experimental data** (generally $N < 30$) more nearly approximates σ if the number of degrees of freedom is substituted for N ($N - 1$ adjusts for the difference between $\bar{x}$ and μ).

$$s = \sqrt{\frac{\Sigma(x_i - \bar{x})^2}{(N - 1)}} \qquad (3.2)$$

The value of s is only an estimate of σ, then, and will more nearly approach σ as the number of measurements increases.

EXAMPLE 3.7 Calculate the mean and the standard deviation of the following set of analytical results: 15.67 g, 15.69 g, 16.03 g.

SOLUTION

x_i	$x_i - \bar{x}$	$(x_i - \bar{x})^2$
15.67	0.13	0.0169
15.69	0.11	0.0121
16.03	0.23	0.0529
Σ 47.39	Σ 0.47	Σ 0.0819

$$\bar{x} = \frac{47.39}{3} = 15.80$$

$$s = \sqrt{\frac{0.0819}{3 - 1}} = 0.20 \text{ g}$$

The standard deviation calculation considered so far is an estimate of the probable error of a single measurement. The arithmetical mean of a series of N measurements taken from an infinite population will show less scatter from the "true value" than will an individual observation. The scatter will decrease as N is increased; as N gets very large the sample average will approach the population average μ, and the scatter approaches zero. The arithmetical mean derived from N measurements can be shown to be $\sqrt{N}$ times more reliable than a single measurement. Hence, the random error in the mean of a series of four observations is one-half that of a single observation. In other words, the **precision of the mean** of a series of N measurements is inversely proportional to the square root of N of the deviation of the individual values. Thus,

$$\textit{standard deviation of the mean} = s \text{ (mean)} = \frac{s}{\sqrt{N}} \qquad (3.3)$$

The standard deviation of the mean is sometimes referred to as the *standard error*.

The standard deviation is sometimes expressed as the **relative standard deviation,** which is just the standard deviation expressed as a percentage of the mean. This term is often called the **coefficient of variation.**

EXAMPLE 3.8 The following set of weights were obtained: 29.8 mg, 30.2 mg, 28.6 mg, and 29.7 mg. Calculate the standard deviation of the individual values and the standard deviation of the mean. Express these as absolute and relative values.

SOLUTION

x_i	$x_i - \bar{x}$	$(x_i - \bar{x})^2$
29.8	0.2	0.04
30.2	0.6	0.36
28.6	1.0	1.00
29.7	0.1	0.01
Σ 118.3	Σ 1.9	Σ 1.41

$$\bar{x} = \frac{118.3}{4} = 29.6$$

$$s = \sqrt{\frac{1.41}{4-1}} = 0.69 \text{ mg (absolute)}; \quad \frac{0.69}{29.6} \times 100\% = 2.3\% \text{ (coefficient of variation)}$$

$$s \text{ (mean)} = \frac{0.69}{\sqrt{4}} = 0.34 \text{ mg (absolute)}; \quad \frac{0.34}{29.6} \times 100\% = 1.1\% \text{ (relative)}$$

The precision of a measurement can be improved by increasing the number of observations. In other words, the spread $\pm s$ of the normal curve in Figure 3.1 becomes smaller as the number of observations is increased and would approach zero as the number of observations approached infinity. However, as seen above (Equation 3.3), the deviation of the mean does not decrease in direct proportion to the number of observations, but instead it decreases as the square root of the number of observations. A point will be reached where a slight increase in precision will require an unjustifiably large increase in the number of observations. For example, to decrease the standard deviation by a factor of 10 requires 100 times as many observations.

The practical limit of useful replication is reached when the standard deviation of the random errors is comparable to the magnitude of the determinate or systematic errors (unless, of course, these can be identified and corrected for). This is because the systematic errors in a determination cannot be removed by replication.

The significance of s in relation to the normal distribution curve is shown in Figure 3.1. The mathematical treatment from which the curve was derived reveals that 68% of the individual deviations fall within one standard deviation from the mean, 95% are less than twice the standard deviation, and 99% are less than 2.5 times the standard deviation. So, 68% of the individual values will fall within the range $\bar{x} \pm s$, 95% will fall within $\bar{x} \pm 2s$, 99% will fall within $\bar{x} \pm 2.5s$, and so on.

Actually, these percentage ranges were derived assuming an infinite number of measurements. There are then two reasons why the analyst cannot be 95% certain that the true value falls within $\bar{x} \pm 2s$. First, one makes a limited number of measurements, and the fewer the measurements, the less certain one will be. Second, the normal distribution curve assumes no determinate errors, but only random errors. Determinate errors, in effect, shift the normal error curve from the true value.

It is apparent that there are a variety of ways in which the precision of a number can be reported. Whenever a number is reported as $\bar{x} \pm x$, you should always qualify under what conditions this holds, that is, how you arrived at $\pm x$. It may represent s, $2s$, s (mean), the coefficient of variation, and so forth.

A term that is sometimes useful in statistics is the **variance.** This is the square of the standard deviation, s^2. We shall use this in determining the propagation of error and in the F test below (Section 3.12).

3.8 PROPAGATION OF ERROR

When discussing significant figures earlier, we stated that the relative uncertainty in the answer to a multiplication or division operation could be no better than the relative uncertainty in the operator that had the poorest relative uncertainty. Also, the absolute uncertainty in the answer of an addition or subtraction could be no better than the absolute uncertainty in the number with the largest absolute uncertainty. Without specific knowledge of the uncertainties, we assumed an uncertainty of at least ± 1 in the last digit of each number.

From a knowledge of the uncertainties in each number, it is possible to estimate the actual uncertainty in the answer. The errors in the individual numbers will propagate throughout a series of calculations, in either a relative or absolute fashion, depending on whether the operation is a multiplication or division or whether it is an addition or subtraction.

Addition and Subtraction

Consider the addition and subtraction of the following numbers:

$$(65.06 \pm 0.07) + (16.13 \pm 0.01) - (22.68 \pm 0.02) = 58.51 \ (\pm ?)$$

The uncertainties listed represent the random or indeterminate errors associated with each number, expressed as standard deviations of the numbers. The maximum error of the summation, expressed as a standard deviation, would be ± 0.10, that is, it could be either $+0.10$ or -0.10 if all uncertainties happened to have the same sign. The minimum uncertainty would be 0.00 if all combined by chance to cancel. Both of these extremes are not highly likely, and statistically the uncertainty will fall somewhere in between. For addition and subtraction, *absolute uncertainties* are additive. The most probable error is represented by the square root of the sum of the *absolute variances*. That is, the absolute variance of the answer is the sum of the individual variances. For $a = b + c - d$,

$$s_a^2 = s_b^2 + s_c^2 + s_d^2 \tag{3.4}$$

$$s_a = \sqrt{s_b^2 + s_c^2 + s_d^2} \tag{3.5}$$

In the above example,

$$s_a = \sqrt{(\pm 0.07)^2 + (\pm 0.01)^2 + (\pm 0.02)^2}$$

$$= \sqrt{(\pm 49 \times 10^{-4}) + (\pm 1 \times 10^{-4}) + (\pm 4 \times 10^{-4})}$$

$$= \sqrt{\pm 54 \times 10^{-4}} = \pm 7.3 \times 10^{-2}$$

So the answer is 58.51 ± 0.07. The number ±0.07 represents the absolute uncertainty. If we wish to express it as a relative uncertainty, this would be

$$\frac{\pm 0.07}{58.51} \times 100\% = \pm 0.1_2\%$$

EXAMPLE 3.9 You have received three shipments of uranium ore of equal weight. Analysis of the three ores indicated contents of 3.978 ± 0.004%, 2.537 ± 0.003%, and 3.680 ± 0.003%, respectively. What is the average uranium content of the ores and what are the absolute and relative uncertainties?

SOLUTION

$$\bar{x} = \frac{3.978 \pm 0.004\%) + (2.536 \pm 0.003\%) + (3.680 \pm 0.003\%}{3}$$

The uncertainty in the summation is

$$s_a = \sqrt{(\pm 0.004)^2 + (\pm 0.003)^2 + (\pm 0.003)^2}$$

$$= \sqrt{(\pm 16 \times 10^{-6}) + (\pm 9 \times 10^{-6}) + (\pm 9 \times 10^{-6})}$$

$$= \sqrt{\pm 34 \times 10^{-6}} = \pm 5.8 \times 10^{-3}\%$$

Hence,

$$x = \frac{10.194}{3} \pm 0.006\% = 3.398 \pm 0.006\%$$

Note that since there is no uncertainty in the divisor 3, the absolute uncertainty in the mean is the same as for the sum of the individual numbers. If we had divided the uncertainty by 3, it would have been less than in any of the individual measurements, an impossibility.

Multiplication and Division

Consider the following operation:

$$\frac{(13.67 \pm 0.02)(120.4 \pm 0.2)}{4.623 \pm 0.006} = 356.0 \ (\pm ?)$$

Here, the *relative uncertainties* are additive, and the most probable error is represented by the square root of the sum of the relative variances. That is, the relative variance of the answer is the sum of the individual relative variances.

For $a = bc/d$,

$$(s_a^2)_{rel} = (s_b^2)_{rel} + (s_c^2)_{rel} + (s_d^2)_{rel} \tag{3.5}$$

$$(s_a)_{rel} = \sqrt{(s_b^2)_{rel} + (s_c^2)_{rel} + (s_d^2)_{rel}} \tag{3.6}$$

In the above example,

$$(s_b)_{rel} = \frac{\pm 0.02}{13.67} = \pm 0.0015$$

$$(s_c)_{rel} = \frac{\pm 0.2}{120.4} = \pm 0.0017$$

$$(s_d)_{rel} = \frac{\pm 0.006}{4.623} = \pm 0.0013$$

$$(s_a)_{rel} = \sqrt{(\pm 0.0015)^2 + (\pm 0.0017)^2 + (\pm 0.0013)^2}$$

$$= \sqrt{(\pm 2.2 \times 10^{-6}) + (\pm 2.9 \times 10^{-6}) + (\pm 1.7 \times 10^{-6})}$$

$$= \sqrt{(\pm 6.8 \times 10^{-6})} = \pm 2.6 \times 10^{-3}$$

The absolute uncertainty is given by

$$s_a = 356.0 \times (\pm 2.6 \times 10^{-3}) = \pm 0.93$$

So the answer is 356.0 ± 0.9.

EXAMPLE 3.10 Calculate the uncertainty in the number of millimoles of chloride contained in 250.0 mL of a sample when three equal aliquots of 25.00 mL are titrated with silver nitrate with the following results: 36.78 mL, 36.82 mL, 36.75 mL. The molarity of the $AgNO_3$ solution is 0.1167 ± 0.0002 M.

SOLUTION
The mean volume is

$$\frac{36.78 + 36.82 + 36.75}{3} = 36.78 \text{ mL}$$

The standard deviation is

x_i	$x_i - \bar{x}$	$(x_i - \bar{x})^2$
36.78	0.00	0.0000
36.82	0.04	0.0016
36.75	0.03	0.0009
		Σ 0.0025

$$s = \sqrt{\frac{0.0025}{3 - 1}} = 0.035; \text{ mean volume} = 36.78 \pm 0.04 \text{ mL}$$

mmol Cl^- titrated $= (0.1167 \pm 0.0002 \text{ mmol/mL})(36.78 \pm 0.04 \text{ mL}) = 4.292 \, (\pm?)$

$$(s_b)_{rel} = \frac{\pm 0.0002}{0.1167} = \pm 0.0017$$

$$(s_c)_{rel} = \frac{\pm 0.035}{36.78} = \pm 0.00095$$

$$(s_a)_{rel} = \sqrt{(\pm 0.0017)^2 + (\pm 0.00095)^2}$$

$$= \sqrt{(\pm 2.9 \times 10^{-6}) + (\pm 0.90 \times 10^{-6})}$$

$$= \sqrt{\pm 3.8 \times 10^{-6}}$$

$$= \pm 1.9 \times 10^{-3}$$

The absolute uncertainty in the millimoles of Cl^- is

$$4.292 \times (\pm 0.0019) = \pm 0.0082$$

$$\text{mmol } Cl^- \text{ in 25 mL} = 4.292 \pm 0.0082 \text{ mmol}$$

$$\text{mmol } Cl^- \text{ in 250 mL} = 10(4.292 \pm 0.0082) = 42.92 \pm 0.08 \text{ mmol}$$

Note that we retained one extra figure in computations until the final answer. Here, the absolute uncertainty determined is proportional to the size of the sample; it would not remain constant for twice the sample size, for example.

If there is a combination of multiplication/division and addition/subtraction in a calculation, the uncertainties of these must be combined.

EXAMPLE 3.11 You have received three shipments of iron ore of the following weights: 2852 lb, 1578 lb, and 1877 lb. There is an uncertainty in the weights of ± 5 lb. Analysis of the ores gives $36.28 \pm 0.04\%$, $22.68 \pm 0.03\%$, and $49.23 \pm 0.06\%$, respectively. You are to pay \$300 per ton of iron. What should you pay for these three shipments and what is the uncertainty in the payment?

SOLUTION
We need to calculate the weight of iron in each shipment, with the uncertainties, and then add these together to obtain the total weight of iron and the uncertainty in this. The relative uncertainties in the weights are

$$\frac{\pm 5}{2852} = \pm 0.0017; \quad \frac{\pm 5}{1578} = \pm 0.0032; \quad \frac{\pm 5}{1877} = \pm 0.0027$$

The relative uncertainties in the analyses are

$$\frac{\pm 0.04}{36.28} = \pm 0.0011; \quad \frac{\pm 0.03}{22.68} = \pm 0.0013; \quad \frac{\pm 0.06}{49.23} = \pm 0.0012$$

The weight of iron in each shipment is

$$\frac{(2852 \pm 5 \text{ lb})(36.28 \pm 0.04\%)}{100} = 1,034.7 \, (\pm ?) \text{ lb Fe}$$

$$(s_a)_{rel} = \sqrt{(\pm 0.0017)^2 + (\pm 0.0011)} = \pm 0.0020$$

$$s_a = 1034.7 \times (\pm 0.0020) = \pm 2.1 \text{ lb}$$

$$\text{lb Fe} = 1034.7 \pm 2.1$$

(We will carry an additional figure throughout.)

$$\frac{(1578 \pm 5 \text{ lb})(22.68 \pm 0.03\%)}{100} = 357.89 \, (\pm ?) \text{ lb Fe}$$

$$(s_a)_{rel} = \sqrt{(\pm 0.0032)^2 + (\pm 0.0013)^2} = \pm 0.0034$$

$$s_a = 357.89 \times (\pm 0.0034) = \pm 1.2 \text{ lb}$$

$$\text{lb Fe} = 357.9 \pm 1.2 \text{ lb}$$

$$\frac{(1877 \pm 5 \text{ lb.})(49.23 \pm 0.06\%)}{100} = 924.05 \, (\pm ?) \text{ lb Fe}$$

$$(s_a)_{rel} = \sqrt{(\pm 0.0027)^2 + (\pm 0.0012)^2} = \pm 0.0030$$

$$s_a = 924.05 \times (\pm 0.0030) = \pm 2.8 \text{ lb}$$
$$\text{lb Fe} = 924.0 \pm 2.8 \text{ lb}$$

$$\text{total Fe} = (1034.7 \pm 2.1 \text{ lb}) + (357.9 \pm 1.2 \text{ lb}) + (924.0 \pm 2.8 \text{ lb})$$
$$= 2{,}316.6 \ (\pm ?) \text{ lb}$$
$$s_a = \sqrt{(\pm 2.1)^2 + (\pm 1.2)^2 + (\pm 2.8)^2} = \pm 3.7 \text{ lb}$$
$$\text{total Fe} = 2317 \pm 4 \text{ lb.}$$
$$\text{price} = (2316.6 \pm 3.7 \text{ lb})(\$0.15/\text{lb}) = \$347.49 \pm 0.56$$

Hence, you should pay 347.50 ± 0.60.

Exponents

Suppose a number is raised to a certain power, for example, $(6.8 \pm 0.3)^2 = 46.2$ $(\pm ?)$. In this case, the *relative error* in the answer is changed by the multiple of the exponent. So if the number is squared, the relative error doubles. If the square root is taken, the relative error is halved, and so forth. For the general case, $a = b^c$,

$$\boxed{(s_a)_{\text{rel}} = c(s_b)_{\text{rel}}} \tag{3.7}$$

This is different from the previous cases because there is no possibility of the error canceling. That is, in multiplying a number by itself $(b \times b)$, the sign of the uncertainty is the same in each, and the relative uncertainty necessarily doubles. In the above example, the relative error in 6.8 is $\pm 0.3/6.8 = \pm 0.044$; so the relative error in the answer is ± 0.088, the absolute error is $(46.2)(\pm 0.088) = \pm 4.1$, and the answer is 46 ± 4.

EXAMPLE 3.12 The radius of a circle is determined to be 34.6 ± 0.3 cm. What is its area?

SOLUTION

$$A = \pi R^2 = \pi (34.6 \pm 0.3)^2$$
$$= 3.1416[1197 \ (\pm ?)] = 3761 \ (\pm ?)$$

The relative error is

$$\frac{\pm 0.3}{34.6} = \pm 0.0087$$

$$(s_a)_{\text{rel}} = 2(\pm 0.0087) = \pm 0.017$$
$$s_a = 3761 \times (\pm 0.017) = \pm 64$$
$$A = 3761 \pm 64 = 3760 \pm 60 \text{ cm}^2 = (3.76 \pm 0.06) \times 10^3 \text{ cm}^2$$

(The uncertainty resides in this position, so we can't know the answer better than this. More on this below.)

Logarithms

For the general expression

$$b = 10^a$$
$$a = \log b = 0.434 \ln b$$

it can be shown by calculus that the *absolute uncertainty* in a is equal to the *relative uncertainty* in b multiplied by 0.434:

$$s_a = 0.434(s_b)_{rel}$$

(3.8)

In other words, the absolute error in a is proportional to the relative error in b.

EXAMPLE 3.13   The hydrogen ion concentration of a solution is determined to be $(3.7 \pm 0.2) \times 10^{-3}$ M. What is the pH?

SOLUTION

$$pH = -\log [H^+] = -\log[(3.7 \pm 0.2) \times 10^{-3}] = 2.432 \, (\pm ?)$$

The relative uncertainty in $[H^+]$ is

$$\frac{\pm 0.2 \times 10^{-3}}{3.7 \times 10^{-3}} = \pm 0.054$$

$$s_a = 0.434(\pm 0.054) = \pm 0.023$$

$$pH = 2.43 \pm 0.02$$

So in pH calculations, the absolute error in pH is proportional to the relative error in $[H^+]$.

EXAMPLE 3.14   The pH of a solution if 8.34 ± 0.03. What is the hydrogen ion concentration?

SOLUTION

$$[H^+] = 10^{-pH} = 10^{-8.34 \pm 0.03} = 10^{0.66 \pm 0.03} \times 10^{-9}$$
$$= (4.57 \, (\pm ?) \times 10^{-9} \, M$$
$$b = 10^a$$

From Equation 3.8,

$$(s_b)_{rel} = \frac{s_a}{0.434} = \frac{\pm 0.03}{0.434} = \pm 0.069$$

$$s_b = 4.57 \times (\pm 0.069) = \pm 0.32$$

$$[H^+] = (4.6 \pm 0.3) \times 10^{-9} \, M$$

Hence, in determining hydrogen ion concentration from pH measurement, the relative error will be proportional to the absolute error in pH.

3.9   SIGNIFICANT FIGURES AND PROPAGATION OF ERROR

We noted earlier that the total uncertainty in a computation determines how accurately we can know the answer. In other words, the uncertainty sets the number of significant figures. Take the following example:

$$(73.1 \pm 0.2)(2.245 \pm 0.008) = 164.1 \pm 0.7$$

We are justified in keeping four figures, even though the key number has three. Here, we don't have to carry the additional figure as a subscript, since we have indicated the actual uncertainty in it. Note that the greatest relative uncertainty in the multipliers is 0.003, while that in the answer is 0.004; so, due to the propagation of error, we know the answer less accurately than the key number. The key number (the one with the greatest uncertainty), when actual uncertainties are known, may not necessarily be the one with the smallest number of digits. For example, the relative uncertainty in 78.1 ± 0.2 is 0.003, while that in 11.21 ± 0.08 is 0.007.

Suppose we have the following calculation:

$$(73.1 \pm 0.9)(2.245 \pm 0.008) = 164.1 \pm 2.1 = 164 \pm 2$$

Now the uncertainty in the answer is in the units place, and so figures beyond that are meaningless. In this instance, the uncertainty in the key number and the answer are similar (± 0.012), since the uncertainty in the other multiplier is significantly smaller.

EXAMPLE 3.15 Provide the answers to the following calculations to the proper number of significant figures:

(a) $$(38.68 \pm 0.07) - (6.16 \pm 0.09) = 32.52$$

(b) $$\frac{(12.18 \pm 0.08)(23.04 \pm 0.07)}{3.247 \pm 0.006} = 86.43$$

SOLUTION
(a) The calculated absolute uncertainty in the answer is ± 0.11. Therefore, the answer is 32.50 ± 0.1.
(b) The calculated relative uncertainty in the answer is 0.0075, so the absolute uncertainty is $0.0075 \times 86.43 = 0.65$. Therefore, the answer is 86.4 ± 0.6, even though we know all the other numbers to four figures; there is substantial uncertainty in the fourth digits, which leads to the uncertainty in the answer. The relative uncertainty in the answer is 0.007, and the largest relative uncertainty in the other numbers is 0.0066, very similar.

3.10 CONTROL CHARTS

A **quality control chart** is a time plot of a measured quantity that is assumed to be constant (with a Gaussian distribution) for the purpose of ascertaining that the measurement remains within a statistically acceptable range. It may be a day-to-day plot of the measured value of a standard that is run intermittently with samples. The control chart consists of a central line representing the known or assumed value of the control and either one or two pairs of limit lines, the **inner** and **outer control limits.** Usually the standard deviation of the procedure is known (a good estimate of σ), and this is used to establish the control limits.

An example of a control chart is illustrated in Figure 3.2, representing a plot of day-to-day results of the analysis of a serum calcium pooled or control sample that is run randomly and blindly with samples each day. A useful inner control limit is two standard deviations since there is only 1 chance in 20 that an individual mea-

FIGURE 3.2
Typical quality control chart.

surement will exceed this purely by chance. This might represent a warning limit. The outer limit might be 2.5 or 3 σ, in which case there is only 1 chance in 100 or 1 chance in 500 a measurement will fall outside this range in the absence of systematic error. Usually, one control is run with each batch of samples (e.g., 20 samples), so several control points may be obtained each day. The mean of these may be plotted each day. The random scatter of this would be expected to be smaller by $\sqrt{N}$, compared to individual points.

Particular attention should be paid to trends in one direction; that is, the points lie largely on one side of the central line. This would suggest that either the control is in error or there is a systematic error in the measurement. A tendency for points to lie outside the control limits would indicate the presence of one or more determinate errors in the determination, and the analyst should check for deterioration of reagents, instrument malfunction, environmental effects and so forth. Trends should signal contamination of reagents, improper calibration or erroneous standards, or change in the control lot.

3.11 THE CONFIDENCE LIMIT

Calculation of the standard deviation for a set of data provides an indication of the precision inherent in a particular procedure or analysis. But unless there is a large number of data, it does not by itself give any information about how close the experimentally determined mean $\bar{x}$ might be to the true mean value μ. Statistical theory, though, allows us to estimate the range within which the true value might fall, within a given probability, defined by the experimental mean and the standard deviation. This range is called the **confidence interval,** and the limits of this range are called the **confidence limit.** The likelihood that the true value falls within the range is called the **probability,** or **confidence level,** usually expressed as a percent. The confidence limit is given by

$$\text{confidence limit} = \bar{x} \pm \frac{ts}{\sqrt{N}} \qquad (3.9)$$

where t is a statistical factor that depends on the number of degrees of freedom and the confidence level desired. Values of t at different confidence levels and degrees of freedom v are given in Table 3.1. Note that the confidence limit is simply the product of t and the standard deviation of the mean $(s/\sqrt{N})$. **(The confidence limit for a single observation is given by $\bar{x} \pm ts$.)**

TABLE 3.1
Values of t for v Degrees of Freedom for Various Confidence Levels[a]

v	Confidence Level, 90%	95%	99%	99.5%
1	6.314	12.706	63.657	127.32
2	2.920	4.303	9.925	14.089
3	2.353	3.182	5.841	7.453
4	2.132	2.776	4.604	5.598
5	2.015	2.571	4.032	4.773
6	1.943	2.447	3.707	4.317
7	1.895	2.365	3.500	4.029
8	1.860	2.306	3.355	3.832
9	1.833	2.262	3.250	3.690
10	1.812	2.228	3.169	3.581
15	1.753	2.131	2.947	3.252
20	1.725	2.086	2.845	3.153
25	1.708	2.060	2.787	3.078
∞	1.645	1.960	2.576	2.807

[a] $v = N - 1$ = degrees of freedom.

EXAMPLE 3.16   A soda ash sample is analyzed in the analytical chemistry laboratory by titration with standard hydrochloric acid. The analysis is performed in triplicate with the following results: 93.50, 93.58, and 93.43% Na_2CO_3. Within what range are you 95% confident that the true value lies?

SOLUTION
The mean is 93.50%. The standard deviation s is calculated to be 0.075% (absolute). At the 95% confidence level and two degrees of freedom, $t = 4.303$ and

$$\text{confidence limit} = \bar{x} \pm \frac{ts}{\sqrt{N}}$$

$$= 93.50 \pm \frac{4.303 \times 0.075}{\sqrt{3}}$$

$$= 93.50 \pm 0.19\%$$

So you are 95% confident that, in the absence of a determinate error, the true value falls within 93.31–93.69%.

As the number of measurements increases, both t and $s/\sqrt{N}$ decrease, with the result that the confidence interval is narrowed. So the more measurements you make, the more confident you will be that the true value lies within a given range or, conversely, that the range will be narrowed at a given confidence level. However, t decreases exponentially with an increase in N, just as the standard deviation of the mean does (see Table 3.1), so a point of diminishing returns is eventually reached in which the increase in confidence is not justified by the increase in the multiple of samples analyses required.

3.12   TESTS OF SIGNIFICANCE

In developing a new analytical method, it is often desirable to compare the results of that method with those of an accepted (perhaps standard) method. How, though, can one tell if there is a significant difference between the new method and the accepted one? Again, we resort to statistics for the answer.

The *F* Test

This is a test designed to indicate whether there is a significant difference between two methods based on their standard deviations. *F* is defined in terms of the variances of the two methods, where the **variance** is the square of the standard deviation:

$$F = \frac{s_1^2}{s_2^2}$$

(3.10)

where $s_1^2 > s_2^2$. There are two different degrees of freedom, ν_1 and ν_2, where degrees of freedom is defined as $N - 1$ for each case.

If the calculated *F* value from Equation 3.10 exceeds a tabulated *F* value at the selected confidence level, then there is a significant difference between the variances of the two methods. A list of *F* values at the 95% confidence level is given in Table 3.2.

TABLE 3.2
Values of *F* at the 95% Confidence Level

$\nu_1 = 2$	3	4	5	6	7	8	9	10	15	20	30
$\nu_2 = 2$ 19.0	19.2	19.2	19.3	19.3	19.4	19.4	19.4	19.4	19.4	19.4	19.5
3 9.55	9.28	9.12	9.01	8.94	8.89	8.85	8.81	8.79	8.70	8.66	8.62
4 6.94	6.59	6.39	6.26	6.16	6.09	6.04	6.00	5.96	5.86	5.80	5.75
5 5.79	5.41	5.19	5.05	4.95	4.88	4.82	4.77	4.74	4.62	4.56	4.50
6 5.14	4.76	4.53	4.39	4.28	4.21	4.15	4.10	4.06	3.94	3.87	3.81
7 4.74	4.35	4.12	3.97	3.87	3.79	3.73	3.68	3.64	3.51	3.44	3.38
8 4.46	4.07	3.84	3.69	3.58	3.50	3.44	3.39	3.35	3.22	3.15	3.08
9 4.26	3.86	3.63	3.48	3.37	3.29	3.23	3.18	3.14	3.01	2.94	2.86
10 4.10	3.71	3.48	3.33	3.22	3.14	3.07	3.02	2.98	2.85	2.77	2.70
15 3.68	3.29	3.06	2.90	2.79	2.71	2.64	2.59	2.54	2.40	2.33	2.25
20 3.49	3.10	2.87	2.71	2.60	2.51	2.45	2.39	2.35	2.20	2.12	2.04
30 3.32	2.92	2.69	2.53	2.42	2.33	2.27	2.21	2.16	2.01	1.93	1.84

EXAMPLE 3.17   You are developing a new colorimetric procedure for determining the glucose content of blood serum. You have chosen the standard Folin-Wu procedure with which to compare your results. From the following two sets of replicate analyses on the same sample, determine whether the variance of your method differs significantly from that of the standard method.

	Your Method, mg/dL		Folin-Wu Method, mg/dL
	127		130
	125		128
	123		131
	130		129
	131		127
	126		125
	129		
mean ($\bar{x}_1$)	127	mean ($\bar{x}_2$)	128

SOLUTION

$$s_1^2 = \frac{\Sigma(x_{i_1} - \bar{x}_1)^2}{N_1 - 1} = \frac{50}{7 - 1} = 8.3$$

$$s_2^2 = \frac{\Sigma(x_{i_2} - \bar{x}_2)^2}{N_2 - 1} = \frac{24}{6 - 1} = 4.8$$

$$F = \frac{8.3}{4.8} = 1.7_3$$

The variances are arranged so that the F value is > 1. The tabulated F value for $\nu_1 = 6$ and $\nu_2 = 5$ is 4.95. Since the calculated value is less than this, we conclude that there is no significant difference in the precision of the two methods, or that the standard deviations are from random error alone and don't depend on the sample.

The Student t Test

Frequently, the analyst wishes to decide whether there is a statistical difference between the results obtained using two different procedures, that is, whether they both indeed measure the same thing. The t test is very useful for such comparisons.

In this method, comparison is made between two sets of replicate measurements made by two different methods; one of them will be the test method, and the other will be an accepted method. A statistical t value is calculated and compared with a tabulated value for the given number of tests at the desired confidence level (Table 3.1). If the calculated t value *exceeds* the tabulated t value, then there is a significant difference between the results by the two methods at that confidence level. If it does not exceed the tabulated value, then we can predict that there is no significant difference between the methods. This in no way implies that the two results are identical.

Three ways in which a t test can be used will be described. (See Reference 2 for a discussion of different t tests.) If an accepted value of μ is available (from other measurements), then the test can be used to determine if a particular analysis method gives results statistically equal to μ at a given confidence level. If an accepted value is not available, then a series of replicate analyses on a single sample may be performed using two methods, or a series of analyses may be performed on a set of different samples by the two methods. One method should be an accepted method. We will describe these various uses of the t test.

1. t TEST WHEN AN ACCEPTED VALUE IS KNOWN. Note that Equation 3.9 is a representation of the true value μ. We can write that

$$\mu = \bar{x} \pm \frac{ts}{\sqrt{N}} \qquad (3.11)$$

It follows that

$$\pm t = (\bar{x} - \mu)\frac{\sqrt{N}}{s} \qquad (3.12)$$

If a good estimate of the "true" value is available from other analyses, for example, from a National Bureau of Standards standard reference material (or the ultimate in chemical analysis, an atomic weight), then Equation 3.12 can be used to determine whether the value obtained from a test method is statistically equal to the accepted value.

EXAMPLE 3.18   You are developing a procedure for determining traces of copper in biological materials using a wet digestion followed by measurement by atomic absorption spectrophotometry. In order to test the validity of the method, you obtain an NBS orchard leaves standard reference material and analyze this material. Five replicas are sampled and analyzed, and the mean of the results is found to be 10.8 ppm with a standard deviation of ±0.7 ppm. The listed value is 11.7 ppm. Does your method give a statistically correct value at the 95% confidence level?

SOLUTION

$$\pm t = (\bar{x} - \mu)\frac{\sqrt{N}}{s}$$

$$= (10.8 - 11.7)\frac{\sqrt{5}}{0.7}$$

$$= 2._9$$

There are five measurements, so there are four degrees of freedom ($N - 1$). From Table 3.1, we see that the tabulated value of t at the 95% confidence level is 2.776. This is *less* than the calculated value, so there is a determinate error in the new procedure. That is, there is a 95% probability that the difference between the reference value and the measured value is not due to chance.

Note from Equation 3.12 that as the precision is improved, that is, as s becomes smaller, the calculated t becomes larger. Thus, there is a greater chance that the tabulated t value will be less than this. That is, as the precision improves, it is easier to distinguish nonrandom differences. Looking again at Equation 3.12, this means as s decreases, so must the difference between the two methods ($\bar{x} - \mu$) in order for the difference to be ascribed only to random error.

2. PAIRED t TEST.   When the t test is applied to two sets of data, μ in Equation 3.12 is replaced by the mean of the second set. The reciprocal of the standard deviation of the mean ($\sqrt{N}/s$) is replaced by that of the differences between the two, which is readily shown to be

$$\sqrt{\frac{N_1 N_2}{N_1 + N_2}}/s_p$$

where s_p is the pooled standard deviation of the individual measurements of two sets:

$$\pm t = \frac{\bar{x}_1 - \bar{x}_2}{s_p} \sqrt{\frac{N_1 N_2}{N_1 + N_2}} \qquad (3.13)$$

The pooled standard deviation, defined below, is sometimes used to obtain an improved estimate of the precision of a method, and it is used for calculating the precision of the two sets of data in a paired t test. That is, rather than relying on a single set of data to describe the precision of a method, it is sometimes preferable to perform several sets of analyses, for example, on different days, or on different samples with slightly different compositions. If the indeterminate (random) error is assumed to be the same for each set, then the precision data of the different sets can be pooled. This provides a more reliable estimate of the precision of a method than is obtained from a single set. The **pooled standard deviation** s_p is given by

$$s_p = \sqrt{\frac{\Sigma(x_{i1} - \bar{x}_1)^2 + \Sigma(x_{i2} - \bar{x}_2)^2 + \cdots + \Sigma(x_{ik} - \bar{x}_k)^2}{N - k}} \qquad (3.14)$$

where $\bar{x}_1, \bar{x}_2, \ldots, \bar{x}_k$ are the means of each of k sets of analyses, and $x_{i1}, x_{i2}, \ldots, x_{ik}$ are the individual values in each set. N is the total number of measurements and is equal to $(N_1 + N_2 + \cdots + N_k)$. If five sets of 20 analyses each are performed, $k = 5$ and $N = 100$. (The number of samples in each set need not be equal.) $N - k$ is the degrees of freedom obtained from $(N_1 - 1) + (N_2 - 1) + \cdots + (N_k - 1)$; one degree of freedom is lost for each subject. This equation represents a combination of the equations for the standard deviations of each set of data.

In applying the t test between two methods, it is assumed that both methods have essentially the same standard deviation, that is, each represents the precision of the population (the same σ). This can be verified using the F test above.

EXAMPLE 3.19 A new gravimetric method is developed for iron (III) in which the iron is precipitated in crystalline form with an organoboron "cage" compound. The accuracy of the method is checked by analyzing the iron in an ore sample and comparing with the results using the standard precipitation with ammonia and weighing of Fe_2O_3. The results, reported as % Fe for each analysis, were as follows.

Test Method	Reference Method
20.10%	18.89%
20.50	19.20
18.65	19.00
19.25	19.70
19.40	19.40
19.99	
	$\bar{x}_2 = 19.24\%$
$\bar{x}_1 = 19.65\%$	

Is there a significant difference between the two methods?

SOLUTION

x_{i1}	$x_{i1} - \bar{x}_1$	$(x_{i1} - \bar{x}_1)^2$	x_{i2}	$x_{i2} - \bar{x}_2$	$(x_{i2} - \bar{x}_2)^2$
20.10	0.45	0.202	18.89	0.35	0.122
20.50	0.85	0.722	19.20	0.04	0.002
18.65	1.00	1.000	19.00	0.24	0.058
19.25	0.40	0.160	19.70	0.46	0.212
19.40	0.25	0.062	19.40	0.16	0.026
19.99	0.34	0.116			

$$\Sigma(x_{i2} - \bar{x}_2)^2 = 0.420$$

$$\Sigma(x_{i1} - \bar{x}_1)^2 = 2.262$$

$$F = \frac{s_1^2}{s_2^2} = \frac{2.262/5}{0.420/4} = 4.31$$

This is less than the tabulated value (6.26), so the two methods have comparable standard deviations and the t test can be applied.

$$s_p = \sqrt{\frac{\Sigma(x_{i1} - \bar{x}_i)^2 + \Sigma(x_{i2} - \bar{x}_2)^2}{N_1 + N_2 - 2}}$$

$$= \frac{2.262 + 0.420}{(6 + 5 - 2)} = 0.546$$

$$\pm t = \frac{19.65 - 19.24}{0.546}\sqrt{\frac{(6)(5)}{(6 + 5)}} = 1.2_3$$

The tabulated t for nine degrees of freedom ($N_1 + N_2 - 2$) at the 95% confidence level is 2.262, so there is no statistical difference in the results by the two methods.

Rather than comparing two methods using one sample, two samples could be compared for comparability using a single analysis method in a manner identical to the above examples.

3. t TEST WITH MULTIPLE SAMPLES. In the clinical laboratory, a new method is frequently tested against an accepted method by analyzing several different samples of slightly varying composition (within physiological range). In this case, the t value is calculated in a slightly different form. The difference between each of the paired measurements on each sample is computed. An average difference $\bar{D}$ is calculated and the individual deviations of each from $\bar{D}$ are used to compute a standard deviation, s_d. The t value is calculated from

$$t = \frac{\bar{D}}{s_d}\sqrt{N} \tag{3.15}$$

$$s_d = \sqrt{\frac{\Sigma(D_i - \bar{D})^2}{N - 1}} \tag{3.16}$$

where D_i is the individual difference between the two methods for each sample, with regard to sign; and $\bar{D}$ is the mean of all the individual differences.

EXAMPLE 3.20 You are developing a new analytical method for the determination of blood urea nitrogen (BUN). You want to determine whether your method differs significantly from a standard one for analyzing a range of sample concentrations expected to be found in the routine laboratory. It has been ascertained that the two methods have comparable precisions. Following are two sets of results for a number of individual samples.

Sample	Your Method, mg/dL	Standard Method, mg/dL	D_i	$D_i - \overline{D}$	$(D_i - \overline{D})^2$
A	10.2	10.5	−0.3	−0.6	0.36
B	12.7	11.9	0.8	0.5	0.25
C	8.6	8.7	−0.1	−0.4	0.16
D	17.5	16.9	0.6	0.3	0.09
E	11.2	10.9	0.3	0.0	0.00
F	11.5	11.1	0.4	0.1	0.01
			Σ 1.7		Σ 0.87
			$\overline{D} =$ 0.28		

SOLUTION

$$s_d = \sqrt{\frac{0.87}{6-1}} = 0.42$$

$$t = \frac{0.28}{0.42} \times \sqrt{6} = 1.6_3$$

The tabulated t value at the 95% confidence level for five degrees of freedom is 2.571. Therefore, $t_{calc} < t_{table}$, and there is no significant difference between the two methods at this confidence level.

Usually, a test at the 95% confidence level is considered significant, while one at the 99% level is highly significant. That is, the smaller the calculated t value, the more confident you are that there is no significant difference between the two methods. If you employ too low a confidence level (e.g., 80%), you are likely to conclude erroneously that there is a significant difference between two methods (type I error). On the other hand, too high a confidence level will require too large a difference to detect (type II error). If a calculated t value is near the tabular value at the 95% confidence level, more tests should be run to ascertain definitely whether the two methods are significantly different.

3.13 REJECTION OF A RESULT

Frequently, when a series of replicate analyses is performed, one of the results will appear to differ markedly from the others. A decision will have to be made whether to reject the result or to retain it. Unfortunately, there are no uniform criteria that can be used to decide if a suspect result can be ascribed to accidental error rather than chance variation. The only reliable basis for rejection occurs when it can be decided that some specific error may have been made in obtaining the doubtful result. No result should be retained in cases where a known error has occurred in its collection.

Experience and common sense may serve as just as practical a basis for judging the validity of a particular observation as a statistical test would be. Frequently, the experienced analyst will gain a good idea of the precision to be expected in a particular method and will recognize when a particular result is suspect.

Frequently, an analyst who knows the standard deviation expected of a method will reject a data point that falls outside $2s$ or $2.5s$ of the mean, because there is about 1 chance in 20 or 1 chance in 100 this will occur.

A wide variety of statistical tests have been suggested and used to determine whether an observation should be rejected. In all of these, a range is established within which statistically significant observations should fall. The difficulty with all of them is determining what the range should be. If it is too small, then perfectly good data will be rejected; and if it is too large, then erroneous measurements will be retained too high a proportion of the time. The **Q test** is, among the several suggested tests, one of the most statistically correct for a fairly small number of observations and is recommended when a test is necessary. The ratio Q is calculated by arranging the data in decreasing order of numbers. The difference between the suspect number and its nearest neighbor is divided by the range, that is, the difference between the highest number and the lowest number. Referring to Figure 3.3, $Q = a/w$. This ratio is compared with tabulated values of Q. If it is equal to or greater than the tabulated value, the suspected observation can be rejected. The tabulated values of Q at the 90% confidence level are given in Table 3.3. If Q exceeds the tabulated value for a given number of observations, the questionable measurement may be rejected with 90% confidence that some definite error is in this measurement.

EXAMPLE 3.21   The following set of chloride analyses on separate aliquots of a pooled serum were reported: 103, 106, 107, and 114 meq/L. One value appears suspect. Determine if it can be ascribed to accidental error.

SOLUTION
The suspect result is 114 meq/L. It differs from its nearest neighbor, 107 meq/L, by 7 meq/L. The range is 114 to 103, or 11 meq/L. Q is therefore $7/11 = 0.64$. The tabulated value for four observations is 0.76. Since the calculated Q is less than the tabulated Q, the suspected number may not be rejected.

For a small number of measurements (e.g., three to five), the discrepancy of the measurement must be quite large before it can be rejected by this criterion, and it is likely that erroneous results may be retained. This would cause a significant

$$Q = \frac{a}{w}$$

FIGURE 3.3
Illustration of the calculation of Q.

TABLE 3.3
**Rejection Quotient, Q at 90 Percent
Confidence Limit[a]**

Number of Observations	Q
3	0.94
4	0.76
5	0.64
6	0.56
7	0.51
8	0.47
9	0.44
10	0.41
∞	0.00

[a]Adapted from R. B. Dean and W. J. Dixon,
Anal. Chem., **23** (1951) 636.

change in the arithmetic mean, because the mean is greatly influenced by a discordant value. For this reason, it has been suggested that the median rather than the mean be reported when a discordant number cannot be rejected from a small number of measurements. The **median** is the middle result of an odd number of results, or the average of the central pair for an even number, when they are arranged in order of magnitude. The median has the advantage of not being unduly influenced by an outlying value. In the above example, the median could be taken as the average of the two middle values [= (106 + 107)/2 = 106]. This compares with a mean of 108, which is influenced more by the suspected number.

The following procedure is suggested for interpretation of the data of three to five measurements if the precision is considerably poorer than expected and if one of the observations is considerably different from the others of the set.

1. Estimate the precision that can reasonably be expected for the method in deciding whether a particular number actually is questionable. Note that for three measurements with two of the points very close, the Q test is likely to fail. (See the statement at the end of this section, below.)
2. Check the data leading to the suspected number to see if a definite error can be identified.
3. If new data cannot be collected, run a Q test.
4. If the Q test indicates retention of the outlying number, consider reporting the median rather than the mean for a small set of data.
5. As a last resort, run another analysis. Agreement of the new result with the apparently valid data previously collected will lend support to the opinion that the suspected result should be rejected. You should avoid, however, continually running experiments until the "right" answer is obtained.

The Q test should not be applied to three data points if two are identical. In that case, the test *always* indicates rejection of the third value, regardless of the magnitude of the deviation, because a is equal to w and Q_{calc} is always equal to 1. The same obviously applies for three identical data points in four measurements, and so forth.

3.14 LINEAR LEAST SQUARES

The analyst is frequently confronted with plotting data which fall on a straight line, as in an analytical calibration curve. This is often done intuitively, that is, by simply "eyeballing" the best straight line by placing a ruler through the points, which invariably have some scatter. A better approach is to apply statistics to define the most probable straight-line fit of the data.

If a straight-line relationship is assumed, then the data fit the equation

$$y = mx + b \tag{3.17}$$

where y is the dependent variable, x is the independent variable, m is the slope of the curve, and b is the intercept on the ordinate (y axis); y is usually the measured variable, plotted as a function of changing x. In a spectrophotometric calibration curve, y would represent the measured absorbances and x would be the concentrations of the standards. Our problem, then, is to establish values for m and b.

It can be shown statistically that the best straight line through a series of experimental points is that line for which the sum of the squares of the deviations of the points from the line is minimum. This is known as the **method of least squares.** If x is the fixed variable (e.g., concentration) and y is the measured variable (absorbance in a spectrophotometric measurement, fluorescence intensity in a spectrofluorometric measurement, etc.), then the deviation of y vertically from the line at a given value of x (x_i) is of interest. If y_l is the value *on the line*, it is equal to $mx_i + b$. The square of the sum of the differences, S, is then

$$S = \Sigma(y_i - y_l)^2 = \Sigma[y_i - (mx_i + b)]^2 \tag{3.18}$$

This equation assumes no error in x, the independent variable.

The best straight line occurs when S goes through a minimum. This is obtained by use of differential calculus by setting the derivatives of S with respect to m and b equal to zero and solving for m and b. The result is

$$m = \frac{\Sigma(x_i - \bar{x})(y_i - \bar{y})}{\Sigma(x_i - \bar{x})^2} \tag{3.19}$$

$$b = \bar{y} - m\bar{x} \tag{3.20}$$

where $\bar{x}$ is the mean of all the values of x_i and $\bar{y}$ is the mean of all the values of y_i. The use of differences in calculations is cumbersome, and Equation 3.19 can be transformed into an easier to use form, especially if a calculator is available:

$$m = \frac{\Sigma x_i y_i - [(\Sigma x_i \Sigma y_i)/n]}{\Sigma x_i^2 - [(\Sigma x_i)^2/n]} \tag{3.21}$$

where n is the number of data points.

EXAMPLE 3.22 Riboflavin (vitamin B_2) is determined in a cereal sample by measuring its fluorescence intensity in 5% acetic acid solution. A calibration curve was prepared by measuring the fluorescence intensities of a series of standards of increasing concentrations. The following data were obtained. Use the method of least squares to obtain the best straight line for the calibration curve and to calculate the concentration of riboflavin in the sample solution. The sample fluorescence intensity was 15.4

Riboflavin µg/mL (x_i)	Fluorescence Intensity, Arbitrary Units (y_i)	x_i^2	$x_i y_i$
0.000	0.0	0.0000	0.00
0.100	5.8	0.0100	0.58
0.200	12.2	0.0400	2.44
0.400	22.3	0.160_0	8.92
0.800	43.3	0.640_0	34.6_4
$\Sigma x_i = 1.500$	$\Sigma y_i = 83.6$	$\Sigma x_i^2 = 0.850_0$	$\Sigma x_i y_i = 46.5_8$

$(\Sigma x_i)^2 = 2.250$ 　　　　　　　　　　$n = 5$

$\bar{x} = \dfrac{\Sigma x_i}{n} = 0.300_0$ 　　$\bar{y} = \dfrac{\Sigma y_i}{n} = 16.7_2$

SOLUTION

Using Equations 3.21 and 3.20,

$$m = \frac{46.5_8 - [(1.500 \times 83.6)/5]}{0.850_0 - 2.250/5} = 53.7_5$$

$$b = 16.7_2 - (53.7_5 \times 0.300_0) = 0.6_0$$

We have retained the maximum number of significant figures in computation. Since the experimental values of y are obtained to only the first decimal place, we can round m and b to the first decimal. The equation of the straight line is

$$y = 53.8x + 0.6$$

The sample concentration is

$$15.4 = 53.8x + 0.6$$

$$x = 0.275 \ \mu g/mL$$

To prepare an actual plot of the line, take two arbitrary values of x sufficiently far apart and calculate the corresponding y values (or vice versa) and use these as points to draw the line. The intercept $y = 0.6$ (at $x = 0$) could be used as one point. At 0.500 µg/mL, $y = 27.5$. A plot of the experimental data and the least-squares line drawn through them is shown in Figure 3.4.

3.15 THE CORRELATION COEFFICIENT

The **correlation coefficient** is used as a measure of the correlation between two variables. When variables x and y are correlated rather than being functionally related (i.e., are not directly dependent upon one another), we do not speak of the

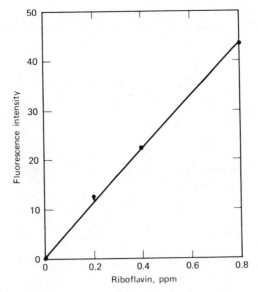

FIGURE 3.4
Least-squares plot of data from Example 3.22.

"best" y value corresponding to a given x value, but only of the most "probable" value. The closer the observed values are to the most probable values, the more definite is the relationship between x and y. This postulate is the basis for various numerical measures of the degree of correlation.

The **Pearson correlation coefficient** is one of the most convenient to calculate. This is given by

$$r = \Sigma \frac{(x_i - \bar{x})(y_i - \bar{y})}{n s_x s_y} \tag{3.22}$$

where r is the correlation coefficient; n is the number of observations; s_x is the standard deviation of x; s_y is the standard deviation of y; x_i and y_i are the individual values of the variables x and y, respectively; and $\bar{x}$ and $\bar{y}$ are their means. The use of differences in the calculation is frequently cumbersome, and the equation can be transformed to a more convenient form:

$$
\begin{aligned}
r &= \frac{\Sigma x_i y_i - n\bar{x}\bar{y}}{\sqrt{(\Sigma x_i^2 - n\bar{x}^2)(\Sigma y_i^2 - n\bar{y}^2)}} \\
&= \frac{n\Sigma x_i y_i - \Sigma x_i \Sigma y_i}{\sqrt{[n\Sigma x_i^2 - (\Sigma x_i)^2][n\Sigma y_i^2 - (\Sigma y_i)^2]}}
\end{aligned}
\tag{3.23}
$$

Despite its formidable appearance, Equation 3.23 is probably the most convenient for calculating r, particularly if a calculator is available.

The maximum value of r is 1. When this occurs, there is exact correlation between the two variables. When the value of r is zero (this occurs when xy is equal to zero), there is complete independence of the variables. The minimum value of r is -1. A negative correlation coefficient indicates that the assumed dependence is opposite to what exists and is therefore a positive coefficient for the reversed relation.

EXAMPLE 3.23   Calculate the correlation coefficient for the data in Example 3.20.

SOLUTION
We calculate that $\Sigma x_i^2 = 903.2$, $\Sigma y_i^2 = 855.2$ $\bar{x} = 12.0$, $\bar{y} = 11.7$, and $\Sigma x_i y_i = 878.5$. Therefore, from Equation 3.23,

$$r = \frac{878.5 - (6)(12.0)(11.7)}{\sqrt{[903.2 - (6)(12.0)^2][855.2 - (6)(11.7)^2]}} = 0.99$$

A correlation coefficient can be calculated for a calibration curve to ascertain the degree of correlation between the measured instrumental variable and the sample concentration. As a general rule, $0.90 < r < 0.95$ indicates a fair curve, $0.95 < r < 0.99$ a good curve, and $r > 0.99$ indicates excellent linearity. An $r > 0.999$ can sometimes be obtained with care. It is common practice in the clinical chemistry literature when comparing two methods to analyze a series of samples over a range of physiological concentrations by both methods and to make a correlation plot of the results of one method against the other. A least-squares plot is prepared and a correlation coefficient is calculated. A typical plot is illustrated in Figure 3.5, in which the results for blood samples analyzed for the enzyme alkaline phosphatase by a new electrochemical method using an oxygen electrode are compared with those using the standard Bessey colorimetric procedure.

It should be mentioned that it is possible to have a high degree of correlation between two methods (r near unity) but to have a statistically significant difference

FIGURE 3.5
Typical correlation plot. (O) 100-μL serum samples, (Δ) 50-μL serum samples. (Reprinted with permission from A. Kumar and G. D. Christian, *Anal Chem.*, **48** (1976) 1283. Copyright by the American Chemical Society.)

between the results of each according to the t test. This would occur, for example, if there were a constant determinate error in one method. This would make the differences significant (not due to chance), but there would be a direct correlation between the results (r would be near unity, but the slope (m) may not be near unity or the intercept (b) not near zero. In principle, an empirical correction factor (a constant) could be applied to make the results by each method the same over the concentration range analyzed.

QUESTIONS

1. Distinguish between accuracy and precision.
2. What is a determinate error? An indeterminate error?
3. The following is a list of common errors encountered in research laboratories. Categorize each as a determinate or an indeterminate error, and further categorize determinate errors as instrumental, operative, or methodic: (a) An unknown being weighed is hygroscopic. (b) One component of a mixture being analyzed quantitatively by gas chromatography reacts with the column packing. (c) A radioactive sample being counted repeatedly without any change in conditions yields a slightly different count at each trial. (d) The tip of the pipet used in the analysis is broken. (e) In measuring the same peak heights of a chromatogram, two technicians each report different heights.

PROBLEMS

Significant Figures

4. How many significant figures does each of the following numbers have? (a) 200.06; (b) 6.030×10^{-4}; (c) 7.80×10^{10}.
5. How many significant figures does each of the following numbers have? (a) 0.02670; (b) 328.0; (c) 7000.0; (d) 0.00200.
6. Calculate the formula weight of $LiNO_3$ to the correct number of significant figures.
7. Calculate the formula weight of $PdCl_2$ to the correct number of significant figures.
8. Give the answer to the following problem to the maximum number of significant figures: $50.00 \times 27.8 \times 0.1167$.
9. Give the answer of the following to the maximum number of significant figures: $(2.776 \times 0.0050) - (6.7 \times 10^{-3}) + (0.036 \times 0.0271)$.
10. An analyst wishes to analyze spectrophotometrically the copper content in a bronze sample. If the sample weighs about 5 g and if the absorbance (A) is to be read to the nearest 0.001 absorbance unit, how accurately should he weigh the sample? Assume he will adjust the volume of the measured solution to obtain minimum error in the absorbance, that is, so that $0.1 < A < 1$.

Expressions of Results

11. A standard serum sample containing 102 meq/L chloride was analyzed by coulometric titration with silver ion. Duplicate results of 101 and 98 meq/L were obtained. Calculate (a) the mean value, (b) the absolute error, (c) the relative error in percent.
12. A batch of nuclear fuel pellets was weighed to determine if they fell within control guidelines. The weights were 127.2, 128.4, 127.1, 129.0 and 128.1 g. Calculate (a) the mean, (b) the median, and (c) the range.

13. Calculate the absolute error and the relative error in percent and in parts per thousand in the following:

	Measured Value	Accepted Value
(a)	22.62 g	22.57 g.
(b)	45.02 mL	45.31 mL
(c)	2.68%	2.71%
(d)	85.6 cm.	85.0 cm.

Standard Deviation

14. The tin and zinc contents of a brass sample are analyzed with the following results. (a) Zn: 33.27, 33.37, and 33.34%; (b) Sn: 0.022, 0.025, and 0.026%. Calculate the standard deviation and the coefficient of variation for each analysis.

15. Replicate water samples are analyzed for water hardness with the following results: 102.2, 102.8, 103.1, and 102.3 ppm $CaCO_3$. Calculate (a) the standard deviation, (b) the relative standard deviation, (c) the standard deviation of the mean, and (d) the relative standard deviation of the mean.

16. Replicate samples of a silver alloy are analyzed and determined to contain 95.67, 95.61, 95.71, and 95.60% Ag. Calculate (a) the standard deviation, (b) the standard deviation of the mean, and (c) the relative standard deviation of the mean (in percent) of the individual results.

Propagation of Error

17. Calculate the uncertainty in the answers of the following: (a) $(128 \pm 2) + (1025 \pm 8) - (636 \pm 4)$; (b) $(16.25 \pm 0.06) - (9.43 \pm 0.03)$; (c) $(46.1 \pm 0.4) + (935 \pm 1)$.

18. Calculate the absolute uncertainty in the answers of the following: (a) $(2.78 \pm 0.04)(0.00506 \pm 0.00006)$; (b) $36.2 \pm 0.04/(27.1 \pm 0.06)$; (c) $(50.23 \pm 0.07)(27.86 \pm 0.05)/(0.1167 \pm 0.0003)$.

19. Calculate the absolute uncertainty in the answer of the following: $[(25.0 \pm 0.1)(0.0215 \pm 0.0003) - (1.02 \pm 0.01)(0.112 \pm 0.001)](17.0 \pm 0.2)/(5.87 \pm 0.01)$.

20. Calculate the uncertainty in the area of a circle whose radius is 125.3 ± 0.4 cm.

21. Calculate the uncertainty in the area of a circle whose diameter is 250.6 ± 0.8 cm.

22. The solubility product for $AB \rightleftharpoons A + B$ is $(4.0 \pm 0.4) \times 10^{-10}$. Calculate the molar solubility of AB and its uncertainty. [*Note:* $K_{sp} = [A][B]$. See Chapter 6 for a review of solubility product calculations.]

23. The equilibrium constant K_a of a weak acid is determined to be $(5.6 \pm 0.6) \times 10^{-6}$. What is pK_a and its uncertainty? (*Note:* $pK_a = -\log K_a$.)

24. The calcium content of a serum sample is measured with an ion-selective electrode, which gives a pCa reading $(= -\log [Ca^{2+}])$ of 2.65 ± 0.05. What is the molar calcium concentration and its uncertainty?

Confidence Limit

25. The following molarities were calculated from replicate standardization of a solution: 0.5026, 0.5029, 0.5023, 0.5031, 0.5025, 0.5032, 0.5027, and 0.5026 M. Assuming no determinate errors, within what range are you 95% certain that the true mean value of the molarity falls:

26. Determination of the sodium level in separate portions of a blood sample by ion-selective electrode measurement gave the following results: 139.2, 139.8, 140.1, and 139.4 meq/L. What is the range within which the true value falls, assuming no determinate error (a) at the 90% confidence level, (b) at the 95% confidence level, and (c) at the 99% confidence level?

27. Lead on leaves by a roadside was measured spectrophotometrically by reaction with

dithizone. The standard deviation for a triplicate analysis was 2.3 ppm. What is the 90% confidence limit?

28. The standard deviation established for the determination of blood chloride by coulometric titration is 0.5 meq/L. What is the 95% confidence limit for a triplicate determination?

29. Estimate the range of the true molarity of the solution at the 90% confidence level from the standardization in Problem 36 below.

Tests of Significance

30. A study is being performed to see if there is a correlation between the concentration of chromium in the blood and a suspected disease. Blood samples from a series of volunteers with a history of the disease and other indicators of susceptibility are analyzed and compared with the results from the analysis of samples from healthy control subjects. From the following results, determine whether the differences between the two groups can be ascribed to chance or whether they are real. Control group (ppb Cr): 15, 23, 12, 18, 9, 28, 11, 10. Disease group: 25, 20, 35, 32, 15, 40, 16, 10, 22, 18.

31. An enzymatic method for determining alcohol in wine is evaluated by comparison with a gas-chromatographic method (G.C.). The same sample is analyzed several times by both methods with the following results (% ethanol). Enzymatic method: 13.1, 12.7, 12.6, 13.3, 13.3. G.C. method: 13.5, 13.3, 13.0, 12.9. Does the enzymatic method give the same value as the G.C. method at the 95% confidence level?

32. Your laboratory is evaluating the precision of a colorimetric method for creatinine in serum in which the sample is reacted with alkaline picrate to produce a color. Rather than perform one set of analyses, several sets with different samples are performed over several days, in order to get a better estimate of the precision of the method. From the following absorbance data, calculate the pooled standard deviation.

Day 1 (Sample A)	Day 2 (Sample B)	Day 3 (Sample C)
0.826	0.682	0.751
0.810	0.655	0.702
0.880	0.661	0.699
0.865		0.724
$\bar{x}_A = 0.845$	$\bar{x}_B = 0.666$	$\bar{x}_C = 0.719$

33. The following replicate calcium determinations on a blood sample using atomic absorption spectrophotometry (AAS) and a new colorimetric method were reported. Is there a significant difference in the precision of the two methods?

AAS, mg/dL	Colorimetric, mg/dL
10.9	9.2
10.1	10.5
10.6	9.7
11.2	11.5
9.7	11.6
10.0	9.3
——	10.1
mean 10.4	11.2
	mean 10.4

34. Potassium dichromate is an oxidizing agent that is used for the volumetric determination of iron by titrating iron(II). Although potassium dichromate is a high-purity material that can be used for the direct preparation of a standard solution of known concentration, the solution is frequently standardized by titrating a known amount of iron(II) prepared from high-purity iron wire or electrolytic iron, using the same procedure as for the sample. This is because the color of the iron(III) product of the titration tends to mask the indicator color (used to detect the end of the titration), causing a slight error. A solution prepared to be 0.1012 N was standardized with the following results: 0.1017, 0.1019, 0.1016, 0.1015 N. Is the supposition that the titration values are statistically different from the actual prepared concentration valid?

35. In the nuclear industry, detailed records are kept of the quantity of plutonium received, transported, or used. Each shipment of plutonium pellets received is carefully analyzed to check that the purity and hence the total quantity is as the supplier claims. A particular shipment is analyzed with the following results: 99.93, 99.87, 99.91, and 99.86%. The listed purity as received from the supplier is 99.95%. Is the shipment acceptable?

Q Test

36. The following replicate molarities were obtained when standardizing a solution: 0.1067, 0.1071, 0.1066, and 0.1050. Can one of the results be discarded as due to accidental error?

37. Can any of the data in Problem 14 be rejected?

38. The precision of a method is being established, and the following data are obtained: 22.23, 22.18, 22.25, 22.09, and 22.15%. Is 22.09% a valid measurement?

Least Squares

39. Calculate the slope of the line in Example 3.22, using Equation 3.19. Compare with the value calculated using Equation 3.21.

40. A calibration curve for the colorimetric determination of phosphorous in urine is prepared by reacting standard solutions of phosphate with molybdenum(VI) and reducing the phosphomolybdic acid complex to produce the characteristic blue color. The measured absorbance A is plotted against the concentration of phosphorous. From the following data, determine the linear least-squares line and calculate the phosphorous concentration in the urine sample:

ppm P	A
1.00	0.205
2.00	0.410
3.00	0.615
4.00	0.820
Urine sample	0.625

Correlation Coefficient

41. From the data given below, determine the correlation coefficient between the amount of toxin produced by a fungus and the percent of yeast extract in the growth medium.

Sample	% Yeast Extract	Toxin, mg
(a)	1.000	0.487
(b)	0.200	0.260
(c)	0.100	0.195
(d)	0.010	0.007
(e)	0.001	0.002

42. The cultures described in Problem 41 had the following fungal dry weights: sample (a) 116 mg, (b) 53 mg, (c) 37 mg, (d) 8 mg, (e) 1 mg. Determine the correlation coefficient between the dry weight and the amount of toxin produced.

43. A new method for the determination of cholesterol in serum is being developed in which the rate of depletion of oxygen is measured with an oxygen electrode upon reaction of the cholesterol with oxygen, when catalyzed by the enzyme cholesterol oxidase. The results for several samples are compared with those of the standard Lieberman colorimetric method. From the following data, determine by the *t* test if there is a statistically significant difference between the two methods and calculate the correlation coefficient. Assume the two methods have similar precisions.

Sample	Enzyme Method, mg/dL	Colorimetric Method, mg/dL
1	305	300
2	385	392
3	193	185
4	162	152
5	478	480
6	455	461
7	238	232
8	298	290
9	408	401
10	323	315

RECOMMENDED REFERENCES

Statistics

1. G. Hoel, *Elementary Statistics*. New York: John Wiley & Sons, 1971.

2. *Industrial Hygiene Service Laboratory Manual, Technical Report No. 78*. Cincinnati: Department of Health, Education and Welfare, PHS, CDC, Washington, D.C.

3. H. A. Laitinen and W. E. Harris, "Statistics in Quantitative Analysis," and "Sampling," *Chemical Analysis,* 2nd ed. New York: McGraw-Hill, 1975, Chapters 26 and 27. The latter chapter discusses the statistics of sampling.

4. M. G. Natrella, *Experimental Statistics*. NBS Handbook 91. Washington, D.C.: U.S. Government Printing Office, 1963. Reprinted 1966.

5. G. W. Snedecor, *Statistical Methods*. Ames, Iowa: Iowa State University Press, 1965.

6. W. J. Youdan and E. H. Steiner, *Statistical Manual of the AOAC*. Washington, D.C.: American Association of Official Analytical Chemists, 1975.

Q Test

7. W. J. Blaedel, V. W. Meloche, and J. A. Ramsay, "A Comparison of Criteria for the Rejection of Measurements," *J. Chem. Educ.,* **28** (1951) 643.

8. R. B. Dean and W. J. Dixon, "Simplified Statistics for Small Numbers of Observations," *Anal. Chem.,* **23** (1951) 636.

Quality Control

9. J. K. Taylor, "Quality Assurance of Chemical Measurements," *Anal. Chem.*, **53** (1981) 1588A.

10. J. K. Taylor, "Validation of Analytical Methods," *Anal. Chem.*, **55** (1983) 600A.

Least Squares

11. S. D. Christian and E. E. Tucker, "Least Squares Analysis With the Microcomputer. Part One: Linear Relationships," *Amer. Lab.*, **14** (8) (1982) 36; "Part Two: Nonlinear Relationships," *Am. Lab.*, **14** (9) (1982) 31.

chapter 4

GENERAL CONCEPTS OF CHEMICAL EQUILIBRIUM

Even though chemical reactions may go far to completion, the reactions never go in only one direction, but actually reach an equilibrium in which the rates of reactions in both directions are equal. In this chapter we review the equilibrium concept and the equilibrium constant and describe general approaches for calculations using equilibrium constants. Activity of ionic species is described along with the calculation of activity coefficients. These are required for calculations using thermodynamic equilibrium constants, that is, for the diverse ion effect, described at the end of the chapter.

4.1 CHEMICAL REACTIONS: THE RATE CONCEPT

Consider the chemical reaction

$$aA + bB \rightleftharpoons cC + dD \qquad (4.1)$$

The rate of the forward reaction depends on the number of collisions per unit time between the reacting species A and B. If the coefficients a and b are greater than unity, the rate is proportional to the concentration of A multiplied by itself a times and the concentration of B multiplied by itself b times, since the number of collisions would increase proportionally. We can write, then, that[1]

$$\text{rate}_f = k_f[A]^a[B]^b \qquad (4.2)$$

rate $_f$ constant

---

[1] [ ] represents moles/liter and here represents the effective concentration. The effective concentration will be discussed under the diverse ion effect.

where rate$_f$ is the rate of the forward reaction and k_f is the **rate constant,** which is a measure of the fraction of collisions that result in reaction (and is dependent on such factors as the temperature and the presence of catalysts). [A] and [B] represent the molar concentrations of A and B. Similarly, for the backward reaction, we can write

$$\text{rate}_b = k_b[C]^c[D]^d \tag{4.3}$$

When the reaction between A and B is initiated, the rate of the forward reaction is large because the concentrations of A and B are large, whereas the reverse reaction is slow since the concentrations of C and D are small (that rate is initially zero). As the reaction progresses, A and B decrease and C and D increase, so that the rate of the forward reaction diminishes while that for the backward reaction increases (Figure 4.1). Eventually, the two rates become equal and the system is in a state of equilibrium. At this point, the individual concentrations of A, B, C, and D remain constant (the relative values will depend on the reaction and how far the equilibrium lies to the right). However, the system remains in dynamic equilibrium, with the forward and backward reactions continuing at equal rates.

When the system is at equilibrium, we can equate the two rates:

$$k_f[A]^a[B]^b = k_b[C]^c[D]^d \tag{4.4}$$

It is not convenient to evaluate the individual rate constants k_f and k_b. We can rearrange the equation to obtain the **molar equilibrium constant** (which holds for dilute solutions) for the reaction, K:

$$\boxed{\frac{[C]^c[D]^d}{[A]^a[B]^b} = \frac{k_f}{k_b} = K} \tag{4.5}$$

K can be evaluated by measuring the concentrations of A, B, C, and D at equilibrium. Since the individual rate constants are dependent on the temperature, it follows that the equilibrium constant is temperature dependent. Note also that the more favorable the rate constant of the forward reaction relative to the backward reaction, the larger will be the equilibrium constant and the farther to the right the reaction will be at equilibrium.

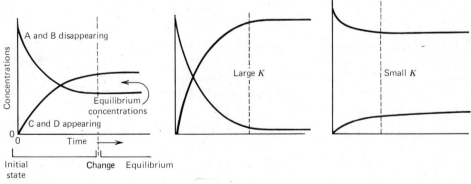

FIGURE 4.1
Progress of a chemical reaction.

You will notice that the equilibrium constant expression is the ratio in which the concentrations of the products appear in the numerator and the concentrations of the reactants appear in the denominator. This is quite arbitrary but is the accepted convention. Hence, a large equilibrium constant indicates the equilibrium lies far to the right.

We should point out that although a particular reaction may have a rather large equilibrium constant, the reaction may proceed from *left* to *right* if sufficiently large concentrations of the products are initially present. The equilibrium constant tells us nothing about how fast a reaction will proceed toward equilibrium. Some reactions, in fact, may be so slow as to be unmeasurable. The equilibrium constant merely tells us the tendency of a reaction to occur and in what direction, not whether it is fast enough to be feasible in practice. (See the chapter on kinetic methods of analysis for the measurement of reaction rates and their application to analyses.)

For the reaction depicted in Equation 4.1, the rate at which equilibrium will be approached will likely be different for either the forward or the reverse reaction. That is, if we start with a mixture of C and D, the rate at which equilibrium is approached may be much slower or faster than for the converse reaction.

4.2 GIBBS FREE ENERGY AND THE EQUILIBRIUM CONSTANT

The tendency for a reaction to occur is defined thermodynamically from its change in **enthalpy** (ΔH) and **entropy** (ΔS). Enthalpy is the heat absorbed when an endothermic reaction occurs under constant pressure. When heat is given off (exothermic reaction), ΔH is negative. Entropy is a measure of the disorder, or randomness, of a substance or system.

A system will always tend toward lower energy and increased randomness, that is, lower enthalpy and higher entropy. For example, a stone on a hill will tend to roll spontaneously down the hill (lower energy state), and a box of marbles ordered by color will tend to become randomly ordered when shaken. The combined effect of these is given by the **Gibbs free energy** G:

$$G = H - TS \tag{4.6}$$

where T is the absolute temperature in K. G is a measure of the energy of the system, and a system spontaneously tends toward lower energy states. The change in energy of a system at a constant temperature is

$$\Delta G = \Delta H - T\,\Delta S \tag{4.7}$$

So a process will be *spontaneous when ΔG is negative,* will be spontaneous in the reverse direction when ΔG is positive, and will be at equilibrium when ΔG is zero. Hence, a reaction is favored by heat given off (negative ΔH), as in endothermic reactions, and by increased entropy (positive ΔS). Both ΔH and ΔS can be either positive or negative, and the relative magnitudes of each and the temperature will determine whether ΔG will be negative so that the reaction will be spontaneous.

Standard enthalpy H^0, standard entropy S^0, and standard free energy G^0 represent the thermodynamic quantities for formation of a substance from its elements at standard state (1 atm, 298 K, unit concentration). Then,

$$\Delta G^0 = \Delta H^0 - T\,\Delta S^0 \tag{4.8}$$

ΔG^0 is related to the equilibrium constant of a reaction by

$$K = e^{-\Delta G^0/RT} \tag{4.9}$$

or

$$\Delta G^0 = -RT \ln K = -2.303RT \log K \tag{4.10}$$

where R is the gas constant (8.314 J deg K^{-1} mol^{-1}). Hence, from a knowledge of the standard free energy of a reaction, the equilibrium constant can be calculated. Obviously, the larger ΔG^0 (when negative), the larger will be K. Note that while ΔG^0 and ΔG give information about the spontaneity of a reaction, they say nothing of the *rate* at which it will occur.

4.3   LE CHÂTELIER'S PRINCIPLE

The equilibrium concentrations of reactants and products can be altered by applying stress to the system, for example, by changing the temperature, the pressure, or the concentration of one of the reactants. The effects of such changes can be predicted from **Le Châtelier's principle,** which states that when stress is applied to a system at chemical equilibrium, the equilibrium will shift in a direction that tends to relieve or counteract that stress. The effects of the above mentioned variables on chemical equilibria are considered below.

4.4   TEMPERATURE EFFECTS ON EQUILIBRIUM CONSTANTS

As we have mentioned, temperature influences the individual rate constants for the forward and backward reactions and therefore the equilibrium constant. An increase in temperature will displace the equilibrium in that direction which results in absorption of heat, since this removes the source of the stress. So an endothermic forward reaction which absorbs heat will be displaced to the right with an increase in the equilibrium constant. The reverse will be true for an exothermic forward reaction which releases heat. The extent of the displacement will depend on the magnitude of the heat of reaction for the system.

   In addition to influencing the position of equilibrium, temperature has a pronounced effect on the rates of the forward and backward reactions involved in the equilibrium, and so it influences the rate at which equilibrium is approached. This is because the number and the energy of collisions between the reacting species increase with increasing temperature. The rates of many reactions increase about two- to threefold for every 10°C rise in temperature.

4.5   PRESSURE EFFECTS ON EQUILIBRIA

Pressure can have a large influence on the position of chemical equilibrium for reactions occurring in the gaseous phase. An increase in pressure favors a shift in the direction that results in a reduction in the volume of the system. But for reactions occurring in solutions, normal pressure changes have a negligible effect on the equilibrium because of the small compressibility of liquids.

4.6 EFFECT OF CONCENTRATIONS ON EQUILIBRIA

The value of an equilibrium constant is independent of the concentrations of the reactants and products. However, the position of equilibrium is very definitely influenced by the concentrations. The direction of change is readily predictable from Le Châtelier's principle. Consider the reaction of iron(III) with iodide:

$$3I^- + 2Fe^{3+} \rightleftharpoons I_3^- + 2Fe^{2+}$$

If the four components are in a state of equilibrium, as determined by the equilibrium constant, addition or removal of one of the components would cause the equilibrium to reestablish itself. For example, suppose we add more iron(II) to the solution. According to Le Châtelier's principle, the reaction will shift to the left to relieve the stress. Equilibrium will eventually be reestablished and its position will still be defined by the same equilibrium constant.

4.7 CATALYSTS

Catalysts alter (speed or retard) the rate at which an equilibrium is attained by affecting the rates of both the forward and reverse reactions. But both rates are affected to the same extent, and so catalysts have no effect on the value of an equilibrium constant.

Catalysts are very important to the analytical chemist in a number of reactions that are normally too slow to be analytically useful. An example is the use of an osmium tetroxide catalyst to speed up the titration reaction between arsenic(III) and cerium(IV), whose equilibrium is very favorable but whose rate is normally too slow to be useful for titrations. The measurement of the change in the rate of a kinetically slow reaction in the presence of a catalyst can actually be used for determining the catalyst, as we will see in Chapter 18.

4.8 COMPLETENESS OF REACTIONS

If the equilibrium of a reaction lies sufficiently to the right that the remaining amount of the substance being determined (reacted) is too small to be measured by the measurement technique, we say the reaction has gone to completion. If the equilibrium is not so favorable, then Le Châtelier's principle may be applied to make it so. We may either increase the concentration of a reactant or decrease the concentration of a product. Product may be decreased by (1) allowing a gaseous product to escape, (2) by precipitating the product, (3) by forming a stable ionic complex of the product in solution, or (4) by preferential extraction.

4.9 EQUILIBRIUM CONSTANTS FOR DISSOCIATING SPECIES

When a substance dissolves in water, it will often partially or completely dissociate or ionize. Partially dissociated electrolytes are called weak electrolytes and completely dissociated ones are strong electrolytes. For example, acetic acid only partially ionizes in water and is a weak electrolyte. But hydrochloric acid is completely ionized and is a strong electrolyte. Some substances completely ionize in water but have limited solubility (we call these slightly soluble substances).

The dissociation of weak electrolytes or the solubility of slightly soluble substances can be quantitatively described by equilibrium constants. Equilibrium constants for completely dissolved and dissociated electrolytes are effectively infinite. Consider the dissociating species AB:

$$AB \rightleftharpoons A + B \tag{4.11}$$

The equilibrium constant for such a dissociation can be written generally as

$$\frac{[A][B]}{[AB]} = K_{eq} \tag{4.12}$$

The larger K_{eq}, the greater will be the dissociation. For example, the larger the equilibrium constant of an acid, the stronger will be the acid.

Some species dissociate stepwise, and an equilibrium constant can be written for each dissociation step. A compound A_2B, for example, may dissociate as follows:

$$A_2B \rightleftharpoons A + AB \quad K_1 = \frac{[A][AB]}{[A_2B]} \tag{4.13}$$

$$AB \rightleftharpoons A + B \quad\quad K_2 = \frac{[A][B]}{[AB]} \tag{4.14}$$

The overall dissociation of the compound is the sum of these two equilibria:

$$A_2B \rightleftharpoons 2A + B \quad K_{eq} = \frac{[A]^2[B]}{[A_2B]} \tag{4.15}$$

If we multiply Equations 4.13 and 4.14 together, we arrive at the overall equilibrium constant:

$$K_{eq} = K_1K_2 = \frac{[A][AB]}{[A_2B]} \cdot \frac{[A][B]}{[AB]}$$
$$= \frac{[A]^2[B]}{[A_2B]} \tag{4.16}$$

When chemical species dissociate in a stepwise manner like this, the successive equilibrium constants generally become progressively smaller. Equilibrium constants for acids, complexes, and precipitates will be described in succeeding chapters.

4.10 CALCULATIONS USING EQUILIBRIUM CONSTANTS

Equilibrium constants are useful for calculating the concentrations of the various species in equilibrium, for example, the hydrogen ion concentration from the dissociation of a weak acid. In this section we present the general approach for calculations using equilibrium constants. The applications to specific equilibria are treated in later chapters dealing with these equilibria.

Chemical Reactions

It is sometimes useful to know the concentrations of reactants and products in equilibrium in a chemical reaction. For example, we may need to know the amount of reactant for the construction of a titration curve or for calculating the potential of an electrode in the solution. These are, in fact, applications we consider in later chapters. Some example calculations will illustrate the general approach to solving equilibrium problems.

EXAMPLE 4.1 The chemicals A and B react as follows to produce C and D:

$$A + B \rightleftharpoons C + D \quad K = \frac{[C][D]}{[A][B]}$$

The equilibrium constant K has a value of 0.30. Assume 0.20 mol of A and 0.50 mol of B are dissolved in 1.00 L, and the reaction proceeds. Calculate the concentrations of reactants and products at equilibrium.

SOLUTION
The initial concentration of A is 0.20 M and that of B is 0.50 M, while C and D are initially 0 M. After the reaction has reached equilibrium, the concentrations of A and B will be decreased and C and D will be increased. Let x represent the equilibrium concentration of C or the moles/liter of A and B reacting. Since we get one mole of D with each mole of C, its concentration will also be x. The concentrations of A and B will each be diminished by x. So the equilibrium concentrations are:

	[A]	[B]	[C]	[D]
Initial	0.20	0.50	0	0
Change (x = mmol/mL reacting)	$-x$	$-x$	$+x$	$+x$
Equilibrium	$0.20 - x$	$0.50 - x$	x	x

We can substitute these values in the equilibrium constant expression and solve for x:

$$\frac{(x)(x)}{(0.20 - x)(0.50 - x)} = 0.30$$

$$x^2 = (0.10 - 0.70x + x^2)0.30$$

$$0.70x^2 + 0.21x - 0.030 = 0$$

This is a quadratic equation and can be solved algebraically for x using the quadratic formula given in Appendix B:

$$x = \frac{-b \pm \sqrt{b^2 - 4ac}}{2a}$$

$$= \frac{-0.21 \pm \sqrt{(0.21)^2 - 4(0.70)(-0.030)}}{2(0.70)}$$

$$= \frac{-0.21 \pm \sqrt{0.044 + 0.084}}{1.40} = 0.11 \ M$$

$$[A] = 0.20 - x = 0.09 \ M$$
$$[B] = 0.50 - x = 0.39 \ M$$
$$[C] = [D] = x \ 0.11 \ M$$

In the above example, appreciable amounts of A remained, even though B was in excess, because the equilibrium constant was not very large. In fact, the equilibrium was only about halfway to the right since C and D were about the same concentration as A. In most reactions of analytical interest, the equilibrium constants are large and the equilibrium lies far to the right. In these cases, the equilibrium concentrations of reactants that are not in excess are generally very small compared to the other concentrations. This simplifies our calculations.

EXAMPLE 4.2   Assume that in Example 4.1 the equilibrium constant was 2.0×10^{16}. Calculate the equilibrium concentrations of A, B, C, and D.

SOLUTION

Since K is very large, the reaction of A with B will be virtually complete, leaving only traces of A at equilibrium. Let x represent the equilibrium concentration of A. An amount of B equal to A will have reacted to form an equivalent amount of C and D (about 0.20 M for each). We can summarize the equilibrium concentrations as follows:

$$[A] = x$$
$$[B] = (0.50 - 0.20) + x = 0.30 + x$$
$$[C] = 0.20 - x$$
$$[D] = 0.20 - x$$

Or, looking at the equilibrium,

A	+	B	$\rightleftharpoons$	C	+	D
x		$0.30 + x$		$0.20 - x$		$0.20 - x$

Basically, we have said that all of A is converted to a like amount of C and D, except for a small amount x. Now x will be very small compared to 0.20 and 0.30 and can be neglected. So we can say

$$[A] = x$$
$$[B] \approx 0.30$$
$$[C] \approx 0.20$$
$$[D] \approx 0.20$$

The only unknown concentration is [A]. Substituting these in the equilibrium constant expression, we have

$$\frac{(0.20)(0.20)}{(x)(0.30)} = 2.0 \times 10^{16}$$

$$x = [A] = 6.7 \times 10^{-18} \ M \text{ (analytically undetectable)}$$

So the calculation was considerably simplified by neglecting x in comparison to other concentrations. If x should turn out to be significant compared to these concentrations, then the solution should be reworked using the quadratic formula. **Generally, if it is less than about** 10% **of the assumed concentration, it can be neglected.** In this case, the error in x itself is usually 5% or less. **This simplification will generally hold if the initial concentration is** $\geq 100\ xK_{eq}$.

EXAMPLE 4.3   A and B react as follows:

$$A + 2B \rightleftharpoons 2C \quad K = \frac{[C]^2}{[A][B]^2}$$

Assume 0.10 mol of A is reacted with 0.20 mol of B in a volume of 1000 mL; $K = 1.0 \times 10^{10}$. What are the equilibrium concentrations of A, B, and C?

SOLUTION

We have stoichiometrically equal amounts of A and B, so both are virtually all reacted, with trace amounts remaining. Let x represent the equilibrium concentration of A. At equilibrium, we have

$$
\begin{array}{ccc}
A & + \quad 2B & \rightleftharpoons \qquad 2C \\
x & 2x & 0.20 - 2x \approx 0.20
\end{array}
$$

For each mole of A that either reacts or is produced, we produce or remove two moles of C. (The same is true for B.) Substituting into the equilibrium constant expression,

$$\frac{(0.20)^2}{(x)(2x)^2} = 1.0 \times 10^{10}$$

$$\frac{0.040}{4x^3} = 1.0 \times 10^{10}$$

$$x = [A] = \sqrt[3]{\frac{4.0 \times 10^{-2}}{4.0 \times 10^{10}}} = \sqrt[3]{1.0 \times 10^{-12}} = 1.0 \times 10^{-4}\ M$$

$$B = 2x = 2.0 \times 10^{-4}\ M \text{ (analytically detectable)}$$

Dissociation Equilibria

Calculations involving dissociating species are not much different from the examples just given for chemical reactions.

EXAMPLE 4.4   Calculate the equilibrium concentrations of A and B in a 0.10 M solution of a weak electrolyte AB with an equilibrium constant of 3.0×10^{-6}.

SOLUTION

$$AB \rightleftharpoons A + B \quad K_{eq} = \frac{[A][B]}{[AB]}$$

Both [A] and [B] are unknown and equal. Let x represent their equilibrium concentrations. The concentration of AB at equilibrium is equal to its initial concentration minus x.

$$AB \quad \rightleftharpoons \quad A \; + \; B$$
$$0.10 - x \qquad x \qquad x$$

K_{eq} is quite small, so we are probably justified in neglecting x compared to 0.10. Otherwise, a quadratic equation will have to be solved. Substituting into the K_{eq} expression,

$$\frac{(x)(x)}{0.10} = 3.0 \times 10^{-6}$$

$$x = [A] = [B] = \sqrt{3.0 \times 10^{-7}} = 5.5 \times 10^{-4} \, M$$

We can now write some general rules for the solving of chemical equilibrium problems. These should be applicable to acid–base dissociation, complex formation, oxidation–reduction reactions, and so forth. That is, all equilibria can be treated similarly.

1. Write down the equilibria involved.
2. Write the equilibrium constant expressions and the numerical values.
3. From a knowledge of the chemistry involved, let x represent the equilibrium concentration of the species that will be unknown and small compared to other equilibrium concentrations; other species of unknown and small concentrations will be multiples of this.
4. List the equilibrium concentrations of all species, adding or subtracting the appropriate multiple of x where needed.
5. Make suitable approximations by neglecting x compared to finite equilibrium concentrations. This is generally valid if the finite concentration is about $100 \times K_{eq}$ or more. Also, if the calculated x is less than approximately 10% of the finite concentration, the assumption was valid.
6. Substitute the approximate representation of individual concentrations into the equilibrium constant expression and solve for x.
7. If the approximations in step 5 are invalid, use the quadratic formula to solve for x.

The application of these rules will become more apparent in subsequent chapters when we deal with specific equilibria in detail.

4.11 THE COMMON ION EFFECT

Equilibria can be markedly affected by adding one or more of the species present, as is predicted from Le Châtelier's principle. A calculation illustrates this principle.

EXAMPLE 4.5 Recalculate the concentration of A in Example 4.4, assuming that the solution also contains 0.20 M B.

SOLUTION
We can represent the equilibrium concentrations as follows:

	[AB]	[A]	[B]
Initial	0.10	0	0.20
Change (x = mmol/mL of AB dissociated)	$-x$	$+x$	$+x$
Equilibrium	$0.10 - x$	x	$0.20 + x$
	≈ 0.10		≈ 0.20

The value of x will be smaller now than before because of the common ion effect of B, so we can certainly neglect it compared to the initial concentrations. Substituting in the equilibrium constant expression,

$$\frac{(x)(0.20)}{(0.10)} = 3.0 \times 10^{-6}$$

$$x = 1.5 \times 10^{-6} \, M$$

The concentration of A was decreased by nearly 400-fold.

4.12   ACTIVITY AND ACTIVITY COEFFICIENTS

Generally, the presence of diverse salts (not containing ions common to the equilibrium involved) will cause an increase in dissociation of a weak electrolyte or in the solubility of a precipitate. Cations attract anions, and vice versa, and so the cations of the analyte attract anions of the diverse electrolyte and the anions of the analyte attract the cations. The attraction of the ions of the equilibrium reaction by the dissolved electrolyte effectively shields them, decreasing their effective concentration and shifting the equilibrium. As the charge on either the diverse salt or the ions of the equilibrium reaction is increased, the diverse salt effect generally increases. This effect on the equilibrium is not predicted by Le Châtelier's principle; but if one thinks in terms of the effective concentrations being changed, it is analogous in some ways to the common ion effect.

This "effective concentration" of an ion in the presence of an electrolyte is called the **activity** of the ion. This can be used to describe quantitatively the effects of salts on equilibrium constants (see the diverse salt effect below). Activity is also important in potentiometric measurements (see Chapter 11). In this section we describe how to estimate activity.

The Activity Coefficient

The **activity** of an ion a_i is defined by

$$\boxed{a_i = C_i f_i}$$

(4.17)

where C_i is the concentration of the ion i and f_i is its **activity coefficient.** The concentration is usually expressed as molarity, and the activity has the same units as the concentration. The activity coefficient varies with the total number of ions in the solution and with their charge, and it is a correction for interionic attraction. In dilute solution, less than $10^{-4} \, M$, the activity coefficient of a simple electrolyte is near unity, and activity is approximately equal to the concentration. As the concentration of an electrolyte increases, or as an extraneous salt is added, the activity coefficient generally decreases and the activity becomes less than the concentration.

The Ionic Strength

From the above discussion, it is apparent that the activity coefficient is a function of the total electrolyte concentration of the solution. The **ionic strength** is a measure of total electrolyte concentration and is defined by

$$\mu = 1/2 \Sigma C_i Z_i^2$$ (4.18)

where μ is the ionic strength and Z_i is the charge on each individual ion. All cations and anions present in solution are included in the calculation. Obviously, for each positive charge there will be a negative charge.

EXAMPLE 4.6   Calculate the ionic strength of a 0.2 M solution of KNO_3 and a 0.2 M solution of K_2SO_4.

SOLUTION
For KNO_3,

$$\mu = \frac{C_{K^+} Z_{K^+}^2 + C_{NO_3^-} Z_{NO_3^-}^2}{2}$$

$$[K^+] = 0.2\ M;\quad [NO_3^-] = 0.2\ M$$

$$\mu = \frac{0.2 \times (1)^2 + 0.2 \times (1)^2}{2} = 0.2$$

For K_2SO_4,

$$\mu = \frac{C_{K^+} Z_{K^+}^2 + C_{SO_4^{2-}} Z_{SO_4^{2-}}^2}{2}$$

$$[K^+] = 0.4\ M;\quad [SO_4^{2-}] = 0.2\ M$$

So,

$$\mu = \frac{0.4 \times (1)^2 + 0.2 \times (2)^2}{2} = 0.6$$

Note that due to the doubly charged SO_4^{2-}, the ionic strength of the K_2SO_4 is three times that of the KNO_3.

If more than one salt is present, then the ionic strength is calculated from the total concentration and charges of all the different ions. For any given electrolyte, the ionic strength will be proportional to the concentration. Strong acids that are completely ionized are treated in the same manner as salts. If the acids are partially ionized, then the concentration of the ionized species must be estimated from the ionization constant before computing the ionic strength. Very weak acids can usually be considered to be nonionized and do not contribute to the ionic strength.

EXAMPLE 4.7   Calculate the ionic strength of a solution consisting of 0.30 M NaCl and 0.20 M Na_2SO_4.

SOLUTION

$$\mu = \frac{C_{Na^+}Z_{Na^+}^2 \ + \ C_{Cl^-}Z_{Cl^-}^2 \ + \ C_{SO_4^{2-}}Z_{SO_4^{2-}}^2}{2}$$

$$= \frac{0.70 \times (1)^2 + 0.30 \times (1)^2 + 0.20 \times (2)^2}{2}$$

$$= 0.90$$

Calculation of Activity Coefficients

In 1923, Debye and Hückel derived a theoretical expression for calculating activity coefficients. The equation, known as the **Debye–Hückel equation,** is

$$-\log f_i = \frac{0.51 Z_i^2 \sqrt{\mu}}{1 + 0.33 \alpha_i \sqrt{\mu}} \tag{4.19}$$

The numbers 0.51 and 0.33 are constants for water at 25°C, and the former includes the $-3/2$ power of both the dielectric constant of the solvent and the absolute temperature; α_i is the **ion size parameter,** which is the effective diameter of the hydrated ion in å angstrom units. A limitation of the Debye–Hückel equation is the accuracy to which α_i can be evaluated. For most singly charged ions, α_i is generally about 3 Å, and for practical purposes Equation 4.19 simplifies to

$$-\log f_i = \frac{0.51 \ Z_i^2 \sqrt{\mu}}{1 + \sqrt{\mu}} \tag{4.20}$$

For common multiply charged ions, α_i may become as large as 11 Å. But at ionic strengths less than 0.01, the second term of the denominator becomes small with respect to 1, so uncertainties in α_i become relatively unimportant and Equation 4.20 can be applied at ionic strengths of 0.01 or less. Equation 4.19 can be applied up to ionic strengths of about 0.2. Reference 7 at the end of the chapter lists values for α_i for different ions, and also includes a table of calculated activity coefficients using Equation 4.20, at ionic strengths ranging from 0.0005 to 0.1.

EXAMPLE 4.8   Calculate the activity coefficients for K^+ and SO_4^{2-} in a 0.0020 M solution of potassium sulfate.

SOLUTION
The ionic strength is 0.0060, so we can apply Equation 4.20:

$$-\log f_{K^+} = \frac{0.51(1)^2 \sqrt{0.0060}}{1 + \sqrt{0.0060}} = 0.037$$

$$f_{K^+} = 10^{-0.037} = 10^{-1} \times 10^{0.963} = 0.918$$

$$-\log f_{SO_4^{2-}} = \frac{0.51(2)^2\sqrt{0.0060}}{1 + \sqrt{0.0060}} = 0.14_7$$

$$f_{SO_4^{2-}} = 10^{-0.14_7} = 10^{-1} \times 10^{0.853} = 0.71_3$$

EXAMPLE 4.9 Calculate the activity coefficients for K^+ and SO_4^{2-} in a 0.020 M solution of potassium sulfate.

SOLUTION
The ionic strength is 0.060, so we would use Equation 4.19. From Reference 7, we find that $\alpha_{K^+} = 3$ Å and $\alpha_{SO_4^{2-}} = 4.0$ Å. For K^+, we can use Equation 4.20:

$$-\log f_{K^+} = \frac{0.51(1)^2\sqrt{0.060}}{1 + \sqrt{0.060}} = 0.10_1$$

$$f_{K^+} = 10^{-0.101} = 10^{-1} \times 10^{0.899} = 0.79_2$$

For SO_4^{2-}, use Equation 4.19:

$$-\log f_{SO_4^{2-}} = \frac{0.51(2)^2\sqrt{0.060}}{1 + 0.33 \times 4.0\sqrt{0.060}} = 0.37_8$$

$$f_{SO_4^{2-}} = 10^{-1} \times 10^{0.622} = 0.41_9$$

This latter compares with a calculated value of 0.39_6 using Equation 4.20. Note the decrease in the activity coefficients compared to 0.002 M K_2SO_4, especially for the SO_4^{2-} ion.

For higher ionic strengths, a number of empirical equations have been developed. Perhaps one of the more useful is the **Davies modification** (see Reference 6):

$$-\log f_i = \frac{0.51Z_i^2\sqrt{\mu}}{1 + 0.33\alpha_i\sqrt{\mu}} - 0.10Z_i^2\,\mu \qquad (4.21)$$

This is identical to the Debye–Hückel equation, but with the correction term on the end. It is valid up to ionic strengths of about 0.6.

At very high electrolyte concentrations, activity coefficients may actually increase and become greater than unity. This is because the activity of the solvent, water, is decreased and solvated ionic species become partially desolvated. This increases their reactivity and hence their activity.

1. The activity coefficients of ions of a given charge type are approximately the same in solutions of a given ionic strength, and this activity coefficient is the same regardless of their concentrations.
2. The behavior of ions becomes less ideal as the charge type increases, resulting in less confidence in calculated activity coefficients.
3. The calculated activity coefficient of an ion in a mixed electrolyte solution will be less accurate than in a single electrolyte solution.
4. The activity coefficients of nonelectrolytes (uncharged molecules) can generally be considered equal to unity in ionic strengths up to 0.1, and deviations from this

> approximation are only moderate in ionic strengths as high as 1. Undissociated acids, HA, are nonelectrolytes whose activity coefficients can be taken as unity.

We can draw some general conclusions about the estimation of activity coefficients.

4.13 THE DIVERSE ION EFFECT: THE THERMODYNAMIC EQUILIBRIUM CONSTANT

We mentioned at the beginning of the last section on activity that the presence of diverse salts will generally increase the dissociation of weak electrolytes due to a shielding (or decrease in the activity) of the ionic species produced upon dissociation. We can quantitatively predict the extent of the effect on the equilibrium by taking into account the activities of the species in the equilibrium.

In our consideration of equilibrium constants thus far, we have assumed no diverse ion effect, that is, an ionic strength of zero and an activity coefficient of 1. Equilibrium constants should more exactly be expressed in terms of activities rather than concentrations. Consider the dissociation of AB. The **thermodynamic equilibrium constant** K_{eq}^0 is

$$K_{eq}^0 = \frac{a_A \cdot a_B}{a_{AB}} = \frac{[A]f_A \cdot [B]f_B}{[AB]f_{AB}} \tag{4.22}$$

Since the **concentration equilibrium constant** $K_{eq} = [A][B]/[AB]$,

$$K_{eq}^0 = K_{eq} \frac{f_A \cdot f_B}{f_{AB}} \tag{4.23}$$

or

$$K_{eq} = K_{eq}^0 \frac{f_{AB}}{f_A \cdot f_B} \tag{4.24}$$

The numberical value of K_{eq}^0 holds for all activities. $K_{eq} = K_{eq}^0$ at zero ionic strength, but at appreciable ionic strengths, a value for K_{eq} must be calculated for each ionic strength using Equation 4.24. The equilibrium constants listed in the appendix are for zero ionic strength, that is, they are really thermodynamic equilibrium constants. (Experimental K_{eq} values are available at different ionic strengths and can be used for equilibrium calculations at the listed ionic strength, using molar concentrations without having to calculate activity coefficients.)

EXAMPLE 4.10 The weak electrolyte AB dissociates to A^+ and B^-, with a thermodynamic equilibrium constant K_{eq}^0 of 2×10^{-8}. (a) Calculate the molar equilibrium constant K_{eq}. (b) Calculate the percent dissociation of a 1.0×10^{-4} M solution of AB in water and in the presence of a diverse salt of ionic strength 0.1, if the activity coefficients of A^+ and B^- are 0.6 and 0.7, respectively, at $\mu = 0.1$.

SOLUTION
(a) $AB \rightleftharpoons A^+ + B^-$

$$K_{eq} = \frac{[A^+][B^-]}{[AB]}$$

$$K_{eq}^0 = \frac{a_{A^+} \cdot a_{B^-}}{a_{AB}} = \frac{[A^+]f_{A^+} \cdot [B^-]f_{B^-}}{[AB]f_{AB}}$$

The activity of a neutral species is unity, so

$$K_{eq}^0 = \frac{[A^+][B^-]}{[AB]} \cdot f_{A^+} \cdot f_{B^-} = K_{eq}f_{A^+} \cdot f_{B^-}$$

$$K_{eq} = \frac{K_{eq}^0}{f_{A^+} \cdot f_{B^-}} = \frac{2 \times 10^{-8}}{(0.6)(0.7)} = 5 \times 10^{-8}$$

(b)

$$\begin{array}{ccccc} AB & \rightleftharpoons & A^+ & + & B^- \\ 1 \times 10^{-4} - x & & x & & x \end{array}$$

In water, $f_{A^+} = f_{B^-} \approx 1$ (since $\mu < 10^{-4}$), $x \ll 10^{-4}$

$$\frac{[A^+][B^-]}{[AB]} = 2 \times 10^{-8}$$

$$\frac{(x)(x)}{1.0 \times 10^{-4}} = 2 \times 10^{-8}$$

$$x = 1._4 \times 10^{-6}\ M$$

$$\% \text{ dissociated} = \frac{1._4 \times 10^{-6}\ M}{1.0 \times 10^{-4}\ M} \times 100\% = 1._4\%$$

For 0.1 M salt,

$$\frac{[A^+][B^-]}{[AB]} = 5 \times 10^{-8}$$

$$\frac{(x)(x)}{1.0 \times 10^{-4}} = 5 \times 10^{-8}$$

$$x = 2._2 \times 10^{-6}$$

$$\% \text{ dissociated} = \frac{2._2 \times 10^{-6}}{1.0 \times 10^{-4}} \times 100\% = 2._2\%$$

which represents a 57% increase in dissociation.

Calculations using the diverse ion effect are illustrated in Chapter 5 for acid dissociation and in Chapter 6 for precipitate solubilities. For illustrative purposes throughout this book, we will in general neglect the diverse ion effects on equilibria. In most cases, we are interested in *relative* changes in equilibrium concentrations, and the neglect of activities will not change our arguments.

PROBLEMS

Equilibrium Calculations

1. A and B react as follows: $A + B \rightleftharpoons C + D$. The equilibrium constant is 2.0×10^3. If 0.30 mol of A and 0.80 mol of B are mixed in 1 L, what are the concentrations of A, B, C, and D after reaction?

2. A and B react as follows: $A + B \rightleftharpoons 2C$. The equilibrium constant is 5.0×10^6. If 0.40 mol of A and 0.70 mol of B are mixed in 1 L, what are the concentrations of A, B, and C after reaction?

3. The dissociation constant for salicylic acid, $C_6H_4(OH)COOH$, is 1.0×10^{-3}. Calculate the percent dissociation of a 1.0×10^{-3} M solution. There is one dissociable proton.

4. The dissociation constant for hydrocyanic acid, HCN, is 7.2×10^{-10}. Calculate the percent dissociation of a 1.0×10^{-3} M solution.

5. Calculate the percent dissociation of the salicyclic acid in Problem 3 if the solution also contained 1.0×10^{-2} M sodium salicylate (the salt of salicylic acid).

6. Hydrogen sulfide, H_2S, dissociates stepwise, with dissociation constants of 9.1×10^{-8} and 1.2×10^{-15}, respectively. Write the overall dissociation reaction and the overall equilibrium constant.

7. Fe^{2+} and $Cr_2O_7^{2-}$ react as follows: $6Fe^{2+} + Cr_2O_7^{2-} + 14H^+ \rightleftharpoons 6Fe^{3+} + 2Cr^{3+} + 7H_2O$. The equilibrium constant for the reaction is 1×10^{57}. Calculate the equilibrium concentrations of the iron and chromium species if 10 mL each of 0.02 M $K_2Cr_2O_7$ in 1.14 M HCl and 0.12 M $FeSO_4$ in 1.14 M HCl are reacted.

Ionic Strength

8. Calculate the ionic strengths of the following solutions. (a) 0.30 M NaCl; (b) 0.30 M Na_2SO_4; (c) 0.30 M NaCl and 0.20 M K_2SO_4; (d) 0.20 M $Al_2(SO_4)_3$ and 0.10 M Na_2SO_4.

9. Calculate the ionic strengths of the following solutions: (a) 0.20 M $ZnSO_4$; (b) 0.40 M $MgCl_2$; (c) 0.50 M $LaCL_3$; (d) 1.0 M $K_2Cr_2O_7$; (e) 1.0 M $Tl(NO_3)_3$ + 1.0 M $Pb(NO_3)_2$.

Activity.

10. Calculate the activity coefficients of the sodium and chloride ions for a 0.00100 M solution of NaCl.

11. Calculate the activity coefficients of each ion in a solution containing 0.0020 M Na_2SO_4 and 0.0010 M $Al_2(SO_4)_3$.

12. Calculate the activity of the NO_3^- ion in a solution of 0.0020 M KNO_3.

13. Calculate the activity of the CrO_4^{2-} ion in a 0.020 M solution of Na_2CrO_4.

Thermodynamic Equilibrium Constants

14. Write thermodynamic equilibrium constant expressions for the following:
    (a) $HCN \rightleftharpoons H^+ + CN^-$
    (b) $NH_3 + H_2O \rightleftharpoons NH_4^+ + OH^-$

15. Calculate the pH of a solution of 5.0×10^{-3} M benzoic acid in (a) water and (b) in the presence of 0.05 M K_2SO_4.

RECOMMENDED REFERENCES

Equilibria

1. A. J. Bard, *Chemical Equilibrium*. New York: Harper & Row, 1966.
2. T. R. Blackburn, *Equilibrium: A Chemistry of Solutions*. New York: Holt, Rinehart and Winston, 1969.

3. J. N. Butler, *Ionic Equilibrium. A Mathematical Approach.* Reading, Mass.: Addison-Wesley, 1964.

4. G. M. Fleck, *Ionic Equilibria in Solution.* New York: Holt, Rinehart and Winston, 1966.

5. H. Freiser and Q. Fernando, *Ionic Equilibria in Analytical Chemistry.* New York: John Wiley & Sons, Inc., 1963.

Activity

6. C. W. Davies, *Ion Association.* London: Butterworth, 1962.

7. J. Kielland, "Individual Activity Coefficients of Ions in Aqueous Solutions," *J. Am. Chem. Soc.,* **59** (1937) 1675.

chapter 5

ACID–BASE EQUILIBRIA

The acidity or basicity of a solution is frequently an important factor in chemical reactions. The use of buffers of a given pH to maintain the solution pH at a desired level is very important. In addition, fundamental acid–base equilibria are important in understanding acid–base titrations and the effects of acids on chemical species and reactions, for example, the effects on complexion or precipitation. In the last chapter, we described the fundamental concept of equilibrium constants. In this chapter, we consider in more detail various acid–base equilibrium calculations, including weak acids and bases, hydrolysis of salts of weak acids and bases, buffers, polyprotic acids and their salts, and physiological buffers. The basic pH concept is reviewed first.

5.1 AQUEOUS ACID–BASE THEORIES

Several theories have been used to describe the acid–base properties of substances in water as well as other solvents. For aqueous systems, either the **Arrhenius theory** or the **Brønsted theory** adequately describes acids and bases. In the Arrhenius theory, an acid is any substance that ionizes (partially or completely) in water to give hydrogen ions (which associate with the solvent to give hydronium ions, H_3O^+):

$$HA + H_2O \rightleftharpoons H_3O^+ + A^-$$

A base ionizes in water to give hydroxyl ions. Weak (partially ionized) bases generally ionize as follows:

$$B + H_2O \rightleftharpoons BH^+ + OH^-$$

while strong bases such as metal hydroxides (e.g. NaOH) dissociate as

$$M(OH)_n \rightarrow M^{n+} + nOH^-$$

In the Brønsted theory, **an acid is any substance that can donate a proton, and a base is any substance that can accept a proton.** Hence, an acid ionizes to give a proton and a base:

$$\boxed{acid = H^+ + base} \tag{5.1}$$

The acid and base pairs in the ionization reaction are called **conjugate pairs,** that is, the base is the conjugate base of the acid and, vice versa, the acid is the conjugate acid of the base.

5.2   ACID–BASE EQUILIBRIA IN WATER

We see from the above that when an acid or base is dissolved in water, it will dissociate, or **ionize,** the amount of ionization being dependent on the strength of the acid. A "strong" electrolyte is completely dissociated, while a "weak" electrolyte is partially dissociated. Table 5.1 lists some common electrolytes, some strong and some weak. Other weak acids and bases are listed in the Appendix.

Hydrochloric acid is a strong acid, and its ionization is complete:

$$HCl + H_2O \rightarrow H_3O^+ + Cl^- \tag{5.2}$$

The proton H^+ exists in water as a hydrated ion, the **hydronium ion,** H_3O^+. Higher hydrates probably exist, particularly $H_9O_4^+$. The hydronium ion is written as H_3O^+ for convenience and to emphasize Brønsted behavior.

Acetic acid[1] is a weak acid, which ionizes only partially:

$$HOAc + H_2O \rightleftharpoons H_3O^+ + OAc^- \tag{5.3}$$

We can write an **equilibrium constant** for this reaction:

$$K = \frac{a_{H_3O^+} \cdot a_{OAc^-}}{a_{HOAc} \cdot a_{H_2O}} \tag{5.4}$$

where K is the **acidity constant** and a is the **activity** of the indicated species. An equilibrium constant for Equation 5.2 would have a value of infinity. Salt cations or anions may also react with water after they are dissociated, for example, acetate ion from dissociated acetate salt, to give HOAc.

The activity can be thought of as representing the effective concentration of an ion (described in Chapter 4). The effects of protons in reactions are often governed by their activities, and it is the activity that is measured by the widely used pH meter (Chapter 11). Methods for predicting numerical values of activity coefficients were described in Chapter 4.

In dilute solutions, the activity of water remains essentially constant. Therefore, it can be incorporated in K, and Equation 5.4 can be written as

---

[1]We shall use the symbol OAc^- to represent the acetate ion $CH_3-\overset{\overset{O}{\|}}{C}-O^-$.

TABLE 5.1
Some Strong Electrolytes and Some Weak Electrolytes

Strong	Weak
HCl	$HC_2H_3O_2$ (acetic acid)
$HClO_4$	NH_3
$H_2SO_4^a$	C_6H_5OH (phenol)
HNO_3	$HCHO_2$ (formic acid)
NaOH	$C_6H_5NH_2$ (aniline)
$NaC_2H_3O_2$	

^{a}The first proton is completely ionized in dilute solution, but the second proton is partially ionized ($K_i = 10^{-2}$).

$$K_a = \frac{a_{H_3O^+} \cdot a_{OAc^-}}{a_{HOAc}} \tag{5.5}$$

Pure water ionizes slightly, or undergoes **autoprotolysis:**

$$2H_2O \rightleftharpoons H_3O^+ + OH^- \tag{5.6}$$

The equilibrium constant for this is

$$K = \frac{a_{H_3O^+} \cdot a_{OH^-}}{a_{H_2O}^2} \tag{5.7}$$

Again, the activity of water is constant in dilute solutions ($\sim$55.3 M), so

$$K_w = a_{H_3O^+} \cdot a_{OH^-} \tag{5.8}$$

where K_w is the **autoprotolysis,** or **self-ionization, constant.**

Calculations are simplified if we neglect activity coefficients. This simplification results in only slight errors for dilute solutions, and we shall use molar concentrations in all our calculations. This will satisfactorily illustrate the equilibria involved. Most of the solutions we will be concerned with are rather dilute, and we will frequently be interested in relative changes in pH (and large ones) in which case small errors are insignificant. We will simplify our expressions by using H^+ in place of H_3O^+. This is not inconsistent, since the waters of solvation associated with other ions or molecules (e.g., metal ions) are not generally written and H_3O^+ is not an entirely accurate representation of the species present anyway.

Molar concentration will be represented by a square bracket [ ] around the species. Simplified equations for the above reactions are

$$HCl \rightarrow H^+ + Cl^- \tag{5.9}$$

$$HOAc \rightleftharpoons H^+ + OAc^- \tag{5.10}$$

$$\boxed{K = \frac{[H^+][OAc^-]}{[HOAc]}} \tag{5.11}$$

$$H_2O \rightleftharpoons H^+ + OH^- \tag{5.12}$$

$$\boxed{K_w = [H^+][OH^-]} \tag{5.13}$$

At 25°C, $K_w = 1.0 \times 10^{-14}$. The product of the hydrogen ion concentration and the hydroxyl ion concentration in aqueous solution is **always** equal to 1.0×10^{-14} at room temperature:

$$[H^+][OH^-] = 1.0 \times 10^{-14} \tag{5.14}$$

In pure water, then, the concentrations of these two species are equal, since there are no other sources of H^+ or OH^- except H_2O dissociation:

$$[H^+] = [OH^-]$$

Therefore,

$$[H^+][H^+] = 1.0 \times 10^{-14}$$

$$[H^+] = 1.0 \times 10^{-7} \, M \equiv [OH^-]$$

If an acid is added to water, we can calculate the hydroxyl ion concentration if we know the hydrogen ion concentration from the acid. Except when the hydrogen ion concentration from the acid is very small, 10^{-6} M or less, any contribution to $[H^+]$ from the ionization of water can be neglected.

EXAMPLE 5.1 A 1.0×10^{-3} M solution of hydrochloric acid is prepared. What is the hydroxyl ion concentration?

SOLUTION

Since hydrochloric acid is a strong electrolyte and is completely ionized, the H^+ concentration is 1.0×10^{-3} M. Thus,

$$(1.0 \times 10^{-3})[OH^-] = 1.0 \times 10^{-14}$$
$$[OH^-] = 1.0 \times 10^{-11} \, M$$

5.3 THE pH SCALE

The concentration of H^+ or OH^- in aqueous solutions can vary over extremely wide ranges, from 1 M or greater to 10^{-14} M or less. To construct a plot of H^+ concentration against some variable would be very difficult if the concentration changed from, say 10^{-1} M to 10^{-13} M. This range is common in a titration. It is more convenient to compress the acidity scale by placing it on a logarithm basis. The **pH** of a solution was defined by Sørenson as

$$pH = -\log [H^+] \tag{5.15}$$

The minus sign is used because most of the concentrations encountered are less than 1 M, and so this designation gives a positive number. (More strictly, pH is now defined as $-\log a_{H^+}$, but we will use the simpler definition of Equation 5.15.) In general, pAnything $= -\log$ Anything, and this method of notation will be used later for other numbers that can vary by large amounts.

EXAMPLE 5.2   Calculate the pH of a 2.0×10^{-3} M solution of HCl.

SOLUTION
HCl is completely ionized, so

$$[H^+] = 2.0 \times 10^{-3}\ M$$

$$pH = -\log (2.0 \times 10^{-3}) = 3 - \log 2.0 = 3 - 0.30 = 2.70$$

A similar definition is made for the hydroxyl ion concentration:

$$\boxed{pOH = -\log [OH^-]} \qquad (5.16)$$

Equation 5.13 can be used to calculate the hydroxyl ion concentration if the hydrogen ion concentration is known, and vice versa. The equation in logarithm form for a more direct calculation of pH or pOH is

$$-\log K_w = -\log [H^+][OH^-] = -\log [H^+] - \log [OH^-], \qquad (5.17)$$

$$\boxed{pK_w = pH + pOH} \qquad (5.18)$$

At 25°C,

$$\boxed{14.00 = pH + pOH} \qquad (5.19)$$

EXAMPLE 5.3   Calculate the pOH and the pH of a 5.0×10^{-2} M solution of NaOH.

SOLUTION

$$[OH^-] = 5.0 \times 10^{-2}\ M$$

$$pOH = -\log (5.0 \times 10^{-2}) = 2 - \log 5.0 = 2 - 0.70 = 1.30$$

$$pH + 1.30 = 14.00$$

$$pH = 12.70$$

or

$$[H^+] = \frac{1.0 \times 10^{-14}}{5.0 \times 10^{-2}} = 2.0 \times 10^{-13}\ M$$

$$pH = -\log (2.0 \times 10^{-13}) = 13 - \log 2.0 = 13 - 0.30 = 12.70$$

EXAMPLE 5.4   Calculate the pH of a solution prepared by mixing 2.0 mL of a strong acid solution of pH 3.00 and 3.0 mL of a strong base of pH 10.00.

SOLUTION

$$[H^+] \text{ of acid solution} = 1.0 \times 10^{-3}\ M$$

$$\text{mmol } H^+ = 1.0 \times 10^{-3}\ M \times 2.0\ mL = 2.0 \times 10^{-3}\ mmol$$

$$pOH \text{ of base solution} = 14.00 - 10.00 = 4.00$$

$$[OH^-] = 1.0 \times 10^{-4} \, M$$

$$\text{mmol } OH^- = 1.0 \times 10^{-4} \, M \times 3.0 \, mL = 3.0 \times 10^{-4} \, mmol$$

There is an excess of acid.

$$\text{mmol } H^+ = 0.0020 - 0.0003 = 0.0017 \, mmol$$

$$[H^+] = 0.0017 \, mmol/5.0 \, mL = 3.4 \times 10^{-4} \, M$$

$$pH = -\log 3.4 \times 10^{-4} = 4 - 0.53 = 3.47$$

EXAMPLE 5.5   The pH of a solution is 9.67. Calculate the hydrogen ion concentration in the solution.

SOLUTION

$$-\log [H^+] = 9.67$$
$$[H^+] = 10^{-9.67} = 10^{-10} \times 10^{0.33}$$
$$[H^+] = 2.1 \times 10^{-10} \, M$$

The hydrogen ion concentration in pure water is 10^{-7} M, and the pH of water is 7. A pH of 7 is therefore **neutral.** Values of pH that are greater than this are **alkaline,** and pH values less than this are **acidic.** The reverse is true of pOH values. A pOH of 7 is also neutral. Note that the product of $[H^+]$ and $[OH^-]$ is always 10^{-14} at 25°C, and the sum of pH and pOH is always 14.

Students are often under the illusion that it is impossible to have a **negative pH.** There is no theoretical basis for this. A negative pH only means that the hydrogen ion concentration is greater than 1 M. In actual practice, a negative pH is uncommon because of two reasons. First, even strong acids may become partially undissociated at high concentrations. For example, 100% H_2SO_4 is so weakly dissociated that it can be stored in iron containers; more dilute H_2SO_4 solutions would contain sufficient protons from dissociation to attack and dissolve the iron. The second reason has to do with the **activity,** which we have chosen to neglect for dilute solutions. Since pH is really $-\log a_{H^+}$ (this is what a pH meter reading is a measure of), a solution that is 1.1 M in H^+ may actually have a positive pH because the activity of the H^+ is less than 1.0 M.[2] This is because at these high concentrations, the activity coefficient is less than unity (although at still higher concentrations the activity coefficient may become greater than unity—see Chapter 4). Nevertheless, there is mathematically no basis for not having a negative pH (or a negative pOH), although it may be rare in analytical solutions encountered.

If the concentration of an acid or base is much less than 10^{-7} M, then its contribution to the acidity or basicity will be negligible compared with the contribution

---

[2]As will be seen in Chapter 11, it is also difficult to *measure* the pH of a solution having a negative pH or pOH because high concentrations of acids or bases tend to introduce an error in the measurement by adding a significant and unknown liquid-junction potential in the measurements.

from water. The pH of a 10^{-8} M sodium hydroxide solution would therefore not differ significantly from 7. If the concentration of the acid or base is around 10^{-7} M, then its contribution is not negligible and neither is that from water, hence the sum of the two contributions must be taken.

EXAMPLE 5.6   Calculate the pH and pOH of a 1.0×10^{-7} M solution of HCl.

SOLUTION
Equilibria:

$$HCl \rightarrow H^+ + Cl^-$$

$$H_2O \rightleftharpoons H^+ + OH^-$$

$$[H^+][OH^-] = 1.0 \times 10^{-14}$$

$$[H^+]_{H_2O\ diss.} = [OH^-]_{H_2O\ diss.} = x$$

Since the hydrogen ions contributed from the ionization of water are not negligible compared to the HCl added,

$$[H^+] = C_{HCl} + [H^+]_{H_2O\ diss.}$$

Then,

$$([H^+]_{HCl} + x)(x) = 1.0 \times 10^{-14}$$

$$(1.00 \times 10^{-7} + x)(x) = 1.0 \times 10^{-14}$$

$$x^2 + 1.00 \times 10^{-7}\,x - 1.0 \times 10^{-14} = 0$$

Using the quadratic equation to solve (see the Appendix),

$$x = \frac{-1.00 \times 10^{-7} \pm \sqrt{1.0 \times 10^{-14} + 4(1.0 \times 10^{-14})}}{2} = 6.2 \times 10^{-8}\ M$$

Therefore, the *total* H^+ concentration $= (1.00 \times 10^{-7} + 6.2 \times 10^{-8}) = 1.62 \times 10^{-7}$ M:

$$pH = -\log 1.62 \times 10^{-7} = 7 - 0.21 = 6.79$$

$$pOH = 14.00 - 6.79 = 7.21$$

or, since $[OH^-] = x$,

$$pOH = -\log (6.2 \times 10^{-8}) = 8 - 0.79 = 7.21$$

Note that, owing to the presence of the added H^+, the ionization of water is suppressed by 38% by the common ion effect (Le Châtelier's principle). At higher acid (or base) concentrations, the suppression is even greater and the contribution from the water becomes negligible. This contribution can be considered negligible if the concentration of protons or hydroxyl ions from an acid or base is **10^{-6} M or greater.**

The calculation in this example is more academic than practical, because *carbon dioxide from the air dissolved in water exceeds these concentrations.* Since carbon dioxide in water forms an acid, extreme care would have to be taken to remove and keep this from the water.

5.4   pH AT ELEVATED TEMPERATURES: BLOOD pH

It is a convenient fact of nature for students and chemists who deal with acidity calculations and pH scales in aqueous solutions at room temperature that K_w is an integral number. At 100°C, for example, $K_w = 5.5 \times 10^{-13}$, and a neutral solution has

$$[H^+] = \sqrt{5.5 \times 10^{-13}} = 7.4 \times 10^{-7}\ M$$
$$pH = pOH = 6.13$$
$$pK_w = 12.26 = pH + pOH$$

Not all measurements or interpretations are done at room temperature, however, and the temperature dependence of K_w must be taken into account (recall from Chapter 4 that equilibrium constants are temperature dependent). An important example is the pH of the body. The pH of blood at body temperature (37°C) is 7.35 to 7.45. This value represents a slightly more alkaline solution relative to neutral water than the same value would be at room temperature. At 37°, $K_w = 2.5 \times 10^{-14}$ and $pK_w = 13.60$. The pH (and pOH) of a neutral solution is 13.60/2 = 6.80. The hydrogen ion (and hydroxide ion) concentration is $\sqrt{2.5 \times 10^{-14}}$ = $1.6 \times 10^{-7}\ M$. Since a neutral blood solution at 37° would have pH 6.8, a blood pH of 7.4 is more alkaline at 37° by 0.2 pH units than it would be at 25°. This is important when one considers that a change of 0.3 pH units in the body is extreme.

The hydrochloric acid concentration in the stomach is about 0.1 to 0.02 M. Since pH = $-\log [H^+]$, the pH at 0.02 M would be 1.7. It will be the same regardless of the temperature, since the hydrogen ion concentration is the same (neglecting solvent volume changes), and the same pH would be measured at either temperature. But, while the pOH would be 14.0 − 1.7 = 12.3 at 25°, it is 13.6 − 1.7 = 11.9 at 37°.

Not only does the temperature affect the ionization of water in the body and therefore change the pH of a neutral solution, it also affects the ionization constants of the acids and bases from which the buffer systems in the body are derived. As we shall see later in the chapter, this influences the pH of the buffers, and so a blood pH of 7.4 measured at 37°C will not be the same when measured at room temperature, in contrast to the stomach pH whose value was determined by the concentration of a strong acid. For this reason, measurement of blood pH for diagnostic purposes is generally done at 37° (see Chapter 11). (Neglecting changes in equilibrium constants of the blood buffer systems, the measured pH would be the same at 25° or 37°—remembering to readjust the acidity scale at 37°—but this is purely academic, since the equilibrium constants do indeed change.)

5.5   WEAK ACIDS AND BASES

We have limited our calculations so far to strong acids and bases in which ionization is complete. Since the concentration of H^+ or OH^- is readily determined from the concentration of the acid or base, the calculations are straightforward. As seen in Equation 5.3, weak acids (or bases) are only partially ionized. The ionization constant can be used to calculate the amount ionized and, from this, the pH. The acidity constant for acetic acid at 25°C is 1.75×10^{-5}:

$$\frac{[H^+][OAc^-]}{[HOAc]} = 1.75 \times 10^{-5} \tag{5.20}$$

When acetic acid ionizes, it dissociates to equal portions of H^+ and OAc^- by such an amount that the computation on the left side of Equation 5.20 will always be equal to 1.75×10^{-5}:

$$HOAc \rightleftharpoons H^+ + OAc^- \tag{5.21}$$

If the original concentration of acetic acid is C and the concentration of ionized acetic acid species (H^+ and OAc^-) is x, then the final concentration for each species at equilibrium is given by

$$\begin{array}{ccc} HOAc & \rightleftharpoons H^+ + & OAc^- \\ (C - x) & x & x \end{array} \tag{5.22}$$

EXAMPLE 5.7 Calculate the pH and pOH of a 1.00×10^{-3} M solution of acetic acid.

SOLUTION

$$HOAc \rightleftharpoons H^+ + OAc^-$$

The concentrations of the various species are as follows:

	[HOAc]	[H$^+$]	[OAc$^-$]
Initial	1.00×10^{-3}	0	0
Change (x = mmol/mL HOAc ionized)	$-x$	$+x$	$+x$
Equilibrium	$1.00 \times 10^{-3} - x$	x	x

From Equation 5.20

$$\frac{(x)(x)}{(1.00 \times 10^{-3} - x)} = 1.75 \times 10^{-5}$$

The solution is that of a quadratic equation. The expression may be simplified if less than about 10 or 15% of the acid is ionized by neglecting x compared with C (10^{-3} M in this case). This is an arbitrary (and not very demanding) criterion. The simplification applies if **K_a is smaller than about 0.01C,** that is, smaller than 10^{-4} at C = 0.01 M, 10^{-3} at C = 0.1 M, and so forth. Under these conditions, the error in calculation is 5% or less (results come out too high), and within the probable accuracy of the equilibrium constant. Our calculation simplifies to

$$\frac{x^2}{1.00 \times 10^{-3}} = 1.75 \times 10^{-5}$$

$$x = 1.32 \times 10^{-4} M \equiv [H^+]$$

Therefore,

$$pH = -\log 1.32 \times 10^{-4} = 4 - \log 1.32 = 4 - 0.12 = 3.88$$

$$pOH = 14.00 - 3.88 = 10.12$$

The simplification in the calculation is not serious, particularly since equilibrium constants are not known to a high degree of accuracy (frequently no better than $\pm 10\%$). In the above example, solution of the quadratic equation results in $[H^+]$ $= 1.26 \times 10^{-4}\ M$ (5% less) and pH $= 3.90$. This pH is within 0.02 unit of that calculated using the simplification, which is near the limit of accuracy to which pH measurements can be made, and almost certainly as close a calculation as is justified in view of the experimental errors in K_a or K_b values and the fact that we are using concentrations rather than activities in the calculations. In our calculations, we also neglected the contribution of hydrogen ions from the ionization of water (which was obviously justified); this is generally permissible except for very dilute ($< 10^{-6}\ M$) or very weak ($K_a < 10^{-12}$) acids.

Similar equations and calculations hold for weak bases.

EXAMPLE 5.8   The basicity constant K_b for ammonia is 1.75×10^{-5} at 25°C. (It is only coincidental that this is equal to K_a for acetic acid.) Calculate the pH and pOH for a $1.00 \times 10^{-3}\ M$ solution of ammonia.

SOLUTION

$$NH_3 \qquad + H_2O \rightleftharpoons NH_4^+ + OH^-$$
$$(1.00 \times 10^{-3} - x) \qquad\qquad x \qquad x$$

$$\frac{[NH_4^+][OH^-]}{[NH_3]} = 1.75 \times 10^{-5}$$

The same rule applies for the approximation applied for a weak acid. Thus,

$$\frac{(x)(x)}{1.00 \times 10^{-3}} = 1.75 \times 10^{-5}$$

$$x = 1.32 \times 10^{-4}\ M = [OH^-]$$

$$pOH = -\log 1.32 \times 10^{-4} = 3.88$$

$$pH = 14.00 - 3.88 = 10.12$$

5.6   SALTS OF WEAK ACIDS AND BASES

The salt of a weak acid, for example, NaOAc, is a strong electrolyte like (almost) all salts, and completely ionizes. In addition, the anion of the **salt of a weak acid** is a **Brønsted base,** which will accept protons. It ionizes in water (a Brønsted acid) to form hydroxide ion and the corresponding undissociated acid. For example,

$$OAc^- + H_2O \rightleftharpoons HOAc + OH^- \tag{5.23}$$

The HOAc here is undissociated and therefore does not contribute to the pH. This ionization is also known as **hydrolysis** of the salt ion. Because it hydrolyzes, sodium acetate is a weak base (the conjugate base of acetic acid). The ionization constant for Equation 5.23 is equal to the basicity constant of the salt. The weaker the conjugate acid, the stronger the conjugate base, that is, the more strongly the salt will combine with a proton, as from the water, to shift the ionization in Equation 5.23 to the right. **Equilibria for these Brønsted bases are treated identically**

to the weak bases we have just considered. We can write an equilibrium constant:

$$K_H = K_b = \frac{[HOAc][OH^-]}{[OAc^-]} \quad \text{Rxn product / initial} \qquad (5.24)$$

K_H is called the **hydrolysis constant** of the salt and is the same as the basicity constant. We will use K_b to emphasize that these salts are treated the same as for any other weak base.

The value of K_b can be calculated from K_a of acetic acid and K_w if we multiply both the numerator and denominator by $[H^+]$:

$$K_b = \frac{[HOAc]}{[OAc^-]} \frac{[OH^-]}{} \cdot \frac{[H^+]}{[H^+]} \qquad (5.25)$$

The quantity inside the dashed line is K_w and the remainder is $1/K_a$. Hence,

$$K_b = \frac{K_w}{k_a} = \frac{1.0 \times 10^{-14}}{1.75 \times 10^{-5}} = 5.7 \times 10^{-10} \qquad (5.26)$$

We see from the small K_b that the acetate ion is quite a weak base with only a small fraction of ionization. **The product of K_a of any weak acid and K_b of its conjugate base is always equal to K_w:**

$$\boxed{K_a K_b = K_w} \qquad (5.27)$$

For any salt of a weak acid HA that ionizes in water,

$$A^- + H_2O \rightleftharpoons HA + OH^- \qquad (5.28)$$

$$\boxed{\frac{[HA][OH^-]}{[A^-]} = \frac{K_w}{K_a} = K_b} \qquad (5.29)$$

The pH of such a salt (a Brønsted base) is calculated in the same manner as for any other weak base. When the salt hydrolyzes, it forms an equal amount of HA and OH^-. If the original concentration of A^- is C_{A^-}, then

$$\underset{(C_{A^-} - x)}{A^-} + H_2O \rightleftharpoons \underset{x}{HA} + \underset{x}{OH^-} \qquad (5.30)$$

The quantity x can be neglected compared to C_{A^-} if $C_{A^-} > 100K_b$, which will generally be the case for such weakly ionized bases.

We can solve for the OH^- concentration using Equation 5.29:

$$\frac{[OH^-][OH^-]}{C_{A^-}} = \frac{K_w}{K_a} = K_b \qquad (5.31)$$

Compare this with the algebraic setup in Example 5.8. They are identical:

$$\boxed{[OH^-] = \sqrt{\frac{K_w}{K_a} \cdot C_{A^-}} = \sqrt{K_b \cdot C_{A^-}}} \qquad (5.32)$$

This equation holds only if $C_{A^-} > 100K_b$, and x can be neglected compared to C_{A^-}. If this is not the case, then the quadratic formula must be solved as for other bases in this situation.

EXAMPLE 5.9   Calculate the pH of a 0.10 M solution of sodium acetate.

SOLUTION
Write the equilibria

$$NaOAc \rightarrow Na^+ + OAc^-$$
$$OAc^- + H_2O \rightleftharpoons HOAc + OH^-$$

Write the equilibrium constant

$$\frac{[HOAc][OH^-]}{[OAc^-]} = K_b = \frac{K_w}{K_a} = \frac{1.0 \times 10^{-14}}{1.75 \times 10^{-5}} = 5.7 \times 10^{-10}$$

Let x represent the concentration of HOAc and OH^- at equilibrium. Then, at equilibrium,

$$[HOAc] = [OH^-] = x$$
$$[OAc^-] = C_{OAc^-} - x = 0.10 - x$$

Since $C_{OAc^-} \gg K_b$, neglect x compared to C_{OAc^-}. Then,

$$\frac{(x)(x)}{0.10} = 5.7 \times 10^{-10}$$
$$x = \sqrt{5.7 \times 10^{-10} \times 0.10} = 7.6 \times 10^{-6}\, M$$

Compare this last step with Equation 5.32. Also, compare the entire setup and solution with those in Example 5.8. The HOAc formed is undissociated and does not contribute to the pH:

$$[OH^-] = 7.6 \times 10^{-6}\, M$$

$$[H^+] = \frac{1.0 \times 10^{-14}}{7.6 \times 10^{-6}} = 1.3 \times 10^{-9}\, M$$

$$pH = -\log 1.3 \times 10^{-9} = 9 - 0.11 = 8.89$$

Similar equations can be derived for the cations of **salts of weak bases** (the salts are completely dissociated). These are **Brønsted acids** and ionize (hydrolyze) in water:

$$BH^+ + H_2O \rightleftharpoons B + H_3O^+ \tag{5.33}$$

The B is undissociated and does not contribute to the pH. The acidity constant is

$$K_H = K_a = \frac{[B][H_3O^+]}{[BH^+]} \tag{5.34}$$

The acidity constant (hydrolysis constant) can be derived by multiplying the numerator and denominator by $[OH^-]$:

$$K_a = \frac{[B]}{[BH^+]} \frac{[H_3O^+]}{} \frac{[OH^-]}{[OH^-]} \tag{5.35}$$

Again, the quantity inside the dashed line is K_w, while the remainder is $1/K_b$. Therefore,

$$\boxed{\frac{[B][H_3O^+]}{[BH^+]} = \frac{K_w}{K_b} = K_a} \tag{5.36}$$

and for NH_4^+,

$$K_a = \frac{K_w}{K_b} = \frac{1.0 \times 10^{-14}}{1.75 \times 10^{-5}} = 5.7 \times 10^{-10} \tag{5.37}$$

We could, of course, have derived K_a from Equation 5.27. It is again coincidence that the numerical value of K_a for NH_4^+ equals K_b for OAc^-.

The salt of a weak base ionizes to form equal amounts of B and H_3O^+ (H^+ if we disregard hydronium ion formation as was done previously). We can therefore solve for the hydrogen ion concentration (by assuming $C_{BH^+} > 100 \, K_a$):

$$\frac{[H^+][H^+]}{C_{BH^+}} = \frac{K_w}{K_b} = K_a \tag{5.38}$$

$$\boxed{[H^+] = \sqrt{\frac{K_w}{K_b} \cdot C_{BH^+}} = \sqrt{K_a \cdot C_{BH^+}}} \tag{5.39}$$

Again, this equation only holds if $C_{BH^+} > 100 \, K_a$. Otherwise, the quadratic formula must be solved.

Concentration of $[H^+] = \sqrt{C_A K_A}$

EXAMPLE 5.10 Calculate the pH of a 0.25 M solution of ammonium chloride.

SOLUTION
Write the equilibria

$$NH_4Cl \rightarrow NH_4^+ + Cl^-$$
$$NH_4^+ + H_2O \rightleftharpoons NH_4OH + H^+$$
$$(NH_4^+ + H_2O \rightleftharpoons NH_3 + H_3O^+)$$

Write the equilibrium constant

$$\frac{[NH_4OH][H^+]}{[NH_4^+]} = K_a = \frac{K_w}{K_b} = \frac{1.0 \times 10^{-14}}{1.75 \times 10^{-5}} = 5.7 \times 10^{-10}$$

Let x represent the concentration of $[NH_4OH]$ and $[H^+]$ at equilibrium. Then, at equilibrium,

$$[NH_4OH] = [H^+] = x$$
$$[NH_4^+] = C_{NH_4^+} - x = 0.25 - x$$

Since $C_{NH_4^+} \gg K_a$, neglect x compared to $C_{NH_4^+}$. Then,

$$\frac{(x)(x)}{0.25} = 5.7 \times 10^{-10}$$

$$x = \sqrt{5.7 \times 10^{-10} \times 0.25} = 1.2 \times 10^{-5} \, M$$

Compare this last step with Equation 5.39. Also, compare the entire setup and solution with those in Example 5.7. The NH_4OH formed is undissociated and does not contribute to the pH:

$$[H^+] = 1.2 \times 10^{-5} M$$
$$pH = -\log (1.2 \times 10^{-5}) = 5 - 0.08 = 4.92$$

5.7 BUFFERS

A **buffer** is defined as a solution that resists change in pH when a small amount of an acid or base is added or when the solution is diluted. This is very useful for maintaining the pH for a reaction at an optimum value. A buffer solution consists of a **mixture of a weak acid and its conjugate base or a weak base and its conjugate acid** at predetermined concentrations or ratios. That is, we have a mixture of a weak acid and its salt or a weak base and its salt. Consider an acetic acid–acetate buffer. The acid equilibrium that governs this system is

$$HOAc \rightleftharpoons H^+ + OAc^-$$

But now, since we have added a supply of acetate ions to the system (from sodium acetate, for example), the hydrogen ion concentration is no longer equal to the acetate ion concentration. The hydrogen ion concentration is

$$[H^+] = K_a \frac{[HOAc]}{[OAc^-]} \tag{5.40}$$

Taking the negative logarithm of each side of this equation, we have

$$-\log [H^+] = -\log K_a - \log\frac{[HOAc]}{[OAc^-]} \tag{5.41}$$

$$pH = pK_a - \log\frac{[HOAc]}{[OAc^-]} \tag{5.42}$$

Upon inverting the last log term, it becomes positive:

$$pH = pK_a + \log\frac{[OAc^-]}{[HOAc]} \tag{5.43}$$

This form of the ionization constant equation is called the **Henderson–Hasselbalch equation.** It is useful for calculating the pH of a weak acid solution containing its salt. A general form can be written for a weak acid HA that ionizes to its salt, A^-, and H^+:

$$HA \rightleftharpoons H^+ + A^- \tag{5.44}$$

$$pH = pK_a + \log \frac{[A^-]}{[HA]} \tag{5.45}$$

$$pH = pK_a + \log \frac{[conjugate\ base]}{[acid]} \tag{5.46}$$

$$pH = pK_a + \log \frac{[proton\ acceptor]}{[proton\ donor]} \tag{5.47}$$

EXAMPLE 5.11 Calculate the pH of a buffer prepared by adding 10 mL of 0.10 M acetic acid to 20 mL of 0.10 M sodium acetate.

SOLUTION
We need to calculate the concentration of the acid and salt in the solution. The final volume is 30 mL:

$$M_1 \times mL_1 = M_2 \times mL_2$$

For HOAc,

$$0.10 \text{ mmol/mL} \times 10 \text{ mL} = M_{HOAc} \times 30 \text{ mL}$$

$$M_{HOAc} = 0.033 \text{ mmol/mL}$$

For OAc$^-$,

$$0.10 \text{ mmol/mL} \times 20 \text{ mL} = M_{OAc^-} \times 30 \text{ mL}$$

$$M_{OAc^-} = 0.067 \text{ mmol/mL}$$

Some of the HOAc dissociates to $H^+ + OAc^-$, and the equilibrium concentration of HOAc would be the amount added (0.033 M) minus the amount dissociated, while that of OAc$^-$ would be the amount added (0.067 M) plus the amount of HOAc dissociated. However, *the amount of acid dissociated is very small*, particularly in the presence of the added salt (ionization suppressed by the common ion effect), and can be neglected. Hence, we can assume the added concentrations to be the equilibrium concentrations:

$$pH = -\log K_a + \log \frac{[\text{proton acceptor}]}{[\text{proton donor}]}$$

$$pH = -\log (1.75 \times 10^{-5}) + \log \frac{0.067 \text{ mmol/mL}}{0.033 \text{ mmol/mL}}$$

$$= 4.76 + \log 2.0$$

$$= 5.06$$

We could have shortened the calculation by recognizing that in the log term the volumes cancel. So we can take the ratio of millimoles only:

$$mmol_{HOAc} = 0.10 \text{ mmol/mL} \times 10 \text{ mL} = 1.0 \text{ mmol}$$

$$mmol_{OAc^-} = 0.10 \text{ mmol/mL} \times 20 \text{ mL} = 2.0 \text{ mmol}$$

$$pH = 4.76 + \log \frac{2.0 \text{ mmol}}{1.0 \text{ mmol}} = 5.06$$

The mixture of a weak acid and its salt may also be obtained by mixing an excess of weak acid with some strong base to produce the salt by neutralization or by mixing an excess of salt with strong acid to produce the weak acid component of the buffer.

EXAMPLE 5.12 Calculate the pH of a solution prepared by adding 25 mL of 0.10 M sodium hydroxide to 30 mL of 0.20 M acetic acid (this would actually be a step in a typical titration).

SOLUTION

$$mmol\ HOAc = 0.20\ M \times 30\ mL = 6.0\ mmol$$

$$mmol\ NaOH = 0.10\ M \times 25\ mL = 2.5\ mmol$$

These react as follows:

$$HOAc + NaOH \rightleftharpoons NaOAc + H_2O$$

After reaction,

$$mmol\ NaOAc = 2.5\ mmol$$

$$mmol\ HOAc = 6.0 - 2.5 = 3.5\ mmol$$

$$pH = 4.76 + \log\frac{2.5}{3.5} = 4.61$$

The **buffering mechanism** for a mixture of a weak acid and its salt can be explained as follows. The pH is governed by the logarithm of the ratio of the salt and acid:

$$pH = constant + \log\frac{[A^-]}{[HA]} \tag{5.48}$$

If the solution is diluted, the ratio remains constant, and so the pH of the solution does not change.[3] If a small amount of a strong acid is added, it will combine with an equal amount of the A^- to convert it to HA. That is, in the equilibrium $HA \rightleftharpoons H^+ + A^-$, Le Châtelier's principle dictates added H^+ will combine with A^- to form HA, with the equilibrium lying far to the left if there is an excess of A^-. The change in the ratio $[A^-]/[HA]$ is small and hence the change in pH is small. If the acid had been added to an unbuffered solution (for example, a solution of NaCl), the pH would have decreased markedly. If a small amount of a strong base is added, it will combine with part of the HA to form an equilvalent amount of A^-. Again, the change in the ratio is small.

The amount of acid or base that can be added without causing a large change in pH is governed by the **buffering capacity** of the solution. This is determined by the concentrations of HA and A^-. The higher their concentrations, the more acid or base the solution can tolerate. The buffering capacity is also governed by the *ratio* of HA to A^-. It is **maximum** when the ratio is unity, that is, when the **pH = pK_a**:

$$pH = pK_a + \log\frac{1}{1} = pK_a \tag{5.49}$$

In general, the buffering capacity is satisfactory over a **pH range of pK_a $\pm$ 1.** We will discuss the buffering capacity on a pictorial basis in Chapter 8, when the *titration curves* of weak acids are discussed.

---

[3]In actuality, the pH will *increase* slightly because the activity coefficient of the salt has been increased by decreasing the ionic strength. The activity of an uncharged molecule (that is, undissociated acid) is equal to its molarity (see Chapter 4), and so the ratio increases, causing a slight increase in pH. See Section 5.11.

EXAMPLE 5.13   A buffer solution is 0.20 M in acetic acid and in sodium acetate. Calculate the change in pH upon adding 1.0 mL of 0.10 M hydrochloric acid to 10 mL of this solution.

SOLUTION
Initially, the pH is equal to pK_a, because the ratio [OAc$^-$]/[HOAc] is unity. The pH is 4.76. To calculate the new pH, we need to determine the new concentrations of HOAc and OAc$^-$. We started with $10 \times 0.20 = 2.0$ mmol OAc$^-$ per 10 mL. We added $1.0 \times 0.10 = 0.10$ mmol of H$^+$ and therefore converted 0.10 mmol OAc$^-$ to HOAc:

$$\text{mmol HOAc} = 2.0 + 0.1 = 2.1 \text{ mmol}$$

$$\text{mmol OAc}^- = 2.0 - 0.1 = 1.9 \text{ mmol}$$

These new amounts of acid and salt are contained in 11 mL but, again, the volumes cancel in our calculations:

$$\text{pH} = 4.76 + \log \frac{(1.9 \text{ mmol/11 mL})}{(2.1 \text{ mmol/11 mL})}$$

$$= 4.76 + \log \frac{1.9 \text{ mmol}}{2.1 \text{ mmol}} = 4.76 + \log 0.90$$

$$= 4.76 + (0.95 - 1) = 4.71$$

   The change in pH is -0.05. This is rather small, especially if we consider that had the HCl been added to an unbuffered neutral solution, the final concentration would have been approximately 10^{-2} M, and the pH would be 2.0.

Note that a buffer can resist a pH change, even when there is added an amount of strong acid or base greater (in moles) than the equilibrium amount of H$^+$ or OH$^-$ (in moles) in the buffer. For example, in Example 5.13, the pH of the buffer is 4.76 and [H$^+$] $= 1.7 \times 10^{-5}$ M, and millimoles H$^+$ $= (1.7 \times 10^{-5}$ mmol/mL)(10 mL) $= 1.7 \times 10^{-4}$ mmol (in equilibrium with the buffer components). We added 0.10 mmol H$^+$, well in excess of this. However, due to the reserve of buffer components (OAc$^-$ to react with H$^+$ in this case), the added H$^+$ is consumed so that the pH remains relatively constant, *so long as we do not exceed the amount of buffer reserve*.

Similar calculations apply for mixtures of a weak base and its salt. We can consider the equilibrium between the base B and its conjugate acid BH$^+$ and write a K_a for the conjugate (Brønsted) acid:

$$BH^+ = B + H^+ \tag{5.50}$$

$$K_a = \frac{[B][H^+]}{[BH^+]} = \frac{K_w}{K_b} \tag{5.51}$$

The logarithmic Henderson–Hasselbalch form is derived exactly as above:

$$[H^+] = K_a \cdot \frac{[BH^+]}{[B]} = \frac{K_w}{K_b} \cdot \frac{[BH^+]}{[B]} \tag{5.52}$$

$$-\log [H^+] = -\log K_a - \log \frac{[BH^+]}{[B]} = -\log \frac{K_w}{K_b} - \log \frac{[BH^+]}{[B]} \tag{5.53}$$

$$pH = pK_a + \log \frac{[B]}{[BH^+]} = (pK_w - pK_b) + \log \frac{[B]}{[BH^+]} \quad (5.54)$$

$$pH = pK_a + \log \frac{[\text{proton acceptor}]}{[\text{proton donor}]} = (pK_w - pK_b) + \log \frac{[\text{proton acceptor}]}{[\text{proton donor}]}$$

$$(5.55)$$

Since $pOH = pK_w - pH$, we can also write, by subtracting either Equation 5.54 or 5.55 from pK_w,

$$pOH = pK_b + \log \frac{[BH^+]}{[B]} = pK_b + \log \frac{[\text{proton donor}]}{[\text{proton acceptor}]} \quad (5.56)$$

A mixture of a weak base and its salt acts as a buffer in the same manner as a weak acid and its salt. When a strong acid is added, it combines with some of the base B to form the salt BH^+. Conversely, a base combined with BH^+ to form B. Since the change in the ratio will be small, the change in pH will be small. Again, the **buffering capacity is maximum at a pH equal to $pK_a = 14 - pK_b$ (or at pOH $= pK_b$), with a useful range of $pK_a \pm 1$.**

EXAMPLE 5.14 Calculate the volume of concentrated ammonia and the weight of ammonium chloride you would have to take to prepare 100 mL of a buffer at pH 10.00 if the final concentration of salt is to be 0.200 M.

SOLUTION

We want 100 mL of 0.200 M NH_4Cl. Therefore, mmol NH_4Cl = 0.200 mmol/mL $\times$ 100 mL = 20.0 mmol

$$mg\ NH_4Cl = 20.0\ mmol \times 53.5\ mg/mmol = 1.07 \times 10^3\ mg$$

Therefore, we need 1.07 g NH_4Cl. We calculate the concentration of NH_3 by

$$pH = pK_a + \log \frac{[\text{proton acceptor}]}{[\text{proton donor}]}$$

$$= (14.00 - pK_b) + \log \frac{[NH_3]}{[NH_4^+]}$$

$$10.00 = (14.00 - 4.76) + \log \frac{[NH_3]}{(0.200\ mmol/mL)}$$

$$\log \frac{[NH_3]}{(0.200\ mmol/mL)} = 0.76$$

$$\frac{[NH_3]}{(0.200\ mmol/mL)} = 10^{0.76} = 5.8$$

$$[NH_3] = (0.200)(5.8) = 1.1_6\ mmol/mL$$

The molarity of concentrated ammonia is 14.8 M. Therefore,

$$100\ mL \times 1.1_6\ mmol/mL = 14.8\ mmol/mL \times mL\ NH_3$$

$$mL\ NH_3 = 7.8\ mL$$

EXAMPLE 5.15 How many grams ammonium chloride and how many milliliters 3.0 M sodium hydroxide should be added to 200 mL water and diluted to 500 mL to prepare a buffer of pH 9.50 with a salt concentration of 0.10 M?

SOLUTION

We need the ratio of $[NH_3]/[NH_4^+]$. From Example 5.14,

$$pH = pK_a + \log \frac{[NH_3]}{[NH_4^+]} = 9.24 + \log \frac{[NH_3]}{[NH_4^+]}$$

$$9.50 = 9.24 + \log \frac{[NH_3]}{[NH_4^+]}$$

$$\log \frac{[NH_3]}{[NH_4^+]} = 0.26$$

$$\frac{[NH_3]}{[NH_4^+]} = 10^{0.26} = 1.8$$

The final concentration of NH_4^+ is 0.10 M, so

$$[NH_3] = (1.8)(0.10) = 0.18 \ M$$

$$\text{mmol } NH_4^+ \text{ in final solution} = 0.10 \ M \times 500 \text{ mL} = 50 \text{ mmol}$$

$$\text{mmol } NH_3 \text{ in final solution} = 0.18 \ M \times 500 \text{ mL} = 90 \text{ mmol}$$

The NH_3 is formed from an equal number of millimoles of NH_4Cl. Therefore, a total of 50 + 90 = 140 mmol NH_4Cl must be taken:

$$\text{mg } NH_4Cl = 140 \text{ mmol} \times 53.5 \text{ mg/mmol} = 7.49 \times 10^3 \text{ mg} = 7.49 \text{ g}$$

The millimoles of NaOH needed is equal to the millimoles of NH_3:

$$3.0 \ M \times x \text{ mL} = 90 \text{ mmol}$$

$$x = 30 \text{ mL NaOH}$$

We see that a buffer solution for a given pH is prepared by choosing a weak acid (or a weak base) and its salt, with a pK_a **value near the pH that we want.** There are a number of such acids and bases, and any pH region can be buffered by a proper choice of these. A weak acid and its salt give the best buffering in acid solution and a weak base and its salt give the best buffering in alkaline solution. Some useful buffers for measurements in physiological solutions are described below. National Bureau of Standards buffers used for calibrating pH electrodes are described in Chapter 11.

You may have wondered why, in buffer mixtures, the salt does not react with water to ionize as an acid or base. This is because the reaction is suppressed by the presence of the acid or base. In Equation 5.28, the presence of appreciable amounts of either HA or OH^- will suppress the ionization to a negligible amount. In Equation 5.33, the presence of either B or H_3O^+ will suppress the ionization.

5.8 POLYPROTIC ACIDS AND THEIR SALTS

Many acids or bases are polyfunctional, that is, have more than one ionizable proton or hydroxide ion. These substances ionize stepwise, and an equilibrium constant can be written for each step. Consider, for example, the ionization of phosphoric acid:

$$H_3PO_4 \rightleftharpoons H^+ + H_2PO_4^- \quad K_{a1} = 1.1 \times 10^{-2} = \frac{[H^+][H_2PO_4^-]}{[H_3PO_4]} \quad (5.57)$$

$$H_2PO_4^- \rightleftharpoons H^+ + HPO_4^{2-} \quad K_{a2} = 7.5 \times 10^{-8} = \frac{[H^+][HPO_4^{2-}]}{[H_2PO_4^-]} \quad (5.58)$$

$$HPO_4^{2-} \rightleftharpoons H^+ + PO_4^{3-} \quad K_{a3} = 4.8 \times 10^{-13} = \frac{[H^+][PO_4^{3-}]}{[HPO_4^{2-}]} \quad (5.59)$$

Recall from Chapter 4 that the overall ionization is the sum of these individual steps and the overall ionization constant is the product of the individual ionization constants:

$$H_3PO_4 \rightleftharpoons 3H^+ + PO_4^{3-}$$
$$K_a = K_{a1}K_{a2}K_{a3} = 4.0 \times 10^{-22} = \frac{[H^+]^3[PO_4^{3-}]}{[H_3PO_4]} \quad (5.60)$$

The individual pK_a values are 1.96, 7.12, and 12.32, respectively, for pK_{a1}, pK_{a2}, and pK_{a3}. In order to make precise pH calculations, the contributions of protons from each ionization step must be taken into account. Exact calculation is difficult and requires a tedious iterative procedure, since $[H^+]$ is unknown in addition to the various phosphoric acid species. See, for example, References 6 and 9 for calculations.

In most cases, approximations can be made so that each ionization step can be considered individually. **If the difference between successive ionization constants is at least 10^4, each proton can be differentiated in a titration,** that is, each is titrated separately to give stepwise pH breaks in the titration curve. (If an ionization constant is less than about 10^{-8}, then the ionization is too small for a pH break to be exhibited in the titration curve—for example, the third proton for H_3PO_4.) Under these conditions, calculations are simplified because **the system can be considered as simply a mixture of three weak acids.**

Buffer Calculations for Polyprotic Acids

The anion on the right side in each ionization step can be considered the salt (conjugate base) of the acid from which it is derived. That is, in Equation 5.57, $H_2PO_4^-$ is the salt of the acid H_3PO_4. In Equation 5.58, HPO_4^{2-} is the salt of the acid $H_2PO_4^-$, and in Equation 5.59, PO_4^{3-} is the salt of the acid HPO_4^{2-}. So each of these pairs constitutes a buffer system, and orthophosphate buffers can be prepared over a wide pH range. The optimum buffering capacity of each pair occurs at a pH corresponding to its pK_a. The $HPO_4^{2-}/H_2PO_4^-$ couple is an effective buffer system in the blood (see below).

EXAMPLE 5.16 The pH of blood is 7.40. What is the ratio of $[HPO_4^{2-}]/[H_2PO_4^-]$ in the blood (assume 25°C)?

SOLUTION

$$pH = pK_a + \log \frac{[\text{proton acceptor}]}{[\text{proton donor}]}$$

$$pK_{a2} = 7.12$$

Therefore,

$$pH = 7.12 + \log \frac{[HPO_4^{2-}]}{[H_2PO_4^-]}$$

$$7.40 = 7.12 + \log \frac{[HPO_4^{2-}]}{[H_2PO_4^-]}$$

$$\frac{[HPO_4^{2-}]}{[H_2PO_4^-]} = \frac{1.9}{1}$$

Dissociation Calculations for Polyprotic Acids

Because the individual ionization constants are sufficiently different, **the pH of a solution of H_3PO_4 can be calculated by treating it just as we would any weak acid.** The H^+ from the first ionization step effectively suppresses the other two ionization steps, so that the H^+ contribution from them is negligible compared to the first ionization. The quadratic equation must, however, be solved because K_{a1} is relatively large.

EXAMPLE 5.17   Calculate the pH of a 0.100 M H_3PO_4 solution.

SOLUTION

$$H_3PO_4 \rightleftharpoons H^+ + H_2PO_4^-$$
$$0.100 - x \quad x \quad x$$

From Equation 5.57,

$$\frac{(x)(x)}{0.100 - x} = 1.1 \times 10^{-2}$$

In order to neglect x, C should be $\geq 100\ K_a$. Here, it is only 10 times as large. Therefore, use the quadratic equation to solve:

$$x^2 + 0.011x - 1.1 \times 10^{-3} = 0$$

$$x = \frac{-0.011 \pm \sqrt{(0.011)^2 - 4(-1.1 \times 10^{-3})}}{2}$$

$$x = [H^+] = 0.028\ M$$

The acid is 28% ionized:

$$pH = -\log 2.8 \times 10^{-2} = 2 - 0.45 = 1.55$$

We can determine if our assumption that the only important source of protons is H_3PO_4 was a realistic one. $H_2PO_4^-$ would be the next most likely source of protons. From Equation 5.58, $[HPO_4^{2-}] = K_{a2}\ [H_2PO_4^-]/[H^+]$. Assuming the concentrations of $H_2PO_4^-$ and H^+ as a first approximation are 0.028 M as calculated, then $[HPO_4^{2-}] \approx K_{a2} = 7.5 \times 10^{-8}\ M$. This is very small compared to 0.028 M $H_2PO_4^-$, and so further dissociation is indeed insignificant. We were justified in our approach.

Fractions of Dissociating Species at a Given pH: α-Values

Often, it is of interest to know the distribution of the different species of a polyprotic acid as a function of pH, that is, at known hydrogen ion concentrations as in a buffered solution.

Consider, for example, the dissociation of phosphoric acid. The equilibria are given in Equations 5.57–5.59. At any given pH, all the four phosphoric acid species will coexist in equilibrium with one another, although the concentrations of some may be diminishingly small. By changing the pH, the equilibria shift, and the relative concentrations change. It is possible to derive general equations for calculating the fraction of the acid that exists in a given form, from the given hydrogen ion concentration.

For a given *total* concentration of phosphoric acid, $C_{H_3PO_4}$, we can write

$$C_{H_3PO_4} = [PO_4^{3-}] + [HPO_4^{2-}] + [H_2PO_4^-] + [H_3PO_4] \tag{5.61}$$

where the terms on the right-hand side of the equation represent the **equilibrium concentrations** of the individual species. We presumably know the initial total concentration $C_{H_3PO_4}$ and wish to find the fractions or concentrations of the individual species at equilibrium.

We define

$$\alpha_0 = \frac{[H_3PO_4]}{C_{H_3PO_4}}; \quad \alpha_1 = \frac{[H_2PO_4^-]}{C_{H_3PO_4}}; \quad \alpha_2 = \frac{[HPO_4^{2-}]}{C_{H_3PO_4}}$$

$$\alpha_3 = \frac{[PO_4^{3-}]}{C_{H_3PO_4}}; \quad \alpha_0 + \alpha_1 + \alpha_2 + \alpha_3 = 1$$

where the α's are the **fractions** of each species present at equilibrium. Note that the subscripts denote the number of dissociated protons or the charge on the species. We can use Equation 5.61 and the equilibrium constant expressions 5.57 through 5.59 to obtain an expression for $C_{H_3PO_4}$ in terms of the desired species. This is substituted into the appropriate equation to obtain α in terms of $[H^+]$ and the equilibrium constants. In order to calculate α_0, for example, we can rearrange Equations 5.57 through 5.59 to solve for all the species except $[H_3PO_4]$ and substitute into Equation 5.61:

$$[PO_4^{3-}] = \frac{K_{a3}[HPO_4^{2-}]}{[H^+]} \tag{5.62}$$

$$[HPO_4^{2-}] = \frac{K_{a2}[H_2PO_4^-]}{[H^+]} \tag{5.63}$$

$$[H_2PO_4^-] = \frac{K_{a1}[H_3PO_4]}{[H^+]} \tag{5.64}$$

We want all these to contain only $[H_3PO_4]$ (and $[H^+]$, the variable). We can substitute Equation 5.64 for $[H_2PO_4^-]$ in Equation 5.63:

$$[HPO_4^{2-}] = \frac{K_{a1}K_{a2}[H_3PO_4]}{[H^+]^2} \tag{5.65}$$

And we can substitute Equation 5.65 into Equation 5.62 for $[HPO_4^{2-}]$:

$$[PO_4^{3-}] = \frac{K_{a1}K_{a2}K_{a3}[H_3PO_4]}{[H^+]^3} \qquad (5.66)$$

Finally, we can substitute 5.64 through 5.66 in Equation 5.61:

$$C_{H_3PO_4} = \frac{K_{a1}K_{a2}K_{a3}[H_3PO_4]}{[H^+]^3} + \frac{K_{a1}K_{a2}[H_3PO_4]}{[H^+]^2} + \frac{K_{a1}[H_3PO_4]}{[H^+]} + [H_3PO_4] \quad (5.67)$$

We can either divide each side of this expression by $[H_3PO_4]$ to obtain $1/\alpha_0$, or we can substitute the expression into the denominator of the above α_0 expression to obtain a value for α_0 ($[H_3PO_4]$ cancels). Doing the former, we have

$$\frac{1}{\alpha_0} = \frac{K_{a1}K_{a2}K_{a3}}{[H^+]^3} + \frac{K_{a1}K_{a2}}{[H^+]^2} + \frac{K_{a1}}{[H^+]} + 1 \qquad (5.68)$$

or doing the latter, we have

$$\alpha_0 = \frac{1}{(K_{a1}K_{a2}K_{a3}/[H^+]^3) + (K_{a1}K_{a2}/[H^+]^2) + (K_{a1}/[H^+]) + 1} \qquad (5.69)$$

either of which can be rearranged to

$$\boxed{\alpha_0 = \frac{[H^+]^3}{[H^+]^3 + K_{a1}[H^+]^2 + K_{a1}K_{a2}[H^+] + K_{a1}K_{a2}K_{a3}}} \qquad (5.70)$$

Similar approaches can be taken to obtain expressions for the other α's. For α_1, for example, the equilibrium constant expressions would be solved for all species in terms of $[H_2PO_4^-]$ and substituted into Equation 5.61 to obtain an expression for $C_{H_3PO_4}$ containing only $[H_2PO_4^-]$ and $[H^+]$, from which α_1 is calculated. The results for the other α's are

$$\boxed{\alpha_1 = \frac{K_{a1}[H^+]^2}{[H^+]^3 + K_{a1}[H^+]^2 + K_{a1}K_{a2}[H^+] + K_{a1}K_{a2}K_{a3}}} \qquad (5.71)$$

$$\alpha_2 = \frac{K_{a1}K_{a2}[H^+]}{[H^+]^3 + K_{a1}[H^+]^2 + K_{a1}K_{a2}[H^+] + K_{a1}K_{a2}K_{a3}} \qquad (5.72)$$

$$\alpha_3 = \frac{K_{a1}K_{a2}K_{a3}}{[H^+]^3 + K_{a1}[H^+]^2 + K_{a1}K_{a2}[H^+] + K_{a1}K_{a2}K_{a3}} \qquad (5.73)$$

Note that all have the same denominator and that *the sum of the numerators equals the denominator*. See Problem 52 for a more detailed derivation of the other α's.

EXAMPLE 5.18 Calculate the equilibrium concentrations of the different species in a 0.10 M phosphoric acid solution at pH 3.00 ($[H^+] = 1.0 \times 10^{-3}$ M).

SOLUTION

Substituting into Equation 5.70,

$$\alpha_0 = \frac{(1.0 \times 10^{-3})^3}{(1.0 \times 10^{-3})^3 + (1.1 \times 10^{-2})(1.0 \times 10^{-3})^2 + (1.1 \times 10^{-2})(7.5 \times 10^{-8})(1.0 \times 10^{-3})}$$

$$\frac{}{+ (1.1 \times 10^{-2})(7.5 \times 10^{-8})(4.8 \times 10^{-13})} = \frac{1.0 \times 10^{-9}}{1.2 \times 10^{-8}} = 8.3 \times 10^{-2}$$

$[H_3PO_4] = C_{H_3PO_4}\, \alpha_0 = 0.10 \times 8.3 \times 10^{-2} = 8.3 \times 10^{-3}$ M

Similarly,

$$\alpha_1 = 0.92$$

$$[H_2PO_4^-] = C_{H_3PO_4}\,\alpha_1 = 0.10 \times 0.92 = 9.2 \times 10^{-2}\ M$$

$$\alpha_2 = 6.9 \times 10^{-5}$$

$$[HPO_4^{2-}] = C_{H_3PO_4}\,\alpha_2 = 0.10 \times 6.9 \times 10^{-5} = 6.9 \times 10^{-6}\ M$$

$$\alpha_3 = 3.3 \times 10^{-14}$$

$$[PO_4^{3-}] = C_{H_3PO_4}\,\alpha_3 = 0.10 \times 3.3 \times 10^{-14} = 3.3 \times 10^{-15}\ M$$

We see that at pH 3, the majority (91%) of the phosphoric acid exists as $H_2PO_4^-$ and 8.3% exists as H_3PO_4. Only 3.3×10^{-12}% exists as PO_4^{3-}!

A plot of the fractions of each phosphoric acid species as a function of pH is given in Figure 5.1.

Equation 5.67 could be used for a rigorous calculation of the hydrogen ion concentration from dissociation of a phosphoric acid solution at a given H_3PO_4 concentration (no other added H^+), but this would involve tedious iterations. As a first approximation, $[H^+]$ could be calculated from K_{a1} as in Example 5.17, assuming that only the first dissociation step of phosphoric acid was significant. (This is, in fact, what we did in that example.) The first calculated $[H^+]$ could then be substituted in Equation 5.67 to calculate a second approximation of $[H_3PO_4]$, which would be used for a second iterative calculation of $[H^+]$ using K_{a1}, and so forth, until the concentration was constant.

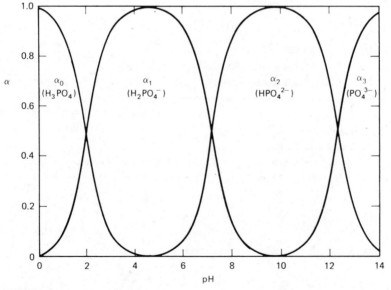

FIGURE 5.1
Fraction of H_3PO_4 species as a function of pH.

Salts of Polyprotic Acids

Salts of acids such as H_3PO_4 may be acidic or basic. The protonated salts possess both acidic and basic properties ($H_2PO_4^-$, HPO_4^{2-}), while the unprotonated salt is simply a Brønsted base that hydrolyzes (PO_4^{3-}).

1. AMPHOTERIC SALTS. $H_2PO_4^-$ possesses both acidic and basic properties. That is, it is **amphoteric.** It ionizes as a weak acid and it also is a Brønsted base that hydrolyzes:

$$H_2PO_4^- \rightleftharpoons H^+ + HPO_4^{2-} \quad K_{a2} = \frac{[H^+][HPO_4^{2-}]}{[H_2PO_4^-]} = 7.5 \times 10^{-8} \quad (5.74)$$

$$H_2PO_4^- + H_2O \rightleftharpoons H_3PO_4 + OH^- \quad K_b = \frac{K_w}{K_{a1}} = \frac{[H_3PO_4][OH^-]}{[H_2PO_4^-]}$$

$$= \frac{1.00 \times 10^{-14}}{1.1 \times 10^{-2}} = 9.1 \times 10^{-13} \quad (5.75)$$

The solution could, hence, be either alkaline or acidic, depending on which ionization is more extensive. Since K_{a2} for the first ionization is nearly 10^5 greater than K_b for the second ionization, the solution in this case will obviously be acidic.

An expression for the hydrogen ion concentration in a solution of an ion such as $H_2PO_4^-$ can be obtained as follows. The total hydrogen ion concentration is equal to the amounts produced from the ionization equilibrium in Equation 5.74 and the ionization of water, less the amount of OH^- produced from the hydrolysis in Equation 5.75. We can write, then,

$$C_{H^+} = [H^+]_{total} = [H^+]_{H_2O} + [H^+]_{H_2PO_4^-} - [OH^-]_{H_2PO_4^-} \quad (5.76)$$

or

$$[H^+] = [OH^-] + [HPO_4^{2-}] - [H_3PO_4] \quad (5.77)$$

We have included the contribution from water, since it will not be negligible if the pH of the salt solution happens to be near 7—although in this particular case, the solution will be acid, making the water ionization negligible.

We can solve for $[H^+]$ by substituting expressions in the right-hand side of Equation 5.77 from the equilibrium constant expressions 5.57 and 5.58 and K_w to eliminate all but $[H_2PO_4^-]$, the experimental variable, and $[H^+]$:

$$[H^+] = \frac{K_w}{[H^+]} + \frac{K_{a2}[H_2PO_4^-]}{[H^+]} - \frac{[H_2PO_4^-][H^+]}{K_{a1}} \quad (5.78)$$

from which (by multiplying each side of the equation by $[H^+]$, collecting the terms containing $[H^+]^2$ on the left side, and solving for $[H^+]^2$)

$$[H^+]^2 = \frac{K_w + K_{a2}[H_2PO_4^-]}{1 + \dfrac{[H_2PO_4^-]}{K_{a1}}} \quad (5.79)$$

$$\boxed{[H^+] = \sqrt{\frac{K_{a1}K_w + K_{a1}K_{a2}[H_2PO_4^-]}{K_{a1} + [H_2PO_4^-]}}} \quad (5.80)$$

That is, for the general case HA^-,

$$[H^+] = \sqrt{\frac{K_{a1}K_w + K_{a1}K_{a2}[HA^-]}{K_{a1} + [HA^-]}} \qquad (5.81)$$

This equation is valid for any salt HA^- derived from an acid H_2A (or for HA^{2-} derived from H_2A^-, etc.) where $[H_2PO_4^-]$ is replaced by $[HA^-]$.

If we assume that the equilibrium concentration $[HA^-]$ is equal to the concentration of salt added, that is, that the extent of ionization and hydrolysis is fairly small, then this value along with the constants can be used for the calculation of $[H^+]$. This assumption is reasonable if the two equilibrium constants (K_{a1} and K_b) involving the salt HA^- are small and the solution is not too dilute. In many cases, $K_{a1}K_w \ll K_{a1}K_{a2}[HA^-]$ in the numerator and can be neglected. This is the equation we would have obtained if we had neglected the dissociation of water. Furthermore, if $K_{a1} \ll [HA^-]$ in the denominator, the equation simplifies to

$$[H^+] = \sqrt{K_{a1}K_{a2}} \qquad (5.82)$$

Therefore, if the assumptions hold, the pH of a solution of $H_2PO_4^-$ is independent of its concentration! This approximation is adequate for our purposes. The equation generally applies if there is a large difference between K_{a1} and K_{a2}. For the case of $H_2PO_4^-$, then,

$$[H^+] \approx \sqrt{K_{a1}K_{a2}} = \sqrt{1.1 \times 10^{-2} \times 7.5 \times 10^{-8}} = 2.9 \times 10^{-5}\ M \qquad (5.83)$$

and the pH is approximately independent of the salt concentration (pH ≈ 4.54). This would be the approximate pH of a NaH_2PO_4 solution.

Similarly, HPO_4^{2-} is both an acid and a base. The K values involved here are K_{a2} and K_{a3} of H_3PO_4 ($H_2PO_4^- \equiv H_2A$ and $HPO_4^{2-} \equiv HA^-$). Since $K_{a2} \gg K_{a3}$, the pH of a Na_2HPO_4 solution can be calculated from

$$[H^+] \approx \sqrt{K_{a2}K_{a3}} = \sqrt{7.5 \times 10^{-8} \times 4.8 \times 10^{-13}} = 1.9 \times 10^{-10} \qquad (5.84)$$

Because the pH of amphoteric salts of this type is essentially independent of concentration, the salts are useful for preparing solutions of known pH for standardizing pH meters. For example, potassium acid phthalate, $KHC_8H_4O_2$, gives a solution of pH 4.0 at 25°C. However, these salts are poor buffers against acids or bases; their pH does not fall in the buffer region but occurs at the end point of a titration curve, where the pH changes markedly, i.e., when a proton has just been neutralized.

2. UNPROTONATED SALT. Unprotonated phosphate is a fairly strong Brønsted base in solution and ionizes as follows:

$$PO_4^{3-} + H_2O \rightleftharpoons HPO_4^{2-} + OH^- \qquad K_b = \frac{K_w}{K_{a3}} \qquad (5.85)$$

The constant K_{a3} is very small, and so the equilibrium lies significantly to the right. Because $K_{a3} \ll K_{a2}$, hydrolysis of HPO_4^{2-} is suppressed by the OH^- from the first step, and the pH of PO_4^{3-} can be calculated just as for a salt of a monoprotic weak acid. However, because K_{a3} is so small, K_b is relatively large, and the amount of OH^- is not negligible compared with the initial concentration of PO_4^{3-} (C_{B-}), and the quadratic equation must be solved.

EXAMPLE 5.19   Calculate the pH of 0.100 M Na$_3$PO$_4$.

SOLUTION

$$PO_4^{3-} + H_2O \rightleftharpoons HPO_4^{2-} + OH^-$$

$$0.100 - x \qquad\qquad x \qquad x$$

$$\frac{[HPO_4^{2-}][OH^-]}{[PO_4^{3-}]} = K_b = \frac{K_w}{K_a} = \frac{1.0 \times 10^{-14}}{4.8 \times 10^{-13}} = 0.020$$

$$\frac{(x)(x)}{0.100 - x} = \frac{1.0 \times 10^{-14}}{4.8 \times 10^{-13}} = 0.020$$

The concentration is only five times K_b, so the quadratic equation is used:

$$x^2 + 0.020x - 2.0 \times 10^{-3} = 0$$

$$x = \frac{-0.020 \pm \sqrt{(0.020)^2 - 4(-2.0 \times 10^{-3})}}{2}$$

$$x = [OH^-] = 0.036 \; M$$

$$pH = 12.56$$

The dissociation (hydrolysis) is 36% complete, and phosphate is quite a strong base.

EXAMPLE 5.20   EDTA is a polyprotic acid with four protons (H$_4$Y). Calculate the hydrogen ion concentration of a 0.0100 M solution of Na$_2$EDTA (Na$_2$H$_2$Y).

SOLUTION
The equilibria are

$$H_2Y^{2-} \rightleftharpoons H^+ + HY^{3-}; \quad K_{a3} = 6.9 \times 10^{-7}$$

and

$$H_2Y^{2-} + H_2O \rightleftharpoons H_3Y^- + OH^-; \quad K_b = \frac{K_w}{K_{a2}} = \frac{1.0 \times 10^{-14}}{2.2 \times 10^{-3}}$$

H_2Y^{2-} is the equivalent of HA$^-$, and H_3Y^- is the equivalent of H$_2$A. The equilibrium constants involved are K_{a2} and K_{a3} (the former for the conjugate acid H_3Y^- of the hydrolyzed salt). Thus,

$$[H^+] = \sqrt{K_{a2}K_{a3}} = \sqrt{(2.2 \times 10^{-3})(6.9 \times 10^{-7})}$$

$$= 3.9 \times 10^{-5} \; M$$

5.9   PHYSIOLOGICAL BUFFERS

The pH of the blood in a healthy individual remains remarkably constant at 7.35 to 7.45. This is because the blood contains a number of buffers that protect against pH change due to the presence of acidic or basic metabolites. From a physiological viewpoint, a change of ±0.3 pH unit is extreme. Acid metabolites are ordinarily produced in greater quantities than basic metabolites, and carbon dioxide is the principal one. The buffering capacity of blood for handling CO_2 is estimated to be distributed among various buffer systems as follows: hemoglobin and oxyhemoglobin, 62%; $H_2PO_4^-/HPO_4^{2-}$, 22%; plasma protein, 11%; bicarbonate, 5%. Proteins contain carboxylic and amino groups, which are weak acids and bases, respectively. They are, therefore, effective buffering agents. The combined buffering capacity of blood to neutralize acids is designated by clinicians as the "alkali reserve," and this is frequently determined in the clinical laboratory. Certain diseases cause disturbances in the acid balance of the body. For example, diabetes may give rise to "acidosis," which can be fatal.

An important diagnostic analysis is the CO_2/HCO_3^- balance in blood. This ratio is related to the pH of the blood by the Henderson–Hasselbalch equation (5.45):

$$pH = 6.10 + \log \frac{[HCO_3^-]}{[H_2CO_3]} \qquad (5.86)$$

where H_2CO_3 can be considered equal to the concentration of dissolved CO_2 in the blood; 6.10 is pK_{a1} of carbonic acid in blood at body temperature (37°C). Normally, the bicarbonate concentration in blood is about 26.0 mmol/L, while the concentration of carbon dioxide is 1.3 mmol/L. Accordingly, for the blood,

$$pH = 6.10 + \log \frac{26 \text{ mmol/L}}{1.3 \text{ mmol/L}} = 7.40$$

The HCO_3^- concentration may be determined by titrimetry (Experiment 8), or the total carbon dioxide content (HCO_3^- + dissolved CO_2) can be determined by acidification and measurement of the evolved gas.[4] If both analyses are performed, the ratio of HCO_3^-/CO_2 can be calculated, and hence conclusions can be drawn concerning acidosis or alkalosis in the patient. Alternatively, if the pH is measured (at 37°), only HCO_3^- or total CO_2 need be measured for a complete knowledge of the carbonic acid balance because the ratio of $[HCO_3^-]/[H_2CO_3]$ can be calculated from Equation 5.86.

The partial pressure, p_{CO_2}, of CO_2 may also be measured (e.g., using a CO_2 electrode), in which case $[H_2CO_3] \approx 0.30 \, p_{CO_2}$. Then, only pH or $[HCO_3^-]$ need be determined.

Note that these equilibria and Equation 5.86 hold although there are other buffer systems in the blood. The pH is the result of all the buffers and the $[HCO_3^-]/[H_2CO_3]$ ratio is set by this pH.

The HCO_3^-/H_2CO_3 buffer system is the most important one in buffering blood in the lung (alveolar blood). As oxygen from inhaled air combines with hemoglobin, the oxygenated hemoglobin ionizes, releasing a proton. This excess acid is removed by reacting with HCO_3^-:

$$H^+ + HCO_3^- \rightarrow H_2CO_3$$

But note that the $[HCO_3^-]/[H_2CO_3]$ ratio at pH 7.4 is 26/1.3 = 20:1. This is not a very effective buffering ratio; and as significant amounts of HCO_3^- are converted

---

[4]The volume of CO_2 is measured, but from the temperature and atmospheric pressure, the number of millimoles of CO_2 and hence its concentration in mmol/L in the solution it originated from can be calculated. At standard temperature and pressure, 22.4 L contain one mole gas.

to H_2CO_3, the pH would have to decrease to maintain the new ratio. But, fortunately, the H_2CO_3 produced is rapidly decomposed to CO_2 and H_2O by the enzyme decarboxylase, and the CO_2 is exhaled by the lungs. Hence, the ratio of HCO_3^-/H_2CO_3 remains constant at 20:1.

EXAMPLE 5.21 The total carbon dioxide content ($HCO_3^- + CO_2$) in a blood sample is determined by acidifying the sample and measuring the volume of CO_2 evolved with a Van Slyke manometric apparatus (see the clinical chemistry chapter). The total concentration was determined to be 28.5 mmol/L. The blood pH at 37°C was determined to be 7.48. What are the concentrations of HCO_3^- and CO_2 in the blood?

SOLUTION

$$pH = 6.10 + \log \frac{[HCO_3^-]}{[CO_2]}$$

$$7.48 = 6.10 + \log \frac{[HCO_3^-]}{[CO_2]}$$

$$\log \frac{[HCO_3^-]}{[CO_2]} = 1.38$$

$$\frac{[HCO_3^-]}{[CO_2]} = 10^{1.38} = 10^1 \times 10^{.38} = 24$$

$$[HCO_3^-] = 24[CO_2]$$

But

$$[HCO_3^-] + [CO_2] = 28.5 \text{ mmol/L}$$

$$24[CO_2] + [CO_2] = 28.5$$

$$[CO_2] = 1.1_4 \text{ mmol/L}$$

$$[HCO_3^-] = 28.5 - 1.1 = 27.4 \text{ mmol/L}$$

5.10 BUFFERS FOR BIOLOGICAL AND CLINICAL MEASUREMENTS

Many biological reactions of interest occur in the pH range of 6 to 8. A number, particularly specific enzyme reactions that might be used for analyses (see Chapter 18), may occur in the pH range of 4 to 10 or even greater. The proper selection of buffers for the study of biological reactions or for use in clinical analyses can be critical in determining whether or not they influence the reaction.

Phosphate Buffers

One useful series of buffers are phosphate buffers. Biological systems usually contain some phosphate already, and phosphate buffers will not interfere in many cases. By choosing appropriate mixtures of $H_3PO_4/H_2PO_4^-$, $H_2PO_4^-/HPO_4^{2-}$, or HPO_4^{2-}/PO_4^{3-}, solutions over a wide pH range can be prepared. Table 5.2 lists the compositions of a series of phosphate buffers at a constant ionic strength of 0.2. Ionic strength is a measure of the total salt content of a solution (see Chapter 4),

TABLE 5.2
The Composition of Phosphate Buffer Solutions[a]

85% H_3PO_4, mL	KH_2PO_4, g	Na_2HPO_4, g	Na_3PO_4, g	Measured pH
0.49	5.44			2.00
0.79	5.44			2.50
0.27	5.44			3.02
0.098	5.44			3.40
0.04	5.44			3.77
	5.44			4.18
	5.386	0.0352		4.68
	5.168	0.0858		5.05
	4.855	0.2014		5.30
	4.189	0.4374		5.72
	3.680	0.805		6.00
	2.170	1.13		6.50
	0.942	1.57		7.00
	0.4285	1.747		7.40
	0.122	1.86		8.00
	0.0181	1.894		8.60
		1.894		9.10
		1.849	0.06232	9.80
		1.784	0.1505	10.00
		1.633	0.3473	10.60
			2.531	11.90

Adapted from G. D. Christian and W. C. Purdy, *J. Electroanal. Chem.*, **3** (1962) 363.
[a]The final volume of all buffer solutions is 200 mL, and the ionic strength is 0.2.

and it frequently influences reactions, particularly in kinetic studies. Hence, these buffers could be used in cases where the ionic strength must be constant. However, the buffering capacity decreases markedly as the pH approaches the values for the single salts listed, and the single salts are not buffers at all; the best buffering capacity is within ± 1 pH unit of the respective pK_a value, that is, 1.96 ± 1, 7.12 ± 1, and 12.32 ± 1. The other solutions are most useful in establishing a given pH when little or no acid or base is to be added to or generated in the solution. The pH 7.40 mixture is very good for buffering at physiological pH values, although above pH 7.5 its buffering capacity begins to diminish.

EXAMPLE 5.22   What weights of NaH_2PO_4 and Na_2HPO_4 would be required to prepare 1 liter of a buffer solution of pH 7.45 that has an ionic strength of 0.100?

SOLUTION
Let $x = [Na_2HPO_4]$ and $y = [NaH_2PO_4]$. There are two unknowns, and two equations are needed. Our first equation is the ionic strength equation:

$$\mu = 1/2 \ \Sigma \ C_i Z_i^2$$

$$0.100 = 1/2 \ [Na^+](1)^2 + [HPO_4^{2-}](2)^2 + [H_2PO_4^-](1)^2$$

$$0.100 = 1/2 \ [(2x + y)(1)^2 + x(2)^2 + y(1)^2)] \tag{1}$$

$$0.100 = 3x + y$$

Our second equation is the Henderson–Hasselbalch equation.

$$pH = pK_{a2} + \log \frac{[HPO_4^{2-}]}{[H_2PO_4^-]}$$

$$7.45 = 7.12 + \log\frac{x}{y} \tag{2}$$

$$\frac{x}{y} = 10^{0.33} = 2.1_4$$

$$x = 2.1_4 y \tag{3}$$

Substitute in (1):

$$0.100 = 3(2.1_4)y + y$$

$$y = 0.013_5 \ M = [NaH_2PO_4]$$

Substitute in (3):

$$x = (2.1_4)(0.013_5) = 0.028_9 \ M = [Na_2HPO_4]$$

$$g \ NaH_2PO_4 = 0.013_5 \ mol/L \times 120 \ g/mol = 1.6_2 \ g/L$$

$$g \ Na_2HPO_4 = 0.028_9 \ mol/L \times 142 \ g/mol = 4.1_0 \ g/L$$

The use of phosphate buffers is limited in certain applications. Besides the limited buffering capacity at certain pH values, phosphate will precipitate or complex many polyvalent cations, and it frequently will participate in or inhibit a reaction. It should not be used, for example, when calcium is present if its precipitation would affect the reaction of interest.

Tris Buffers

A buffer that is widely used in the clinical laboratory and in biochemical studies in the physiological pH range is that prepared from *tris*(hydroxymethyl)aminomethane [(HOCH$_2$)$_3$CNH$_2$—**Tris,** or **THAM**] and its conjugate acid (the amino group is protonated). It is a primary standard and has good stability, has a high solubility in physiological fluids, is nonhygroscopic, does not absorb CO_2 appreciably, does not precipitate calcium salts, does not appear to inhibit many enzyme systems, and is compatible with biological fluids. It has a pK close to physiological pH (pK_a = 8.08 for the conjugate acid), but its buffering capacity below pH 7.5 does begin to diminish, a disadvantage. Other disadvantages are that the primary aliphatic amine has considerable potential reactivity and it is reactive with linen fiber junctions, as found in saturated calomel reference electrodes used in pH measurements (Chapter 11); a reference electrode with a ceramic, quartz, or sleeve junction should be used. These buffers are usually prepared by adding an acid such as hydrochloric acid to a solution of Tris to adjust the pH to the desired value.

5.11 THE DIVERSE ION EFFECT ON ACIDS AND BASES: K_a^0 and K_b^0

In Chapter 4, we discussed the thermodynamic equilibrium constant based on activities rather than on concentrations. Diverse salts affect the activities and therefore the extent of dissociation of weak electrolytes such as weak acids or bases. The

activity coefficient of the undissociated acid or base is essentially unity if it is uncharged. Then, for the acid HA,

$$K_a^0 = \frac{a_{H^+} \cdot a_{A^-}}{a_{HA}} \approx \frac{a_{H^+} \cdot a_{A^-}}{[HA]} \tag{5.87}$$

$$K_a^0 = \frac{[H^+]f_{H^+} \cdot [A^-]f_{A^-}}{[HA]} = K_a f_{H^+} f_{A^-} \tag{5.88}$$

$$K_a = \frac{K_a^0}{f_{H^+} f_{A^-}} \tag{5.89}$$

Therefore, we would predict an increase in K_a and in the dissociation with increased ionic strength, as the activity coefficients decrease. See Example 4.10 and
Problem 15 in Chapter 4. A similar relationship holds for weak bases (see Problem
54).

Since the ionic strength affects the dissociation of weak acids and bases, it will
have an effect on the pH of a buffer. We can write the Henderson–Hasselbalch
equation in terms of either K_a or K_a^0:

$$pH = pK_a + \log \frac{[A^-]}{[HA]} \tag{5.90}$$

$$pH = pK_a^0 + \log \frac{a_{A^-}}{[HA]} \tag{5.91}$$

If the ionic strength increases, K_a will increase and pK_a will decrease. So Equation
5.90 predicts a decrease in pH. Likewise, the activity of A^- will decrease, and a
similar decrease in pH is predicted by Equation 5.91. If a buffer solution is diluted,
its ionic strength will decrease, and so there would be a slight *increase* in pH, even
though the ratio of $[A^-]/[HA]$ remains constant. See Footnote 3.

QUESTION

1. Explain the difference between a strong electrolyte and a weak electrolyte. Is an "insoluble" salt a weak or a strong electrolyte?

PROBLEMS

Strong Acids and Bases

2. Calculate the pH and pOH of the following strong acid solutions: (a) 0.020 M $HClO_4$;
   (b) 1.3×10^{-4} M HNO_3; (c) 1.2 M HCl; (d) 1.2×10^{-9} M HCl; (e) 2.4×10^{-7} M
   HNO_3.

3. Calculate the pH and pOH of the following strong base solutions: (a) 0.050 M NaOH;
   (b) 0.14 M $Ba(OH)_2$; (c) 2.4 M NaOH; (d) 3.0×10^{-7} M KOH; (e) 3.7×10^{-3} M
   KOH.

4. Calculate the hydroxyl ion concentration of the following solutions: (a) 2.6×10^{-5} M
   HCl; (b) 0.20 M HNO_3; (c) 2.7×10^{-9} M $HClO_4$; (d) 1.9 M $HClO_4$.

5. Calculate the hydrogen ion concentration of the solutions with the following pH values: (a) 3.47; (b) 0.20; (c) 8.60; (d) −0.60; (3) 14.35; (f) −1.25.

6. Calculate the pH and pOH of a solution obtained by mixing equal volumes of 0.10 M H_2SO_4 and 0.30 M NaOH.

7. Calculate the pH of a solution obtained by mixing equal volumes of a strong acid solution of pH 3.00 and a strong base solution of pH 12.00.

Temperature Effect

8. Calculate the hydrogen ion concentration and pH of a neutral solution at 50°C (K_w = 5.5 × 10^{-14} at 50°C).

9. Calculate the pOH of a blood sample whose pH is 7.40 at 37°C.

Weak Acids and Bases

10. The pH of an acetic acid solution is 3.26. What is the concentration of acetic acid and what is the percent acid ionized?

11. The pH of a 0.20 M solution of a primary amine, RNH_2, is 8.42. What is the pK_b of the amine?

12. A monoprotic organic acid with a K_a of 6.7 × 10^{-4} is 3.5% ionized when 100 g of it is dissolved in 1 liter. What is the formula weight of the acid?

13. Calculate the pH of a 0.25 M solution of propanoic acid.

14. Calculate the pH of a 0.10 M solution of aniline, a weak base.

15. Calculate the pH of a 0.1 M solution of iodic acid, HIO_3.

16. The first proton of sulfuric acid is completely ionized, but the second proton is only partially dissociated, with an acidity constant K_{a2} of 1.2 × 10^{-2}. Calculate the hydrogen ion concentration in a 0.0100 M H_2SO_4 solution.

17. Calculate the hydrogen ion concentration in a 0.100 M solution of trichloroacetic acid.

18. An amine, RNH_2, has a pK_b of 4.20. What is the pH of a 0.20 M solution of the base?

19. What is the concentration of a solution of acetic acid if it is 3.0% ionized?

20. By how much should a 0.100 M solution of a weak acid HA be diluted in order to double its percent ionization? Assume $C > 100K_a$.

Salts of Weak Acids and Bases

21. If 25 mL of 0.20 M NaOH is added to 20 mL of 0.25 M boric acid, what is the pH of the resultant solution?

22. Calculate the pH of a 0.010 M solution of NaCN.

23. Calculate the pH of a 0.050 M solution of sodium benzoate.

24. Calculate the pH of a 0.25 M solution of pyridinium hydrochloride (pyridine · HCl, $C_6H_5NH^+Cl$).

25. Calculate the pH of the solution obtained by adding 12.0 mL of 0.25 M H_2SO_4 to 6.0 mL of 1.0 M NH_3.

26. Calculate the pH of the solution obtained by adding 20 mL of 0.10 M HOAc to 20 mL of 0.10 M NaOH.

27. Calculate the pH of the solution prepared by adding 0.10 mol each of hydroxylamine and hydrochloric acid to 500 mL water.

28. Calculate the pH of a 0.0010 M solution of sodium salicylate, $C_6H_4(OH)COONa$.

29. Calculate the pH of a 1.0 × 10^{-4} M solution of NaCN.

Polyprotic Acids and Their Salts

30. What is the pH of 0.0100 M solution of phthalic acid?

31. What is the pH of a 0.0100 M solution of potassium phthalate?

32. What is the pH of a 0.0100 M solution of potassium acid phthalate (KHP)?

33. Calculate the pH of a 0.600 M solution of Na_2S.

34. Calculate the pH of a 0.500 M solution of Na_3PO_4.
35. Calculate the pH of a 0.250 M solution of $NaHCO_3$.
36. Calculate the pH of a 0.600 M solution of NaHS.
37. Calculate the pH of a 0.050 M solution of the trisodium salt of EDTA (ethylenedi-aminetetraacetic acid), Na_3HY.

Buffers

38. Calculate the pH of a solution that is 0.050 M in formic acid and 0.10 M in sodium formate.
39. Calculate the pH of a solution prepared by mixing 5.0 mL of 0.10 M NH_3 with 10.0 mL of 0.020 M HCl.
40. An acetic acid-sodium acetate buffer of pH 5.00 is 0.100 M in NaOAc. Calculate the pH after the addition of 10 mL of 0.1 M NaOH to 100 mL of the buffer.
41. A buffer solution is prepared by adding 20 mL of 0.10 M sodium hydroxide solution to 50 mL of 0.10 M acetic acid solution. What is the pH of the buffer?
42. A buffer solution is prepared by adding 25 mL of 0.050 M sulfuric acid solution to 50 mL of 0.10 M ammonia solution. What is the pH of the buffer?
43. Aspirin (acetylsalicylic acid) is absorbed from the stomach in the free (un-ionized) acid form. If a patient takes an antacid that adjusts the pH of the stomach contents to 2.95 and then takes two 5-grain aspirin tablets (total 0.65 g), how many grams of aspirin are available for immediate absorption from the stomach, assuming immediate dissolution. Also assume that aspirin does not change the pH of the stomach contents. The pK_a of aspirin is 3.50, and its formula weight is 180.2.
44. *Tris*(hydroxymethyl)aminomethane [$(HOCH_2)_3CNH_2$—Tris, or THAM] is a weak base frequently used to prepare buffers in biochemistry. Its K_b is 1.2×10^{-6} and pK_b is 5.92. The corresponding pK_a is 8.08, which is near the pH of the physiological buffers, and so it exhibits good buffering capacity at physiological pH. What weight of THAM must be taken with 100 mL of 0.50 M HCl to prepare 1 liter of a pH 7.40 buffer?
45. Calculate the hydrogen ion concentration for Problem 17 if the solution contains also 0.100 M sodium trichloroacetate.

Buffers from Polyprotic Acids

46. What is the pH of a solution that is 0.20 M in phthalic acid (H_2P) and 0.10 M in potassium acid phthalate (KHP)?
47. What is the pH of a solution that is 0.25 M each in potassium acid phthalate (KHP) and potassium phthalate (K_2P)?
48. The total phosphate concentration in a blood sample is determined by spectrophoto-metry to be 3.0×10^{-3} M. If the pH of the blood sample is 7.45, what are the concentrations of $H_2PO_4^-$ and HPO_4^{2-} in the blood?

Constant-Ionic-Strength Buffers

49. What weight of Na_2HPO_4 and KH_2PO_4 would be required to prepare 200 mL of a buffer solution of pH 7.40 that has an ionic strength of 0.20? (See Chapter 4 for a definition of ionic strength.)
50. What volume of 85% (wt/wt) H_3PO_4 (sp. gr. 1.69) and what weight of KH_2PO_4 are required to prepare 200 mL of a buffer of pH 3.00 that has an ionic strength of 0.20?

α Calculations

51. Calculate the equilibrium concentrations of the different species in a 0.0100 M solution of sulfurous acid, H_2SO_3, at pH 4.00 ($[H^+] = 1.0 \times 10^{-4}$ M).
52. Derive Equations 5.71, 5.72, and 5.73 for α_1, α_2, and α_3 of phosphoric acid.

Diverse Salt Effect

53. Calculate the hydrogen ion concentration for a 0.0200 M solution of HCN in 0.100 M NaCl (diverse ion effect).

54. Derive the equivalent of Equation 5.89 for the diverse salt effect on an uncharged weak base B.

RECOMMENDED REFERENCES

Acid–Base Theories, Buffers

1. R. G. Bates, "Concept and Determination of pH," in I. M. Kolthoff and P. J. Elving, eds., *Treatise on Analytical Chemistry,* Part I, Vol. **1.** New York: Wiley-Intersience 1959, pp. 361–401.

2. N. W. Good, G. D. Winget, W. Winter, T. N. Connally, S. Izawa, and R. M. M. Singh, "Hydrogen Ion Buffers for Biological Research," *Biochem.,* **5** (1966) 467.

3. D. E. Gueffroy, ed., *A Guide for the Preparation and Use of Buffers in Biological Systems.* La Jolla, Cal.: Calbiochem, 1975.

4. I. M. Kolthoff, "Concepts of Acids and Bases," in I. M. Kolthoff and P. J. Elving, eds., *Treatise on Analytical Chemistry,* Part I, Vol. **1.** New York: Wiley-Interscience, 1959, pp. 405–420.

Equilibrium Calculations

5. S. Brewer, *Solving Problems in Analytical Chemistry.* New York: John Wiley & Sons, 1980. Describes iterative approach for solving equilibrium calculations.

6. J. N. Butler, *Ionic Equilibria. A Mathematical Approach.* Reading, Mass.: Addison-Wesley, 1964.

7. D. D. DeFord, "The Reliability of Calculations Based on the Law of Chemical Equilibrium," *J. Chem. Ed.,* **31** (1954) 460.

8. H. A. Laitinen and W. E. Harris, "Acid–Base Equilibria in Water," in *Chemical Analysis,* 2nd ed. New York: McGraw-Hill, 1975. Chapter 3.

9. E. R. Nightingale, "The Use of Exact Expressions in Calculating H^+ Concentrations," *J. Chem. Educ.,* **34** (1957) 277.

chapter
6

GRAVIMETRIC ANALYSIS

Gravimetric analysis is one of the most accurate and precise methods of macro-quantitative analysis. The analyte is selectively converted to an insoluble form. The separated precipitate is dried or ignited, possibly to another form, and is accurately weighed. From the weight of the precipitate and a knowledge of its chemical composition, the weight of analyte in the desired form is calculated.

Before describing the principles and operations of gravimetric analysis, a brief discussion of precipitation equilibria will be given. The specific steps of gravimetric analysis will then be described, including preparing the solution in proper form for precipitation, the precipitation process and how to obtain the precipitate in pure and filterable form, the filtration and washing of the precipitate to prevent losses and impurities, and heating of the precipitate to convert it to a weighable form. Finally, calculation procedures are given for computing the quantity of analyte from the weight of precipitate, followed by some common examples of gravimetric analysis.

6.1 PRECIPITATION EQUILIBRIA: THE SOLUBILITY PRODUCT

When substances have limited solubility and their solubility is exceeded, the ions of the dissolved portion exist in equilibrium with the solid material. So-called "insoluble" compounds generally exhibit this property.

When a compound is referred to as insoluble, it is not completely insoluble but is **slightly soluble.** For example, if solid AgCl is added to water, a small portion of it will dissolve:

$$\underline{AgCl} \rightleftharpoons (AgCl)_{aq} \rightleftharpoons Ag^+ + Cl^- \tag{6.1}$$

The precipitate will have a definite solubility (i.e., a definite amount that will dissolve) in g/L, or mol/L, at a given temperature (a saturated solution). A small amount of undissociated compound usually exists in equilibrium in the aqueous phase (e.g., on the order of 0.1%) and its concentration is constant. It is difficult to measure the undissociated molecule, and we are interested in the ionized form as a measure of a compound's solubility and chemical availability. Hence, the presence of any undissociated species can generally be neglected.

An overall equilibrium constant can be written for the above stepwise equilibrium, called the **solubility product** K_{sp}. $(AgCl)_{aq}$ cancels when the two stepwise equilibrium constants are multiplied together.

$$K_{sp} = [Ag^+][Cl^-] \tag{6.2}$$

The "concentration" of any solid such as AgCl is constant and is combined in the equilibrium constant to give K_{sp}. The above relationship holds regardless of the presence of any undissociated intermediate, that is, the concentrations of free ions are rigorously defined by Equation 6.2, and we will take these as a measure of a compound's solubility. From a knowledge of the value of the solubility product at a specified temperature, the solubility of the compounds can be calculated. (The solubility product is determined in the reverse order, by measuring the solubility.)

The amount of a slightly soluble salt that dissolves does not depend on the amount of the solid in equilibrium with the solution, so long as there is enough to saturate the solution. Instead, the amount that dissolves depends on the *volume* of the solvent. A nonsymmetric salt such as Ag_2CrO_4 would have a K_{sp} as follows:

$$Ag_2CrO_4 \rightleftharpoons 2Ag^+ + CrO_4^{2-} \tag{6.3}$$

$$K_{sp} = [Ag^+]^2[CrO_4^{2-}] \tag{6.4}$$

Such electrolytes do not dissolve or dissociate in steps because they are really strong electrolytes. That portion that dissolves ionizes completely. Therefore, we do not have stepwise K_{sp} values. As with any equilibrium constant, the K_{sp} product holds under all conditions at the specified temperature.

EXAMPLE 6.1   The K_{sp} of AgCl at 25°C is 1.0×10^{-10}. Calculate the concentrations of Ag^+ and Cl^- in a saturated solution of AgCl, and the molar solubility of AgCl.

SOLUTION
When AgCl ionizes, equal amounts of Ag^+ and Cl^- are formed: $AgCl \rightleftharpoons Ag^+ + Cl^-$ and $K_{sp} = [Ag^+][Cl^-]$. Let s represent the molar solubility of AgCl. Since each mole of AgCl that dissolves gives one mole of either Ag^+ of Cl^-, then

$$[Ag^+] = [Cl^-] = s$$

$$s^2 = 1.0 \times 10^{-10}$$

$$s = 1.0 \times 10^{-5} \, M$$

The solubility of AgCl is $1.0 \times 10^{-5} \, M$.

If there is an excess of one ion over the other, the concentration of the other is suppressed **(common ion effect),** and the solubility of the precipitate is decreased. The concentration can still be calculated from the solubility product.

EXAMPLE 6.2   Ten milliliters of 0.20 M $AgNO_3$ is added to 10 mL of 0.10 M NaCl. Calculate the concentration of Cl^- remaining in solution at equilibrium, and the solubility of the AgCl.

SOLUTION
The final volume is 20 mL. The millimoles Ag^+ added equals $0.20 \times 10 = 2.0$ mmol. The millimoles Cl^- taken equals $0.10 \times 10 = 1.0$ mmol. Therefore, the millimoles excess Ag^+ equals $(2.0 - 1.0) = 1.0$ mmol. From Example 6.1, we see that the Ag^+ concentration contributed from the precipitate is small, that is, on the order of 10^{-5} mmol/mL in the absence of a common ion. This will be even smaller in the presence of excess Ag^+ since the solubility is suppressed. Therefore, we can neglect the amount of Ag^+ contributed from the precipitate compared to the excess Ag^+. Hence, the final concentration of Ag^+ is 1.0 mmol/20 mL = 0.050M, and

$$(0.050)[Cl^-] = 1.0 \times 10^{-10}$$

$$[Cl^-] = 2.0 \times 10^{-9} \, M$$

The Cl^- concentration again equals the solubility of the AgCl, and so the solubility is $2.0 \times 10^{-9} \, M$.

Because the K_{sp} product always holds, **precipitation will not take place unless the product of $[Ag^+]$ and $[Cl^-]$ exceeds the K_{sp}.** If the product is just equal to K_{sp}, all the Ag^+ and Cl^- would remain in solution.

EXAMPLE 6.3   What must be the concentration of added Ag^+ to just start precipitation of AgCl in a $1.0 \times 10^{-3} \, M$ solution of NaCl?

SOLUTION

$$[Ag^+](1.0 \times 10^{-3}) = 1.0 \times 10^{-10}$$

$$[Ag^+] = 1.0 \times 10^{-7} \, M$$

The concentration of Ag^+ must, therefore, just exceed $10^{-7} \, M$ to begin precipitation.

EXAMPLE 6.4   What is the solubility of PbI_2, in g/L, if the solubility product is 7.1×10^{-9}?

SOLUTION
The equilibrium is $PbI \rightleftharpoons Pb^{2+} + 2I^-$, and $K_{sp} = [Pb^{2+}][I^-]^2 = 7.1 \times 10^{-9}$. Let s represent the molar solubility of PbI_2. Then

$$[Pb^{2+}] = s \text{ and } [I^-] = 2s$$

$$(s)(2s)^2 = 7.1 \times 10^{-9}$$

$$s = \sqrt[3]{\frac{7.1 \times 10^{-9}}{4}} = 1.2 \times 10^{-3} \, M$$

Therefore, the solubility, in g/L, is

$$1.2 \times 10^{-3} \text{ mol/L} \times 461.0 \text{ g/mol} = 0.55 \text{ g/L}$$

Note that the concentration of I^- was *not* doubled before squaring; $2s$ represented its actual equilibrium concentration, not twice its concentration. We could have let s represent the concentration of I^-, instead of the molar solubility of PbI_2, in which case $[Pb^{2+}]$ and the solubility of PbI_2 would have been $1/2s$. The calculated s would have been twice as great, but the concentrations of each species would have been the same. You try this calculation!

EXAMPLE 6.5   Calculate the molar solubility of $PbSO_4$ and compare it with that of PbI_2.

SOLUTION

$$PbSO_4 \rightleftharpoons Pb^{2+} + SO_4^{2-}$$

$$[Pb^{2+}][SO_4^{2-}] = 1.6 \times 10^{-8}$$

$$(s)(s) = 1.6 \times 10^{-8}$$

$$s = 1.3 \times 10^{-4}\ M$$

Although the K_{sp} of PbI_2 is smaller than that of $PbSO_4$, the solubility of PbI_2 is greater (see Example 6.4).

For electrolytes of the same valence type, the order of solubility will be the same as the order of the corresponding solubility products. But when comparing salts of different valence type, the order may be different. Compound AB will have a smaller molar solubility than compound AC_2 when both have identical K_{sp} values.

We take advantage of the common ion effect to decrease the solubility of a precipitate in gravimetric analysis. For example, sulfate ion is determined by precipitating $BaSO_4$ with added barium chloride solution. Figure 6.1 illustrates the effect of excess barium ion on the solubility of $BaSO_4$.

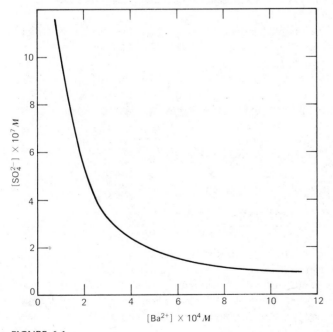

FIGURE 6.1
Predicted effect of excess barium ion on solubility of $BaSO_4$. The sulfate concentration is the amount in equilibrium and is equal to the $BaSO_4$ solubility. In the absence of excess barium ion, the solubility is $10^{-5}\ M$.

6.2 THE DIVERSE ION EFFECT ON SOLUBILITY: K_{sp}^0

In Chapter 4 we defined the thermodynamic equilibrium constant written in terms of activities to account for the effects of inert electrolytes on equilibria. The presence of diverse salts will generally increase the solubility of precipitates due to the shielding of the dissociated ionic species. (Their activity is decreased.) Consider the solubility of AgCl. The thermodynamic solubility product K_{sp}^0 is

$$K_{sp}^0 = a_{Ag^+}a_{Cl^-} = [Ag^+]f_{Ag^+}[Cl^-]f_{Cl^-} \tag{6.5}$$

Since the *concentration* solubility product K_{sp} is $[Ag^+][Cl^-]$,

$$K_{sp}^0 = K_{sp}f_{Ag^+}f_{Cl^-} \tag{6.6}$$

or

$$K_{sp} = \frac{K_{sp}^0}{f_{Ag^+}f_{Cl^-}} \tag{6.7}$$

The numerical value of K_{sp}^0 holds at all activities. K_{sp} equals K_{sp}^0 at zero ionic strength, but at appreciable ionic strengths, a value must be calculated for each ionic strength using Equation 6.7. Note that this equation predicts, as we predicted qualitatively, that decreased activity of the ions will result in an increased K_{sp} and, therefore, increased molar solubility.

EXAMPLE 6.6   Calculate the solubility of silver chloride in 0.10 M NaNO$_3$.

SOLUTION
The equilibrium constants listed in the Appendix are for zero ionic strength, that is, they are really thermodynamic equilibrium constants.[1] Therefore, from Table D.3 in Appendix D, $K_{sp}^0 = 1.0 \times 10^{-10}$.

We need the activity coefficients of Ag$^+$ and Cl$^-$. The ionic strength is 0.10. From Reference 6 in Chapter 4, we find that $f_{Ag^+} = 0.75$ and $f_{Cl^-} = 0.76$. (You could also have used the values of α_{Ag^+} and α_{Cl^-} in the reference to calculate the activity coefficients using Equation 4.19.) From Equation 6.7,

$$K_{sp} = \frac{1.0 \times 10^{-10}}{(0.75)(0.76)} = 1.8 \times 10^{-10} = [Ag^+][Cl^-] = s^2$$

$$s = \sqrt{1.8 \times 10^{-10}} = 1.3 \times 10^{-5} \, M$$

This is 30% greater than at zero ionic strength ($s = 1.0 \times 10^{-5} \, M$).

Figure 6.2 illustrates the increase in solubility of BaSO$_4$ in the presence of NaNO$_3$ due to the diverse ion effect.

The increase in solubility is greater with precipitates containing multiply charged ions. At very high ionic strengths, where activity coefficients may become greater than unity, the solubility is decreased. In gravimetric analysis, a sufficiently large

---

[1]Experimental K_{sp} values are available at different ionic strengths and can be used to calculate molar solubilities at the listed ionic strengths without having to calculate activity coefficients.

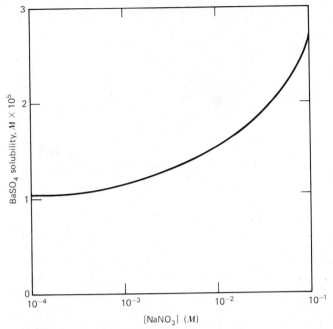

FIGURE 6.2
Predicted effect of increased ionic strength on solubility of $BaSO_4$. Solubility at zero ionic strength is 1.0×10^{-5} M.

excess of precipitating agent is added so that the solubility is reduced to such a small value that we do not need to worry about this effect.

Acids frequently affect the solubility of a precipitate. As the H^+ concentration increases, it competes more effectively with the metal ion of interest for the precipitating agent (which may be the anion of a weak acid). With less free reagent available, and a constant K_{sp}, the solubility of the salt must increase:

$$
\begin{array}{ll}
M^{+n} + nR^- \rightleftharpoons MR_n & \text{(desired reaction)} \\
R^- + H^+ \rightleftharpoons HR & \text{(competing reaction)} \\
MR_n + nH^+ \rightleftharpoons M^{n+} + nHR & \text{(overall reaction)}
\end{array}
$$

Similarly, a complexing agent that reacts with the metal ion of the precipitate will increase the solubility, for example, when ammonia reacts with silver chloride:

$$
AgCl + 2NH_3 \rightleftharpoons Ag(NH_3)_2^+ + Cl^-
$$

The quantitative treatment of these effects in solubility calculations will be treated in Chapter 9, after we have covered acid–base equilibria and complex equilibria.

6.3 UNIT OPERATIONS IN GRAVIMETRIC ANALYSIS

A successful gravimetric analysis consists of a number of important operations in order to obtain a pure and filterable precipitate suitable for weighing.

Steps of a Gravimetric Analysis

The steps required in a gravimetric analysis, after the sample has been dissolved, can be summarized as follows:

1. Preparation of the solution.
2. Precipitation.
3. Digestion.
4. Filtration.
5. Washing.
6. Drying or igniting.
7. Weighing.
8. Calculation.

These operations and the reasons for them are described in the remainder of this chapter.

Preparation of the Solution

Some form of preliminary separation may be necessary to eliminate interfering materials. Also, the solution conditions must be adjusted to maintain low solubility of the precipitate and to obtain it in a form suitable for filtration. Proper adjustment of the solution conditions prior to precipitation may also mask potential interferences. Factors that must be considered include the volume of the solution during precipitation, the concentration range of the test substance, the presence and concentrations of other constituents, the temperature, and the pH.

Although preliminary separations may be required, in other instances the precipitation step in gravimetric analysis is sufficiently selective that other separations are not required. The pH is important because often it influences the solubility of the analytical precipitate and the possibility of interferences from other substances. For example, calcium oxalate is insoluble in basic medium; but at low pH, the oxalate ion combines with the hydrogen ions to form a weak acid. 8-Hydroxyquinoline (oxine) can be used to precipitate a large number of elements; but by control of pH, elements can be precipitated selectively. Aluminum ion can be precipitated at pH 4, but the concentration of the anion form of oxine is too low at this pH to precipitate magnesium ion.

A higher pH is required to shift the ionization step to the right in order to precipitate magnesium. If the pH is too high, magnesium hydroxide will precipitate, causing interference.

The effects of the other factors mentioned above will become apparent as we discuss the precipitation step.

Conditions for Analytical Precipitation

The precipitate should first be **sufficiently insoluble** that the amount lost due to solubility will be negligible. It should consist of **large crystals** so that they can be easily filtered. All precipitates tend to carry some of the other constituents of the solution with them. This contamination should be negligible; it is minimized also by keeping the crystals large.

We can achieve an appreciation of the proper conditions for precipitation by first looking at the **precipitation process:** When a solution of precipitating agent is added to a test solution to form a precipitate, such as in the addition of $AgNO_3$ to a chloride solution to precipitate $AgCl$, the actual precipitation occurs in a series of steps. First, **supersaturation** occurs, that is, the solution phase contains more of the dissolved salt than occurs at equilibrium. This is a metastable condition, and the driving force will be for the system to approach equilibrium (saturation). This is started by **nucleation.** For nucleation to occur, a minimum number of particles must come together to produce microscopic nuclei of the solid phase. The higher the degree of supersaturation, the greater the rate of nucleation. The formation of a greater number of nuclei per unit time will ultimately result in more total crystals of smaller size. The total crystal surface area will be larger, with increased danger of adsorption of impurities (see below).

Although nucleation should theoretically occur spontaneously, it is usually induced, for example, on dust particles, scratches on the vessel surface, or added seed crystals of the precipitate (not in quantitative analysis).

Following nucleation, the initial nucleus will grow by deposition of other precipitate particles to form a crystal of a certain geometric shape. Again, the greater the supersaturation, the more rapid the crystal growth rate. With increased growth rate, there is increased chance of imperfections in the crystal and trapping of impurities.

Von Weimarn discovered that the particle size of precipitates is inversely proportional to the relative supersaturation of the solution during the precipitation process:

$$\text{relative supersaturation} = \frac{Q - S}{S}$$

where Q is the concentration of the mixed reagents *before* precipitation occurs and is the **degree of supersaturation,** and S is the **solubility** of the precipitate at equilibrium. This ratio is also called the **von Weimarn ratio.**

As mentioned above, when a solution is supersaturated, it is in a state of metastable equilibrium, and this favors rapid nucleation to form a large number of small particles. That is,

> high relative supersaturation → many small crystals
>
> (high surface area)
>
> low relative supersaturation → fewer, larger crystals
>
> (low surface area)

Obviously, then, we want to keep Q low and S high during precipitation. Several steps are commonly taken to maintain **favorable conditions for precipitation:**

---

1. Precipitate from **dilute solution.** This keeps Q low.
2. Add dilute precipitating reagents **slowly,** with effective **stirring.** This also keeps Q low. Local excesses of the reagent are prevented by stirring.
3. Precipitate from **hot solution.** This increases S. The solubility should not be too great or the precipitation will not be quantitative (with less than 1 ppt remaining). The bulk of the precipitation will not be quantitative (with less than 1 ppt remaining). The bulk of the precipitation may be performed in the hot solution, and then the solution may be cooled to make the precipitation quantitative.
4. Precipitate at as **low** a **pH** as is possible to still maintain quantitative precipitation. As shown above, many precipitates are more soluble in acid medium, and this slows the rate of precipitation. They are more soluble because the anion of the precipitate combines with protons in the solution.

Most of these operations also often decrease the degree of contamination. The concentration of impurities is kept lower and their solubility is increased, and the slower rate of precipitation decreases their chance of being trapped. The larger crystals have a smaller specific surface area (i.e., a smaller surface area relative to the mass) and so have less chance of adsorbing impurities.

When the precipitation is performed, a slight excess of precipitating reagent is added to decrease the solubility by mass action (common ion effect) and to assure complete precipitation. If the approximate amount of analyte is known, a 10 percent excess of reagent is generally added. Completeness of precipitation is checked by waiting until the precipitate has settled and then adding a few drops of precipitating reagent to the clear solution above it. If no new precipitate forms, precipitation is complete.

Digestion of the Precipitate

It is known that very small crystals with a large specific surface area have a higher surface energy and a higher apparent solubility than large crystals. This is an initial rate phenomenon and does not represent the equilibrium condition. When a precipitate is allowed to stand in the presence of the mother liquor (the solution from which it was precipitated), the large crystals grow at the expense of the small ones. This process is called **digestion,** or **Ostwald ripening.** This process is illustrated in Figure 6.3. The small particles tend to dissolve and reprecipitate on the surfaces of the larger crystals. In addition, individual particles **agglomerate** to effectively share a common counterion layer (see below), and the agglomerated particles finally **cement** together through formation of connecting bridges. This results in an appreciable decrease in surface area.

Also, imperfections of the crystals tend to disappear, and adsorbed or trapped impurities tend to go into solution. Digestion is usually done at elevated temperatures to speed the process, although in some cases it is done at room temperature. It improves the filterability of the precipitate and its purity.

Many precipitates do not give a favorable von Weimarn ratio, especially very insoluble ones. Hence, it is impossible to yield a crystalline precipitate (small number of large particles), and the precipitate is first **colloidal** (large number of small particles).

Colloidal particles are very small (1 to 100 μm) and have a very large surface-to-mass ratio, which promotes surface adsorption. They are formed by virtue of the

FIGURE 6.3
Ostwald ripening.

precipitation mechanism. As a precipitate forms, the ions are arranged in a fixed pattern. In AgCl, for example, there will be alternating Ag^+ and Cl^- ions on the surface (see Figure 6.4). While there are localized $+$ and $-$ charges on the surface, the net surface charge is zero. However, the surface does tend to adsorb the ion of the precipitate particle that is in excess in the solution, for example, Cl^- if precipitating Cl^- with Ag^+; this imparts a charge. (With crystalline precipitates, the degree of such adsorption will generally be small in comparison with particles that tend to form colloids.) The adsorption creates a **primary layer** that is strongly adsorbed and is an integral part of the crystal. It will attract ions of the opposite charge in a **counterlayer** or secondary layer and give an overall neutral particle. There will be solvent molecules interspersed between the layers. Normally, the counterlayer completely neutralizes the primary layer and is close to it, so the particles will collect together to form larger-size particles, that is, they will **coagulate.** However, if the secondary layer is loosely bound, the primary surface charge will tend to repel like particles, maintaining a colloidal state.

When coagulated particles are filtered, they retain the adsorbed primary and secondary ion layers along with solvent. If washed with water, the extent of solvent (water) molecules between the layers is increased, causing the secondary layer to be loosely bound, and the particles revert to the colloidal state. This process is called **peptization** and is discussed in more detail below in describing washing of the precipitate. An added electrolyte will result in a closer secondary layer and will promote coagulation. Heating tends to decrease adsorption and the effective charge in the adsorbed layers, thereby aiding coagulation. Stirring will also help.

$$Na^+ \qquad K^+$$
$$K^+ \qquad H^+ \Big\} \text{ Counter layer}$$

$$Cl^- \qquad Cl^- \qquad Cl^- \qquad Cl^- \Big\} \text{ Primary adsorptive layer}$$

$$\overline{\begin{matrix} Ag^+Cl^-Ag^+Cl^-Ag^+Cl^- \\ Cl^-Ag^+Cl^-Ag^+Cl^-Ag^+ \end{matrix}} \Big\} \text{ Crystal}$$

FIGURE 6.4
Representation of silver chloride crystal and adsorptive layers when Cl^- is in excess.

While all colloidal precipitates cause difficulties in analytical precipitates, some are worse than others. There are two types of colloids, **hydrophilic** and **hydrophobic.** Hydrophilic means "water-loving," and these colloids have a strong affinity for water. A solution of a hydrophilic colloid is therefore viscous. A hydrophobic colloid has little attraction for water. A solution of this type of colloid is called a **sol.**

Coagulation of a hydrophobic colloid is fairly easy and results in a curdy precipitate. An example is silver chloride. Coagulation of a hydrophilic colloid, such as hydrous ferric oxide, is more difficult, and it produces a gelatinous precipitate that is difficult to filter because it tends to clog the pores of the filter. In addition, gelatinous precipitates adsorb impurities readily because of their very large surface area. Sometimes a *reprecipitation* of the filtered precipitate is required. During the reprecipitation, the concentration of impurities in solution (from the original sample matrix) has been reduced to a low level, and adsorption will be very small.

Despite the colloidal nature of silver chloride, the gravimetric determination of chloride is one of the most accurate determinations.

Impurities in Precipitates

Precipitates tend to carry down from the solution other constituents that are normally soluble, causing contamination of the precipitate. This process is called **coprecipitation.** There are a number of ways in which a foreign material may be coprecipitated.

1. OCCLUSION. In the process of **occlusion,** foreign ions are trapped inside the crystal as it is formed. As the crystal grows, it will contain a primary adsorbed layer of the lattice ion *that is in excess.* For example, when $BaSO_4$ is precipitated by the slow addition of Ba^{2+} (as $BaCl_2$) to a solution of SO_4^{2-}, the sulfate ion will be in excess throughout the precipitation, and the primary adsorbed layer will be sulfate ions. This is neutralized by a secondary layer of counterions from the solution. For example, Na^+ may be the counterion (two per SO_4^{2-}). As more barium chloride is added, the loosely bound Na^+ is replaced by Ba^{2+}, and the crystal continues to grow. However, if the growth takes place too rapidly, all the sodium ions may not be replaced by barium ions.

Occlusion may also take place by mechanical entrapment of solvent in imperfections (holes) in the crystal. The solvent will contain dissolved impurities.

Occluded impurities are difficult to remove. Digestion may help some, but this is not completely effective. The impurities cannot be removed by washing. Purification by dissolving and reprecipitating may be helpful.

2. SURFACE ADSORPTION. As we mentioned above, the surface of the precipitate will have a primary adsorbed layer of the lattice ions in excess. This results in **surface adsorption.** After the barium sulfate is completely precipitated, the lattice ion in excess will be barium, and this will form the primary layer. The counterion will be a foreign anion, say, a nitrate anion, two for each barium ion. The net effect then is an adsorbed layer of barium nitrate. These adsorbed layers can often be removed by washing, or they can be replaced by ions that are readily volatilized. Gelatinous precipitates are especially troublesome, though. Digestion reduces the surface area and the amount of adsorption.

3. POSTPRECIPITATION. Sometimes, when the precipitate is allowed to stand in contact with the mother liquor, a second substance will slowly form a precipitate with the precipitating reagent. This is called **postprecipitation.** For example, when calcium oxalate is precipitated in the presence of magnesium ions, magnesium oxalate does not immediately precipitate because it tends to form supersaturated solutions. But it will come down if the solution is allowed to stand too long before being filtered. Similarly, copper sulfide will precipitate in acid solution in the presence of zinc ions, but eventually zinc sulfide will precipitate.

Two compounds are said to be *isomorphous* if they have the same type of formula and crystallize in similar geometric forms. When their lattice dimensions are about the same, one ion can replace another in a crystal, resulting in a **mixed crystal.** This process is called **isomorphous replacement.** For example, in the precipitation of Mg^{2+} as magnesium ammonium phosphate, K^+ has nearly the same ionic size as NH_4^+ and can replace it to form magnesium potassium phosphate. Isomorphous replacement, when it occurs, is very serious and little can be done about it. Precipitates in which it occurs are seldom used analytically.

Washing and Filtering the Precipitates

Coprecipitated impurities, especially those on the surface, can be removed by washing the precipitate after filtering. The precipitate will be wet with the mother liquor, which is also removed by washing. Many precipitates cannot be washed with pure water, because **peptization** occurs. This is the reverse of coagulation, as mentioned above.

The process of coagulation discussed above is at least partially reversible. As mentioned, coagulated particles have a neutral layer of adsorbed primary and counterions. It was also pointed out that the presence of another electrolyte will cause the counterions to be forced into closer contact with the primary layer, promoting coagulation. These foreign ions are carried along in the coagulation. Washing with water will dilute and remove foreign ions and the counterion will occupy a larger volume, with more solvent molecules between it and the primary layer. The result is that the repulsive forces between particles become strong again, and the particles partially revert to the colloidal state and pass through the filter. Prevention consists of adding an electrolyte to the wash liquid, for example, HNO_3 or NH_4NO_3 for AgCl precipitate (but not KNO_3 since it is nonvolatile—see below).

The electrolyte must be one that is volatile at the temperature to be used for drying or ignition, and it must not dissolve the precipitate. For example, dilute nitric acid is used as the wash solution for silver chloride. The nitric acid replaces the adsorbed layer of $Ag^+|anion^-$, and it is volatilized when dried at 110°C. Ammonium nitrate is used as the wash electrolyte for hydrous ferric oxide. It is decomposed to NH_3, HNO_3, N_2, and oxides of nitrogen when the precipitate is dried by ignition at a high temperature.

When a precipitate is washed, a test should be made to determine when the washing is complete. This is usually done by testing the filtrate for the presence of an ion of the precipitating reagent. After several washings with small volumes of the wash liquid, a few drops of the filtrate are collected in a test tube for the testing. For example, if chloride ion is determined by precipitating with silver nitrate reagent, the filtrate is tested for silver ion by adding sodium chloride or dilute HCl.

The technique of filtering was described in Chapter 2.

Drying or Igniting the Precipitate

If the collected precipitate is in a form suitable for weighing, it must be heated to remove water and to remove the adsorbed electrolyte from the wash liquid. This drying can usually be done by heating at 110 to 120°C for 1 to 2 hours. **Ignition** at a much higher temperature is usually required if a precipitate must be converted to a more suitable form for weighing. For example, magnesium ammonium phosphate, $MgNH_4PO_4$, is decomposed to the pyrophosphate, $Mg_2P_2O_7$, by heating at 900°C. Hydrous ferric oxide, $Fe_2O_3 \cdot xH_2O$, is ignited to the anhydrous ferric oxide. Many metals that are precipitated by organic reagents (for example, 8-hydroxyquinoline) or by sulfide can be ignited to their oxides.

If a precipitate is to be ignited in a porcelain filter crucible, the moisture should be driven off first at a low heat. The ignition may be done in a muffle furnace or by heating with a burner. If a burner is to be used, the filter crucible should be placed in a porcelain or platinum crucible to prevent reducing gases of the flame diffusing through the pores of the filter.

When precipitates are collected on filter paper, the cone-shaped filter containing the precipitate is removed from the funnel, the upper edge is flattened, and the corners are folded in. Then, the top is folded over and the paper and contents are placed in a crucible with the bulk of the precipitate on the bottom. The paper must now be dried and charred off. The crucible is placed at an angle on a triangle support with the crucible cover slightly ajar, as illustrated in Figure 6.5. The moisture is removed by low heat from the burner, with care taken to avoid splattering. The heating is gradually increased as moisture is evolved and the paper begins to char. Care should be taken to avoid directing the reducing portion of the flame into the crucible. A sudden increase in the volume of smoke evolved indicates that the paper is about to burst into flame, and the burner should be removed. If it does burst into flame, it should be smothered quickly by replacing the crucible cover. Carbon particules will undoubtedly appear on the cover, and these will ultimately have to be ignited. Finally, when no more smoke is detected, the charred paper is burned off by gradually increasing the flame temperature. The carbon residue should glow but should not flame. Continue heating until all the carbon and tars on the crucible and its cover are burned off. The crucible and precipitate are now ready for igniting. The ignition can be continued with the burner used at highest temperature or with the muffle furnace.

Before a precipitate is collected in a filter crucible or transferred to a crucible, the crucible should be dried to constant weight (e.g., 1 hour of heating, followed by $\frac{1}{2}$-hour heatings) if the precipitate is to be dried, or it should be ignited to constant weight if the precipitate is to be ignited. Constant weight is considered to have been achieved with an analytical student balance when successive weighings

FIGURE 6.5
Crucible and cover supported on a wire triangle for charring off paper.

agree within about 0.3 or 0.4 mg. The crucible plus the precipitate are heated to constant weight in a similar manner. After the first heating, the time of heating can be reduced by half. The crucible should be allowed to cool in a desiccator for at least $\frac{1}{2}$-hour before weighing. Red-hot crucibles should be allowed to cool below redness before placing them in the desiccator (use crucible tongs—usually nickel-plated or stainless steel to minimize contamination from rust, etc.). Before weighing a covered crucible, check for any radiating heat by placing your hand near it (don't touch).

Precipitation from Homogeneous Solution

We have seen that the most favorable conditions for precipitation involve precipitating from dilute solution and adding the precipitating agent slowly with effective stirring. These operations maintain a low degree of supersaturation (low Q). In spite of these operations, local excesses of precipitating reagent are unavoidable when the reagent is added to the sample solution. A process of precipitation from homogeneous solution avoids such difficulties. In this technique, the precipitating reagent is generated *in situ* by a chemical reaction which occurs uniformly throughout the solution.

An example is the generation of hydroxyl ions for the precipitation of hydrous iron oxide, aluminium oxide, and so forth. The precipitating reagent is formed by hydrolysis of urea in an initially acid solution:

$$(NH_2)_2CO + 3H_2O \rightarrow CO_2 \uparrow + 2NH_4^+ + 2OH^-$$

This hydrolysis reaction takes place slowly and occurs at temperatures just below the boiling point of water. Hydrous oxide precipitates formed using this method of reagent addition have higher densities than the corresponding precipitates formed by external addition of reagent. Hence, the volume of the precipitate is about one-tenth that produced using external addition of reagent, and the precipitates are much more easily filtered. In addition, because of the decreased specific surface area of the precipitates, the degree of coprecipitation is decreased and a purer precipitate is obtained.

There are many examples of homogeneous precipitation reactions. The above reaction for generating hydroxyl ions, for example, may be used for homogeneous pH control. Thus, calcium may be precipitated homogeneously as the oxalate by adding the oxalate reagent to an acidic solution and then generating hydroxyl ions homogeneously to bring the pH up to a value at which calcium oxalate precipitates. Sulfate ion can be generated homogeneously by heating a solution containing sulfamic acid. The sulfamic acid hydrolyzes as follows:

$$HSO_3NH_2 + H_2O \rightarrow H^+ + SO_4^{2-} + NH_4^+$$

Thus, barium sulfate or lead sulfate may be homogeneously precipitated. In the presence of nitric acid, sulfate ion is also formed homogeneously by the following reaction:

$$HSO_3NH_2 + HNO_3 \rightarrow 2H^+ + SO_4^{2-} + N_2O + H_2O$$

Figures 6.6 and 6.7 compare the lead sulfate precipitates obtained using the conventional sulfuric acid precipitation method and the homogeneous precipitation method in the presence of nitric acid. Note the increase in regularity of the shape of the crystals of lead sulfate formed by homogeneous precipitation. Creeping of the precipitate does not occur, and contamination from coprecipitation is small.

FIGURE 6.6
Lead sulfate precipitated by conventional sulfuric acid method. (From J. E. Koles, P. A. Shinners, and W. F. Wagner, *Talanta,* **12** (1965) 297. Reproduced by permission of Pergamon Press, Ltd.)

FIGURE 6.7
Lead sulfate precipitated from homogeneous solution. (From J. E. Koles, P. A. Shinners, and W. F. Wagner, *Talanta,* **12** (1965) 297. Reproduced by permission of Pergamon Press, Ltd.)

6.4   CALCULATIONS

Calculations are usually made on a percentage basis:

$$\% \ A = \frac{g_A}{g_{sample}} \times 100\% \tag{6.8}$$

where g_A represents the grams of analyte (the desired test substance) and g_{sample}

represents the grams of sample taken for analysis. We have seen that the analyte is almost always weighed in a form different from the desired form of the results. We must, therefore, calculate the weight of the desired substance from the weight of the gravimetric precipitate. This can be done by using a direct proportion. For example, if we are analyzing for the percentage of chloride in a sample by weighing it as AgCl, we can write

$$Cl^- \xrightarrow{\text{precipitating reagent}} AgCl$$

We derive one mole AgCl from one mole Cl^-, so

$$\frac{g\ Cl^-}{g\ AgCl} = \frac{\text{at. wt. Cl (g Cl/mol Cl)}}{\text{f. wt. AgCl (g AgCl/mol AgCl)}} \times \frac{1\ mol\ Cl}{1\ mol\ AgCl}$$

or

$$g\ Cl^- = g\ AgCl \times \frac{\text{at. wt. Cl}}{\text{f. wt. AgCl}}\ \text{(g Cl/g AgCl)}$$

In other words, the weight of Cl contained in or derived from AgCl is equal to the weight of AgCl times the **fraction** of Cl in it.

Calculation of the corresponding weight of Cl_2 that would be contained in the sample would proceed thus:

$$Cl_2 \xrightarrow{\text{precipitating reagent}} 2AgCl$$

We derive two moles of AgCl from each mole of Cl_2, so

$$\frac{g\ Cl_2}{g\ AgCl} = \frac{\text{f. wt. } Cl_2\ (g\ Cl_2/mol\ Cl_2)}{\text{f. wt. AgCl (g AgCl/mol AgCl)}} \times \frac{1\ mol\ Cl_2}{2\ mol\ AgCl}$$

and

$$g\ Cl_2 = g\ AgCl \times \frac{\text{f. wt. } Cl_2}{2\ (\text{f. wt. AgCl})}\ \text{(g } Cl_2\text{/g AgCl)}$$

Note that we have calculated the amount of Cl_2 gas *derivable* from the sample instead of the amount of Cl^- ion, the form in which it probably exists in the sample and the form in which it is weighed. If we precipitate the chloride as $PbCl_2$,

$$2Cl^- \xrightarrow{\text{precipitating agent}} PbCl_2$$

and

$$Cl_2 \rightarrow PbCl_2$$

Then,

$$\frac{g\ Cl^-}{g\ PbCl_2} = \frac{2(\text{f. wt. Cl})}{\text{f. wt. } PbCl_2}$$

or

$$g\ Cl^- = g\ PbCl_2 \times \frac{2(\text{f. wt. Cl})}{\text{f. wt. } PbCl_2}\ \text{(g Cl/g } PbCl_2\text{)}$$

and

$$\frac{g\ Cl_2}{g\ PbCl_2} = \frac{\text{f. wt. } Cl_2}{\text{f. wt. } PbCl_2}$$

or

$$g\ Cl_2 = g\ PbCl_2 \times \frac{f\ wt.\ Cl_2}{f.\ wt.\ PbCl_2}\ (g\ Cl_2/g\ PbCl_2)$$

From the above examples, it is apparent that the weight of the precipitate is always multiplied by a certain fraction, which we will define as the **gravimetric factor.** The gravimetric factor is the appropriate ratio of the formula weight of the substance **sought** to that of the substance **weighed:**

$$\boxed{\text{gravimetric factor} = \frac{f.\ wt.\ of\ substance\ sought}{f.\ wt.\ of\ substance\ weighed} \times \frac{a}{b}} \qquad (6.9)$$

where a and b are integers that make the formula weights in the numerator and denominator chemically equivalent. In the above examples, the gravimetric factors were $(Cl/AgCl) \times 1/1$, $(Cl_2/AgCl) \times 1/2$, $(Cl/PbCl_2) \times 2/1$, and $(Cl_2/PbCl_2) \times 1/1$. Note that one or both of the formula weights may be multiplied by an integer in order to keep the same number of atoms of the key element in the numerator and denominator.

The weight of the substance sought is obtained by multiplying the weight of the precipitate by the gravimetric factor:

$$\boxed{\text{weight (g)} \times \frac{f.\ wt.\ of\ substance\ sought}{f.\ wt.\ of\ substance\ weighed} \times \frac{a}{b} = \text{sought (g)}} \qquad (6.10)$$

Note that the *species* and the *units* of the equation can be checked by dimensional analysis (canceling of like species and units). For example,

$$AgCl\ (g) \times \frac{Cl_2\ (f.\ wt.)}{2AgCl\ (f.\ wt.)} = Cl_2\ (g)$$

Following are some other examples of gravimetric factors:

Sought	Weighed	Gravimetric Factor
SO_3	$BaSO_4$	$\dfrac{SO_3\ f.\ wt.}{BaSO_4\ f.\ wt.}$
Fe_3O_4	Fe_2O_3	$\dfrac{2Fe_3O_4\ f.\ wt.}{3Fe_2O_3\ f.\ wt.}$
Fe	Fe_2O_3	$\dfrac{2Fe\ f.\ wt.}{Fe_2O_3\ f.\ wt.}$
MgO	$Mg_2P_2O_7$	$\dfrac{2MgO\ f.\ wt.}{Mg_2P_2O_7\ f.\ wt.}$
P_2O_5	$Mg_2P_2O_7$	$\dfrac{P_2O_5\ f.\ wt.}{Mg_2P_2O_7\ f.\ wt.}$

We can now write a general formula for calculating the percentage of composition of the substance sought:

$$\boxed{\%\ \text{sought} = \frac{\text{weight of precipitate (g)} \times \text{gravimetric factor}}{\text{weight of sample (g)}} \times 100\%} \qquad (6.11)$$

EXAMPLE 6.7   Orthophosphate (PO_4^{3-}) is determined by weighing as ammonium phosphomolybdate, $(NH_4)_3PO_4 \cdot 12MoO_3$. Calculate the percent P in the sample and the percent P_2O_5 if 1.1682 g precipitate were obtained from a 0.2711-g sample.

SOLUTION

$$\% \text{ P} = \frac{1.1682 \text{ g ppt} \times \dfrac{P}{(NH_4)_3PO_4 \cdot 12MoO_3} \text{ (g P/g ppt)}}{0.2711 \text{ g sample}} \times 100\%$$

$$= \frac{1.1682 \text{ g} \times (30.97/1876.5)}{0.2711 \text{ g}} \times 100\% = 7.111\%$$

$$\% \text{ P}_2O_5 = \frac{1.1682 \text{ g ppt} \times \dfrac{P_2O_5}{2(NH_4)_3PO_4 \cdot 12MoO_3} \text{ (g P}_2O_5\text{/g ppt)}}{0.2711 \text{ g sample}} \times 100\%$$

$$= \frac{1.1682 \text{ g} \times [141.95/(2 \times 1876.5)]}{0.2711 \text{ g}} \times 100\%$$

$$= 16.30\%$$

EXAMPLE 6.8   An ore is analyzed for the manganese content by converting the manganese to Mn_3O_4 and weighing it. If a 1.52-g sample yields Mn_3O_4 weighing 0.126 g, what would be the percent Mn_2O_3 in the sample? The percent Mn?

SOLUTION

$$\% \text{ Mn}_2O_3 = \frac{0.126 \text{ g Mn}_3O_4 \times \left(\dfrac{3Mn_2O_3}{2Mn_3O_4}\right) \text{ (g Mn}_2O_3\text{/g Mn}_3O_4\text{)}}{1.52 \text{ g sample}} \times 100\%$$

$$= \frac{0.126 \text{ g} \times [3(157.9)/2(228.8)]}{1.52 \text{ g}} \times 100\% = 8.58\%$$

$$\% \text{ Mn} = \frac{0.126 \text{ g Mn}_3O_4 \times \dfrac{3 \text{ Mn}}{Mn_3O_4} \text{ (g Mn/g Mn}_3O_4\text{)}}{1.52 \text{ g sample}} \times 100\%$$

$$= \frac{0.126 \text{ g} \times [3(54.94)/228.8]}{1.52 \text{ g}} \times 100\% = 5.97\%$$

The following two examples illustrate some special additional capabilities of gravimetric computations.

EXAMPLE 6.9   What weight of pyrite ore (impure FeS_2) must be taken for analysis so that the $BaSO_4$ precipitate weight obtained will be equal to one-half that of the percent S in the sample?

SOLUTION
If we have $a\%$ of S, then we will obtain $1/2a$ g of $BaSO_4$. Therefore,

$$A\% \text{ S} = \frac{1/2A(\text{g BaSO}_4) \times \dfrac{\text{S}}{\text{BaSO}_4}\,(\text{g S/g BaSO}_4)}{\text{g sample}} \times 100\%$$

or

$$1\% \text{ S} = \frac{1/2 \times \dfrac{32.064}{233.40}}{\text{g sample}} \times 100\%$$

$$\text{g sample} = 6.869 \text{ g}$$

EXAMPLE 6.10 A mixture containing only $FeCl_3$ and $AlCl_3$ weighs 5.95 g. The chlorides are converted to the hydrous oxides and ignited to Fe_2O_3 and Al_2O_3. The oxide mixture weighs 2.62 g. Calculate the percent Fe and Al in the original mixture.

SOLUTION

There are two unknowns, so two simultaneous equations must be set up and solved. Let x = g Fe and y = g Al. Then,

$$\text{g FeCl}_3 + \text{g AlCl}_3 = 5.95 \text{ g} \tag{1}$$

$$x\left(\frac{FeCl_3}{Fe}\right) + y\left(\frac{AlCl_3}{Al}\right) = 5.95 \text{ g} \tag{2}$$

$$x\left(\frac{162.21}{55.85}\right) + y\left(\frac{133.34}{26.98}\right) = 5.95 \text{ g} \tag{3}$$

$$2.90x + 4.94y = 5.95 \text{ g} \tag{4}$$

$$\text{g Fe}_2O_3 + \text{g Al}_2O_3 = 2.62 \text{ g} \tag{5}$$

$$x\left(\frac{Fe_2O_3}{2Fe}\right) + y\left(\frac{Al_2O_3}{2Al}\right) = 2.62 \text{ g} \tag{6}$$

$$x\left(\frac{159.69}{2 \times 55.85}\right) + y\left(\frac{101.96}{2 \times 26.98}\right) = 2.62 \text{ g} \tag{7}$$

$$1.43x + 1.89y = 2.62 \text{ g} \tag{8}$$

Solving (4) and (8) simultaneously for x and y:

$$x = 1.07 \text{ g}$$

$$y = 0.58 \text{ g}$$

$$\% \text{ Fe} = \frac{1.07 \text{ g}}{5.95 \text{ g}} \times 100\% = 18.0\%$$

$$\% \text{ Al} = \frac{0.58 \text{ g}}{5.95 \text{ g}} \times 100\% = 9.8\%$$

6.5 EXAMPLES OF GRAVIMETRIC ANALYSIS

Some of the most precise and also accurate analyses are gravimetric analyses. There are many examples, and you should be familiar with some of the more common ones. These are summarized in Table 6.1, which lists the substance

TABLE 6.1
Some Commonly Employed Gravimetric Analyses

Substance Analyzed	Precipitate Formed	Precipitate Weighed	Interferences
Fe	$Fe(OH)_3$	Fe_2O_3	Many. Al, Ti, Cr, etc.
	Fe cupferrate	Fe_2O_3	Tetravalent metals
Al	$Al(OH)_3$	Al_2O_3	Many. Fe, Ti, Cr, etc.
	$Al(ox)_3^a$	$Al(ox)_3$	Many. Mg does not interfere in acidic solution
Ca	CaC_2O_4	$CaCO_3$ or CaO	All metals except alkalis and Mg
Mg	$MgNH_4PO_4$	$Mg_2P_2O_7$	All metals except alkalis
Zn	$ZnNH_4PO_4$	$Zn_2P_2O_7$	All metals except Mg
Ba	$BaCrO_4$	$BaCrO_4$	Pb
SO_4^{2-}	$BaSO_4$	$BaSO_4$	NO_3^-, PO_4^{3-}, ClO_3^-
Cl^-	AgCl	AgCl	Br^-, I^-, SCN^-, CN^-, S^{2-}, $S_2O_3^{2-}$
Ag	AgCl	AgCl	Hg(I)
PO_4^{3-}	$MgNH_4PO_4$	$Mg_2P_2O_7$	MoO_4^{2-}, $C_2O_4^{2-}$, K^+
Ni	$Ni(dmg)_2^b$	$Ni(dmg)_2$	Pd

[a] ox = Oxine (8-hydroxyquinoline) with 1 H^+ removed.
[b] dmg = Dimethylglyoxime with 1 H^+ removed.

sought, the precipitate formed, the form in which it is weighed, and the common elements that will interfere and must be absent. You should consult more advanced texts and comprehensive analytical reference books for details on these and other determinations.

6.6 ORGANIC PRECIPITATES

All the precipitating agents we have talked about so far, except for oxine, cupferrate, and dimethylglyoxime (Table 6.1), have been inorganic in nature. There are a large number of organic compounds that are very useful precipitating agents for metals. Some of these are very selective, and others are very broad in the number of elements they will precipitate.

Organic precipitating agents have the advantages of giving precipitates with very low solubility in water and a favorable gravimetric factor. Most of them are **chelating agents** that form slightly soluble, uncharged **chelates** with the metal ions. A chelating agent is a type of complexing agent that has two or more groups capable of complexing with a metal ion. The complex formed is called a chelate. See Chapter 9 for a more thorough discussion of chelates.

Since chelating agents are weak acids, the number of elements precipitated, and thus the selectivity, can usually be regulated by adjustment of the pH. The reactions can be generalized as

$$M^{n+} + nHX \rightleftharpoons \underline{MX_n} + nH^+$$

There may be more than one ionizable proton on the organic reagent. The weaker the metal chelate, the higher the pH needed to achieve precipitation. Some of the commonly used organic precipitants are listed in Table 6.2. Some of these precipitates are not stoichiometric, and more accurate results are obtained by igniting them to form the metal oxides. Some, such as sodium diethyldithiocarbamate, can

TABLE 6.2
Some Organic Precipitating Agents

Reagent	Structure	Metals Precipitated
Dimethylglyoxime	$CH_3—C{=}NOH$ | $CH_3—C{=}NOH$	Ni(II) in NH_3 or bufferedHOAc; Pd(II) in HCl $(M^{2+} + 2HR \rightarrow \underline{MR_2} + 2H^+)$
α-Benzoinoxime (cupron)		Cu(II) in NH_3 and tartrate; Mo(VI) and W(VI) in H^+ ($M^{2+} +$ $H_2R \rightarrow \underline{MR} + 2H^+$; $M^{2+} =$ Cu^{2+}, MoO_2^+, WO_2^{2+}) Metal oxide weighed
Ammonium nitrosophenyl-hydroxylamine (cupferron)		Fe(III), V(V), Ti(IV), Zr(IV), Sn(IV), U(IV) $(M^{n+} + nNH_4R \rightarrow \underline{MR_n} + nNH_4^+)$ Metal oxide weighed
8-Hydroxyquinoline (oxine)		Many metals. Useful for Al(III) and Mg(II) $(M^{n+} + nHR \rightarrow \underline{MR_n} + nH^+)$
Sodium diethyldithio-carbamate	S $\|$ $(C_2H_5)_2N—C—S^-Na^+$	Many metals from acid solution $(M^{n+} + nNaR \rightarrow \underline{MR_n} + nNa^+)$
Sodium tetraphenylboron	$NaB(C_6H_5)_4$	K^+, Rb^+, Cs^+, Tl^+, Ag^+, $Hg(I)$, Cu(I), NH_4^+, RNH_3^+, $R_2NH_2^+$, R_3NH^+, R_4N^+. Acidic solution $(M^+ + NaR \rightarrow \underline{MR} + Na^+)$
Tetraphenylarsonium chloride	$(C_6H_5)_4AsCl$	$Cr_2O_7^{2-}$, MnO_4^-, ReO_4^-, MoO_4^{2-}, WO_4^{2-}, ClO_4^-, I_3^-. Acidic solution $(A^{n-} + nRCl \rightarrow \underline{R_nA} + nCl^-)$

be used to perform group separations, as is done with hydrogen sulfide. Treatises on analytical chemistry should be consulted for applications of these and other organic precipitating reagents. The multivolume treatise by Hollingshead on the uses of oxine and its derivatives is very helpful for applications of this versatile reagent (see references, end of chapter).

QUESTIONS

1. Describe the unit operations commonly employed in gravimetric analysis and briefly indicate the purpose of each.
2. What is the von Weimarn ratio? Define the terms in it.
3. What information concerning optimum conditions of precipitation does the von Weimarn ratio give us?
4. What is digestion of a precipitate and why is it necessary?
5. Outline the optimum conditions for precipitation that will obtain a pure and filterable precipitate.
6. What is coprecipitation? List the different types of coprecipitation and indicate how they may be minimized or treated for.
7. Why must a filtered precipitate be washed?
8. Why must a wash liquid generally contain an electrolyte? What are the requirements for this electrolyte?
9. What are the principles and advantages of precipitation from homogeneous solution?

PROBLEMS

Solubility Product Calculations

10. Write solubility product expressions for the following: (a) $AgSCN$; (b) $La(IO_3)_3$; (c) Hg_2Br_2; (d) $Ag[Ag(CN)_2]$; (e) $Zn_2Fe(CN)_6$; (f) Bi_2S_3.
11. Bismuth iodide, BiI_3, has a solubility of 7.76 mg/L. What is its K_{sp}?
12. What is the concentration of Ag^+ and CrO_4^{2-} in a saturated solution of Ag_2CrO_4?
13. Calculate the concentration of barium in the solution at equilibrium when 15.0 mL of 0.200 M K_2CrO_4 is added to 25.0 mL of 0.100 M $BaCl_2$.
14. What must be the concentration of PO_4^{3-} to just start precipitation of Ag_3PO_4 in a 0.10 M $AgNO_3$ solution?
15. What must be the concentration of Ag^+ to just start precipitating 0.10 M PO_4^{3-}? 0.10 M Cl^-?
16. What weight of Ag_3AsO_4 will dissolve in 250 mL water?
17. What is the solubility of Ag_2CrO_4 in 0.10 M K_2CrO_4?
18. Compounds AB and AC_2 each have solubility products equal to 4×10^{-18}. Which is more soluble, as expressed in moles per liter?
19. The solubility product of Bi_2S_3 is 1×10^{-97} and that of HgS is 4×10^{-53}. Which is the least soluble?
20. A student proposes to analyze barium gravimetrically by precipitating BaF_2 with NaF. Assuming a 200-mg sample of Ba^{2+} in 100 mL is to be precipitated and that the precipitation must be 99.9% complete for quantitative results, comment on the feasibility of the analysis.

Diverse Ion Effect on Solubility

21. Write the thermodynamic solubility product expressions for the following:
  (a) $BaSO_4 \rightleftharpoons Ba^{2+} + SO_4^{2-}$
  (b) $\overline{Ag_2CrO_4} \rightleftharpoons 2Ag^+ + CrO_4^{2-}$

22. Calculate the solubility of $BaSO_4$ in 0.0125 M $BaCl_2$. Take into account the diverse ion effect.

23. You are to determine fluoride ion gravimetrically by precipitating CaF_2. $Ca(NO_3)_2$ is added to give an excess of 0.015 M calcium ion after precipitation. The solution also contains 0.025 M $NaNO_3$. How many grams fluoride will be in solution at equilibrium if the volume is 250 mL?

Gravimetric Factor

24. Calculate the weight of sodium present in 50.0 g Na_2SO_4.

25. If the salt in Problem 24 is analyzed by precipitating and weighing $BaSO_4$, what weight of precipitate would be obtained?

26. Calculate the gravimetric factors for:

Substance Sought	Substance Weighed
As_2O_3	Ag_3AsO_4
$FeSO_4$	Fe_2O_3
K_2O	$KB(C_6H_5)_4$
SiO_2	$KAlSi_3O_8$

27. How many grams CuO would 1.00 g Paris green, $Cu_3(AsO_3)_2 \cdot 2As_2O_3 \cdot Cu(C_2H_3O_2)_2$, give? Of As_2O_3?

Quantitative Calculations

28. A 523.1-mg sample of impure KBr is treated with excess $AgNO_3$ and 814.5 mg AgBr is obtained. What is the purity of the KBr?

29. What weight of Fe_2O_3 precipitate would be obtained from a 0.4823 $-$ g sample of iron wire that is 99.89% pure?

30. The aluminum content of an alloy is determined gravimetrically by precipitating it with 8-hydroxyquinoline (oxine) to give $Al(C_9H_6ON)_3$. If a 1.021-g sample yielded 0.1862 g of precipitate, what is the percent aluminum in the alloy?

31. Iron in an ore is to be analyzed gravimetrically by weighing as Fe_2O_3. It is desired that the results be obtained to four significant figures. If the iron content ranges between 11 and 15%, what is the minimum size sample that must be taken to obtain 100.0 mg of precipitate?

32. The chloride in a 0.12-g sample of 95% pure $MgCl_2$ is to be precipitated as AgCl. Calculate the volume of 0.100 M $AgNO_3$ solution required to precipitate the chloride and give a 10% excess.

33. Ammonium ions can be analyzed by precipitating with H_2PtCl_6 as $(NH_4)_2PtCl_6$ and then igniting the precipitate to platinum metal, which is weighed [$(NH_4)_2PtCl_6 \xrightarrow{heat} Pt + 2NH_4Cl \uparrow + 2Cl_2 \uparrow$]. Calculate the percent ammonia in a 1.00-g sample that yields 0.100 g Pt by this method.

34. A sample is to be analyzed for its chloride content by precipitating and weighing silver chloride. What weight of sample would have to be taken so that the weight of precipitate is equal to the percent chloride in the sample?

35. Pyrite ore (impure FeS_2) is analyzed by converting the sulfur to sulfate and precipitating $BaSO_4$. What weight of ore should be taken for analysis so that the grams of precipitate will be equal to 0.1000 times the percent of FeS_2?

See p. 167 Example 6.8

36. A mixture containing only BaO and CaO weighs 2.00 g. The oxides are converted to the corresponding mixed sulfates, which weigh 4.00 g. Calculate the percent Ba and Ca in the original mixture.

37. A mixture containing only $BaSO_4$ and $CaSO_4$ contains one-half as much Ba^{2+} as Ca^{2+} by weight. What is the percentage of $CaSO_4$ in the mixture?

38. A mixture containing only AgCl and AgBr weighs 2.000 g. It is quantitatively reduced to silver metal, which weighs 1.300 g. Calculate the weight of AgCl and AgBr in the original mixture.

RECOMMENDED REFERENCES

General and Inorganic

1. F. E. Beamish and W. A. E. McBryde, "Inorganic Gravimetric Analysis," in C. L. Wilson and D. W. Wilson, eds., *Comprehensive Analytical Chemistry,* Vol. 1A. New York: Elsevier Publishing Company, 1959, Chapter VI.

2. I. M. Kolthoff, E. B. Sandell, E. J. Meehan, and S. Bruckenstein, *Quantitative Chemical Analysis,* 4th ed. London: The Macmillan Company, 1969.

3. "Gravimetric Determination of the Elements," in C. L. Wilson and D. W. Wilson, eds., *Comprehensive Analytical Chemistry,* Vol. 1C. New York: Elsevier Publishing Company, 1960.

Organic Reagents

4. K. L. Cheng, K. Ueno, and T. Imamura, eds., *Handbook of Organic Analytical Reagents.* Boca Raton: CRC Press, 1982.

5. R. G. W. Hollingshead, *Oxine and Its Derivatives.* London: Butterworth Scientific Publications, 1954–56.

6. F. Holmes, "Organic Reagents in Inorganic Analysis," in C. L. Wilson and D. W. Wilson, eds., *Comprehensive Analytical Chemistry,* Vol. 1A. New York: Elsevier Publishing Company, 1959, Chapter II.8.

Precipitation from Homogeneous Solution

7. L. Gordon, M. L. Salulsky, and H. H. Willard, *Precipitation from Homogeneous Solution.* New York: John Wiley & Sons, 1959.

chapter
7

PRINCIPLES OF
VOLUMETRIC ANALYSIS

Volumetric analysis is one of the most useful and accurate analytical techniques, especially for millimole amounts of analyte. It is fairly rapid or can be automated, and it can be applied to smaller amounts of analyte when combined with a sensitive instrumental technique for detection of the completion of the titration reaction, for example, spectrophotometry or pH measurement. Manual titrations nowadays are used in situations requiring high accuracy for relatively small numbers of samples. They are used, for example, to calibrate or validate more routine instrumental methods. Automated titrations are useful when large numbers of samples must be processed. (A titration may be automated, for instance, by means of a color change or a pH change that activates a motor-driven buret to stop delivery; the volume delivered may then be registered on a digital counter.) In this chapter, the types of titrations that can be performed are described and the principles applicable to all are given, including the principles and requirements of a titration and standard solutions. Calculations in volumetric analysis are given, including molarity and normality, titer, and back-titrations.

7.1 GENERAL PRINCIPLES

Titration

In a **titration,** the test substance (analyte) reacts with a reagent added as a solution of known concentration. This is referred to as a **standard solution,** and it is generally added from a buret. The added solution is called the **titrant.** The volume of titrant required to just completely react with the analyte is measured. Since the concentration is known and since the reaction between the analyte and the reagent is known, the amount of analyte can be calculated. The **requirements of a titration** are as follows:

---

1. The reaction should be **stoichiometric.** That is, there should be a well-defined and known reaction between the analyte and the titrant. In the titration of acetic acid in vinegar with sodium hydroxide, for example, a well-defined reaction takes place:

$$HC_2H_3O_2 + NaOH \rightarrow NaC_2H_3O_2 + H_2O$$

2. The reaction should be **rapid.** Most ionic reactions, as above, are very rapid.
3. There should be **no side reactions,** and the reaction should be specific. If there are interfering substances, these must be removed. In the above example, there should be no other acids present.
4. There should be a **marked change in some property of the solution when the reaction is complete.** This may be a change in the color of the solution or in some electrical or other physical property of the solution. In the titration of acetic acid with sodium hydroxide, there is a marked increase in the pH of the solution when the reaction is complete. A color change is usually brought about by addition of an **indicator,** whose color is dependent on the properties of the solution, for example, the pH.
5. The point at which an equivalent or stoichiometric amount of titrant is added is called the **equivalence point.** The point at which the reaction is *observed* to be complete is called the **end point,** that is, when a change in some property of the solution is detected. The end point should coincide with the equivalence point or be at a reproducible interval from it.
6. The reaction should be **quantitative.** That is, the equilibrium of the reaction should be far to the right so that a sufficiently *sharp* change will occur at the end point to obtain the desired accuracy. If the equilibrium does not lie far to the right, then there will be gradual change in the property marking the end point (e.g., pH), and this will be difficult to detect.

Standard Solutions

A standard solution is prepared by dissolving an accurately weighed quantity of a highly pure material called a **primary standard** and diluting to an accurately known volume in a volumetric flask. Alternatively, if the material is not sufficiently pure, a solution is prepared to give approximately the desired concentration, and this is **standardized** by titrating a weighed quantity of a primary standard. For example, since sodium hydroxide is not sufficiently pure to prepare a standard solution directly, it is standardized by titrating a primary standard acid, such as potassium acid phthalate (KHP). Potassium acid phthalate is a solid that can be weighed accurately.

A **primary standard** should fulfill these **requirements:**

> 1. It should be **100.00% pure,** although 0.01 to 0.02% impurity is tolerable if accurately known.
> 2. It should be **stable to drying** temperatures, and it should be stable indefinitely at room temperature. The primary standard is always dried before weighing.[1]
> 3. It should be **readily available.**
> 4. Although not necessary, it should have a **high formula weight.** This is so that a relatively large amount of it will have to be weighed to get enough to titrate. The relative error in weighing a greater amount of material will be smaller than that for a small amount.
> 5. If it is to be used in a titration, it should possess the **properties required for a titration** listed above. In particular, the equilibrium of the reaction should be far to the right so that a very sharp end point will be obtained.

Classification of Volumetric Methods

There are four general classes of volumetric or titrimetric methods.

   1. ACID–BASE.   Many compounds, both inorganic and organic, are either acids or bases, and can be titrated with a standard solution of a strong base or a strong acid. The end points of these titrations are easy to detect, either by means of an indicator or by following the change in pH with a pH meter. The acidity and basicity of many organic acids and bases can be enhanced by titrating in a *non-aqueous solvent* (Chapter 8). The result is a sharper end point, and weaker acids and bases can be titrated.

   2. PRECIPITATION.   In the case of precipitation, the titrant forms an insoluble product with the analyte. An example is the titration of chloride ion with silver nitrate solution. Again, indicators can be used to detect the end point, or the potential of the solution can be monitored electrically.

   3. COMPLEXOMETRIC.   In complexometric titrations, the titrant is a complexing agent and forms a water-soluble complex with the analyte, a metal ion. The titrant is often a **chelating agent.**[2] The reverse titration may be carried out also. Ethylenediaminetetraacetic acid (EDTA) is one of the most useful chelating agents used for titration. It will react with a large number of elements, and the reactions can be controlled by adjustment of the pH. Indicators can be used to form a highly colored complex with the metal ion.

   4. REDUCTION–OXIDATION.   These "redox" titrations involve the titration of an oxidizing agent with a reducing agent, or vice versa. An oxidizing agent gains electrons and a reducing agent loses electrons in a reaction between them. There

---

[1]There are a few exceptions when the primary standard is a hydrate.

[2]A chelating agent (the term is derived from the Greek word for "clawlike") is a type of complexing agent that contains two or more groups capable of complexing with a metal ion. EDTA has six such groups.

must be a sufficiently large difference between the oxidizing and reducing capabilities of these agents for the reaction to go to completion and give a sharp end point, that is, one should be a fairly strong oxidizing agent (strong tendency to gain electrons) and the other a fairly strong reducing agent (strong tendency to lose electrons). Appropriate indicators can be used for these titrations, or various electrometric means can be employed to detect the end point.

These different types of titrations and means of detecting their end points will be treated separately in succeeding chapters.

7.2 MOLARITY CALCULATIONS

We shall use molarity throughout the majority of the text for volumetric calculations. Another useful concentration unit for volumetric calculations is **normality,** using the concept of **equivalents** and **equivalent weights** in place of moles and formula weights. Normal concentration depends on the particular reaction, and the reaction should be specified. Some instructors prefer to introduce the concept of normality, and the student is likely to encounter it in reference books. Therefore, a review of equivalents and normality is given following the discussion of calculations using molarity.

We reviewed some of the ways of expressing concentrations in Chapter 1. Basic concepts are summarized below.

$$\text{moles} = \frac{\text{g}}{\text{f. wt. (g/mol)}}; \quad \text{millimoles} = \frac{\text{mg}}{\text{f. wt. (mg/mmol)}} \tag{7.1}$$

$$\text{molar concentration} = M = \frac{\text{mol}}{\text{L}}; \quad M = \frac{\text{mmol}}{\text{mL}} \tag{7.2}$$

By rearrangement of these equations we obtain the expressions for calculating other quantities.

$$M \text{ (mol/L)} \times \text{L} = \text{mol}; \quad M \text{ (mmol/mL)} \times \text{mL} = \text{mmol} \tag{7.3}$$

$$\text{g} = \text{mol} \times \text{f. wt. (g/mol)} \quad \text{mg} = \text{mmol} \times \text{f. wt. (mg/mmol)} \tag{7.4}$$

$$\text{g} = M \text{ (mol/L)} \times \text{L} \times \text{f. wt. (g/mol)}$$

$$\text{mg} = M \text{ (mmol/mL)} \times \text{mL} \times \text{f. wt. (mg/mmol)} \tag{7.5}$$

We usually work with millimole (mmol) and milliliter (mL) quantities in titrations; therefore, the righthand equations are more useful. Note that the expression for formula weight contains the same numerical value whether it be in g/mol or mg/mmol. Note also that care must be taken in utilization of "milli" quantities (millimoles, milligrams, milliliters, etc.). Incorrect use could result in calculation errors of 1000-fold.

Assume 25.0 mL of 0.100 M $AgNO_3$ is required to titrate a sample containing sodium chloride. The reaction is

$$Cl^- + Ag^+ \rightarrow \underline{AgCl}$$

Since Ag^+ and Cl^- react on a 1:1 molar basis, the number of millimoles of Cl^- is equal to the number of millimoles of Ag^+ needed for the titration. We can calculate the milligrams of NaCl as follows:

$$mmol_{NaCl} = mL_{AgNO_3} \times M_{AgNO_3}$$

$$= 25.0 \text{ mL} \times 0.100 \text{ (mmol/mL)} = 2.50 \text{ mmol}$$

$$mg_{NaCl} = mmol \times f. \text{ wt.}_{NaCl}$$

$$= 2.50 \text{ mmol} \times 58.44 \text{ mg/mmol} = 146 \text{ mg}$$

The percentage of analyte A that reacts on a **1:1 mole basis** with the titrant can be calculated from the following general formula:

$$\% \text{ A} = \frac{mg_{analyte}}{mg_{sample}} \times 100\%$$

$$= \frac{M \text{ (mmol/mL)} \times mL \times f. \text{ wt.}_{analyte} \text{ (mg/mmol)}}{mg_{sample}} \times 100\% \qquad (7.6)$$

General Calculations with Molarity

Many substances do not react on a 1:1 mole basis, and so the simple calculation in the above example cannot be applied to all reactions. A generalized formula for calculations applicable to all reactions can be written from the **balanced equation** for reactions.

Consider the general reaction

$$aA \times tT \rightarrow P \qquad (7.7)$$

where A is the analyte, T is the titrant, reacting in the ratio a/t to give products P. Then,

$$mmol_A = mmol_T \times \frac{a}{t} \text{ (mmol A/mmol T)} \qquad (7.8)$$

$$mmol_A = M_T \text{ (mmol/mL)} \times mL_T \times \frac{a}{t} \text{ (mmol A/mmol T)} \qquad (7.9)$$

$$mg_A = mmol_A \times f. \text{ wt.}_A \text{ (mg/mmol)} \qquad (7.10)$$

$$mg_A = M_T \text{ (mmol/mL)} \times mL_T \times \frac{a}{t} \text{ (mmol A/mmol T)}$$
$$\times f. \text{ wt.}_A \text{ (mg/mmol)} \qquad (7.11)$$

Note that the a/t factor serves much the same function as the gravimetric factor in Chapter 6. To avoid a mistake in setting up the factor, it is helpful to remember that when calculating the amount of analyte, the amount of titrant is multiplied by the a/t ratio (a comes first). Conversely, if the amount of titrant (e.g., molarity) is being calculated from a known amount of analyte titrated, you multiply the amount of analyte by the t/a ratio (t comes first). See below. The best way, of course, to ascertain the correct ratio is to do a dimensional analysis to obtain the correct units.

The percent analyte A in a sample can be determined by titrating a known weight of sample with a standard solution of T:

$$\% \text{ A} = \frac{M \text{ (mmol/mL)} \times \text{mL}_T \times (a/t) \text{ (mmol A/mmol T)} \times \text{f. wt.}_A \text{ (mg/mmol)}}{\text{mg}_{\text{sample}}} \times 100\% \qquad (7.12)$$

Or the concentration of a solution of T standardized by titrating a known number of millimoles of primary standard A can be calculated from

$$M_T \text{ (mmol/mL)} \times \text{mL}_T = \text{mmol}_A \times \frac{t}{a} \text{ (mmol T/mmol A)}$$

$$= \frac{\text{mg}_A}{\text{f. wt.}_A \text{ (mg/mmol)}} \times \frac{t}{a} \text{ (mmol T/mmol A)} \qquad (7.13)$$

Therefore,

$$M_T = \frac{(\text{mg}_A/\text{f. wt.}_A) \times (t/a) \text{ (mmol T/mmol A)}}{\text{mL}_T} \qquad (7.14)$$

EXAMPLE 7.1 A 0.4671-g sample containing sodium bicarbonate was dissolved and titrated with a solution of hydrochloric acid, requiring 40.72 mL. The reaction is

$$\text{HCO}_3^- + \text{H}^+ \rightarrow \text{H}_2\text{O} + \text{CO}_2$$

The hydrochloric acid was standardized by titrating 0.1876 g sodium carbonate by the following reaction:

$$\text{CO}_3^{2-} + 2\text{H}^+ \rightarrow \text{H}_2\text{O} + \text{CO}_2$$

The titration required 37.86 mL acid. Calculate the percent sodium bicarbonate in the sample.

SOLUTION
From Equation 7.14, the molarity of the HCl is

$$M_{\text{HCl}} = \frac{(\text{mg}_{\text{Na}_2\text{CO}_3}/\text{f. wt.}_{\text{Na}_2\text{CO}_3}) \times (2/1) \text{ (mmol HCl/mmol Na}_2\text{CO}_3)}{\text{mL}_{\text{HCl}}}$$

$$= \frac{[187.6 \text{ mg}/105.99 \text{ (mg/mmol)}] \times 2 \text{ (mmol HCl/mmol Na}_2\text{CO}_3)}{37.86 \text{ mL}}$$

$$= 0.09350 \text{ mmol/mL}$$

Note that we multiplied the amount of analyte, Na_2CO_3, by the t/a ratio (mmol titrant/mmol analyte). Note also that although all measurements were to four significant figures, we computed the formula weight of Na_2CO_3 to five figures. This is because with four figures, it would have become the key number with an uncertainty of about one part per thousand compared to 187.6 with an uncertainty of about half that. It is not bad practice, as a matter of routine, to carry the formula weight to one additional figure, particularly if a calculator is available. From Equation 7.12,

$$\% \text{ NaHCO}_3 = \frac{M_{\text{HCl}} \times \text{mL}_{\text{HCl}} \times 1/1 \text{ (mmol NaHCO}_3/\text{mmol HCl}) \times \text{f. wt.}_{\text{NaHCO}_3}}{\text{mg}_{\text{sample}}}$$

$$\times 100\%$$

$$= \frac{0.09350 \text{ mmol/mL} \times 40.72 \text{ mL} \times 1 \times 84.01 \text{ mg/mmol}}{467.1 \text{ mg}} \times 100\%$$

$$= 68.48\%$$

EXAMPLE 7.2   How many milliliters of 0.25 M solution of H_2SO_4 will react with 10 mL of a 0.25 M solution of NaOH?

SOLUTION
The reaction is

$$H_2SO_4 + 2NaOH \rightarrow Na_2SO_4 + 2H_2O$$

One-half as many millimoles of H_2SO_4 as of NaOH will react, or

$$M_{\text{H}_2\text{SO}_4} \times \text{mL}_{\text{H}_2\text{SO}_4} = M_{\text{NaOH}} \times \text{mL}_{\text{NaOH}} \times \tfrac{1}{2}(\text{mmol H}_2\text{SO}_4/\text{mmol NaOH})$$

Therefore,

$$\text{mL}_{\text{H}_2\text{SO}_4} = \frac{0.25 \text{ mmol/mL} \times 10 \text{ mL} \times \tfrac{1}{2}(\text{mmol H}_2\text{SO}_4/\text{mmol NaOH})}{0.25 \text{ mmol/mL}} = 5.0 \text{ mL}$$

Note that, in this case, we multiplied the amount of titrant by the a/t ratio (mmol analyte/mmol titrant).

EXAMPLE 7.3   A sample of impure salicylic acid, $C_6H_4(OH)COOH$ (one titratable proton), is analyzed by titration. What size sample should be taken so that the percent purity is equal to five times the milliliters of 0.0500 M NaOH used to titrate it?

SOLUTION
Let x = mL NaOH; % salicylic acid (HA) = $5x$:

$$\% \text{ HA} = \frac{M_{\text{NaOH}} \times \text{mL}_{\text{NaOH}} \times 1(\text{mmol HA/mmol NaOH}) \times \text{f. wt.}_{\text{HA}} \text{ (mg/mmol)}}{\text{mg}_{\text{sample}}}$$

$$\times 100\%$$

$$5x\% = \frac{0.0500 \, M \times x \text{ mL} \times 1 \times 138 \text{ mg/mmol}}{\text{mg}_{\text{sample}}} \times 100\%$$

$$\text{mg}_{\text{sample}} = 138 \text{ mg}$$

The above examples of acid–base calculations are applicable to the titrations described in Chapter 8.

Dilution Calculations

We can use the relationship $M \times \text{mL} = \text{mmol}$ to calculate the dilution required to prepare a certain concentration of a solution from a more concentrated solution. If we wish to prepare 500 mL of a 0.100 M solution by diluting a more concentrated solution, we can calculate the millimoles of the solution that must be taken. From this we can calculate the volume of the more concentrated solution to be diluted to 500 mL.

EXAMPLE 7.4   You wish to prepare 500 mL of a 0.100 M $K_2Cr_2O_7$ solution from a 0.250 M solution. What volume of the 0.250 M solution must be diluted to 500 mL?

SOLUTION

$$M_{final} \times mL_{final} = M_{original} \times mL_{original}$$

$$0.100 \text{ mmol/mL} \times 500 \text{ mL} = 0.250 \text{ mmol/mL} \times mL_{original}$$

$$mL_{original} = 200 \text{ mL}$$

EXAMPLE 7.5   What volume of 0.40 M $Ba(OH)_2$ must be added to 50 mL of 0.30 M NaOH to give a solution to 0.50 M in OH^-?

SOLUTION
Let x = mL $Ba(OH)_2$. The final volume is $(50 + x)$ mL.

$$\text{mmol } OH^- = \text{mmol NaOH} + 2 \times \text{mmol } Ba(OH)_2$$

$$0.50 \text{ } M \times (50 + x) \text{ mL} = 0.30 \text{ } M \times 50 \text{ mL} + 2 \times 0.40 \text{ } M \times x \text{ mL}$$

$$x = 33 \text{ mL}$$

Often, the analyst is confronted with serial dilutions of a sample or standard solution. Again, obtaining the final concentration simply requires keeping tract of the number of millimoles and the volumes.

EXAMPLE 7.6   You are to determine iron in a sample by spectrophotometry by reacting Fe^{2+} with 1,10-phenanthroline to form an orange color. This requires preparation of a series of standards against which to compare absorbances or color intensities (i.e., to prepare a calibration curve). A stock standard solution of 1.000×10^{-3} M iron is prepared from ferrous ammonium sulfate. Working standards A and B are prepared by adding with pipets 2.000 and 1.000 mL, respectively, of this solution to 100-mL volumetric flasks and diluting to volume. Working standards C, D, and E are prepared by adding 20.00, 10.00, and 5.000 mL of working standard A to 100-mL volumetric flasks and diluting to volume. What are the concentrations of the working solutions?

SOLUTION
solution A: $M_{stock} \times mL_{stock} = M_A \times mL_A$

$$(1.000 \times 10^{-3} \text{ } M)(2.000 \text{ mL}) = M_A \times 100.0 \text{ mL}$$

$$M_A = 2.000 \times 10^{-5} \text{ } M$$

solution B: $(1.000 \times 10^{-3} \text{ } M)(1.000 \text{ mL}) = M_B \times 100.0 \text{ mL}$

$$M_B = 1.000 \times 10^{-5} \text{ } M$$

solution C: $M_A \times mL_A = M_C \times mL_C$

$$(2.000 \times 10^{-5} \text{ } M)(20.00 \text{ mL}) = M_C \times 100.0 \text{ mL}$$

$$M_C = 4.000 \times 10^{-6} \text{ } M$$

solution D: $(2.000 \times 10^{-5}\,M)(10.00\,\text{mL}) = M_D \times 100.0\,\text{mL}$

$\qquad M_D = 2.000 \times 10^{-6}\,M$

solution E: $(2.000 \times 10^{-5}\,M)(5.000\,\text{mL}) = M_E \times 100.0\,\text{mL}$

$\qquad M_E = 1.000 \times 10^{-6}\,M$

The above calculations apply to all types of reactions, including acid–base, redox, precipitation, and complexometric reactions. The primary requirement before making calculations is to know the ratio in which the substances react, that is, start with a balanced reaction.

EXAMPLE 7.7 The iron(II) in an acidified solution is titrated with a 0.0206 M solution of potassium permanganate:

$$5Fe^{2+} + MnO_4^- + 8H^+ \rightarrow 5Fe^{3+} + Mn^{2+} + 4H_2O$$

If the titration required 40.2 mL, how many milligrams iron are in the solution?

SOLUTION
There are five times as many millimoles of iron as there are of permanganate that react with it, so

$$\text{mmol}_{Fe} = \frac{\text{mg}_{Fe}}{\text{f. wt.}_{Fe}} = M_{KMnO_4} \times \text{mL}_{KMnO_4} \times \frac{5}{1}$$

mg_{Fe} = 0.0206 mmol/mL $\times$ 40.2 mL $\times$ 5 (mmol Fe/mmol MnO_4^-) $\times$ 55.8 mg/mmol

$\qquad$ = 231 mg

Calculations of this type are used for the titrations described in Chapter 10.

Following is a list of typical precipitation and complexometric titration reactions and the factors for calculating the milligrams of analyte from millimoles of titrant.[3]

$Cl^- + Ag^+ \rightarrow \underline{AgCl}$ $\qquad\qquad$ $\text{mg}_{Cl^-} = M_{Ag^+} \times \text{mL}_{Ag^+} \times 1(\text{mmol } Cl^-/\text{mmol } Ag^+)$
$\qquad\qquad\qquad\qquad\qquad\qquad\qquad \times$ f. wt.$_{Cl^-}$

$Cl^- + \frac{1}{2}Pb^{2+} \rightarrow \frac{1}{2}PbCl_2$ $\qquad$ $\text{mg}_{Cl^-} = M_{Pb^{2+}} \times \text{mL}_{Pb^{2+}} \times 2\,(\text{mmol } Cl^-/\text{mmol } Pb^{2+})$
$\qquad\qquad\qquad\qquad\qquad\qquad\qquad \times$ f. wt.$_{Cl^-}$

$PO_4^{3-} + 3\,Ag^+ \rightarrow \underline{Ag_3PO_4}$ $\qquad$ $\text{mg}_{PO_4^{3-}} = M_{Ag^+} \times \text{mL}_{Ag^+} \times \frac{1}{3}(\text{mmol } PO_4^{3-}/\text{mmol } Ag^+)$
$\qquad\qquad\qquad\qquad\qquad\qquad\qquad \times$ f. wt.$_{PO_4^{3-}}$

$CN^- + \frac{1}{2}Ag^+ \rightarrow \frac{1}{2}Ag(CN)_2-$ $\qquad$ $\text{mg}_{CN^-} = M_{Ag^+} \times \text{mL}_{Ag^+} \times 2(\text{mmol } CN^-/\text{mmol } Ag^+)$
$\qquad\qquad\qquad\qquad\qquad\qquad\qquad \times$ f. wt.$_{CN^-}$

$CN^- + Ag^+ \rightarrow \frac{1}{2}Ag[Ag(CN)_2]$ $\qquad$ $\text{mg}_{CN^-} = M_{Ag^+} \times \text{mL}_{Ag^+} \times 1(\text{mmol } CN^-/\text{mmol } Ag^+)$
$\qquad\qquad\qquad\qquad\qquad\qquad\qquad \times$ f. wt.$_{CN^-}$

$Ba^{2+} + SO_4^{2-} \rightarrow \underline{BaSO_4}$ $\qquad$ $\text{mg}_{Ba^{2+}} = M_{SO_4^{2-}} \times 1(\text{mmol } Ba^{2+}/\text{mmol } SO_4^{2-}) \times$ f. wt.$_{Ba^{2+}}$

$Ca^{2+} + H_2Y^{2-} \rightarrow CaY^{2-} + 2H^+$ $\quad$ $\text{mg}_{Ca^{2+}}$ $M_{EDTA} \times$ 1(mmol Ca^{2+}/mmol EDTA) $\times$ f.wt.$_{Ca^{2+}}$

---

[3]H_4Y = EDTA in the last equation.

These formulas are useful for calculations involving the titrations described in Chapter 9.

EXAMPLE 7.8   Aluminum is determined by titrating with EDTA:

$$Al^{3+} + H_2Y^{2-} \rightarrow AlY^- + 2H^+$$

A 1.00-g sample requires 20.5 mL EDTA for titration. The EDTA was standardized by titrating 25.0 mL of a 0.100 M $CaCl_2$ solution, requiring 30.0 mL EDTA. Calculate the percent Al_2O_3 in the sample.

SOLUTION
Since Ca^{2+} and EDTA react on a 1:1 mole ratio,

$$M_{EDTA} = \frac{0.100 \text{ mmol/ml} \times 25.0 \text{ mL}}{30.0 \text{ mL}} = 0.0833 \text{ mmol/mL}$$

The millimoles Al^{3+} are equal to the millimoles EDTA used in the sample titration, but there are one-half this number of millimoles Al_2O_3 (since 1 $Al^{3+} \rightarrow \frac{1}{2}Al_2O_3$). Therefore,

$$\% \ Al_2O_3 = \frac{M_{EDTA} \times mL_{EDTA} \times \frac{1}{2}(\text{mmol } Al_2O_3/\text{mmol EDTA}) \times \text{f. wt.}_{Al_2O_3}}{mg_{sample}} \times 100\%$$

$$\% \ Al_2O_3 = \frac{0.0833 \text{ mmol/mL} \times 20.5 \text{ mL} \times \frac{1}{2} \times 101.96 \text{ mg/mmol}}{1000 \text{ mg}} \times 100\% = 8.71\%$$

Variable Reactions in Molarity Calculations

Some substances can undergo reaction to different products. The factor used in calculating millimoles of a substance from the millimoles of titrant reacted with it will depend on the reaction. Sodium carbonate, for example, can react as a diprotic or a monoprotic base:

$$CO_3^{2-} + 2H^+ \rightarrow H_2O + CO_2$$

or

$$CO_3^{2-} + H^+ \rightarrow HCO_3^-$$

In the first case, mmol $NaCO_3$ = mmol acid $\times \frac{1}{2}$(mmol CO_3^{2-}/mmol H^+). In the second case, mmol Na_2CO_3 = mmol acid. Similarly, phosphoric acid can be titrated as a monoprotic or a diprotic acid:

$$H_3PO_4 + OH^- \rightarrow H_2PO_4^- + H_2O$$

or

$$H_3PO_4 + 2OH^- \rightarrow HPO_4^{2-} + 2H_2O$$

Note that in these examples, if normality and milliequivalents are used in calculations, rather than molarity and millimoles, the equivalent weight and hence the number of milliequivalents calculated will depend on which reaction we are talking about.

EXAMPLE 7.9   In acid solution, potassium permanganate reacts with H_2O_2 to form Mn^{2+}:

$$5H_2O_2 + 2MnO_4^- + 6H^+ \rightarrow 5O_2 + 2Mn^{2+} + 8H_2O$$

In neutral solution, it reacts with $MnSO_4$ to form MnO_2:

$$3Mn^{2+} + 2MnO_4^- + 4OH^- \rightarrow \underline{5MnO_2} + 2H_2O$$

Calculate the number of milliliters of 0.100 M $KMnO_4$ that will react with 50.0 mL of 0.200 M H_2O_2 and with 50.0 mL of 0.200 M $MnSO_4$.

SOLUTION

The number of millimoles of MnO_4^- will be equal to two-fifths of the number of millimoles of H_2O_2 reacted:

$$M_{MnO_4^-} \times mL_{MnO_4^-} = M_{H_2O_2} \times mL_{H_2O_2} \times \tfrac{2}{5}(mmol\ MnO_4^-/mmol\ H_2O_2)$$

$$mL_{MnO_4^-} = \frac{0.200\ mmol/mL \times 50.0\ mL \times \tfrac{2}{5}}{0.100\ mmol/mL} = 40.0\ mL$$

The number of millimoles of MnO_4^- reacting with Mn^{2+} will be equal to two-thirds of the number of millimoles of Mn^{2+}:

$$M_{MnO_4^-} \times mL_{MnO_4^-} = M_{Mn^{2+}} \times \tfrac{2}{3}(mmol\ MnO_4^-/mmol\ Mn^{2+})$$

$$mL_{MnO_4^-} = \frac{0.200\ mmol/mL \times 50.0\ mL \times \tfrac{2}{3}}{0.100\ mmol/mL} = 66.7\ mL$$

EXAMPLE 7.10   Oxalic acid, $H_2C_2O_4$, is a reducing agent that reacts with $KMnO_4$ as follows:

$$5H_2C_2O_4 + 2MnO_4^- + 6H^+ \rightarrow 10CO_2 + 2Mn^{2+} + 8H_2O$$

Its two protons are also titratable with a base. How many milliliters of 0.100 M NaOH and 0.100 M $KMnO_4$ will react with 500 mg $H_2C_2O_4$?

SOLUTION

$$mmol\ NaOH = 2 \times mmol\ H_2C_2O_4$$

$$0.100\ M \times x\ mL = \frac{500\ mg\ H_2C_2O_4}{90.0\ mg/mmol} \times 2(mmol\ OH^-/mmol\ H_2C_2O_4)$$

$$x = 111\ mL\ NaOH$$

$$mmol\ KMnO_4 = \tfrac{2}{5} \times mmol\ H_2C_2O_4$$

$$0.100\ M \times x\ mL = \frac{500\ mg\ H_2C_2O_4}{90.0\ mg/mmol} \times \tfrac{2}{5}(mmol\ MnO_4^-/mmol\ H_2C_2O_4)$$

$$x = 22.2\ mL\ KMnO_4$$

EXAMPLE 7.11   Pure $Na_2C_2O_4$ plus $KHC_2O_4 \cdot H_2C_2O_4$ (three replaceable protons, KH_3A_2) are mixed in such a proportion that each gram of the mixture will react with equal volumes of 0.100 M $KMnO_4$ and 0.100 M NaOH. What is the proportion?

SOLUTION

Assume 10.0 mL titrant, so there is 1.00 mmol NaOH or $KMnO_4$. The acidity is due to $KHC_2O_4 \cdot H_2C_2O_4(KH_3A_2)$:

$$\text{mmol } KH_3A_2 = \text{mmol NaOH} \times \tfrac{1}{3}(\text{mmol } KH_3A_2/\text{mmol } OH^-)$$

$$1.00 \times \tfrac{1}{3} = 0.333 \text{ mmol } KH_3A$$

From Example 7.10, each mmol $Na_2C_2O_4$ (Na_2A) reacts with $\tfrac{2}{5}$ mmol $KMnO_4$:

$$\text{mmol } KMnO_4 = \text{mmol } Na_2A \times \tfrac{2}{5}(\text{mmol } MnO_4^-/\text{mmol } Na_2A) + \text{mmol } KH_3A_2$$

$$\times \tfrac{4}{5}(\text{mmol } MnO_4^-/\text{mmol } KH_3A_2)$$

$$1.00 \text{ mmol} = \text{mmol } Na_2A \times \tfrac{2}{5} + 0.333 \text{ mmol } KH_3A_2 \times \tfrac{4}{5} \text{ mmol } Na_2A = 1.8_2 \text{ mmol}$$

The ratio is 1.8_2 mmol $Na_2A/0.333$ mmol $KH_3A_2 = 5.4_7$ mmol $Na_2A/\text{mmol } KH_3A_2$.

The weight ratio is $\dfrac{5.4_7 \text{ mmol } Na_2A \times 134 \text{ mg/mmol}}{218 \text{ mg } KH_3A_2/\text{mmol}} = 3.36 \text{ g } Na_2A/\text{g } KH_3A_2$

Back-Titrations

Sometimes a reaction is slow to go to completion and a sharp end point cannot be obtained, for example, in the titration of antacid tablets with a strong acid such as HCl. In these cases, a **back-titration** will often yield useful results. In this technique, a measured amount of the reagent, which would normally be the titrant, is added to the sample so that there is a slight excess. After the reaction with the analyte is allowed to go to completion, the amount of excess (unreacted) reagent is determined by titration with another standard solution; the analyte reaction also may be speeded up in the presence of excess reagent. So by knowing the number of millimoles of reagent taken and by measuring the number of millimoles remaining unreacted, we can calculate the number of millimoles of sample that reacted with the reagent:

mmol reagent reacted = mmol taken − mmol back-titrated

mg analyte = mmol reagent reacted × factor (mmol analyte/mmol reagent)
 × f. wt. analyte

EXAMPLE 7.12 A 0.200-g sample of pyrolusite is analyzed as follows. Add 50.0 mL of a 0.100 M solution of ferrous ammonium sulfate to reduce the MnO_2 to Mn^{2+}. After reduction is complete, the excess ferrous ion is titrated in acid solution with 0.0200 M $KMnO_4$, requiring 15.0 mL. Calculate the percent manganese in the sample as Mn_3O_4 (only part or none of the manganese may exist in this form, but we can make the calculations on the assumption that it does).

SOLUTION
The reaction between Fe^{2+} and MnO_4^- is

$$5Fe^{2+} + MnO_4^- + 8H^+ \rightarrow 5Fe^{3+} + Mn^{2+} + 4H_2O$$

and so there are five times as many millimoles of excess Fe^{2+} as of MnO_4^- that reacted with it.
 The reaction between Fe^{2+} and MnO_2 is

$$MnO_2 + 2Fe^{2+} + 4H^+ \rightarrow Mn^{2+} + 2Fe^{3+} + 2H_2O$$

and there are one-half as many millimoles of MnO_2 as millimoles of Fe^{2+} that react with it. There are one-third as many millimoles of Mn_3O_4 as of MnO_2 (1 $MnO_2 \rightarrow \frac{1}{3} Mn_3O_4$). Therefore,

$$\% \ Mn_3O_4 = \{[M_{Fe^{2+}} \times mL_{Fe^{2+}} - M_{MnO_4^-} \times mL_{MnO_4^-} \times 5(mmol/Fe^{2+}/mmol \ MnO_4^-)$$

$$\times \tfrac{1}{2}(mmol \ MnO_2/mmol \ Fe^{2+}) \times \tfrac{1}{3}(mmol \ Mn_3O_4/mmol \ MnO_2)$$

$$\times f. \ wt._{Mn_3O_4}]/mg_{sample}\} \times 100\%$$

$$= \frac{(0.100 \times 50.0 - 0.0200 \times 15.0 \times 5) \times \tfrac{1}{2} \times \tfrac{1}{3} \times 228.8 \ mg/mmol}{200}$$

$$\times 100\%$$

$$= 66.7\%$$

7.3   NORMALITY CALCULATIONS

Many substances do not react on a 1:1 mole basis, and so solutions of equal molar concentration do not react on a 1:1 volume basis. By introducing the concepts of equivalents and normality, it is possible in these cases to make calculations similar to molar calculations for 1:1 mole reactions. We define a new unit of concentration called **normality.** The symbol N stands for **normal,** just as M stands for molar. The normality of a solution is equal to the number of **equivalents** of material per liter of solution:

$$\boxed{N = \frac{eq}{L} = \frac{meq}{mL}} \tag{7.15}$$

where meq represents **milliequivalents.**

Equivalents are based on the same concept as moles, but the number of equivalents will depend on the number of **reacting units** supplied by each molecule or the number with which it will react. For example, if we have one mole of HCl, we have one mole of H^+ to react as an acid. Therefore, we have one equivalent of H^+. If, on the other hand, we have one mole of H_2SO_4, we have two moles of the reacting unit H^+ and two equivalents of H^+. The number of equivalents can be calculated from the number of moles by

$$\boxed{\begin{array}{l} eq = mol \times no. \ of \ reacting \ units \ per \ molecule \\ meq = mmol \times no. \ of \ reacting \ units \ per \ molecule \end{array}}$$

The Equivalent Weight

The **equivalent weight** is that weight of a substance in grams that will furnish one mole of the reacting unit. Thus, for HCl, the equivalent weight is equal to the formula weight. For H_2SO_4, it takes only one-half the number of molecules to furnish one mole of H^+, and so the equivalent weight is one-half the formula weight:

$$\text{eq. wt. HCl} = \frac{\text{f. wt.}_{HCl} \text{ (g/mol)}}{1 \text{ (eq/mol)}}$$

$$\text{eq. wt. H}_2\text{SO}_4 = \frac{\text{f. wt.}_{H_2SO_4} \text{ (g/mol)}}{2 \text{ (eq/mol)}}$$

See Chapter 1 for a discussion of reacting units for precipitation and complexation cases. Just as the number of moles can be calculated from the number of grams by dividing by the formula weight, the number of equivalents of a substance can be calculated by dividing by the equivalent weight:

$$\text{eq} = \frac{\text{g}}{\text{eq. wt. (g/eq)}}; \quad \text{meq} = \frac{\text{mg}}{\text{eq. wt. (mg/meq)}} \qquad (7.16)$$

The normality of a solution, then, is calculated from the number of equivalents and the volume:

$$N = \frac{\text{eq}}{\text{L}} = \frac{\text{g/eq. wt. (g/eq)}}{\text{L}}; \quad N = \frac{\text{meq}}{\text{mL}} = \frac{\text{mg/eq. wt. (mg/meq)}}{\text{mL}} \qquad (7.17)$$

The advantage of expressing concentrations in normality and quantities as equivalents is that one equivalent of substance A will **ALWAYS** *react with one equivalent of substance B.* Thus, one equivalent of NaOH ($\equiv$ 1 mol) will react with one equivalent of HCl ($\equiv$ 1 mol) or with one equivalent of H_2SO_4 ($\equiv \frac{1}{2}$ mol). We can, therefore, calculate the weight of analyte from the number of equivalents of titrant, because the latter is equal to the equivalents of analyte.

Remember that the number of reacting units of a compound, and hence the number of equivalents, will depend on the reaction it is undergoing. If A is the sample and T is the titrant,

$$\text{meq}_A = \text{meq}_T \qquad (7.18)$$

$$\text{meq}_A = \frac{\text{mg}_A}{\text{eq. wt.}_A \text{ (mg/meq)}} = N_T \text{ (meq/mL)} \times \text{mL} \qquad (7.19)$$

$$\text{mg}_A = \text{meq}_T = \text{eq. wt.}_A \text{ (mg/meq)} \qquad (7.20)$$

$$\text{mg}_A = N_T \text{ (meq/mL)} \times \text{mL}_T \times \text{eq. wt.}_A \text{ (mg/meq)} \qquad (7.21)$$

The equivalent weight of A is determined in the same way as it is for T; that is, how many reacting units does A liberate or react with per molecule?

A general equation for calculating the percent of a constituent in the sample can now be written:

$$\% \text{ A} = \frac{N_T \text{ (meq/mL)} \times \text{mL}_T \times \text{eq. wt.}_A \text{ (mg/meq)}}{\text{mg}_{sample}} \times 100\% \qquad (7.22)$$

EXAMPLE 7.13 A 0.4671-g sample containing sodium bicarbonate (a monoacidic base) is dissolved and titrated with a standard solution of hydrochloric acid, requiring 40.72 mL. The hydrochloric acid was standardized by titrating 0.1876 g sodium carbonate, which required 37.86 mL acid. (See Example 7.1 for reactions.) Calculate the percent sodium bicarbonate in the sample.

SOLUTION

Na_2CO_3 is a base that reacts with two H^+ per molecule:

$$N_{HCl} = \frac{meq_{Na_2CO_2}}{mL_{HCl}} = \frac{mg_{Na_2CO_3}/(f.\ wt._{Na_2CO_3}/2)}{mL_{HCl}}$$

$$= \frac{187.6\ mg/(105.99/2\ mg/meq)}{37.86\ mL} = 0.09350\ meq/mL$$

$$\%\ NaHCO_3 = \frac{N_{HCl} \times mL_{HCl} \times (f.\ wt._{NaHCO_3}/1)}{mg_{sample}} \times 100\%$$

$$= \frac{0.09350\ meq/mL \times 40.72\ mL \times (84.01/1\ mg/meq)}{476.1\ mg} \times 100\%$$

$$= 67.18\%$$

It is important to remember that one equivalent of a substance will always react with one equivalent of its counterpart. It is useful to recognize that, since

$$meq_A = meq_B \tag{7.23}$$

one can calculate the volumes of two solutions that will react by

$$\boxed{N_A\ (meq/mL) \times mL_A = N_T\ (meq/mL) \times mL_T} \tag{7.24}$$

EXAMPLE 7.14   How many milliliters of a 0.25 M solution of H_2SO_4 will react with 10 mL of a 0.25 M solution of NaOH?

SOLUTION

Since there are two reacting units per molecule of H_2SO_4, the normality of this solution will be twice its molarity:

$$N_{H_2SO_4} = 2\ (eq/mol) \times 0.25\ (mol/L) = 0.50\ eq/L$$

The normality of the NaOH will be the same as its molarity, since it will consume one reacting unit per molecule:

$$N_{NaOH} = 0.25\ eq/L$$

$$meq_{H_2SO_4} = meq_{NaOH}$$

$$N_{H_2SO_4} \times mL_{H_2SO_4} = N_{NaOH} \times mL_{NaOH}$$

$$\therefore 0.50\ meq/mL \times mL_{H_2SO_4} = 0.25\ meq/mL \times 10\ mL$$

$$mL_{H_2SO_4} = 5.0\ mL$$

Equation 7.24 can also be used to calculate the dilution required to prepare a certain normality of a solution from a more concentrated solution in a manner similar to molarity dilutions (see Example 7.4).

Reacting Units in Normality Calculations

1. ACID–BASE. As we have mentioned, the reacting unit for acids and bases is the proton H^+. If the substance reacts as an acid, we must determine the number of reactive protons it possesses per molecule. If it reacts as a base, we must determine the number of protons it will react with per molecule. Then,

$$\boxed{\text{eq. wt.} = \frac{\text{f. wt.}}{\text{no. of } H^+}} \tag{7.25}$$

This may depend on the particular reaction we choose. With H_2SO_4, both protons are strongly ionized, thus there are two equivalents of reacting units per mole of H_2SO_4. These react together to give a single end point, corresponding to the titration of both protons. In H_3PO_4, the first proton is fairly strongly ionized, the second weakly so, and the third is too weakly ionized to be titrated. So, we can titrate the first and second protons *stepwise* to obtain two separate end points, while the third is too weak to yield a detectable end point. If we choose to titrate only the first proton, then there will be only one reacting unit per molecule of H_3PO_4, **in this particular reaction.** The number of equivalents will, therefore, be equal to the number of moles of H_3PO_4. If we choose, however, to titrate two protons, then the number of equivalents will be twice the number of moles.

Na_2CO_3 is a strong base that will react with an acid to produce $NaHCO_3$. The $NaHCO_3$ is a weak base and can be titrated one step further to carbonic acid:

$$CO_3^{2-} + H^+ \rightleftharpoons HCO_3^-$$

$$HCO_3^- + H^+ \rightleftharpoons H_2O + CO_2.$$

Again, we can have two separate end points, and the number of equivalents (and the equivalent weight) will depend on which reaction we choose. It is then essential to know the reaction taking place before beginning a calculation.

EXAMPLE 7.15 A solution of sodium carbonate is prepared by dissolving 0.212 g Na_2CO_3 and diluting to 100 mL. Calculate the normality of the solution (a) if it is used as a monoacidic base, and (b) if it is used as a diacidic base.

SOLUTION

(a)
$$N = \frac{mg_{Na_2CO_3}/(Na_2CO_3/1)}{mL} = \frac{212 \text{ mg}/(106.0/1 \text{ mg/meq})}{100 \text{ mL}} = 0.0200 \text{ meq/mL}$$

(b)
$$N = \frac{mg_{Na_2CO_3}/(Na_2CO_3/2)}{mL} = \frac{212 \text{ mg}/(106.0/2 \text{ mg/meq})}{100 \text{ mL}} = 0.0400 \text{ meq/mL}$$

---

2. REDUCTION–OXIDATION. The reacting unit here is the **electron.** A reducing agent liberates electrons and is thereby oxidized, and an oxidizing agent takes on electrons and is thereby reduced. For example, in the reaction

$$5Fe^{2+} + MnO_4^- + 8H^+ \rightleftharpoons 5Fe^{3+} + Mn^{2+} + 4H_2O$$

each Fe^{2+} (reducing agent) loses one electron, and each MnO_4^- (oxidizing agent) gains five electrons in being reduced from Mn^{7+} to Mn^{2+}. We treat the number of reacting units just as we did with acids and bases. Thus,

$$\boxed{\text{eq. wt.} = \frac{\text{f. wt.}}{\text{no. of moles of electrons gained or lost}}} \tag{7.26}$$

In this example, the number of equivalents of iron is equal to its number of moles, and the equivalent weight is equal to the atomic weight of iron. There are five times as many equivalents of the permanganate as moles, and its equivalent weight is one-fifth its formula weight. But just as with acids and bases, one equivalent of reducing agent will react with one equivalent of oxidizing agent.

EXAMPLE 7.16   Iodine (I_2) is an oxidizing agent that in reactions with reducing agents is reduced to iodide ion (I^-). How many grams I_2 would you weigh out to prepare 100 mL of a 0.100 N I_2 solution?

SOLUTION

Since each molecule of I_2 consumes two electrons,

$$I_2 + 2e^- \rightarrow 2I^-$$

The equivalent weight is one-half the formula weight:

$$\text{eq. wt.} = \frac{\text{f. wt.}_{I_2} \ (\text{g/mol})}{2 \ (\text{eq/mol})}$$

$$N \times \text{mL} = \text{meq} = \frac{\text{mg}_{I_2}}{\text{eq. wt.}} = \frac{\text{mg}_{I_2}}{\text{f. wt.}_{I_2}/2}$$

$$0.100 \text{ meq/mL} \times 100 \text{ mL} = \frac{\text{mg}_{I_2}}{254/2 \text{ mg/meq}}$$

$$\text{mg}_{I_2} = 0.100 \text{ meq/mL} \times 100 \text{ mL} \times (254/2 \text{ mg/meq}) = 1270 \text{ mg}$$

You would, therefore, weigh out 1.27 g.

EXAMPLE 7.17   Calculate the normality of a solution of 0.25 g/L $H_2C_2O_4$, both as an acid and as a reducing agent.

SOLUTION

The equivalent weight as an acid is half the formula weight;

$$\therefore N_{\text{acid}} = \frac{250 \text{ mg}/(90.04/2 \text{ mg/meq})}{1{,}000 \text{ mL}} = 0.00555 \text{ meq/mL}$$

Each oxalate ion gains two electrons in being oxidized to CO_2 (each carbon is oxidized from a valence of $+3$ to a valence of $+4$). Therefore, the equivalent weight is half the formula weight, and the normality as a reducing agent is the same as it is as an acid.

---

3. PRECIPITATION AND COMPLEXOMETRIC.   In these cases, there is no reacting unit exchanged; the reactants merely combine. The charge on the cation (metal ion) is taken as the reacting unit, and so the equivalent weight is the atomic weight divided by the charge. The equivalent weight of the precipitating or complexing agent is its formula weight divided by the number of equivalents that each mole reacts with:

$$\text{eq. wt.}_{M^{n+}} = \frac{\text{at. wt.}}{\text{charge } (n+)} \tag{7.27}$$

$$\text{eq. wt.}_{\text{reagent}} = \frac{\text{f. wt.}}{\text{eq } M^{n+}/\text{mol reagent}} \tag{7.28}$$

One equivalent of reagent will react with one equivalent of metal ion.

EXAMPLE 7.18 What would be the normality of a 0.1 M K_2CrO_4 solution used as a precipitating agent for (a) Pb^{2+} and (b) Ag^+?

SOLUTION
(a) The reaction is

$$Pb^{2+} + CrO_4^{2-} \rightleftharpoons \underline{PbCrO_4}$$

Each mole of CrO_4^{2-} reacts with two equivalents of Pb^{2+}. Therefore, the equivalent weight of K_2CrO_4 is one-half its formula weight, and its normality is twice its molarity ($N = 0.2$).

(b) The reaction is

$$2Ag^+ + CrO_4^{2-} \rightleftharpoons \underline{Ag_2CrO_4}$$

Again, each mole of CrO_4^{2-} reacts with two equivalents of Ag^+ (since the charge is $1+$ and there are two Ag^+), so the equivalent weight and normality are the same as in (a), that is, $N = 0.2$.

EXAMPLE 7.19 What would be the normality of a 0.1 M EDTA solution used as a complexometric titrating agent for (a) Ca^{2+} and (b) Al^{3+}?

SOLUTION
(a) Let H_4Y represent EDTA. The Y^{4-} species forms the complex. The reaction is

$$Ca^{2+} + Y^{4-} \rightleftharpoons CaY^{2-}$$

Each mole of EDTA reacts with two equivalents of Ca^{2+}. So its equivalent weight is one-half its formula weight, and the normality is twice the molarity ($N = 0.2$).

(b) The reaction is

$$Al^{3+} + Y^{4-} \rightleftharpoons AlY^-$$

In this case, each mole of EDTA reacts with three equivalents of Al^{3+}. So the equivalent weight is one-third the formula weight, and the normality is three times the molarity ($N = 0.3$).

Summary of Normality Calculations

Summarized here are the equations most frequently used in applying the concept of normality to volumetric titrations:

$$\text{meq} = \frac{\text{mg}}{\text{eq. wt. (mg/meq)}} \tag{7.29}$$

$$N = \frac{\text{meq}}{\text{mL}} \tag{7.30}$$

One of the following equations will usually be used to calculate the results of a titration in which substance A is titrated with substance T:

$$N_T(\text{meq/mL}) \times \text{mL}_T \times \text{eq. wt.}_A (\text{mg/meq}) = \text{mg}_A \qquad (7.31)$$

$$\frac{N_T(\text{meq/mL}) \times \text{mL}_T \times \text{eq. wt.}_A (\text{mg/meq})}{\text{mg}_{\text{sample}}} \times 100\% = \% \text{ A} \qquad (7.32)$$

7.4   TITER

For routine titrations, it is often convenient to calculate the **titer** of the titrant. The titer is the weight of analyte that is chemically equivalent to 1 mL of the titrant, usually expressed in milligrams. For example, if a potassium dichromate solution has a titer of 1.267 mg Fe, each milliliter potassium dichromate will react with 1.267 mg iron, and the weight of iron titrated is obtained by simply multiplying the volume of titrant used by the titer. The titer can be expressed in terms of any form of the analyte desired, for example, milligrams FeO or Fe_2O_3.

EXAMPLE 7.20   A standard solution of potassium dichromate contains 5.442 g/L. What is its titer in terms of milligrams Fe_3O_4?

SOLUTION
The iron is titrated as Fe^{2+} and each $Cr_2O_7^{2-}$ will react with $6Fe^{2+}$ or the iron from $2Fe_3O_4$:

$$6Fe^{2+} + Cr_2O_7^{2-} + 14H^+ \rightarrow 6Fe^{3+} + 2Cr^{3+} + 7H_2O$$

The molarity of the $K_2Cr_2O_7$ solution is:

$$M_{Cr_2O_7^{2-}} = \frac{\text{g/L}}{\text{f. wt.}_{K_2Cr_2O_7}} = \frac{5.442 \text{ g/L}}{294.19 \text{ g/mol}} = 0.01850 \text{ mol/L}$$

Therefore the titer is:

$$0.01850\left(\frac{\text{mmol } K_2Cr_2O_7}{\text{mL}}\right) \times \frac{2}{1}\left(\frac{\text{mmol } Fe_3O_4}{\text{mmol } K_2Cr_2O_7}\right) \times 231.54\left(\frac{\text{mg } Fe_3O_4}{\text{mmol } Fe_3O_4}\right)$$

$$= 8.567 \text{ mg } Fe_3O_4/\text{mLK}_2Cr_2O_7$$

QUESTIONS

1. List the requirements for a titration. What are the four classes of titrations?
2. What is the equivalence point of a titration? The end point?
3. What is a standard solution? How is it prepared?
4. What are the requirements of a primary standard?
5. Why should a primary standard have a high formula weight?

PROBLEMS

Dilution Calculations
6. Twelve and five-tenths milliliters of a solution is diluted to 500 mL, and its molarity is determined to be 0.125. What is the molarity of the original solution?

7. What volume of 0.50 M H_2SO_4 must be added to 65 mL of 0.20 M H_2SO_4 to give a final solution of 0.35 M? Assume volumes are additive.

8. How many milliliters of 0.10 M H_2SO_4 must be added to 50 mL of 0.10 M NaOH to give a solution that is 0.050 M in H_2SO_4? Assume volumes are additive.

9. You are required to prepare working standard solutions of 1.00×10^{-5}, 2.00×10^{-5}, 5.00×10^{-5}, and 1.00×10^{-4} M glucose from a 0.100 M stock solution. You have available 100-mL volumetric flasks and pipets of 1.00-, 2.00-, 5.00-, and 10.00-mL volume. Outline a procedure for preparing the working standards.

10. A 0.500-g sample is analyzed spectrophotometrically for manganese by dissolving it in acid and transferring to a 250-mL flask and diluting to volume. Three aliquots are analyzed by transferring 50-mL portions with a pipet to 500-mL Erlenmeyer flasks and reacting with an oxidizing agent, potassium peroxydisulfate, to convert the manganese to permanganate. After reaction, these are quantitatively transferred to 250-mL volumetric flasks, diluted to volume, and measured spectrophotometrically. By comparison with standards, the average concentration in the final solution is determined to be 1.25×10^{-5} M. What is the percent manganese in the sample?

Standardization Calculations

11. A preparation of soda ash is known to contain 98.6% Na_2CO_3. If a 0.678-g sample requires 36.8 mL of a sulfuric acid solution for complete neutralization, what is the molarity of the sulfuric acid solution?

12. A 0.1 M sodium hydroxide solution is to be standardized by titrating primary standard sulfamic acid (NH_2SO_3H). What weight of sulfamic acid should be taken so that the volume of NaOH delivered from the buret is about 40 mL?

Analysis Calculations

13. A sample of U.S.P.-grade citric acid ($H_3C_6H_5O_7$, three titratable protons) is analyzed by titrating with 0.1087 M NaOH. If a 0.2678-g sample requires 38.31 mL for titration, what is the purity of the preparation? (U.S.P. requires 99.5%.)

14. Calcium in a 200-μL serum sample is titrated with 1.87×10^{-4} M EDTA solution, requiring 2.47 mL. What is the calcium concentration in the blood in mg/dL?

15. A 0.372-g sample of impure $BaCl_2 \cdot 2H_2O$ is titrated with 0.100 M $AgNO_3$, requiring 27.2 mL. Calculate (a) the percent Cl in the sample and (b) the percent purity of the compound.

16. An iron ore is analyzed for iron content by dissolving in acid, converting the iron to Fe^{2+}, and then titrating with standard 0.0150 M $K_2Cr_2O_7$ solution. If 35.6 mL is required to titrate the iron in a 1.68-g ore sample, how much iron is in the sample, expressed as percent Fe_2O_3? (See Example 7.20 for the titration reaction.)

17. Calcium in a 2.00-g sample is determined by precipitating CaC_2O_4, dissolving this in acid, and titrating the oxalate with 0.0200 M $KMnO_4$. What percent of CaO is in the sample if 35.6 mL $KMnO_4$ is required for the titration? (The reaction is $5H_2C_2O_4 + 2MnO_4^- + 6H^+ \rightarrow 10CO_2 + 2Mn^{2+} + 8H_2O$.)

18. A potassium permanganate solution is prepared by dissolving 4.68 g $KMnO_4$ in water and diluting to 500 mL. How many milliliters of this will react with the iron in 0.500 g of an ore containing 35.6% Fe_2O_3? (See Example 7.12 for the titration reaction.)

19. A sample contains $BaCl_2$ plus inert matter. What weight must be taken so that when the solution is titrated with 0.100 M $AgNO_3$, the milliliters of titrant will be equal to the percent $BaCl_2$ in the sample?

20. A 0.250-g sample of impure $AlCl_3$ is titrated with 0.100 M $AgNO_3$, requiring 48.6 mL. What volume of 0.100 M EDTA would react with a 0.350-g sample? (EDTA reacts with Al^{3+} in a 1:1 ratio.)

21. A 425.2-mg sample of a purified monoprotic organic acid is titrated with 0.1027 M NaOH, requiring 28.78 mL. What is the formula weight of the acid?

22. The purity of a 0.287-g sample of $Zn(OH)_2$ is determined by titrating with a standard HCl solution, requiring 37.8 mL. The HCl solution was standardized by precipitating AgCl in a 25.0-mL aliquot and weighing (0.462 g AgCl obtained). What is the purity of the $Zn(OH)_2$?

23. A sample of pure $KHC_2O_4 \cdot H_2C_2O_4 \cdot 2H_2O$ (three replaceable hydrogens) requires 46.2 mL of 0.100 M NaOH for titration. How many milliliters of 0.100 M $KMnO_4$ will the same-size sample react with? (See Problem 17 for reaction with $KMnO_4$.)

Back-Titrations

24. A 0.500-g sample containing Na_2CO_3 plus inert matter is analyzed by adding 50.0 mL of 0.100 M HCl, a slight excess, boiling to remove CO_2, and then back-titrating the excess acid with 0.100 M NaOH. If 5.6 mL NaOH is required for the back-titration, what is the percent Na_2CO_3 in the sample?

25. A hydrogen peroxide solution is analyzed by adding a slight excess of standard $KMnO_4$ solution and back-titrating the unreacted $KMnO_4$ with standard Fe^{2+} solution. A 0.587-g sample of the H_2O_2 solution is taken, 25.0 mL of 0.0215 M $KMnO_4$ is added, and the back-titration requires 5.10 mL of 0.112 M Fe^{2+} solution. What is the percent H_2O_2 in the sample? (See Examples 7.9 and 7.12 for the reactions.)

26. The sulfur content of a steel sample is determined by converting it to H_2S gas, absorbing the H_2S in 10.0 mL of 0.00500 M I_2, and then back-titrating the excess I_2 with 0.00200 M $Na_2S_2O_3$. If 2.6 mL $Na_2S_2O_3$ is required for the titration, how many milligrams sulfur are contained in the sample?
    Reactions:

$$H_2S + I_2 \rightarrow S + 2I^- + 2H^+$$

$$I_2 + 2S_2O_3^{2-} \rightarrow 2I^- + S_4O_6^{2-}$$

Titer

27. Express the titer of a 0.100 M EDTA solution in mg BaO/mL.

28. Express the titer of a 0.0500 M $KMnO_4$ solution in mg Fe_2O_3/mL.

29. The titer of a silver nitrate solution is 22.7 mg Cl/mL. What is its titer in mg Br/mL?

Equivalent Weight

30. Calculate the equivalent weight of KHC_2O_4 (a) as an acid, (b) as a reducing agent in reaction with MnO_4^- ($5HC_2O_4^- + 2MnO_4^- + 11H^+ \rightarrow 10CO_2 + 2Mn^{2+} + 8H_2O$), and (c) as a precipitating agent for Ca^{2+} ($\rightarrow CaC_2O_4$).

31. Mercuric oxide, HgO, can be analyzed by reaction with iodide and then titration with an acid: $HgO + 4I^- \rightarrow HgI_4^{2-} + 2OH^-$. What is its equivalent weight?

32. Calculate the grams of one equivalent each of the following for the indicated reaction: (a) $FeSO_4$ ($Fe^{2+} \rightarrow Fe^{3+}$); (b) H_2S ($\rightarrow S^0$); (c) H_2O_2 ($\rightarrow O_2$); (d) H_2O_2 ($\rightarrow H_2O$).

33. $BaCl_2 \cdot 2H_2O$ is to be used to titrate Ag^+ to yield AgCl. How many milliequivalents are contained in 0.5000 g $BaCl_2 \cdot 2H_2O$?

Normality

34. A solution is prepared by dissolving 7.82 g NaOH and 9.26 g $Ba(OH)_2$ in water and diluting to 500 mL. What is the normality of the solution as a base?

35. What weight of arsenic trioxide, As_2O_3, is required to prepare 1 L of 0.1000 N arsenic(III) solution (arsenic $3+$ is oxidized to $5+$ in redox reactions).

36. If 2.73 g $KHC_2O_4 \cdot H_2C_2O_4$ (three ionizable protons) having 2.0% inert impurities and

1.68 g $KHC_8H_4O_4$ (one ionizable proton) are dissolved in water and diluted to 250 mL, what is the normality of the solution as an acid, assuming complete ionization?

37. A solution of $KHC_2O_4 \cdot H_2C_2O_4 \cdot 2H_2O$ (three replaceable hydrogens) is 0.200 N as an acid. What is its normality as a reducing agent? (See Problem 17 for its reaction as a reducing agent.)

38. $Na_2C_2O_4$ and $KHC_2O_4 \cdot H_2C_2O_4$ are mixed in such a proportion by weight that the normality of the resulting solution as a reducing agent is 3.62 times the normality as an acid. What is the proportion? (See Problem 17 for its reaction as a reducing agent.)

39. What weight of $K_2Cr_2O_7$ is required to prepare 1.000 L of 0.1000 N solution? (In reaction, $Cr_2O_7^{2-} + 14H^+ + 6e^- \rightleftharpoons 2Cr^3 + 7H_2O$.)

RECOMMENDED REFERENCES

1. M. R. F. Ashworth, *Titrimetric Organic Analysis*. New York: Interscience, 1964.

2. I. M. Kolthoff, E. B. Sandell, E. J. Meehan, and S. Bruckenstein, *Quantitative Chemical Analysis*, 4th ed. London: The Macmillan Company, 1969.

chapter

8

ACID–BASE TITRATIONS

In Chapter 5, we introduced the principles of acid–base equilibria. These are important for the construction and interpretation of titration curves in acid–base titrations. In this chapter, we discuss the various types of acid–base titrations, including the titration of strong acids or bases and of weak acids or bases. The shapes of titration curves obtained are illustrated. Through a description of the theory of indicators, we discuss the selection of a suitable indicator for detecting the completion of a particular titration reaction. The titrations of weak acids or bases with two or more titratable groups and of mixtures of acids or bases are presented. The important Kjeldahl analysis method is described for analyzing nitrogen in organic and biological samples, and, finally, the preparation of a standard acid or base solution is given.

8.1 STRONG ACID VERSUS STRONG BASE

An acid–base titration involves a **neutralization** reaction in which an acid is reacted with an equivalent amount of base. By constructing a **titration curve,** we can easily explain how the **end points** of these titrations can be detected; the end point signals the completion of the reaction. A titration curve is constructed by plotting the pH of the solution as a function of the volume of titrant added. **The titrant is always a strong acid or a strong base.** The analyte may be either a strong base or acid or a weak base or acid.

In the case of a strong acid versus a strong base, both the titrant and the analyte are completely ionized. An example is the titration of hydrochloric acid with sodium hydroxide:

$$H^+ + Cl^- + Na^+ + OH^- \rightarrow H_2O + Na^+ + Cl^- \qquad (8.1)$$

The H^+ and OH^- combine to form H_2O, and the other ions (Na^+ and Cl^-) remain unchanged, so the net result of neutralization is conversion of the HCl to a neutral solution of NaCl. The titration curve for 100 mL of 0.1 M HCl titrated with 0.1 M NaOH is shown in Figure 8.1.

The calculations of titration curves simply involves computation of the pH from the concentration of the particular species present at the various stages of the titration, using the procedures given in Chapter 5. The volume changes during the titration must be employed for determining the concentration of the species.

At the beginning of the titration, we have 0.1 M HCl, so the initial pH is 1.0. As the titration proceeds, part of the H^+ is removed from solution as H_2O. So the concentration of H^+ gradually decreases. At 90% neutralization (90 mL NaOH), only 10% of the H^+ remains. Neglecting the volume change, the H^+ concentration at this point would be 10^{-2} M, and the pH would have risen by only one pH unit. (If we correct for volume change, it will be slightly higher.) However, as the **equivalence point** is approached (the point at which a stoichiometric amount of base is added), the H^+ concentration is rapidly reduced until *at* the equivalence point, when the neutralization is complete, a neutral solution of NaCl remains and the pH is 7.0. As we continue to add NaOH, the OH^- concentration rapidly increases from 10^{-7} M at the equivalence point and levels off between 10^{-2} and 10^{-1} M; we then have a solution of NaOH plus NaCl. Thus, the pH remains fairly constant on either side of the equivalence point, but it changes markedly very near the equivalence point. This large change allows the determination of the completion of the reaction by measurement of either the pH or some property that changes with pH.

FIGURE 8.1
Titration curve for 100 mL of 0.1 M HCl versus 0.1 M NaOH.

EXAMPLE 8.1   Calculate the pH at 0, 10, 90, 100, and 110% titration for the titration of 50.0 mL of 0.100 M HCl with 0.100 M NaOH.

SOLUTION

At 0%, pH = $-\log 0.100 = 1.00$.

At 10%, 5.0 mL NaOH is added. We start with 0.100 M × 50.0 mL = 5.00 mmol H^+. Calculate the concentration of H^+ after adding the NaOH:

$$
\begin{aligned}
\text{mmol } H^+ \text{ at start} &= 5.00 \text{ mmol } H^+ \\
\text{mmol } OH^- \text{ added} = 0.100\ M \times 5.0 \text{ mL} &= \underline{0.500 \text{ mmol } OH^-} \\
\text{mmol } H^+ \text{ left} &= 4.50 \text{ mmol } H^+ \text{ in } 55.0 \text{ mL}
\end{aligned}
$$

$$
\begin{aligned}
[H^+] &= 4.50 \text{ mmol}/55.0 \text{ mL} = 0.0818\ M \\
\text{pH} &= -\log 0.0818 = 1.09
\end{aligned}
$$

At 90%:

$$
\begin{aligned}
\text{mmol } H^+ \text{ at start} &= 5.00 \text{ mmol } H^+ \\
\text{mmol } OH^- \text{ added} = 0.100\ M \times 45.0 \text{ mL} &= \underline{4.50 \text{ mmol } OH^-} \\
\text{mmol } H^+ \text{ left} &= 0.50 \text{ mmol } H^+ \text{ in } 95.0 \text{ mL}
\end{aligned}
$$

$$
\begin{aligned}
[H^+] &= 0.00526\ M \\
\text{pH} &= -\log 0.00526 = 2.28
\end{aligned}
$$

At 100%: All the H^+ has been reacted with OH^-, and we have a 0.0500 M solution of NaCl. Therefore, the pH is 7.00.

At 110%: We now have a solution consisting of NaCl and excess added NaOH.

$$
\begin{aligned}
\text{mmol } OH^- &= 0.100 M \times 5.00 \text{ mL} = 0.50 \text{ mmol } OH^- \text{ in } 105 \text{ mL} \\
[OH^-] &= 0.00476\ M \\
\text{pOH} &= -\log 0.00476 = 2.32;\ \text{pH} = 11.68
\end{aligned}
$$

The magnitude of the break will depend on both the concentration of the acid and the concentration of the base. Titration curves at different concentrations are shown in Figure 8.2. The reverse titration gives the mirror image of these curves. The titration of 0.1 M NaOH with 0.1 M HCl is shown in Figure 8.3.

8.2   DETECTION OF THE END POINT: INDICATORS

We wish to determine when the equivalence point is reached. The point at which the reaction is **observed to be complete** is called the **end point.** A measurement is chosen such that the end point coincides with or is very close to the equivalence point. The most obvious way of determining the end point is to measure the pH at different points of the titration and make a plot of this versus milliliters of titrant. This is done with a pH meter, which is discussed in Chapter 11.

It is usually more convenient to add an **indicator** to the solution and visually detect a color change. An indicator for an acid–base titration is a weak acid or weak base that is highly colored. The color of the ionized form is markedly different from that of the un-ionized form. One form may be colorless, but the other must be colored. These substances are usually composed of highly conjugated organic constituents that give rise to the color (see Chapter 13).

FIGURE 8.2
Dependence of the magnitude of the end point break on concentration. Curve 1: 100 mL 0.1 M HCl versus 0.1 M NaOH. Curve 2: 100 mL 0.01 M HCl versus 0.01 M NaOH. Curve 3: 100 mL 0.001 M HCl versus 0.001 M NaOH.

FIGURE 8.3
Titration curve for 100 mL 0.1 M NaOH versus 0.1 M HCl.

Assume the indicator is a weak acid, designated HIn, and assume that the un-ionized form is red while the ionized form is blue:

$$\begin{array}{ll} \text{HIn} \rightleftharpoons \text{H}^+ + \text{In}^- & \\ \text{(red)} \hspace{2.5cm} \text{(blue)} & \end{array} \qquad (8.2)$$

We can write a Henderson–Hasselbalch equation for this, just as for other weak acids:

$$\text{pH} = \text{p}K_{\text{In}} + \log\frac{[\text{In}^-]}{[\text{HIn}]} \qquad (8.3)$$

The indicator changes color over a pH **range.** The transition range depends on the ability of the observer to detect small color changes. With indicators in which both forms are colored, generally only one color is observed if the ratio of the concentration of the two forms is 10:1; only the color of the more concentrated form is seen. From this information, we can calculate the pH transition range required to go from one color to the other. When only the color of the un-ionized form is seen, $[\text{In}^-]/[\text{HIn}] = \frac{1}{10}$. Therefore,

$$\text{pH} = \text{p}K_a + \log\tfrac{1}{10} = \text{p}K_a - 1 \qquad (8.4)$$

When only the color of the ionized form is observed, $[\text{In}^-]/[\text{HIn}] = \frac{10}{1}$, and

$$\text{pH} = \text{p}K_a + \log\tfrac{10}{1} = \text{p}K_a + 1 \qquad (8.5)$$

So the pH in going from one color to the other has changed from $\text{p}K_a - 1$ to $\text{p}K_a + 1$. This is a pH change of 2, and **most indicators require a transition range of about two pH units.** During this transition, the observed color is a mixture of the two colors.

Midway in the transition, the concentrations of the two forms are equal, an pH $= \text{p}K_a$. Obviously, then, **the pK_a of the indicator should be close to the pH of the equivalence point.**

Calculations similar to these can be made for weak base indicators, and they reveal the same transition range; the pOH midway in the transition is equal to $\text{p}K_b$, and the pH equals $14 - \text{p}K_b$. Hence, a weak-base indicator should be selected such that pH $= 14 - \text{p}K_b$. Figure 8.4 illustrates the colors and transition ranges of some commonly used indicators. The range may be somewhat less in some cases, depending on the colors; some colors are easier to see than others. The transition is easier to see if one form of the indicator is colorless. For this reason, phenolphthalein is usually used as an indicator for strong acid–base titrations (Figure 8.1.). In dilute solutions, however, phenolphthalein falls outside the steep portion of the titration curve (Figure 8.2), and an indicator such as bromothymol blue must be used.

Since an indicator is a weak acid or base, the amount added should be kept minimal so that it does not contribute appreciably to the pH and so that only a small amount of titrant will be required to cause the color change. That is, the color change will be sharper when the concentration is lower, because less acid or base is required to convert it from one form to the other. Of course, sufficient indicator must be added to impart an easily discernible color to the solution. Gen-

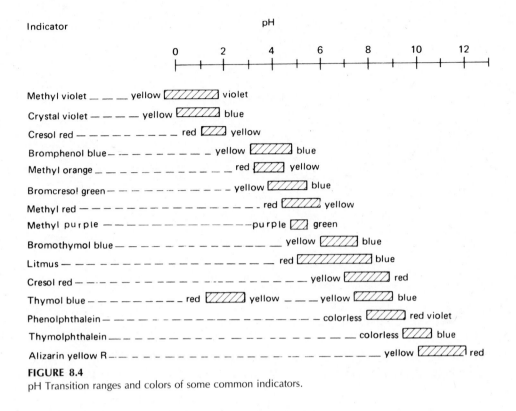

FIGURE 8.4
pH Transition ranges and colors of some common indicators.

erally, a few tenths percent solution (wt/vol) of the indicator is prepared and two or three drops are added to the solution to be titrated.

8.3 WEAK ACID VERSUS STRONG BASE

The titration curve for 100 mL of 0.1 M acetic acid titrated with 0.1 M sodium hydroxide is shown in Figure 8.5. The neutralization reaction is

$$HOAc + Na^+ + OH^- \rightarrow H_2O + Na^+ + OAc^- \tag{8.6}$$

The acetic acid, which is only a few percent ionized, depending on the concentration, is neutralized to water and an equivalent amount of the salt, sodium acetate. Before the titration is started, we have 0.1 M HOAc, and the pH is calculated as described for weak acids in Chapter 5. Table 8.1 summarizes the equations governing the different portions of the titration curve, as developed in Chapter 5. As soon as the titration is started, some of the HOAc is converted to NaOAc, and a buffer system is set up. As the titration proceeds, the pH slowly increases as the ratio [OAc$^-$]/[HOAc] changes. At the **midpoint of the titration,** [OAc$^-$] = [HOAc], and the **pH is equal to pK_a.** At the equivalence point, we have a solution of NaOAc. Since this is a Brønsted base (it hydrolyzes), **the pH at the equivalence point will be alkaline.** The pH will depend on the concentration of NaOAc (see Equation 5.32 and Figure 8.6). The greater the concentration, the higher the pH. As excess NaOH is added beyond the equivalence point, the ionization of the base OAc$^-$ is suppressed to a negligible amount (see Equation 5.23), and the pH is

FIGURE 8.5
Titration curve for 100 mL 0.1 M HOAc versus 0.1 M NaOH.

determined only by the concentration of excess OH^-. Therefore, **the titration curve beyond the equivalence point follows that for the titration of a strong acid.**

TABLE 8.1
Equations Governing a Weak-Acid (HA) or Weak-Base (B) Titration

		Weak Acid		Weak Base
Fraction F Titrated	Present	Equation	Present	Equation
$F = 0$	HA	$[H^+] = \sqrt{K_a \cdot C_{HA}}$ (Eq. 5.20)	B	$[OH^-] = \sqrt{K_b \cdot C_B}$ (Example 5.6)
$0 < F < 1$	HA/A$^-$	$pH = pK_a + \log\dfrac{C_{A^-}}{C_{HA}}$ (Eq. 5.45)	B/BH$^+$	$pH = (pK_w - pK_b) + \log\dfrac{C_B}{C_{BH^+}}$ (Eq. 5.54)
$F = 1$	A$^-$	$[OH^-] = \sqrt{\dfrac{K_w}{K_a} \cdot C_{A^-}}$ (Eq. 5.32)	BH$^+$	$[H^+] = \sqrt{\dfrac{K_w}{K_b} \cdot C_{BH^+}}$ (Eq. 5.39)
$F > 1$	OH$^-$	$[OH^-] = $ [excess titrant]	H$^+$	$[H^+] = $ [excess titrant]

EXAMPLE 8.2   Calculate the pH at 0, 10.0, 25.0, 50.0, and 60.0 mL titrant in the titration of 50.0 mL of 0.100 M acetic acid with 0.100 M NaOH.

SOLUTION
At 0 mL, we have a solution of only 0.100 M HOAc:

$$\frac{(x)(x)}{0.100 - x} = 1.75 \times 10^{-5}$$

$$[H^+] = x = 1.32 \times 10^{-3} M$$

$$pH = 2.88$$

At 10.0 mL, we started with 0.100 M $\times$ 50.0 mL = 5.00 mmol HOAc; part has reacted with OH$^-$ and has been converted to OAc$^-$:

mmol HOAc at start = 5.00 mmol HOAc
mmol OH$^-$ added = 0.100 M $\times$ 10.0 mL = 1.00 mmol OH$^-$
= mmol OAc$^-$ formed in 60.0 mL
mmol HOAc left = 4.00 mmol HOAc in 60.0 mL

We have a buffer. Since volumes cancel, use millimoles:

$$pH = pK_a + \log\frac{[OAc^-]}{[HOAc]}$$

$$pH = 4.76 + \log\frac{1.00}{4.00} = 4.16$$

At 25.0 mL, one-half the HOAc has been converted to OAc$^-$, so pH = pK_a:

mmol HOAc at start = 5.00 mmol HOAc
mmol OH$^-$ = 0.100 M $\times$ 25.0 mL = 2.50 mmol OAc$^-$ formed
mmol HOAc left = 2.50 mmol HOAc

$$pH = 4.76 + \log\frac{2.50}{2.50} = 4.76$$

At 50.0 mL; all the HOAc has been converted to OAc$^-$ (5.00 mmol in 100 mL, or 0.0500 M):

$$[OH^-] = \sqrt{\frac{K_w}{K_a}[OAc^-]}$$

$$= \sqrt{\frac{1.0 \times 10^{-14}}{1.75 \times 10^{-5}} \times 0.0500} = 5.35 \times 10^{-6} M$$

$$pOH = 5.27; \quad pH = 8.73$$

At 60.0 mL, we have a solution of NaOAc and excess added NaOH. The hydrolysis of the acetate is negligible in the presence of added OH$^-$. So the pH is determined by the concentration of excess OH$^-$:

mmol OH$^-$ = 0.100 M $\times$ 10.0 mL = 1.00 mmol in 110 mL
$[OH^-]$ = 0.00909 M
pOH = 2.04; pH = 11.96

The slowly rising region before the equivalence point is called the **buffer region.** It is flattest at the midpoint, and so the **buffering capacity is greatest at a pH corresponding to pK_a.** The buffering capacity also depends on the concentrations of HOAc and OAc⁻, and the **total buffering capacity** increases as the concentration increases. In other words, the distance of the flat portion on either side of pK_a will increase as [HOAc] and [OAc⁻] increase. As the pH deviates to the acid side of pK_a, the buffer will tolerate more base but less acid; the change in pH with a given small amount of added base will be greater, though, than at a pH equal to pK_a, because the curve is not so flat. Conversely, on the alkaline side of pK_a, more acid but less base can be tolerated.

You may have noticed that the corresponding region for a strong acid–strong base titration (Figures 8.1 and 8.2) is much flatter than for the weak acid case. In this respect, a solution of a strong acid or of a strong base is much more resistant to pH change upon addition of H⁺ or OH⁻ than the buffer systems we have discussed. The problem is that they are restricted to a very narrow pH region, either very acid or very alkaline, especially if the acid or base concentration is to be strong enough to have any significant capacity against a pH change. So these are regions that are rarely of practical value for buffering. Also, solutions of strong acids and bases are not resistant to pH change upon dilution, as buffers are. Therefore, we usually use mixtures of weak acids or bases with their salts, so that the desired pH region can be selected. Often, a buffer is used only to give a specified pH, and no extraneous acids or bases are added. A desired pH can be obtained more easily with conventional buffers than with a strong acid or a strong base.

The transition range of the indicator for this titration of a weak acid must fall within a pH range of about 7 to 10 (Figure 8.5). Phenolphthalein fits this nicely. If an indicator such as methyl red were used, it would begin changing color shortly after the titration began and would gradtually change to the alkaline color up to pH 6, before the equivalence point was even reached.

The dependence of the shape of the titration curve and of the equivalence point pH on concentration is shown in Figure 8.6 for different concentrations of HOAc

FIGURE 8.6
Dependence of the titration curve of a weak acid on the concentration. Curve 1: 100 mL of 0.1 *M* HOAc versus 0.1 *M* NaOH. Curve 2: 100 mL of 0.01 *M* HOAc versus 0.01 *M* NaOH. Curve 3: 100 mL of 0.001 *M* HOAc versus 0.001 *M* NaOH.

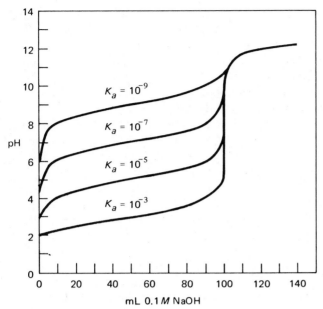

FIGURE 8.7
Titration curves for 100 mL 0.1 M weak acids of different K_a values versus 0.1 M NaOH.

and NaOH. Obviously, phenolphthalein could not be used as an indicator for solutions as dilute as 10^{-3} M (curve 3). Note that the equivalence point pH decreases as the weak-acid system becomes more dilute (which doesn't happen in the strong-acid system).

The equivalence point for the titration of any weak acid with a strong base will be alkaline. **The weaker the acid** (the smaller the K_a), the larger the K_b of the salt and **the more alkaline the equivalence point.**

Figure 8.7 shows the titration curves for 100 mL of 0.1 M solutions of weak acids of different K_a values titrated with 0.1 M NaOH. The sharpness of the end point decreases as K_a decreases. As in Figure 8.6, the sharpness will also decrease as the concentration decreases. Generally, for macro titrations (ca. 0.1 M), acids with K_a values of 10^{-6} can be titrated accurately with a visual indicator; and with suitable color comparisons, those with K_a values approaching 10^{-8} can be titrated with reasonable accuracy. A pH meter can be used to obtain better precision for the very weak acids by plotting the titration curve. Weaker acids can be titrated in nonaqueous solvents that do not possess the acidity or basicity of water (see the end of the chapter).

8.4 WEAK BASE VERSUS STRONG ACID

The titration of a weak base with a strong acid is completely analogous to the above case, but the titration curves are the reverse of those for a weak acid versus a strong base. The titration curve for 100 mL of 0.1 M ammonia titrated with 0.1 M hydrochloric acid is shown in Figure 8.8. The neutralization reaction is

$$NH_3 + H^+ + Cl^- \rightarrow NH_4^+ + Cl^- \qquad (8.7)$$

At the beginning of the titration, we have 0.1 M NH_3, and the pH is calculated as described for weak bases in Chapter 5. See Table 8.1. As soon as some acid is added, some of the NH_3 is converted to NH_4^+, and we are in the buffer region. At the **midpoint of the titration,** $[NH_4^+]$ equals $[NH_3]$, and the **pH is equal to (14 − pK_b).** At the equivalence point, we have a solution of NH_4Cl, a weak Brønsted acid which hydrolyzes to give an **acid solution.** Again, the pH will depend on the concentration; the greater the concentration, the lower the pH (see Equation 5.39). Beyond the equivalence point, the free H^+ suppresses the ionization (see Equation 5.33), and the pH is determined by the concentration of H^+ added in excess. Therefore, the titration curve beyond the equivalence point will follow that for titration of a strong base (Figure 8.3). Because K_b for ammonia happens to be equal to K_a for acetic acid, the titration curve for ammonia versus a strong acid is just the mirror image of the titration curve for acetic acid versus a strong base.

The indicator for the titration in Figure 8.8 must have a transition range within about pH 4 to 7. Methyl red meets this requirement, as shown in the figure. If phenolphthalein had been used as the indicator, it would have gradually lost its color between pH 10 and 8, before the equivalence point was reached.

Titration curves for different concentrations of NH_3 titrated with varying concentrations of HCl would be the mirror images of the curves in Figure 8.6. Methyl red could not be used as an indicator in dilute solutions. The titration curves for weak bases of different K_b values (100 mL, 0.1 M) versus 0.1 M HCl are shown in Figure 8.9. In macro titrations, one can accurately titrate a base with a K_b of 10^{-6} using a visual indicator.

8.5 TITRATION OF SODIUM CARBONATE

Sodium carbonate is a Brønsted base that is used as a primary standard for the standardization of strong acids. It hydrolyzes in two steps:

$$CO_3^{2-} + H_2O \rightleftharpoons HCO_3^- + OH^- K_{H1} = K_{b1} = \frac{K_w}{K_{a2}} = 2.1 \times 10^{-4} \quad (8.8)$$

$$HCO_3^- + H_2O \rightleftharpoons CO_2 + H_2O + OH^- K_{H2} = K_{b2} = \frac{K_w}{K_{a1}} = 2.3 \times 10^{-8} \quad (8.9)$$

where K_{a1} and K_{a2} refer to the K_a values of H_2CO_3; HCO_3^- is the conjugate acid of CO_3^{2-} and H_2CO_3 is the conjugate acid of HCO_3^-; and the K_b values are calculated as described in Chapter 5 for salts of weak acids and bases (i.e., from $K_bK_b = K_w$).

Sodium carbonate can be titrated to give end points corresponding to the step-wise additions of protons to form HCO_3^- and CO_2. **The K_b values should differ by at least 10^4** to obtain good separation of the equivalence point breaks in a case such as this.

A titration curve for Na_2CO_3 versus HCl is shown in Figure 8.10 (solid line). Even though K_{b1} is considerably larger than the 10^{-6} required for a sharp end point, the pH break is decreased by the formation of CO_2 beyond the first equivalence point. The second end point is not very sharp either, because K_{b2} is smaller than the 10^{-6} we would like. Fortunately, this end point can be sharpened, because the CO_2 produced from the neutralization of HCO_3^- is volatile and can be boiled out of the solution. This is described below.

At the start of the titration, the pH is determined by the hydrolysis of the

FIGURE 8.8
Titration curve for 100 mL 0.1 M NH$_3$ versus 0.1 M HCl.

Brønsted base CO_3^{2-}. After the titration is begun, part of the CO_3^{2-} is converted to HCO_3^-, and a CO_3^{2-}/HCO_3^- buffer region is established. At the first equivalence point, there remains a solution of HCO_3^-, and $[H^+] \approx \sqrt{K_{a1}K_{a2}}$. Beyond the first equivalence point, the HCO_3^- is partially converted to $H_2CO_3(CO_2)$ and a second buffer region is established, the pH being established by $[HCO_3^-]/[CO_2]$. The pH at the second equivalence point is determined by the concentration of the weak acid CO_2.

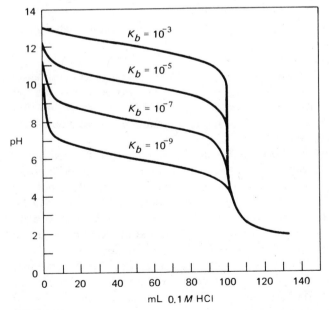

FIGURE 8.9
Titration curves for 100 mL 0.1 M weak bases of different K_b values versus 0.1 M HCl.

FIGURE 8.10
Titration curve for 50 mL 0.1 M Na$_2$CO$_3$ versus 0.1 M HCl. Dashed line represents a boiled solution with the CO$_2$ removed.

   Phenolphthalein is used to detect the first end point, and methyl orange is used to detect the second one. Neither end point, however, is very sharp. In actual practice, the phenolphthalein end point is used only to get an approximation of where the second end point will occur; phenolphthalein is colorless beyond the first end point and does not interfere. The second equivalence point, which is used for accurate titrations, is normally not very accurate with methyl orange indicator because of the gradual change in the color of the methyl orange. This is caused by the gradual decrease in the pH due to the HCO$_3^-$/CO$_2$ buffer system beyond the first end point.

   If beyond the first equivalence point we were to boil the solution after each addition of HCl to remove the CO$_2$ from the solution, the buffer system of HCO$_3^-$/CO$_2$ would be removed, leaving only HCO$_3^-$ in solution. This is both a weak acid and a weak base whose pH ($\approx$ 8.3) is independent of concentration ([H$^+$] = $\sqrt{K_{a1}K_{a2}}$ or [OH$^-$] = $\sqrt{K_{b1}K_{b2}}$; see Chapter 5). In effect, then, the pH would remain essentially constant until the equivalence point when we are left with a neutral solution of water and NaCl (pH = 7). The titration curve would then follow the dashed line in Figure 8.10.

   The following procedure can, therefore, be employed to sharpen the end point, as illustrated in Figure 8.11. Methyl red is used as the indicator, and the titration is continued until the solution just turns from yellow through orange to a definite red color. This will occur just before the equivalence point. The change will be very gradual because the color will start changing at about pH 6.3, well before the equivalence point. At this point, the titration is stopped and the solution is gently boiled to remove CO$_2$. The color should now revert to yellow because we have a dilute solution of only HCO$_3^-$. The solution is cooled and the titration is continued to a sharp color change to red or pink. The equivalence point here does not occur at pH 7 as indicated on the dashed line in Figure 8.10, because there is a small amount of HCO$_3^-$ still remaining to be titrated after boiling. That is, there will still be a slight buffering effect throughout the remainder of the titration, and dilute CO$_2$ will still remain at the equivalence point.

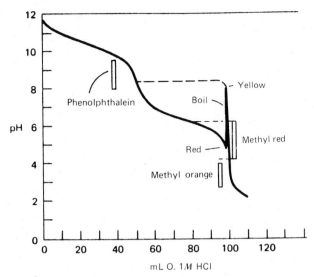

FIGURE 8.11
Titration of 50 mL 0.1 M Na_2CO_3 with 0.1 M HCl using methyl red indicator.

Bromcresol green can be used in a manner similar to methyl red. Its transition range is pH 3.8 to 5.4, with a color change from blue through pale green to yellow (see Experiment 7). Similarly, methyl purple can be used (see Experiment 17).

The methyl orange end point can be used (without boiling) by adding a blue dye, xylene cyanole FF, to the indicator. This mixture is called **modified methyl orange.** The blue color is complementary to the orange color of the methyl orange at about pH 2.8. This imparts a gray color at the equivalence point, the transition range of which is smaller than that of methyl orange. This results in a sharper end point. It is still not so sharp as the methyl red end point. Methyl orange can also be used by titrating to the color of the indicator in a solution of potassium acid phthalate, which has a pH close to 4.0.

8.6 TITRATION OF POLYPROTIC ACIDS

Diprotic acids can be titrated stepwise just as sodium carbonate was. In order to obtain good endpoint breaks for titration of the first proton, K_{a1} **should be at least** $10^4 \times K_{a2}$. If K_{a2} is in the required range of 10^{-7} to 10^{-8} for a successful titration, an endpoint break is obtained for titrating the second proton. Triprotic acids can be titrated similarly, but K_{a3} is generally too small to obtain an end point break for titration of the third proton. Figure 8.12 illustrates the titration curve for a diprotic acid H_2A. The pH at the beginning of the titration is determined from the ionization of the first proton if the solution is not too dilute (see discussion of polyprotic acids in Chapter 5). During titration up to the first equivalent point, an HA^-/H_2A buffer region is established. At the first equivalence point, a solution of HA^- exists, and $[H^+] \approx \sqrt{K_{a1}K_{a2}}$. Beyond this point, an A^{2-}/HA^- buffer exists; and finally at the second equivalence point, the pH is determined from the hydrolysis of A^{2-}.

The titration of polyprotic acids is similar to the titration of mixtures of acids discussed below.

FIGURE 8.12
Titration of a diprotic acid, H_2A, with sodium hydroxide.

8.7 MIXTURES OF ACIDS OR BASES

Mixtures of acids (or bases) can be titrated stepwise if there is an appreciable difference in their strengths. **There must generally be a difference in K_a values of at least 10^4,** unless perhaps a pH meter is used to construct the titration curve. If one of the acids is a strong acid, a separate end point will be observed for the strong acid *only if K_a is about 10^{-5} or smaller.* See, for example, Figure 8.13, where only a small break is seen for the HCl. The stronger acid will titrate first and will give a pH break at its equivalence point. This will be followed by titration of the weaker acid and a pH break at its equivalence point. The titration curve for a mixture of hydrochloric acid and acetic acid versus sodium hydroxide is shown in Figure 8.13. At the equivalence point for HCl, a solution of HOAc and NaCl remains, and so the equivalence point is acidic, as in Figure 8.13. Beyond the equivalence point, the OAc$^-$/HOAc buffer region is established, and this markedly suppresses the pH break for HCl. The remainder of the titration curve is identical to Figure 8.5 for the titration of HOAc.

If two strong acids are titrated together, there will be no differentiation between them, and only one equivalence point break will occur, corresponding to the titration of both acids. The same is true for two weak acids if their K_a values are not too different. For example, a mixture of acetic acid, $K_a = 1.75 \times 10^{-5}$, and propionic acid, $K_a = 1.3 \times 10^{-5}$, would titrate together to give a single equivalence point.

With H_2SO_4, the first proton is completely dissociated and the second proton has a K_a of about 10^{-2}. Therefore, the second proton is ionized sufficiently to titrate as a strong acid, and only one equivalence point break is found. The same is true for a mixture of a strong acid and a weak acid with a K_a in the neighborhood of 10^{-2}.

The first ionization constant of sulfurous acid, H_2SO_3, is 1.3×10^{-2}, and the second ionization constant is 5×10^{-6}. Therefore, in a mixture with HCl, the first proton of H_2SO_3 would titrate along with the HCl, and the pH at the equivalence

FIGURE 8.13
Titration curve for 50 mL of a mixture of 0.1 M HCl and 0.2 M HOAc with 0.1 M NaOH

point would be determined by the HSO_3^- remaining; that is, $[H^+] = \sqrt{K_{a1}K_{a2}}$, since HSO_3^- is both an acid and a base. This would be followed by titration of the second proton to give a second equivalence point. The volume of titrant required to reach the first end point would always be greater than that in going from the first to the second, since the first includes the titration of both acids. The amount of H_2SO_3 could be determined from the amount of base required for the titration of the second proton. The amount of HCl could be found by subtracting from the first end point the volume of base required to titrate the second proton of H_2SO_3, which is equal to the volume required to titrate the first proton. In practice, this titration actually would find little practical use, because H_2SO_3 is volatilized as SO_2 in strong acid solution.

Phosphoric acid in mixture with a strong acid acts in a manner similar to the above example. The first proton titrates with the strong acid, followed by titration of the second proton to give a second equivalence point; the third proton is too weakly ionized to be titrated.

EXAMPLE 8.3 A mixture of HCl and H_3PO_4 is titrated with 0.1000 M NaOH. The first end point (methyl red) occurs at 35.00 mL, and the second end point (bromothymol blue) occurs at a total of 50.00 mL (15.00 mL after the first end point). Calculate the millimoles HCl and H_3PO_4 present in the solution.

SOLUTION
The second end point corresponds to that in the titrating of one proton of H_3PO_4 ($H_2PO_4^- \rightarrow HPO_4^-$). Therefore, the millimoles H_3PO_4 is the same as the millimoles NaOH used in the 15.00 mL for titrating that proton.

$$mmol_{H_3PO_4} = M_{NaOH} \times mL_{NaOH} = 0.1000 \text{ mmol/mL} \times 15.00 \text{ mL}$$
$$= 1.500 \text{ mmol}$$

The HCl and the first proton of H_3PO_4 titrate together. Fifteen milliliters base was used to titrate the first proton of H_3PO_4 (same as for the second proton), leaving 20.00 mL used to titrate the HCl. Therefore,

$$mmol_{HCl} = 0.1000 \text{ mmol/mL} \times (35.00 - 15.00) \text{ mL} = 2.000 \text{ mmol}$$

Similarly, mixtures of bases can be titrated if their strengths are sufficiently different. Again, the difference in K_b values must be at least 10^4. Also, if one base is a strong base, the weak base must have a K_b no greater than about 10^{-5} to obtain separate end points. For example, sodium hydroxide does not give a separate end point from that for the titration of CO_3^{2-} to HCO_3^- when titrated in the presence of Na_2CO_3 ($K_{b1} = 2.1 \times 10^{-4}$).

8.8 TITRATION OF AMINO ACIDS

Amino acids are important in pharmaceutical chemistry and biochemistry. These are amphoteric substances that contain both acidic and basic groups (i.e., they can act as acids or bases). The acid group is a carboxylic acid group ($-CO_2H$), and the basic group is an amine group ($-NH_2$). In aqueous solutions, these substances tend to undergo internal proton transfer from the carboxylic acid group to the amino group because the RNH_2 is a stronger base than RCO_2^-. The result is a **zwitterion:**

$$R-CH-CO_2^-$$
$$|$$
$$NH_3^+$$

Since they are amphoteric, these substances can be titrated with either a strong acid or a strong base. Many amino acids are too weak to be titrated in aqueous solutions, but some will give adequate end points, especially if a pH meter is used to construct a titration curve.

We can consider the conjugate acid of the zwitterion as a *diprotic acid* which ionizes stepwise:

$$\underset{\substack{\text{conjugate acid} \\ \text{of zwitterion}}}{R-\underset{NH_3^+}{\overset{|}{C}}H-CO_2H} \overset{K_{a1}}{\rightleftharpoons} H^+ + \underset{\text{zwitterion}}{R-\underset{NH_3^+}{\overset{|}{C}}H-CO_2^-} \overset{K_{a2}}{\rightleftharpoons} H^+ + \underset{\substack{\text{conjugate base} \\ \text{of zwitterion}}}{R-\underset{NH_2}{\overset{|}{C}}H-CO_2^-}$$

$$(8.10)$$

K_{a1} and K_{a2} values are frequently tabulated for amino acids (see Table D.1 in the Appendix). The values listed represent the successive ionization of the protonated form (i.e., the conjugate acid of the zwitterion; it ionizes to give first the amphoteric zwitterion and second to give the conjugate base which is the same as a salt of a weak acid that hydrolyzes. Acid–base equilibria of amino acids are therefore treated just as for any other diprotic acid. The hydrogen ion concentration of the zwitterion is calculated in the same way as for any amphoteric salt, such as HCO_3^-, as we described in Chapter 5; that is,

$$[H^+] = \sqrt{K_{a1}K_{a2}} \tag{8.11}$$

The titration of amino acids is not unlike the titration of other amphoteric substances, such as HCO_3^-. In the latter case, titration with a base gives CO_3^{2-}, with an intermediate CO_3^{2-}/HCO_3^- buffer region, and titration with an acid gives H_2CO_3, with an intermediate HCO_3^-/H_2CO_3 buffer region.

When the zwitterion of an amino acid is titrated with strong acid, a buffer region is first established, consisting of the zwitterion (the "salt") and the conjugate acid. Halfway to the equivalence point, $pH = pK_{a1}$ (just as with HCO_3^-/H_2CO_3); and at the equivalence point, the pH is determined by the conjugate acid (and K_{a1}, as with H_2CO_3). When the zwitterion is titrated with a strong base, a buffer region of conjugate base (the "salt") and zwitterion (now the "acid") is established. Halfway to the equivalence point, $pH = pK_{a2}$ (as with CO_3^{2-}/HCO_3^-); and at the equivalence point, the pH is determined by the conjugate base (whose $K_b = K_{a2}/K_w$, as with CO_3^{2-}).

Amino acids may contain more than one carboxyl or amine group; in these cases, they may yield stepwise end points like other poylprotic acids (or bases), provided the different groups differ in K's by 10^4 and are still strong enough to be titrated.

8.9 KJELDAHL ANALYSIS

An important method for accurately determining nitrogen in protein and other nitrogen-containing compounds is the **Kjeldahl analysis.** The quantity of protein can be calculated from a knowledge of the percent nitrogen contained in it. Although other more rapid methods for determining proteins exist, the Kjeldahl method is the standard on which all other methods are based.

The material is digested with sulfuric acid to decompose it and convert the nitrogen to ammonium hydrogen sulfate:

$$C_aH_bN_c \xrightarrow[\text{catalyst}]{H_2SO_4} aCO_2 \uparrow + \tfrac{1}{2}bH_2O + cNH_4HSO_4$$

The solution is cooled, concentrated alkali is added to make the solution alkaline, and the volatile ammonia is distilled into a solution of standard acid, which is in excess. Following distillation, the excess acid is back-titrated with standard base.

$$cNH_4HSO_4 \xrightarrow{OH^-} cNH_3 \uparrow + cSO_4^{2-}$$

$$cNH_3 + (c + d)HCl \longrightarrow cNH_4Cl + dHCl$$

$$dHCl + dNaOH \longrightarrow \tfrac{1}{2}dH_2O + dNaCl$$

> mmol N(c) = mmol reacted HCl = mmol HCl taken
>
> $(c + d)$ − mmol NaOH(d)
>
> mmol $C_aH_bN_c$ = mmol N × 1/c

The digestion is speeded up by adding potassium sulfate to increase the boiling point and by a catalyst such as a selenium, mercury, or copper salt. The amount of the nitrogen-containing compound is calculated from the weight of nitrogen analyzed by multiplying it by the gravimetric factor.

EXAMPLE 8.4   A 0.2000-g sample containing urea,

$$O$$
$$\|$$
$$NH_2\!-\!C\!-\!NH_2$$

is analyzed by the Kjedahl method. The ammonia is collected in 50.00 mL of 0.05000 M H_2SO_4, and the excess acid is back-titrated with 0.05000 M NaOH, a procedure requiring 3.40 mL. Calculate the percent urea in the sample.

SOLUTION
The titration reaction is

$$H_2SO_4 + 2NaOH \rightarrow Na_2SO_4 + 2H_2O$$

So there are one-half as many millimoles excess H_2SO_4 as NaOH that reacted with it. The reaction with the NH_3 is

$$2NH_3 + H_2SO_4 \rightarrow (NH_4)_2SO_4$$

There are twice as many millimoles NH_3 as H_2SO_4 that reacted with it. And there are one-half as many millimoles urea as NH_3. Therefore,

% urea $= \{[\text{mmol}_{H_2SO_4} - \text{mmol}_{NaOH} \times \frac{1}{2}(\text{mmol } H_2SO_4/\text{mmol NaOH})]$

$\times\ 2(\text{mmol } NH_3/\text{mmol } H_2SO_4) \times \frac{1}{2}[\text{mmol } (NH_2)_2CO/\text{mmol } NH_3]$

$\times\ (NH_2)_2CO \text{ mg/mmol urea}\}/\text{mg}_{sample} \times 100\%$

$$= \frac{(0.0500\ M \times 50.00\ \text{mL} - 0.0500\ M \times 3.40\ \text{mL} \times \frac{1}{2}) \times 2 \times \frac{1}{2} \times 60.05}{200}$$

$\times\ 100\% = 72.51\%$

Instead of first calculating the millimoles of NH_3, we could have calculated the weight of $(NH_2)_2CO$ directly by multiplying the millimoles of reacted H_2SO_4 by the molecular weight of urea, 60.05; the number of millimoles of $(NH_2)_2CO$ is equal to the number of millimoles of H_2SO_4.

A large number of different **proteins** contain very nearly the same percentage of nitrogen. The gravimetric factor for conversion of weight of N to weight of protein for normal mixtures of serum proteins (globulins and albumin) and protein in feeds is 6.25, (i.e., the proteins contain 16% nitrogen). When the sample is made up almost entirely of gamma globulin, the factor 6.24 is more accurate, while if it contains mostly albumin, 6.27 is preferred.

In the conventional Kjeldahl method, two standard solutions are required, the acid for collecting the ammonia and the base for back-titration. A modification can be employed that requires only standard acid for direct titration. The ammonia is collected in a solution of boric acid. In the distillation, an equivalent amount of ammonium borate is formed:

$$NH_3 + H_3BO_3 \rightleftharpoons NH_4^+ + H_2BO_3^- \tag{8.12}$$

Boric acid is too weak to be titrated, but the borate, which is equivalent to the amount of ammonia, is a fairly strong Brønsted base that can be titrated with a standard acid to a methyl red end point. The boric acid is so weak it does not interfere, and its concentration need not be known accurately.

EXAMPLE 8.5 A 0.300-g feed sample is analyzed for its protein content by the modified Kjeldahl method. If 25.0 mL of 0.100 M HCl is required for titration, what is the percent protein in the sample?

SOLUTION
Since this is a direct titration with HCl which reacts 1:1 with NH_3, the millimoles of NH_3 (and therefore of N) equals the millimoles of HCl. Multiplication by 6.25 gives the milligrams of protein.

$$\% \text{ protein} = \frac{0.10 \text{ mmol/mL} \times 25.0 \text{ mL} \times 14.01 \text{ mg N/mmol} \times 6.25 \text{ mg protein/mgN}}{300 \text{ mg}}$$

$$\times 100\% = 73.0\%$$

The boric acid method (a direct method) is simpler and is usually more accurate, since it requires the standardization and accurate measurement of only one solution. However, the end point break is not so sharp, and the indirect method requiring back-titration is usually preferred for **micro-Kjeldahl analysis.** A macro-Kjeldahl analysis of blood requires about 5 mL blood, while a micro-Kjeldahl analysis requires about 0.1 mL.

We have confined our discussion to those substances in which the nitrogen exists in the 3− valence state, as in ammonia. Such compounds include amines and amides. Compounds containing oxidized forms of nitrogen, such as organic nitro and azo compounds, must be treated with a reducing agent prior to digestion in order to obtain complete conversion to ammonium ion. Reducing agents such as iron(II) or thiosulfate are used.

8.10 PREPARATION OF STANDARD BASE SOLUTIONS

Sodium hydroxide is usually used as the titrant when a base is required. It contains significant amounts of water and sodium carbonate, and so it cannot be used as a primary standard. For accurate work, the sodium carbonate must be removed from the NaOH, because it reacts to form a buffer that decreases the sharpness of the end point. In addition, an error will result if the NaOH is standardized using a phenolphthalein end point (in which case the CO_3^{2-} is titrated only to HCO_3^-), and then a methyl orange end point is used in the titration of a sample (in which case the CO_3^{2-} is titrated to CO_2). In other words, the effective molarity of the base has been increased, owing to further reaction of the HCO_3^-.

Sodium carbonate is essentially insoluble in nearly saturated sodium hydroxide. It is conveniently removed by dissolving the weighed NaOH in a volume (milliliters) of water equal to its weight in grams. The insoluble Na_2CO_3 can be allowed to settle for several days, and then the clear supernatant liquid can be carefully decanted,[1] or it can be filtered in a Gooch crucible with an asbestos mat (do not

---

[1]The solution must be kept in a nonglass-stoppered test tube or other appropriate vessel to keep out atmospheric carbon dioxide, which would continue to react with the sodium hydroxide to produce sodium carbonate. Use a rubber stopper since concentrated alkali causes glass joints to "freeze."

wash the filtered Na_2CO_3). The former procedure is preferred because of the carcinogenic nature of asbestos. This procedure does not work with KOH because K_2CO_3 remains soluble.

Water dissolves CO_2 from the air. In many routine determinations not requiring the highest degree of accuracy, carbonate or CO_2 impurities in the water will result in an error that is small enough to be considered negligible. For the highest accuracy, however, CO_2 should be removed from all water used to prepare solutions for acid–base titrations, particularly the alkaline solutions. This is conveniently done by boiling and then cooling under the cold-water tap.

Sodium hydroxide is usually standardized by titrating a weighed quantity of primary standard potassium acid phthalate (KHP), which is a moderately weak acid ($K_a = 4 \times 10^{-6}$), approximately like acetic acid; a phenolphthalein end point is used. The sodium hydroxide solution should be stored in a rubber-stoppered Pyrex bottle to prevent absorption of CO_2 from the air and "freezing" of glass to glass; or else plastic bottles should be used. If the bottle must be open (e.g., a siphon bottle), the opening is protected with an **Ascarite** (asbestos impregnated with NaOH) tube.

8.11   PREPARATION OF STANDARD ACID SOLUTIONS

Hydrochloric acid is the usual titrant for the titration of bases. Most chlorides are soluble, and few side reactions are possible with this acid. It is convenient to handle. It is not a primary standard (although constant-boiling HCl, which is a primary standard, can be prepared), and an approximate concentration is prepared simply by diluting the concentrated acid. For most accurate work, the water used to prepare the solution should be boiled, although use of boiled water is not so critical as with NaOH; CO_2 will have a low solubility in strongly acidic solutions and will tend to escape during shaking of the solution.

Primary standard sodium carbonate is usually used to standardize HCl solutions. Its disadvantage is that the end point is not sharp unless methyl red, methyl purple, and so forth, is used as the indicator and the solution is boiled at the end point. A modified methyl orange end point may be used without boiling, but this is not so sharp. Another disadvantage is the low formula weight of Na_2CO_3. Tris-(hydroxymethyl)aminomethane (THAM), $(HOCH_2)_3CNH_2$, is another primary standard that is more convenient to use. It is nonhygroscopic, but it is still a fairly weak base ($K_b = 1.3 \times 10^{-6}$) with a low molecular weight. The end point is not complicated by released CO_2, and it is recommended as the primary standard unless the HCl is being used to titrate carbonate samples.

If a standardized NaOH solution is available, the HCl can be standardized by titrating an aliquot with the NaOH. The end point is sharp and the titration is more rapid. The NaOH solution is a **secondary standard.** Any error in standardizing this will be reflected in the accuracy of the HCl solution. The HCl is titrated with the base, rather than the other way around, to minimize absorption of CO_2 in the titration flask. Phenolphthalein or bromothymol blue can be used as indicator.

Acid–base titrations are used to determine the bicarbonate content of blood (see Experiment 8 and Reference 3 at the end of the chapter). An excess of standard HCl solution is added to volatilize the bicarbonate as CO_2, and then the excess is back-titrated with standard base. They are also used in the micro-Kjeldahl determination of blood protein nitrogen and nonprotein nitrogen (see Reference 1).

8.12 NONAQUEOUS SOLVENTS

Nonaqueous solvents are useful for the titration of very weak acids or bases that cannot be titrated in water. The Brønsted theory (Chapter 5) can be applied to nonaqueous solvents. Some acid–base reactions in different solvents are illustrated in Table 8.2. In the first example, acetate ion is the conjugate base of acetic acid, and ammonium ion is the conjugate acid of ammonia. The first four examples represent ionization of an acid or a base in a solvent, while the others represent a neutralization reaction between an acid and a base in the solvent.

A substance cannot act as an acid unless a base is present to accept the protons. Thus, acids will ionize in basic solvents such as water, liquid ammonia, or ethanol. But in neutral or "inert" solvents, ionization will not occur. However, ionization in the solvent is *not* a prerequisite for an acid–base reaction, as in the last example in the table, where picric acid reacts with aniline. Some of the most successful nonaqueous titrations are done in so-called inert solvents. These aprotic solvents (see below) allow the existence of a wide range of acid–base strengths.

Solvents may in principle be classified into three groups:

1. **Amphiprotic,** those which possess both acidic and basic properties, like water, ethanol, and methanol. These are **ionizable** solvents.
2. **Aprotic,** those that are neither appreciably acidic nor basic, the "inert" solvents, like benzene and carbon tetrachloride.
3. **Basic but not acidic**—nonionizable—for example, ether, dioxane, ketones, and pyridine.[2] Most of these are extremely weak bases. There are no known examples of solvents that are acidic but not basic.

Amphiprotic solvents undergo self-ionization, or **autoprotolysis:**

$$2HSolv = H_2Solv^+ + Solv^- \tag{8.13}$$

TABLE 8.2
Brønsted Acid–Base Reactions

Solvent	Acid 1	+	Base 2	→	Acid 2	+	Base 1
NH_3 (liq.)	HOAc		NH_3		NH_4^+		OAc^-
H_2O	HCl		H_2O		H_3O^+		Cl^-
H_2O	NH_4^+		H_2O		H_3O^+		NH_3
H_2O	H_2O		OAc^-		HOAc		OH^-
H_2O	HCO_3^-		OH^-		H_2O		CO_3^{2-}
C_2H_5OH	NH_4^+		$C_2H_5O^-$		C_2H_5OH		NH_3
C_6H_6	H Picrate		$C_6H_5NH_2$		$C_6H_5NH_3^+$		Picrate$^-$

---

[2]These solvents contain an atom or atoms with a pair of free electrons (oxygen or nitrogen) that will accept a proton and are therefore **Lewis bases,** for example, pyridine:

where Hsolv represents the (hydrogen-containing) solvent. The **autoprotolysis constant** is

$$k_s = [H_2Solv^+][Solv^-] \qquad (8.14)$$

Amphiprotic solvents may have acid–base properties comparable to those of water, that is, they are about equally acidic or basic (e.g., ethanol: $2C_2H_5OH \rightleftharpoons C_2H_5OH_2^+ + C_2H_5O^-$). Or they may be much stronger acids and, therefore, much weaker bases than water (e.g., glacial acetic acid: $2HOAc \rightleftharpoons H_2OAc^+ + OAC^-$). And there are those that are much stronger bases and, therefore, much weaker acids than water (e.g., liquid ammonia: $2NH_3 \rightleftharpoons NH_4^+ + NH_2^-$).

  The strongest acid that can exist in an amphiprotic solvent is the cation of the solvent, and the strongest base that can exist is the anion of the solvent. Thus, the strongest acid that can exist in water is the hydronium ion, H_3O^+, and the strongest base is the hydroxyl ion, OH^-. Any strong acid that is completely ionized[3] in water is "leveled" to the strength of the hydronium ion by the reaction

$$H^+ + H_2O \rightarrow H_3O^+$$

Thus, the mineral acids $HClO_4$, HCl, and HNO_3 all transfer their protons completely to water (due to the relative basicity of water) and are all leveled to the same strength in water. This is so even though perchloric acid is inherently a stronger acid than the others. This phenomenon is called the **leveling effect.** Acids that are leveled to the same strength cannot be differentiated.

  In glacial acetic acid, the strongest acid that can exist is H_2OAc^+, but the mineral acids are not completely ionized in this acidic solvent and their strengths can be differentiated. On the other hand, acetic acid is even a stronger leveling solvent than water is for bases. For example, both aliphatic and aromatic amines are completely ionized[4] in acetic acid and are leveled to the strength of the acetate ion, the strongest base that can exist in this solvent:

$$RNH_2 + HOAc \rightarrow RNH_3^+ + OAc^-$$

This reaction proceeds quantitatively to the right. Thus, these amines are "strong" bases in acetic acid while they are weak bases in water. In water, which is a weaker leveling solvent for bases than is acetic acid, amines are only partially leveled (partially ionized). Methylamine (CH_3NH_2) is more than 10^4 times stronger as a base than is the aromatic amine aniline ($C_6H_5NH_2$) in water. The former can be titrated but the latter cannot.

  On the other end of the scale, strongly basic amphiprotic solvents, such as ethylenediamine, will exhibit a stronger leveling effect for weak acids (and make them "stronger" acids) than does water, but they will be more differentiating for strong bases. Benzoic acid is a weak acid in water, but in ethylenediamine its ionization (leveling) is complete:

$$C_6H_5CO_2H + H_2NCH_2CH_2NH_2 \rightleftharpoons C_6H_5CO_2^- + H_2NCH_2CH_2NH_3^+$$

---

[3]The ionization constant of an acid or base in a solvent is not a measure of its intrinsic acidity or basicity, because the basicity or acidity of the **solvent** also determines the extent of ionization; the K_a of an acid in a given solvent can be considered a measure of the ability of the acid to transfer its proton to the solvent.

[4]They actually exist as *ion pairs* rather than completely dissociated ions, because of the low dielectric constant of acetic acid.

In **inert solvents,** an acid (or base) will not ionize unless there is a base (or acid) present to react with it. But because these solvents do not possess the acidic or basic properties of water, weaker acids and bases can be titrated in them (the solvent does not compete for the titrant).

8.13   TITRATIONS IN NONAQUEOUS SOLVENTS

Acids and bases with ionization constants less than about 10^{-7} to 10^{-8} are too weak to be titrated accurately in aqueous solutions using conventional equivalence point detection, due to the acidity and basicity of water. By choosing a solvent less basic than water, it is possible to titrate much weaker bases. Conversely, by choosing a solvent less acidic than water, much weaker acids can be titrated. These may include basic solvents, inert (aprotic) solvents, or amphiprotic solvents.

Glacial acetic acid, an amphiprotic solvent, is widely used for the titration of weak bases such as amines. (Ionized as follows: $RNH_2 + HOAC \rightleftharpoons RNH_3^+ + OAc^-$; the OAc^- is then titrated, as OH^- is in water.) Perchloric acid is intrinsically the strongest mineral acid, so it is usually used as the titrant, dissolved in glacial acetic acid. Even this acid is only partially dissociated in glacial acetic acid—about the same extent as acetic acid in water.

Weak acids such as phenol may be titrated in ethylenediamine, an amphiprotic solvent, using tetrabutylammonium hydroxide, $(C_4H_9)_4NOH$, as the titrant. Many titrations may be carried out in ethanol using sodium ethoxide, $NaOC_2H_5$, as the titrant. This is the anion of the solvent—the strongest base.

Acids or bases of various strengths may be differentiated in inert solvents since they are not leveled in these solvents, especially if a pH meter is used to detect the end points. The inherently stronger ones will titrate first. Methylisobutyl ketone, acetonitrile, and chloroform are typical solvents.

QUESTIONS

1. What is the minimum pH change required for a sharp indicator color change at the end point? Why?
2. What criterion is used in selecting an indicator for a particular acid–base titration?
3. At what pH is the buffering capacity of a buffer the greatest?
4. Is the pH at the end point for the titration of a weak acid neutral, alkaline, or acidic? Why?
5. What would be a suitable indicator for the titration of ammonia with hydrochloric acid? Of acetic acid with sodium hydroxide?
6. Explain why boiling the solution near the end point in the titration of sodium carbonate increases the sharpness of the end point.
7. What is the approximate pK of the weakest acid or base that can be titrated in aqueous solution?
8. What must be the difference in the strengths of two acids in order to differentiate between them in titration?
9. Distinguish between a primary standard and a secondary standard.
10. What is an amphiprotic solvent? An aprotic solvent?
11. What is the leveling effect?
12. What solvent and titrant would you use for titrating phenol? Aniline, $C_6H_5NH_2$?

PROBLEMS

Standardization Calculations

13. A hydrochloric acid solution is standardized by titrating 0.4541 g of primary standard tris(hydroxymethyl)aminomethane. If 35.37 mL is required for the titration, what is the molarity of the acid?

14. A hydrochloric acid solution is standardized by titrating 0.2329 g of primary standard sodium carbonate to a methyl red end point by boiling the carbonate solution near the end point to remove carbon dioxide. If 42.87 mL acid is required for the titration, what is its molarity?

15. A sodium hydroxide solution is standardized by titrating 0.8592 g of primary standard potassium acid phthalate to a phenolphthalein end point, requiring 32.67 mL. What is the molarity of the base solution?

16. A 10.00-mL aliquot of a hydrochloric acid solution is treated with excess silver nitrate, and the silver chloride precipitate formed is determined by gravimetry. If 0.1682 g precipitate is obtained, what is the molarity of the acid?

Indicators

17. Write a Henderson–Hasselbalch equation for a weak-base indicator, B, and calculate the required pH change to go from one color of the indicator to the other. Around what pH is the transition?

Titration Curves

18. Calculate the pH at 0, 10.0, 25.0, and 30.0 mL of titrant in the titration of 50.0 mL of 0.100 M NaOH with 0.200 M HCl.

19. Calculate the pH at 0, 10.0, 25.0, 50.0, and 60.0 mL of titrant in the titration of 25.0 mL of 0.200 M HA with 0.100 M NaOH. $K_a = 2.0 \times 10^{-5}$.

20. Calculate the pH at 0, 10.0, 25.0, 50.0, and 60.0 mL of titrant in the titration of 50.0 mL of 0.100 M NH$_3$ with 0.100 M HCl.

21. Calculate the pH at 0, 25.0, 50.0, 75.0, 100, and 125% titration in the titration of both protons of the diprotic acid H$_2$A with 0.100 M NaOH, starting with 100 mL of 0.100 M H$_2$A. $K_{a1} = 1.0 \times 10^{-3}$, $K_{a2} = 1.0 \times 10^{-7}$.

22. Calculate the pH at 0, 25.0, 50.0, 75.0, 100, and 150% titration in the titration of 100 mL of 0.100 M Na$_2$HPO$_4$ with 0.100 M HCl to H$_2$PO$_4^-$.

Quantitative Determinations

23. A 0.492-g sample of KH$_2$PO$_4$ is titrated with 0.112 M NaOH, requiring 25.6 mL:

$$H_2PO_4^- + OH^- \rightarrow HPO_4^{2-} + H_2O$$

What is the percent purity of the KH$_2$PO$_4$?

24. What volume of 0.155 M H$_2$SO$_4$ is required to titrate 0.293 g of 90.0% pure LiOH?

25. An indication of the average formula weight of a fat is its saponification number, expressed as the milligrams KOH required to hydrolyze (saponify) 1 gram of the fat:

$$
\begin{array}{ll}
\text{CH}_2\text{CO}_2\text{R} & \text{CH}_2\text{OH} \\
| & | \\
\text{CHCO}_2\text{R} + 3\text{KOH} \rightarrow \text{CHOH} + 3\text{RCO}_2\text{K} \\
| & | \\
\text{CH}_2\text{CO}_2\text{R} & \text{CH}_2\text{OH}
\end{array}
$$

where R can be variable. A 1.10-g sample of butter is treated with 25.0 mL of 0.250 M KOH solution. After the saponification is complete, the unreacted KOH is back-titrated with 0.250 M HCl, requiring 9.26 mL. What is the saponification number of the fat and what is its average formula weight (assuming the butter is all fat)?

26. A sample containing the amino acid alanine, $CH_3CH(NH_2)COOH$, plus inert matter is analyzed by the Kjeldahl method. A 2.00-g sample is digested, the NH_3 is distilled and collected in 50.0 mL of 0.150 M H_2SO_4, and a volume of 9.0 mL of 0.100 M NaOH is required for back-titration. Calculate the percent alanine in the sample.

27. A 2.00-mL serum sample is analyzed for protein by the modified Kjeldahl method. The sample is digested, the ammonia is distilled into boric acid solution, and 15.0 mL of standard HCl is required for the titration of the ammonium borate. The HCl is standardized by treating 0.330 g pure $(NH_4)_2SO_4$ in the same manner. If 33.3 mL acid is required in the standardization titration, what is the concentration of protein in the serum in g% (wt/vol)?

Quantitative Determinations of Mixtures

28. A 100-mL aliquot of a solution containing HCl and H_3PO_4 is titrated with 0.200 M NaOH. The methyl red end point occurs at 25.0 mL, and the bromothymol blue end point occurs 10.0 mL later (total 35.0 mL). What are the concentrations of HCl and H_3PO_4 in the solution?

29. A 0.527-g sample of a mixture containing $NaCO_3$, $NaHCO_3$, and inert impurities is titrated with 0.109 M HCl, requiring 15.7 mL to reach the phenolphthalein end point and a total of 43.8 mL to reach the modified methyl orange end point. What is the percent each of Na_2CO_3 and $NaHCO_3$ in the mixture?

30. Sodium hydroxide and Na_2CO_3 will titrate together to a phenolphthalein end point $(OH^- \rightarrow H_2O; CO_3^{2-} \rightarrow HCO_3^-)$. A mixture of NaOH and Na_2CO_3 is titrated with 0.250 M HCl, requiring 26.2 mL for the phenolphthalein end point and an additional 15.2 mL to reach the modified methyl orange end point. How many milligrams NaOH and Na_2CO_3 are in the mixture?

31. Sodium carbonate can coexist with either NaOH or $NaHCO_3$ but not with both simultaneously, since they would react to form Na_2CO_3. Sodium hydroxide and Na_2CO_3 will titrate together to a phenolphthalein end point $(OH^- \rightarrow H_2O; CO_3^{2-} \rightarrow HCO_3^-)$. A mixture of either NaOH and Na_2CO_3 or of Na_2CO_3 and $NaHCO_3$ is titrated with HCl. The phenolphthalein end point occurs at 15.0 mL and the modified methyl orange end point occurs at 50.0 mL (35.0 mL beyond the first end point). The HCl was standardized by titrating 0.477 g Na_2CO_3, requiring 30.0 mL to reach the modified methyl orange end point. What mixture is present and how many millimoles of each constituent are present?

32. What would be the answers to Problem 31 if the second end point had occurred at 25.0 mL (10.0 mL beyond the first end point)?

33. A mixture containing only $BaCO_3$ and Li_2CO_3 weighs 0.150 g. If 25.0 mL of 0.120 M HCl is required for complete neutralization $(CO_3^{2-} \rightarrow H_2CO_3)$, what is the percent $BaCO_3$ in the sample?

34. A sample of P_2O_5 contains some H_3PO_4 impurity. A 0.405-g sample is reacted with water $(P_2O_5 + 3H_2O \rightarrow 2H_3PO_4)$, and the resulting solution is titrated with 0.250 M NaOH $(H_3PO_4 \rightarrow Na_2HPO_4)$. If 42.5 mL is required for the titration, what is the percent H_3PO_4 impurity?

RECOMMENDED REFERENCES

1. R. M. Archibald, "Nitrogen by the Kjeldahl Method," in *Standard Methods of Clinical Chemistry*, Vol. 2, D. Seligson, ed. New York: Academic Press, 1958, pp. 91–99.

2. G. Gorin, "Indicators and the Basis for their Use," *J. Chem. Educ.,* **33** (1956) 318.

3. M. E. Hodes, "Carbon Dioxide Content (Titrimetric)," in *Standard Methods of Clinical Chemistry,* Vol. 1, M. Reiner, ed. New York: Academic Press, 1953, pp. 19–22.

4. R. K. McAlpine, "Changes in pH at the Equivalence Point," *J. Chem. Educ.,* **25** (1948) 694.

chapter 9

PRECIPITATION AND COMPLEXOMETRIC TITRATIONS

Many metal ions form slightly soluble salts or slightly dissociated complexes. The formation of these can serve as the basis of accurate and convenient titrations for metal ions. Conversely, anions can be titrated with metals, such as chloride titrated with silver ion. Complexometric titrations especially are useful for the determination of a large number of metals. These titrations offer ease of determination of metal ions and anions at the millimole level. By use of appropriate masking agents and pH control, needed selectivity can often be achieved.

In this chapter, we discuss complexes of metal ions and their equilibria and titrations with the very useful complexing agent EDTA. The effects of acidity and complexation on precipitation equilibria are described, and then titration curves in precipitation titrations are presented along with the different types of indicators and their theory. You should review fundamental precipitation equilibria described in Chapter 6.

9.1 COMPLEXES: FORMATION CONSTANTS

Many cations will form complexes in solution with a variety of substances that have a pair of unshared electrons (e.g., on N, O, S atoms in the molecule) capable of satisfying the coordination number of the metal. [The metal ion is a Lewis acid (electron pair acceptor) and the complexer is a Lewis base (electron pair donor).] The number of molecules of the complexing agent, called the **ligand,** will depend on the coordination number of the metal and on the number of complexing groups on the ligand molecule.

Ammonia is a simple complexing agent with one pair of unshared electrons that will complex silver ion. Two ammonia molecules complex with each silver ion in a stepwise fashion, and we can write an equilibrium constant for each step, called the **formation constant K_f:**

$$Ag^+ + NH_3 \rightleftharpoons Ag(NH_3)^+ \quad K_{f1} = \frac{[Ag(NH_3)^+]}{[Ag^+][NH_3]}$$

$$= 2.5 \times 10^3 \qquad (9.1)$$

$$Ag(NH_3)^+ + NH_3 \rightleftharpoons Ag(NH_3)_2^+ \quad K_{f2} = \frac{[Ag(NH_3)_2^+]}{[Ag(NH_3)^+][NH_3]} \qquad (9.2)$$

$$= 1.0 \times 10^4$$

The overall reaction is the sum of the two steps and the overall formation constant is the product of the stepwise formation constants:

$$Ag^+ + 2NH_3 = Ag(NH_3)_2^+ \quad \left(K_f\right) = \left(K_{f1} \cdot K_{f2}\right) = \frac{[Ag(NH_3)_2^+]}{[Ag^+][NH_3]^2} \qquad (9.3)$$

$$= 2.5 \times 10^7$$

Note that the products of the reactions are written in the numerators of the equilibrium constant expressions in the usual manner, even though we wrote the reactions as associations rather than dissociations. The formation constant is also called the **stability constant K_s, or K_{stab}.**

We could write the equilibria in the opposite direction, as dissociations. If we do this, the concentration terms are inverted in the equilibrium constant expressions. The equilibrium constants then are simply the reciprocals of the formation constants, and they are called **instability constants K_i, or dissociation constants K_d:**

$$Ag(NH_3)_2^+ \rightleftharpoons Ag^+ + 2NH_3 \quad K_d = \frac{1}{K_f} = \frac{[Ag^+][NH_3]^2}{[Ag(NH_3)_2^+]} \qquad (9.4)$$

Either constant can be used in calculations, as long as it is used with the proper reaction and the correct expression.

EXAMPLE 9.1 A divalent metal M^{2+} reacts with a ligand L to form a 1:1 complex:

$$M^{2+} + L \rightleftharpoons ML^{2+} \quad K_f = \frac{[ML^{2+}]}{[M^{2+}][L]}$$

Calculate the concentration of M^{2+} in a solution prepared by mixing equal volumes of 0.20 M M^{2+} and 0.20 M L. $K_f = 1.0 \times 10^8$.

SOLUTION

We have added stoichiometrically equal amounts of M^{2+} and L. The complex is sufficiently strong that their reaction is virtually complete. Since we added equal volumes, the initial concentrations were halved. Let x represent $[M^{2+}]$. At equilibrium, we have

$$M^{2+} + L \rightleftharpoons \quad ML^{2+}$$
$$x \qquad x \qquad 0.10 - x \approx 0.10$$

Essentially, all the M^{2+} (original concentration 0.20 M) was converted to an equal amount of ML^{2+}, with only a small amount of uncomplexed metal remaining. Substituting into the K_f expression,

$$\frac{0.10}{(x)(x)} = 1.0 \times 10^8$$

$$x = [M^{2+}] = 3.2 \times 10^{-5} \, M$$

EXAMPLE 9.2 Silver ion forms a stable 1:1 complex with triethylenetetramine, called "trien" $[NH_2(CH_2)_2 NH(CH_2)_2 NH(CH_2)_2 NH_2]$. Calculate the silver ion concentration at equilibrium when 25 mL of 0.010 M silver nitrate is added to 50 mL of 0.015 M trien. $K_f = 5.0 \times 10^7$.

SOLUTION

$$Ag^+ + trien \rightleftharpoons Ag(trien)^+ \quad K_f = \frac{[Ag(trien)^+]}{[Ag^+][trien]}$$

Calculate the millimoles Ag^+ and trien added:

$$mmol \, Ag^+ = 25 \, mL \times 0.010 \, mmol/mL = 0.25 \, mmol$$

$$mmol \, trien = 50 \, mL \times 0.015 \, mmol/mL = 0.75 \, mmol$$

The equilibrium lies far to the right, so assume that virtually all the Ag^+ reacts with 0.25 mmol of the trien (leaving 0.50 mmol trien in excess) to form 0.25 mmol complex. Calculate the molar concentrations:

$$[Ag^+] = x = mol/L \, unreacted$$

$$[trien] = (0.50 \, mmol/75 \, mL) + x = 6.7 \times 10^{-3} + x$$

$$\approx 6.7 \times 10^{-3}$$

$$[Ag(trien)^+] = 0.25 \, mmol/75 \, mL - x$$

$$= 3.3 \times 10^{-3} - x \approx 3.3 \times 10^{-3}$$

Try neglecting x compared to the other concentrations:

$$\frac{(3.3 \times 10^{-3})}{(x)(6.7 \times 10^{-3})} = 5.0 \times 10^7$$

$$x = [Ag^+] = 9.8 \times 10^{-9} \, M$$

We were justified in neglecting x.

9.2 CHELATES: EDTA

Simple complexing agents such as ammonia are rarely used as titrating agents because a sharp end point corresponding to a stoichiometric complex is generally difficult to achieve. This is because the stepwise formation constants are frequently close together and not very large, and a single stoichiometric complex cannot be observed. Certain complexing agents that have two or more complexing groups on the molecule, however, do form well-defined complexes and can be used as titrating agents. One such reagent that is widely used is ethylenediaminetetraacetic acid (EDTA).

An organic agent that has two or more groups capable of complexing with a metal ion is called a **chelating agent.** The complex formed is called a **chelate.** The chelating agent is called the *ligand.* Titration with a chelating agent is called a **chelometric titration,** a type of complexometric titration. The formula for EDTA is

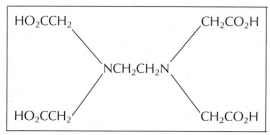

A pair of unshared electrons capable of complexing with a metal ion is contained on each of the two nitrogens and each of the four carboxyl groups. Thus, there are six complexing groups in EDTA. We will represent EDTA by the symbol H_4Y. It is a tetraprotic acid, and the hydrogens in H_4Y refer to the four ionizable hydrogens on the four carboxyl groups. It is the unprotonated ligand Y^{4-} that forms complexes with metal ions.

EDTA Equilibria

EDTA can be represented as having four K_a values corresponding to the stepwise dissociation of the four protons[1]:

$$H_4Y \rightleftharpoons H^+ + H_3Y^- \quad K_{a1} = 1.0 \times 10^{-2} = \frac{[H^+][H_3Y^-]}{[H_4Y]} \tag{9.5}$$

$$H_3Y^- \rightleftharpoons H^+ + H_2Y^{2-} \quad K_{a2} = 2.2 \times 10^{-3} = \frac{[H^+][H_2Y^{2-}]}{[H_3Y^-]} \tag{9.6}$$

$$H_2Y^{2-} \rightleftharpoons H^+ + HY^{3-} \quad K_{a3} = 6.9 \times 10^{-7} = \frac{[H^+][HY^{3-}]}{[H_2Y^{2-}]} \tag{9.7}$$

$$HY^{3-} \rightleftharpoons H^+ + Y^{4-} \quad K_{a4} = 5.5 \times 10^{-11} = \frac{[H^+][Y^{4-}]}{[HY^{3-}]} \tag{9.8}$$

Polyprotic acid equilibria are treated in Section 5.8 in Chapter 5, and you should review this section before proceeding with the following discussion.

The fraction of each form of EDTA as a function of pH is shown in Figure 9.1. Since the anion Y^{4-} is the ligand species in complex formation, the complexation equilibria are affected markedly by the pH. H_4Y has a very low solubility in water, and so the disodium salt $Na_2H_2Y \cdot 2H_2O$ is usually used, in which two of the acid groups are neutralized. This dissociates in solution to give predominantly H_2Y^{2-};

---

[1] Actually, the nitrogens on the EDTA molecule can protonate weakly, and so there are in reality six dissociation steps and six K_a values, the first two being 1.0 and 0.032 for dissociation from the nitrogen atoms. We will use the simplified version above in our discussions. This protonation does affect the solubility of EDTA in acid.

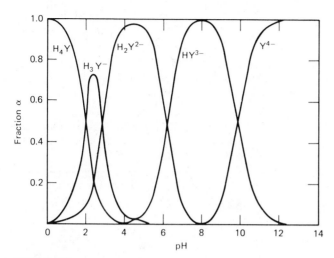

FIGURE 9.1
Fraction of EDTA species as a function of pH.

the pH of the solution is approximately 4 to 5 (theoretically 4.4 from $[H^+] = \sqrt{K_{a2}K_{a3}}$).

The Formation Constant

Consider the formation of the EDTA chelate of Ca^{2+}. This can be represented by

$$Ca^{2+} + Y^{4-} \rightleftharpoons CaY^{2-} \tag{9.9}$$

The formation constant for this is

$$K_f = \frac{[CaY^{2-}]}{[Ca^{2+}][Y^{4-}]} \tag{9.10}$$

The value of some representative EDTA formation constants are given in the Appendix.

Effect of pH on EDTA Equilibria

The equilibrium in Equation 9.9 is shifted to the left as the hydrogen ion concentration is increased, due to competition for the chelating anion by hydrogen ion. From a knowledge of the pH and the equilibria involved, Equation 9.10 can be used to calculate the concentration of free Ca^{2+} under various solution conditions (e.g., in a titration to interpret a titration curve). The Y^{4-} concentration is calculated at different pH values as follows (see Chapter 5, polyprotic acids). If we let C_{H_4Y} represent the total concentration of all forms of uncomplexed EDTA, Then

$$C_{H_4Y} = [Y^{4-}] + [HY^{3-}] + [H_2Y^{2-}] + [H_3Y^-] + [H_4Y] \tag{9.11}$$

Solving for the equilibrium concentrations of H_4Y, H_3Y^-, H_2Y^{2-}, and HY^{3-} in Equations 9.5 through 9.8, substituting these in Equation 9.11 to eliminate all forms but Y^{4-}, and dividing by $[Y^{4-}]$ gives

$$\frac{C_{H_4Y}}{[Y^{4-}]} = \frac{1}{\alpha_4} = 1 + \frac{[H^+]}{K_{a4}} + \frac{[H^+]^2}{K_{a3}K_{a4}} + \frac{[H^+]^3}{K_{a2}K_{a3}K_{a4}} + \frac{[H^+]^4}{K_{a1}K_{a2}K_{a3}K_{a4}} \quad (9.12)$$

where α_4 is the **fraction** of the total EDTA species that exists as Y^{4-} ($\alpha_4 = [Y^{4-}]/C_{H_4Y}$). Similar equations can be derived for the fraction of each of the other EDTA species α_0, α_1, α_2, and α_3, as in Chapter 5 (this is the way Figure 9.1 was constructed).

Equation 9.12 can be used, then, to calculate the fraction of the EDTA that exists as Y^{4-} at a given pH; and from a knowledge of the concentration of uncomplexed EDTA (C_{H_4Y}), the free Ca^{2+} can be calculated using Equation 9.10.

EXAMPLE 9.3   Calculate the fraction of EDTA that exists as Y^{4-} at pH 10, and from this calculate pCa in 100 mL of solution of 0.100 M Ca^{2+} at pH 10 after adding 100 mL of 0.100 M EDTA.

SOLUTION
From Equation 9.12,

$$\frac{1}{\alpha_4} = 1 + \frac{1.0 \times 10^{10}}{5.5 \times 10^{-11}} + \frac{(1.0 \times 10^{-10})^2}{(6.9 \times 10^{-7})(5.5 \times 10^{-11})}$$

$$+ \frac{(1.0 \times 10^{-10})^3}{(2.2 \times 10^{-3})(6.9 \times 10^{-7})(5.5 \times 10^{-11})}$$

$$+ \frac{(1.0 \times 10^{-10})^4}{(1.0 \times 10^{-2})(2.2 \times 10^{-3})(6.9 \times 10^{-7})(5.5 \times 10^{-11})}$$

$$= 1 + 1.82 + 2.6 \times 10^{-4} + 1.2 \times 10^{-11} + 1.2 \times 10^{-19} = 2.82$$

$$\alpha_4 = 0.35$$

Stoichiometric amounts of Ca^{2+} and EDTA are added to produce an equivalent amount of CaY^{2-}, less the amount dissociated:

$$\text{mmol } Ca^{2+} = 0.100 \, M \times 100 \, \text{mL} = 10.0 \, \text{mmol}$$

$$\text{mmol EDTA} = 0.100 \, M \times 100 \, \text{mL} = 10.0 \, \text{mmol}$$

We have formed 10.0 mmol CaY^{2-} in 200 mL, or 0.0500 M:

$$Ca^{2+} + \text{EDTA} \rightleftharpoons \quad CaY^{2-}$$
$$x \qquad x \qquad 0.0500 \, M - x$$
$$\approx 0.0500 \, M \text{ (since } K_f \text{ is large)}$$

where x represents the total equilibrium EDTA concentration in all forms, C_{H_4Y}. $[Y^{-4}]$, needed to apply Equation 9.10, is equal to $\alpha_4 C_{H_4Y}$. Hence, we can write Equation 9.10 as

$$K_f = \frac{[CaY^{2-}]}{[Ca^{2+}]\alpha_4[C_{H_4Y}]}$$

From the Appendix, $K_f = 5.0 \times 10^{10}$. Hence,

$$5.0 \times 10^{10} = \frac{(0.0500)}{(x)(0.35)(x)}$$

$$x = 1.7 \times 10^{-6} \, M$$

$$pCa = 5.77$$

The Conditional Formation Constant

We can substitute $\alpha_4 C_{H_4Y}$ for $[Y^{4-}]$ in Equation 9.10:

$$K_f = \frac{[CaY^{2-}]}{[Ca^{2+}]\alpha_4 C_{H_4Y}}$$ (9.13)

Rearrangement yields

$$K_f\alpha_4 = K_f' = \frac{[CaY^{2-}]}{[Ca^{2+}]C_{H_4Y}}$$ (9.14)

where K_f' is the **conditional formation constant** and is dependent on α_4 and, hence, the pH. We can use this equation to calculate the equilibrium concentrations of the different species at a given pH in place of Equation 9.10.

EXAMPLE 9.4 The formation constant for CaY^{2-} is 5.0×10^{10}. At pH 10, α_4 is calculated (Example 9.3) to be 0.35 to give a conditional constant (from Equation 9.14) of 1.8 $\times 10^{10}$. Calculate pCa in 100 mL of a solution of 0.100 M Ca^{2+} at pH 10 after addition of (a) 0 mL, (b) 50 mL, (c) 100 mL, and (d) 150 mL of 0.100 M EDTA.

SOLUTION

(a)

$$pCa = -\log [Ca^{2+}] = -\log 1.00 \times 10^{-1} = 1.00$$

(b) We started with 0.100 M $\times$ 100 mL = 10.0 mmol Ca^{2+}. The millimoles of EDTA added are 0.100 M $\times$ 50 mL = 5.0 mmol. The conditional constant is large so that the reaction 9.9 will be far to the right. Hence, we can neglect the amount of Ca^{2+} from the dissociation of CaY^{2-} and the number of millimoles of free Ca^{2+} is essentially equal to the number of unreacted millimoles:

$$\text{mmol } Ca^{2+} = 10.0 - 5.0 = 5.0 \text{ mmol}$$

$$[Ca^{2+}] = 5.0 \text{ mmol}/150 \text{ mL} = 0.030 \text{ } M$$

$$pCa = -\log 3.0 \times 10^{-2} = 1.48$$

(c) At the equivalence point, we have converted all the Ca^{2+} to CaY^{2-}. We must, therefore, use Equation 9.14 to calculate the equilibrium concentration of Ca^{2+}. The number of millimoles of CaY^{2-} formed is equal to the number of millimoles of Ca^{2+} started with, and $[CaY^{2-}] = 10.0 \text{ mmol}/200 \text{ mL} = 0.0500 \text{ } M$. From the dissociation of CaY^{2-}, $[Ca^{2+}]$ $= C_{H_4Y} = x$, and the equilibrium $[CaY^{2-}] = 0.050 \text{ } M - x$. But since the dissociation is slight, we can neglect x compared to 0.050 M. Therefore, from Equation 9.14,

$$\frac{(0.050)}{(x)(x)} = 1.8 \times 10^{10}$$

$$x = 1.7 \times 10^{-6} \text{ } M = [Ca^{2+}]$$

$$pCa = -\log 1.7 \times 10^{-6} = 5.77$$

Compare this value with that calculated using K_f in Example 9.3, instead of K_f'.

(d) The concentration of C_{H_4Y} is equal to the concentration of excess EDTA added (neglecting the dissociation of CaY^{2-}, which will be even smaller in the presence of excess EDTA). The millimoles of CaY^{2-} will be as in (c). Hence,

$$[CaY^{2-}] = 10.0 \text{ mmol}/250 \text{ mL} = 0.0400 \ M$$

$$\text{mmol excess } C_{H_4Y} = 0.100 \ M \times 150 \text{ mL} - 0.100 \ M \times 100 \text{ mL} = 5.0 \text{ mmol}$$

$$C_{H_4Y} = 5.0 \text{ mmol}/250 \text{ mL} = 0.020 \ M$$

$$\frac{(0.040)}{[Ca^{2+}](0.020)} = 1.8 \times 10^{10}$$

$$[Ca^{2+}] = 1.1 \times 10^{-10} \ M$$

$$pCa = -\log 1.1 \times 10^{-10} = 9.95$$

The pH can affect stability of the complex (i.e., K_f') by affecting not only the form of the EDTA but also that of the metal ion. For example, hydroxy species may form ($M^{2+} + OH^- \rightarrow MOH^+$). That is, OH^- competes for the metal ion just as H^+ competes for the Y^{4-}. Figure 9.2 shows how K_f' changes with pH for three metal–EDTA chelates with moderate (Ca) to strong (Hg) formation constants. The calcium chelate is obviously too weak to be titrated in acid solution ($K_f' < 1$), while the mercury chelate is strong enough to be titrated in acid. At pH 13, all K_f' values are virtually equal to the K_f values because α_4 is essentially unity; that is, the EDTA is completely dissociated to Y^{4-}. The curves parallel one another because at each pH, each K_f is multiplied by the same α_4 value to obtain K_f'.

9.3   EDTA TITRATION CURVES

A titration is performed by adding the chelating agent to the sample; the reaction occurs as in Equation 9.9. The titration curve for Ca^{2+} titrated with EDTA at pH 10 is shown in Figure 9.3. Before the equivalence point, the Ca^{2+} concentration is nearly equal to the amount of unchelated (unreacted) calcium since the dissociation of the chelate is slight (analogous to the amount of an unprecipitated ion). At the equivalence point and beyond, pCa is determined from the dissociation of the

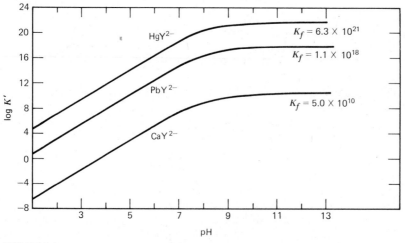

FIGURE 9.2
Effect of pH on K' values for EDTA chelates.

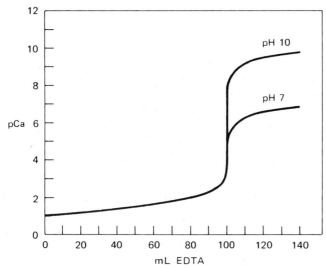

FIGURE 9.3
Titration curves for 100 mL 0.1 M Ca^{2+} versus 0.1 M Na_2EDTA at pH 7 and pH 10.

chelate at the given pH as described in Example 9.3 or 9.4, using K_f or K_f'. The effect of pH on the titration is apparent from the curve in Figure 9.3 for titration at pH 7.

The more stable the chelate (the larger K_f), the farther to the right will be the equilibrium of the reaction (Equation 9.9), and the larger will be the end point break. Also, the more stable the chelate, the lower the pH at which the titration can be performed (Figure 9.3). This is important because it allows the titration of some metals in the presence of others whose EDTA chelates are too weak to titrate at the lower pH.

Figure 9.4 shows the minimum pH at which different metals can be titrated with EDTA. The points on the curve represent the pH at which the conditional formation constant K_f' for each metal is 10^6 ($-pK_f' = 6$), which was arbitrarily chosen as the minimum needed for a sharp end point. Note that the smaller K_f, the more alkaline the solution must be to obtain a K_f' of 10^6 (i.e., the larger α_4 must be). The dashed lines in the figure divide the metals into separate groups according to their formation constants. One group is titrated in quite acid (pH < 4) solution, a second group at pH 4 to 7, and a third group at pH > 7. At the highest pH range, all the metals will titrate. At the intermediate pH range, the third group will not titrate and the second group of metals can be titrated in the presence of the third group. And finally, in the most acidic pH range, only the first group will titrate and can be determined in the presence of the others.

9.4 DETECTION OF THE END POINT: INDICATORS

We can measure the pM potentiometrically if a suitable electrode is available (see Chapter 11), but it is simpler if an indicator can be used. Indicators used for chelometric titrations are themselves chelating agents. They are usually dyes of the o,o'-dihydroxy azo type.

Eriochrome Black T is a typical indicator. It contains three ionizable protons, so

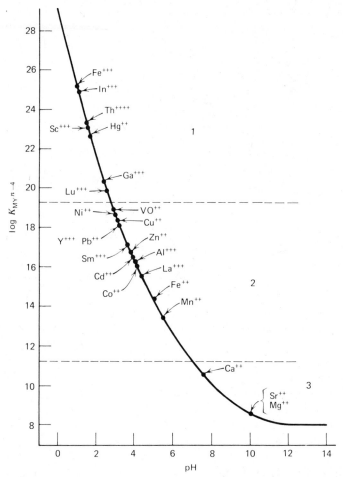

FIGURE 9.4
Minimum pH for effective titration of various metal ions with EDTA. (Reprinted with permission from C. N. Reilley and R. W. Schmid, *Anal. Chem.*, **30** (1958) 947. Copyright by the American Chemical Society.)

we will represent it by H_3In. This can be used for the titration of Mg^{2+} with EDTA. A small amount of indicator is added to the sample solution, and it forms a red complex with part of the Mg^{2+}; the color of the uncomplexed indicator is blue. As soon as all the free Mg^{2+} is titrated, the EDTA displaces the indicator from the magnesium, causing a change in the color from red to blue:

$$\boxed{\begin{array}{l} MgIn^- + H_2Y^{2-} \rightarrow MgY^{2-} + HIn^{2-} + H^+ \\ \text{(red)} \quad \text{(colorless)} \quad \text{(colorless)} \text{ (blue)} \end{array}} \qquad (9.15)$$

This will occur over a pMg range, and the change will be sharper if the indicator is kept as dilute as possible and will still give a good color.

Of course, the metal indicator complex must be less stable than that of the metal–EDTA complex, or else the EDTA will not displace it from the metal. On the other hand, it must not be too weak, or the EDTA will start replacing it at the beginning of the titration, and a diffuse end point will result. In general, the metal indicator complex should be 10 to 100 times less stable than the metal–titrant complex.

The formation constants of the EDTA complexes of calcium and magnesium are too close to differentiate between them in an EDTA titration, even by pH adjustment. So they will titrate together, and the Eriochrome Black T end point can be used as above. This titration is used to determine "total hardness" of water (Ca^{2+} plus Mg^{2+}—see Experiment 12. Eriochrome Black T cannot be used to indicate the direct titration of calcium alone with EDTA, however, because the indicator forms too weak a complex with calcium to give a sharp end point. Therefore, a small measured amount of Mg^{2+} is added to the Ca^{2+} solution; and as soon as the Ca^{2+} and the small amount of free Mg^{2+} are titrated, the end point color change occurs as above. (The Ca^{2+} titrates first since its EDTA chelate is more stable.) A correction is made for the amount of EDTA used for titration of the Mg^{2+} by performing a "blank" titration on the same amount of Mg^{2+} added to the buffer.

It is more convenient to add, instead, about 2 mL of 0.005 M Mg–EDTA instead of $MgCl_2$. This is prepared by adding together equal volumes of 0.01 M $MgCl_2$ and 0.01 M EDTA solutions and adjusting the ratio with dropwise additions until a portion of the reagent turns a dull violet when treated with pH 10 buffer and Eriochrome Black T indicator. When this is true, one drop of 0.01 M EDTA will turn the solution blue and one drop of 0.01 M EDTA will turn it red.

When Mg–EDTA is added to the sample, the Ca^{2+} in the sample displaces the EDTA from the Mg^{2+} (since the Ca–EDTA is more stable) so that the Mg^{2+} is free to react with the indicator. At the end point, an equivalent amount of EDTA displaces the indicator from the Mg^{2+}, causing the color change, and no correction is required for the added Mg–EDTA. This procedure is used in Experiment 12.

An alternative method is to add a small amount of Mg^{2+} to the EDTA solution. This immediately reacted with EDTA to form MgY^{2-} with very little free Mg^{2+} in equilibrium. This, in effect, reduces the molarity of the EDTA. So the EDTA is standardized **after** adding the Mg^{2+} by titrating primary standard calcium carbonate (dissolved in HCl and pH adjusted). When the indicator is added to the calcium solution, it is blue. But as soon as the titration is started, the indicator is complexed by the magnesium and it turns red. At the end point, it changes back to blue, as the indicator is displaced from the magnesium. No correction is required for the Mg^{2+} added, because it is accounted for in the standardization. This solution should not be used to titrate metals other than calcium.

High-purity EDTA can be prepared from $Na_2H_2Y \cdot 2H_2O$ by drying at 80°C for 2 hours. The waters of hydration remain intact.

The titration of calcium and magnesium with EDTA is done at pH 10, using an ammonia–ammonium chloride buffer. The pH must not be too high or the metal hydroxide may precipitate, causing the reaction with EDTA to be very slow. Calcium can actually be titrated in the presence of magnesium by raising the pH to 12 with strong alkali; $Mg(OH)_2$ precipitates and does not titrate.

Since Ericochrome Black T and other indicators are weak acids, their color will depend somewhat on the pH, because the different ionized species of the indicator have different colors. For example, with Eriochrome Black T, H_2In^- is red (pH < 6), HIn^{2-} is blue (pH 6 to 12), and In^{3-} is yellow orange (pH > 12). Thus, indicators will be usable over definite pH ranges. The pH also affects the stability of the complex formed between the indicator and the metal ion, as well as between EDTA and the metal ion.

An indicator is useful for indication of titrations of only those metals that form a more stable complex with the titrant than with the indicator at the given pH. This all sounds rather complex but, fortunately, suitable indicators have been described for many titrations with several different chelating agents.

Calmagite indicator gives a somewhat improved end point over Eriochrome Black T for the titration of calcium and magnesium with EDTA. It also has a longer shelf life. Xylenol orange is useful for titration of metal ions that form very strong EDTA complexes and are titrated at pH 1.5 to 3.0. Examples are the direct titration of thorium(IV) and bismuth(III), and the indirect determination of zirconium(IV) and iron(III) by back-titration with one of the former two metals. There are many other indicators for EDTA titrations. The treatise by Wilson and Wilson (Reference 3 at the end of the chapter) gives many examples of EDTA titrations, and you are referred to this excellent source for detailed descriptions of procedures for different metals.

There are a number of other useful reagents for complexometric titrations. A notable example is ethyleneglycol bis(β-aminoethyl ether)-N,N,N',N'-tetraacetic acid **(EGTA)**. This is an ether analogue of EDTA that will selectively titrate calcium in the presence of magnesium:

$$\text{HOOCCH}_2 \diagdown \atop \text{HOOCCH}_2 \diagup \text{NCH}_2\text{CH}_2\text{OCH}_2\text{CH}_2\text{OCH}_2\text{CH}_2\text{N} \diagup \atop \diagdown \text{CH}_2\text{COOH} \atop \text{CH}_2\text{COOH}$$

Complexing agents having ether linkages have a strong tendency to complex the alkaline earths heavier than magnesium. Log K_f for calcium–EGTA is 11.0 while that for magnesium–EGTA is only 5.2. For other chelating agents and their applications, see References 3 and 6 at the end of the chapter.

With the exception of the alkali metals, nearly every metal can be determined with high precision and accuracy by chelometric titration. These methods are much more rapid and convenient than gravimetric procedures and have largely replaced them, except in those few instances when greater accuracy may be offered and required.

Precipitation and complexometric titrations in the clinical laboratory are limited to those substances that occur in fairly high concentrations, since volumetric methods are generally not too sensitive. The only common precipitation titration is the titration of chloride in biological fluids with silver ion. This serves as the basis for the Cotlove chloridimeter, a coulometric titrator (Chapter 12). The chloride can also be titrated with mercury(II), but here the titration product is a slightly dissociated molecule. See Experiment 11 and Reference 5. The most important complexometric titration is the determination of calcium in blood (see Reference 1). Chelating agents such as EDTA are used in the treatment of heavy metal poisoning, for example, when children ingest chipped paint that contains lead. The calcium chelate (as Na$_2$CaY) is administered to prevent complexation and removal of calcium in the bones. Heavy metals such as lead form more stable EDTA chelates than calcium does and will displace the calcium from the EDTA. The chelated lead is excreted via the kidneys.

A table of formation constants of some EDTA chelates appears in the Appendix.

9.5 FRACTION OF DISSOCIATING SPECIES IN POLYLIGAND COMPLEXES: β-VALUES

Complexes such as $Ag(NH_3)_2^+$ dissociate stepwise, just as polyprotic acids do. As with the acids, calculation of the equilibrium concentrations of the individual complex species is difficult unless excess ligand is present. Otherwise, an iterative procedure must be used. When excess ligand is present, treatment of the equilibria is similar to the calculation of the equilibrium concentration of the various species of a polyprotic acid at a given hydrogen ion concentration. See Chapter 5, Equations 5.61 through 5.73. Suppose, for example, we wish to calculate the equilibrium concentration of the various silver species in Equations 9.1 and 9.2 at a given ammonia concentration. We can define the fractions of each species as follows:

$$\beta_0 = \frac{[Ag^+]}{C_{Ag}} \quad \beta_1 = \frac{[Ag(NH_3)^+]}{C_{Ag}}$$

$$\beta_2 = \frac{[Ag(NH_3)_2^+]}{C_{Ag}} \quad \beta_0 + \beta_1 + \beta_2 = 1$$

where C_{Ag} is the total silver ion concentration in all forms. We use β to represent the various fractions for complexes to avoid confusion with α used for the fractions of acid species. The subscript denotes the number of ligands associated with the metal ion. The total concentration of silver species is given by

$$C_{Ag} = [Ag(NH_3)_2^+] + [Ag(NH_3)^+] + [Ag^+] \tag{9.16}$$

If we wish to calculate β_0, the fraction of Ag^+, then we can use the equilibrium constant expressions in Equations 9.1 and 9.2 to substitute in Equation 9.16 to obtain an expression containing only $[Ag^+]$ as the silver species. From Equation 9.1,

$$[Ag(NH_3)^+] = K_{f1}[Ag^+][NH_3] \tag{9.17}$$

From Equations 9.2 and 9.17,

$$[Ag(NH_3)_2^+] = K_{f2}[Ag(NH_3)^+][NH_3] = K_{f1}K_{f2}[Ag^+][NH_3]^2 \tag{9.18}$$

Substituting Equations 9.17 and 9.18 in 9.16,

$$C_{Ag} = K_{f1}K_{f2}[Ag^+][NH_3]^2 + K_{f1}[Ag^+][NH_3] + [Ag^+] \tag{9.19}$$

Substituting in the denominator of the β_0 expression, we have

$$\beta_0 = \frac{1}{K_{f1}K_{f2}[NH_3]^2 + K_{f1}[NH_3] + 1} \tag{9.20}$$

We can derive similar equations for the other β's by substituting from Equations 9.1 and 9.2 to obtain expressions in terms of either $[Ag(NH_3)^+]$ for β_1 or $[Ag(NH_3)_2^+]$ for β_2. Or, since we already have an expression for the denominators of β_1 and β_2, we can substitute 9.17 and 9.18 in the numerators of β_1 and β_2, respectively. The result is

$$\beta_1 = \frac{K_{f1}[NH_3]}{K_{f1}K_{f2}[NH_3]^2 + K_{f1}[NH_3] + 1} \tag{9.21}$$

$$\beta_2 = \frac{K_{f1}K_{f2}[NH_3]^2}{K_{f1}K_{f2}[NH_3]^2 + K_{f1}[NH_3] + 1} \tag{9.22}$$

These equations may be compared with Equations 5.70 through 5.73 in Chapter 5 for α's. They are somewhat different in form because the equilibria were written as associations rather than dissociations. If dissociation constants were used in place of formation constants, the equations would be identical to the α equations for a diprotic acid, except that $[H^+]$ would be replaced by $[NH_3]$ and the K_a's would be replaced by the K_d's, and the order of β's would be reversed from the order of α's ($\beta_0 \leftrightharpoons \alpha_2$) since they were defined in reverse order.

EXAMPLE 9.5 Calculate the equilibrium concentrations of the different silver ion species for 0.010 M silver(I) in 0.10 M NH_3.

SOLUTION
From Equations 9.20 through 9.22,

$$\beta_0 = \frac{1}{(2.5 \times 10^3)(1.0 \times 10^4)(0.10)^2 + (2.5 \times 10^3)(0.10) + 1} = 4.0 \times 10^{-6}$$

Similarly,

$$\beta_1 = 1.0 \times 10^{-3}; \qquad \beta_2 = 1.0$$

$$[Ag^+] = C_{Ag}\beta_0 = (0.010)(4.0 \times 10^{-6}) = 4.0 \times 10^{-8}\ M$$

$$[Ag(NH_3)^+] = C_{Ag}\beta_1 = (0.010)(1.0 \times 10^{-3}) = 1.0 \times 10^{-5}\ M$$

$$[Ag(NH_3)_2^+] = C_{Ag}\beta_2 = (0.010)(1.0) = 0.010\ M$$

Essentially, all the silver exists as the diammine complex in 0.10 M ammonia.

We neglected in the above calculation any consumption of ammonia in forming the complexes. We see that 20% of it was consumed, as a first approximation. A recalculation of the β's at 0.08 M NH_3 still results in β_2 equal to 1.0 and most of the silver still exists as $Ag(NH_3)_2^+$. The relative values of β_0 and β_1 would change on recalculation, however. This is an **iterative procedure** or **method of successive approximations.** It can be used in any equilibrium calculation in which assumptions are made to simplify the calculations, including simple acid–base equilibria where the amount of acid dissociated is assumed negligible compared to the initial concentration. Usually, two or at most three iterative calculations are adequate.

9.6 EFFECT OF ACIDITY ON SOLUBILITY OF PRECIPITATES

Before describing precipitation titrations, we shall consider the effects of competing equilibria on the solubility of precipitates. A review of polyprotic acid equilibria and the calculation of α's, the fractions of each acid species in equilibrium at a given pH, will be helpful (Chapter 5).

The solubility of a precipitate whose anion is derived from a weak acid will be increased in the presence of added acid because the acid will tend to combine with the anion and thus remove the anion from solution. For example, the precipitate MA that partially dissolves to give M^+ and A^- ions will exhibit the following equilibria:

$$MA \rightleftharpoons M^+ + A^-$$
$$+$$
$$H^+$$
$$\upharpoonleft\downharpoonright$$
$$HA$$

The anion A^- can combine with protons to increase the solubility of the precipitates. By suitable application of the equilibrium constants for the equilibria involved, the solubility of the precipitate at a given acidity can be calculated.

Consider, for example, the solubility of CaC_2O_4 in the presence of a strong acid. The equilibria are

$$\underline{CaC_2O_4} \rightleftharpoons Ca^{2+} + C_2O_4^{2-} \quad K_{sp} = [Ca^{2+}][C_2O_4^{2-}] = 2.6 \times 10^{-9} \quad (9.23)$$

$$C_2O_4^{2-} + H^+ \rightleftharpoons HC_2O_4^- \quad K_{a2} = \frac{[H^+][C_2O_4^{2-}]}{[HC_2O_4^-]} = 6.1 \times 10^{-5} \quad (9.24)$$

$$HC_2O_4^- + H^+ \rightleftharpoons H_2C_2O_4 \quad K_{a1} = \frac{[H^+][HC_2O_4^-]}{[H_2C_2O_4]} = 6.5 \times 10^{-2} \quad (9.25)$$

The solubility s of CaC_2O_4 is equal to $[Ca^{2+}] = C_{H_2C_2O_4}$, where $C_{H_2C_2O_4}$ represents the concentrations of all the oxalate species in equilibrium ($= [H_2C_2O_4] + [HC_2O_4^-] + [C_2O_4^{2-}]$). We can substitute $C_{H_2C_2O_4}\alpha_2$ for $[C_2O_4^{2-}]$ in the K_{sp} expression:

$$\boxed{K_{sp} = [Ca^{2+}]C_{H_2C_2O_4}\alpha_2} \quad (9.26)$$

where α_2 is the fraction of the oxalate species present as $C_2O_4^{2-}$ ($\alpha_2 = [C_2O_4^{2-}]/C_{H_2C_2O_4}$). Using the approach described in Chapter 5 for H_3PO_4 to calculate α's, we find that

$$\alpha_2 = \frac{K_{a1}K_{a2}}{[H^+]^2 + K_{a1}[H^+] + K_{a1}K_{a2}} \quad (9.27)$$

We can write, then, that

$$\boxed{\frac{K_{sp}}{\alpha_2} = K'_{sp} = [Ca^{2+}]C_{H_2C_2O_4} = s^2} \quad (9.28)$$

where K'_{sp} is the **conditional solubility product,** similar to the conditional formation constant described earlier.

EXAMPLE 9.6 Calculate the solubility of CaC_2O_4 in a solution containing 0.0010 M hydrochloric acid.

SOLUTION

$$\alpha_2 = \frac{(6.5 \times 10^{-2})(6.1 \times 10^{-5})}{(1.0 \times 10^{-3})^2 + (6.5 \times 10^{-2})(1.0 \times 10^{-3}) + (6.5 \times 10^{-2})(6.1 \times 10^{-5})}$$

$$= 5.7 \times 10^{-2}$$

$$s = \sqrt{K_{sp}/\alpha_2} = \sqrt{2.6 \times 10^{-9}/5.7 \times 10^{-2}} = 2.1 \times 10^{-4} \, M$$

This compares with a calculated solubility in water using Equation 9.23 of $5.1 \times 10^{-5} \, M$ (a 400% increase in solubility!). Note that both $[Ca^{2+}]$ and $C_{H_2C_2O_4} = 2.1 \times 10^{-4} \, M$. The concentrations of the other oxalate species in equilibrium can be obtained by multiplying this number by α_0, α_1, and α_2 for oxalic acid at 0.0010 M H^+ to obtain $[H_2C_2O_4]$, $[HC_2O_4^-]$, and $[C_2O_4^{2-}]$, respectively. We will not derive α_0 and α_1 here, but the results would be $[C_2O_4^{2-}] = 1.2 \times 10^{-5} \, M$, $[HC_2O_4^-] = 2.0 \times 10^{-4} \, M$, and $[H_2C_2O_4] = 3.1 \times 10^{-6} \, M$. (You can try the calculations.)

We neglected in the above calculations the fact that some of the acid was consumed by reaction with oxalate. We see that one-fifth of it reacted to form $HC_2O_4^-$. The amount reacted to form $H_2C_2O_4$ is negligible. If a more exact solution is desired, the amount of acid reacted, as calculated above, can be subtracted from the initial acid concentration and then the calculation repeated using the new acid concentration. This process is repeated until the change in the final answer is within the desired accuracy, an iterative procedure. Recalculation using $0.8 \times 10^{-3} \, M$ acid gives a calcium concentration of $1.9 \times 10^{-4} \, M$.

It should be emphasized that, when dealing with multiple equilibria, the validity of a given equilibrium expression is in no way compromised by the existence of additional competing equilibria. Thus, in the above example, the solubility product expression for CaC_2O_4 describes the relationship between Ca^{2+} and $C_2O_4^{2-}$ ions, whether or not acid is added. In other words, the product $[Ca^{2+}][C_2O_4^{2-}]$ is always a constant as long as there is solid CaC_2O_4 present. The quantity of $Ca_2C_2O_4$ that dissolves is increased, however, because part of the $C_2O_4^{2-}$ in solution is converted to $HC_2O_4^-$ and $H_2C_2O_4$.

9.7   EFFECT OF COMPLEXATION ON SOLUBILITY

Complexing agents can compete for the metal ion in a precipitate, just as acids compete for the anion. A precipitate MA that dissociates to give M^+ and A^- and whose metal complexes with the ligand L to form ML^+ would have the following equilibria:

$$MA \rightleftharpoons M^+ + A^-$$
$$+$$
$$L$$
$$\updownarrow$$
$$ML^+$$

Calculations for such a situation are handled in a manner completely analogous to those for the effects of acids on solubility.

Consider the solubility of AgBr in the presence of NH_3. The equilibria are

$$AgBr \rightleftharpoons Ag^+ + Br^- \qquad (9.29)$$

$$Ag^+ + NH_3 \rightleftharpoons Ag(NH_3)^+ \qquad (9.30)$$

$$Ag(NH_3)^+ + NH_3 \rightleftharpoons Ag(NH_3)_2^+ \qquad (9.31)$$

The solubility s of AgBr is equal to $[Br^-] = C_{Ag}$, where C_{Ag} represents the concentrations of all the silver species in equilibrium $(= [Ag^+] + [Ag(NH_3)^+] + [Ag(NH_3)_2^+])$. As before, we can substitute $C_{Ag}\beta_0$ for $[Ag^+]$ in the K_{sp} expression, where β_0 is the fraction of silver species present as Ag^+ (Equation 9.20):

$$K_{sp} = [Ag^+][Br^-] = C_{Ag}\beta_0 [Br^-] = 4 \times 10^{-13} \qquad (9.32)$$

Then,

$$\boxed{\frac{K_{sp}}{\beta_0} = K'_{sp} = C_{Ag} [Br^-] = s^2} \qquad (9.33)$$

where K'_{sp} is again the conditional solubility product, whose value depends on the concentration of ammonia.

EXAMPLE 9.7   Calculate the molar solubility of silver bromide in a 0.10 M ammonia solution.

SOLUTION
From Example 9.5, β_0 for 0.10 M NH_3 = 4.0×10^{-6}:

$$s = \sqrt{\frac{K_{sp}}{\beta_0}} = \sqrt{4 \times 10^{-13}/4.0 \times 10^{-6}} = 3._2 \times 10^{-4} \, M$$

This compares with a solubility in water of 6×10^{-7} M (530 times more soluble). Note again that both $[Br^-]$ and $C_{Ag} = 3._2 \times 10^{-4}$ M. The concentrations of the other silver species in equilibrium can be obtained by multiplying this number by β_0, β_1, and β_2 at 0.10 M NH_3 to obtain $[Ag^+]$, $[Ag(NH_3)^+]$, and $[Ag(NH_3)_2^+]$, respectively. Taking the β-values from Example 9.5, the results are $[Ag^+] = 1._3 \times 10^{-9}$ M, $[Ag(NH_3^+)] = 3._2 \times 10^{-7}$ M, and $[Ag(NH_3)_2^+] = 3._2 \times 10^{-4}$ M. Note that the majority of the dissolved silver exists in the $Ag(NH_3)_2^+$ form.

We neglected the amount of ammonia consumed in the reaction with the silver. We see that it was indeed negligible compared to 0.10 M [6×10^{-4} M was used in forming $Ag(NH_3)_2^+$, even less in forming $Ag(NH_3)^+$]. Had the amount of ammonia consumed been appreciable, an iterative procedure could have been used to obtain a more exact solution; that is, the amount of ammonia consumed could have been subtracted from the original concentration and then the new concentration used to calculate new β's and a new solubility.

9.8   PRECIPITATION TITRATIONS

Titrations with precipitating agents are useful for determining certain analytes, provided the equilibria are rapid and a suitable means of detecting the end point is

available. A consideration of titration curves will aid in the understanding of indicator selection, precision, and the titration of mixtures.

Titration Curves

Consider the titration of Cl^- with a standard solution of $AgNO_3$. A titration curve
can be prepared by plotting pCl ($-\log [Cl^-]$) against the volume of $AgNO_3$, in a
manner similar to that used for acid–base titrations. A typical titration curve is
illustrated in Figure 9.5; pX in the figure refers to the negative logarithm of the
halide concentration. At the beginning of the titration, we have 0.1 M Cl^-, and
pCl is 1. As the titration continues, part of the Cl^- is removed from solution by
precipitation as AgCl, and the pCl is determined by the concentration of the remaining Cl^-; the contribution of Cl^- from dissociation of the precipitate is negligible, except near the equivalence point. At the equivalence point, we have a
saturated solution of AgCl, pCl $= 5$, and $[Cl^-] = \sqrt{K_{sp}} = 10^{-5}$ M (see Chapter
6). Beyond the equivalence point, there is excess Ag^+, and the Cl^- concentration
is determined from the concentration of Ag^+ and K_{sp} as in Example 6.2 in Chapter
6.

The smaller the K_{sp}, the larger the break at the equivalence point. This is illustrated by a comparison of the titration curves for Cl^-, Br^-, and I^- versus Ag^+ in
Figure 9.5. The K_{sp} values of AgCl, AgBr, and AgI are 1×10^{-10}, 4×10^{-13}, and
1×10^{-16}, respectively. The concentration of each anion has been chosen to be
the same at the beginning of the titration, and so up to near the equivalence point
the concentration of each remains the same, since the same fraction is removed
from solution. At the equivalence point, $[X^-]$ is smaller for the smaller K_{sp} values,
hence pX is larger for a saturated solution of the salt. Beyond the equivalence
point, $[X^-]$ is smaller when K_{sp} is smaller, also resulting in a larger jump in pX. So
the overall effect is a larger pX break at the equivalence point when the compound
is more insoluble.

If the chloride titration were performed in reverse, that is, Ag^+ titrated with Cl^-,
the titration curve would be the reverse of the curve in Figure 9.5 if pCl were
plotted against the volume of Cl^-. Before the equivalence point, the Cl^- concentration would be governed by the concentration of excess Ag^+ and K_{sp}; and beyond the equivalence point, it would be governed merely by the excess Cl^-. pAg
could be plotted, instead, against the volume of chloride solution, and the curve
would look the same as that of Figure 9.5.

Detection of the End Point: Indicators

The end point can be detected by measuring either pCl or pAg with an appropriate
electrode and a potentiometer. We discuss this in Chapter 11. It is more convenient
if an indicator can be employed. The indicator theory for these titrations is different
from that for acid–base indicators. The properties of the indicators do not necessarily depend on the concentration of some ion in solution, that is, on pCl or pAg.

Two types of indicators are commonly employed. The first type forms a colored
compound with the titrant when it is in excess. The second type, called an **adsorption indicator,** suddenly becomes adsorbed on the precipitate at the equivalence
point owing to a property of the precipitate at the equivalence point, and the color
of the indicator changes when it is adsorbed. The mechanism will be discussed
below.

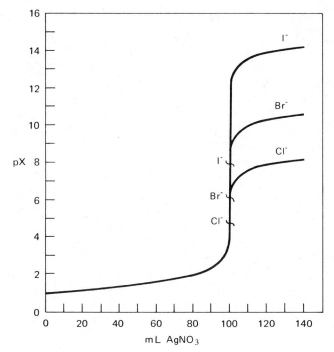

FIGURE 9.5
Titration curves for 100 mL 0.1 *M* chloride, bromide, and iodide solutions versus 0.1 *M* AgNO₃.

1. INDICATORS REACTING WITH THE TITRANT. There are several examples of this type of indicator. The **Mohr method** for determining chloride serves as an example. The chloride is titrated with standard silver nitrate solution. A soluble chromate salt is added as the indicator. This produces a yellow solution. When the precipitation of the chloride is complete, the first excess of Ag^+ reacts with the indicator to precipitate red silver chromate:

$$CrO_4^{2-} + 2Ag^+ \rightarrow Ag_2CrO_4$$

(yellow)   (red)

(9.34)

The concentration of indicator is important. The Ag_2CrO_4 should just start precipitating at the equivalence point, where we have a saturated solution of AgCl. From K_{sp}, the concentration of Ag^+ at the equivalence point is 10^{-5} M. (It is less than this before the equivalence point.) So, Ag_2CrO_4 should precipitate just when $[Ag^+] = 10^{-5}$ M. The solubility product of Ag_2CrO_4 is 1.1×10^{-12}. By inserting the Ag^+ concentration in the K_{sp} equation for Ag_2CrO_4, we calculate that, for this to occur, $[CrO_4^{2-}]$ should be 0.011 *M*:

$$(10^{-5})^2[CrO_4^{2-}] = 1.1 \times 10^{-12}$$

$$[CrO_4^{2-}] = 1.1 \times 10^{-2} M$$

If the concentration is greater than this, Ag_2CrO_4 will begin precipitation when $[Ag^+]$ is less than 10^{-5} M (before the equivalence point). If it is less than 0.011 *M*, then the $[Ag^+]$ will have to exceed 10^{-5} M (beyond the equivalence point) before precipitation of Ag_2CrO_4 begins.

In actual practice, the indicator concentration is kept at 0.002 to 0.005 M. If it is much higher than this, the intense yellow color of the chromate ion obscures the red Ag_2CrO_4 precipitate color, and an excess of Ag^+ is required to produce enough precipitate to be seen. An indicator blank should always be run and subtracted from the titration to correct for errors.

The Mohr titration must be performed at a pH of about 8. If the solution is too acid (pH < 6) part of the indicator is present as $HCrO_4^-$, and more Ag^+ will be required to form the Ag_2CrO_4 precipitate. Above pH 8, silver hydroxide may be precipitated (at pH > 10). The pH is properly maintained by adding solid calcium carbonate to the solution. (While the carbonate ion is a fairly strong Brønsted base, the concentration in a saturated calcium carbonate solution is just sufficient to give a pH about 8.)

The Mohr titration is useful for determining chloride in neutral or unbuffered solutions, such as drinking water.

A second example of this type of indicator is illustrated in the **Volhard titration.** This is an indirect titration procedure for determining anions that precipitate with silver (Cl^-, Br^-, I^-, SCN^-), and it is performed in acid (HNO_3) solution. A measured excess of $AgNO_3$ is added to precipitate the anion, and the excess Ag^+ is determined by back-titration with standard potassium thiocyanate solution:

$$\boxed{\begin{array}{l} X^- + Ag^+ \rightarrow \underline{AgX} + \text{excess } Ag^+ \\ \text{excess } Ag^+ + SCN^- \rightarrow \underline{AgSCN} \end{array}} \tag{9.35}$$

The end point is detected by adding iron(III) as a ferric alum (ferric ammonium sulfate), which forms a soluble red complex with the first excess of titrant:

$$\boxed{Fe^{3+} + SCN^- \rightarrow Fe(SCN)^{2+}} \tag{9.36}$$

If the precipitate, AgX, is less soluble than AgSCN, it is not necessary to remove the precipitate before titrating. Such is the case with I^-, Br^-, and SCN^-. In the case of I^-, the indicator is not added until all the I^- is precipitated, since it would be oxidized by the iron(III). If the precipitate is more soluble than AgSCN, it will react with the titrant to give a high and diffuse end point. Such is the case with AgCl:

$$\underline{AgCl} + SCN^- \rightarrow \underline{AgSCN} + Cl^- \tag{9.37}$$

Therefore, the precipitate is removed by filtration before titrating.

Obviously, these indicators must not form a compound with the titrant that is more stable than the precipitate, or the color reaction would occur on addition of the first drop of titrant.

2. ADSORPTION INDICATORS.   With adsorption indicators, the indicator reaction takes place on the surface of the precipitate. The indicator, which is a dye, exists in solution as the ionized form, usually an anion, In^-. To explain the mechanism of the indicator action, we must recall the mechanism occurring during precipitation (Chapter 6).

Consider the titration of Cl^- with Ag^+. Before the equivalence point, Cl^- is in excess and the primary adsorbed layer is Cl^-. This repulses the indicator anion, and the more loosely held secondary (counter) layer of adsorbed ions is cations, such as Na^+:

$$AgCl:Cl^-::Na^+$$

Beyond the equivalence point, Ag^+ is in excess and the surface of the precipitate becomes positively charged, with the primary layer being Ag^+. This will now attract the indicator anion and adsorb it in the counter layer:

$$AgCl:Ag^+::In^-$$

The color of the adsorbed indicator is different from that of the unadsorbed indicator, and this difference signals the completion of the titration. A possible explanation is that the indicator forms a colored complex with Ag^+, which is too weak to exist in solution, but whose formation is facilitated by adsorption on the surface of the precipitate (it becomes "insoluble").

The pH is important. If it is too low, the indicator, which is usually a weak acid, will dissociate too little to allow it to be adsorbed as the anion. Also, the indicator must not be too strongly adsorbed at the given pH, or it will displace the anion of the precipitate (e.g., Cl^-) in the primary layer before reaching the equivalence point. This will, of course, depend on the degree of adsorption of the anion of the precipitate. For example, Br^- forms a more insoluble precipitate with Ag^+ and is more strongly adsorbed. A more strongly adsorbed indicator can therefore be used.

The degree of adsorption of the indicator can be decreased by increasing the acidity. The stronger an acid the indicator is, the wider the pH range over which it can be adsorbed. In the case of Br^-, since a more acidic (more strongly adsorbed) indicator can be used, the pH of the titration can be more acid than with Cl^-.

A list of some adsorption indicators is given in Table 9.1. Fluorescein can be used as an indicator for any of the halides at pH 7, because it will not displace any of them. Dichlorofluoroscein will displace Cl^- at pH 7 but not at pH 4. Hence, results tend to be low when titrations are performed at pH 7. Titration of chloride using these indicators is called **Fajans' method.** The former indicator was the original one described by Fajans, but the latter indicator is now preferred. Eosin cannot be used for the titration of chloride at any pH because it is too strongly adsorbed.

Because most of these end points do not coincide with the equivalence point, the titrant should be standardized by the same titration as used for the sample. In this way, the errors will nearly cancel if about the same amount of titrant is used for both the standardization and analysis.

TABLE 9.1
Adsorption Indicators

Indicator	Titration	Solution
Fluorescein	Cl^- with Ag^+	pH 7–8
Dichlorofluorescein	Cl^- with Ag^+	pH 4
Bromcresol green	SCN^- with Ag^+	pH 4–5
Eosin	Br^-, I^-, SCN^- with Ag^+	pH 2
Methyl violet	Ag^+ with Cl^-	Acid solution
Rhodamine 6 G	Ag^+ with Br^-	HNO_3 ($\leq 0.3\ M$)
Thorin	SO_4^{2-} with Ba^{2+}	pH 1.5–3.5
Bromphenol blue	Hg^{2+} with Cl^-	0.1 M solution
Orthochrome T	Pb^{2+} with CrO_4^{2-}	Neutral, 0.02 M solution

A chief source of error in titrations involving silver is photodecomposition of AgX, which is catalyzed by the adsorption indicator. By proper standardization, accuracies of one part per thousand can be achieved, however.

The precipitate is uncharged at the equivalence point (neither ion is in excess). Colloidal precipitates, such as silver chloride, therefore tend to coagulate at this point, especially if the solution is shaken. This is just what we want for gravimetry, but the opposite of what we want here. It decreases the surface area for absorption of the indicator, which decreases the sharpness of the end point. Coagulation of silver chloride is prevented by adding some dextrin to the solution.

Titration of Sulfate with Barium

One of the experiments deals with the precipitation titration of sulfate in urine by titration with barium ion. As in the gravimetric determination of sulfate by precipitation of barium sulfate, this titration is subject to errors by coprecipitation. Cations such as K^+, Na^+, and NH_4^+ (especially the first) coprecipitate as sulfates:

$$BaSO_4 : SO_4^{2-} : 2M^+$$

This results in requiring less barium ion to complete the precipitation of the sulfate ion, and calculated results are low. Some metal ions will complex the indicator and interfere. Foreign anions may coprecipitate as the barium salts to cause high results. Errors from chloride, bromide, and perchlorate are small, but nitrate must be absent.

Cation interferences are readily removed with a strong cation exchange resin in the hydrogen form:

$$2Rz^-H^+ + (M)_2SO_4 \rightarrow 2Rz^-M^+ + H_2SO_4$$

The cations are replaced by protons. The principles of ion exchange chromatography are covered in Chapter 16.

The titration is carried out in an aqueous–nonaqueous solvent mixture. The organic solvent decreases the dissociation of the indicator and thereby hinders formation of a barium–indicator complex. It also results in a more flocculant precipitate with better adsorption properties for the indicator.

QUESTIONS

1. Explain the Volhard titration of chloride. The Fajan titration. Which is used for acid solutions? Why?
2. Explain the principles of adsorption indicators.
3. Distinguish between a complexing agent and a chelating agent.
4. Explain the principles of chelation titration indicators.
5. Why is a small amount of magnesium salt added to the EDTA solution used for the titration of calcium with an Eriochrome Black T indicator?

PROBLEMS

Complex Equilibrium Calculations (K_f)
6. Calcium ion forms a weak 1:1 complex with nitrate ion with a formation constant of 2.0. What would be the equilibrium concentrations of Ca^{2+} and $Ca(NO_3)^+$ in a solution

prepared by adding 10 mL each of 0.010 M $CaCl_2$ and 2.0 M $NaNO_3$? Neglect diverse ion effects.

7. The formation constant of the silver–ethylenediamine complex, $Ag(NH_2CH_2CH_2NH_2)^+$, is 5.0×10^4. Calculate the concentration of Ag^+ in equilibrium with a 0.10 M solution of the complex. (Assume no higher-order complexes.)

8. What would be the concentration of Ag^+ in Problem 7 if the solution contained also 0.10 M ethylenediamine, $NH_2CH_2CH_2NH_2$?

9. Silver ion forms stepwise complexes with thiosulfate ion, $S_2O_3^{2-}$, with $K_{f1} = 6.6 \times 10^8$ and $K_{f2} = 4.4 \times 10^4$. Calculate the equilibrium concentrations of all silver species for 0.0100 M $AgNO_3$ in 1.00 M $Na_2S_2O_3$. Neglect diverse ion effects.

Conditional Formation Constants

10. The formation constant for the lead–EDTA chelate (PbY^{2-}) is 1.10×10^{18}. Calculate the conditional formation constant (a) at pH 3 and (b) at pH 10.

11. Using the conditional constants calculated in Problem 10, calculate the pPb ($-\log$ $[Pb^{2+}]$) for 50.0 mL of a solution of 0.0250 M Pb^{2+} (a) at pH 3 and (b) at pH 10 after the addition of (1) 0 mL, (2) 50 mL, (3) 125 mL, and (4) 200 mL of 0.0100 M EDTA.

12. The conditional formation constant for the calcium–EDTA chelate was calculated for pH 10 in Example 9.4 to be 1.8×10^{10}. Calculate the conditional formation constant at pH 3. Compare this with that calculated for lead at pH 3 in Problem 10. Could lead be titrated with EDTA at pH 3 in the presence of calcium?

Effect of Acidity on Solubility

13. Calculate the solubility of $AgIO_3$ in 0.100 M HNO_3. Also calculate the equilibrium concentrations of IO_3^- and HIO_3.

14. Calculate the solubility of CaF_2 in 0.100 M HCl. Also calculate the equilibrium concentrations of F^- and HF.

15. Calculate the solubility of PbS in 0.0100 M HCl. Also calculate the equilibrium concentrations of S^{2-}, HS^-, and H_2S.

Effect of Complexation on Solubility

16. Silver ion forms a stepwise 1:2 complex with ethylenediamine (en) with formation constants of $K_{f1} = 5.0 \times 10^4$ and $K_{f2} = 1.4 \times 10^3$. Calculate the solubility of silver chloride in 0.100 M ethylenediamine. Also calculate the equilibrium concentrations of $Ag(en)^+$ and $Ag(en)_2^+$.

Standard Solutions

17. Calculate the weight of $Na_2H_2Y \cdot 2H_2O$ required to prepare 500.0 mL of 0.05000 M EDTA.

18. An EDTA solution is standardized against high-purity $CaCO_3$ by dissolving 0.3982 g $CaCO_3$ in hydrochloric acid, adjusting the pH to 10 with an ammoniacal buffer, and titrating. If 38.26 mL was required for the titration, what is the molarity of the EDTA?

19. Calculate the titer of 0.1000 M EDTA in mg $CaCO_3$/mL.

20. If 100.0 mL of a water sample is titrated with 0.01000 M EDTA, what is the titer of the EDTA in terms of water hardness/mL?

Quantitative Complexometric Determinations

21. Calcium in powdered milk is determined by ashing a 1.50-g sample and then titrating the calcium with EDTA solution, 12.1 mL being required. The EDTA was standardized by titrating 10.0 mL of a zinc solution prepared by dissolving 0.632 g zinc metal in acid and diluting to 1 L (10.8 mL EDTA required for titration). What is the concentration of calcium in the powdered milk in parts per million?

22. Calcium is determined in serum by microtitration with EDTA. A 100-μL sample is treated with two drops of 2 M KOH, Cal-Red indicator is added, and the titration is performed with 0.00122 M EDTA, using a microburet. If 0.203 mL EDTA is required for titration, what is the level of calcium in the serum in mg/dL and in meq/L?

23. In the Liebig titration of cyanide ion, a soluble complex is formed; and at the equivalence point, solid silver cyanide is formed, signaling the end point:

$$2CN^- + Ag^+ \rightarrow Ag(CN)_2^- \quad \text{(titration)}$$

$$Ag(CN)_2^- + Ag^+ \rightarrow Ag[Ag(CN)_2] \quad \text{(end point)}$$

A 0.4723-g sample of KCN was titrated with 0.1025 M AgNO$_3$, requiring 34.95 mL. What is the percent purity of the KCN?

24. Copper in salt water near the discharge of a sewage treatment plant is determined by first separating and concentrating it by solvent extraction of its dithizone chelate at pH 3 into methylene chloride and then evaporating the solvent, ashing the chelate to destroy the organic portion, and titrating the copper with EDTA. Three 1-L portions of the sample are each extracted with 25-mL portions of methylene chloride and the extracts are combined in a 100-mL volumetric flask and diluted to volume. A 50-mL aliquot is evaporated, ashed, and titrated. If the EDTA solution has a CaCO$_3$ titer of 2.69 mg/mL and 2.67 mL is required for titration of the copper, what is the concentration of copper in the seawater in parts per million?

25. Chloride in serum is determined by titration with Hg(NO$_3$)$_2$: $2Cl^- + Hg^{2+} \rightleftharpoons HgCl_2$. The Hg(NO$_3$)$_2$ is standardized by titrating 2.00 mL of a 0.0108 M NaCl solution, requiring 1.12 mL to reach the diphenylcarbazone end point. A 0.500-mL serum sample is treated with 3.50 mL water, 0.50 mL 10% sodium tungstate solution, and 0.50 mL of 0.33 M H$_2$SO$_4$ solution to precipitate proteins. After the proteins are precipitated, the sample is filtered through a dry filter into a dry flask. A 2.00-mL aliquot of the filtrate is titrated with the Hg(NO$_3$)$_2$ solution, requiring 1.23 mL. Calculate the meq/L chloride in the serum.

Quantitative Precipitation Determinations

26. Chloride in a brine solution is determined by the Volhard method. A 10.00-mL aliquot of the solution is treated with 15.00 mL of standard 0.1182 M AgNO$_3$ solution. The excess silver is titrated with standard 0.101 M KSCN solution, requiring 2.38 mL to reach the red Fe(SCN)$^{2+}$ end point. Calculate the concentration of chloride in the brine solution, in g/L.

27. In a Mohr titration of chloride with silver nitrate, an error is made in the preparation of the indicator. Instead of 0.011 M chromate indicator in the titration flask at the end point, there is only 0.0011 M. If the flask contains 100 mL at the end point, what is the error in the titration in milliliters of 0.100 M titrant? Neglect errors due to the color of the solution.

RECOMMENDED REFERENCES

1. F. W. Fales, "Calcium (Complexometric)," in *Standard Methods of Clinical Chemistry*, Vol. 2, D. Seligson, ed., New York: Academic Press, 1958, pp. 1–11.
2. H. Flaschka, *EDTA Titrations*. New York: Pergamon Press, 1959.
3. H. A. Flaschka and A. J. Barnard, Jr., "Titrations with EDTA and Related Compounds," in *Comprehensive Analytical Chemistry*, Vol. 1B, C. L. Wilson and D. W. Wilson, eds. New York: Elsevier Publishing Co., 1960.

4. J. Kragten, *Atlas of Metal-Ligand Equilibria in Aqueous Solution*. London: Ellis Horwood, 1978.

5. O. Schales, "Chloride," in *Standard Methods of Clinical Chemistry,* Vol. 1, M. Reiner, ed., 1955, pp. 37–42.

6. G. Schwarzenbach, *Complexometric Titrations*. New York: Interscience Publishers, 1957.

7. F. J. Welcher, *The Analytical Uses of Ethylenediaminetetraacetic Acid*. Princeton: D. Van Nostrand Co., 1958.

chapter 10

REDUCTION–OXIDATION TITRATIONS

An **oxidation** is defined as a loss of electrons to give a higher oxidation state (more positive), and **reduction** is defined as a gain of electrons to give a lower oxidation state (more negative). Titrations involving oxidizing and reducing agents, in which electrons are transferred from one to the other, are useful for the determination of many substances. The calculations in volumetric analysis require that the balanced reaction be known, and so you should review the balancing of reduction–oxidation reactions in the Appendix. Practice problems for balancing reduction–oxidation reactions are given at the end of the chapter. In this chapter, we introduce the principles of electrochemical cells and how electrode potentials can be used to predict which oxidizing and reducing agents might react and for constructing titration curves. Various means of end point detection will be described, and then some representative reduction–oxidation titrations will be given, plus the necessary procedures to get the sample in the correct oxidation state for titration.

10.1 PRINCIPLES

A reduction–oxidation reaction (commonly called a **"redox"** reaction) is one that occurs between a reducing and an oxidizing agent:

$$Ox_1 + Red_2 \rightleftharpoons Red_1 + Ox_2 \qquad (10.1)$$

Ox_1 is reduced to Red_1 and Red_2 is oxidized to Ox_2. Ox_1 is the oxidizing agent and Red_2 is the reducing agent. The reducing or oxidizing tendency of a substance will depend on its reduction potential, described below. An oxidizing substance will tend to take on an electron or electrons and be reduced to a lower oxidation state:

$$M^{a+} + ne^- \rightarrow M^{(a-n)+} \tag{10.2}$$

Conversely, a reducing substance will tend to give up an electron or electrons and be oxidized:

$$M^{a+} \rightarrow M^{(a+n)+} + ne^- \tag{10.3}$$

If the oxidized form of a metal ion is complexed, it is more stable and will be more difficult to reduce; so its tendency to take on electrons will be decreased.

An understanding of the oxidizing or reducing tendency of substances is available from electrochemical cells and electrode potentials.

10.2 ELECTROCHEMICAL CELLS

There are two kinds of electrochemical cells, **galvanic** (voltaic) and **electrolytic.** In the former, a chemical reaction spontaneously occurs to **produce electrical energy.** The lead storage battery and the ordinary flashlight battery are common examples. In the latter, **electrical energy is used** to force a nonspontaneous chemical reaction to occur, that is, to go in the reverse direction it would in a galvanic cell. An example is the electrolysis of water. In both types of these cells, the electrode at which oxidation occurs is the **anode,** and that at which reduction occurs is the **cathode.** Galvanic cells will be of importance in our discussions in the next two chapters, while electrolytic cells are important in the electrochemical methods discussed in Chapter 12.

Consider the following redox reaction:

$$Fe^{2+} + Ce^{4+} \rightleftharpoons Fe^{3+} + Ce^{3+} \tag{10.4}$$

If a solution containing Fe^{2+} is mixed with one containing Ce^{4+}, there is a certain tendency for the ions to transfer electrons. Assume the Fe^{2+} and Ce^{4+} are in separate beakers connected by a **salt bridge,** as shown in Figure 10.1. (A salt bridge allows charge transfer through the solutions but prevents mixing of the solutions.) No reaction can occur, since the solutions do not make contact. Now put an inert platinum wire in each solution and connect the two wires. The setup now constitutes a galvanic cell. If a microammeter is connected in series, it indicates that a current is flowing. The Fe^{2+} is being oxidized at the platinum wire (the anode):

$$Fe^{2+} \rightarrow Fe^{3+} + e^- \tag{10.5}$$ ½ Rxn

The released electrons flow through the wire to the other beaker where the Ce^{4+} is reduced (at the cathode):

$$Ce^{4+} + e^- \rightarrow Ce^{3+} \tag{10.6}$$ ½ Rxn

This occurs because of the tendency of these ions to transfer electrons. The net result is the reaction written in Equation 10.4 which would occur if Fe^{2+} and Ce^{4+} were added together in a single beaker. The platinum wires can be considered **electrodes.** Each will adopt an electrical **potential** that is determined by the tendency of the ions to give off or take on electrons, and this is called the **electrode potential.** A voltmeter placed between the electrodes will indicate the *difference* in the potentials between the two electrodes. The larger the potential difference,

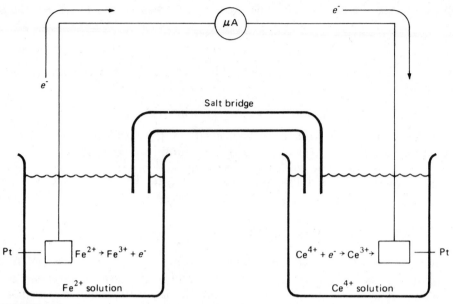

FIGURE 10.1
Galvanic cell.

the greater the tendency for the reaction between Fe^{2+} and Ce^{4+}. The driving force of the chemical reaction (the potential difference) can be used to perform work such as lighting a light bulb or running a motor, as in a battery.

Equations 10.5 and 10.6 are **half-reactions.** No half-reaction can occur by itself. There must be an **electron donor** (a reducing agent) and an **electron acceptor** (an oxidizing agent). In this case, Fe^{2+} is the reducing agent and Ce^{4+} is the oxidizing agent. Each half-reaction will generate a definite potential that would be adopted by an inert electrode dipped in the solution.

If the potentials of all half-reactions could be measured, then we could determine which oxidizing and reducing agents will react. Unfortunately, there is no way to measure individual electrode potentials. But, as described above, the *difference* between two electrode potentials can be measured. The electrode potential of the half-reaction[1]

$$2H^+ + 2e^- \rightleftharpoons H_2 \qquad\qquad (10.7)$$

has arbitrarily been assigned a value of 0.000 V. This is called the **normal hydrogen electrode** (NHE), or the **standard hydrogen electrode** (SHE). The potential differences between this half-reaction and other half-reactions have been measured and arranged in decreasing order. Some of these are listed in Table 10.1. Potentials are dependent on concentrations and all standard potentials refer to conditions of unit activity for all species (or 1 atmosphere partial pressure in the case of gases, as for hydrogen in the NHE). The effects of concentrations on potentials are described below. A more complete listing of potentials is given in the Appendix.

---

[1]The reaction could have been written $H^+ + e^- = \frac{1}{2} H_2$. The way it is written does not affect its potential.

TABLE 10.1
Some Standard Potentials

Half-Reaction	E^0, V
$H_2O_2 + 2H^+ + 2e^- = 2H_2O$	1.77
$MnO_4^- + 4H^+ + 3e^- = MnO_2 + 2H_2O$	1.695
$Ce^{4+} + e^- = Ce^{3+}$	1.61
$MnO_4^- + 8H^+ + 5e^- = Mn^{2+} + 4H_2O$	1.51
$Cr_2O_7^{2-} + 14H^+ + 6e^- = 2Cr^{3+} + 7H_2O$	1.33
$MnO_2 + 4H^+ + 2e^- = Mn^{2+} + 2H_2O$	1.23
$2IO_3^- + 12H^+ + 10e^- = I_2 + 6H_2O$	1.20
$H_2O_2 + 2e^- = 2OH^-$	0.88
$Cu^{2+} + I^- + e^- = CuI$	0.86
$Fe^{3+} + e^- = Fe^{2+}$	0.771
$O_2 + 2H^+ + 2e^- = H_2O_2$	0.682
$I_2(aq) + 2e^- = 2I^-$	0.6197
$H_3AsO_4 + 2H^+ + 2e^- = HAsO_2 + 2H_2O$	0.559
$I_3^- + 2e^- = 3I^-$	0.5355
$Sn^{4+} + 2e^- = Sn^{2+}$	0.154
$S_4O_6^{2-} + 2e^- = S_2O_3^{2-}$	0.08
$2H^+ + 2e^- = H_2$	0.000
$Zn^{2+} + 2e^- = Zn$	-0.763
$2H_2O + 2e^- = H_2 + 2OH^-$	-0.828

The potentials are for the half-reaction written as a *reduction*, and so they represent **reduction potentials.** We will use the Gibbs–Stockholm electrode potential convention, adopted at the 17th Conference of the International Union of Pure and Applied Chemistry in Stockholm, 1953. In this convention, the half-reaction is written as a reduction, and the potential increases as the tendency for reduction (of the oxidized form of the half-reaction) increases.

The electrode potential for $Sn^{4+} + 2e^- \rightleftharpoons Sn^{2+}$ is +0.15 V. In other words, the potential of this half reaction relative to the NHE in a cell like that in Figure 10.1 would be 0.15 V. Since the above couple has a larger (more positive) reduction potential than the NHE, Sn^{4+} has a stronger tendency to be reduced than H^+ has. We can draw some general conclusions from the electrode potentials:

1. The more *positive* the electrode potential, the greater the tendency of the oxidized form to be reduced. In other words, **the more positive the electrode potential, the stronger an oxidizing agent the oxidized form is and the weaker a reducing agent the reduced form is.**
2. The more *negative* the electrode potential, the greater the tendency of the reduced form to be oxidized. In other words, **the more negative the reduction potential, the weaker an oxidizing agent the oxidized form is and the stronger a reducing agent the reduced form is.**

The reduction potential for $Ce^{4+} + e^- \rightleftharpoons Ce^{3+}$ is very positive, so Ce^{4+} is a strong oxidizing agent, while Ce^{3+} is a very weak reducing agent. On the other hand, the potential for $Zn^{2+} + 2e^- \rightleftharpoons Zn$ is very negative, and so Zn^{2+} is a very weak oxidizing agent, while metallic zinc is a very strong reducing agent.

The oxidized form of any half-reaction is capable of oxidizing the reduced form of any half-reaction whose reduction potential is more *negative* than its own, and vice versa: The reduced form of any half-reaction is capable of reducing the oxidized form of any half-reaction whose reduction potential is more *positive* than its own.

For example, consider the two half-reactions

$$Fe^{3+} + e^- = Fe^{2+} \quad E^0 = 0.771 \text{ V} \tag{10.8}$$

$$Sn^{4+} + 2e^- = Sn^{2+} \quad E^0 = 0.154 \text{ V} \tag{10.9}$$

There are two combinations for possible reaction between an oxidizing and a reducing agent in these two half-reactions, which we arrive at by *subtracting* one from the other (multiply the first half-reaction by 2 so the electrons cancel):

$$2Fe^{3+} + Sn^{2+} = 2Fe^{2+} + Sn^{4+} \tag{10.10}$$

and

$$Sn^{4+} + 2Fe^{2+} = Sn^{2+} + 2Fe^{3+} \tag{10.11}$$

[There is no possibility of reaction between Fe^{3+} and Sn^{4+} (both oxidizing agents) or between Fe^{2+} and Sn^{2+} (both reducing agents).] Perusal of the potentials tells us that Reaction 10.10 will take place, that is, the reduced form Sn^{2+} of Reaction 10.9 (with the more negative potential) will react with the oxidized form of Reaction 10.8 (with the more positive potential). Note that the number of electrons donated and accepted must be equal (see the Appendix on balancing redox reactions).

EXAMPLE 10.1   For the following substances, list the oxidizing agents in decreasing order of oxidizing capability and the reducing agents in decreasing order of reducing capability: MnO_4^-, Ce^{3+}, Cr^{3+}, IO_3^-, Fe^{3+}, I^-, H^+, Zn^{2+}.

SOLUTION
Looking at Table 10.1, the following must be oxidizing agents (are in the oxidized forms) and are listed from the most positive E^0 to the least positive: MnO_4^-, IO_3^-, Fe^{3+}, H^+, Zn^{2+}. MnO_4^- is a very good oxidizing agent, Zn^{2+} is very poor. The remainder are in the reduced form and their reducing power is in the order I^-, Cr^{3+}, Ce^{3+}. I^- is a reasonably good reducing agent, Ce^{3+} is poor.

   If the potentials are subtracted in the same manner as the half-reactions to give the net reaction, the result is the **cell voltage** that would be observed in a galvanic cell (Equation 10.8 minus Equation 10.9, or 0.771 V − 0.154 V = +0.620 V in the above).[2] **If this calculated cell voltage is positive, the reaction goes as written.** If it is negative, the reaction will occur in the reverse reaction. This is the result of the convention that, for a spontaneous reaction, the free energy is negative. The free energy at standard conditions is given by

$$\Delta G^0 = -nF \Delta E^0 \tag{10.12}$$

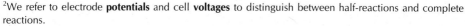

[2]We refer to electrode **potentials** and cell **voltages** to distinguish between half-reactions and complete reactions.

and so a positive potential difference provides the necessary negative free energy. Hence, we can tell from the relative standard potentials for two reactions, and from their signs, which reaction combination will produce a negative free-energy change and be spontaneous. For example, for the Ce^{4+}/Ce^{3+} half-reaction, E^0 is $+1.61$ V (Table 10.1); and for the Fe^{3+}/Fe^{2+} half-reaction, E^0 is $+0.771$ V. ΔG^0 for the former is more negative than for the latter, and subtraction of the latter half-reaction from the former will provide the spontaneous reaction that would occur to give a negative free energy. That is, Ce^{4+} would spontaneously oxidize Fe^{2+}.

By convention, a cell is written with the anode on the left:

$$\boxed{\text{anode/solution/cathode}} \tag{10.13}$$

The single lines represent a boundary between either an electrode phase and a solution phase or two solution phases. In Figure 10.1, the cell would be written as

$$Pt/Fe^{2+}(C_1),\ Fe^{3+}(C_2)//Ce^{4+}(C_3),\ Ce^{3+}(C_4)/Pt \tag{10.14}$$

C_1, C_2, C_3, and C_4 represent the concentrations of the different species. The double line represents the salt bridge. If a galvanic cell were constructed for the above iron and tin half-reactions, it would be written as

$$Pt/Sn^{2+}(C_1),\ Sn^{4+}(C_2)//Fe^{3+}(C_3),\ Fe^{2+}(C_4)/Pt \tag{10.15}$$

Since oxidation occurs at the anode and reduction occurs at the cathode, the stronger reducing agent is placed on the left and the stronger oxidizing agent is placed on the right. The potential of the cell is given by

$$\boxed{E_{cell} = E_{right} - E_{left} = E_{cathode} - E_{anode} = E_+ - E_-} \tag{10.16}$$

When the cell is set up properly, **the calculated voltage will always be positive,** and the cell reaction is written correctly, that is, the cathode half-reaction is written as a reduction and the anode half-reaction is written as an oxidation. In cell (10.15), we would have at standard conditions

$$E^0_{cell} = E^0_{Fe3+,Fe2+} - E^0_{Sn4+,Sn2+} = 0.771 - 0.154 = 0.617\ V$$

To take some more examples of possible redox reactions, Fe^{3+} will not oxidize Mn^{2+}. Quite the contrary, MnO_4^- will oxidize Fe^{2+}. I_2 is a moderate oxidizing agent and will oxidize Sn^{2+}. On the other hand, I^- is a fairly good reducing agent and will reduce Fe^{3+}, $Cr_2O_7^{2-}$, and so on. For a reaction to be complete enough to obtain a sharp end point in a titration, there should be **at least 0.2 to 0.3 V difference between the two electrode potentials.**

EXAMPLE 10.2   From the potentials listed in Table 10.1, determine the reaction between the following half-reactions and calculate the corresponding cell voltage:

$$Fe^{3+} + e^- = Fe^{2+} \quad E^0 = 0.771\ V$$

$$I_3^- + 2e^- = 3I^- \quad E^0 = 0.5355\ V$$

SOLUTION

Since the Fe^{3+}/Fe^{2+} potential is the more positive, Fe^{3+} is a better oxidizing agent than I_3^-. Hence, Fe^{3+} will oxidize I^- and $E^0_{cell} = E_{cathode} - E_{anode} = E^0_{Fe3+/Fe2+} - E^0_{I_3-/I^-}$. In the same fashion, the second half-reaction must be subtracted from the first (multiplied by 2) to give the overall cell reaction:

$$2Fe^{3+} + 3I^- = 2Fe^{2+} + I_3^- \quad E^0_{cell} = 0.771 \text{ V} - 0.536 \text{ V} = +0.235 \text{ V}$$

Note that multiplying a half-reaction by any number does not change its potential.

10.3   THE NERNST EQUATION

The potentials listed in Table 10.1 were determined for the case when the concentrations of both the oxidized and reduced forms (and all other species) were at **unit activity**, and they are called the **standard potentials**, designated by E^0. This potential is dependent on the concentration of the species. This potential dependence is described by the **Nernst equation**[3]:

$$aOx + ne^- \rightleftharpoons bRed \tag{10.17}$$

$$E = E^0 - \frac{2.3026RT}{nF} \log \frac{[Red]^b}{[Ox]^a} \tag{10.18}$$

where E is the reduction potential at the specific concentration, n is the number of electrons involved in the half-reaction (equivalents per mole), R is the gas constant (8.3143 V coul deg^{-1} mol^{-1}), T is the absolute temperature, and F is the Faraday constant (96,487 coul eq^{-1}). At 25°C (298.16 K), the value of $2.3026RT/F$ is 0.05916, or 1.9842×10^{-4}(°C + 273.16). The concentration of pure substances such as precipitates and liquids (H_2O) is taken as unity.

EXAMPLE 10.3   A solution is 10^{-3} M in $Cr_2O_7^{2-}$ and 10^{-2} M in Cr^{3+}. If the pH is 2.0, what is the potential of the half-reaction?

SOLUTION

$$Cr_2O_7^{2-} + 14H^+ + 6e^- \rightleftharpoons 2Cr^{3+} + 7H_2O$$

$$E = E^0_{Cr_2O_7^{2-},Cr^{3+}} - \frac{0.059}{6} \log \frac{[Cr^{3+}]^2}{[Cr_2O_7^{2-}][H^+]^{14}}$$

$$= 1.33 - \frac{0.059}{6} \log \frac{(10^{-2})^2}{(10^{-3})(10^{-2})^{14}}$$

$$= 1.33 - \frac{0.059}{6} \log 10^{27} = 1.33 - 27\left(\frac{0.059}{6}\right)$$

$$= 1.06 \text{ V}$$

This calculated potential is the potential an electrode would adopt, relative to the NHE, if it were placed in the solution, and it is a measure of the oxidizing or reducing

[3]More correctly, activities, rather than concentrations, should be used; but we will use concentrations for this discussion. In the next chapter involving potential measurements for direct calculation of concentrations, we will use activities.

power of that solution. Theoretically, the potential would be infinite if there were no Cr^{3+} at all in solution. In actual practice, the potential is always finite (but impossible to calculate from the simple Nernst equation). Either there will be a small amount of impurity of the oxidized or reduced form present or, more probably, the potential will be limited by another half-reaction, such as the oxidation or reduction of water, that prevents it from going to infinity.

The potential of a solution containing the ions of two half-reactions at equilibrium (e.g., at different points in a titration) can be calculated using the Nernst equation for *either* half-reaction. This is because when the reaction comes to equilibrium, the potentials of the two half-reactions become identical; otherwise, the reaction would still be going on. An electrode dipped in the solution will adopt the **equilibrium potential.** The equilibrium potential is dictated by the equilibrium concentrations of either half-reaction and the Nernst equation.

EXAMPLE 10.4   5.0 mL of 0.10 M Ce^{4+} solution is added to 5.0 mL of 0.30 M Fe^{2+} solution. Calculate the potential of a platinum electrode dipping in the solution.

SOLUTION
We start with $0.30 \times 5.0 = 1.5$ mmol Fe^{2+} and add $0.10 \times 5.0 = 0.50$ mmol Ce^{4+}. So we form 0.50 mmol each of Fe^{3+} and Ce^{3+} and have 1.0 mmol Fe^{2+} remaining. The reaction lies far to the right at equilibrium if there is at least 0.2 V difference between the standard electrode potentials of two half-reactions. But a small amount of $Ce^{4+}(=x)$ will exist at equilibrium and an equal amount of Fe^{2+} will be formed:

$$Fe^{2+} \quad + \quad Ce^{4+} \rightleftharpoons \quad Fe^{3+} \quad + \quad Ce^{3+}$$
$$1.0 + x \qquad x \qquad 0.50 - x \quad 0.50 - x$$

where the numbers and x represent millimoles. This is analogous to "ionization" of the product in precipitation or acid–base reactions; a slight shift of the equilibrium here to the left would be the "ionization." The quantity x is very small compared with 0.50 or 1.0 and can be neglected. Either half-reaction can be used to calculate the potential. Since the concentrations of both species in the Fe^{3+}/Fe^{2+} couple are known, we will use this:

$$Fe^{3+} \quad + \quad e^- \rightleftharpoons Fe^{2+}$$
$$0.50 \qquad\qquad 1.0$$

$$E = 0.771 - 0.059 \log \frac{[Fe^{2+}]}{[Fe^{3+}]}$$

The final volume is 10 mL, so

$$E = 0.771 - 0.059 \log \frac{(1.0 \text{ mmol}/10 \text{ mL})}{(0.50 \text{ mmol}/10 \text{ mL})} = 0.771 - 0.059 \log 2.0$$

$$= 0.771 - 0.059(0.30)$$

$$= 0.753 \text{ V}$$

A Nernst equation can also be written for a **complete cell reaction:**

$$E_{cell} = E_{cell}^0 - \frac{2.303RT}{nF} \log \frac{[\text{products}]}{[\text{reactants}]} \qquad (10.19)$$

In Example 10.2,

$$E_{cell} = E_{cathode} - E_{anode}$$

$$= \left(E^0_{Fe^{3+},Fe^{2+}} - \frac{0.059}{2} \log \frac{[Fe^{2+}]^2}{[Fe^{3+}]^2} \right) - \left(E^0_{I_3^-,I^-} - \frac{0.059}{2} \log \frac{[I^-]^3}{[I_3^-]} \right) \quad (10.20)$$

$$= E^0_{Fe^{3+},Fe^{2+}} - E^0_{I_3^-,I^-} - \frac{0.059}{2} \log \frac{[Fe^{2+}]^2[I_3^-]}{[Fe^{3+}]^2[I^-]^3}$$

Note that it was necessary to multiply the Fe^{3+}/Fe^{2+} half-reaction by 2 (as when subtracting the two half-reactions) in order to combine the two log terms (with $n = 2$), and the final equation is the same as we would have written from the cell reaction. Note also that $E^0_{Fe^{3+},Fe^{2+}} - E^0_{I_3^-,I^-}$ is the cell standard potential, E^0_{cell}.

The term on the right of the log sign is the **equilibrium constant expression** for the reaction.

$$2Fe^{3+} + 3I^- \rightleftharpoons 2Fe^{2+} + I_3^- \quad (10.21)$$

The cell voltage represents the tendency of a reaction to occur when the reacting species are put together (just as it does in a battery, that is, it represents the potential for work). After the reaction has reached equilibrium, the cell voltage necessarily becomes zero and the reaction is complete (i.e., no more work can be derived from the cell). That is, the potentials of the two half-reactions are equal at equilibrium.

EXAMPLE 10.5 One beaker contains a solution of 0.0200 M $KMnO_4$, 0.00500 M $MnSO_4$, and 0.500 M H_2SO_4; and a second beaker contains 0.150 M $FeSO_4$ and 0.00150 M $Fe_2(SO_4)_3$. The two beakers are connected by a salt bridge and platinum electrodes are placed in each. The electrodes are connected via a wire with a voltmeter in between. What would be the potential of each half-cell (a) before reaction and (b) after reaction? What would be the measured cell voltage (c) at the start of the reaction and (d) after the reaction reaches equilibrium? Assume H_2SO_4 to be completely ionized and equal volumes in each beaker.

SOLUTION
The cell reaction is

$$5Fe^{2+} + MnO_4^- + 8H^+ \rightleftharpoons 5Fe^{3+} + Mn^{2+} + 4H_2O$$

and the cell is

$$Pt/Fe^{2+}(0.150\ M),\ Fe^{3+}(0.00300\ M)//MnO_4^-(0.0200\ M),\ Mn^{2+}(0.00500\ M),\ H^+(1.00\ M)/Pt$$

(a) $$E_{Fe} = E^0_{Fe^{3+},Fe^{2+}} - 0.059 \log \frac{[Fe^{2+}]}{[Fe^{3+}]}$$

$$= 0.771 - 0.059 \log \frac{(0.150)}{(0.00300)} = 0.671\ V$$

$$E_{Mn} = E^0_{MnO_4^-,Mn^{2+}} - \frac{0.059}{5} \log \frac{[Mn^{2+}]}{[MnO_4^-][H^+]^8}$$

$$= 1.51 - \frac{0.059}{5} \log \frac{(0.00500)}{(0.0200)(1.00)^8} = 1.52 \text{ V}$$

(b) At equilibrium, $E_{Fe} = E_{Mn}$. We can calculate E from either half-reaction. First, calculate the equilibrium concentrations. Five moles of Fe^{2+} will react with each mole of MnO_4^-. The Fe^{2+} is in excess. It will be decreased by $5 \times 0.0200 = 0.100$ M, so 0.050 M Fe^{2+} remains and 0.100 M Fe^{3+} is formed (total now is $0.100 + 0.003 = 0.103$ M). Virtually all the MnO_4^- is converted to Mn^{2+} (0.0200 M) to give a total of 0.0250 M. A small unknown amount of MnO_4^- remains at equilibrium, and we would need the equilibrium constant to calculate it; this can be obtained from $E_{cell} = 0$ at equilibrium—see Equation 10.20—and will be treated below, but we need not go to this trouble, since $[Fe^{2+}]$ and $[Fe^{3+}]$ are known:

$$E_{Mn} = E_{Fe} = 0.771 - 0.059 \log \frac{(0.050)}{(0.103)} = 0.790 \text{ V}$$

Note that the equilibrium half-cell potentials are in between the values for the two half-cells before reaction.

(c) $E_{cell} = E_{Mn} - E_{Fe} = 1.52 - 0.671 = 0.85$ V

(d) At equilibrium, $E_{Mn} = E_{Fe}$, and so E_{cell} is zero volts.

Note that if one of the species had not been initially present in a half-reaction, we could not have calculated an initial potential for that half-reaction.

10.4 CALCULATION OF THE EQUILIBRIUM CONSTANT OF A REACTION

Since a cell voltage is zero at reaction equilibrium, the difference between the two half-reaction potentials is zero (or the two potentials are equal), and the Nernst equations for the half-reactions can be equated. On combining the equations, the log term is that of the equilibrium constant expression for the reaction (see Equation 10.20), and a numerical value can be calculated for the equilibrium constant. This is a consequence of the relationship between the free energy and the equilibrium constant of a reaction. Recall from Chapter 4, Equation 4.9, that $\Delta G^0 = -RT \ln K$. Since $\Delta G^0 = -nFE^0$ for the reaction, then

$$-RT \ln K = -nFE^0 \qquad (10.22)$$

or

$$E^0 = \frac{RT}{nF} \ln K$$

For the spontaneous reaction, ΔG^0 is negative and E^0 is positive.

EXAMPLE 10.6 Calculate the potential of the solution in Example 10.4 using the cerium half-reaction.

SOLUTION

In order to use this half-reaction, we need to solve for x. This can only be done using the equilibrium constant, which is obtained by equating the two half-reaction potentials. The

Ce^{4+}/Ce^{3+} half-reaction is

$$Ce^{4+} + e^- \rightleftharpoons Ce^{3+}$$

$$E = 1.61 - 0.059 \log \frac{[Ce^{3+}]}{[Ce^{4+}]}$$

Therefore,

$$1.61 - 0.059 \log \frac{[Ce^{3+}]}{[Ce^{4+}]} = 0.771 - 0.059 \log \frac{[Fe^{2+}]}{[Fe^{3+}]}$$

$$0.84 = 0.059 \log \frac{[Ce^{3+}][Fe^{3+}]}{[Ce^{4+}][Fe^{2+}]} = 0.059 \log K_{eq}$$

$$\frac{[Ce^{3+}][Fe^{3+}]}{[Ce^{4+}][Fe^{2+}]} = 10^{0.84/0.059} = 10^{14.2} = 1.6 \times 10^{14} = K_{eq}$$

Note that the large magnitude of K_{eq} indicates the reaction lies far to the right at equilibrium. Now, since the volumes cancel, we can use millimoles instead of millimoles/milliliters (molarity) and

$$[Ce^{3+}] = 0.50 - x \approx 0.50 \text{ mmol}$$

$$[Ce^{4+}] = x \text{ mmol}$$

$$[Fe^{3+}] = 0.50 - x \approx 0.50 \text{ mmol}$$

$$[Fe^{2+}] = 1.0 + x \simeq 1.0 \text{ mmol}$$

Therefore,

$$\frac{(0.50 \text{ mmol})(0.50 \text{ mmol})}{(x \text{ mmol})(1.0 \text{ mmol})} = 1.6 \times 10^{14}$$

$$x = 1.6 \times 10^{-15} \text{ mmol } (=1.6 \times 10^{-16} \text{ M})$$

We see how very small $[Ce^{4+}]$ is. Nevertheless, it is finite, and knowing its concentration, we calculate the potential from the Nernst equation, using millimoles:

$$E = 1.61 - 0.059 \log \frac{[Ce^{3+}]}{[Ce^{4+}]} = 1.61 - 0.059 \log \frac{(0.50 \text{ mmol})}{(1.6 \times 10^{-15} \text{ mmol})}$$

$$= 0.75 \text{ V}$$

Obviously, it is easier to to make the calculations using the half-reaction we have the most information about; in essence, the potential of that half-reaction must be calculated anyway during the calculation using the other half-reaction. The calculations illustrate that in a mixture, the concentrations of all species **at equilibrium** are such that the potential of each half-reaction is the same. Note that the potential will be close to the standard potential (E^0) of the half-reaction in which there is an excess of the reactant; in this case, there is an excess of Fe^{2+}.

It should be pointed out here that the n values in the two half-reactions do not have to be equal in order to equate the Nernst equations. For convenience, the two half-reactions are generally adjusted to the same n value before equating the Nernst equations.

When there are stoichiometric amounts of reactants, for example, at the equivalence point of a titration, the equilibrium concentrations of the species in neither half-reaction is known and an approach similar to the calculation in Example 10.6 is required.

EXAMPLE 10.7 Calculate the potential of a solution obtained by reacting 10 mL each of 0.20 M Fe^{2+} and 0.20 M Ce^{4+}.

SOLUTION

The reactants are essentially quantitatively converted to equivalent quantities of Fe^{3+} and Ce^{3+} and the concentration of each of the products is 0.10 M (neglecting the amount of the reverse reaction):

$$Fe^{2+} + Ce^{4+} \rightleftharpoons Fe^{3+} + Ce^{3+}$$
$$\phantom{Fe^{2+}} x x 0.10 - x 0.10 - x$$

where x represents the molar concentration of Fe^{2+} and Ce^{4+}. We can solve for x as in Example 10.6 and then plug it in the Nernst equation for either half-reaction to calculate the potential (do this for practice). Another approach follows.

The potential is given by either Nernst equation:

$$E = E^0_{Fe^{3+},Fe^{2+}} - \frac{0.059}{n_{Fe}} \log \frac{[Fe^{2+}]}{[Fe^{3+}]}; \quad n_{Fe}E = n_{Fe}E^0_{Fe^{3+},Fe^{2+}} - 0.059 \log \frac{x \text{ mmol/mL}}{0.10 \text{ mmol/mL}}$$

$$E = E^0_{Ce^{4+},Ce^{3+}} - \frac{0.059}{n_{Ce}} \log \frac{[Ce^{3+}]}{[Ce^{4+}]}; \quad n_{Ce}E = n_{Ce}E^0_{Ce^{4+},Ce^{3+}} - 0.059 \log \frac{0.10 \text{ mmol/mL}}{x \text{ mmol/mL}}$$

Note that Nernst equations for both species are written for reductions even though one of the species, here Fe, is actually being oxidized in the reaction. We can add these equations together and solve for E, the potential of each half-reaction, and hence the potential of the solution at equilibrium:

$$n_{Fe}E + n_{Ce}E = n_{Fe}E^0_{Fe^{3+},Fe^{2+}} + n_{Ce}E^0_{Ce^{4+},Ce^{3+}} - 0.059 \log \frac{x \text{ mmol/mL}}{0.10 \text{ mmol/mL}} \cdot \frac{0.10 \text{ mmol/mL}}{x \text{ mmol/mL}}$$

$$E = \frac{n_{Fe}E^0_{Fe^{3+},Fe^{2+}} + n_{Ce}E^0_{Ce^{4+},Ce^{3+}}}{n_{Fe} + n_{Ce}} = \frac{(1)0.77 + (1)1.61}{1 + 1} = 1.19 \text{ V}$$

The above approach is general, that is, E for stoichiometric quantities of reactants (E at the equivalence point of a titration) is given by

$$\boxed{E = \frac{n_1 E^0_1 + n_2 E^0_2}{n_1 + n_2}} \tag{10.23}$$

where n_1 and E^0_1 are the n value and standard potential for one half-reaction, and n_2 and E^0_2 are the values for the other half-reaction. In other words, E is the weighted average of the E^0 values. In the above example, it was the simple average since the n values were each unity. This equation holds only for reactions in which there are no polyatomic species (e.g., $Cr_2O_7^{2-}$) and no hydrogen ion dependence (or when the pH is zero). The equation contains additional terms if pH and concentration factors must be considered (see Problem 30). It can be applied if **formal potentials** are used, that is, for the specified conditions of acidity (see Section 10.7).

10.5 TITRATION CURVES

We can use our understanding of redox equilibria to describe titration curves for redox titrations. The shape of a titration curve can be predicted from the E^0 values of the analyte half-reaction and the titrant half-reaction. Roughly, the potential change in going from one side of the equivalence point to the other will be equal to the difference in the two E^0 values; the potential will be near E^0 for the analyte half-reaction before the equivalence point and near that of the titrant half-reaction beyond the equivalence point. Consider the titration of 100 mL of 0.1 M Fe^{2+} with 0.1 M Ce^{4+} in 1 M HNO_3. Each millimole Ce^{4+} will oxidize one millimole Fe^{2+}, and so the end point will occur at 100 mL. The titration curve is shown in Figure 10.2. This is actually a plot of the potential of the titration solution relative to the NHE, whose potential by definition is zero. In Chapter 11, we will see that the potential difference between redox half-cells can be measured with inert electrodes such as platinum in a cell similar to that in Figure 10.1. The electrode dipped in the titration or test solution is called the **indicator electrode,** and the other is called the **reference electrode,** whose potential remains constant. Hence, the potential of the indicator electrode will change relative to that of the reference electrode as indicated in Figure 10.2. The potential of the solution is plotted against the volume of titrant. This is analogous to plotting the pH of a solution versus the volume of titrant in an acid–base titration, or pM against volume of titrant in a precipitation or complexometric titration. In a redox titration, it is the potential of the solution rather than the pH that changes with concentration. At the beginning of the titration, we have only a solution of Fe^{2+}, and so we cannot calculate the potential. As soon as the first drop of titrant is added, a known amount of Fe^{2+} is converted to Fe^{3+}, and we know the ratio of $[Fe^{2+}]/[Fe^{3+}]$. So the potential can be determined from the Nernst equation of this couple. It will be near the E^0 value for this couple (the sample) **before** the end point.

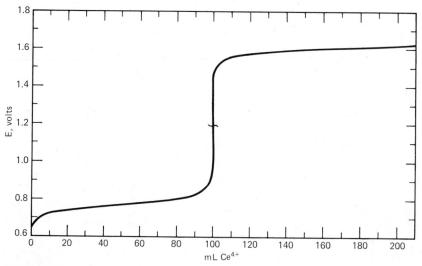

FIGURE 10.2
Titration curve for 100 mL 0.1 M Fe^{2+} versus 0.1 M Ce^{4+}.

Note that, since $[Fe^{2+}]/[Fe^{3+}]$ is equal to unity at the midpoint of the titration and log 1 = 0, the potential is equal to E^0 at this point in the titration. This will only be true if the half-reaction is symmetrical. For example, in the half-reaction I_2 + $2e^- \rightleftharpoons 2I^-$, the $[I^-]$ would be twice $[I_2]$ at midway in a titration, and the ratio would be $[I^-]^2/[I_2] = (2)^2/(1) = 4$. So the potential would be less than E^0 by $(-0.059/2) \log 4$, or $-0.018V$.

At the equivalence point of our titration, we have the following conditions:

$$Fe^{2+} + Ce^{4+} \rightleftharpoons Fe^{3+} + Ce^{3+}$$
$$\phantom{Fe^{2+}}x x C - x C - x$$

where C is the concentration of Fe^{3+}, which we know since all the Fe^{2+} is converted to Fe^{3+} (x is negligible compared to C). Now, we have an unknown quantity in both half-reactions, and so we must solve for x by equating the two Nernst equations, as was done in Example 10.6. Then, we can calculate the potential from either half-reaction. Alternatively, Equation 10.23 could be used since this is a symmetrical reaction.

Beyond the equivalence point, we have an excess of Ce^{4+} and an unknown amount of Fe^{2+}. Since we now have more information about the Ce^{4+}/Ce^{3+} half-reaction, it is easier to calculate the potential from its Nernst equation. Note that here, with an excess of titrant, the potential is near the E^0 value of the *titrant*. At 200% of the titration, $[Ce^{4+}]/[Ce^{3+}] = 1$, and E is E^0 of the cerium couple.

The example below illustrates that the magnitude of the end point break is directly related to the difference in the E^0 values of the sample and the titrant half-reactions. At least a 0.2-V difference is required for a sharp end point.

The equivalence point for this titration is indicated in Figure 10.2. Because the reaction is symmetrical, the equivalence point (inflection point of the curve—that point at which it is steepest) occurs at the midpoint of the rising part of the curve. In nonsymmetrical titrations, the inflection point will not occur at the midpoint. For example, in the titration of Fe^{2+} with MnO_4^-, the steepest portion occurs near the top of the break because of the consumption of protons in the reaction, causing it to be nonsymmetrical.

EXAMPLE 10.8   Calculate the potential of the solution in the above titration of 100 mL of 0.100 M Fe^{2+} at 10.0, 50.0, 100, and 200 mL of 0.100 M Ce^{4+}.

SOLUTION
The reaction is

$$Fe^{2+} + Ce^{4+} \rightleftharpoons Fe^{3+} + Ce^{3+}$$

10.0 mL: mmol Ce^{4+} added = 0.100 M × 10.0 mL = 1.00 mmol
     mmol Fe^{2+} reacted = 1.00 mmol = mmol Fe^{3+} formed
     mmol Fe^{2+} left = 0.100 M × 100 mL − 1.00 mmol = 9.0 mmol Fe^{2+}

$$E = 0.771 - 0.059 \log \frac{9.0}{1.00} = 0.715 \text{ V}$$

50.0 mL: One-half the Fe^{2+} is converted to Fe^{3+} (5.00 mmol each)

$$E = 0.771 - 0.059 \log \frac{5.00}{5.00} = 0.771 \text{ V}$$

$$100 \text{ mL: mmol Fe}^{3+} = 10.0 - x \approx 10.0$$
$$\text{mmol Fe}^{2+} = x$$
$$\text{mmol Ce}^{3+} = 10.0 - x \approx 10.0$$
$$\text{mmol Ce}^{4+} = x$$

We must solve for x. Since our calculations are for when equilibrium is achieved between the two half-reactions, the two Nernst equations are equal:

$$0.771 - \frac{0.059}{1} \log \frac{[Fe^{2+}]}{[Fe^{3+}]} = 1.61 - \frac{0.059}{1} \log \frac{[Ce^{3+}]}{[Ce^{4+}]}$$

$$-0.84 = -0.059 \log \frac{[Fe^{3+}][Ce^{3+}]}{[Fe^{2+}][Ce^{4+}]} = -0.059 \log K_{eq}$$

$$K_{eq} = 1.7 \times 10^{14}$$

Substituting into K_{eq} and solving for x (use millimoles, since volumes cancel):

$$\frac{(10.0)(10.00)}{(x)(x)} = 1.7 \times 10^{14}$$

$$x = 7.7 \times 10^{-7} \text{ mmol Fe}^{2+} = \text{mmol Ce}^{4+}$$

Use either half-reaction to calculate the potential:

$$E = 0.771 - 0.059 \log \frac{7.7 \times 10^{-7}}{10.0} = 1.19 \text{ V}$$

Compare this with the potential calculated in Example 10.7. Try calculating this potential using the Ce^{4+}/Ce^{3+} Nernst equation. Note that this potential is halfway between the two E^0 potentials.

200 mL: We have 100 mL excess titrant (Ce^{4+}). It is now easier to use the Ce^{4+}/Ce^{3+} half-reaction:

$$\text{mmol Ce}^{3+} = 10.0 - x \approx 10.0$$
$$\text{mmol Ce}^{4+} = 0.100 \text{ M} \times 100 \text{ mL} + x \approx 10.0 \text{ mmol}$$
$$E = 1.61 - 0.059 \log \frac{10.0}{10.0} = 1.61 \text{ V}$$

(We could have used the Fe^{3+}/Fe^{2+} half-reaction to calculate this potential by calculating x ($[Fe^{2+}]$) as above from K_{eq}.)

For nonsymmetrical reactions, we must keep track of the ratio in which the chemicals react. Also, if protons are consumed or produced in the reaction, the change in $[H^+]$ must be calculated.

EXAMPLE 10.9 Calculate the potential of the solution at the equivalence point in the titration of 100 mL of 0.100 M Fe^{2+} in 0.500 M H$_2$SO$_4$ with 100 mL of 0.0200 M MnO$_4^-$.[4]

---

[4]In 0.5 M H$_2$SO$_4$, the second proton is only about 2% dissociated (calculate it from the ionization constant and see!); but for simplicity, we will assume that it is completely ionized to give a H$^+$ concentration of 1 M.

SOLUTION
The reaction is

$$5Fe^{2+} + MnO_4^- + 8H^+ \rightleftharpoons 5Fe^{3+} + Mn^{2+} + 4H_2O$$
$$\quad x \quad\quad 1/5x \quad\quad\quad C - x \quad 1/5C - 1/5x$$

$$mmol\ Fe^{3+} = 0.100\ M \times 100\ mL - x \approx 10.0$$
$$mmol\ Fe^{2+} = x$$
$$mmol\ Mn^{2+} = 1/5(10.0) - 1/5x \approx 2.00$$
$$mmol\ MnO_4^- = 1/5x$$

Solve for x by equating the two Nernst equations (since we are at equilibrium, they are equal). Multiply the Fe^{2+}/Fe^{3+} half-reaction by 5 to equate the electrons:

$$0.771 - \frac{0.059}{5} \log \frac{[Fe^{2+}]^5}{[Fe^{3+}]^5} = 1.51 - \frac{0.059}{5} \log \frac{[Mn^{2+}]}{[MnO_4^-][H^+]^8}$$

$$-0.74 = -\frac{0.059}{5} \log \frac{[Mn^{2+}][Fe^{3+}]^5}{[MnO_4^-][Fe^{2+}]^5[H^+]^8} = -\frac{0.059}{5} \log K_{eq}$$

$$K_{eq} = 5._0 \times 10^{62}$$

We started with $1.00 \times 100 = 100$ mmol H^+. We consumed 8 mmol per 5 mmol Fe^{2+} reacted ($= 8/5 \times 10.0 = 16.0$ mmol). Therefore, we have 84 mmol H^+ left in 200 mL, or 0.42 M. For the other species, we can use millimoles since the volumes all cancel:

$$\frac{(2.00)(10.0)^5}{(1/5x)(x)^5(0.42)^8} = 5._0 \times 10^{62}$$

$$x = 1.1 \times 10^{-9}\ mmol\ Fe^{2+};\ mmol\ MnO_4^- = 1/5(1.1 \times 10^{-9})$$
$$= 2.2 \times 10^{-10}$$

Use either half-reaction to calculate the potential:

$$E = 0.771 - 0.059 \log \frac{1.1 \times 10^{-9}}{10} = 1.35_9\ V$$

E^0 for the Mn^{2+}/MnO_4^- couple is 1.51 V. Note that the potential halfway between the two E^0 potentials is 1.14 V. The equivalence point (inflection point) for this unsymmetrical titration reaction is therefore closer to the titrant couple and the titration curve is unsymmetrical.

10.6 DETECTION OF THE END POINT

Obviously, the end point can be determined by measuring potential of the solution with an indicating electrode relative to a reference and plotting this against the volume of titrant. This will be described in the next chapter. But as in other titrations, it is usually more convenient to use a visual indicator. There are three methods used for visual indication.

Self-Indication

If the titrant is highly colored, this color may be used to detect the end point. For example, a 0.02 M solution of potassium permanganate is deep purple. A dilute solution of potassium permanganate is pink. The product of its reduction, Mn^{2+}, is nearly colorless, being a very faint pink. During a titration with potassium permanganate, the purple color of the MnO_4^- is removed as soon as it is added because it is reduced to Mn^{2+}. As soon as the titration is complete, a fraction of a drop of excess MnO_4^- solution imparts a definite pink color to the solution, indicating that the reaction is complete. Obviously, the end point does not occur at the equivalence point, but at a fraction of a drop beyond. The error is small and can be corrected for by running a blank titration, or it is accounted for in standardization.

Starch Indicator

This indicator is used for titrations involving iodine. Starch forms a not very reversible complex with I_2 that is a very dark-blue color. The color reaction is sensitive to very small amounts of iodine. In titrations of reducing agents with iodine, the solution remains colorless up to the equivalence point. A fraction of a drop of excess titrant turns the solution a definite blue.

Redox Indicators

The above two methods of indication do not depend on the potential of the solution, although the completeness of the titration reaction and hence the sharpness of the end point do. Examples of these first two methods of visual indication are few, and most types of redox titrations are detected using **redox indicators.** These are highly colored dyes that are weak reducing or oxidizing agents that can be oxidized or reduced; the colors of the oxidized and reduced forms are different. The oxidation state of the indicator and hence its color will depend on the potential of the solution. A half-reaction and Nernst equation can be written for the indicator:

$$Ox_{In} + ne^- \rightleftharpoons Red_{In} \tag{10.24}$$

$$E_{In} = E_{In}^0 - \frac{0.059}{n} \log \frac{[Red_{In}]}{[Ox_{In}]} \tag{10.25}$$

The potential of the solution determines E_{In} and hence the ratio of $[Red_{In}]/[Ox_{In}]$. This is analogous to the ratio of the different forms of a pH indicator being determined by the pH of the solution. So the ratio, and therefore the color, will change as the potential of the solution changes. If we assume, as with acid–base indicators, that the ratio must change from 10/1 to 1/10 in order that a sharp color change can be seen, then **a potential equal to 2 × (0.059/n) V is required.** If n for the indicator is equal to 1, then a 0.12-V change is required. If E_{In}^0 **is near the equivalence point potential** of the titration, where there is a rapid change in potential in excess of 0.12 V, then the color change occurs at the equivalence point. Again, this is analogous to the requirement that the pK_a value of an acid–base indicator be near the pH of the equivalence point.

   If there is a hydrogen ion dependence in the indicator reaction, Equation 10.24, then this will appear in the corresponding Nernst equation, Equation 10.25, and the potential at which the indicator changes color will be displaced from E_{In}^0 by the hydrogen ion term.

So, redox indicators will have a transition range over a certain potential, and this transition range must fall within the steep equivalence point break of the titration curve. The redox indicator reaction must be *rapid,* and to use terms of the electrochemist, it must be *reversible.* If the reaction is slow or is *irreversible* (slow rate of electron transfer), the color change will be gradual and a sharp end point will not be detected.

There are not many good redox indicators. Table 10.2 lists some of the common indicators arranged in order of decreasing standard potentials. Ferroin [tris(1,10-phenanthroline) iron(II) sulfate] is one of the best indicators. It is useful for many titrations with cerium(IV). It is oxidized from the red color to a pale blue at the equivalence point. Other phenanthroline-type indicators are listed in the table. Diphenylaminesulfonic acid is used as an indicator for titrations with dichromate. The potential of the $Cr_2O_7^{2-}/Cr^{3+}$ couple is lower than that of the cerium couple, and so this indicator with a lower E^0 is required. The color at the end point is purple. The indicator used may depend on the sample titrated, since the magnitude of the end point break is also dependent on the potential of the sample half-reaction.

10.7 FORMAL POTENTIAL

The E^0 values listed in Table 10.1 are for the case where *all* species are at an activity of 1 M. However, the potential of a half-reaction may depend on the conditions of the solution. For example, the E^0 value for $Ce^{4+} + e^- \rightleftharpoons Ce^{3+}$ is 1.61 V. However, this potential can be changed by changing the acid used to acidify the solution. (See Table D.5 in the Appendix.) This happens because the anion of the acid complexes with the cerium, and the concentration of free cerium ion is thereby reduced.

If we know the form of the complex, we could write a new half-reaction involving the acid anion and determine an E^0 value for this reaction, keeping the acid and all other species at unit activity. However, the complexes are frequently of unknown composition. So we define the **formal potential** and designate this as $E^{0\prime}$. This is the standard potential of a redox couple with the oxidized and reduced forms at 1 M concentrations and **with the solution conditions specified.** For example, the formal potential of the Ce^{4+}/Ce^{3+} couple in 1 M HCl is 1.28 V. The Nernst equation is written as usual, using the formal potential in place of the standard potential. Some formal potentials are listed in the table in the Appendix.

TABLE 10.2
Redox Indicators

| | Color | | | |
Indicator	Reduced Form	Oxidized Form	Solution	E^0, V
Nitroferroin	Red	Pale blue	1 M H$_2$SO$_4$	1.25
Ferroin	Red	Pale blue	1 M H$_2$SO$_4$	1.06
Diphenylaminesulfonic acid	Colorless	Purple	Diluted acid	0.84
Diphenylamine	Colorless	Violet	1 M H$_2$SO$_4$	0.76
Methylene blue	Blue	Colorless	1 M acid	0.53
Indigo tetrasulfonate	Colorless	Blue	1 M acid	0.36

Dependence of Potential on pH

Hydrogen or hydroxyl ions are involved in many redox half-reactions. The potential of these redox couples can be changed by changing the pH of the solution. Consider the As(V)/As(III) couple:

$$H_3AsO_4 + 2H^+ + 2e^- \rightleftharpoons H_3AsO_3 + H_2O \tag{10.26}$$

$$E = E^0 - \frac{0.059}{2} \log \frac{[H_3AsO_3]}{[H_3AsO_4][H^+]^2} \tag{10.27}$$

This can be rearranged to[5]

$$E = E^0 + 0.059 \log [H^+] - \frac{0.059}{2} \log \frac{[H_3AsO_3]}{[H_3AsO_4]} \tag{10.28}$$

or

$$E = E^0 - 0.059 \, pH - \frac{0.059}{2} \log \frac{[H_3AsO_3]}{[H_3AsO_4]} \tag{10.29}$$

The term $E^0 - 0.059 \, pH$, where E^0 is the standard potential for the half-reaction, can be considered as equal to a formal potential $E^{0\prime}$, which can be calculated from the pH of the solution.[6]

In strongly acid solution, H_3AsO_4 will oxidize I^- to I_2. But in neutral solution, the potential of the As(V)/As(III) couple ($E^{0\prime} = 0.146$ V) is less than that for I_2/I^-, and the reaction goes in the reverse; that is, I_2 will oxidize H_3AsO_3.

Dependence of Potential on Complexation

If one or both of the ions in a redox couple are complexed, the concentration of the free ion is reduced. This causes the potential of the couple to change. For example, E^0 for the Fe^{3+}/Fe^{2+} couple is 0.771 V. In HCl solution, the Fe^{3+} is complexed, probably to a variety of species. This reduces the concentration of Fe^{3+} and so the potential is decreased. In 1 M HCl, the formal potential is 0.70 V. If we assume that the complex is $FeCl_4^-$, then the half-reaction would be

$$FeCl_4^- + e^- \rightleftharpoons Fe^{2+} + 4Cl^- \tag{10.30}$$

and if we assume that [HCl] is constant at 1 M,

$$E = 0.70 - 0.059 \log \frac{[Fe^{2+}]}{[FeCl_4^-]} \tag{10.31}$$

In effect, we have stabilized the Fe^{3+} by complexing it, making it more difficult to reduce. So the reduction potential is decreased. If we complexed the Fe^{2+}, the reverse effect would be observed.

---

[5]The H^+ term in the log term can be separated as $(-0.059/2) \log (1/[H^+]^2) = (+0.059/2) \log [H^+]^2$. The squared term can be brought to the front of the log term to give $0.059 \log [H^+]$.

[6]Actually, this is an oversimplification of the effect of pH in this case, because H_3AsO_4 and H_3AsO_3 are also weak acids, and the effect of their ionization, that is, their K_a values, should be taken into account as well.

10.8   LIMITATIONS OF ELECTRODE POTENTIALS

Electrode potentials (E^0 or $E^{0\prime}$) will predict whether a given reaction will occur, but they indicate nothing about the **rate** of the reaction. If a reaction is reversible, it will occur fast enough for a titration. But if the rate of the electron transfer step is slow, the reaction may be so slow that equilibrium will be reached only after a very long time. We say that such a reaction is **irreversible.**

Some reactions in which one half-reaction is irreversible do occur rapidly. Several oxidizing and reducing agents containing oxygen are reduced or oxidized irreversibly but may be speeded up by addition of an appropriate catalyst. The oxidation of arsenic(III) by cerium(IV) is slow, but it is catalyzed by a small amount of osmium tetroxide, OsO_4.

These slow reactions can serve as the basis for determining the trace amounts of substances that catalyze them. This will be discussed in Chapter 18.

So, while electrode potentials are useful for predicting many reactions, they do not assure the success of a given reaction. They are useful in that they will predict that a reaction will **not** occur if the potential differences are not sufficient.

10.9   TITRATIONS INVOLVING IODINE: IODIMETRY AND IODOMETRY

Redox titrations are among the most important types of analyses performed in many areas of application. Examples in clinical laboratories are rare, since most analyses are for traces, but these titrations are still extremely useful for standardizing reagents. You should be familiar with some of the more commonly used titrants.

Iodine is an oxidizing agent that can be used to titrate fairly strong reducing agents. On the other hand, iodide ion is a mild reducing agent and serves as the basis for determining strong oxidizing agents.

Iodimetry

Iodine is a moderately strong oxidizing agent and can be used to titrate reducing agents. **Titrations with I_2 are called iodimetric methods.** These titrations are usually performed in neutral or mildly alkaline (pH 8) to weakly acid solutions. If the pH is too alkaline, I_2 will disproportionate to hypoiodate and iodide:

$$I_2 + 2OH^- = IO^- + H_2O \tag{10.32}$$

There are three reasons for keeping the solution from becoming strongly acidic. First, the starch used for the end point detection tends to hydrolyze or decompose in strong acid, and so the end point may be affected. Second, the reducing power of several reducing agents is increased in neutral solution. For example, consider the reaction of I_2 with As(III):

$$H_3AsO_3 + I_2 + H_2O = H_3AsO_4 + 2I^- + 2H^+ \tag{10.33}$$

This equilibrium is affected by the hydrogen ion concentration. At low hydrogen ion concentration, the equilibrium is shifted to the right. We have already seen in Equation 10.29 that in neutral solution the potential of the As(V)/As(III) couple is decreased sufficiently that arsenic(III) will reduce I_2. But in acid solution, the equilibrium is shifted the other way, and the reverse reaction occurs. The third reason for avoiding acid solutions is that the I^- produced in the reaction tends to be oxidized by dissolved oxygen in acid solution:

$$4I^- + O_2 + 4H^+ \rightarrow 2I_2 + 2H_2O \tag{10.34}$$

The pH for the titration of arsenic(III) with I_2 can be maintained neutral by adding $NaHCO_3$. The bubbling action of the CO_2 formed also removes the dissolved oxygen and maintains a blanket of CO_2 over the solution to prevent air oxidation of the I^-.

Because I_2 is not a strong oxidizing agent, the number of reducing agents that can titrated is limited. Nevertheless, several examples exist, and the moderate oxidizing power of I_2 makes it a more selective titrant than the strong oxidizing agents. Some commonly determined substances are listed in Table 10.3. Antimony behaves similarly to arsenic, and the pH is critical for the same reasons. Tartrate is added to complex the antimony and keep it in solution to prevent hydrolysis.

Although high-purity I_2 can be obtained by sublimation, iodine solutions are usually standardized against a primary standard reducing agent such as As_2O_3. Arsenious oxide is not soluble in acid, and so it is dissolved in sodium hydroxide. The solution is neutralized after dissolution is complete. If arsenic(III) solutions are to be kept for any length of time, they should be neutralized or acidified, because arsenic(III) is slowly oxidized in alkaline solution.

Iodine has a low solubility in water but the complex I_3^- is very soluble. So iodine solutions are prepared by dissolving I_2 in a concentrated solution of potassium iodide:

$$I_2 + I^- \rightarrow I_3^- \tag{10.35}$$

I_3^- is therefore the actual species used in the titration.

EXAMPLE 10.10 The purity of a hydrazine, N_2H_4 (note: a violent poison!), sample is determined by titration with iodine. A sample of the oily liquid weighing 1.4286 g is dissolved in water and diluted to 1 L in a volumetric flask. A 50.00-mL aliquot is taken with a pipet and titrated with standard iodine solution, requiring 42.41 mL. The iodine was standardized against 0.4123 g primary standard As_2O_3 by dissolving the As_2O_3 in a small amount of NaOH solution, adjusting the pH to 8, and titrating, requiring 40.28 mL iodine solution. What is the percent purity by weight of the hydrazine?

SOLUTION
Standardization:

$$H_2AsO_3^- + I_2 + H_2O \rightarrow HAsO_4^{2-} + 2I^- + 3H^+$$

Each As_2O_3 gives $2H_2AsO_3^-$, so mmol I_2 = 2 × mmol As_2O_3.

$$M_{I_2} \times 40.28 \text{ mL} = \frac{412.3 \text{ mg}}{197.85 \text{ mg/mmol As}_2O_3} \times 2 \text{ mmol I}_2/\text{mmol As}_2O_3$$

$$M_{I_2} = 0.1034_7 \text{ mmol/mL}$$

Analysis:

$$N_2H_4 + 2I_2 \rightarrow N_2 + 4H^+ + 4I^-$$

$$\text{mmol } N_2H_4 = \tfrac{1}{2} \times \text{mmol } I_2$$

$$\text{weight of } N_2H_4 \text{ titrated} = 1.4286 \text{ g} \times \frac{50.00}{1000.0} = 0.07143 \text{ g}$$

$$\% \ N_2H_4 = [0.1034_7 \ M \times 42.41 \text{ mL} \times \tfrac{1}{2}(\text{mmol } N_2H_4/\text{mmol } I_2)$$

$$\times \ 32.045 \text{ mg/mmol } N_2H_4]/71.43 \text{ mg} \times 100\% = 98.43\%$$

The problem can, of course, also be worked using equivalents and normality. To illustrate this approach, the calculation is given. The equivalent weight of As_2O_3 is one-fourth its formula weight, since each arsenic is oxidized from $+3$ to $+5$ valence and there are two arsenics per molecule. Therefore,

$$N_{I_2} \times 40.28 \text{ mL} = \frac{412.3 \text{ mg}}{197.85/4 \ (\text{mg/meq } As_2O_3)}$$

$$N_{I_2} = 0.2069_4 \text{ meq/mL}$$

Each nitrogen in hydrazine is oxidized from -2 to 0 valence for a total valence change of 4 electrons per molecule. Therefore, its equivalent weight is one-fourth its formula weight. Hence,

$$\% \ N_2H_4 = \frac{0.2069_4 \ N \times 42.41 \text{ mL} \times 32.045/4 \ (\text{mg/meq } N_2H_4)}{71.43 \text{ mg}} \times 100\% = 98.43\%$$

Note that because of the low molecular weight of hydrazine, it would have been difficult to weigh out the required sample to four significant figures, and by titrating an accurately measured aliquot, a larger sample can be weighed.

Iodometry

Iodide ion is a weak reducing agent and will reduce strong oxidizing agents. It is not used, however, as a titrant mainly because of lack of convenient visual indicator system, as well as other factors such as speed of the reaction.

When an excess of iodide is added to a solution of an oxidizing agent, **I_2 is produced in an amount equivalent to the oxidizing agent present.** This I_2 can, therefore, be titrated with a reducing agent, and the result will be the same as if the oxidizing agent were titrated directly. The titrating agent used is sodium thiosulfate.

TABLE 10.3
Some Substances Determined by Iodimetry

Substance Determined	Reaction with Iodine	Solution Conditions
H_2S	$H_2S + I_2 \rightarrow S + 2I^- + 2H^+$	Acid solution
SO_3^{2-}	$SO_3^{2-} + I_2 + H_2O \rightarrow SO_4^{2-} + 2I^- + 2H^+$	
Sn^{2+}	$Sn^{2+} + I_2 \rightarrow Sn^{4+} + 2I^-$	Acid solution
As(III)	$H_2AsO_3^- + I_2 + H_2O \rightarrow HAsO_4^{2-} + 2I^- + 3H^+$	pH 8
N_2H_4	$N_2H_4 + 2I_2 \rightarrow N_2 + 4H^+ + 4I^-$	

Analysis of an oxidizing agent in this way is called an **iodometric method.** Consider, for example, the determination of dichromate:

$$Cr_2O_7^{2-} + 6I^- \text{ (excess)} + 14H^+ \rightarrow 2Cr^{3+} + 3I_2 + 7H_2O \qquad (10.36)$$

$$\boxed{I_2 + 2S_2O_3^{2-} \rightarrow 2I^- + S_4O_6^{2-}} \qquad (10.37)$$

Each $Cr_2O_7^{2-}$ produces $3I_2$, which in turn react with $6S_2O_3^{2-}$. The millimoles of $Cr_2O_7^{2-}$ are equal to one-sixth the millimoles of $S_2O_3^{2-}$ used in the titration.

Iodate can be determined iodometrically:

$$IO_3^- + 5I^- + 6H^+ \rightarrow 3I_2 + 3H_2O \qquad (10.38)$$

Each IO_3^- produces $3I_2$, which again react with $6S_2O_3^{2-}$, and the millimoles of IO_3^- are obtained by multiplying the millimoles of $S_2O_3^{2-}$ used in the titration by $\frac{1}{6}$.

EXAMPLE 10.11   0.200-g sample containing copper is analyzed iodometrically. Copper(II) is reduced to copper(I) by iodide:

$$2Cu^{2+} + 4I^- \rightarrow 2\underline{CuI} + I_2$$

What is the percent copper in the sample if 20.0 mL of 0.100 M $Na_2S_2O_3$ is required for titration of the liberated I_2?

SOLUTION
One-half mole of I_2 is liberated per mole of Cu^{2+}, and since each I_2 reacts with $2S_2O_3^{2-}$, each Cu^{2+} is equivalent to one $S_2O_3^{2-}$, and mmol Cu^{2+} = mmol $S_2O_3^{2-}$:

$$\% \text{ Cu} = \frac{0.100 \text{ mmol/mL} \times 20.0 \text{ mL} \times \text{Cu}}{200 \text{ mg}} \times 100\%$$

$$= \frac{0.100 \text{ mmol/mL} \times 20.0 \text{ mL} \times 63.54 \text{ mg/mmol}}{200 \text{ mg}} \times 100\% = 63.5\%$$

Why not titrate the oxidizing agents directly with the thiosulfate? Because strong oxidizing agents oxidize thiosulfate to oxidation states higher than that of tetrathionate (e.g., to SO_4^{2-}), but the reaction is generally not stoichiometric. Also, several oxidizing agents form mixed complexes with thiosulfate (e.g., Fe^{3+}). By reaction with iodide, the strong oxidizing agent is destroyed and an equivalent amount of I_2 is produced, which will react stoichiometrically with thiosulfate and for which a satisfactory indicator exists. The titration can be considered a direct titration.

The end point for iodometric titrations is detected with starch. The disappearance of the blue starch–I_2 color indicates the end of the titration. The starch is not added at the beginning of the titration when the iodine concentration is high. Instead, it is added just before the end point when the dilute iodine color becomes pale yellow. There are two reasons for such timing. One is that the iodine–starch complex is only slowly dissociated, and a diffuse end point would result if a large amount of the iodine were adsorbed on the starch. The second reason is that most iodometric titrations are performed in strongly acid medium and the starch has a tendency to hydrolyze in acid solution. The reason for using acid solutions is that reactions between many oxidizing agents and iodide are promoted by high acidity. Thus,

$$2MnO_4^- + 10I^- + 16H^+ \rightarrow 5I_2 + 2Mn^{2+} + 8H_2O \qquad (10.39)$$

$$H_2O_2 + 2I^- + 2H^+ \rightarrow I_2 + 2H_2O \qquad (10.40)$$

and so on.

The titration should be performed rapidly to minimize air oxidation of the iodide. Stirring should be efficient in order to prevent local excesses of thiosulfate, because it is decomposed in acid solution:

$$S_2O_3^{2-} + 2H^+ \rightarrow H_2SO_3 + S \qquad (10.41)$$

Indications of such excess is the presence of colloidal sulfur, which makes the solution cloudy. In iodometric methods, a large excess of iodide is added to promote the reaction (common ion effect). The unreacted iodide does not interfere, but it may be air-oxidized if the titration is not performed immediately.

Sodium thiosulfate solution is standardized iodometrically against a pure oxidizing agent such as $K_2Cr_2O_7$, KIO_3, $KBrO_3$, or metallic copper (dissolved to give Cu^{2+}). With potassium dichromate, the deep green color of the resulting chromic ion makes it a little more difficult to determine the iodine–starch end point. When titrating copper(II) iodometrically, the end point is diffuse unless thiocyanate ion is added. The primary reaction is given in Example 10.11. But iodine is adsorbed on the surface of the cuprous iodide precipitate and only slowly reacts with the thiosulfate titrant. The thiocyanate coats the precipitate with CuSCN and displaces the iodine from the surface. The potassium thiocyanate should be added near the end point, since it is slowly oxidized by iodine to sulfate. The pH must be buffered to around 3. If it is too high, copper(II) hydrolyzes and cupric hydroxide will precipitate. If it is too low, air oxidation of iodide becomes appreciable, because it is catalyzed in the presence of copper. Copper metal is dissolved in nitric acid, with oxides of nitrogen being produced. These oxides will oxidize iodide, and they are removed by addition of urea.

Some examples of iodometric determinations are listed in Table 10.4.

TABLE 10.4
Iodometric Determinations

Substance Determined	Reaction with Iodide
MnO_4^-	$2MnO_4^- + 10I^- + 16H^+ \rightleftharpoons 2Mn^{2+} + 5I_2 + 8H_2O$
$Cr_2O_7^{2-}$	$Cr_2O_7^{2-} + 6I^- + 14H^+ \rightleftharpoons 2Cr^{3+} + 3I_2 + 7H_2O$
IO_3^-	$IO_3^- + 5I^- + 6H^+ \rightleftharpoons 3I_2 + 3H_2O$
BrO_3^-	$BrO_3^- + 6I^- + 6H^+ \rightleftharpoons Br^- + 3I_2 + 3H_2O$
Ce^{4+}	$2Ce^{4+} + 2I^- \rightleftharpoons 2Ce^{3+} + I_2$
Fe^{3+}	$2Fe^{3+} + 2I^- \rightleftharpoons 2Fe^{2+} + I_2$
H_2O_2	$H_2O_2 + 2I^- + 2H^+ \underset{\text{[Mo(VI) catalyst]}}{\rightleftharpoons} 2H_2O + I_2$
As(V)	$H_3AsO_4 + 2I^- + 2H^+ \rightleftharpoons H_3AsO_3 + I_2 + H_2O$
Cu^{2+}	$2Cu^{2+} + 4I^- \rightleftharpoons 2CuI + I_2$
HNO_2	$2HNO_2 + 2I^- \rightleftharpoons I_2 + 2NO + H_2O$
SeO_3^{2-}	$SeO_3^{2-} + 4I^- + 6H^+ \rightleftharpoons Se + 2I_2 + 3H_2O$
O_3	$O_3 + 2I^- + 2H^+ \rightleftharpoons O_2 + I_2 + H_2O$
	(can determine in presence of O_2 above pH 7)
Cl_2	$Cl_2 + 2I^- \rightleftharpoons 2Cl^- + I_2$
Br_2	$Br_2 + 2I^- \rightleftharpoons 2Br^- + I_2$
HClO	$HClO + 2I^- + H^+ \rightleftharpoons Cl^- + I_2 + H_2O$

EXAMPLE 10.12 A solution of $Na_2S_2O_3$ is standardized iodometrically against 0.1262 g of high-purity $KBrO_3$, requiring 44.97 mL $Na_2S_2O_3$. What is the molarity of the $Na_2S_2O_3$?

SOLUTION
The reactions are

$$BrO_3^- + 6I^- + 6H^+ \rightarrow Br^- + 3I_2 + 3H_2O$$
$$3I_2 + 6S_2O_3^{2-} \rightarrow 6I^- + 3S_4O_6^{2-}$$

So mmol $S_2O_3^{2-} = 6 \times$ mmol BrO_3^-:

$$M_{S_2O_3^{2-}} \times 44.97 \text{ mL} = \frac{126.2 \text{ mg}}{167.01 \text{ (mg/mmol } KBrO_3)} \times 6 \text{ (mmol } S_2O_3^{2-}/\text{mmol } BrO_3^-)$$

$$M_{S_2O_3^{2-}} = 0.1008_2 \text{mmol/mL}$$

10.10 TITRATIONS WITH OXIDIZING AGENTS

We have already mentioned some oxidizing agents that can be used as titrants. The titrant should be fairly stable and should be convenient to prepare and to handle. If it is too strong an oxidizing agent, it will be so reactive that its stability will not be great. Thus, fluorine is one of the strongest oxidizing agents known, but it is certainly not convenient to use in the analytical laboratory ($E^0 = 3.06$ V). Chlorine would make a good titrant, except, that it is volatile from aqueous solution, and to prepare and maintain a standard solution would be difficult.

Potassium permanganate is a widely used oxidizing titrant. It acts as a self-indicator for end point detection and is a very strong oxidizing agent ($E^0 = 1.51$ V). The solution is stable if precautions are taken in its preparation. When the solution is first prepared, small amounts of reducing impurities in the solution reduce a small amount of the MnO_4^-. In neutral solution, the reduction product of this permanganate is MnO_2, rather than Mn^{2+} produced in acid medium. The MnO_2 acts as a catalyst for further decomposition of the permanganate, which produces more MnO_2, and so on. This is called **autocatalytic decomposition.** The solution can be stabilized by removing the MnO_2. So, before standardizing, the solution is boiled to hasten oxidation of all impurities and is allowed to stand overnight. The MnO_2 is then removed by filtering through a sintered-glass filter. Potassium permanganate can be standardized by titrating primary standard sodium oxalate, $Na_2C_2O_4$, which, dissolved in acid, forms oxalic acid:

$$5H_2C_2O_4 + 2MnO_4^- + 6H^+ = 10CO_2 + 2Mn^{2+} + 8H_2O \qquad (10.42)$$

The solution must be heated for rapid reaction. The reaction is catalyzed by the Mn^{2+} product and it goes very slowly at first until some Mn^{2+} is formed. Pure electrolytic iron metal can also be used as the primary standard. It is dissolved in acid and reduced to Fe^{2+} for titration (see below, under Preparation of the Analyte Solution).

A difficulty arises when permanganate titrations of iron(II) are performed in the presence of chloride ion. The oxidation of chloride ion to chlorine by permanganate at room temperature is normally slow. However, the oxidation is catalyzed by the presence of iron. If an iron sample has been dissolved in hydrochloric acid, or if stannous chloride has been used to reduce it to iron(II) (see next page), the titration can be performed by adding the **Zimmermann–Reinhardt reagent.** This contains manganese(II) and phosphoric acid. The manganese(II) reduces the potential of the MnO_4^-/Mn^{2+} couple sufficiently so that permanganate will not oxidize chloride ion; the formal potential is less than E^0, due to the large concentration of Mn^{2+}. This decrease in the potential decreases the magnitude of the endpoint break. Therefore, phosphoric acid is added to complex the iron(III) and decrease the potential of the Fe^{3+}/Fe^{2+} couple also; the iron(II) is not complexed. In other words, iron(III) is removed from the solution as it is formed to shift the equilibrium of the titration reaction to the right and give a sharp end point. The overall effect is still a large potential break in the titration curve, but the entire curve has been shifted to a lower potential.

An added effect of complexing the iron(III) is that the phosphate complex is nearly colorless, while the chloro complex (normally present in chloride medium) is deep yellow. A sharper end point color change results.

Potassium dichromate, $K_2Cr_2O_7$, is a slightly weaker oxidizing agent than potassium permanganate. The great advantage of this reagent is its availability as a primary standard, and the solution need not be standardized in most cases. In the titration of iron(II), standardizing potassium dichromate against electrolytic iron is preferable, however, because the green color of the chromic ion introduces a small error in the end point (diphenylamine sulfonate indicator). Standardization is necessary only for the most accurate work.

Oxidation of chloride ion is not a problem with dichromate. However, the formal potential of the $Cr_2O_7^{2-}/Cr^{3+}$ couple is reduced from 1.33 V to 1.00 V in 1 M hydrochloric acid, and phosphoric acid must be added to reduce the potential of the Fe^{3+}/Fe^{2+} couple. Such addition is also necessary because it decreases the equivalence point potential to near the standard potential for the diphenylamine sulfonate indicator (0.84 V). Otherwise, the end point would occur too soon.

Cerium(IV) is a powerful oxidizing agent. Its formal potential depends on the acid used to keep it is solution (it hydrolyzes to form ceric hydroxide if the solution is not acid). Titrations are usually performed in sulfuric acid or perchloric acid. In the former acid, the formal potential is 1.44 V, and in the latter acid, it is 1.70 V. So cerium(IV) is a stronger oxidizing agent in perchloric acid. Cerium(IV) can be used for most titrations in which permanganate is used, and it possesses a number of advantages. It is a very strong oxidizing agent and its potential can be varied by choice of the acid used. The rate of oxidation of chloride ion is slow, even in the presence of iron, and titrations can be carried out in the presence of moderate amounts of chloride without the use of a Zimmermann–Reinhardt-type preventive solution. The solution can be heated but should not be boiled, or chloride ion will be oxidized. Sulfuric acid solutions of cerium(IV) are stable indefinitely. Nitric acid and perchloric acid solutions, however, do decompose, but only slowly. An added advantage of cerium is that a salt of cerium(IV), ammonium hexanitratocerate, $(NH_4)_2Ce(NO_3)_6$, can be obtained as a primary standard, and the solution does not have to be standardized. The main disadvantage of cerium(IV) is its increased cost over potassium permanganate, although this should not be a serious factor if a saving in time is achieved. Ferroin is a suitable indicator for many cerate titrations.

Cerium(IV) solutions can be standardized against primary standard As_2O_3, $Na_2C_2O_4$, or electrolytic iron. The reaction with arsenic(III) is slow, and it must be catalyzed by adding either osmium tetroxide (OsO_4) or iodine monochloride (ICl). Ferroin is used as the indicator. The reaction with oxalate is also slow at room temperature, and the same catalyst can be used. The reaction is rapid, however, at room temperature in the presence of 2 M perchloric acid. Nitroferroin is used as the indicator.

Cerium(IV) solutions to be standardized are usually prepared from ammonium sulfatocerate, $(NH_4)_4Ce(SO_4)_4 \cdot 2H_2O$, ammonium nitratocerate, $(NH_4)_2Ce(NO_3)_6$ (*not* the high-purity primary standard variety though), or hydrous ceric oxide, $CeO_2 \cdot 4H_2O$. Primary standard ammonium nitratocerate is used only if the solution is not to be standardized, because of its increased expense.

10.11   TITRATIONS WITH REDUCING AGENTS

Standard solutions of reducing agents are not used as widely as oxidizing agents are, because most of them are oxidized by dissolved oxygen. They are, therefore, less convenient to prepare and use. **Thiosulfate** is the only common reducing agent that is stable to air oxidation and that can be kept for long periods of time. This is the reason that iodometric titrations are so popular for determining oxidizing agents. However, stronger reducing agents than iodide ion are sometimes required.

Iron(II) is only slowly oxidized by air in sulfuric acid solution and is a common titrating agent. It is not a strong reducing agent ($E^0 = 0.771$ V) and can be used to titrate strong oxidizing agents such as cerium(IV), chromium(VI) (dichromate), and vanadium(V) (vanadate). Ferroin is a good indicator for the first two titrations, and oxidized diphenylamine sulfonate is used for the last titration. The iron(II) standardization should be checked daily.

Chromium(II) and **titanium(III)** are very powerful reducing agents, but they are readily air-oxidized and difficult to handle. The standard potential of the former is -0.41 V (Cr^{3+}/Cr^{2+}) and that of the latter is 0.04 V (TiO^{2+}/Ti^{3+}). The oxidized forms of copper, iron, silver, gold, bismuth, uranium, tungsten, and other metals have been titrated with chromium(II). The principal use of Ti^{3+} is in the titration of iron(III) as well as copper(II), tin(IV), chromate, vanadate, and chlorate.

10.12   PREPARATION OF THE ANALYTE SOLUTION

When samples are dissolved, the element to be analyzed is usually in a mixed oxidation state or is in an oxidation state other than that required for titration. There are various oxidizing and reducing agents that can be used to convert different metals to certain oxidation states prior to titration. The excess preoxidant or pre-reductant must generally be removed before titrating the metal ion.

Reduction of the Sample Prior to Titration

The reducing agent should not interfere in the titration or, if it does, unreacted reagent should be readily removable. Most reducing agents will, of course, react with oxidizing titrants, and they must be removable. **Sodium sulfite,** Na_2SO_3, and **sulfur dioxide** are good reducing agents in acid solution ($E^0 = 0.17$ V), and the excess can be removed by bubbling with CO_2 or in some cases by boiling. If SO_2 is not available, sodium sulfite or bisulfite can be added to an acidified solution. Thallium(III) is reduced to the $+1$ state, arsenic(V) and antimony(V) to the $+3$ state, vanadium(V) to the $+4$ state, and selenium and tellurium to the elements.

Iron(III) and copper(II) can be reduced to the $+2$ and $+1$ states, respectively, if thiocyanate is added to catalyze the reaction.

Stannous chloride, $SnCl_2$, is usually used for the reduction of iron(III) to iron(II) for titrating with cerium(IV) or dichromate. The reaction is rapid in the presence of chloride (hot HCl). When iron samples (for example, ores) are dissolved (usually in hydrochloric acid), part or all of the iron is in the $+3$ oxidation state and must be reduced. The reaction with stannous chloride is

$$Fe^{3+} + SnCl_2 \rightarrow Fe^{2+} + SnCl_4 \qquad (10.43)$$

The reaction is complete when the yellow color of the iron(III)–chloro complex disappears. The excess tin(II) is removed by addition of mercuric chloride:

$$SnCl_2 + 2HgCl_2 \text{ (excess)} \rightarrow SnCl_4 + \underline{Hg_2Cl_2} \text{ (calomel)} \qquad (10.44)$$

A large excess of cold $HgCl_2$ must be added rapidly with stirring. If too little is added, or if it is added slowly, some of the mercury will be reduced, by local excesses of $SnCl_2$, to elemental mercury, a gray precipitate. The calomel, Hg_2Cl_2, which is a milky-white precipitate, does not react at an appreciable rate with dichromate or cerate, but mercury will. In order to prevent a large excess of tin(II) and subsequent danger of formation of mercury, the stannous chloride is added dropwise until the yellow color of iron(III) just disappears. If a gray precipitate is noted after adding the $HgCl_2$, the sample must be discarded. Stannous chloride can also be used to reduce As(V) to As(III), Mo(VI) to Mo(V), and, with $FeCl_3$ catalyst, U(VI) to U(IV).

Metallic reductors are widely used for preparing samples. These are usually used in a granular form in a column through which the sample solution is passed. The sample is eluted from the column by slowly passing dilute acid through it. The oxidized metal ion product does not interfere in the titration and no excess reductant is present, since the metal is insoluble. For example, lead can be used to reduce tin(IV):

$$\underline{Pb} + Sn^{4+} \rightarrow Pb^{2+} + Sn^{2+} \qquad (10.45)$$

The solution eluted from the column will contain Pb^{2+} and Sn^{2+}, but no Pb. Table 10.5 lists several commonly used metallic reductors and some elements they will reduce. The reductions are carried out in acid solution. In the case of zinc, metallic zinc is amalgamated to prevent attack by acid to form hydrogen:

$$Zn + 2H^+ \rightarrow Zn^{2+} + H_2 \qquad (10.46)$$

Sometimes, the reduced sample is rapidly air-oxidized and the sample must be titrated under an atmosphere of CO_2, by the addition of sodium bicarbonate, to an acid solution. Air must be excluded from tin(II) and titanium(III) solutions. Sometimes, elements rapidly air-oxidized are eluted from the column into an iron(III) solution, with the end of the column immersed in the solution. The iron(III) is reduced by the sample to give an equivalent amount of iron(II), which can be titrated with dichromate. Molybdenum(III), which is oxidized to molybdenum(VI) by the iron, and copper(I) are determined in this way.

Oxidation of the Sample Prior to Titration

Very strong oxidizing agents are required to oxidize most elements. We mentioned in Chapter 3 that hot anhydrous **perchloric acid** is a strong oxidizing agent. Per-

TABLE 10.5
Metallic Reductors

Reductor	Element Reduced
Zn (Hg) (Jones reductor)	Fe(III) → Fe(II), Cr(VI) → Cr(II), Cr(III) → Cr(II), Ti(IV) → Ti(III), V(V) → V(II), Mo(VI) → Mo(III), Ce(IV) → Ce(III), Cu(II) → Cu
Ag (1 M HCl) (Walden reductor)	Fe(III) → Fe(II), U(VI) → U(IV), Mo(VI) → Mo(V) (2 M HCl), Mo(VI) → Mo(III) (4 M HCl), V(V) → V(IV), Cu(II) → Cu(I)
Al	Ti(IV) → Ti(III)
Pb	Sn(IV) → Sn(II), U(VI) → U(IV)
Cd	$ClO_3^- \rightarrow Cl^-$

chloric acid can be used to oxidize chromium(III) to dichromate. Dilute perchloric acid is not a strong oxidizing agent, and the solution needs only to be diluted following the oxidation. Chlorine is a product of perchloric acid reduction, and this must be removed by boiling the diluted solution.

Potassium persulfate, $K_2S_2O_8$, is a powerful oxidizing agent that can be used to oxidize chromium(III) to dichromate, vanadium(IV) to vanadium(V), cerium(III) to cerium(IV), and manganese(II) to permanganate. The oxidations are carried out in hot acid solution, and a small amount of silver(I) catalyst must be added. The excess persulfate is destroyed by boiling. This boiling will always reduce some permanganate.

Bromine can be used to oxidize several elements, such as Tl(I) to Tl(III) and iodide to iodate. The excess is removed by adding phenol, which is brominated. **Chlorine** is an even stronger oxidizing agent. **Permanganate** oxidizes V(IV) to V(V) and Cr(III) to Cr(VI). The latter reaction is rapid only in alkaline solution. It has been used to oxidize trace quantities of Cr(III) in acid solution, however, by heating. Excess permanganate is destroyed by adding hydrazine, the excess of which is destroyed by boiling. **Hydrogen peroxide** will oxidize Fe(II) to Fe(III), Co(II) to Co(III) in mildly alkaline solution, and Cr(II) to Cr(VI) in strongly alkaline solution.

For most redox determinations, specified procedures have been described for the preparation of different elements in various types of samples. You should be able to recognize the reasoning behind the operations from the discussions in this chapter.

The only redox titration commonly applied in the clinical laboratory is for the analysis of calcium in biological fluids. Calcium oxalate is precipitated and filtered, the precipitate is dissolved in acid, and the oxalate, which is equivalent to the calcium present, is titrated with standard potassium permanganate solution. This method is largely replaced now by more convenient techniques such as complexometric titration with EDTA (Chapter 9) or measurement by atomic absorption spectrophotometry (Chapter 14).

QUESTIONS

1. What is an oxidizing agent? A reducing agent?

2. What is the Nernst equation?

3. Describe the ways in which the end points of redox titrations may be detected visually.

4. What is the standard potential? The formal potential?

5. Distinguish between iodimetry and iodometry.
6. Why are iodimetric titrations usually done in neutral solution and iodometric titrations in acid solution?
7. What is the function of a salt bridge in an electrochemical cell?
8. What is the NHE? SHE?
9. The standard potential for the half-reaction $M^{4+} + 2e^- = M^{2+}$ is $+0.98$ V. Is M^{2+} a good or a poor reducing agent?
10. What should be the minimum potential difference between two half-reactions so that a sharp end point will be obtained in a titration involving the two half-reactions?
11. Does the end point in a permanganate titration coincide with the equivalence point? Explain and suggest how any discrepancies might be corrected.
12. Why cannot standard or formal electrode potentials always be used to predict whether a given titration will work?
13. Explain the function of the Zimmermann–Reinhardt reagent in the titration of iron(II) with permanganate.

PROBLEMS

Balancing Redox Reactions

14. Balance the following reactions:
    (a) $IO_3^- + I^- \rightarrow I_2$ (acid solution)
    (b) $Se_2Cl_2 \rightarrow H_2SeO_3 + Se + HCl$
    (c) $H_3PO_3 + HgCl_2 \rightarrow Hg_2Cl_2 + H_3PO_4 + HCl$
15. Balance the following reactions:
    (a) $MnO_4^{2-} \rightarrow MnO_2 + MnO_4^-$ (alkaline solution)
    (b) $MnO_4^- + H_2S \rightarrow Mn^{2+} + S$
    (c) $SbH_3 + Cl_2O \rightarrow H_4Sb_2O_7 + HCl$
    (d) $FeS + NO_3^- \rightarrow Fe^{3+} + NO_2 + S$ (acid solution)
    (e) $Al + NO_3^- \rightarrow AlO_2^- + NH_3$
    (f) $FeAsS + ClO_2 \rightarrow Fe^{3+} + AsO_4^{3-} + SO_4^{2-} + Cl^-$ (acid solution)
    (g) $K_2NaCo(NO_2)_6 + MnO_4^- \rightarrow K^+ + Na^+ + Co^{3+} + NO_3^- + Mn^{2+}$ (acid solution)

Redox Strengths

16. Arrange the following substances in decreasing order of oxidizing strengths: H_2SeO_3, H_3AsO_4, Hg^{2+}, Cu^{2+}, Zn^{2+}, O_3, $HClO$, K^+, Co^{2+}.
17. Arrange the following substances in decreasing order of reducing strengths: I^-, V^{3+}, Sn^{2+}, Co^{2+}, Cl^-, Ag, H_2S, Ni, HF.
18. Which of the following pairs would be expected to give the largest end point break in a titration of one component with the other in each pair?
    (a) $Fe^{2+} - MnO_4^-$ or $Fe^{2+} - Cr_2O_7^{2-}$
    (b) $Fe^{2+} - Ce^{4+}$ (H_2SO_4) or $Fe^{2+} - Ce^{4+}$ ($HClO_4$)
    (c) $H_3AsO_3 - MnO_4^-$ or $Fe^{2+} - MnO_4^-$
    (d) $Fe^{3+} - Ti^{2+}$ or $Sn^{2+} - I_3^-$

Galvanic Cells

19. Write the equivalent galvanic cells for the following reactions (assume all concentrations are 1 M):
    (a) $6Fe^{2+} + Cr_2O_7^{2-} + 14H^+ \rightleftharpoons 6Fe^{3+} + 2Cr^{3+} + 7H_2O$
    (b) $IO_3^- + 5I^- + 6H^+ \rightleftharpoons 3I_2 + 3H_2O$
    (c) $Zn + Cu^{2+} \rightleftharpoons Zn^{2+} + Cu$
    (d) $Cl_2 + H_2SeO_3 + H_2O \rightleftharpoons 2Cl^- + SeO_4^{2-} + 4H^+$

20. For each of the following cells, write the cell reactions:
    (a) $Pt/V^{2+}, V^{3+}//PtCl_4^{2-}, PtCl_6^{2-}, Cl^-/Pt$
    (b) $Ag/AgCl(s)/Cl^-//Fe^{3+}, Fe^{2+}/Pt$
    (c) $Cd/Cd^{2+}//ClO_3^-, Cl^-, H^+/Pt$
    (d) $Pt/I^-, I_2//H_2O_2, H^+/Pt$

Potential Calculations

21. What is the potential of a solution containing 0.50 M $KBrO_3$ and 0.20 M Br_2 at pH 2.5?
22. What is the potential of the solution prepared by adding 90 mL of 5.0 M KI to 10 mL of 0.10 M H_2O_2 buffered at pH 2.0?
23. Equal volumes of 0.20 M $TlNO_3$ and 0.20 M Br_2 in 1 M HCl are mixed. What is the potential of the solution?
24. A solution of a mixture of Pt^{4+} and Pt^{2+} is 3.0 M in HCl, which produces the chloro complexes of the Pt ions (see Problem 31). If the solution is 0.015 M in Pt^{4+} and 0.025 M in Pt^{2+}, what is the potential of the half-reaction?
25. Equal volumes of 0.100 M UO_2^{2+} and 0.100 M V^{2+} in 0.10 M H_2SO_4 are mixed. What would the potential of a platinum electrode dipped in the solution be at equilibrium? Assume H_2SO_4 is completely ionized.
26. Calculate the potential of the solution in the titration of 50.0 mL of 0.100 M Fe^{2+} in 1.00 M $HClO_4$ with 0.0167 M $Cr_2O_7^{2-}$ at 10, 25, 50, and 60 mL titrant added.
27. Calculate the potential of the solution in the titration of 100 mL of 0.100 M Fe^{2+} in 0.500 M H_2SO_4 with 0.0200 M $KMnO_4$ at 10.0, 50.0, 100, and 200 mL of 0.0200 M $KMnO_4$ titrant. Assume the H_2SO_4 is completely ionized.

Equivalence Point Potentials

28. What would be the potential at the equivalence point in the titration of Fe^{3+} with Sn^{2+}?
29. Equation 10.23 was derived using two half-reactions with equal n values. Derive a similar equation for the following reaction (used in Problem 28), using the numerical n values:

$$2Fe^{3+} + Sn^{2+} \rightleftharpoons 2Fe^{2+} + Sn^{4+}$$

30. Derive an equation similar to Equation 10.23 for the following reaction, remembering to include a hydrogen ion term:

$$5Fe^{2+} + MnO_4^- + 8H^+ \rightleftharpoons 5Fe^{3+} + Mn^{2+} + 4H_2O$$

Use the derived equation to calculate the end point potential in the titration in Example 10.9 and compare with the value obtained in that example by calculating equilibrium concentrations.

Cell Voltages

31. From the standard potentials of the following half-reactions, determine the reaction that will occur, and calculate the cell voltage from the reaction:

$$PtCl_6^{2-} + 2e^- = PtCl_4^{2-} + 2Cl^-$$

$$E^0 = 0.68 \text{ V}$$

$$V^{3+} + e^- = V^{2+}$$

$$E^0 = -0.255 \text{ V}$$

32. Calculate the voltages of the following cells:
    (a) Pt/I^- (0.100 M), I_3^- (0.0100 M)//IO_3^- (0.100 M), I_2 (0.0100 M), H^+ (0.100 M)/Pt
    (b) $Ag/AgCl(s)/Cl^-$ (0.100 M)//UO_2^{2+} (0.200 M), U^{4+} (0.0500 M), H^+ (1.00 M)/Pt

    (c) Pt/Tl^+ (0.100 M), Tl^{3+} (0.0100 M)//MnO_4^- (0.0100 M), Mn^{2+} (0.100 M), H^+ (pH 2.00)/Pt

33. For the following cells, calculate the cell voltage before reaction and each half-cell potential after reaction. Also calculate the equilibrium constants of the reactions:

    (a) Zn/Zn^{2+} (0.250 M)//Cd^{2+} (0.0100 M)/Cd

    (b) Pb/Pb^{2+} (0.0100 M)//I_3^- (0.100 M), I^- (1.00 M)/Pt

34. From the standard potentials, determine the reaction between the following half-reactions and calculate the corresponding standard cell voltage:

$$VO_2^+ + 2H^+ + e^- = VO^{2+} + H_2O$$

$$UO_2^{2+} + 4H^+ + 2e^- = U^{4+} + 2H_2O$$

Quantitative Calculations

35. Selenium in a 10.0-g soil sample is distilled as the tetrabromide, which is collected in aqueous solution where it is hydrolyzed to SeO_3^{2-}. The SeO_3^{2-} is determined iodometrically, requiring 4.5 mL standard thiosulfate solution for the titration. If the thiosulfate titer is 0.049 mg $K_2Cr_2O_7$/mL, what is the concentration of selenium in the soil in ppm?

36. The calcium in a 5.00-mL serum sample is precipitated as CaC_2O_4 with ammonium oxalate. The filtered precipitate is dissolved in acid, the solution is heated, and the oxalate is titrated with 0.00100 M $KMnO_4$, requiring 4.94 mL. Calculate the concentration of calcium in the serum in meq/L (equivalents based on charge).

37. A 2.50-g sample containing As_2O_5, Na_2HAsO_3, and inert material is dissolved and the pH is adjusted to neutral with excess $NaHCO_3$. The As(III) is titrated with 0.150 M I_2 solution, requiring 11.3 mL to just reach the end point. Then, the solution (all the arsenic in the +5 state now) is acidified with HCl, excess KI is added, and the liberated I_2 is titrated with 0.120 M $Na_2S_2O_3$, requiring 41.2 mL. Calculate the percent As_2O_5 and Na_2HAsO_3 in the sample.

38. If 1.00 mL $KMnO_4$ solution will react with 0.125 g Fe^{2+} and if 1.00 mL $KHC_2O_4 \cdot H_2C_2O_4$ solution will react with 0.175 mL of the $KMnO_4$ solution, how many milliliters of 0.200 M NaOH will react with 1.00 mL of the textroxalate solution? (All three protons on the tetroxalate are titratable.)

RECOMMENDED REFERENCES

Nernst Equation
1. L. Meites, "A 'Derivation' of the Nernst Equation for Elementary Quantitative Analysis," *J. Chem. Ed.*, **29** (1952) 142.

Redox Equations
2. C. A. Vanderhoff, "A Consistent Treatment of Oxidation–Reduction," *J. Chem. Educ.*, **25** (1948) 547.

3. R. G. Yolman, "Writing Oxidation–Reduction Equations," *J. Chem. Educ.*, **36** (1959) 215.

Equivalence Point Potential
4. A. J. Bard and S. H. Simpsonsen, "The General Equation for the Equivalence Point Potential in Oxidation–Reduction Titrations," *J. Chem. Educ.*, **37** (1960) 364.

Electrode Sign Conventions
5. F. C. Anson, "Electrode Sign Convention," *J. Chem. Educ.*, **36** (1959) 394.

6. T. S. Light and A. J. de Bethune, "Recent Developments Concerning the Signs of Electrode Conventions," *J. Chem. Educ.*, **34** (1957) 433.

chapter 11

POTENTIOMETRY AND POTENTIOMETRIC TITRATIONS

In Chapter 10, we mentioned measurement of the potential of a solution and described a platinum electrode whose potential was determined by the half-reaction of interest. This was a special case, and there are a number of electrodes available for measuring solution potentials. In this chapter, we list the various types of electrodes that can be used for measuring solution potentials and how to select the proper one for measuring a given analyte. The apparatus for making potentiometric measurements is described along with limitations and accuracies of potentiometric measurements. The important glass pH electrode is described, as well as standard buffers required for its calibration. The various kinds of ion-selective electrodes are discussed and, finally, the use of electrodes in potentiometric titrations is considered.

Electrodes measure activity rather than concentration, and we will use activities in this chapter in describing electrode potentials. An understanding of activity and the factors that affect it are important for direct potentiometric measurements, as in pH or ion-selective electrode measurements. You should, therefore, review the material on activity and activity coefficients in Chapter 4.

11.1 ELECTRODES OF THE FIRST KIND

An electrode of this type is a metal in contact with a solution containing its cation. An example is a silver metal electrode dipping in a solution of silver nitrate.

For all electrode systems, an electrode half-reaction can be written from which the potential of the electrode is determined. The electrode system can be represented by $M|M^{n+}$, in which the line represents an electrode–solution interface. For the silver electrode, we have

$$Ag|Ag^+ \tag{11.1}$$

and the half-reaction is

$$Ag^+ + e^- = Ag \tag{11.2}$$

The potential of the electrode is described by the Nernst equation:

$$\boxed{E = E^0_{Ag^+,Ag} - \frac{2.303RT}{F} \log \frac{1}{a_{Ag^+}}} \tag{11.3}$$

where a_{Ag^+} represents the **activity** of the silver ion (see Chapter 4). We will use the more correct unit of activity in discussions in this chapter because, in the interpretation of direct potentiometric measurements, significant errors would result if concentrations were used in calculations.

The potential calculated from Equation 11.3 is the potential *relative to the normal hydrogen electrode (NHE)*, and the potential increases with increasing Ag^+. That is, in a cell measurement using the NHE as the second half-cell, the voltage is

$$E_{measd.} = E_{cell} = E_{ind\ vs.\ NHE} = E_{ind} - E_{NHE} \tag{11.4}$$

where E_{ind} is the potential of the **indicator electrode** (the one that responds to the test solution, Ag^+ ions in this case). Since E_{NHE} is zero,

$$E_{cell} = E_{ind} \tag{11.5}$$

corresponds to writing the cell as

$$E_{ref}|solution|E_{ind} \tag{11.6}$$

and

$$E_{cell} = E_{right} - E_{left} = E_{ind} - E_{ref} = E_{ind} - constant \tag{11.7}$$

where E_{ref} is the potential of the **reference electrode,** whose potential is constant. Note that E_{cell} (or E_{ind}) may be positive or negative, depending on the activity of the silver ion or the relative potentials of the two electrodes. This is in contrast to the convention used in Chapter 10 for a galvanic cell, in which a cell was always set up to give a positive voltage and thereby indicate what the spontaneous cell reaction would be. Here, we are interested in how the potential of the test electrode (indicating electrode) changes with analyte concentration, as measured against some constant reference electrode. Equation 11.7 is arranged so that changes in E_{cell} reflect the same changes in E_{ind}, *including sign*. This point is discussed more below when we talk about cells and measurement of electrode potentials.

The activity of silver metal above, as with other pure substances, is taken as unity. So an electrode of this kind can be used to monitor the activity of a metal ion in solution. There are few reliable electrodes of this type, because many metals tend to form an oxide coating that changes the potential. A mercury electrode can be used to measure the activity of mercury ions:

$$Hg|Hg_2^+ \ or \ Hg|Hg^{2+}$$

11.2 ELECTRODES OF THE SECOND KIND

The general form of this type of electrode is $M|MX|X^{n-}$, where MX is a slightly soluble salt. An example is the silver–silver chloride electrode:

$$Ag|AgCl(s)|Cl^- \tag{11.8}$$

The (s) indicates a solid, (g) is used to indicate a gas, and (l) is used to indicate a pure liquid. The half-reaction is

$$AgCl + e^- = Ag + Cl^- \tag{11.9}$$

and the potential is defined by

$$\boxed{E = E^0_{AgCl,Ag} - \frac{2.303RT}{F} \log a_{Cl^-}} \tag{11.10}$$

This electrode, then, can be used to measure the activity of chloride ion in solution. Note that, as the activity of chloride increases, the potential decreases. A silver wire is coated with silver chloride precipitate (e.g., by electrically oxidizing it in a solution containing chloride ion, the reverse reaction of Equation 11.9). Actually, as soon as a silver wire is dipped in a chloride solution, it adopts a thin layer of silver chloride and pretreatment is usually not required.

Note that this electrode can be used to monitor either a_{Cl^-} or a_{Ag^+}. It really measures ("sees") only silver ion, and the activity of this is determined by the solubility of the slightly soluble salt. Since $a_{Cl^-} = K_{sp}/a_{Ag^+}$, Equation 11.10 can be rewritten

$$E = E^0_{AgCl,Ag} - \frac{2.303RT}{F} \log \frac{K_{sp}}{a_{Ag^+}} \tag{11.11}$$

$$E = E^0_{AgCl,Ag} - \frac{2.303RT}{F} \log K_{sp} - \frac{2.303RT}{F} \log \frac{1}{a_{Ag^+}} \tag{11.12}$$

Comparing this with Equation 11.3, we see that

$$\boxed{E^0_{Ag^+,Ag} = E^0_{AgCl,Ag} - \frac{2.303RT}{F} \log K_{sp}} \tag{11.13}$$

K_{sp} here is really the thermodynamic activity coefficient K^0_{sp} (see Chapter 4), since we defined it in terms of activity. We could have arrived at an alternative form of Equation 11.10 by substituting K_{sp}/a_{Cl^-} for a_{Ag^+} in Equation 11.3 (see Example 11.1 below).

In a solution containing a mixture of Ag^+ and Cl^- (for instance, a titration of Cl^- with Ag^+), the concentrations of each **at equilibrium** will be such that the potential of a silver wire dipping in the solution can be calculated by either Equation 11.3 or Equation 11.10. This is completely analogous to the statement in Chapter 10 that the potential of one half-reaction must be equal to the potential of the other in a chemical reaction at equilibrium. Equations 11.2 and 11.9 are the two half-reactions in this case, and when one is subtracted from the other, the result is the **overall chemical reaction**

$$Ag^+ + Cl^- = AgCl \tag{11.14}$$

Note that as Cl^- is titrated with Ag^+, the former decreases and the latter increases. Equation 11.10 predicts an increase in potential as Cl^- decreases; and, similarly, Equation 11.12 predicts the same increase as Ag^+ increases.

The silver electrode can also be used to monitor other anions that form slightly soluble salts with silver, such as I^-, Br^-, and S^{2-}. The E^0 in each case would be that for the particular half-reaction $AgX + e^- = Ag + X^-$.

Another widely used electrode of this type is the **calomel electrode,** Hg, $Hg_2Cl_2(s)|Cl^-$. This will be described in more detail when we talk about *reference electrodes.*

EXAMPLE 11.1 Given that the standard potential of the calomel electrode is 0.268 V and that of the Hg/Hg_2^{2+} electrode is 0.789 V, calculate K_{sp} for calomel (Hg_2Cl_2).

SOLUTION
For $Hg_2^{2+} + 2e^- = Hg$,

$$E = 0.780 - \frac{0.05916}{2} \log \frac{1}{a_{Hg_2^{2+}}} \tag{1}$$

For $Hg_2Cl_2 + 2e^- = 2Hg + 2Cl^-$,

$$E = 0.268 - \frac{0.05916}{2} \log (a_{Cl^-})^2 \tag{2}$$

Since $K_{sp} = a_{Hg_2^{2+}} \cdot (a_{Cl^-})^2$,

$$E = 0.268 - \frac{0.05916}{2} \log \frac{K_{sp}}{a_{Hg_2^{2+}}} \tag{3}$$

$$E = 0.268 - \frac{0.05916}{2} \log K_{sp} - \frac{0.05916}{2} \log \frac{1}{a_{Hg_2^{2+}}} \tag{4}$$

From (1) and (4),

$$0.789 = 0.268 - \frac{0.05916}{2} \log K_{sp}$$

$$K_{sp} = 2.4_4 \times 10^{-18}$$

11.3 REDOX ELECTRODES

In the redox electrode, an inert metal is in contact with a solution containing the soluble oxidized and reduced forms of the redox half-reaction. This type of electrode was mentioned in Chapter 10.

The inert metal used is usually platinum. The potential of such an inert electrode is determined by the ratio of the reduced and oxidized species in the half-reaction:

$$M^{a+} + ne^- = M^{(a-n)+} \tag{11.15}$$

$$E = E^0_{M^{a+},M^{(a-n)+}} - \frac{2.303RT}{nF} \log \frac{a_{M^{(a-n)+}}}{a_{M^{a+}}} \tag{11.16}$$

An example is the measurement of the ratio of MnO_4^-/Mn^{2+}:

$$MnO_4^- + 8H^+ + 5e^- = Mn^{2+} + 4H_2O \tag{11.17}$$

$$E = E^0_{MnO_4^-,Mn^{2+}} - \frac{2.303RT}{5F} \log \frac{a_{Mn^{2+}}}{a_{MnO_4^-} \cdot (a_{H^+})^8} \tag{11.18}$$

The pH is usually held constant, and so the ratio $a_{Mn^{2+}}/a_{MnO_4^-}$ is measured, as in a redox titration.

A very important example of this type of electrode is the **hydrogen electrode,** $Pt|H_2, H^+$:

$$H^+ + e^- = \tfrac{1}{2}H_2 \tag{11.19}$$

$$E = E^0_{H^+,H_2} - \frac{2.303RT}{F} \log \frac{(p_{H_2})^{1/2}}{a_{H^+}} \tag{11.20}$$

The construction of the hydrogen electrode is shown in Figure 11.1. The pressure of gases, in atmospheres, is used in place of activities. If the hydrogen pressure is held at 1 atmosphere, then, since E^0 for Equation 11.19 is defined as zero,

$$E = -\frac{2.303RT}{F} \log \frac{1}{a_{H^+}} = -\frac{2.303RT}{F} pH \tag{11.21}$$

EXAMPLE 11.2   Calculate the pH of a solution whose potential at 25°C measured with a hydrogen electrode at an atmospheric pressure of 1.012 atm (corrected for the vapor pressure of water at 25°) is -0.324 V (relative to the NHE).

SOLUTION

$$-0.324 = -0.05916 \log \frac{(1.012)^{1/2}}{a_{H^+}}$$

$$= -0.05916 \log (1.012)^{1/2} - 0.05916 \, pH$$

$$pH = 5.48$$

While the hydrogen electrode is very important for specific applications (e.g., establishing standard potentials or the pH of standard buffers—see below), its use for routine pH measurements is limited. First, it is inconvenient to prepare and use. The partial pressure of hydrogen must be established at the measurement temperature. A layer of platinum black must be placed on the surface of the platinum electrode by cathodically electrolyzing in a H_2PtCl_6 solution. The platinum black provides a larger surface area for adsorption of hydrogen molecules and catalyzes their oxidation. Too much platinum black, however, can adsorb traces of other substances such as organic molecules, H_2S, and so on, causing erratic behavior of the electrode. The solution should not contain other oxidizing or reducing agents since these will alter the potential of the electrode.

FIGURE 11.1
The hydrogen electrode.

11.4 CELLS WITHOUT LIQUID JUNCTION

To make potential measurements, a complete cell consisting of two half-cells must be set up, as was described in Chapter 10. One half-cell usually comprises the test solution and an electrode whose potential is determined by the analyte we wish to measure. This electrode is the **indicator electrode.** The other half-cell is any arbitrary half-cell whose potential is not dependent on the analyte. This half-cell electrode is designated the **reference electrode.** Its potential is constant, and the measured cell voltage reflects the indicator electrode potential relative to that of the reference electrode. Since the reference electrode potential is constant, any **changes** in potential of the indicator electrode will be reflected by an equal change in the cell voltage.

There are two basic ways a cell may be set up, either without or with a salt bridge. The first is called a **cell without liquid junction.** An example of a cell of this type would be

$$Pt|H_2(g), HCl \text{ (solution)}|AgCl(s)|Ag \qquad (11.22)$$

The solid line represents an electrode–solution interface. An electrical cell such as this is a galvanic cell, and the cell is set up for the spontaneous cell reaction (positive E_{cell}). The hydrogen electrode is the anode, since its potential is the more negative (see Chapter 10 for a review of cell voltage conventions for galvanic cells). The potential of the left electrode would be given by Equation 11.20, and that for the right electrode would be given by Equation 11.10, and the cell voltage would be equal to the difference in these two potentials:

$$E_{cell} = \left(E^0_{AgCl,Ag} - \frac{2.303RT}{F} \log a_{Cl^-} \right)$$

$$- \left(E^0_{H^+,H_2} - \frac{2.303RT}{F} \log \frac{(p_{H_2})^{\frac{1}{2}}}{a_{H^+}} \right) \qquad (11.23)$$

$$E_{cell} = E^0_{AgCl,Ag} - E^0_{H^+,H_2} - \frac{2.303RT}{F} \log \frac{a_{H^+} a_{Cl^-}}{(p_{H_2})^{\frac{1}{2}}} \qquad (11.24)$$

The **cell reaction** would be (half-reaction)$_{right}$ − (half-reaction)$_{left}$ (to give a positive E_{cell} and the spontaneous reaction), or

$$AgCl + e^- = Ag + Cl^- \qquad (11.25)$$

$$-(H^+ + e^- = \tfrac{1}{2}H_2) \qquad (11.26)$$

$$AgCl + \tfrac{1}{2}H_2 = Ag + Cl^- + H^+ \qquad (11.27)$$

Equation 11.23 would also represent the cell voltage if the right half-cell were used as an indicating electrode in a potentiometric measurement of chloride ion and the left cell were the reference electrode (see Equations 11.6 and 11.7). That is, the voltage (and hence the indicator electrode potential) would decrease with increasing chloride ion. If we were to use the hydrogen electrode as the indicating electrode to measure hydrogen ion activity or pH, we would reverse the cell setup, Cell 11.22, and Equation 11.23, and the voltage (and indicator electrode potential) would increase with increasing acidity or decreasing pH.

A cell without liquid junction is always used for the most accurate measurements because there are no uncertain potentials to account for. However, there are few examples of cells without liquid junction (sometimes called **cells without transference**), and these are inconvenient to use. Therefore, the more convenient but less accurate cells with liquid junction are most commonly used.

11.5   CELLS WITH LIQUID JUNCTION

An example of this type of cell is

$$Hg|Hg_2Cl_2(s)|KCl \text{ (saturated)}||HCl \text{ (solution)}, H_2(g)|Pt \qquad (11.28)$$

The double line represents the **liquid junction** between two dissimilar solutions and is usually in the form of a **salt bridge.** The purpose of this is to prevent mixing of the two solutions. In this way, the potential of one of the electrodes will be constant, independent of the composition of the test solution, and determined by the solution in which it dips. The electrode on the left of cell 11.28 is the **saturated calomel electrode,** which is a commonly used reference electrode (see below). The cell is set up using the hydrogen electrode as the indicating electrode to measure pH.

The disadvantage of a cell of this type is that there is a potential associated with the liquid junction, called the **liquid-junction potential.** The potential of the above cell is

$$E_{cell} = (E_{right} - E_{left}) + E_j \qquad (11.29)$$

where E_j is the liquid-junction potential; E_j may be positive or negative. The liquid-junction potential results from the unequal diffusion of the ions on each side of the boundary.

A typical boundary might be a fine-porosity sintered-glass frit with two different solutions on either side of it; the frit prevents appreciable mixing of the two solutions. The simplest type of liquid junction occurs between two solutions containing the same electrolyte at different concentrations. An example is HCl (0.1 M)||HCl (0.01 M), illustrated in Figure 11.2. Both hydrogen ions and chloride ions will migrate across the boundary in both directions, but the net migration will be from the more concentrated to the less concentrated side of the boundary, the driving force for this migration being proportional to the concentration difference. Hydrogen ions migrate about five times faster than chloride ions. Therefore, a net positive charge is built up on the right side of the boundary, leaving a net negative charge on the left side; that is, there is a separation of charge, and this represents a potential. A steady state is rapidly achieved by the action of this built-up positive charge in repulsing the further migration of hydrogen ions; the converse applies to the negative charge on the left-hand side. Hence, a constant potential difference is quickly attained between the two solutions.

The E_j for this junction is $+40$ mV, and $E_{cell} = (E_{right} - E_{left}) + 40$ mV. This E_j is very large, owing to the rapid mobility of the hydrogen ion. As the concentration of HCl on the left side of the boundary is decreased, the net charge built up will be less and the liquid junction potential will be decreased until, at equal concentration, it will be zero, because equal amounts of HCl diffuse in each direction.

A second example of this type of liquid junction is $0.1\ M$ KCl/$0.01\ M$ KCl. This situation is completely analogous to that above, except that in this case the K^+ and Cl^- ions migrate at nearly the same rate, with the chloride ion moving only about 4% faster. So a net *negative* charge is built up on the right side of the junction. Thus, E_j will be negative and is equal to -1.0 mV.

The nearly equal migration of potassium and chloride ions makes it possible to decrease significantly the liquid-junction potential. Decreasing is possible because, if an electrolyte on one side of a boundary is in large excess over that on the other side, the flux of the migration of the ions of this electrolyte will be much greater than that of the more dilute electrolyte and the liquid-junction potential will be determined largely by the migration of this more concentrated electrolyte. Thus, E_j of the junction KCl $(3.5\ M)\|H_2SO_4\ (0.05\ M)$ is only -4 mV, even though the hydrogen ions diffuse at a much more rapid rate than sulfate.

Some examples of different liquid-junction potentials are given in Table 11.1. (The signs are for those as set up, and they would be the sign in a potentiometric measurement if the solution on the left were used for the salt bridge and the one on the right were the test solution. If solutions on each side of the junction were reversed, the signs of the junction potentials would be reversed.) It is apparent that the liquid-junction potential can be minimized by keeping a high concentration of a salt on one side of the boundary whose ions have nearly the same mobility, such as KCl. Ideally, the same high concentration of such a salt should be on both sides

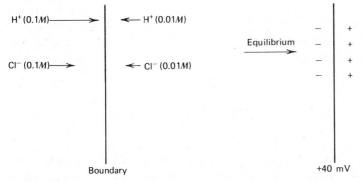

FIGURE 11.2
Representation of a liquid-junction potential.

TABLE 11.1
Some Liquid-Junction Potentials at 25°C[a]

Boundary	E_j, mV
0.1 M KCl‖0.1 M NaCl	+6.4
3.5 M KCl‖0.1 M NaCl	+0.2
3.5 M KCl‖1 M NaCl	+1.9
0.01 M KCl‖0.01 M HCl	−26
0.1 M KCl‖0.1 M HCl	−27
3.5 M KCl‖0.1 M HCl	+3.1
0.1 M KCl‖0.1 M NaOH	+18.9
3.5 M KCl‖0.1 M NaOH	+2.1
3.5 M KCl‖1 M NaOH	+10.5

[a]Adapted from G. Milazzo, *Electrochemie*. Vienna: Springer-Verlag, 1952; and D. A. MacInnes and Y. L. Yeh, *J. Am. Chem. Soc.*, **43** (1921) 2563.

of the junction. This is generally not possible for the test solution side of a salt bridge. However, the solution in the other half-cell in which the other end of the salt bridge forms a junction can often be made high in KCl to minimize that junction potential. As noted before, this half-cell, which is connected via the salt bridge to form a complete cell, is the reference electrode. See the discussion of the saturated calomel electrode below.

As the concentration of the (dissimilar) electrolyte on the other side of the boundary (in the test solution) increases, or as the ions are made different, the liquid-junction potential will get larger. Very rarely can the liquid-junction potential be considered to be negligible. The liquid-junction potential with neutral salts is less than when a strong acid or base is involved. The variation is due to the unusually high mobilities of the hydrogen ion and the hydroxyl ion. Therefore, **the liquid-junction potential will vary with the pH of the solution,** an important fact to remember in potentiometric pH measurements. A potassium chloride salt bridge, at or near saturation, is ually employed, except when these ions may interfere in a determination. Ammonium chloride or potassium nitrate may be used if the potassium or chloride ion interferes.

A commonly used salt bridge is a 3% agar-saturated potassium chloride salt bridge, which is prepared by adding 100 mL cold water to 3 g granulated agar. The mixture is heated on a steam bath and shaken until a homogeneous solution is obtained. Then, 25 g solid potassium chloride is added and the solution is stirred until the salt dissolves. The mixture will gel on cooling. Various other types of electrolyte junctions or salt bridges have been designed, such as a ground-glass joint, a wick of asbestos sealed into glass, a porous glass or ceramic plug, or a fine capillary drip.

11.6 REFERENCE ELECTRODES: THE SATURATED CALOMEL ELECTRODE

A commonly used reference electrode is the **saturated calomel electrode** (SCE). The term "saturated" refers to the concentration of potassium chloride; and at 25°C, the potential of the SCE is 0.242 V versus NHE.

An SCE is easily prepared and the proper construction is shown in Figure 11.3. A small amount of mercury is triturated (mixed) with some solid Hg_2Cl_2 (calomel), solid KCl, and enough saturated KCl to moisten the mixture. This is poured into the vessel containing sufficient mercury to cover a platinum wire electrode for making electrical contact. More solid KCl is added and the vessel is filled with saturated KCl solution. The salt bridge makes contact with the test solution, and the lead from the platinum wire electrode goes to one terminal of the potential measuring device. The salt bridge should contain the same salt used in the reference electrode to prevent contamination with foreign salts. That is, a potassium chloride salt bridge should be used with the SCE. If it is necessary to use a different salt bridge to prevent contamination of the test solution, then one end of this second salt bridge should be dipped in a beaker containing saturated potassium chloride. The reference electrode is dipped in this same solution to make electrical contact. The obvious disadvantage in this operation is that a second liquid-junction potential is introduced. This second E_j will be constant, and if calibration of the indicating electrode is performed, it is of no consequence.

A commercial probe-type SCE is shown in Figure 11.4. This contains a porous fiber as the salt bridge in the tip that allows very slow leakage of the saturated potassium chloride solution. It has a small mercury pool area and so the current it can pass without its potential being affected is more limited (as will be seen below, a small current is usually drawn during potential measurements). Also, the fiber salt bridge *has a high resistance* (about 2500 ohms), compared with the agar salt bridge. This limits the sensitivity in measurements with a potentiometer, in which the reading is made as close as possible at zero current flow; with increased external resistance, a given current deflection from zero will result in an increased error in the potential reading. This is no serious problem in many potential measurements (e.g., in titrations), but a lower-resistance electrode is still preferred. The

FIGURE 11.3
Bottle-type saturated calomel electrode.

FIGURE 11.4
Commercial saturated calomel electrode. (Courtesy of Arthur H. Thomas Company.)

fiber SCE is perfectly satisfactory for use with a pH meter, though, which is designed to make measurements with high-resistance electrodes.

EXAMPLE 11.3   Calculate the potential of the cell consisting of a silver electrode dipping in a silver nitrate solution with a_{Ag^+} 0.0100 M and an SCE reference electrode.

SOLUTION
Neglecting the liquid junction potential,

$$E_{cell} = E_{ind} - E_{ref}$$

$$E_{cell} = \left(E^0_{Ag^+, Ag} - 0.0592 \log \frac{1}{a_{Ag^+}} \right) - E_{SCE}$$

$$= 0.799 - 0.0592 \log \frac{1}{0.0100} - 0.242$$

$$= 0.439 \text{ V}$$

EXAMPLE 11.4   A cell voltage measured using an SCE reference electrode is -0.774
V. (The indicating electrode is the more negative half-cell.) What would the cell voltage
be with a silver/silver chloride reference electrode (1 M KCl; $E = 0.228$ V) or with a
NHE?

SOLUTION
The potential of the Ag/AgCl electrode is more negative than that of the SCE by 0.242 −
0.228 = 0.014 V. Hence, the cell voltage using the former electrode is less negative by
this amount:

$$E_{\text{vs. Ag/AgCl}} = E_{\text{vs. SCE}} + 0.014$$

$$= -0.774 + 0.014 = -0.760 \text{ V}$$

Similarly, the cell voltage using the NHE is 0.242 V less negative:

$$E_{\text{vs. NHE}} = E_{\text{vs. SCE}} + 0.242 \text{ V}$$

$$= -0.774 + 0.242 = -0.532 \text{ V}$$

11.7 MEASUREMENT OF POTENTIAL

The Potentiometer and pH Meter

There are two commonly used instruments for making potential measurements. One is the **potentiometer** and the other is the **pH meter** (a voltmeter). pH measurements with a glass (or other) electrode involve the measurement of potentials (see below).

The potentiometer can be used for measurements of low-resistance circuits. The pH meter is a voltage measuring device designed for use with high-resistance glass electrodes and can be used with both low- and high-resistance circuits. **Electrometers** can also be used with high-resistance circuits.

The principle of operation of the potentiometer is schematically represented in Figure 11.5. The potentiometer consists of a known voltage source connected to a movable slidewire (variable resistance). By adjusting the points on the slidewire, a known fraction of the standard voltage is bucked against the unknown voltage. When the two voltages are equal, no current flows through the galvanometer connected between the two voltage sources, and the voltage of the unknown can be read from the position of the slidewire. The sensitivity of the potentiometer is gov-

FIGURE 11.5
Principle of operation of a potentiometer.

erned by the sensitivity of the galvanometer and the cell resistance. Most commercial potentiometers will measure to ± 0.1 or ± 0.01 mV, and more sensitive ones are available. This is more than adequate for nearly all analytical purposes; and even with most fiber junction SCE reference electrodes, the sensitivity is sufficient. The sensitivity required for potentiometric titrations is less than that for direct potentiometry.

A pH meter is a voltmeter that converts the unknown voltage to a current that is amplified and read out, and it is not normally a null-type device. It draws less current than a potentiometer does. Within the latter, an appreciable current must actually be drawn in order that the null point may be determined (ca. 10^{-6} amp, depending on the galvanometer sensitivity). This will upset the chemical equilibrium at the electrode surface. However, if the redox system is **reversible,** it will reestablish equilibrium quickly and a stable equilibrium reading is obtained readily. *If the system is sluggish* **(irreversible),** *slow in attaining equilibrium, or has a high resistance, then a pH meter or electrometer should be used for the potential measurement.* These are "high-input impedance" devices. (Impedance in an ac circuit is comparable to resistance in a dc circuit. These devices convert the signal to an ac signal for amplification.) Because of their high-input resistance, very little current is drawn, typically 10^{-13} to 10^{-15} amp, and so chemical equilibrium is not greatly disturbed. High-input impedance circuits must be used with high-resistance electrodes (e.g., several megohms—10^6 MΩ) because adequate current cannot be drawn through them with a potentiometer to detect the null point, and also because the current drawn must be very small for the voltage drop across the cell ($=iR$ or current $\times$ cell resistance) to be low enough not to cause error in the measurement.

Expanded-scale pH meters are available that will measure the potential to a few tenths of a millivolt, about 10 times more closely than conventional pH meters. They are well suited for direct potentiometric measurements with ion-selective electrodes.

The Cell for Potential Measurements

In potentiometric measurements, a cell of the type in Figure 11.6 is set up. For direct potentiometric measurements in which the activity of one ion is to be calculated from the potential of the indicating electrode, the potential of the reference electrode will have to be known or determined. The voltage of the cell is described by Equation 11.7 in the previous chapter, and when a salt bridge is employed, the liquid junction potential must be included. Then,

$$E_{cell} = (E_{ind} - E_{ref}) + E_j \qquad (11.30)$$

The E_j can be combined with the other constants in Equation 11.30 into a single constant, assuming that the liquid-junction potential does not differ significantly from one solution to the next. We are forced to accept this assumption, since E_j cannot be evaluated under most circumstances. E_{ref}, E_j, and E_{ind}^0 are lumped together into a constant k:

$$k = E_{ind}^0 - E_{ref} + E_j \qquad (11.31)$$

Then,

$$\boxed{E_{cell} = k - \frac{2.303RT}{nF} \log \frac{a_{red}}{a_{ox}}} \qquad (11.32)$$

To potentiometer

Reference electrode

Indicating electrode

FIGURE 11.6
Cell for potentiometric measurements.

The constant k is determined by measuring the potential of a standard solution in which the activities are known.

11.8   DETERMINATION OF CONCENTRATIONS FROM POTENTIAL MEASUREMENTS

Usually, we are interested in determining the concentration of a test substance rather than its activity. Activity coefficients are not generally available, and it is inconvenient to calculate activities of solutions used to standardize the electrode.

If the ionic strength of all solutions is held constant at the same value, the activity coefficient of the test substance remains constant for all concentrations of the substance. We can then write for the log term in the Nernst equation:

$$\frac{2.303RT}{nF} \log f_i C_i = \frac{2.303RT}{nF} \log f_i + \frac{2.303RT}{nF} \log C_i \qquad (11.33)$$

Under the prescribed conditions, the first term on the right-hand side of this equation is constant and can be included in k, so that at constant ionic strength,

$$E_{cell} = k - \frac{2.303RT}{nF} \log \frac{C_{red}}{C_{ox}} \qquad (11.34)$$

In other words, the electrode potential undergoes a change of $2.303RT/nF$ volts for each 10-fold change in *concentration* of the oxidized or reduced form.

It is best to determine a **calibration curve** of potential versus log concentration; this should have a slope of $2.303RT/nF$. In this way, any deviation from this theoretical response will be accounted for in the calibration curve.

Since the ionic strength of an unknown solution is usually not known, a high concentration of an electrolyte is added both to the standards and to the samples to maintain about the same ionic strength. The standard solutions should contain any species in the test solution that will change the activity of the analyte, such as complexing agents.

11.9 THE RESIDUAL LIQUID-JUNCTION POTENTIAL

We have assumed above in Equations 11.32 and 11.34 that k is the same in measurements of both standards and samples. This is so only if the liquid-junction potential at the reference electrode is the same in both solutions. But the test solution will usually have a somewhat different composition from the standard solution, and the magnitude of the liquid-junction potential will vary from solution to solution. The difference in the two liquid-junction potentials is called the *residual liquid-junction potential,* and it will remain unknown. The difference can be kept to a minimum by keeping the pH of the test solution and pH of the standard solution as close as possible, and by keeping the ionic strength of both solutions as close as possible. *The former is particularly important.*

11.10 ACCURACY OF DIRECT POTENTIOMETRIC MEASUREMENTS

We can get an idea of the accuracy required in potentiometric measurements from the percent error caused by a 1-mV error in the reading at 25°C. For an electrode responsive to a monovalent ion such as silver,

$$E_{cell} = k - 0.05915 \log \frac{1}{a_{Ag^+}} \qquad (11.35)$$

and

$$a_{Ag^+} = \text{antilog} \frac{E_{cell} - k}{0.05915} \qquad (11.36)$$

A ± 1-mV error in the terms in parentheses results in an error in a_{Ag^+} of **±4 percent.** This is quite significant in direct potentiometric measurements. The same percent error in activity will result for all activities of silver ion with a 1-mV error in the measurement. **The error is doubled when n is doubled** to 2. So, a 1-mV error for a copper/copper(II) electrode would result in an 8% error in the activity of copper(II). It is obvious, then, that the residual liquid-junction potential can have an appreciable effect on the accuracy.

The accuracy and precision of potentiometric measurements are also limited by the **poising capacity** of the redox couple being measured. This is analogous to the buffering capacity in pH measurements. If the solution is very dilute, the solution is poorly poised and potential readings will be sluggish. Also, the potential of the electrode may be governed by other electrode reactions. In a very dilute silver solution, for example, $-\log (1/a_{Ag^+})$ becomes very negative and the potential of the electrode is very reducing. Under these conditions, an oxidizing agent in solution (such as oxygen) may be reduced *at the electrode surface,* setting up a second redox couple (O_2/OH^-); the potential will be a **mixed potential.**

Usually, the lower limit of concentration that can be measured with a degree of certainty is 10^{-5} to 10^{-6} M, although the actual range should be determined experimentally. As the solution becomes more dilute, a longer time should be taken to establish the equilibrium potential reading, because of the sluggishness. An exception to this limit is in pH measurements in which the hydrogen ion concentration of the solution is well poised, either by a buffer or by excess acid or base. At pH 10, the hydrogen ion concentration is 10^{-10} M, and this can be measured with a glass pH electrode (see Section 11.11). A neutral, unbuffered solution is poorly poised, however, and pH readings are sluggish.

11.11   THE GLASS pH ELECTRODE

Principle

Although there are other more conventional electrode systems for measuring pH, sometimes more accurately, the glass electrode, because of its convenience, is used almost universally for pH measurements today. Its potential is essentially not affected by the presence of oxidizing or reducing agents, and it is operative over a wide pH range. It is fast responding and functions well in physiological systems. No other pH-measuring electrode possesses all these properties.

A typical construction of a pH glass electrode is shown in Figure 11.7. For measurement, only the bulb need be submerged. There is an internal reference electrode and electrolyte $(Ag|AgCl|Cl^-)$ for making electrical contact with the glass membrane; its potential is necessarily constant and is set by the concentration of HCl. A complete cell, then, can be represented by

$$\left.\begin{array}{c}\text{reference}\\\text{electrode}\\\text{(external)}\end{array}\right\|\ H^+ \text{ (unknown)}\ \left|\ \begin{array}{c}\text{glass}\\\text{membrane}\end{array}\ \right|\ H^+ \text{ (internal)}\ \left|\ \begin{array}{c}\text{reference}\\\text{electrode}\\\text{(internal)}\end{array}\right.$$

The potential of the glass membrane is given by

$$E_{glass} = \text{constant} - \frac{2.303RT}{F} \log \frac{a_{H^+ \ int}}{a_{H^+ \ unk}} \tag{11.37}$$

and the voltage of the cell is given by

$$E_{cell} = k + \frac{2.303RT}{F} \log a_{H^+ \ unk} \tag{11.38}$$

where k is a constant that includes the potentials of the two reference electrodes,

Glass membrane

Internal filling solution (HCl)

Ag/AgCl reference electrode

FIGURE 11.7
Glass pH electrode.

the liquid-junction potential, a potential at the glass membrane due to H^+ (internal), and a term known as the **asymmetry potential.**

The asymmetry potential is a small potential across the membrane that is present even when the solutions on both sides of the membrane are identical. It is associated with nonuniform composition of the membrane, strains within the membrane, mechanical and chemical attack of the external surface, the degree of hydration of the membrane, and so forth. It slowly changes with time, especially if the membrane is allowed to dry out, and is unknown. For this reason, **a glass pH electrode should be calibrated from day to day.** The asymmetry potential will vary from one electrode to another, owing to differences in construction of the membrane.

Since $pH = -\log a_{H^+}$, Equation 11.38 can be rewritten[1]

$$E_{cell} = k - \frac{2.303RT}{F} pH_{unk} \tag{11.39}$$

or

$$pH_{unk} = \frac{k - E_{cell}}{2.303RT/F} \tag{11.40}$$

It is apparent that the glass electrode will undergo a $2.303RT/F$-volt response for each change of 1 pH unit (10-fold change in a_{H^+}). k must be determined by calibration with a **standard buffer** (see below) of known pH:

$$k = E_{cell} + \frac{2.303RT}{F} pH_{std} \tag{11.41}$$

Substitution of Equation 11.40 into Equation 11.39 yields

$$pH_{unk} = pH_{std} + \frac{E_{cell\ std} - E_{cell\ unk}}{2.303RT/F} \tag{11.42}$$

Note that since the determination involves potential measurements with a very high-resistance membrane electrode (1 to 100 megohms), it is very important to minimize the iR drop by using a pH meter that draws very little current (see before, measurement of potential).

EXAMPLE 11.5   A glass electrode-SCE pair is calibrated at 25°C with a pH 4.01 standard buffer, the measured voltage being 0.814 V. What voltage would be measured in a 1.00 $\times$ 10^{-3} M acetic acid solution? Assume $a_{H^+} = [H^+]$.

SOLUTION
From Example 5.7 in Chapter 5, the pH of a 1.00 $\times$ 10^{-3} M acetic solution is 3.88;

$$\therefore 3.88 = 4.01 + \frac{0.814 - E_{cell\ unk}}{0.0592}$$

$$E_{cell\ unk} = 0.822\ V$$

---

[1]We will assume the proper definition of pH as $-\log a_{H^+}$ in this chapter, since this is what the glass electrode measures.

Theory of the Glass Membrane Potential

The pH glass electrode functions as a result of ion exchange on the surface of a hydrated layer. The membrane of a pH glass electrode consists of chemically bonded Na_2O and SiO_2. The surface of a new glass electrode contains fixed silicate groups associated with sodium ions, $—SiO^-Na^+$. It is known that for the electrode to become operative, it must be soaked in water. During this process, the outer surface of the membrane becomes *hydrated*. The inner surface is already hydrated. The glass membrane is usually 0.03 to 0.1 mm thick, and the hydrated layers are 10^{-5} to 10^{-4} mm thick.

When the outer layer becomes hydrated, the sodium ions are exchanged for protons in the solution:

$$—SiO^-Na^+ + \quad H^+ \quad \rightleftharpoons —SiO^-H^+ + \quad Na^+$$
$$\text{solid} \qquad \text{solution} \qquad \text{solid} \qquad \text{solution} \qquad (11.43)$$

Other ions in the solution can exchange for the Na^+ (or H^+) ions, but the equilibrium constant for the above exchange is very large because of the large affinity of the glass for protons. Thus, the surface of the glass is made up almost entirely of silicic acid, except in very alkaline solution, where the proton concentration is small. The $—SiO^-$ sites are fixed, but the protons are free to move and exchange with other ions. (By varying the glass composition, the exchange for other ions becomes more favorable, and this forms the basis of electrodes selective for other ions—see below.)

The potential of the membrane consists of two components, the boundary potential and the diffusion potential. The former is almost the sole hydrogen ion activity-determining potential. The **boundary potential** resides at the surface of the glass membrane, that is, at the interface between the hydrated gel layer and the external solution. When the electrode is dipped in an aqueous solution, a boundary potential is built up, which is determined by the activity of hydrogen ions in the external solution and the activity of hydrogen ions on the surface of the gel. One explanation of the potential is that the ions will tend to migrate in the direction of lesser activity, much as at a liquid junction. The result is a microscopic layer of charge built up on the surface of the membrane that represents a potential. Hence, as the solution becomes more acidic (the pH decreases), protons migrate to the surface of the gel, building up a positive charge, and the potential of the electrode increases, as indicated by Equations 11.37 and 11.38. The reverse is true as the solution becomes more alkaline.

The **diffusion potential** results from a tendency of the protons in the inner part of the gel layer to diffuse toward the dry membrane, which contains $—SiO^-Na^+$, and a tendency of the sodium ions in the dry membrane to diffuse to the hydrated layer. The ions migrate at a different rate, creating a type of liquid-junction potential. But a similar phenomenon occurs on the other side of the membrane, only in the opposite direction. These in effect cancel each other, and so the potential of the membrane is determined largely by the boundary potential. (Small differences in boundary potentials may occur due to differences in the glass across the membrane—these represent a part of the asymmetry potential.)

The Alkaline Error

Two types of error do occur that result in non-Nernstian behavior (deviation from the theoretical response). The first is called the **"alkaline error."** Such error is due to the capability of the membrane for responding to other cations besides the hydrogen ion. As the hydrogen ion activity becomes very small, these other ions can compete successfully in the potential-determining mechanism. Although the hydrated gel layer prefers protons, sodium ions will exchange with the protons in the layer when the hydrogen ion activity in the external solution is very low (reverse of Equation 11.43). The potential then depends partially on the ratio of $a_{Na^+ \; external}/a_{Na^+ \; gel}$; that is, the electrode becomes a sodium ion electrode.

The error is negligible at pH less than about 9; but at pH values above this, the H^+ concentration is very small relative to that of other ions, and the electrode response to the other ions such as Na^+, K^+, and so on, becomes appreciable. In effect, the electrode appears to "see" more hydrogen ions than are present, and the pH reading is too low. The magnitude of this negative error is illustrated in Figure 11.8. Sodium ion causes the largest errors, which is unfortunate, because many analytical solutions, especially alkaline ones, contain significant amounts of sodium. Commercial general-purpose glass electrodes will usually be supplied with a nomogram for correcting the alkaline error if the sodium ion concentration is known, and these electrodes are useful up to pH about 11.

By a change in the composition of the glass, the affinity of the glass for sodium ion can be reduced. If the Na_2O in the glass membrane is largely replaced by Li_2O, then the error due to sodium ions is markedly decreased. This is the so-called lithium glass electrode, high-pH electrode, or full-range electrode (0 to 14 pH

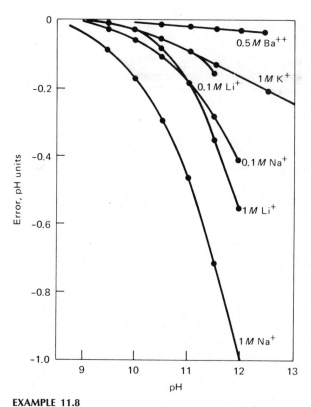

EXAMPLE 11.8

Error of the Corning 015 glass electrode in strongly alkaline solutions containing various cations. (From L. Meites and L. C. Thomas, *Advanced Analytical Chemistry*. Copyright © 1958, McGraw-Hill, Inc., New York. Used with permission of McGraw-Hill Book Company.)

range). A general-purpose electrode is preferred for use below pH 11 because it will provide faster response and greater stability due to its lower-resistance glass. As mentioned before, it was the discovery that variation in the glass composition could change its affinity for different ions that led to the development of glasses selective for ions other than protons, that is, of ion-selective electrodes.

The Acid Error

The **"acid error,"** which is more aptly described as the **"water activity" error,** is the second type causing non-Nernstian response. Such error occurs because the potential of the membrane depends on the activity of the water with which it is in contact. If the activity is unity, the response is Nernstian. In very acid solutions, the activity of water is less than unity (an appreciable amount is used in solvating the protons), and a positive error in the pH reading results (Figure 11.9). A similar type of error will result if the activity of the water is decreased by a high concentration of dissolved salt or by addition of nonaqueous solvent such as ethanol. In these cases, a large liquid-junction potential may also be introduced and another error will thereby result, although this is not very large with small amounts of ethanol.

11.12   STANDARD BUFFERS

The National Bureau of Standards (NBS) has developed a series of certified standard buffers for use in calibrating pH measurements. The pH of the buffers were determined by measuring their pH using a hydrogen-indicating electrode in a cell without liquid junction (similar to the cell given by Equation 11.22). A silver/silver chloride reference electrode was used. From Equation 11.24, we see that the activity of the chloride ion must be calculated (to calculate the potential of the reference electrode) using the Debye–Hückel theory; **this ultimately limits the accuracy of the pH of the buffers to about ± 0.01 pH unit.** The partial pressure of hydrogen is determined from the atmospheric pressure at the time of the measurement (minus the vapor pressure of the water at the temperature of the solution).

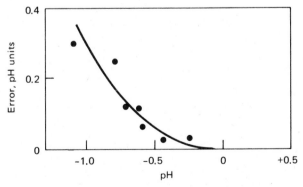

EXAMPLE 11.9
Error of the glass electrode in hydrochloric acid solutions. (From L. Meites and L. C. Thomas, *Advanced Analytical Chemistry*. Copyright © 1958, McGraw-Hill, Inc., New York. Used with permission of McGraw-Hill Book Company.)

The compositions and pH of NBS standard buffers are given in Table 11.2. Although the absolute value of the pH is no better than 0.01 unit, the buffers have been measured **relative to one another** to 0.001 pH. The potentials used in calculating the pH can be measured reproducibly this closely, and the discrimination of differences of thousandths of pH units is sometimes important (i.e., an electrode may have to be calibrated to a thousandth of a pH unit). The pH of the buffers is temperature dependent because of the dependence of the ionization constants of the parent acids or bases on temperature.

It should be pointed out that if a glass electrode/S.C.E. cell is calibrated with one standard buffer and is used to measure the pH of another, the new reading will not correspond exactly to the standard value of the second because of the residual liquid-junction potential.

The KH_2PO_4–Na_2HPO_4 buffer (pH 7.384 at 38°C) is particularly suited for calibration for blood pH measurements. Many blood pH measurements are made at 38°C, which is near body temperature; thus, the pH of the blood in the body is indicated.

TABLE 11.2
pH Values of NBS Buffer Solutions[a]

	Buffer						
t, °C	Tetroxalate[b]	Tartrate[c]	Phthalate[d]	Phosphate[e]	Phosphate[f]	Borax[g]	Calcium Hydroxide[h]
0	1.666	—	4.003	6.984	7.534	9.464	13.423
5	1.668	—	3.999	6.951	7.500	9.395	13.207
10	1.670	—	3.998	6.923	7.472	9.332	13.003
15	1.672	—	3.999	6.900	7.448	9.276	12.810
20	1.675	—	4.002	6.881	7.429	9.225	12.627
25	1.679	3.557	4.008	6.865	7.413	9.180	12.454
30	1.683	3.552	4.015	6.853	7.400	9.139	12.289
35	1.688	3.549	4.024	6.844	7.389	9.102	12.133
38	1.691	3.549	4.030	6.840	7.384	9.081	12.043
40	1.694	3.547	4.035	6.838	7.380	9.068	11.984
45	1.700	3.547	4.047	6.834	7.373	9.038	11.841
50	1.707	3.549	4.060	6.833	7.367	9.011	11.705
55	1.715	3.554	4.075	6.834	—	8.985	11.574
60	1.723	3.560	4.091	6.836	—	8.962	11.449
70	1.743	3.580	4.126	6.845	—	8.921	—
80	1.766	3.609	4.164	6.859	—	8.885	—
90	1.792	3.650	4.205	6.877	—	8.850	—
95	1.806	3.674	4.227	6.886	—	8.833	—

[a]From R. G. Bates, *J. Res. Natl. Bur. Std.*, **A66** (1962) 179. (Reproduced by permission of the U.S. Government Printing Office.)

[b]0.05 *m* potassium textroxalate (*m* refers to molality, but only small errors result if molarity is used instead).

[c]Satd. (25°C) potassium hydrogen tartrate.

[d]0.05 *m* potassium hydrogen phthalate.

[e]0.025 *m* potassium dihydrogen phosphate, 0.025 *m* disodium monohydrogen phosphate.

[f]0.008695 *m* potassium dihydrogen phosphate, 0.03043 *m* disodium hydrogen phosphate.

[g]0.01 *m* borax.

[h]Satd. (25°C) calcium hydroxide.

11.13   ACCURACY OF pH MEASUREMENTS

The accuracy of pH measurements is governed by the accuracy to which the hydrogen ion activity of the standard buffer is known. As mentioned above, this accuracy is not better than ±0.01 pH unit because of limitations in calculating the activity coefficient of a single ion.

A second limitation in the accuracy is the residual liquid-junction potential. The cell is standardized in one solution, and then the unknown pH is measured in a solution of a different composition. We have mentioned that this residual liquid-junction potential is minimized by keeping the pH and compositions of the solutions as near as possible. Because of this, **the cell should be standardized at a pH close to that of the unknown.** The error in standardizing at a pH far removed from that of the test solution is generally within 0.01 to 0.02 pH unit but can be as large as 0.05 pH unit for very alkaline solutions.

The residual liquid junction potential, combined with the uncertainty in the standard buffers, limits the **absolute accuracy of measurement of pH of an unknown solution to about ±0.02 pH unit.** It may be possible, however, to **discriminate** between the pH of two similar solutions with differences as small as ±0.004 or even ±0.002 pH units, although their accuracy is no better than ±0.02 pH units. Such discrimination is possible because the liquid-junction potentials of the two solutions will be virtually identical in terms of true a_{H^+}. For example, if the pH values of two blood solutions are close, we can measure the difference between them accurately to ±0.004 pH. If the pH difference is fairly large, however, then the residual liquid-junction potential will increase and the difference cannot be measured as accurately. For discrimination of 0.02 pH unit, large changes in the ionic strength may not be serious, but it is important for smaller changes than this.

An error of ±0.02 pH unit corresponds to an error in a_{H^+} of ±4.8% (±1.2 mV),[2] and a discrimination of ±0.04 pH unit would correspond to a discrimination of ±1.0% in a_{H^+} (±0.2 mV).

If pH measurements are made at a temperature other than that at which the standardization is made, other factors being equal, the liquid-junction potential will change with temperature. For example, in a rise from 25°C to 38°C, a change of +0.76 mV has been reported for blood and −0.55 mV for buffer solutions. Thus, for very accurate work, the cell should be standardized at the same temperature as the test solution.

11.14   MEASUREMENTS WITH THE pH METER

We have already mentioned that owing to the high resistance of the glass electrode, an electrometer or pH meter must be used to make the potential measurements. If voltage is measured directly, Equation 11.40 or 11.42 is applied to calculate the pH. The value of $2.303RT/F$ at 298.16 K (25°C) is 0.05916; if a different temperature is used, this value should be corrected in direct proportion to the temperature.

A diagram of a pH meter is shown in Figure 11.10. The potential scale is calibrated in pH units, with each pH unit equal to 59.16 mV at 25°C (Equation 11.39).

---

[2]The electrode response is 59 mV/pH at 25°C.

FIGURE 11.10
Typical pH meter.

The pH meter is adjusted with the calibrate knob to indicate the pH of the standard buffer. Then, the standard buffer is replaced by the unknown solution and the pH is read from the scale. This procedure, in effect, sets the constant k in Equation 11.40 and adjusts for the asymmetry potential as well as the other constants included in k.

The pH meter contains a temperature adjustment dial, which changes the sensitivity response (mV/pH) so that it will be equal to $2.303RT/R$. For example, it is 54.1 mV at 0°C and 66.0 mV at 60°C. Note that this does **not** compensate for the change in the pH of standard buffers with temperature, and the pH value of the buffer at the given temperature is used.

Ordinary pH meters are precise to ±0.1 to ±0.01 pH unit, with a full-meter scale of 14 pH units (about 840 mV). The meters can be set to read millivolts directly (usually with a sensitivity of 1400 mV full scale). Expanded-scale pH meters, which amplify the potential signal, are capable of reading to ±0.001 pH unit with a full-scale reading typically of 1.4 pH units (140 mV on the millivolt-scale setting); to accomplish this, the potential must be read to closer than 0.1 mV.

When the pH of an unbuffered solution near neutrality is measured, readings will be sluggish because the solution is poorly poised and a longer time will be required to reach a stable reading. The solution should be stirred because a small amount of the glass tends to dissolve, making the solution at the electrode surface alkaline.

11.15   pH MEASUREMENT OF BLOOD

Recall from Chapter 5 that, because the equilibrium constants of the blood buffer systems change with temperature, the pH of blood at the body temperature of 37°C is different than at room temperature. Hence, to obtain meaningful blood pH measurements that can be related to actual physiological conditions, the measurements should be made at 37°C and the samples should not be exposed to the atmosphere. (Also recall that the pH of a neutral aqueous solution at 37° is 6.80, and so the acidity scale is changed by 0.20 pH unit.)

Some useful rules in making blood pH measurements are as follows:

1. Calibrate the electrodes using a standard buffer at 37°C, making sure to select the proper pH of the buffer at 37° and to set the temperature knob on the pH meter at 37° (slope = 61.5 mV/pH). It is a good idea to use two standards for calibration, narrowly bracketing the sample pH; this assures that the electrode is functioning properly. Also, the electrodes must be equilibrated at 37° before calibration and measurement. The potential of the internal reference electrode inside the glass electrode is temperature dependent, as may be the potential-determining mechanism at the glass membrane interface; and the potentials of the SCE reference electrode and the liquid junction are temperature dependent. (We should note here that if pH or other potential measurements are made at less than room temperature, the salt bridge or the reference electrode should not contain saturated KCl, but somewhat less concentrated, because solid KCl crystals will precipitate in the bridge and increase its resistance.)
2. Blood samples must be kept anaerobically to prevent loss or absorption of CO_2. Make pH measurements within 15 minutes after sample collection, if possible, or else keep the sample on ice and make the measurements within 2 hours. The sample is equilibrated to 37°C before measuring. (If a pCO_2 measurement is to be performed also, do this within 30 minutes.)
3. To prevent coating of the electrode, flush the sample from the electrode with saline solution after each measurement. A residual blood film can be removed by dipping for *only* a few minutes in 0.1 M NaOH, followed by 0.1 M HCl and water or saline.

Generally, venous blood is taken for pH measurement, although arterial blood may be required for special applications. The 95% confidence limit range (see Chapter 3) for arterial blood pH is 7.31 to 7.45 (mean 7.40) for all ages and sexes. A range of 7.37 to 7.42 has been suggested for subjects at rest. Venous blood may differ from arterial blood by up to 0.03 pH unit and may vary with the vein sampled. Intracellular erythrocyte pH is about 0.15 to 0.23 unit lower than that of the plasma.

11.16   pH MEASUREMENTS IN NONAQUEOUS SOLVENTS

Measurement of pH in a nonaqueous solvent when the electrode is standardized with an aqueous solution has little significance in terms of possible hydrogen ion activity because of the unknown liquid junction potential, which can be rather large, depending on the solvent. Measurements made in this way are usually referred to as "apparent pH." pH scales and standards for nonaqueous solvents are being developed using an approach similar to the one for aqueous solutions. These scales have no relation to the aqueous pH scale, however. You are referred to the book by Bates (Reference 2) for a discussion of this topic.

11.17   ION-SELECTIVE ELECTRODES

Various types of membrane electrodes have been developed in which the membrane potential is selective toward a given ion or ions, just as the potential of the glass membrane of a conventional glass electrode is selective toward hydrogen ions. These electrodes are finding importance in the measurement of ions, especially in small concentrations. Generally, they are not "poisoned" by the presence of proteins, as some other electrodes are, and so they are ideally suited to measurements in biological media. This is especially true for the glass membrane ion-selective electrodes.

None of these electrodes is *specific* for a given ion, but each will possess a certain *selectivity* toward a given ion or ions. So they are properly referred to as **ion-selective electrodes,** rather than specific-ion electrodes, a misnomer touted by some electrode manufacturers for obvious reasons.

Glass Membrane Electrodes

These are similar in construction to the pH glass electrode. Varying the composition of the glass membrane can cause the hydrated glass to acquire an increased affinity for various monovalent cations, with a much lower affinity for protons than the pH glass electrode has. The membrane potential becomes dependent on these cations, probably through an ion exchange mechanism similar to that presented for the glass pH electrode; that is, a boundary potential is produced, determined by the relative activities of the cations on the surface of the gel and in the external solution. Increased cation activity results in increased positive charge on the membrane and a positive increase in electrode potential.

The construction is similar to Figure 11.7. The internal filling solution will usually be the chloride salt of the cation to which the electrode is most responsive. The composition of the membrane will vary from manufacturer to manufacturer, but we can classify three general types of glass electrodes:

1. pH type. This is the conventional pH glass electrode, and it has a selectivity order of $H^+ \ggg Na^+ > K^+, Rb^+, Cs^+ \ldots \gg Ca^{2+}$. The response to ions other than H^+ is the "alkaline error" we talked about above.
2. Cation-sensitive type. This responds in general to monovalent cations, and the order of selectivity is $H^+ > K^+ > Na^+ > NH_4^+, Li^+ \ldots \gg Ca^{2+}$.
3. "Sodium-sensitive" type. The selectivity order is $Ag^+ > H^+ > Na^+ \gg K^+$, $Li^+, \ldots \gg Ca^{2+}$.

Note that all the electrodes are responsive to hydrogen ion, but the second two are much less so than the first type. Because of this response, the electrodes must be used at a pH sufficiently high that the hydrogen ion activity is so low that the response will be determined primarily from the ion of interest. This lower pH limit will vary from electrode to electrode and ion to ion.

The "sodium-sensitive" type of electrode can be used to determine the activity of sodium ion in the presence of appreciable amounts of potassium ion. Its selectivity for sodium over potassium is on the order of 3000 or more. Glass electrodes can be obtained that show a selectivity ratio for silver over sodium of greater than 1000. Glass electrodes can be used in liquid ammonia and molten salt media.

Precipitate Electrodes

The construction of typical precipitate electrodes is shown in Figure 11.11. They are used primarily for anion measurements, but are generally slow responding and subject to poisoning. A sparingly soluble inorganic salt having as its anion the anion to be measured is suspended in an inert, semiflexible matrix to hold the precipitate in place. Such a membrane is called a heterogeneous or **precipitate-impregnated membrane.** The inert support material can be silicone rubber, poly(vinyl chloride), and so on. For example, in an iodide-selective electrode, the membrane consists of 50 weight-% silver iodide in silicone rubber, and the filling solution is potassium iodide. The internal reference electrode is then a silver/silver iodide electrode. Potassium sulfate plus a halide salt can be used as the filling solution, the halide salt being used to establish a constant reference electrode potential.

These electrodes are suitable for measuring iodide, bromide, chloride, and sulfide ions. Ion-selective electrodes are advantageous over the silver/silver halide wire electrode in that they are insensitive to redox interferences and surface poisoning, and they possess better selectivity for one halide ion over another.

The sensitivity of precipitate electrodes is limited by the solubility of the membrane precipitates. The solubility of silver iodide is about 10^{-8} M, and the iodide ion can be measured to concentrations as low as 10^{-7} M. The solubility of silver chloride is 10^{-5} M, and so the limit of determination is about 1000 times higher than for iodide. The selectivity for the given anion over other anions is also increased as the solubility is decreased.

Solid-State Electrodes

The construction of these electrodes is shown in Figure 11.12. The most successful example is the fluoride electrode. The membrane consists of a single crystal of lanthanum fluoride doped with some europium(II) to increase the conductivity of the crystal. Lanthanum fluoride is very insoluble, and this electrode exhibits Nernstian response to fluoride down to 10^{-5} M and non-Nernstian response down to 10^{-6} M (19 ppb!). This electrode has at least a 1000-fold selectivity for fluoride ion over chloride, bromide, iodide, nitrate, sulfate, monohydrogen phosphate, and bicarbonate anions and a 10-fold selectivity over hydroxide ion. Hydroxide ion

FIGURE 11.11
Precipitate-impregnated membrane electrode.

Internal filling solution

Ag/AgCl reference electrode

Synthetic single–crystal membrane

FIGURE 11.12
Crystal membrane electrode. (Reproduced by permission of Orion Research, Inc.)

appears to be the only serious interference. The pH range is limited by the formation of hydrofluoric acid at the acid end and by hydroxide ion response at the
alkaline end; a pH range of 4 to 9 is claimed.

A useful solution for minimizing interferences with the fluoride electrode consists of a mixture of an acetate buffer at pH 5.0 to 5.5, 1 M NaCl, and cyclohexy-
lenedinitrilo tetraacetic acid (CDTA). This solution is commercially available as
TISAB (total ionic strength adjustment buffer). A 1:1 dilution of samples and standards with the solution provides a high ionic-strength background, swamping out
moderate variations in ionic strength between solutions. This keeps both the junction potential and the activity coefficient of the fluoride ion constant from solution
to solution. The buffer provides a pH at which appreciable HF formation is avoided
and hydroxide response is not present. CDTA is a chelating agent, similar to EDTA,
that complexes with polyvalent cations such as Al^{3+}, Fe^{3+}, and Si^{4+}, which would
otherwise complex with F^- and change the fluoride activity.

A useful solid-state electrode is based on an Ag_2S membrane. By itself, this
electrode responds to either Ag^+ or S^{2-} ions, down to about 10^{-8} M. This lower
limit of detection is too high to be caused by solubility of Ag_2S ($K_{sp} = 10^{-51}$). It
probably reflects difficulties in preparing extremely dilute solutions and ionic adsorption on, and desorption from, the surfaces of the electrodes and the vessel
containing the solution. This membrane is a good ionic conductor with low resistance; and by mixing the Ag_2S with other silver or sulfide salts (whose resistances
might be high), it will become responsive to other ions. For example, a mixed
membrane of AgI/Ag_2S will respond to I^- ion in addition to Ag^+ and S^{2-}. A mixed
CuS/Ag_2S membrane will respond to Cu^{2+} in addition to the other membrane ions.

The chief restriction of these mixed-salt electrodes is that the solubility of the
second salt must be much larger than that of Ag_2S; but, on the other hand, it must
be sufficiently insoluble that its dissolution does not limit to relatively high values
the test ion concentration that can be detected. As long as the membrane contains
sufficient silver sulfide to provide silver ion conducting pathways through the membrane, it will function as a silver electrode. The potential, then, is related to the
test ion through a series of equilibria similar to those described for an electrode of
the second kind or an electrode of the third kind (Problem 12); that is, the available
Ag^+ is governed by the solubility equilibria. Other Ag_2S mixed-crystal electrodes
are available for Cl^-, Br^-, SCN^-, CN^-, Pb^{2+}, and Cd^{2+}. These electrodes, of

course, are subject to interference from other ions that react with Ag^+ (in the case of anion electrodes) or with S^{2-} (in the case of cation electrodes).

Liquid–Liquid Electrodes

The basic construction of these electrodes is shown in Figure 11.13. Here, the potential-determining "membrane" is a layer of a water-immiscible liquid ion exchanger held in place by an inert, porous membrane. The porous membrane allows contact between the test solution and the ion exchanger but minimizes mixing. It is either a synthetic, flexible membrane or a porous glass frit, depending on the manufacturer. The internal filling solution contains the ion for which the ion exchanger is specific plus a halide ion for the internal reference electrode.

An example of this electrode is the calcium-selective electrode. The ion exchanger is a calcium organophosphorous compound. The sensitivity of the electrode is governed by the solubility of the ion exchanger in the test solution. A Nernstian response is obtained down to about 5×10^{-5} M. The selectivity of this electrode is about 3000 for calcium over sodium or potassium, 200 over magnesium, and 70 over strontium. It can be used over the pH range of 5.5 to 11. Above pH 11, calcium hydroxide precipitates. A phosphate buffer should not be used for calcium measurements because the calcium activity will be lowered by complexation or precipitation. Experience has shown that these and other liquid-membrane electrodes are often subject to poisoning, for example, in biological fluids.

A "divalent cation" ion exchange electrode that responds to several cations is available. Its response is nearly equal for calcium and magnesium, and it is useful for measuring water hardness. A copper and a lead electrode are also available. Anion-selective electrodes of this type are available for nitrate, perchlorate, and chloride. They are the same in principle, except that a liquid anion exchanger is used instead of a cation exchanger.

Table 11.3 summarizes the characteristics of some commercial ion-selective electrodes.

Enzyme Electrodes

Ion-selective electrodes used in conjunction with immobilized enzymes can serve as the basis of electrodes that are selective for specific enzyme substrates. Enzymes are proteins that catalyze specific reactions to a high degree of specificity. The

Ag/AgCl reference electrode

Internal aqueous filling solution

Membrane or salt bridge

Liquid ion exchanger

Porous membrane

FIGURE 11.13
Liquid-membrane electrode.

TABLE 11.3
Typical Properties of Some Commercial Ion-Selective Electrodes

Electrode	Concentration Range, M	Principal Interferences[a]
Liquid–Liquid Ion-Exchange Electrodes		
Ca^{2+}	$10°–10^{-5}$	$Zn^{2+}(3)$; $Fe^{2+}(0.8)$; $Pb^{2+}(0.6)$; $Mg^{2+}(0.1)$; $Na^+(0.003)$
Cl^-	$10^{-1}–10^{-5}$	$I^-(17)$; $NO_3^-(4)$; $Br^-(2)$; $HCO_3^-(0.2)$; SO_4^{2-}, $F^-(0.1)$
Divalent Cation	$10°–10^{-8}$	Fe^{2+}, $Zn^{2+}(3.5)$; $Cu^{2+}(3.1)$; $Ni^{2+}(1.3)$; Ca^{2+}, $Mg^{2+}(1)$; $Ba^{2+}(0.94)$; $Sr^{2+}(0.54)$; $Na^+(0.015)$
BF_4^-	$10^{-1}–10^{-5}$	$NO_3^-(0.1)$; $Br^-(0.04)$; OAc^-, $HCO_3^-(0.004)$; $Cl^-(0.001)$
NO_3^-	$10^{-1}–10^{-5}$	$ClO_4^-(1000)$; $I^-(20)$; $Br^-(0.1)$; $NO_2^-(0.04)$; $Cl^-(0.004)$; $CO_3^{2-}(0.0002)$; $F^-(0.00006)$; $SO_4^{2-}(0.00003)$
ClO_4^-	$10^{-1}–10^{-5}$	$I^-(0.01)$; NO_3^-, $OH^-(0.0015)$; $Br^-(0.0006)$; F^-, $Cl^-(0.0002)$
K^+	$10°–10^{-5}$	$Cs^+(1)$; $NH_4^+(0.03)$; $H^+(0.01)$; $Na^+(0.002)$; Ag^+, $Li^+(0.001)$
Solid-State Electrodes[b]		
F^-	$10°–10^{-6}$	Maximum level: $OH^- < 0.1\ F^-$
Ag^+ or S^{2-}	$10°–10^{-7}$	$Hg^{2+} < 10^{-7}\ M$

[a]Number in parentheses is the relative selectively for the interfering ion over the test ion (see "The Selectivity Ratio" below.

[b]Interference concentrations given represent maximum tolerable concentrations.

reactants are the substrates. Enzymes and their properties are discussed in more detail in Chapter 18.

Consider the hydrolysis of the substrate urea in the presence of the enzyme urease:

$$NH_2CONH_2 + 2H_2O + H^+ \xrightleftharpoons{urease} 2NH_4^+ + HCO_3^-$$

A urea electrode can be prepared by immobilizing urease in a gel and coating it on the surface of a cation-sensitive-type glass electrode (that responds to monovalent cations). When the electrode is dipped in a solution containing urea, the urea diffuses into the gel layer and the enzyme catalyzes its hydrolysis to form ammonium ions. The ammonium ions diffuse to the surface of the electrode where they are sensed by the cation-sensitive glass to give a potential reading. After about 30 to 60 seconds, a steady-state reading is reached which, over a certain working range, is a linear function of the logarithm of the urea concentration. By appropriate choice of immobilized enzyme and electrode, a number of other substrate-selective enzyme electrodes have been described. Some are based on amperometric rather than potentiometric measurements (Chapter 12).

Mechanism of Membrane Response

The mechanisms of ion-selective electrode membrane response have not been as extensively studied as the glass pH electrode, and even less is known about how their potentials are determined. Undoubtedly, there is a similarity of mechanisms. The active membrane generally contains the ion of interest selectively bound to a reagent in the membrane, either as a precipitate or a complex. Or else the electrode must be equilibrated in solution of the test ion, in which case the ion also binds selectively to the membrane reagent. This can be compared to the

—SiO^-H^+ sites on the glass pH electrode. When the ion-selective electrode is immersed in a solution of the test ion, a boundary potential is established at the interface of the membrane and the external solution. The possible mechanism again is due to the tendency of the ions to migrate in the direction of lesser activity to produce a "liquid-junction" type potential. Positive ions will result in a positive charge and a change in the potential in the positive direction, while negative ions will result in a negative charge and a change in the potential in the negative direction.

The secret in constructing ion-selective electrodes, then, is to find a material with sites that show strong affinity for the ion of interest. Thus, the calcium liquid ion exchange electrode exhibits high selectivity for calcium over magnesium and sodium ions because the organic phosphate cation exchanger (in the calcium form) has a high chemical affinity for calcium ions. The ion exchange equilibria at the membrane–solution interface involves calcium ions, and the potential depends on the ratio of the activity of calcium ions in the external solution to that of calcium ions in the membrane phase.

The Selectivity Ratio

The potential of an ion-selective electrode in the presence of a single ion follows an equation similar to Equation 11.38 for the pH glass membrane electrode:

$$E_{ISE} = k + \frac{S}{n} \log a_{ion} \tag{11.44}$$

where S represents the slope (theoretically $2.303RT/F$) and n is the ion charge, *including sign*. Often, the slope is less than Nernstian; but for monovalent ion electrodes, it is usually close. The constant k depends on the nature of the internal reference electrode, the filling solution, and the construction of the membrane. It is determined by measuring the potential of a solution of the ion of known activity.

EXAMPLE 11.6 A fluoride electrode is used to determine fluoride in a water sample. Standards and samples are diluted 1:10 with TISAB solution. For a 1.00×10^{-3} M (before dilution) standard, the potential reading relative to the reference electrode is -211.3 mV; and for a 4.00×10^{-3} M standard, it is -238.6 mV. The reading with the unknown is -226.5 mV. What is the concentration of fluoride in the sample?

SOLUTION
Since the ionic strength remains constant due to dilution with the ionic-strength adjustment solution, the response is proportional to log [F^-]:

$$E = k + \frac{S}{n} \log [F^-] = k - S \log [F^-]$$

where n is -1. First calculate S:

$$-211.3 = k - S \log (1.00 \times 10^{-3}) \tag{1}$$
$$-238.6 = k - S \log (4.00 \times 10^{-3}) \tag{2}$$

Subtract (2) from (1):

$$27.3 = S \log (4.00 \times 10^{-3}) - S \log (1.00 \times 10^{-3}) = S \log \frac{4.00 \times 10^{-3}}{1.00 \times 10^{-3}}$$

$$27.3 = S \log 4.00$$

$$S = 45.3 \text{ mV (somewhat sub-Nernstian)}$$

Calculate k:

$$-211.3 = k - 45.3 \log (1.00 \times 10^{-3})$$
$$k = -347.2 \text{ mV}$$

For the unknown:

$$-226.5 = -347.2 - 45.3 \log [F^-]$$
$$[F^-] = 2.16 \times 10^{-3} \text{ M}$$

If the electrode is in a solution containing a mixture of cations (or anions, if it is an anion-responsive electrode), it may respond to the other cations. Suppose, for example, we have a mixture of sodium and potassium ions and an electrode that responds to both. The Nernst equation must include an additive term for the potassium activity:

$$E_{NaK} = k_{Na} + S \log (a_{Na^+} + K_{NaK}a_{K^+}) \tag{11.45}$$

The constant k_{Na} corresponds to that in the Nernst equation for the primary ion, sodium, alone. E_{NaK} is the potential of the electrode in a mixture of sodium and potassium. K_{NaK} is the **selectivity ratio** of the electrode for **potassium over sodium.** It is equal to the reciprocal of K_{KNa}, which is the selectivity ratio of *sodium over potassium.*

K_{NaK} and k_{Na} are determined by measuring the potential of two different standard solutions containing sodium and potassium and then solving the two simultaneous equations for the two constants. Alternatively, one of the solutions may contain only sodium, and k_{Na} is determined from Equation 11.44.

We can write general equations for mixtures of two cations:

1. For A^+ and B^+:

$$E_{AB} = k_A + S \log (a_{A^+} + K_{AB}a_{B^+}) \tag{11.46}$$

2. For A^{2+} and B^+:

$$E_{AB} = k_A + \frac{S}{2} \log (a_{A2^+} + K_{AB}a_{B^+}^2) \tag{11.47}$$

3. For A^{2+} and B^{2+}:

$$E_{AB} = k_A + \frac{S}{2} \log (a_{A2^+} + K_{AB}a_{B2^+}) \tag{11.48}$$

Similar equations can be written for anion-selective electrodes.

In principle, we can write equations for multicomponent mixtures. Consider an electrode in a mixture of Ca^{2+}, Mg^{2+}, and H^+:

$$E_{CaMgH} = k_{Ca} + \frac{S}{2} \log (a_{Ca2^+} + K_{CaMg}a_{Mg2^+} + K_{CaH}a_{H^+}^2) \tag{11.49}$$

In actual practice, equations expanded beyond two components do not hold strictly, and so this technique is rarely used. Since all electrodes respond more or less to hydrogen ions, the practice is to keep the activity of the hydrogen ion low enough that the product $K_{CaH}a_{H^+}^2$ in Equation 11.49, for example, is negligible compared with the other two terms in the parentheses.

One problem with selectivity ratios should be mentioned. They sometimes vary with the relative concentrations of ions and are not constant. For this reason, it is difficult to use the selectivity ratio in calculations involving mixtures of ions. They are useful for predicting conditions under which interfering ions can be neglected. That is, in practice, conditions are generally adjusted so the product $K_{AB}a_{B^{n+}}^n$ is negligible and a simple Nernst equation applies for the test ion. Usually, a calibration curve is prepared, and if an interfering ion is present, this can be added to standards at the same concentrations as in the unknowns; the result would be a nonlinear but corrected calibration curve. This technique can obviously only be used if the concentration of the interfering ion remains nearly constant in the samples.

An example of how the selectivity ratio might be used in a calculation is given in the following problem.

EXAMPLE 11.7 A cation-sensitive electrode is used to determine the activity of calcium in the presence of sodium. The potential of the electrode in 0.0100 M $CaCl_2$ measured against an SCE is + 195.5 mV. In a solution containing 0.0100 M $CaCl_2$ and 0.0100 M NaCl, the potential is 201.8 mV. What is the activity of calcium ion in an unknown solution if the potential of the electrode in this is 215.6 mV versus SCE and the sodium ion activity has been determined with a sodium ion-selective electrode to be 0.0120 M? Assume Nernstian response.

SOLUTION
The ionic strength of 0.0100 M $CaCl_2$ is 0.0300, and that of the mixture is 0.0400. Therefore, from Equation 4.20, the activity coefficient of calcium ion in the pure $CaCl_2$ solution is 0.55, and for calcium and sodium ions in the mixture, it is 0.51 and 0.83, respectively. Therefore,

$$k_{Ca} = E_{Ca} - 29.58 \log a_{Ca^{2+}}$$
$$= 195.5 - 29.58 \log (0.55 \times 0.0100)$$
$$= 262.3 \text{ mV}$$
$$E_{CaNa} = k_{Ca} + 29.58 \log (a_{Ca^{2+}} + K_{CaNa}a_{Na^+}^2)$$
$$201.8 = 262.3 + 29.58 \log [0.51 \times 0.0100 + K_{CaNa}(0.83 \times 0.0100)^2]$$
$$K_{CaNa} = 47$$
$$215.6 = 262.3 + 29.58 \log (a_{Ca^{2+}} + 47 \times 0.0120^2)$$
$$a_{Ca^{2+}} = 0.0196 \text{ M}$$

Use of Ion-Selective Electrodes

As with pH glass electrodes, most ion-selective electrodes have high resistance, and an electrometer or pH meter must be used to make measurements. An expanded-scale pH meter is generally used. It is often necessary to pretreat ion-selective electrodes by soaking in a solution of the ion to be determined.

The problems and accuracy limitations discussed under pH and other direct potentiometric measurements apply to ion-selective electrodes.

A calibration curve of potential versus log activity is usually prepared. If concentrations are to be measured, then the technique of maintaining a constant ionic

strength as described earlier is used (Equation 10.32). For example, the concentration of unbound calcium ion in serum is determined by diluting samples and standards with 0.15 M NaCl. Only the *unbound* calcium is measured and not the
fraction that is complexed.

The activity coefficient of sodium ion in normal human serum has been estimated, using ion-selective electrodes, to be 0.780 ± 0.001, and in serum water
to be 0.747 (serum contains about 96% water by volume). Standard solutions of
sodium chloride and potassium chloride are usually used to calibrate electrodes for
the determination of sodium and potassium in serum. Concentrations of 1.0, 10.0,
and 100.0 mmol/L can be prepared with respective activities of 0.965, 9.03, and
77.8 mmol/L for sodium ion in pure sodium chloride solution and 0.965, 9.02,
and 77.0 mmol/L for potassium ion in pure potassium chloride solution.

The response of ion-selective electrodes is frequently slow, and considerable
time must be taken to establish an equilibrium reading. The response becomes
more sluggish as the concentration is decreased. Some electrodes, on the other
hand, respond sufficiently rapidly that they can be used to monitor reaction rates.

We can summarize some of the advantages and disadvantages of ion-selective
electrodes and some precautions and limitations in their use as follows:

1. They measure activities rather that concentrations, a unique advantage but a factor that must be considered in obtaining concentrations from measurements. Interference can occur from ionic strength effects.
2. They measure "free" ions (i.e., the portion that is not associated with other species). Chemical interference can occur from complexation, protonation, and the
   like.
3. They are not specific but merely more selective toward a particular ion. Hence,
   they are subject to interference from other ions. They respond to hydrogen ions
   and are, therefore, pH-limited.
4. They function in turbid or colored solutions, where photometric methods fail.
5. They have a logarithmic response, which results in a wide dynamic working
   range, generally from four to six orders of magnitude. This logarithmic response
   also results in an essentially constant, albeit relatively large, error over the working range where the Nernst relation holds.
6. Except in dilute solution, their response is fairly rapid, often requiring less than 1
   minute for a measurement. Electrode response is frequently rapid enough to allow
   process stream monitoring (see Chapter 17).
7. The response is temperature dependent, by RT/nF.
8. The required measuring equipment can be made portable for field operations, and
   small samples (e.g., 1 mL) can be measured.
9. The sample is not destroyed in measurement.
10. While certain electrodes may operate down to 10^{-6} M, many will not, and elec-
    trodes are not ultrasensitive tools as some other techniques that may approach
    10^{-9} M sensitivities or lower.
11. Frequent calibration is generally required.
12. Few primary activity standards are available, as there are for pH measurements,[3]
    and the standard solutions that are used are not "buffered" in the ion being tested.
    Impurities, especially in dilute standards, may cause erroneous results.

---

[3]Activity standards similar to pH standards are available from the National Bureau of Standards for some
salts, such as sodium chloride.

In spite of some limitations, ion-selective electrodes are becoming more and more important because they represent an approach to the analytical chemist's dream of a probe that is specific for the test substance and, therefore, requires essentially no sample preparation for fluids. Research is producing more and better electrodes, and their importance will undoubtedly increase with time.

Electrode Hookup to Potential Measuring Device

In making potentiometric measurements, the reference electrode is connected to the common (ground), or black, terminal of the potential measuring device and the indicating electrode is connected to the red terminal. In this manner, the sign (+ or −) on the voltage scale of the meter (electrometer, pH meter, etc.) denotes the relative polarity of the indicator electrode with respect to the reference electrode of the cell. Under these conditions, since $E_{cell} = E_{ind} - E_{ref}$, the voltage with a cation ISE will increase in a positive direction with increased analyte concentration (Na^+, H^+, etc.); the voltage with an anion ISE will increase in a negative direction (F^-, Cl^-, etc.) and will be in agreement with the equations presented in this chapter. (It should be noted that with modern pH meters, the glass electrode or ISE is actually connected to the common lead because high-impedance electrodes such as glass electrodes are less susceptible to noise pickup if they are held near the ground potential of the meter power supply. However, the potential displayed on the meter is reversed, thus preserving the readout of the indicating electrode relative to the reference electrode.)

11.18 POTENTIOMETRIC TITRATIONS (INDIRECT POTENTIOMETRY)

Volumetric titrations are usually most conveniently performed with a visual indicator. In cases where a visual indicator is unavailable, potentiometric indication of the end point can often be used. Potentiometric titrations are among the most accurate known, because the potential follows the actual change in activity and, therefore, the end point will often coincide directly with the equivalence point. And, as we have mentioned in our discussions of volumetric titrations, they are more sensitive than visual indicators. Potentiometry is, therefore, often employed for dilute solutions.

Potentiometric titrations are straightforward. They involve measurement of an indicating electrode potential against a convenient reference electrode and plotting the change of this potential difference against volume of titrant. A large potential break will occur at the equivalence point. Since we are interested only in the potential *change,* the *correct* potential of the indicating electrode need not be known. For example, in pH titrations, the glass electrode does not have to be calibrated with a standard buffer; it will still give the same *shape* of titration curve that may be shifted up or down on the potential axis. It is a good idea, however, to have some indication of the correct value so the end point can be anticipated and any anomalous difficulties can be detected.

Because we are not interested in "absolute" potentials, the liquid-junction potential becomes unimportant. It will remain somewhat constant throughout the titration, and small changes will be negligible compared to the change in potential at the end point. Also, the potential need not be read very closely, and so a conventional pH meter, the scale of which is divided to the nearest 10 mV and can be estimated to the nearest 1 mV, can be used for most titrations.

pH Titrations

We have shown in Chapter 8 that in acid–base titrations the pH of the solution exhibits a large break at the equivalence point. This pH change can easily be monitored with a pH glass electrode. By plotting the measured pH against volume of titrant, one can obtain titration curves similar to those shown in Chapter 8. The end point is taken as the **inflection point** of the large pH break occurring at the equivalence point; this is the steepest part of the curve.

A glass indicating electrode with a saturated calomel electrode as the reference electrode can be used for most titrations of weak acids or bases in aqueous solvents. A sleeve-type SCE is used in nonaqueous solutions, rather than a conventional fiber junction type, in order to decrease the resistance of the cell. This has a tapered ground-glass sleeve that fits over the end of the electrode; saturated KCl seeps out of a hole in the side of the electrode to give a thin film of solution between the sleeve and electrode that makes contact with the nonaqueous solution. For solvents of low dielectric constant, the electrodes must be kept very close together to decrease the resistance. A silver wire coated with silver chloride offers an advantage as a reference electrode in some cases. The glass electrode for these nonaqueous titrations responds to the solvated proton concentration, but there is no simple relationship between this and the "pH." Thus, the readings are taken on the millivolt scale of the pH meter.

The glass/SCE pair can be used for the titration of acids in alcohols, but the glass electrode fails to function in more basic solvents. An antimony electrode is usually used as the indicating electrode in these solvents. In some solvents such as in butylamine, the glass electrode can serve as a reference electrode, since its potential remains constant, the advantage being that there is no salt bridge.

Precipitation Titrations

The indicating electrode in precipitation titrations is used to follow the change in pM or pA, where M is the cation of the precipitate and A is the anion. In the titration of chloride ion with silver ion, for example, either Equation 11.3 or 11.10 will hold. In the former equation, the term $\log (1/a_{Ag^+})$ is equal to pAg; and in the latter equation, the term $\log a_{Cl^-}$ is equal to $-$pCl. Therefore, the potential of the silver electrode will vary in direct proportion to pAg or pCl, changing $2.30RT/F$ V (ca. 59 mV) for each 10-fold change in a_{Ag^+} or a_{Cl^-}. A plot of the potential versus volume of titrant will give a curve identical in shape to that in Figure 9.5. (Note that since $a_{Ag^+} a_{Cl^-} =$ constant, a_{Cl^-} is proportional to $1/a_{Ag^+}$ and pCl is proportional to $-$pAg, so the same shaped curve results if we plot or measure either pCl or pAg.)

Redox Titrations

Because there is generally no difficulty in finding a suitable indicator electrode, redox titrations are widely used; an inert metal such as platinum is usually satisfactory for the electrode. Both the oxidized and reduced forms are usually soluble and their ratio varies throughout the titration. The potential of the indicating electrode will vary in direct proportion to $\log (a_{red}/a_{ox})$, as in the calculated potential for the titration curves shown in Chapter 10. See, for example, Figure 10.2 for the titration of Fe^{2+} with Ce^{4+}. As pointed out in Chapter 10, the potential is determined by either half-reaction. Generally, the pH in these titrations is held nearly constant, and any H^+ term in the Nernst equation will drop out of the log term.

A potentiometric plot, as in Figure 10.2, is useful for evaluating or selecting a suitable visual indicator for the titration, particularly for a new titration. From a knowledge of the transition potential, it is often possible to select an indicator whose color transition occurs within this potential range. Or the potential can actually be measured during the visual titration and the color transition range on the potentiometric curve noted to see whether it corresponds to the equivalence point.

Ion-Selective Electrodes in Titrations

The term $\log a_{ion}$ in Equation 11.44 is equal to $-pIon$, and so ion-selective electrodes can be used to monitor changes in pM during a titration. For example, a cation-selective glass electrode that is sensitive to silver ion can be used to follow changes in pAg in titrations with silver nitrate. A calcium-sensitive electrode can be used for the titration of calcium with EDTA, and so forth. The electrode should not respond to sodium ion, since the disodium salt of EDTA is usually used. If the electrode responds to a second ion in the solution whose activity remains approximately constant throughout the titration, then Equation 11.46, 11.47, or 11.48 will hold, and the titration curve will be distorted; this is so because the electrode potential is determined by $\log (a_{ion} + \text{constant})$ and not $\log a_{ion}$. If the contribution from the second ion is not too large, then the distortion will not be too great and a good break will still occur at the end point. Titrations involving anions can also be monitored with anion-selective electrodes. For example, fluoride ion can be precipitated with lanthanum(III), and a fluoride electrode can be used to mark the end point of the titration.

Potentiometric titrations are always more accurate than direct potentiometry because of the uncertainties involved in potential measurements. Whereas accuracies of better than a few percent are rarely possible in direct potentiometry, accuracies of a few tenths of a percent are common by potentiometric titration.

We can make some general statements concerning potentiometric titrations:

1. The potential readings are usually sluggish in dilute solutions and near the end point because the solution is poorly poised.
2. It is necessary to plot the potential only near the end point. Small increments of titrant are added near the end point, 0.1 or 0.05 mL, for example. The exact end point volume need not be added, but it is determined by interpolation on the E-versus-volume plot.
3. The polarity of the indicating electrode relative to the reference electrode may change during the titration. That is, the potential difference may go from one polarity to zero and then to the reverse polarity; hence, the polarity of the potential measuring device may have to be changed.

Derivative Titrations

1. **FIRST DERIVATIVE PLOT.** We noted above that at the end point the slope of the titration curve was maximum. In other words, the rate of change of potential with addition of titrant is maximum at the end point. So, if we could plot the rate of change of potential with change in volume ($\Delta E/\Delta V$) against volume, then a "spiked" curve should result and the peak of this spike should occur at the end point. This is conveniently done by adding equal increments of titrant near the end point. Consider the data collected during a titration, as shown in Table 11.4. Disregard the last four columns for the time being. We want to plot $\Delta E/\Delta V$ against the volume to get the first derivative. Such a plot is shown in Figure 11.14. The volume

TABLE 11.4
Potentiometric Titration Data

I V, mL	II $E_{vs\ ref}$, Volts	III ΔE	IV ΔV_{1st}	V $\Delta E/\Delta V_{1st}$	VI V_{ave} for 1st Deriv. Plot	VII $\Delta(\Delta E/$ $\Delta V_{1st})$	VIII ΔV_{2nd}	IX $\Delta^2 E/\Delta V^2$ $[\Delta(\Delta E/$ $\Delta V_{1st})/$ $\Delta V_{2nd}]$	X V_{ave} for 2nd Deriv. Plot
35.45	0.630								
		0.020	0.05	0.4	35.475				
35.50	0.650					+0.2	0.05	+4	35.500
		0.030	0.05	0.6	35.525				
35.55	0.680					+0.8	0.05	+16	35.550
		0.070	0.05	1.4	35.575				
35.60	0.750					−0.2	0.05	−4	35.600
		0.060	0.05	1.2	35.625				
35.65	0.810					−0.6	0.05	−12	35.650
		0.030	0.05	0.06	35.675				
35.70	0.840					−0.2	0.05	−4	35.700
		0.020	0.05	0.4	35.725				
35.75	0.860								

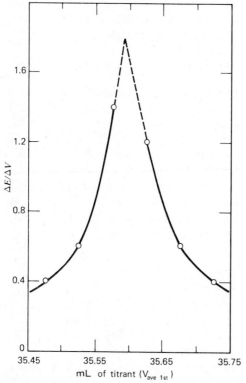

FIGURE 11.14
First derivative plot of a potentiometric titration curve.

used is the **average** of the two volumes used to calculate ΔE (column VI in Table 11.4). So, the volume for $\Delta E/\Delta V = 0.4$ is 35.475 mL, and so on. The end point is theoretically the maximum of this plot, which, extrapolated, occurs at 35.58 mL. This extrapolation leads to an uncertainty that can be partially avoided by a second derivative plot (see below).

Note that we have used equal volume increments here, and so ΔE could have been plotted in place of $\Delta E/\Delta V$. These equal increments are not necessary but do shorten the calculations. Although the average volume may be calculated to 0.001 mL for plotting, experimentally, we are not justified in reporting the end point to more than 0.01 mL.

2. SECOND DERIVATIVE PLOT. Mathematically, the second derivative of a titration curve should pass through zero at the equivalence point. The last four columns in Table 11.4 illustrate how such a plot can be accomplished. The second derivative is the rate of change of the first (column VII) with respect to the change in the average volume (column VIII). Division of column VII by column VIII, then, gives the second derivative, $\Delta^2 E/\Delta V^2$ (column IX). The average of the two successive volumes used for the first derivative plot (column VI) is used for the second derivative plot (column X). See Figure 11.15. Again, there is some extrapolation, but it is less significant than in the first derivative plot. The end point is taken as

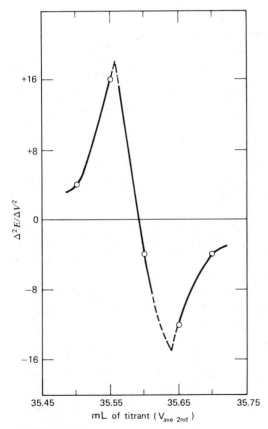

FIGURE 11.15
Second derivative plot of a potentiometric titration curve.

35.58 mL. As before, we are experimentally justified in reporting it to the nearest 0.01 mL. Again, since equal volume increments were added, V_{2nd} was constant, and we could have plotted column VII rather than column IX to save calculations. This will not work, however, if equal volume increments are not added. See Reference 13 by Carter and Huff for a discussion of second derivative plots.

In both these methods, the volume increment should not be too large or there will not be sufficient points near the end point. If the increments are small enough, then extrapolation of the second derivative plots may not be necessary at all, because there will be two or more points on the straight-line portion of the plot that passes through zero. On the other hand, the increments should not be small enough to be tedious and to fall within experimental error of the volume measurement. Usually, more points are taken than we have illustrated here. Of course, these small volume increments are taken only near the end point. In some titrations, the potential break is sufficiently large that the magnitude of potential change can be noted with equal added volume increments and the end point taken as that point where the change is largest. Also, it is convenient sometimes merely to titrate to an end point potential, which has been determined by calculations or empirically from a measured titration curve.

A word of caution should be mentioned with respect to derivative methods. The derivatives tend to emphasize noise or scatter in the data points, being worse for the second derivative. Hence, if a particular titration is subject to noise, a direct plot may be preferred.

Gran's Plots for End Point Detection

Assume that instead of plotting the electrode potential (which is a logarithmic function of concentration) against volume of titrant, we plotted the concentration of sample remaining at each point in the titration. A straight-line plot would in principle result (neglecting volume changes) in which the concentration would decrease to zero at the equivalence point (assuming the equilibrium for the titration reaction lies far to the right). This is because at 20% titrated, 80% of the sample will remain, at 50% titrated, 50% will remain, at 80% titrated, 20% will remain, and so on. (In practice, a plot in the region of the end point is made.) Similarly, a plot of titrant concentration beyond the equivalence point would be a linear plot of increasing concentration that would extrapolate to zero concentration at the end point. A plot like this can be obtained in a number of ways: a calibration curve can be used to convert potential readings to concentrations, or the log scale on the potential measuring device (e.g., pH meter) can be used to read concentration values directly (each $59/n$ mV being equal to a 10-fold change in concentration—an initial calibration of at least one point would be required, and preferably more points to establish the slope). Alternatively, the antilogarithm of the potential or pH reading can be calculated and plotted against volume of titrant ($E \propto \log C$, antilog $E \propto C$). Such a plot is a **Gran's plot** (see References 14 to 16).

A typical Gran's plot is shown in Figure 11.16 for the titration of small amounts of chloride with silver ion. A plot proportional to the titrant concentration is shown (right-hand ordinate) along with the usual S-shaped potentiometric plot (left-hand ordinate); a small potentiometric inflection point occurs due to the small concentrations involved. The straight-line plot is extrapolated back to the horizontal axis to determine the end point (a blank titration is performed and the linear plot is extrapolated to zero milliliters to accurately determine the horizontal axis). Curva-

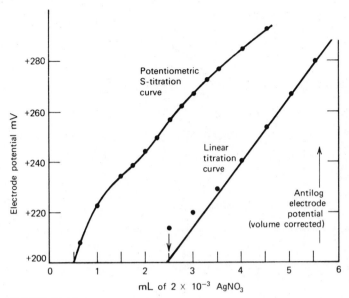

FIGURE 11.16
Gran's plot for titration of 100 mL 5×10^{-5} M Cl^- with $AgNO_3$ using Ag_2S electrode. (Courtesy of Orion Research, Inc.)

ture of the straight line around the end point generally indicates appreciable solubility of a precipitate, dissociation of a complex, and so on.

Several advantages accrue from the linear plots. It is only necessary to obtain a few points to define the straight line, and the end point is easily identified by extrapolating the line to the horizontal axis. Points only need be accurately determined a bit away from the equivalence point, where the titrant is in sufficient excess to suppress dissociation of the titration product and where electrode response is rapid because one of the ions is at relatively high levels compared to the levels at the equivalence point. In case of small inflection points (Figure 11.16), the end point is more readily defined by a Gran's plot.

A Gran's type plot is convenient in **"standard additions"** or **"known additions"** procedures. Standard additions methods are useful ways of calibration when the sample matrix affects the analyte signal. In these methods, a signal is recorded for the sample, and then a known amount of standard is added to the sample and the change in signal is measured. This latter measurement provides calibration in the same matrix as the unknown analyte, and the matrix should have the same effect on both unknown and standard. In this case, it is the electrode response that is calibrated. Most analytical methods give a linear response to analyte; but in potentiometry, it is a logarithmic response. By employing a Gran's type plot, a linear graph can be obtained, simplifying calculation. Here, the potential of the sample is initially recorded and then known amounts of standard are added to the sample. The antilog values are plotted as a function of the amount standard added, and the best straight line is drawn through them (e.g. least-squares analysis). Extrapolation to the horizontal axis (determined from similar measurements on a "blank" with extrapolation to zero concentration) gives the equivalent amount of analyte in the sample (Figure 11.17).

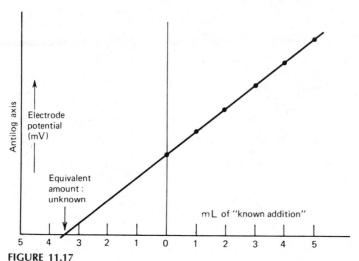

FIGURE 11.17
Known addition using Gran's plot paper. (Courtesy of Orion Research, Inc.)

The standard additions method can also be applied mathematically, as illustrated in the following example.

EXAMPLE 11.8 The calcium ion concentration in serum is determined using an ion-selective electrode. The potential measured with the electrode in the sample is $+217.6$ mV. Addition of 100 μL of a 2000 ppm standard to 2.00 mL of sample and measurement of the potential gives $+226.8$ mV. Assuming a Nernstian response (59.2/2 mV per 10-fold change in activity), what is the concentration of calcium ion in the sample?

SOLUTION
Since the analyte and the standard are subjected to the same matrix and ionic strength, the electrode responds in a Nernstian fashion to concentration (see Section 11.8). We can write

$$E = k + 29.6 \log [Ca^{2+}]$$

The standard (0.100 mL) is diluted in the sample (2.00 mL) about 1:20 to give an added concentration of 100 ppm or, more precisely, correcting for the 5% volume change:

$$C = 2000 \text{ ppm} \times \frac{0.100 \text{ mL}}{2.10 \text{ mL}} = 95.2 \text{ ppm}$$

Let x equal the unknown concentration in parts per million:

$$217.6 \text{ mV} = k + 29.6 \log x \tag{1}$$
$$226.8 \text{ mV} = k + 29.6 \log (x + 95.2) \tag{2}$$

Subtracting (2) from (1):

$$-9.2 \text{ mV} = 29.6 \log x - 29.6 \log (x + 95.2)$$

$$-9.2 \text{ mV} = 29.6 \log \frac{x}{x + 95.2}$$

$$\log \frac{x}{x + 95.2} = -0.31_1$$

$$\frac{x}{x + 95.2} = 0.467$$

$$x = 83.5 \text{ ppm}$$

If the actual slope for the electrode is not known, then multiple additions of the standard should be made to determine the actual slope.

The antilog concentration values must be corrected for volume changes. Corrected values are obtained by multiplying the observed values by $(V + v)/V$, where V is the initial volume and v is the volume added. Semiantilog paper (one axis scaled to antilogarithm values and the other a linear scale) is commercially available (e.g., from Orion Research Incorporated) in which potential readings are plotted directly on the antilog axis; these may be constructed in a manner so that volume correction over a significant range is automatically made. The antilog axis must be divided according to the electrode "slope"—for example, in 59-mV increments—or else an error in the intercept with the horizontal axis will result. A "Gran ruler" has been described that can be used to prepare Gran's plots directly from potentiometric titration data or standard addition measurements using linear graph paper. (See Reference 16.) The antilogarithmic functions for Nernstian potential changes of monovalent and divalent ion electrodes are built into the ruler, with volume corrections.

QUESTIONS

1. What is the liquid-junction potential? Residual liquid-junction potential? How can these be minimized?
2. Describe the different types of ion-selective electrodes. Include in your discussion the construction of the electrodes, differences in membranes, and their usefulness.
3. What is the selectivity ratio? Discuss its significance and how you would determine its value.
4. Discuss the pH limitations of ion-selective electrodes.

PROBLEMS

Standard Potentials

5. The standard potential of the silver/silver bromide electrode is 0.073 V. Calculate the solubility product of silver bromide.
6. A sample of thiocyanate is titrated with silver solution. The potential at the end point is 0.202 V versus SCE. Calculate the standard potential for $Ag^+ + e^- = Ag$. The K_{sp} for silver thiocyanate is 1.00×10^{-12}.

Galvanic Cells

7. For each of the following reactions, (1) separate the reaction into its component half-reactions; (2) write a schematic representation of a cell in which the reaction would occur in the direction as written; (3) calculate the standard potential of the cell; (4)

assign the polarity of each electrode under conditions that the reaction would occur as written.

(a) $Ag + Fe^{3+} = Ag^+ + Fe^{2+}$

(b) $VO_2^+ + V^{3+} = 2VO^{2+}$

(c) $Ce^{4+} + Fe^{2+} = Ce^{3+} + Fe^{3+}$

8. For the following cells, write the half-reactions occurring at each electrode and the complete cell reaction, and calculate the cell potential:

(a) Pt, H_2(0.2 atm)$\|$HCl(0.5 M)$|Cl_2$(0.2 atm), Pt

(b) Pt$|Fe^{2+}$(0.005 M), Fe^{3+}(0.05 M), $HClO_4$(0.1 M)$\|$HClO$_4$(0.1 M), VO_2^+(0.001 M), VO^{2+}(0.002) $M|$Pt

Redox Potentiometric Measurements

9. What would the potentials of the following half-cells at standard conditions be versus a saturated calomel electrode? (a) Pt/Br$_2$(aq), Br$^-$; (b) Ag/AgCl/Cl$^-$; (c) Pt/V^{3+}, V^{2+}.

10. What would be the observed potential if the following half-cell were connected with a saturated calomel electrode?

$$Fe^{3+}(0.00200\ M),\ Fe^{2+}(0.0500\ M)|Pt(\text{numbers represent activities})$$

11. A 50-mL solution that is 0.10 M in chloride and iodide ions is titrated with 0.10 M silver nitrate. (a) Calculate the percent iodide remaining unprecipitated when silver chloride begins to precipitate. (b) Calculate the potential of a silver electrode versus the SCE when silver chloride begins to precipitate and compare this with the theoretical potential corresponding to the end point for the titration of iodide. (c) Calculate the potential at the end point for chloride. Use concentrations in calculations.

12. The potential of the electrode Hg$|$Hg–EDTA, N–EDTA, N^{n+} is a function of the metal in N^{n+} and can be shown as

$$E = E^0_{Hg^{2+},Hg} - \frac{2.303RT}{2F}\log\frac{K_{f(HgEDTA)}}{K_{f(NEDTA)}} - \frac{2.303RT}{2F}\log\frac{a_{NEDTA}}{a_{HgEDTA}} - \frac{2.303RT}{2F}\log\frac{1}{a_{N^{n+}}}$$

The stability of N–EDTA must be less than that of Hg–EDTA (a very stable chelate; $K_{f(Hg-EDTA)} = 10^{22}$). An Hg$|$Hg–EDTA electrode can be used to monitor N^{n+} during the course of a titration with EDTA. Starting with the Hg$|$Hg^{2+} electrode, derive the above equation. This represents an electrode of the third kind.

13. The potential of a hydrogen electrode in an acid solution is -0.465 V when measured against an SCE reference electrode. What would the potential be measured against a normal calomel electrode (1 M KCl)?

pH Potentiometric Measurements

14. (a) How accurately can the pH of an unknown solution generally be measured? What limits this? What is this (calculate it) in terms of millivolts measured? In terms of percent error in the hydrogen ion activity? (b) How precisely can the pH of a solution be measured? How much is this in terms of millivolts measured? In terms of percent variation in the hydrogen ion activity?

15. A glass electrode was determined to have a potential of 0.395 V when measured against the SCE in a standard pH-7.00 buffer solution. Calculate the pH of the unknown solution for which the following potential readings were obtained (the potential decreases with increasing pH):

(a) 0.467 V

(b) 0.209 V

(c) 0.080 V

(d) -0.013 V

16. Calculate the potential of the cell consisting of a hydrogen electrode (P_{H_2} = 1 atm) and a saturated calomel reference electrode (a) in a solution of 0.00100 M HCl, (b) in a solution of 0.00100 M acetic acid, and (c) in a solution containing equal volumes of 0.100 M acetic acid and 0.100 M sodium acetate. Assume that activities are the same as concentrations.

17. The quinhydrone electrode can be used for the potentiometric determination of pH. The solution to be measured is saturated with quinhydrone, an equimolar mixture of quinone (Q) and hydroquinone (HQ), and the potential of the solution is measured with a platinum electrode. The half-reaction and its standard potential are as follows:

quinone (Q)　　　　　　hydroquinone (HQ)

What is the pH of a solution saturated with quinhydrone if the potential of a platinum electrode in the solution, measured against a saturated calomel electrode, is −0.205 V? Assume the liquid junction potential to be zero.

Ion Selective Electrode Measurements

18. It can be shown from Equations 11.44 and 11.46 that, for monovalent ions, log K_{AB} = (k_B − k_A)/S. Derive this equation.

19. A potassium ion-selective electrode is used to measure the concentration of potassium ion in a solution that contains 6.0 × 10^{-3} M cesium (activity). From Table 11.3, the electrode responds equally to either ion (K_{KCs} = 1). If the potential versus a reference electrode is −18.3 mV for a 5.0 × 10^{-3} M KCl solution and +20.9 mV in the sample solution, what is the activity of K^+ in the sample? Assume Nernstian response.

20. The nitrate concentration in an industrial effluent is determined using a nitrate ion-selective electrode. Standards and samples are diluted 20-fold with 0.1 M K_2SO_4 to maintain constant ionic strength. Nitrate standards of 0.0050 and 0.0100 M give potential readings of −108.6 and −125.2 mV, respectively. The sample gives a reading of −119.6 mV. What is the concentration of nitrate in the sample?

21. The perchlorate concentration in a sample containing 0.015 M iodide is determined using a perchlorate ion-selective electrode. All samples and standards are diluted 1:10 with 0.2 M KCl to maintain constant ionic strength. A 0.00100 M $KClO_4$ standard gives a reading of −27.2 mV, and a 0.0100 M KI standard gives a reading of +32.8 mV. The sample solution gives a reading of −15.5 mV. Assuming Nernstian response, what is the concentration of perchlorate in the sample?

22. The potential of a glass cationic-sensitive electrode is measured against an SCE. In a sodium chloride solution of activity 0.100 M, this potential is 113.0 mV, and in a potassium chloride solution of the same activity, it is 67.0 mV. (a) Calculate the selectivity ratio of this electrode for potassium over sodium, using the relationship derived in Problem 18. (b) What would be the expected potential in a mixture of sodium (a = 1.00 × 10^{-3} M) and potassium (a = 1.00 × 10^{-2} M) chlorides? Assume Nernstian response, 59.2 mV/decade.

23. The selectivity ratio for a cation-selective electrode for B^+ with respect to A^+ is deter-

mined from measurements of two solutions containing different activities of the two ions. The following potential readings were obtained: (1) 2.00×10^{-4} M A^+ + 1.00×10^{-3} M B^+, +237.8 mV; and (2) 4.00×10^{-4} M A^+ + 1.00×10^{-3} M B^+, +248.2 mV. Calculate K_{AB}. The electrode slope is 56.7 mV/decade.

24. The sulfide content in a pulp plant effluent is determined with a sulfide ion-selective electrode using the method of standard additions for calibration. A 10.0-mL sample is diluted to 25.0 mL with water and gives a potential reading of −216.4 mV. A similar 10.0-mL sample plus 1.00 mL of 0.030 M sulfide standard diluted to 25.0 mL gives a reading of −224.0 mV. Calculate the concentration of sulfide in the sample.

RECOMMENDED REFERENCES

Activity Coefficients
1. J. Kielland, "Individual Activity Coefficients of Ions in Aqueous Solutions," *J. Am. Chem. Soc.,* **59** (1937) 1675.

pH Electrodes and Measurements
2. R. G. Bates, *Determination of pH.* New York: John Wiley & Sons, 1964.

3. J. E. Nelson, *pH in Plain Language.* Hillsboro, Oregon: Chemtrix, Inc.

4. J. V. Straumford, Jr., "Determination of Blood pH," in *Standard Methods of Clinical Chemistry,* Vol. 2, D. Seligson, ed., New York: Academic Press, 1958, pp. 107–121.

5. C. C. Westcott, "Biomedical pH Measurements," *Am. Lab.,* (March 1978) 131.

6. C. C. Westcott, "Selection and Care of pH Electrodes," *Am. Lab.* (August 1978) 71.

Ion-Selective Electrodes and Measurements
7. R. G. Bates, "Approach to Conventional Scales of Ionic Activity for the Standardization of Ion-Selective Electrodes," *Pure Appl. Chem.,* **37** (1974) 573.

8. H. J. Berman and N. C. Herbert, eds., *Ion-Selective Microelectrodes.* New York: Plenum Publishing Corp., 1974.

9. R. A. Durst, ed., *Ion Selective Electrodes.* Washington, D.C.: National Bureau of Standards Special Publication 314, U.S. Government Printing Office, 1969.

10. E. Pungor, "Theory and Applications of Anion Selective Membrane Electrodes," *Anal. Chem.,* **39** (13) (1967) 28A.

11. G. A. Rechnitz, "New Directions for Ion Selective Electrodes," *Anal. Chem.,* **41** (12) (1969) 109A.

12. G. A. Rechnitz, "Ion Selective Electrodes," *Chem. Eng. News,* **45** (1967) (June 12).

Derivative Titrations
13. K. N. Carter and R. B. Huff, "Second Derivative Curves and End-Point Determination," *J. Chem. Educ.,* **56** (1979) 26.

Gran's Plots
14. G. Gran, "Determination of Equivalent Point in Potentiometric Titrations," *Acta Chem. Scand.,* **4** (1950) 559.

15. G. Gran, "Determination of the Equivalence Point in Potentiometric Titrations. Part II," *Analyst,* **77** (1952) 661.

16. C. C. Westcott, "Ion-Selective Measurements by Gran Plots With a Gran Ruler," *Anal. Chim. Acta,* **86** (1976) 269.

chapter
12

ELECTROLYTIC METHODS

Electrolytic methods include some of the most accurate, as well as most sensitive, instrumental techniques. In these methods, an analyte is oxidized or reduced at an appropriate electrode and the amount of electricity (quantity or current) involved in the electrolysis is related to the amount of analyte. The electrolysis may be complete, as in electrogravimetry or coulometric methods. These methods are usually suited for larger quantities of analyte, for example, millimole amounts, although smaller amounts can sometimes be measured. The fraction of analyte electrolyzed may be very small, in fact negligible, as in the current–voltage techniques of voltammetry or polarography. Micromolar or smaller concentrations can be measured. Since the potential at which a given analyte will be oxidized or reduced is dependent on the particular substance, selectivity can be achieved in electrolytic methods by appropriate choice of electrolysis potential. Owing to the specificity of the methods or the fact that a separation may be performed during the measurement, prior separations are often unnecessary. These methods can therefore be rapid. In this chapter, we discuss a variety of techniques. In electrogravimetry, a metal (or metal oxide) is deposited on an electrode by electroreduction (or electrooxidation) and then weighed. Controlled-potential coulometry is a related technique, but the quantity of electricity required for complete electrolysis is measured and related to the amount of analyte using Faraday's law; hence, the electrolysis product can remain in solution. Controlled-current coulometry is a technique in which a titrant is generated by electrolysis of an appropriate reagent at constant current and reacted with the analyte; the quantity of titrant used to reach the end point is calculated from the quantity of electricity used to generate it, using Faraday's law. Voltammetric techniques use a microelectrode for microelectrolysis. Here, the potential is scanned and a dilute solution of the analyte produces a limiting current at a given potential, in the microampere range or less, which is proportional to the analyte concentration. Polarography is a special type of voltammetry in which the working microelectrode is a dropping mercury electrode. Variations of this include dc polarography and pulse polarography, the latter capable of excellent sensitivity and resolution. Anodic-stripping voltammetry is a technique in which a metal is concentrated on a microelectrode by electrodeposition followed by anodically reoxidizing it to produce a large current; this is among the most sensitive of analytical techniques for the determination of heavy metals. Amperometry is the application of voltammetry (or polarography) at a fixed potential to follow, via the current, changes in concentration of a given species, for example, during a titration.

We describe in detail each of these techniques in this chapter. It will be helpful to review Chapter 11 on potentiometry before reading this material.

12.1   ELECTROGRAVIMETRY

In electrogravimetry, generally the metal to be determined is quantitatively plated onto a platinum cathode, and from the gain in weight of the cathode, the amount of metal in the sample is calculated. Because a weight must be accurately measured, the main use of electrogravimetry is in macroanalysis.

Principles

A typical setup is shown in Figure 12.1. A fraction of the voltage from the battery, as measured by the voltmeter V, is applied to electrodes by moving the pointer on the slidewire (see the operation of a potentiometer, Figure 11.5). A platinum gauze electrode is used as the cathode and a platinum spiral is used as the anode. The anode is placed concentric to the cathode to allow a uniform electric field and thus

FIGURE 12.1
Setup for electrogravimetry: B, battery; V, voltmeter; A, ammeter.

a more uniform deposit. A direct-current (dc) source is used, either a battery or a rectified alternating-current (ac) line source (ac converted to dc).

The determination of copper is one of the most important applications of electrogravimetry. The possible electrode reactions in a solution containing copper sulfate and sulfuric acid are:

Cathode:

$$Cu^{2+} + 2e^- \rightarrow \underline{Cu} \qquad E^0 = 0.337 \text{ V} \qquad (12.1)$$

$$H^+ + e^- \rightarrow \tfrac{1}{2}H_2 \qquad E^0 = 0.000 \text{ V} \qquad (12.2)$$

Anode:

$$H_2O \rightarrow \tfrac{1}{2}O_2 + 2H^+ + 2e^- \qquad E^0 = 1.229 \text{ V} \qquad (12.3)$$

The E^0 value for the oxidation of sulfate ion $(2SO_4^{2-} \rightarrow S_2O_8^{2-} + 2e^-)$ is 21.0 V, which is more positive than the potential required for the oxidation of the solvent, water. Hence, sulfate ion is not electroactive in aqueous solutions. The potential for each half-reaction can be calculated from the Nernst equation. If the copper(II) concentration is 0.01 M and the sulfuric acid concentration is 0.05 M ($[H^+] = 0.1$ M), then at 25°C (assuming gases are at atmospheric pressure),

$$E_{Cu^{2+},Cu} = 0.337 - \frac{0.0592}{2} \log \frac{1}{10^{-2}} = 0.278 \text{ V} \qquad (12.4)$$

$$E_{H^+,H_2} = 0.000 - 0.0592 \log \frac{1}{10^{-1}} = -0.059 \text{ V} \qquad (12.5)$$

$$E_{O_2,H_2O} = 1.229 - \frac{0.0592}{2} \log \frac{1}{(10^{-1})^2} = 1.170 \text{ V} \qquad (12.6)$$

Since the potential for the copper half-reaction is more positive than that for the hydrogen half-reaction, copper(II) will be reduced more easily than hydrogen ions, and the cathode reaction will be Equation 12.1. The overall electrolytic **cell reaction** will be

$$Cu^{2+} + H_2O \rightarrow \underline{Cu} + \tfrac{1}{2}O_2 + 2H^+ \qquad (12.7)$$

The actual voltage that must be supplied by the dc source will depend on a number of variables.

Back-emf

Note that the potential for Equation 12.3 is more positive than that for Equation 12.1. You will recall from Chapter 10 that the redox couple with the more positive potential will be the more strongly oxidizing couple. Therefore, the reaction in Equation 12.7 will spontaneously go in the opposite direction in a **galvanic cell** (Chapter 10), where the electrodes are merely shorted together. In an **electrolytic cell** (here), then, we must supply a voltage equal to and opposite that of the galvanic cell in order to reverse the direction of the reaction.[1] This voltage is called

---

[1] Actually, a slightly greater voltage is required; otherwise, there would be an equal tendency of the reaction to go in either direction.

the **back-emf** (E_{back}). The back-emf in this case is $1.170 - 0.278 = 0.892$ V, and the battery must supply this voltage to the cell, with the negative terminal connected to the platinum gauze electrode (the cathode).

Overvoltage

Often, when a gas product is involved in an electrode reaction, a certain potential in excess of the reversible potential calculated from the Nernst equation is required to make the reaction go. This is called the **overpotential.**[2] The combined overpotential of each electrode in a cell is called the **overvoltage.** Because of this overvoltage, a total voltage equal to the overvoltage, in addition to the back emf, must be supplied to the cell.

The overvoltage will depend on the gas, the electrode characteristics, and the current density. Table 12.1 lists the overvoltage for hydrogen and oxygen on a bright platinum at different current densities. The overvoltage can be decreased on a platinized platinum surface, but this is rarely used for electrogravimetry because of a tendency of the deposit to flake off. It is seen that the overvoltage can be minimized by keeping the electrode area large and the current small, so the current density will be low.

Concentration Polarization

The reversible potential required for electrolysis is determined by the concentration of the electroactive species **at the electrode surface** rather than that in the bulk of the solution. When an electrolysis is started, the concentration at the electrode surface is decreased. **Concentration polarization** results, and causes the potential of the half-reaction at the cathode to become more negative, by the Nernst equation. Therefore, a somewhat larger back emf must be applied to maintain the electrode reaction at a sufficient rate. [At the anode, there is generally a larger supply of the substance being oxidized (e.g., the solvent) and there is only a small or no corresponding increase in the potential of its half-reaction. Any such increase would add to the total required back emf.] Concentration polarization can be minimized by efficient stirring and low current density. Aside from concentration polarization, the concentration in the bulk of the solution decreases as the electrolysis proceeds, and the applied voltage must be increased to assure a high percentage of electrolysis of the metal.

TABLE 12.1
Overvoltage of Hydrogen and Oxygen on Bright Platinum

Current Density, amp/cm²	Overvoltage, Volts	
	Hydrogen	Oxygen
0.001	0.024	0.72
0.01	0.068	0.85
0.10	0.29	1.28
1.0	0.68	1.49

[2]Again, as in Chapter 10, we use **potential** in referring to a single electrode and **voltage** for a complete cell.

iR Drop

When current is passed through a cell, a certain voltage drop, called the iR *drop,* will develop in the circuit. This will be equal to the current in amperes times the resistance in ohms. Most of the iR drop occurs between the electrodes in the cell, and it can be minimized by adding a high concentration of an inert electrolyte to decrease the resistance of the solution. Solution resistance can be very significant in nonaqueous solvents.

The total cell voltage, then, required for electrolysis is

$$E = E_{back} + E_{overvoltage} + E_{polarization} + E_{iR} \qquad (12.8)$$

EXAMPLE 12.1   A 1.00×10^{-3} *M* solution of copper(II) is to be analyzed by electrogravimetry. What potential versus the NHE would have to be applied to the electrode to deposit 99.9% of the copper? What voltage would have to be applied to a cell containing 0.0500 *M* H_2SO_4 and a platinum anode? (Assume no overvoltage, concentration polarization, or iR drop.)

SOLUTION

The concentration of Cu^{2+} when 0.1% remains is 1.00×10^{-6} *M*.

$$E = 0.337 - \frac{0.0592}{2} \log \frac{1}{1.00 \times 10^{-6}} = 0.159 \text{ V versus NHE}$$

From Equation 12.6, the potential of the anode is 1.170 V versus NHE. (The amount of H^+ produced during the electrolysis was twice the amount of copper—see cell reaction 12.7—which represents only 2% of the initial concentration and can be neglected.)

$$E_{cell} = 1.170 - 0.159 = 1.011 \text{ V}$$

Desired Characteristics of the Deposited Metal

The metal should be deposited smoothly and uniformly on the cathode and should adhere tightly. If the current is too large, the metal does not have time to plate uniformly. If a gas is evolved simultaneously, the gas bubbles prevent a smooth deposit. Current and gas evolution are controlled by adjusting either the current or the voltage applied to the cell.

Typical Determinations

Because hydrogen reduction occurs at fairly small potentials on a platinum cathode, copper is the only element commonly deposited from acid solution. Silver can be deposited from acid solutions, but a more adhesive deposit is obtained from an alkaline cyanide solution in which the silver is complexed. More electronegative metals such as zinc can be deposited from ammoniacal or strong alkali solution. The potential of the hydrogen evolution is more negative owing to the decreased hydrogen ion concentration. The metal must not precipitate in alkaline solution, however. Zinc forms an ammine complex, $Zn(NH_3)_4^{2+}$, in ammonia and the zincate ion, ZnO_2^{2-}, in strong alkali. This also helps in obtaining an adhesive deposit, as in the case of the complexed silver; the complexes require a somewhat more negative potential for reduction, but the change is less than that for the reduction of hydrogen ions.

For the deposition of electronegative metals, the more electropositive metals must be absent. For example, zinc cannot be deposited in the presence of copper. The copper can be removed by constant-current deposition in acid solution (hydrogen reduction prevents deposition of zinc), or by controlled-potential deposition in the ammoniacal solution at a potential at which zinc is not deposited. Lead is unusual in that it is deposited at the anode as PbO_2, by oxidation of lead(II).

Quantitative electrodeposition of analytical samples typically takes $\frac{1}{2}$ to 1 hour. At the completion of electrolysis, to prevent the metal from spontaneously going back into solution, **the electrode is removed from the solution before breaking the circuit;** remember, the reaction wants to go in the other direction.

12.2 CONTROLLED-POTENTIAL COULOMETRY

Constant-potential coulometry is an electrolysis technique related to electrogravimetry at controlled potential, but electrodeposition is not required. The cell used is illustrated in Figure 12.2. The **working electrode** is either a mercury pool cathode for the negative potential region to prevent hydrogen evolution, or a platinum gauze for more positive potentials and may be either the cathode or the anode. It is depicted as the cathode in the figure. Its potential is controlled relative to a **reference elctrode** (E versus ref) by adjusting the applied voltage (V) between the working electrode and an **auxiliary electrode.** The potential control can be accomplished automatically in modern-day **potentiostats.**

Because relatively large currents flow during electrolysis ($> 100 \mu A$), the current cannot be allowed to flow through the reference electrode or else it would become polarized, causing its potential to change, and this is the reason a third electrode is used; the current flows between it and the working electrode. While the current

FIGURE 12.2
Setup for constant potential coulometry: Q, coulometer; V, volts.

flows between the auxiliary electrode and the working electrode, the potentiostat measures the potential *(E)* between the working and reference electrodes; it adjusts the applied voltage between the working and auxiliary electrodes as necessary to maintain the desired working electrode potential relative to the reference electrode. This potential is nominally measured under zero current flow between the reference electrode and the working electrode. This is analogous to measuring the voltage drop across a resistor through which current is flowing with a simple voltmeter through which no current is flowing. The auxiliary electrode is separated from the test solution by a salt bridge or a glass tube with a sintered-glass frit end to prevent its electrolysis products from undergoing side reactions at the working electrode.

The potential of the working electrode is adjusted to a value at which the substance is either oxidized or reduced (corresponding to the plateau of a voltammetric wave of the substance—see below) and the electrolysis is allowed to go to completion. Instead of weighing a deposited substance, however, a **coulometer** is placed in series with the electrolysis circuit. This is a device for measuring the *quantity* of current that flows. Since it is in series with the electrolysis cell, the same amount of current will flow through the coulometer as through the cell. The amount of test substance electrolyzed is calculated from **Faraday's law:**

$$\text{weight electrolyzed (g)} = \frac{\text{coulombs} \times \text{f. wt. (g/mol)}}{96{,}487 \ (\text{coul/mol e}^-) \times n(\text{mol e}^-/\text{mol analyte})}$$

$$(12.9)$$

The number 96,487 is the **Faraday constant** and n is the electron change the substance undergoes during electrolysis. Note that the Faraday constant (1 faraday) is one mole of electrons. The Faraday constant relates the moles of electrons consumed to the measured coulombs. From moles of electrons, the moles of analyte is calculated. Note also that the quantity of current is proportional to the number of *equivalents* of analyte [g/f. wt./n), where f. wt./n is the equivalent weight].

Either chemical, mechanical, or electronic coulometers are used to measure the coulombs of current passed. An example of a chemical coulometer is the following cell (called a titration coulometer) placed in series in the electrolysis circuit:

$$\text{Ag/KBr}(0.03 \ M), \ \text{K}_2\text{SO}_4(0.2 \ M)/\text{Pt} \qquad (12.10)$$

Potassium sulfate is the supporting electrolyte, and the anode reaction is

$$\underline{\text{Ag}} + \text{Br}^- \rightarrow \underline{\text{AgBr}} + \text{e}^- \qquad (12.11)$$

while the cathode reaction is

$$\text{H}_2\text{O} + \text{e}^- \rightarrow \tfrac{1}{2}\text{H}_2 + \text{OH}^- \qquad (12.12)$$

The hydroxide ion produced at the cathode (one mole per mole electrons) is titrated with standard hydrochloric acid to bring the pH back to its initial value. The coulombs of current are then calculated by use of Equation 12.9, where f. wt. is the formula weight of hydrochloric acid, n is 1, and g is the weight of hydrochloric acid used. You can see that g/f. wt. is equal to the number of moles (and the number of equivalents) of acid used in the titration, and the weight of acid need not be calculated.

Mechanical or electronic coulometers automatically measure the coulombs of current passed.

EXAMPLE 12.2   A titration coulometer is placed in series in a controlled-potential electrolysis circuit. At the completion of electrolysis, a total of 25.72 mL of 0.01000 M HCl
is required to bring the pH back to its initial value. How many coulombs of current were
used in the electrolysis?

SOLUTION

mol HCl = 0.01000 M × 0.02572 L

       = 2.572×10^{-4} mol

coulombs = 2.572×10^{-4} mol HCl × 96,487 (coul/mol e^{-}) × 1 (mol e^{-}/mol HCl)

       = 24.82 coul

12.3   CONTROLLED-CURRENT COULOMETRY

This technique, known as **coulometric titration,** is essentially a titration in which
the buret and the titrant solution are replaced by a constant-current source and a
timer plus an appropriate pair of electrodes and electrolyte. The titrant is then
electrogenerated. For example, I_2 can be generated by oxidation of I^- at a platinum anode. The generated I_2 can be used for titrations instead of being added from
a buret, the amount used being calculated from Faraday's law (Equation 12.13
below).

    A typical coulometric titration cell is shown in Figure 12.3. The auxiliary elec-
trode is again isolated to prevent its (the auxiliary electrode's) electrolysis products

FIGURE 12.3
Constant-current coulometry titration cell.

such as H_2 from producing side reactions at the working electrode or with the generated titrant (such as I_2). This also prevents the generated titrant from reacting at the auxiliary electrode surface (e.g., $I_2 + 2e^- \rightarrow 2I^-$) and maintains 100% current efficiency at the working electrode. The working electrode may be either the cathode or the anode, depending on the titrant being generated. There are only two electrodes required for titrant generation. The other two electrodes are separate from the generating electrodes and are indicating electrodes, used to detect the end point of the titration. These are not needed if a visual indicator is employed.

Consider the titration of dichromate with Fe^{2+}, in which it is desired to generate Fe^{2+} as the titrant. This can be done by reducing a solution of Fe^{3+} at a platinum cathode. If an acidified solution of 0.1 M $FeCl_3$ is placed in the cell with the working electrode as the cathode and a constant current is passed through the cell, Fe^{2+} will be generated at a constant rate. The anode reaction will be the oxidation of chloride ion to chlorine, but this is isolated from the test solution and will not interfere. The Fe^{2+} reacts with the dichromate as if added from a buret. The end point of the titration is detected in any conventional fashion, and the time of generation required to reach the end point is measured.

The amount of material titrated is calculated from Faraday's law as follows. Since the current is maintained constant, the coulombs of electricity passed during the titration can be calculated from the product of the current in amperes and the time in seconds. Thus,

$$\text{weight titrated (g)} = \frac{\text{amperes (coul/second)} \times \text{seconds} \times \text{f. wt.(g/mol)}}{96{,}487 \text{ (coul/mol } e^-\text{)} \times n\text{(mol } e^-\text{/mol analyte)}} \quad (12.13)$$

If we wish to calculate the weight of $K_2Cr_2O_7$ reacted in the above titration, f. wt. is equal to the formula weight of $K_2Cr_2O_7$ and n is equal to 6, since each chromium is reduced from +6 to +3 and there are two Cr atoms per molecule of $K_2Cr_2O_7$. That is, six electrons are required to produce six ions of Fe^{2+} to react with each molecule of $Cr_2O_7^{2-}$. In a calculation of the weight of Cr from the titration, f. wt. would be replaced by the atomic weight of Cr, and n would be 3.

EXAMPLE 12.3 Thallium(I) is titrated with coulometrically generated bromine ($Tl^+ + Br_2 \rightarrow Tl^{3+} + 2Br^-$). If a generating current of 10.00 mA is used and 102.0 seconds are required to reach the end point, how many grams thallium were present in the solution?

SOLUTION

$$W = \frac{0.01000 \text{ A} \times 102.0 \text{ seconds (A} \cdot \text{s} = \text{coul)} \times 204.4 \text{ g/mol Tl}}{96{,}487 \text{ (coul/mol)} \times 2 \text{ (mol } e^-\text{/mol Tl)}}$$

$$= 1.080 \times 10^{-3} \text{ g}$$

Note that each Br_2 is generated by $2\,Br^- \rightarrow Br_2 + 2e^-$, and therefore two electrons are consumed in the reaction. The titrant just acts as a transfer source for the electrons and the titrant can be considered the electron.

The generating current can be set to give readout in either microequivalents of sample titrated or in micrograms of sample titrated.

EXAMPLE 12.4 Small amounts of water can be determined (e.g., in organic solvents) by Karl Fischer titration. The sample is dissolved in anhydrous methanol, and the titrant consists of a mixture of iodine, sulfur dioxide, and pyridine dissolved in methanol. The I_2 and SO_2 exist as addition compounds with pyridine (C_5H_5N), and the overall reaction is (where R = pyridine)

$$H_2O + R \cdot I_2 + R \cdot SO_2 + R + CH_3OH \rightarrow 2R \cdot HI + R \cdot CH_3OSO_3H$$

The water can be titrated coulometrically by replacing the I_2 with I^- and generating the I_2 at a platinum anode. What generating current should be used so that the seconds of titration will equal the micrograms of water titrated?

SOLUTION
Each H_2O consumes one I_2, which requires two electrons for generation.

$$1 \; \mu g = \frac{i(A) \times 1 \text{ second (coul)} \times 18.02 \times 10^6 \; \mu g/mol}{96,487 \text{ (coul/mol } e^-) \times 2 \text{ (mol } e^-/\text{mol } H_2O)}$$

$$i = 0.01071 \text{ A}$$

In coulometric titrations, the titrant is added in increments, as with a buret (generated at a constant rate for each increment). Or, since the titrant is added at a very constant rate (generally ±0.1%), it can be generated continuously and the end point signal automatically recorded if instrumental detection of the end point is employed.

While, in principle, visual indicators can be used to detect the end point, coulometric titrations are generally applied on a microscale, and visual indicators are not sensitive enough for sharp end points. Therefore, potentiometry or amperometry (below) is usually employed.

An important advantage of coulometric titrations over other instrumental techniques for microanalysis is that very high precision and accuracy can be obtained. The current is usually known to four significant figures, and so, if the time of the titration is known to four significant figures, the amount of sample titrated can be calculated to four significant figures. The time is generally measured to the nearest 0.1 second, and so, if the titration time exceeds 100 seconds, four-significant-figure accuracy is achieved. This is limited by the precision with which the end point can be detected, but with currents of 10 mA or above, this is usually not a problem.

An added advantage of coulometric titrations over volumetric titrations is that many unstable titrants can be generated and used, since they are reacted immediately with the sample. Examples include Br_2, Cl_2, OBr^-, Cu(I), Mo(V), Ag(II), and Cr(II). A summary of some of the titrations that have been performed coulometrically is given in Table 12.2. The generating electrolyte is generally about 0.1 M and the concentration need be only approximate. The current can range from 100 mA down to a few microamperes. Many volumetric titrations that are performed today could probably be done coulometrically. Other advantages include no need to prepare standard solutions and no need to worry about shelflife of the solutions.

The most important application of constant-current coulometry in the clinical laboratory is in the determination of chloride in biological fluids. The titration is performed with the "Cotlove chloridimeter," an automatic constant-current coulometer. This is illustrated in Figure 12.4. The titrant, Ag^+, is generated from a silver wire anode. The end point is detected biamperometrically with a pair of silver wire

TABLE 12.2
Some Coulometric Titrations[a]

Substance Determined	Titrant	Generating Electrolyte
m-Aminophenol	Br_2	0.2 M NaBr in 0.1 M H_2SO_4
Anthracene	Biphenyl radical anion	Biphenyl and tetra-n-butylammonium bromide in dimethylformamide
As(III)	Br_2	
As(III)	I_2	0.2 M KI in pH 8 buffer
Br^-	Ag(I)	Ag anode in HNO_3
Ca^{2+}	EDTA	0.02 M Hg(II)–EDTA in ammoniacal buffer, pH 8.5; mercury cathode
Ca^{2+} (in presence of Mg^{2+})	EGTA	0.1 M Hg(II)–EGTA in triethanolamine solution, pH 8.6; mercury cathode
Ce(IV)	Fe(II)	Acid solution of ferric ammonium sulfate
Cl^-	Ag(I)	
$Cr_2O_7^{2-}$	Mo(V)	0.7 M Mo(VI) in 4 M H_2SO_4
Fe(II)	Ce(IV)	0.1 M cerous sulfate in 3 M H_2SO_4
HCl	OH^-	0.1 M Na_2SO_4 (electrolysis of water)
H_2O	I_2	Pyridine, SO_2, MeOH, KI (Karl Fisher titration)
H_2O_2	Mn(III)	$MnSO_4$ in 2 M H_2SO_4
8-Hydroxyquinoline	Br_2	
Mercaptans (RSH)	Ag(I)	
Olefins	Br_2	
NH_3	OBr^-	1 M NaBr in borate buffer, pH 8.6
O_2	Cr(II)	$Cr_2(SO_4)_3$ in H_2SO_4
Pyridine	H^+	0.1 M Na_2SO_4 (electrolysis of water)
S^{2-}	I_2	
Tl(I)	Br_2	
V(V)	Cu(I)	0.1 M $CuSO_4$ in M HCl
V(V)	Ti(III)	3.6 M $TiCl_4$ in 7 M HCl

[a]From G. D. Christian, in *Advances in Biomedical and Medical Physics,* Vol. IV, S. N. Levine, ed. New York: Wiley-Interscience Div., 1971. (Reproduced by permission of John Wiley & Sons, Inc.)

electrodes with about 100 millivolts impressed between them; no current flows between the biamperometric electrodes until beyond the end point when both electrodes are depolarized, that is, excess Ag^+ is reduced at the cathode (the anode reaction is oxidation of Ag to Ag^+). The titration is performed automatically, with the sudden increase in amperometric current triggering a relay circuit that shuts the instrument off. The time of the titration is then read off the timer. The auxiliary generating electrode (cathode) is not isolated and, consequently, standard chloride solutions are titrated to calibrate the instrument. A 0.1-mL or 0.01-mL serum sample, depending on the titration current selected, is taken for analysis.

12.4 VOLTAMMETRY

Voltammetry is essentially an electrolysis on a microscale, using a micro working electrode (e.g., a platinum wire). As the name implies, it is a current–voltage technique. The potential of the micro working electrode is varied (scanned slowly) and

FIGURE 12.4
Cotlove chloridimeter. (Courtesy of the American Instrument Company.)

the resulting current is recorded as a function of applied potential. The recording is called a **voltammogram.** If an electroactive (reducible or oxidizable) species is present, a current will be recorded when the applied potential becomes sufficiently negative or positive for it to be electrolyzed. (By convention, a cathodic (reduction) current is + and an anodic (oxidation) current is −). If the solution is dilute, the current will reach a limiting value, because the analyte can only diffuse to the electrode and be electrolyzed at a finite rate, depending on its concentration. We will see below that the limiting current is proportional to the concentration of the species. The microelectrode restricts the current to a few microamperes or less, and so in most applications the concentration of the test substance in solution remains essentially unchanged after recording the voltammogram.

The Voltammetric Cell

A voltammetric cell consists of the micro working electrode, the auxiliary electrode, and a reference electrode, usually an SCE. A potentiostat is employed as described in Section 12.2. The current of the working electrode is recorded as a function of its potential measured against the reference electrode, but the voltage is applied between and the current passes between the working and auxiliary electrodes, as in Figure 12.2.

The Current–Voltage Curve

Potentials in voltammetry are by convention referred to the SCE. The following relationship can be used to convert potentials versus SCE to the corresponding potentials versus NHE, and vice versa:

$$\boxed{E_{\text{vs SCE}} = E_{\text{vs NHE}} - 0.242}$$

(12.14)

This relationship can be used to calculate the applied potential required for the electrolysis of the test ion at the microelectrode. Suppose, for example, we place a 10^{-3} M solution of cadmium nitrate in the test cell with a carbon microelectrode and impress a voltage difference between the working and auxiliary electrodes, making the microelectrode negative relative to the SCE. The minimum working electrode potential (back emf) required to begin reducing the cadmium will be

$$E_{vs\ SCE} = -0.403 - \frac{0.0592}{2} \log \frac{1}{10^{-3}} - 0.242 = -0.556\ V \quad (12.15)$$

This is called the decomposition potential. As the potential is increased beyond the decomposition potential, the current will increase linearly in accordance with Ohm's law,

$$i = \frac{E}{R_{circuit}} = kE$$

As the electrolysis proceeds, the ions in the vicinity of the electrode are depleted by being reduced, creating a concentration gradient between the surface of the electrode and the bulk solution. As long as the applied potential is small, the ions from the bulk of the solution can diffuse rapidly enough to the electrode surface to maintain the electrolysis current. But as the potential is increased, the current is increased, creating a larger concentration gradient. Hence, the ions must diffuse at a more rapid rate in order to maintain the current. The concentration gradient, and hence the rate of diffusion, is proportional to the bulk concentration. If the solution is dilute, a potential will eventually be reached at which the rate of diffusion reaches a maximum and all the ions are reduced as fast as they can diffuse to the electrode surface. Hence, a **limiting current** value, i_l, is reached, and further increased potential will not result in increased current.

A typical recorded current–voltage curve is illustrated in Figure 12.5. If the solution is stirred or the electrode is rotated, an S-shaped plot (curve A) is obtained, that is, the limiting current remains constant once it is established. This occurs because the **diffusion layer** or thickness of the concentration gradient across which the analyte must diffuse remains small and constant since the analyte is continually brought to near the electrode surface by mass transfer (stirring). But if the electrode is unstirred and in a quiet solution, the diffusion layer will extend farther out into the solution with time, with the result that the limiting current decreases exponen-

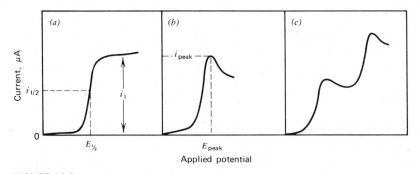

FIGURE 12.5
Different types of voltammetric curves. (a) Stirred solution or rotated electrode, (b) Unstirred solution, (c) Stepwise reduction (or oxidation) of analyte or of a mixture of two electroactive substances (unstirred solution).

tially with time and a "peaked" wave is recorded (curve B). For this reason and others, the voltage scan using stationary, microelectrodes is usually fairly rapid, for example, 50 mV seconds. (Actually, even with stirred solutions, the waves tend to be "peaked" to some extent.)

Although the decomposition potential required to initiate the electrolysis will vary slightly with concentration, the potential at which the current is one-half the limiting current is independent of concentration. This is called the **half-wave potential** $E_{1/2}$. It is a constant related to the standard or formal potential of the redox couple, and so voltammetry serves as a qualitative tool for identifying the reducible or oxidizable species.

An electrode whose potential is dependent on the current flowing is called a **polarizable electrode.** If the electrode area is small and a limiting current is reached, then the electrode is said to be **depolarized.** Therefore, a substance that is reduced or oxidized at a microelectrode is referred to as a **depolarizer.**

If a depolarizer is reduced at the working electrode, a **cathodic current** is recorded at potentials more negative than the decomposition potential. If the depolarizer is oxidized, then an **anodic current** is recorded at potentials more positive than the decomposition potential.

Stepwise Reduction or Oxidation

An electroactive substance may be reduced to a lower oxidation state at a certain potential and then be reduced to a still lower oxidation state when the potential reaches another more negative value. For example, copper(II) in ammonia solution is reduced at a graphite electrode to a stable Cu(I)–ammine complex at -0.2 V versus SCE, which is then reduced to the metal at -0.5 V, each a one-electron reduction step. In such cases, two successive voltammetric waves will be recorded as in curve C in Figure 12.5. The relative heights of the waves will be proportional to the number of electrons involved in the reduction or oxidation. In this case, the two waves would be of equal height.

When a solution contains two or more electroactive substances that are reduced at different potentials, then a similar stepwise reduction will occur. For example, lead is reduced at potentials more negative than -0.4 V versus SCE (Pb^{2+} + $2e^- \rightarrow Pb$) and cadmium is reduced at potentials more negative than -0.6 V (Cd^{2+} + $2e^- \rightarrow Cd$). So a solution containing a mixture of these would exhibit two voltammetric waves at a graphite electrode, one for lead at -0.4 V, followed by another stepwise wave for cadmium at -0.6 V. The relative heights will be proportional to the relative concentrations of the two substances, as well as the relative n values in their reduction or oxidation.

Mixtures of electroactive substances can be determined by their stepwise voltammetric waves. There should be at least 0.2 V between the $E_{1/2}$ values for good resolution. If the $E_{1/2}$ values are equal, then a single composite wave will be seen, equal in height to the sum of the individual waves. If a major component in much higher concentration is reduced (or oxidized for anodic scans) before the test substance(s) of interest, its wave will mask succeeding waves and it may not even reach a limiting current. In such cases, most of the interfering substance will have to be removed before the analysis can be performed. A common procedure is to preelectrolyze it at a macroelectrode using a setup similar to Figure 12.2, at a potential corresponding to its limiting current plateau but not sufficient to electrolyze the test substance.

Stepwise oxidations may also occur to give stepwise anodic waves.

The Supporting Electrolyte

It was assumed above that when a concentration gradient existed in a quiescent solution, the only way the reducible ion could get to the electrode surface was by diffusion. It can also get to the electrode surface by electrical (coulombic) attraction or repulsion.

The electrode surface will be either postively or negatively charged, depending on the applied potential, and this surface charge will either repulse or attract the ion diffusing to the electrode surface. This will cause an increase or decrease in the limiting current, which is called the **migration current.** The migration current can be prevented by adding a high concentration, at least 100-fold greater than the test substance, of an inert **supporting electrolyte** such as potassium nitrate. Potassium ion is reduced only at a very negative potential and will not interfere. The high concentration of inert ions essentially eliminates the attraction or repulsion forces between the electrode and the analyte, and the inert ions are attracted or repulsed instead. The inert ions are not electrolyzed, however.

A second reason for adding a supporting electrolyte is to decrease the iR drop of the cell. For this reason, at least 0.1 M supporting electrolyte is commonly added. This is true for nearly all electrochemical techniques, potentiometry being an exception. The supporting electrolyte is frequently chosen to provide optimum conditions for the particular analysis, such as buffering at the proper pH or elimination of interferences by selective complexation of some species. When a metal ion is complexed, it is generally stabilized against electrolysis and its voltammetric half-wave is shifted to more negative reduction potentials, or it may even become nonelectroactive. Commonly used complexing agents include tartrate, citrate, cyanide, ammonia, and EDTA.

Irreversible Reduction or Oxidation

If a substance is reduced or oxidized reversibly, then its half-wave potential will be near the standard potential for the redox reaction. If it is reduced or oxidized irreversibly, the mechanism of electron transfer at the electrode surface involves a slow step with a high energy of activation. Therefore, extra energy must be applied to the electrode for the electrolysis to occur at an appreciable rate. This is in the form of increased applied potential and is called the **activation overpotential.** $E_{1/2}$ will therefore be more negative than the standard potential in the case of a reduction, or it will be more positive in the case of an oxidation. An irreversible wave is more drawn out than a reversible wave. Nevertheless, an S-shaped wave is still obtained, and its diffusion current will be the same as if it were reversible, because i_l is limited only by the rate of diffusion of the substance to the electrode surface.

The Working Potential Range

The potential range over which voltammetric techniques can be used will depend on the electrode material, the solvent, the supporting electrolyte, and the acidity of the solution. If a platinum electrode is used in aqueous solution, the limiting positive potential would be oxidation of water ($H_2O \rightarrow \frac{1}{2} O_2 + 2e^-$), unless the supporting electrolyte contains a more easily oxidizable ion (e.g., Cl^-). E^0 for the water half-reaction is $+ 1.0$ V versus SCE, and so the limiting positive potential is about $+1$ V versus SCE, depending on the pH. The negative limiting potential will be from the reduction of hydrogen ions. Platinum has a low hydrogen overvoltage at low current densities, and so this will occur at about -0.1 V versus SCE. Because oxygen is not reduced at these potentials, it need not be removed from the solution, unless oxygen interferes chemically.

Carbon electrodes are frequently used for voltammetry. Their positive potential limit will be essentially the same as with platinum electrodes, but more negative potentials can be reached because hydrogen has a rather high overvoltage on carbon. Potentials of about -1 V versus SCE or more can be used, again depending on the pH. With potentials more negative than -0.1 V versus SCE, oxygen must be removed from solutions. An advantage of carbon electrodes is that they are not troubled by oxide formation on the surface, as platinum electrodes are. While carbon electrodes can be used at fairly negative potentials, a dropping mercury electrode (see below, polarography) is usually preferred because better reproducibility can be achieved.

Solid electrode voltammetry is used largely for the oxidation of substances at fairly positive potentials, although for very easily reducible substances, it is also useful. However, reproducibility frequently suffers because the surface characteristics of the electrodes are not reproducible and the surface becomes contaminated. For this reason, the related technique of polarography is preferred when applicable.

12.5 POLAROGRAPHY

Polarography is a class of voltammetry in which the working electrode is a **dropping mercury electrode** (DME). The fact that its discoverer, J. Heyrovsky, received the 1952 Nobel prize in chemistry for its development (1921) attests the importance of this technique.

A typical polarographic cell with three electrodes is shown in Figure 12.6. The reference electrode is a saturated calomel electrode. A salt bridge separates the two compartments.

The Dropping Mercury Electrode

In principle, any microelectrode could be used for the working electrode. But many metal cations are reduced at fairly negative potentials up to -2 V or more versus SCE. As mentioned previously, a platinum electrode cannot be used for these because of competing hydrogen ion reduction. For this reason, a dropping mercury electrode is used. This consists of a capillary through which mercury flows, causing drops to form and fall. It has a number of advantages:

1. Mercury has a high hydrogen overvoltage.
2. The surface is continually renewable and so it does not become contaminated or changed by the deposited metal.
3. The surface is very reproducible.

The term **polarography** is applied to current–voltage curves when the dropping mercury electrode is used.

Oxygen Removal

Dissolved oxygen in water is reduced in two steps, first to H_2O_2 and then to H_2O. The first step occurs at about -0.1 V versus SCE, which severely limits the useful potential range. Oxygen is removed from the solution by bubbling high-purity nitrogen through the solution for 10 to 15 minutes prior to the sample run. A nitrogen atmosphere is maintained above the solution while the polarogram is run.

FIGURE 12.6
Polarographic H-cell. (Adapted from J. J. Lingane and H. A. Laitinen, *Ind. Eng. News, Anal. Ed.,* **11** (1939) 504. Copyright © 1939 by The American Chemical Society. Reprinted by permission of the copyright owner.)

Current Oscillations

A polarogram[3] using a DME is shown in Figure 12.7. Instead of a smooth current–voltage curve, it has current oscillations. This occurs because the area of the electrode surface is not constant but continually grows until the drop falls. The current is directly proportional to the area, and it will grow as the drop grows and then decrease suddenly when the drop falls. A new drop immediately begins to form, and the current begins to grow again. Damping or limited response time of the recorder keeps the recorded current from actually falling back to zero. The limiting current at the DME is called the **diffusion current** i_d.

---

[3]A current–voltage curve using a DME is called a polarogram.

The Residual Current

If there is no electrolysis, for example, at the foot of the polarographic wave as shown in Figure 12.7, the current should be zero and there should be no oscillations. However, there are oscillations, indicating a finite current. This current is called the **residual current.**

If a polarogram is run with only the supporting electrolyte in solution, a current–voltage curve such as that of Figure 12.8 is obtained. The positive potential is limited by oxidation of the electrode. Mercury is oxidized fairly easily and this limits the useful potential range to about $+0.3$ or 0.2 V versus SCE.[4] The oxidation of the electrode is called the "mercury dissolution current."

The negative working potential is limited by the reduction of the supporting electrolyte cation, in this case, potassium ion. Since it is present in high concentration, no limiting current is reached. More negative potentials can be reached by using a quarternary ammonium salt such as tetraethylammonium iodide, and po-

FIGURE 12.7
Polarogram of 1 mM $Cd(NO_3)_2$ in 0.1 M KNO_3.

---

[4]Note that "positive" potential is just a relative term and will depend on the arbitrary reference electrode chosen. Positive *currents* in Figure 12.8 are reductions, and negative *currents* are oxidations occurring at the working electrode. Polarographic waves can occur at any of the potentials, with either cathodic or anodic electrode reactions.

FIGURE 12.8
Residual current in 0.1 M KNO$_3$.

tassium ion can actually be measured polarographically. If the solution is acidic, then the negative working potential is limited by reduction of hydrogen ions on mercury at a somewhat more positive potential, depending on the acidity.

A finite current flows at potentials between the limiting potentials, and at only one potential is the current equal to zero. This residual current is due to two factors. The most important one is the **charging current,** required for the charging of the so-called **electrical double layer** at the surface of the mercury electrode. The electrode acts as a capacitor. When an external potential is applied to the electrode, the electrode adopts either a positive or a negative charge as electrons move toward or away from the mercury surface, and this movement of electrons represents a flow of current. When the surface of the drop is positive, an anodic current flows; when it is neutral, no current flows; and when it is negative, a cathodic current flows. The greater the charge, the larger the charging current and the larger the oscillations. The charge increases as the potential is moved farther from zero charge.

The magnitude of this charging current limits the sensitivity of polarography. Obviously, the farther away from zero current, the poorer the sensitivity. The potential of zero current will depend on the supporting electrolyte and other ions in solution. If no adsorbable or complexing species are in solution to disturb the electrical double layer, it will occur at about -0.56 V versus SCE.

The other factor contributing to the residual current is the presence of impurities in the supporting electrolyte. These will usually be small compared to the charging current.

A residual current is also observed at solid electrodes. But it will generally be smaller because once the electrode is initially charged (when the potential is first applied), the charging current decays exponentially with time as the double layer becomes fully charged (just as current flow into a capacitor decreases as the capacitor becomes charged). With a DME, the double layer must be charged each time a new drop is formed and as the area increases. This is a disadvantage of polarography, and various means have been employed to minimize the residual current (see pulse polarography below).

As mentioned above, the sensitivity of polarography is limited by the magnitude of the residual current. For most cases, the lower limit is about 10^{-5} M and the upper limit, about 10^{-2} M. In favorable cases, 10^{-6} M can be detected. Cadmium represents one of the most ideal cases, because it is reduced at a potential near zero residual current. (The sensitivity for voltammetry at solid electrodes is typically an order of magnitude better than polarography, due to the lower residual current, depending on the potential region; but as mentioned, the reproducibility is not as good.)

Maxima

Sometimes, instead of being S-shaped, the wave will have a large peak, as in Figure 12.9. This is called a **maximum.** Maxima are not completely understood but have been shown by high-speed photography to be associated with swirling or streaming of solution around the mercury electrode. They are believed to be related to two phenomena, electrostatic stirring of the electrode surface and adsorption on the electrode. The first may be associated with imperfections in the capillary. Fortunately, maxima can usually be suppressed by adding a small amount of a surface-active substance such as gelatin or Triton X-100, about 0.005%. Care must be taken not to add too much or the diffusion current may also be suppressed.

Maxima are more likely to occur as the charge on the mercury drop surface increases, and so the farther the applied potential is from zero current on the residual curve, the more probable are maxima. Cadmium does not exhibit a maximum, because it is reduced near the potential of zero charge on the drop.

The Ilkovic Equation

The diffusion current at a dropping mercury electrode is defined by the **Ilkovic equation:**

$$i_d = 706 \, nD^{1/2}Cm^{2/3}t^{1/6} \tag{12.16}$$

where

i_d = the maximum diffusion current during the life of the drop, in microamperes

FIGURE 12.9
Polarogram of 2 mM $Pb(NO_3)_2$ in 0.1 M KNO_3, illustrating maximum.

706 = a constant including π, the Faraday constant, and the density of mercury

n = the number of electrons involved in the electrode reaction

D = the diffusion coefficient of the test substance, in cm^2/second

C = the concentration of the electroactive substance, in mmol/L

m = the mass of mercury flowing through the capillary, in mg/second

t = the drop time, in seconds

All these except C and i_d are held constant in an analysis, so $i_d \propto C$. D is a constant for a given electroactive species in a given solution; it will vary with the viscosity of the solution, the size of the ion, the temperature, and so forth. However, for most metal ions, the diffusion coefficient will be close to 10^{-5} cm^2/second, and so the height of the wave for a given concentration will be about the same for all metals with the same n value.

The quantities m and t will vary with the capillary and its length, and they will vary with the pressure of mercury, that is, the height of the mercury column above the capillary. It is important, then, that the height of the mercury column be constant for quantitative work, since the usual practice is not to measure parameters in the Ilkovic equation but to hold them constant and prepare a calibration curve of i_d versus C. The unknown concentration is then read from the calibration graph, using the measured i_d for the unknown.

The Ilkovic equation holds for drop times of about 2 to 6 seconds, and the usual practice is to adjust the capillary length and mercury pressure to bring the drop time within this range.

Although with carefully controlled conditions, accuracy approaching $\pm 0.2\%$ can be achieved in polarography, under usual conditions an accuracy of $\pm 2\%$ is attained. This is certainly acceptable in most cases, because usually trace constituents are being determined. This is also within the realm of other instrumental methods.

Time Required to Run a Polarogram

The potential is usually scanned automatically. Because of the relatively long drop time, the potential must be scanned fairly slowly. About 0.2 V/minute is usual. So, if the entire potential scan of about 2 V is run, 10 minutes would be required to obtain the polarogram, exclusive of the time required for deaeration. Of course, the entire polarogram need not be run if only one element is being determined.

Inorganic and Organic Polarography

Voltammetric techniques, especially polarography, have been widely used for the determination of trace metals. Table 12.3 lists the $E_{1/2}$ values of some representative metals in different media. By appropriate choice of background electrolyte, pH, or complexing agent, it is possible to simultaneously resolve and determine several elements. However, the more recently developed flame-spectroscopic techniques, which are generally more rapid and often more sensitive (Chapter 14), have largely replaced conventional polarography as the preferred methods for trace

TABLE 12.3
Polarographic Half-Wave Potentials of Some Metal Ions Versus SCE

Metal Ion	0.1M KCl	Supporting Electrolyte		
		1M NH_3/1M NH_4Cl	1M NaOH	Other
Bi^{3+}				−0.09 (1M HCl)
Cd^{2+}	−0.60	−0.81	−0.78	
Co^{2+}	−1.20	−1.29	−1.46	
Cr^{3+}		−1.43 (Cr(III) → Cr(II))		
		−1.71 (Cr(II) → Cr⁰)		
Cu^{2+}	+0.04	−0.24 (Cu(II) → Cu(I))	−0.41	
		−0.51 (Cu(I) → Cu⁰)		
Fe^{3+}				−0.17
				(Fe(III) → Fe(II))
				−1.50
				(Fe(II) → Fe⁰)
				(0.1M EDTA/2M NaOAc)
Fe^{2+}	−1.3			
Ni^{2+}	−1.1	−1.1	−0.76	
Pb^{2+}	−0.40		−0.76	
Sn^{4+}			−0.75	
Sn^{2+}				−0.47 (1M HCl)
Tl^+	−0.48	−0.48	−0.48	
Zn^{2+}	−1.00	−1.35	−1.50	

metal analysis (but pulse polarography and anodic-stripping voltammetry are very competitive—see below).

However, voltammetric techniques are not limited to metals, and they are very important for the determination of organic substances. A few that are of biological interest include drugs such as streptomycin, chloramphenicol, and derivatives of the sulfanilamides. The reduction of the first compound probably involves reduction of the aliphatic aldehyde group to the alcohol. The second involves reduction of the nitro group on the molecule to an amine, and reduction of the acid hydrogen is involved with the sulfanilamides.

Most of the vitamins give polarographic waves. Exceptions are the oil-soluble vitamins of the A and D groups. The electrode process depends on the vitamin. Enzyme reactions have been studied polarographically, many involving sulfhydryl groups. The references by Brezina and Zuman (Reference 6) and by Purdy (Reference 2) at the end of the chapter are recommended for a more thorough treatment of the polarography of organic and biochemical substances.

12.6 ANODIC-STRIPPING VOLTAMMETRY

There are a number of voltammetric techniques that are more sensitive than conventional dc polarography. The most sensitive one for trace metals is **anodic-stripping voltammetry.** Most commonly, a hanging mercury drop electrode (HMDE) is used, where a mercury drop is suspended from a mercury capillary tube, or from the end of a mercury-plated platinum wire or a gold wire sealed in a glass tube. First, the solution is stirred at a reproducible and constant rate, while a constant potential negative enough to be on the diffusion current portion of the polarographic wave of a metal ion is applied. The reduction product of the metal ion must form an amalgam with the electrode. During this step, the metal ion reaches the electrode surface by diffusion and mass convection (stirring), where it is reduced and concentrated in the drop. After a preselected time, the stirring is stopped, the solution is allowed to quiet, and the potential is scanned anodically (toward the more positive potentials). If the electrode process is reversible, the amalgamated metal will be anodically stripped out of the electrode, that is, reoxidized, giving rise to an anodic wave. The wave will be peaked, and because the concentration of the amalgam in the electrode is much greater than that of the metal ion in solution, the peak height will be greatly magnified over that of the normal polarographic reduction wave.

The peak height is proportional to the square root of the rate of potential scan, and so a fairly rapid scan is employed, about 50 mV/second, and the voltammogram is thus recorded in about 20 seconds. The peak current is also directly proportional to the time of preelectrolysis, and the time used will depend on the concentration. The upper concentration limit is about 10^{-6} to 10^{-5} M, and the lower concentration limit is 10^{-9} to 10^{-10} M! The lower concentration corresponds to about 0.01 ppb. It is not difficult to use 1-mL volumes in these techniques, and so 10^{-10} g can be determined. This compares favorably with almost any other analytical technique.

For 10^{-6} M, preelectrolysis times of 60 seconds or less are sufficient, and proportionately longer times are used as the concentration decreases. For 10^{-9} M, electrolysis times of 1 hour are usually required. The problem of electrolyte impurities becomes serious in this technique. Therefore, blank determinations are important.

Most biological tissues and fluids, including skin, liver, blood, muscle, urine, brain, lung, and kidney, contain sufficient salts that if a 10-g sample is dry-ashed and the residue dissolved in 10 mL deionized water, no supporting electrolyte need be added for anodic-stripping analysis. See the reference by Newberry and Christian (Reference 8) at the end of this chapter.

12.7   PULSE POLAROGRAPHY

Another technique that offers improved sensitivity over conventional polarography is **pulse polarography.** With the advent of commercial instrumentation, this technique has largely supplanted dc polarography in the laboratory.

We mentioned before that the sensitivity of polarography is limited by the charging current. This charging current is transient; that is, it is large initially in the drop life, and as soon as the surface becomes charged, the current needed to charge it decays away. The decay is exponential with time. After the drop falls, the new drop must be charged, hence the reason for the continuing residual current.

In pulse polarography, advantage is taken of the decaying nature of the charging current. Instead of an increasing dc voltage ramp, a square-wave **voltage pulse** (⊓) is applied to the electrode during the last quarter of the drop life (when the surface area changes little with time). The duration of the pulse is about 40 ms. At the instant the voltage pulse is applied, the charging current is large, but it decays rapidly. The current is then measured during the second half of the pulse when the charging current is small. A new voltage pulse is applied to each drop and the amplitude of the applied voltage increases linearly with time (with each drop) to give a voltage scan. The shape of the resulting current–voltage curve resembles the S-shaped wave in dc polarography, except that it has a "staircase" appearance since the current is sampled once during each drop life and is stored until the next sample period.

A variation is **differential pulse polarography.** Here, the pulses have *constant amplitude* (5 to 100 mV) and are superimposed on a slowly increasing linear (dc) voltage ramp. Two measuring periods are used, one just before the pulse is applied and one near the end of the pulse. The difference in the two currents is plotted. The magnitude of Δi is maximum when the rate of current change with voltage change (ΔV, the pulse) is maximum, that is, on the steep portion of the polarographic wave, and it diminishes to zero on the plateau of the wave and on the foot of the wave. The result is a peak-shaped plot of current (Δi) as a function of potential. A typical differential-pulse polarogram is shown in Figure 12.10. The "derivative" type wave has the advantage that resolution of polarographic waves is improved and much larger concentrations of substances whose electrolysis precedes that of the test substances can be tolerated.

The detection limit of pulse polarography is about 10^{-6} to 10^{-7} M, while that for differential pulse polarography is about 10^{-7} to 10^{-8} M, depending on the substance. Arsenic(III), for example, has been detected at the 4×10^{-9} M level. Differential-pulse polarography has been used to determine lead in microliter volumes of blood, following an ashing step to remove organic matter.

Anodic-stripping voltammetry can be applied to pulse polarography, and sensitivites are at least an order of magnitude improved over dc methods.

FIGURE 12.10
Typical differential-pulse polarogram. Mixture of maleic and fumaric acids in 0.2 M phosphate buffer and 1 M NH₄Cl, pH adjusted to 8.2. (Courtesy of Princeton Applied Research Corp.)

12.8   THE STATIC MERCURY DROP ELECTRODE (SMDE)

A commercially available DME assembly, known as the static mercury drop elec-trode, or SMDE, offers some convenience for polarographic and stripping voltam-metry measurements. By means of a micrometer arrangement, it pushes a fixed volume of mercury from a reservoir through the capillary and then stops for a preselected time to provide a reproducible and constant drop size. The current can then be measured at a constant drop area without the usual oscillation. Also, the charging current can be allowed to decay for a fixed time before measurement, for improved sensitivity. A hammer dislodges the drop at the initiation of a new cycle. For pulse polarography, this is synchronized with the timing cycle of the pulse. This apparatus is convenient for stripping voltammetry. The hammer is inactivated, and a fixed-area hanging mercury drop is produced that remains during the depo-sition and stripping cycles. Then, a new drop can be produced for the next mea-surement.

12.9   AMPEROMETRY

Amperometry is the application of voltammetric measurements at a fixed potential to detect changes in currents as a function of concentration of electroactive spe-cies, for example, in a titration for detection of the end point. We will indicate some representative examples in the following discussions.

Amperometric Titrations

Amperometry is in principle identical to voltammetry. In titrations, the limiting current of a voltammetric wave is measured at different points in a titration.

Consider the titration of silver ion with chloride. As chloride is added, the silver ion is removed from solution, and consequently the limiting current of the silver voltammetric wave decreases when a voltammogram is run. This is illustrated in Figure 12.11. The height of the wave decreases to zero at the end point of the titration, when all the silver has been removed from solution. Beyond the end point, then, no silver wave is observed. A plot of i_l of the silver wave versus amount of titrant added is shown in Figure 12.12. There is a slight residual current when all the silver has been removed.

The end point need not be pinpointed in the titration. The current on each side of the end point is extrapolated and the intersection is taken as the end point. The current change will not be linear with the volume of titrant added, though, due to dilution, and a curvature will result. Hence, the measured current must be corrected for volume changes during the titration, in order to obtain a straight-line plot:

$$i_{corrected} = i_{observed} \left(\frac{V + v}{V} \right) \qquad (12.17)$$

where V is the original volume and v is the added volume. There will usually be some rounding at the end point, especially with dilute solutions.

In practice, it is inconvenient to run a complete voltammogram after each addition of titrant. Instead, a constant potential is applied that resides on the plateau (limiting current) of the wave and the current at that potential is measured.

Different shaped titration curves are possible, depending on the voltammetric behavior of the analyte, titrant, their product, and the applied potential. If the above titration is performed in reverse, the current remains near zero until the end

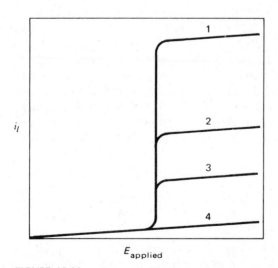

$E_{applied}$

FIGURE 12.11
Voltammograms of silver solution titrated with chloride. Curve 1, 0% titrated; curve 2, 50% titrated; curve 3, 75% titrated; curve 4, 100% titrated.

FIGURE 12.12
Amperometric titration curve.

point and then increases linearly when excess titrant (silver) is added. If both sample and product give a voltammetric reduction wave at the applied potential, then a V-shaped titration curve will be obtained (inverted V for anodic waves). For example, lead ion may be titrated with chromate ion ($Pb^{2+} + CrO_4^{2-} \rightarrow PbCrO_4$). Chromate ion is reduced at 0 V versus SCE, but lead is reduced only at potentials more negative than -0.4 V. If an amperometric titration is performed with a DME at 0 V, then titration curve A in Figure 12.13 will be obtained. The current rise beyond the end point is due to the excess chromate titrant. But if the applied potential is -0.5 V, curve B is obtained. The current before the end point is due to the lead ion, whose concentration is reduced throughout the titration, and that beyond the end point is due to the excess chromate titrant, which is still reduced at -0.5 V. A sharper end point is obtained at -0.5 V; but at 0 V, the solution need not be deaerated since oxygen is not reduced at that potential, and the DME and SCE need only be shorted together in series with a galvanometer (i.e., no external potential source is required at 0 V).

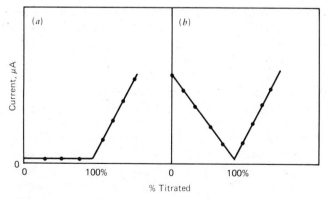

FIGURE 12.13
Amperometric titration of chromate with lead ion (a) at 0 V versus SCE, (b) at -0.5 V versus SCE.

If the analyte gives a cathodic wave but the titrant gives an anodic wave, then a break will not occur. A decreasing cathodic current will be recorded as the analyte is titrated, reaching zero at the equivalence point. As excess titrant is added, current will continue changing in the same direction, that is, an increasing anodic current will be recorded. For example, when Fe^{3+} is titrated with Ti^{3+} (Fe^{3+} + $Ti^{3+} \rightarrow Fe^{2+} + Ti^{4+}$), the Ti^{3+} titrant is oxidized at the same potential at which Fe^{3+} is reduced, and a titration curve like the one just described will be obtained at this potential. In cases like this, the end point must be determined from the titrant volume at the zero current reading (actually, the residual current reading at the applied potential is used), unless the diffusion coefficients of the electroactive analyte and titrant species are sufficiently different to result in a change in the slope beyond the equivalence point.

As in voltammetry, a supporting electrolyte should be present in amperometric titrations; and if a DME is used as the indicating electrode, the solution must be deaerated. A rotating platinum electrode (RPE) is often used for more positive potentials, and frequently oxygen does not have to be removed.

Amperometric titrations are quite sensitive and can be used to detect the end points of titrations at concentrations of 10^{-5} M or less. An advantage over potentiometric titrations is that electrode response is more rapid, especially in dilute solutions and near the end point, and the plots are linear, making selection of the end point easier. Also, poisoning of the electrode is not so serious a problem. (It should be noted that potentiometric Gran's plots, discussed in the previous chapter, are similar to amperometric titration plots in that a linear function of concentration is plotted versus volume of titrant.)

Biamperometry

Biamperometry is a technique in which two identical electrodes are used with a small potential, such as 50 to 200 mV, impressed between them. If both the oxidized and reduced forms of a reversible electroactive couple are present in solution, a current will flow. (They are reduced and oxidized at the same potential, so only a small applied potential difference is required for electrolysis of each to occur.) Thus, if Br_2 and Br^- are in solution together, the Br^- is oxidized to Br_2 at the anode and the Br_2 is reduced to Br^- at the cathode, that is, both electrodes are depolarized. But if either Br_2 or Br^- is absent, no current will flow. The current is limited by the species which is at the smaller concentration. If As(III) is titrated to As(V) (irreversible couple) with Br_2 using a pair of platinum electrodes, no current flows before the equivalence point, since Br_2 is removed by reaction with As(III) and converted to Br^-. Beyond the equivalence point, excess Br_2 is added and a biamperometric current flows in proportion to the Br_2 concentration. The technique of titrating to a sudden current surge is called a *"dead-stop"* titration.

If the sample and its product are reversibly electroactive, then a current is recorded as soon as the titration begins, increases to a maximum at 50% titration, and decreases to zero at the eqiivalence point. For example, if the above titration is performed in reverse, current begins as soon as some As(III) is added and some of the Br_2 is converted to Br^- and is proportional to the amount of Br^- produced. The current reaches a maximum when half the Br_2 is converted to Br^- (when there are equivalent quantities of each). Then, it decreases as the Br_2 concentration becomes limiting and reaches zero at the equivalence point and remains zero beyond.

If both the titrant and sample couples are reversibly electroactive, then the titration curve is a combination of the above two, that is, the current goes through a maximum, drops to zero at the equivalence point, and increases beyond.

Enzyme Electrodes

Amperometry is a general-purpose detection method and is useful for other applications than in titrations. For example, amperometric electrodes have been used to prepare enzyme electrodes similar to those described in Chapter 11. A glucose electrode can be prepared by immobilizing the enzyme glucose oxidase on a platinum anode. The glucose oxidase catalyzes the air oxidation of glucose:

$$\text{glucose} + O_2 + H_2O \xrightleftharpoons{\text{glucose oxidase}} \text{gluconic acid} + H_2O_2$$

The hydrogen peroxide diffuses to the electrode where it is oxidized at $+0.6$ V versus SCE to give a steady-state anodic current after several seconds. A linear relationship is obtained between the steady-state current and the glucose concentration.

The Oxygen Electrode

An important amperometric electrode is the oxygen electrode. It consists of a thin plastic film such as Teflon stretched over a platinum or gold cathode that allows diffusion of gases but is impermeable to ions in solution (Figure 12.14). Oxygen diffuses through the membrane and is reduced at the cathode, producing an amperometric current. A potential suitable to cause reduction of the oxygen is applied between the oxygen-indicating electrode and the reference electrode, usually a silver/silver chloride electrode built into the probe. A gel containing an electrolyte is usually placed between the membrane and the glass insulator to provide electrical contact between the reference electrode and the indicating electrode.

The rate of diffusion of oxygen to the cathode is proportional to the partial pressure of the oxygen in the sample to which the electrode is exposed, and the amperometric current is proportional to this. Measurements are read at atmospheric pressure. Halogens and other gases (SO_2, etc.) that are also reduced at the fixed polarization potential interfere. Hydrogen sulfide poisons the electrode.

The meter is precalibrated by exposure of the probe to samples with known oxygen content, for example, air with assumed 20.9% O_2 content or water saturated with either oxygen or air. At 37°C and sea level (P_{O_2} of 159 torr), air-saturated water contains 5.6 μL O_2 per milliliter and oxygen-saturated water contains 28 μL

"O" Ring

Membrane

KCl Solution

Anode coil

Cathode ring

FIGURE 12.14
Construction of an oxygen electrode. (Courtesy of Arthur H. Thomas Company.)

O_2 per milliliter. See Reference 10 at the end of the chapter for a discussion of the calibration of electrodes and the calculation of P_{O_2} and oxygen concentration.

A polarographic oxygen analyzer is often used by biochemists to follow the consumption or release of oxygen in biochemical and enzymatic reactions in order to determine the kinetics of the reactions, and in the clinical laboratory for analytical measurements of enzymes or substrates whose reactions involve consumption of oxygen.

Chromatography Detectors

Amperometric electrodes are finding important applications as microdetectors in liquid chromatography for following the elution of electroactive substances. This application is described in Chapter 16.

QUESTIONS

1. Define back-emf, overpotential, overvoltage, concentration polarization, and iR drop.
2. List the advantages and disadvantages (limitations, etc.) of coulometric titrations as compared with volumetric titrations.
3. Define half-wave potential, diffusion current, depolarizer, DME, residual current, and voltammetry.
4. Give two reasons for using a supporting electrolyte in voltammetry.
5. Give three reasons for using a dropping mercury electrode in polarography.
6. A solution contains about 10^{-2} M Fe^{3+} and 10^{-5} M Pb^{2+}. It is desired to analyze for the lead content polarographically. Fe^{3+} is reduced to Fe^{2+} at all potentials accessible with the DME up to -1.5 V versus SCE, and is reduced along with Fe^{2+} to the metal at potentials more negative than -1.5 V. Pb^{2+} is reduced at -0.4 V. Suggest a scheme for measuring the lead polarographically.
7. Describe the principles of anodic stripping voltammetry, and compare its sensitivity with that of conventional dc polarography.
8. What effect does complexation have on the polarographic reduction of a metal ion?
9. Describe the principles of pulse polarography. Compare its sensitivity with that of conventional dc polarography. Why is it more sensitive?

PROBLEMS

Electrolysis/Electrogravimetry

10. A solution contains 0.10 M silver ($E^0 = 0.799$ V) and 0.10 M copper ($E^0 = 0.337$ V). We wish to separate the silver from the copper by plating it out on a platinum cathode at controlled current. Calculate the percent silver remaining in solution when its decomposition potential due to its decreasing concentration becomes equal to that required for the deposition of copper. Assume no overvoltage, concentration polarization, or iR drop; that is, decomposition potential = back emf.
11. What was the current efficiency in plating zinc on a platinum cathode if 0.50 g was deposited in 10 minutes, with a current of 17 A used?
12. Copper (0.010 M) is to be analyzed by electrogravimetry. What potential is required if 99.999% of the copper is to be plated out? Assume no overvoltage, concentration polarization, or iR drop.

Coulometry

13. An alloy is analyzed for its copper content by constant-potential coulometry. A 0.500-g sample is dissolved and electrolyzed at a mercury pool cathode (-0.3 V versus SCE). A titration coulometer is placed in series with the electrolysis circuit. At the end of the electrolysis, a total of 40.0 mL of 0.0500 M HCl is required to bring the pH in the coulometer back to its initial value. What is the percent copper in the alloy?

14. Iron(II) in a sample is determined by controlled-potential coulometry by oxidizing it to iron(III) at a platinum anode. A silver coulometer in series is used to determine the number of coulombs of current required. In this, silver in solution is deposited on a platinum cathode and the amount deposited is determined by weighing. (The cathode must be removed from the solution before interrupting the current, or else the silver will spontaneously go back into solution.) The cathode in the coulometer gains 0.2878 g during complete electrolysis of the sample. A blank electrolysis on the supporting electrolyte for the same time results in a weight gain of 0.0022 g. What is the weight of iron in the sample?

15. A protein sample is analyzed by a Kjedahl procedure by digesting it with sulfuric acid to convert protein nitrogen to ammonium sulfate. The ammonia produced is determined by adjusting the pH to 8.6 and titrating coulometrically with electrogenerated hypobromite:

$$Br^- + 2OH^- \rightarrow OBr^- + H_2O + 2e^-$$

$$2NH_3 + 3OBr^- \rightarrow N_2 + 3Br^- + 3H_2O$$

If the titration is performed using 19.30 milliamperes current and the end point occurs at 120.0 seconds, how many milligrams protein were present in the sample? One gram of nitrogen is contained in each 6.25 g protein.

16. For the determination in Problem 15, you are asked to design a constant-current source that will read out directly in micrograms of protein titrated. What current must it supply so that seconds of titration will be equal to micrograms of protein?

17. A common clinical analysis is the coulometric titration of chloride in serum. The results are expressed in meq/L. If a 0.100-mL sample is titrated, what current should be used so that seconds of titration equals meq/L of chloride in the serum?

Polarography/Amperometry

18. The diffusion current of lead in an unknown solution is 5.60 μA. One milliliter of a 1.00×10^{-3} M lead solution is added to 10.0 mL of the unknown solution and the diffusion current of the lead is increased to 12.2 μA. What is the concentration of lead in the unknown solution?

19. An organic substance is reduced polarographically. At a concentration of 2.0×10^{-4} M, it gives a wave with a diffusion current of 23.7 μA when a capillary with a flow rate of 3.4 mg/second and a drop time of 2.7 seconds is used. If the diffusion coefficient of the compound in the supporting electrolyte has been determined by other means to be 0.90×10^{-5} cm²/second, what is the n value for the polarographic reduction of the compound?

20. Iron(III) is polarographically reduced to iron(II) at potentials more negative than about $+0.4$ V versus SCE and is further reduced to iron(0) at -1.5 V versus SCE. Iron(II) is also reduced to the metal at -1.5 V. A polarogram is run (using a DME) on a solution containing Fe^{3+} and/or Fe^{2+}. A current is recorded at zero applied volts, and its magnitude is 12.5 μA. A wave is also recorded with $E_{1/2}$ equal to -1.5 V versus SCE, and its height is 30.0 μA. Identify the iron species in solution ($3+$ and/or $2+$) and calculate the relative concentration of each.

21. Consider the titration of B^{2+} with A^{3+}: $B^{2+} + A^{3+} \rightarrow B^{3+} + A^{2+}$. Polarograms of A and B are as follows:

Sketch the shape of the amperometric titration curves for the cases when the potential of the indicating electrode is at a, b, and c, respectively. Be sure to indicate whether the current measured is cathodic ($+$) or anodic ($-$). Mark where the equivalence point occurs with an X. Neglect volume changes and assume that all species have the same diffusion coefficient. Explain the shapes of the curves.

RECOMMENDED REFERENCES

General

1. J. J. Lingane, *Electroanalytical Chemistry*, 2nd ed. New York: Interscience Publishers, 1958. A classic. Excellent general text.
2. W. C. Purdy, *Electroanalytical Methods in Biochemistry*. New York: McGraw-Hill, 1965.

Conductivity

3. E. A. Cotlove, H. C. Trantham, and R. L. Bowman, "An Instrument and Method for Rapid, Accurate, and Sensitive Titration of Chloride in Biologic Samples," *J. Lab. Clin. Med.*, **51** (1958) 461.
4. G. W. Milner and G. Phillips, *Coulometry in Analytical Chemistry*. Oxford: Pergamon Press, 1967.

Polarography

5. M. Brezina and P. Zuman, *Polarography in Medicine, Biochemistry, and Pharmacy*. New York: Interscience Publishers, 1958.
6. I. M. Kolthoff and J. J. Lingane, *Polarography*, Vols. 1 and 2, 2nd ed. New York: Interscience Publishers, 1952. The bible of polarography.

Anodic-Stripping Voltammetry

7. Kh. Z. Brainina, *Stripping Voltammetry in Chemical Analysis*. New York: John Wiley & Sons, 1974.
8. C. L. Newberry and G. D. Christian, "Anodic Stripping Voltammetry of Biological Samples," *J. Electroanal. Chem.*, **9** (1965) 468.

Amperometry

9. I. M. Kolthoff and Y. Pan, "The Amperometric Titration of Lead with Dichromate or Chromate," *J. Am. Chem. Soc.*, **61** (1939) 3402.
10. M. A. Lessler and G. P. Brierley, in *Methods of Biochemical Analysis*, Vol. 17, D. Glick, ed. New York: Interscience Publishers, 1969, p. 1. Describes the oxygen electrode.
11. J. T. Stock, *Amperometric Titrations*. New York: John Wiley & Sons, 1965.

chapter
13

SPECTROMETRY

Spectrometry, particularly in the visible region of the electromagnetic spectrum, is one of the most widely used methods of analysis. It is very widely used in clinical chemistry and environmental laboratories because many substances can be selectively converted to a colored derivative. The instrumentation is readily available and generally fairly easy to operate. In this chapter, we (1) describe the absorption of radiation by molecules and its relation to molecular structure, (2) make quantitative calculations, relating the amount of radiation absorbed to the concentration of an absorbing analyte, and (3) describe the instrumentation required for making measurements. Measurements can be made in the infrared, visible, and ultraviolet regions of the spectrum. The wavelength region of choice will depend upon factors such as availability of instruments, whether the analyte is colored or can be converted to a colored derivative, whether it contains functional groups that absorb in the ultraviolet or infrared regions, and whether other absorbing species are present in the solution. Infrared spectrometry is generally less suited for quantitative measurements but better suited for qualitative or fingerprinting information than are ultraviolet (UV) and visible spectrometry. Visible spectrometers are generally less expensive and more available than UV spectrometers.

We also describe a related technique, fluorescence spectrometry, in which the amount of light emitted upon excitation is related to the concentration. This is an extremely sensitive analytical technique.

13.1 INTERACTION OF ELECTROMAGNETIC RADIATION WITH MATTER

In spectrometric methods, the sample solution absorbs electromagnetic radiation from an appropriate source, and the amount absorbed is related to the concentration of the analyte in the solution. A solution of copper is blue because it absorbs the complementary color yellow from white light and transmits the remaining blue light (see Table 13.1). The more concentrated the copper solution, the more yellow light is absorbed and the deeper the resulting blue color of the solution. In a spectrometric method, the amount of this yellow light absorbed would be measured and related to the concentration. We can obtain a better understanding of absorption spectrometry from a consideration of the electromagnetic spectrum and how molecules absorb radiation.

The Electromagnetic Spectrum

Electromagnetic radiation, for our purposes, can be considered a form of radiant energy that is propagated as a transverse wave. It vibrates perpendicular to the direction of propagation and this imparts a wave motion to the radiation, as illustrated in Figure 13.1. The wave is described either in terms of its **wavelength,** the distance of one complete cycle, or in terms of the **frequency,** the number of cycles passing a fixed point per unit time. The reciprocal of the wavelength is called the **wavenumber** and is the number of waves in a unit length or distance per cycle.

The relationship between the wavelength and frequency is

$$\lambda = \frac{c}{\nu} \qquad (13.1)$$

where λ is the wavelength in centimeters (cm),[1] ν is the frequency in reciprocal seconds (s^{-1}), or hertz (Hz), and c is the velocity of light (3×10^{10} cm/s). The wavenumber is represented by $\bar{\nu}$, in cm^{-1}:

$$\bar{\nu} = \frac{1}{\lambda} = \frac{\nu}{c} \qquad (13.2)$$

The wavelength of electromagnetic radiation varies from a few angstroms to several meters. The units used to describe the wavelength are as follows:

$Å$ = angstrom = 10^{-10} meter = 10^{-8} centimeter = 10^{-4} micrometer

nm = nanometer = 10^{-9} meter = 10 angstroms = 10^{-3} micrometer

μm = micrometer = 10^{-6} meter = 10^4 angstroms

The wavelength unit preferred for the **ultraviolet** and **visible** regions of the spectrum is nanometer, while the unit micrometer is preferred for the **infrared** region.[2] In this last case, wavenumbers are often used in place of wavelength, and the unit is cm^{-1}. See below for a definition of the ultraviolet, visible, and infrared regions of the spectrum.

Electromagnetic radiation possesses a certain amount of energy. The energy of a unit of radiation, called the **photon,** is related to the frequency or wavelength by

---

[1]More correctly, the units are centimeters per cycle for wavelength and cycles per second for frequency, but the cycles unit is often assumed. In place of cycles/s, the unit **hertz** (Hz) is now commonly used.

[2]Nanometer (nm) is the preferred term over millimicron (mμ), the unit used extensively prior to this. In the infrared region, micrometer (μm) is the preferred term in place of the previously used term micron (μ).

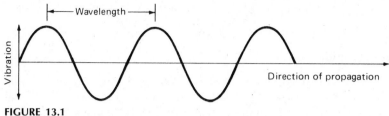

FIGURE 13.1
Wave motion of electromagnetic radiation.

$$E = h\nu = \frac{hc}{\lambda}$$ (13.3)

where E is the energy of the photon in ergs and h is Planck's constant, 6.62×10^{-27} erg-s. It is apparent, then, that *the shorter the wavelength or the greater the frequency, the greater the energy.*

As indicated above, the electromagnetic spectrum is arbitrarily broken down into different regions according to wavelength. The various regions of the spectrum are shown in Figure 13.2. We will not be concerned with the gamma-ray and X-ray regions in this chapter, although these high-energy radiations can be used in principle in the same manner as lower-energy radiations. The *ultraviolet* region extends from about 10 to 380 nm, but the most analytically useful region is from 200 to 380 nm, called the **near-ultraviolet region.** Below 200 nm, the air absorbs appreciably and so the instruments are operated under a vacuum; hence, this wavelength region is called the **vacuum-ultraviolet region.** The **visible region** is actually a very small part of the electromagnetic spectrum, and it is the region of wavelengths that can be seen by the human eye, that is, the light appears as a color. The visible region extends from the near-ultraviolet region (380 nm) to about 780 nm. The **infrared region** extends from about 0.78 μm (780 nm) to 300 μm, but the range from 2.5 to 15 μm is the most frequently used for analysis. The 0.8- to 2.5-μ range is known as the **near-infrared region,** the 2.5- to 16-μm region as the **NaCl-infrared region,** and longer wavelengths as the **far-infrared region.** We shall not be concerned with lower-energy radiation (radio or microwave) in this

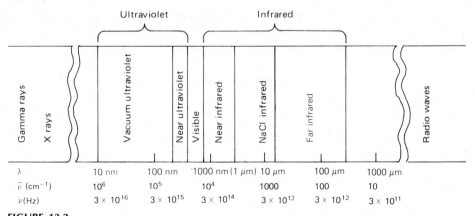

FIGURE 13.2
The electromagnetic spectrum.

chapter. Nuclear magnetic resonance spectroscopy involves the interaction of low-energy microwave radiation with the nuclei of atoms.

The Absorption of Radiation

A qualitative picture of the absorption of radiation can be obtained by considering the absorption of light in the visible region. We "see" objects because they transmit or reflect only a portion of the light in this region. When polychromatic light (white light), which contains the whole spectrum of wavelengths in the visible region, is passed through an object, the object will absorb certain of the wavelengths, leaving the unabsorbed wavelengths to be transmitted. These residual transmitted wavelengths will be seen as a color. This color is **complementary** to the absorbed colors. In a similar manner, opaque objects will absorb certain wavelengths, leaving a residual color to be reflected and "seen."

Table 13.1 summarizes the approximate colors associated with different wavelengths in the visible spectrum. As an example, a solution of potassium permanganate absorbs light in the green region of the spectrum with an absorption maximum at 525 nm, and the solution is purple.

There are three basic processes by which a molecule can absorb radiation; all involve raising the molecule to a higher internal energy level, the increase in energy being equal to the energy of the absorbed radiation ($h\nu$). The three types of internal energy are **quantized;** that is, they exist at discrete levels. First, the molecule rotates about various axes, the energy of rotation being at definite energy levels, so the molecule may absorb radiation and be raised to a higher rotational energy level, in a **rotational transition.** Second, the atoms or groups of atoms within a molecule vibrate relative to each other, and the energy of this vibration occurs at definite quantized levels. The molecule may then absorb a discrete amount of energy and be raised to a higher vibrational energy level, in a **vibrational transition.** Third, the electrons of a molecule may be raised to a higher electron energy, corresponding to an **electronic transition.**

Since each of these internal energy transitions is quantized, they will occur only at *definite wavelengths* corresponding to an energy $h\nu$ equal to the quantized jump in the internal energy. There are, however, many *different* possible energy levels for each type of transition, and several wavelengths may be absorbed. The transitions can be illustrated by an energy level diagram like that in Figure 13.3. The relative energy levels of the three transition processes are in the order electronic > vibrational > rotational, each being about an order of magnitude different in its energy level. Rotational transitions thus can take place at very low energies (long wavelengths, that is, the microwave or far-infrared region), but vibrational transi-

TABLE 13.1
Colors of Different Wavelength Regions

Wavelength Absorbed, nm	Absorbed Color	Transmitted Color (Complement)
380 – 450	Violet	Yellow-green
450 – 495	Blue	Yellow
495 – 570	Green	Violet
570 – 590	Yellow	Blue
590 – 620	Orange	Green-blue
620 – 750	Red	Blue-green

FIGURE 13.3

Energy level diagram illustrating energy changes associated with absorption of electromagnetic radiation: A, pure rotational changes (far infrared); B, rotational-vibrational changes (near infrared); C, rotational-vibrational-electronic transitions (visible and ultraviolet). E_0 is electronic ground state and E_1 is first electronic excited state.

tions require higher energies in the near-infrared region, while electronic transitions require still higher energies (in the visible and ultraviolet regions).

Purely rotational transitions can occur in the **far-infrared** and **microwave** regions (ca. 100 μm to 10 cm), where the energy is insufficient to cause vibrational or electronic transitions. The molecule, at room temperature, is usually in its lowest electronic energy state, called the **ground state** (E_0). Thus, the pure rotational transition will occur at the ground-state electronic level (A in Figure 13.3), although it is also possible to have an appreciable population of **excited states** of the molecule. When only rotational transitions occur, discrete absorption *lines* will occur in the spectrum, the wavelength of each line corresponding to a particular transition. Hence, fundamental information can be obtained about rotational energy levels of molecules. This region has been of little use analytically, however.

As the energy is increased (the wavelength decreased), vibrational transitions occur *in addition* to the rotational transitions, with different combinations of vibrational-rotational transitions. *Each* rotational level at the lowest vibrational level can be excited to different rotational levels of the excited vibrational level (B in Figure 13.3). In addition, there may be several different excited vibrational levels, each with a number of rotational levels. This leads to numerous discrete transitions. The result is a spectrum of *peaks* or "envelopes" of unresolved fine structure. The wavelengths at which these peaks occur can be related to vibrational modes within the molecule. These occur in the mid- and far-infrared regions. Some typical infrared spectra are shown in Figure 13.4.

At still higher energies (visible and ultraviolet wavelengths), different levels of electronic transitions take place, and rotational and vibrational transitions are superimposed on these (C in Figure 13.3). This results in an even larger number of possible transitions. Although all the transitions occur in quantized steps corre-

FIGURE 13.4

Typical infrared spectra. (From 26 *Frequently Used Spectra for the Infrared Spectroscopist* Standard Spectra-Midget Edition. Copyright © Sadtler Research Laboratories, Inc. Permission for the publication herein of Sadtler Standard Spectra ® has been granted, and all rights are reserved by Sadtler Research Laboratories, Inc.)

FIGURE 13.5.
Typical visible absorption spectrum. Tartaric acid reacted with β-naphthol in sulfuric acid. 1, Sample; 2, Blank. From G. D. Christian, *Talanta,* **16** (1969) 255. (Reproduced by permission of Pergamon Press, Ltd.)

sponding to discrete wavelengths, these individual wavelengths are too numerous and too close to resolve into the individual lines or vibrational peaks, and the net result is a spectrum of broad *bands* of absorbed wavelengths. Typical visible and ultraviolet spectra are shown in Figures 13.5 and 13.6.

Not all molecules can absorb in the infrared region. For absorption to occur, there must be *a change in the dipole moment (polarity) of the molecule.* A diatomic molecule must have a permanent dipole (polar covalent bond in which a pair of electrons is shared unequally) in order to absorb, but larger molecules do not. For example, nitrogen, N≡N, cannot exhibit a dipole and will not absorb in the in-

FIGURE 13.6
Typical ultraviolet spectrum. 5-Methoxy-6-(p-methoxyphenyl)-4-phenyl-2(1H)-pyridone in methanol. From *Sadtler Standard Spectra-u.v.* Copyright© Sadtler Research Laboratories, Inc., 1963. (Permission for the publication herein of Sadtler Standard Spectra® has been granted and all rights are reserved by Sadtler Research Laboratories, Inc.)

frared region. An unsymmetrical diatomic molecule such as carbon monoxide does have a permanent dipole and hence will absorb. Carbon dioxide, $O{=}C{=}O$, does not have a permanent dipole, but by vibration it may exhibit a dipole moment. Thus, in the vibration mode $O{\Longrightarrow}C{\Longleftarrow}O$, there is symmetry and no dipole moment. But in the mode $O{\Longleftarrow}C{\Longleftarrow}O$, there is a dipole moment and the molecule can absorb infrared radiation, that is, via an induced dipole. The types of absorbing groups and molecules for the infrared and other wavelength regions will be discussed below.

Our discussions have been confined to molecules, since nearly all absorbing species in solution are molecular in nature. In the case of single atoms (which occur in a flame or an electric arc) that do not vibrate or rotate, only electronic transitions occur. These occur as sharp lines corresponding to definite transitions and will be the subject of discussion in the next chapter.

The lifetimes of excited states of molecules are rather short, and the molecules will lose their energy of excitation and drop back down to the ground state. However, rather than emitting this energy as a photon of the same wavelength as absorbed, most of them will be deactivated by collisional processes in which the energy is lost as heat; the heat will be too small to be detected in most cases. This is the reason for a solution or a substance being colored. If the light were reemitted, then it would appear colorless.[3] In some cases, light will be emitted, usually at longer wavelengths; we discuss this more under the topic "Fluorescence."

13.2　ELECTRONIC SPECTRA AND MOLECULAR STRUCTURE

The electronic transitions that take place in the visible and ultraviolet regions of the spectrum are due to the absorption of radiation by specific types of groups, bonds, and functional groups within the molecule. The wavelength of absorption and the intensity are dependent on the type. The wavelength of absorption is a measure of the energy required for the transition. Its intensity is dependent on the probability of the transition occurring when the electronic system and the radiation interact and on the polarity of the excited state.

Kinds of Transitions

Electrons in a molecule can be classified into four different types. (1) Closed-shell electrons that are not involved in bonding. These have very high excitation energies and do not contribute to absorption in the visible or UV regions. (2) Covalent single-bond electrons (σ, or sigma, electrons). These also possess too high an excitation energy to contribute to absorption of visible or UV radiation (e.g., single-valence bonds in saturated hydrocarbons, $-CH_2-CH_2-$). (3) Paired nonbonding outer-shell electrons (n electrons), such as those on N, O, S, and halogens. These are less tightly held than σ electrons and can be excited by visible or UV radiation. (4) Electrons in π (pi) orbitals, for example, in double or triple bonds. These are the most readily excited and are responsible for a majority of electronic spectra in the visible and UV regions.

---

[3]With unidirectional parallel radiation, the solution should still appear colored, however, because the emitted light would be emitted as a point source in all directions.

Electrons reside in orbitals. A molecule also possesses *unoccupied orbitals* called **antibonding orbitals;** these correspond to excited-state energy levels and are either σ^* or π^* orbitals. Hence, absorption of radiation results in an electronic transition to an antibonding orbital. The most common transitions are from π or n orbitals to antibonding π^* orbitals, and these are represented by $\pi \to \pi^*$ and $n \to \pi^*$ transitions, indicating a transition to an excited π^* state. The nonbonding n electron can also be promoted, at very short wavelengths, to an antibonding σ^* state: $n \to \sigma^*$. These occur at wavelengths less than 200 nm.

Examples of $\pi \to \pi^*$ and $n \to \pi^*$ transitions occur in ketones (R—C—R'), where the carbonyl carbon is double-bonded to O. Representing the electronic transitions by valence bond structures, we can write

$$\begin{array}{c} \diagdown \\ \diagup \end{array}\!C\!\!=\!\!O \longrightarrow \begin{array}{c} \diagdown \\ \diagup \end{array}\!\overset{+}{C}\!\!-\!\!\overset{-}{O}$$

$\pi \to \pi^*$ transition

$$\begin{array}{c} \diagdown \\ \diagup \end{array}\!C\!\!=\!\!O \longrightarrow \begin{array}{c} \diagdown \\ \diagup \end{array}\!\overset{-}{C}\!\!\equiv\!\!\overset{+}{O}$$

$n \to \pi^*$ transition

Acetone, for example, exhibits a high-intensity $\pi \to \pi^*$ transition and a low-intensity $n \to \pi^*$ transition in its absorption spectrum. An example of $n \to \pi^*$ transition occurs in ethers (R—O—R'). Since this occurs below 200 nm, ethers as well as thioethers (R—S—R'), disulfides (R—S—S—R), alkyl amines (R—NH$_2$), and alkyl halides (R—X) are transparent in the visible and UV regions, that is, they have no absorption bands in these regions.

The relative intensity of an absorption band can be represented by its **molar absorptivity** ϵ, which is really a measure of the probability of the electron transition taking place. Molar absorptivity is proportional to the fraction of radiation absorbed at a given wavelength and will be described quantitatively below when we discuss Beer's law. For our purposes now, we can simply state that it represents the *absorbance* of radiation passing through a 1M solution of 1-cm depth, where absorbance is $-\log$ fraction of radiation transmitted.

The probability of $\pi \to \pi^*$ transitions is greater than for $n \to \pi^*$ transitions, and so the intensities of the absorption bands are greater for the former. Molar absorptivities at the band maximum for $\pi \to \pi^*$ transitions are typically 1000 to 100,000 while for $n \to \pi^*$ transitions they are less than 1000; ϵ is a direct measure of the intensities of the bands.

Absorption by Isolated Chromophores

The absorbing groups in a molecule are called **chromophores.** A molecule containing a chromophore is called a **chromogen.** An **auxochrome** does not itself absorb radiation, but, if present in a molecule, it can enhance the absorption by a chromophore or shift the wavelength of absorption when attached to the chromophore. Examples are hydroxy groups, amino groups, and halogens. These possess unshared (n) electrons that can interact with the π electrons in the chromophore (n–π conjugation).

Spectral changes can be classed as follows: (1) **bathochromic shift**—absorption maximum shifted to longer wavelength, (2) **hypsochromic shift**—absorption maximum shifted to shorter wavelength, (3) **hyperchromism**—an increase in absorption intensity, and (4) **hypochromism**—a decrease in absorption intensity.

In principle, the spectrum due to a chromophore is not markedly affected by minor structural changes elsewhere in the molecule. For example, acetone,

$$CH_3{-}\underset{\underset{O}{\|}}{C}{-}CH_3,$$

and 2-butanone, $CH_3\underset{\underset{O}{\|}}{C}{-}CH_2CH_3$, give spectra similar in shape and intensity. If the alteration is major or is very close to the chromophore, then changes can be expected.

Similarly, the spectral effects of two isolated chromophores in a molecule (separated by at least two single bonds) are, in principle, independent and are additive. Hence, in the molecule CH_3CH_2CNS, an absorption maximum due to the CNS group occurs at 245 nm with an ϵ of 800. In the molecule $SNCCH_2CH_2CH_2CNS$, an absorption maximum occurs at 247 nm, with approximately double the intensity ($\epsilon = 2000$). Interaction between chromophores may perturb the electronic energy levels and alter the spectrum.

Table 13.2 lists some common chromophores and their approximate wavelengths of maximum absorption.

Absorption by Conjugated Chromophores

Where multiple (e.g., double, triple) bonds are separated by just one single bond each, they are said to be conjugated. The π orbitals overlap, which decreases the energy gap between adjacent orbitals. The result is a bathochromic shift in the

TABLE 13.2
Electronic Absorption Bands for Representative Chromophores[a]

Chromophore	System	λ_{max}	ϵ_{max}
Amine	$-NH_2$	195	2,800
Ethylene	$-C{=}C-$	190	8,000
Ketone	$\diagdown{C}{=}O$ $\diagup$	195	1,000
		270–285	18–30
Aldehyde	$-CHO$	210	Strong
		280–300	11–18
Nitro	$-NO_2$	210	Strong
Nitrite	$-ONO$	220–230	1,000–2,000
		300–400	10
Azo	$-N{=}N-$	285–400	3–25
Benzene		184	46,700
		202	6,900
		255	170
Naphthalene		220	112,000
		275	5,600
		312	175
Anthracene		252	199,000
		375	7,900

[a]From M. M. Willard, L. L. Merritt, and J. A. Dean, *Instrumental Methods of Analysis*, 4th ed. Copyright© 1948, 1951, 1958, 1965, by Litton Educational Publishing, Inc., by permission of Van Nostrand Reinhold Company.

absorption spectrum and generally an increase in the intensity. The greater the degree of conjugation (i.e., several alternating double, or triple, and single bonds), the greater the shift. Conjugation of multiple bonds with nonbonding electrons (n–

π conjugation) also results in spectral changes, for example, $\overset{\displaystyle\diagdown}{\underset{\displaystyle\diagup}{C}}={CH}-{NO_2}$.

Absorption by Aromatic Compounds

Aromatic systems (containing phenyl or benzene groups) exhibit conjugation. The spectra are somewhat different, however, than in other conjugated systems, being more complex. Benzene, ⬡ , absorbs strongly at 200 nm ($\epsilon_{max} = 6900$) with a weaker band at 230–270 nm ($\epsilon_{max} = 170$); see Figure 13.7. The weaker band exhibits considerable fine structure, each peak being due to the influence of vibrational sublevels on the electronic transitions.

As substituted groups are added to the benzene ring, a smoothing of the fine structure generally results, with a bathochromic shift and an increase in intensity. Hydroxy (—OH), methoxy (—OCH_3), amino (—NH_2), nitro (—NO_2), and aldehydic (—CHO) groups, for example, increase the absorption about 10-fold; this large effect is due to n–π conjugation. Halogens and methyl (—CH_3) groups act as auxochromes.

Polynuclear aromatic compounds (fused benzene rings), for example, naphthalene, ⬡⬡ , have increased conjugation and so absorb at longer wavelengths.

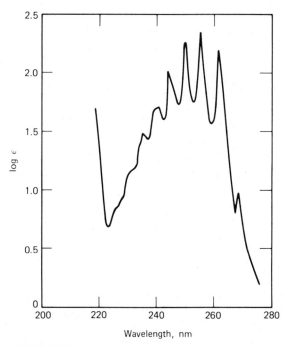

FIGURE 13.7
Ultraviolet spectrum of benzene.

Naphthacene (four rings) has an absorption maximum at 470 nm (visible) and is yellow, and pentacene (five rings) has an absorption maximum at 575 nm and is blue (see Table 13.1).

In polyphenyl compounds, ⬡ (⬡)$_n$ ⬡ , para-linked molecules (1,6-positions, as shown) are capable of resonance interactions (conjugation) over the entire system, and increased numbers of para-linked rings result in bathochromic shifts (e.g., from 250 nm to 320 nm in going from $n = 0$ to $n = 4$).In meta-linked molecules (1,3-positions), however, such conjugation is not possible and no appreciable shift occurs up to $n = 16$. The intensity of absorption increases, however, due to the additive effects of the identical chromophores.

Many heterocyclic aromatic compounds, for example, pyridine, ⬡N , absorb in the UV region, and added substituents will cause spectral changes as for phenyl compounds.

Indicator dyes used for acid–base titrations and redox titrations (Chapters 8 and 10) are extensively conjugated systems and therefore absorb in the visible region. Loss or addition of a proton or an electron will markedly change the electron distribution and hence the color.

If a compound (organic or inorganic) does not absorb in the ultraviolet or visible region, it may be possible to prepare a derivative of it that does. For example, proteins will form a colored complex with copper(II) (biuret reagent). Metals form highly colored chelates with many of the organic precipitating reagents listed in Table 6.2 in Chapter 6, as well as with others. These may be dissolved or extracted (Chapter 15) in an organic solvent such as ethylene chloride and the color of the solution measured spectrometrically. The mechanism of absorption of radiation by inorganic compounds is described below.

Spectrometric measurements in the visible region or the ultraviolet region (particularly the former) are widely employed in clinical chemistry, frequently by forming a derivative or reaction product that is colored and can be related to the test substance. For example, creatinine in blood is reacted with picrate ion in alkaline solution to form a colored product that absorbs at 490 nm. Iron is reacted with bathophenanthroline and measured at 535 nm; inorganic phosphate is reacted with molybdenum(VI) and the complex formed is reduced to form "molybdenum blue" (a +5 species) that absorbs at 660 nm; and uric acid is oxidized with alkaline phosphotungstate, and the blue reduction product of phosphotungstate is measured at 680 nm. Ultraviolet measurements include the determination of barbiturates in alkaline solution at 252 nm, and the monitoring of many enzyme reactions by following the change in absorbance at 340 nm due to changes in the reduced form of nicotinamide adenine dinucleotide (NADH), a common reactant or product in enzyme reactions. Clinical measurements are discussed in more detail in Chapter 19.

Inorganic Compounds

The absorption of ultraviolet or visible radiation by a metal complex can be ascribed to one or more of the following transitions: (1) *excitation of the metal ion,* (2) *excitation of the ligand,* or (3) *charge transfer transition.* Excitation of the metal ion in a complex usually has a very low molar absorptivity (ϵ), on the order of 1 to 100, and is not useful for quantitative analysis. Most ligands used are organic chelating agents that exhibit the absorption properties discussed above, that is, can undergo $\pi \rightarrow \pi^*$ and $n \rightarrow \pi^*$ transitions. Complexation with a metal ion is similar to protonation of the molecule and will result in a change in the wavelength and intensity of absorption. These changes are slight in most cases.

The intense color of metal chelates is frequently due to charge transfer transitions. This is simply the movement of electrons from the metal ion to the ligand, or vice versa. Such transitions include promotion of electrons from π levels in the ligand or from σ bonding orbitals to the unoccupied orbitals of the metal ion, or promotion of σ-bonded electrons to unoccupied π orbitals of the ligand.

When such transitions occur, a redox reaction actually occurs between the metal ion and the ligand. Usually, the metal ion is reduced and the ligand is oxidized, and the wavelength (energy) of maximum absorption is related to the ease with which the exchange occurs. A metal ion in a lower oxidation state, complexed with a high electron affinity ligand, may be oxidized without destroying the complex. An important example is the 1,10-phenanthroline chelate of iron(II).

Charge transfer transitions are extremely intense, with ϵ values typically 10,000 to 100,000; they occur in either the visible or UV regions. The intensity (ease of charge transfer) is increased by increasing the extent of conjugation in the ligand. Metal complexes of this type are intensely colored due to their high absorption and are well suited for the detection and measurement of trace concentration of metals.

13.3 INFRARED ABSORPTION AND MOLECULAR STRUCTURE

Absorbing (vibrating) groups in the infrared region absorb within a certain wavelength region, and the exact wavelength will be influenced by neighboring groups. The absorption peaks are much sharper than in the ultraviolet or visible regions, however, and easier to identify. In addition, each molecule will have a complete absorption spectrum unique to that molecule, and so a "fingerprint" of the molecule is obtained. See, for example, the top spectrum in Figure 13.4. Catalogues of infrared spectra are available for a large number of compounds for comparison purposes. See the references at the end of the chapter. Mixtures of absorbing compounds will, of course, exhibit the combined spectra of compounds. Even so, it is often possible to identify the individual compounds from the absorption peaks of specific groups on the molecules. Figure 13.8 summarizes regions where certain types of groups absorb. Absorption in the 6- to 15-μm region is very dependent on the molecular environment, and this is called the **fingerprint region.** A molecule can be identified by a comparison of its unique absorption in this region with catalogued known spectra.

Although the most important use of infrared spectroscopy is in identification and structure analysis, it is useful for quantitative analysis of complex mixtures of similar compounds because some absorption peaks for each compound will occur at a definite and selective wavelength, with intensities proportional to the concentration of absorbing species.

13.4 SOLVENTS FOR SPECTROMETRY

Obviously, the solvent used to prepare the sample must not absorb appreciably in the wavelength region where the measurement is being made. In the visible region, this is no problem. There are many colorless solvents and, of course, water is used for inorganic substances. Water can be used in the ultraviolet region. Many sub-

FIGURE 13.8
Simple correlations of group vibrations to regions of infrared absorption. From R. T. Conley, *Infrared Spectroscopy*, 2nd Ed. Boston: Allyn and Bacon, Inc., 1972. (Reproduced by permission of Allyn and Bacon, Inc.)

stances measured in the ultraviolet region are organic compounds that are insoluble in water and so an organic solvent must be used. Table 13.3 lists a number of solvents for use in the ultraviolet region. The cutoff point is the lowest wavelength at which the absorbance (see below) approaches unity, using a 1-cm cell with water as the reference. These solvents can all be used at least up to the visible region.

The choice of solvent will sometimes affect the spectrum in the ultraviolet region due to solvent–solute interactions. In going from a nonpolar to a polar solvent, loss of fine structure may occur and the wavelength of maximum absorption may shift (either bathochromic or hypsochromic, depending on the nature of the transition and the type of solute–solvent interactions).

TABLE 13.3
Lower Transparency Limit of Solvents in the Ultraviolet Region

Solvent	Cutoff Point, nm[a]	Solvent	Cutoff Point, nm[a]
Water	200	Dichloromethane	233
Ethanol (95%)	205	Butyl ether	235
Acetonitrile	210	Chloroform	245
Cyclohexane	210	Ethyl proprionate	255
Cyclopentane	210	Methyl formate	260
Heptane	210	Carbon tetrachloride	265
Hexane	210	N,N-Dimethylformamide	270
Methanol	210	Benzene	280
Pentane	210	Toluene	285
Isopropyl alcohol	210	m-Xylene	290
Isooctane	215	Pyridine	305
Dioxane	220	Acetone	330
Diethyl ether	220	Bromoform	360
Glycerol	220	Carbon disulfide	380
1,2-Dichloroethane	230	Nitromethane	380

[a]Wavelength at which the absorbance is unity for a 1-cm cell, with water as the reference.

The problem of finding a suitable solvent is more serious in the infrared region, where it is difficult to find one that is completely transparent. The use of either carbon tetrachloride or carbon disulfide (health effects aside) will cover the most widely used region of 2.5 to 15 μm (see Figure 13.4). Water exhibits strong absorption bands in the infrared region, and it can be employed only for certain portions of the spectrum. Also, special cell materials compatible with water must be used; rock salt is usually used in cells for infrared measurements because glass absorbs the radiation, but rock salt would dissolve in water. The solvents must be moisture-free if rock salt cells are used.

13.5 QUANTITATIVE CALCULATIONS

The fraction of radiation absorbed by a solution of an absorbing analyte can be quantitatively related to its concentration. Here, we present calculations for single species and for mixtures of absorbing species.

Beer's Law

The amount of monochromatic radiation absorbed by a sample is described by the Beer–Bouguer–Lambert law, commonly called **Beer's law.** Consider the absorption of monochromatic radiation as in Figure 13.9. Incident radiation of radiant power P_0 passes through a solution of an absorbing species at concentration c and path length b, and the emergent (transmitted) radiation has radiant power P. This radiant power is the quantity measured by spectrometric detectors. Bouguer in 1729 and Lambert in 1760 recognized that when electromagnetic energy is absorbed, the power of the transmitted energy decreases geometrically (exponentially). Assume, for example, that 25% of the radiant energy in Figure 13.9 is absorbed in a path length of b. Twenty-five percent of the reamaining energy (25% of $0.75P_0$) will be absorbed in the next path length b, leaving 56.25% as the emergent radiation. Twenty-five percent of this would be absorbed in another path length of b, and so on, so that an infinite path length would be required to absorb all the radiant energy. Since the fraction of radiant energy transmitted decays exponentially with path length, we can write it in exponential form:

$$T = \frac{P}{P_0} = 10^{-kb} \tag{13.4}$$

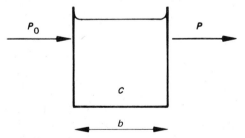

FIGURE 13.9
Absorption of radiation. P_0 = power of incident radiation, P = power of transmitted radiation, C = concentration, b = path length.

where k is a constant and T is called the **transmittance,** the fraction of radiant energy transmitted. Putting this in logarithmic form,

$$\log T = \log \frac{P}{P_0} = -kb \tag{13.5}$$

In 1852, Beer and Bernard each stated that a similar law holds for the dependence of T on the concentration:

$$T = \frac{P}{P_0} = 10^{-k'c} \tag{13.6}$$

where k' is a new constant, or

$$\log T = \frac{P}{P_0} = -k'c \tag{13.7}$$

Combining these two laws, we have "Beer's" law, which describes the dependence of T on both the path length and the concentration?

$$T = \frac{P}{P_0} = 10^{-abc} \tag{13.8}$$

where a is a combined constant of k and k', and

$$\log T = \log \frac{P}{P_0} = -abc \tag{13.9}$$

It is more convenient to omit the negative sign on the right-hand side of the equation and to define a new term, **absorbance:**

$$A = -\log T = \log \frac{1}{T} = \log \frac{P_0}{P} = abc \tag{13.10}$$

where A is the absorbance. This is the common form of Beer's law. Note that it is the *absorbance* that is directly proportional to the concentration.

The **percent transmittance** is given by

$$\% \, T = \frac{P}{P_0} \times 100 \tag{13.11}$$

Equation 13.10 can be rearranged. Since $T = \% \, T/100$,

$$A = \log \frac{100}{\% \, T} = \log 100 - \log \% \, T$$

Or

$$A = 2.00 - \log \% \, T$$

and

$$\% \, T = \text{antilog} \, (2.00 - A) \tag{13.12}$$

The path length b in Equation 13.10 is expressed in centimeters and the concentration c in grams per liter. The constant a is called the **absorptivity** and is dependent on the wavelength and the nature of the absorbing material. In an absorption spectrum, the absorbance varies with wavelength in direct proportion to a (b and c are held constant). The product of the absorptivity and the molecular weight of the absorbing species is called the **molar absorptivity** ϵ. Thus,

$$A = \epsilon bc \qquad (13.13)$$

where c is now in *moles per liter*. The cell path length in ultraviolet and visible spectrophotometry is often 1 cm; ϵ has the units $cm^{-1} mol^{-1} L$, while a has the units $cm^{-1} g^{-1} L$. Beer's law holds strictly for monochromatic radiation, since the absorptivity varies with wavelength.

We have used the symbols and terminology recommended by the journal *Analytical Chemistry*. Other terms such as optical density (OD) in place of absorbance, extinction coefficient in place of absorptivity, and so forth, may appear, especially in the older literature, but their use is not now recommended.

EXAMPLE 13.1 A sample in a 1.0-cm cell is determined with a spectrometer to transmit 80% light at a certain wavelength. If the absorptivity of this substance at this wavelength is 2.0, what is the concentration of the substance?

SOLUTION
The percent transmittance is 80%, and so $T = 0.80$:

$$\log \frac{1}{0.80} = 2.0 \times 1.0 \times c$$

$$\log 1.2_5 = 2.0 \times c$$

$$c = \frac{0.10}{2.0} = 0.050 \text{ g/L}$$

EXAMPLE 13.2 A solution containing 1.00 mg iron (as the thiocyanate complex) in 100 mL was observed to transmit 70.0% of the incident light compared to an appropriate blank. (a) What is the absorbance of the solution at this wavelength? (b) What fraction of light would be transmitted by a solution of iron four times as concentrated?

SOLUTION

(a) $\qquad\qquad\qquad T = 0.700$

$$A = \log \frac{1}{0.700} = \log 1.43 = 0.155$$

(b) $\qquad\qquad\qquad 0.155 = ab(0.0100 \text{ g/L})$

$$ab = 15.5 \text{ L/g}$$

Therefore,

$$A = 15.5 \text{ L/g } (4 \times 0.0100 \text{ g/L}) = 0.620$$

$$\log \frac{1}{T} = 0.620$$

$$T = 0.240$$

The absorbance of the new solution could have been calculated more directly:

$$\frac{A_1}{A_2} = \frac{abc_1}{abc_2} = \frac{c_1}{c_2}$$

$$A_2 = A_1 \times \frac{c_2}{c_1} = 0.155 \times \frac{4}{1} = 0.620$$

EXAMPLE 13.3   Amines, RNH_2, react with picric acid to form amine picrates which absorb strongly at 359 nm ($\epsilon = 1.25 \times 10^4$). An unknown amine (0.1155 g) is dissolved in water and diluted to 100 mL. If this solution exhibits an absorbance of 0.454 at 359 nm using a 1.00-cm cell, what is the formula weight of the amine? What is a probable formula?

SOLUTION

$$A = \epsilon bc$$

$$0.454 = 1.25 \times 10^4 \times 1.00 \times c$$

$$c = 3.63 \times 10^{-5} \text{ mol/L}$$

$$\frac{(3.63 \times 10^{-5} \text{ mol/L})(0.250 \text{ L})}{1.00 \text{ mL}} \times 100 \text{ mL} = 9.08 \times 10^{-4} \text{ mol in original flask}$$

$$\frac{0.1155 \text{ g}}{9.08 \times 10^{-4} \text{ mol}} = 127._2 \text{ g/mol}$$

The formula weight of chloroaniline, $ClC_6H_4NH_2$, is 127.6, and so this is the probable amine.

EXAMPLE 13.4   Chloroaniline in a sample is determined as the amine picrate as described in Example 13.3. A 0.0265-g sample is reacted with picric acid and diluted to 1 L. The solution exhibits an absorbance of 0.368 in a 1-cm cell. What is the percentage chloraniline in the sample?

SOLUTION

$$A = \epsilon bc$$

$$0.368 = 1.2 \times 10^4 \times 1.00 \times c$$

$$c = 2.94 \times 10^{-5} \text{ mol/L}$$

$$(2.94 \times 10^{-5} \text{ mol/L})(127.6 \text{ g/mol}) = 3.75 \times 10^{-3} \text{ g chloroaniline}$$

$$\frac{3.75 \times 10^{-3} \text{ g}}{2.65 \times 10^{-2} \text{ g}} \times 100\% = 15.0\%$$

Mixtures

It is possible to make quantitative calculations when two absorbing species in solution have overlapping spectra. It is apparent from Beer's law that the total absorbance A at given wavelength will be equal to the sum of the absorbances of all absorbing species. For two absorbing species, then, if c is in grams per liter,

$$A = a_x bc_x + a_y bc_y \tag{13.14}$$

$$A = \epsilon_x bc_x + \epsilon_x bc_y \tag{13.15}$$

where the subscripts refer to sbustances x and y, respectively.

Consider, for example, the determination of substances x and y whose individual absorption spectra at their given concentration would appear as the solid curves in Figure 13.10, and the combined spectrum of the mixture is the dashed curve. Since there are two unknowns, two measurements will have to be made. The technique is to choose two wavelengths for measurement, one occurring at the absorption maximum for x (λ_1 in the figure) and the other at the maximum for y (λ_2 in the figure). We can write, then,

$$A_1 = A_{x1} + A_{y1} = \epsilon_{x1} bc_x + \epsilon_{y1} bc_y \tag{13.16}$$

$$A_2 = A_{x2} + A_{y2} = \epsilon_{x2} bc_x + \epsilon_{y2} bc_y \tag{13.17}$$

where A_1 and A_2 are the absorbances at wavelengths 1 and 2, respectively (for the mixture); A_{x1} and A_{y1} are the absorbances contributed by x and y, respectively, at wavelength 1; and A_{x2} and A_{y2} are the absorbances contributed by x and y, respectively, at wavelength 2. Similarly, ϵ_{x1} and ϵ_{y1} are the molar absorptivities of x and y, respectively, at wavelength 1; while ϵ_{x2} and ϵ_{y2} are the molar absorptivities of x and y, respectively, at wavelength 2. These molar absorptivities are determined by making absorbance measurements on pure solutions (known molar concentrations) of x and y at wavelengths 1 and 2. So C_x and C_y become the only two unknowns

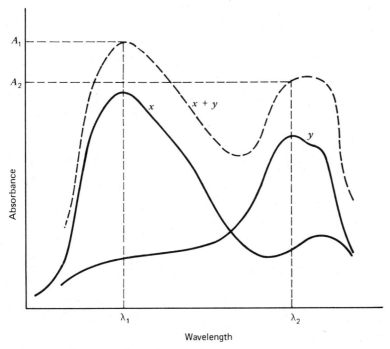

FIGURE 13.10
Absorption spectra of pure substances x and y and of a mixture of x and y at the same concentrations.

in Equations 13.16 and 13.17, and they can be calculated from the solution of the two simultaneous equations.

EXAMPLE 13.5 Potassium dichromate and potassium permanganate have overlapping absorption spectra in 1 M H_2SO_4. $K_2Cr_2O_7$ has an absorption maximum at 440 nm, and $KMnO_4$ has a band at 545 nm (the maximum is actually at 525 nm, but the longer wavelength is generally used where interference from $K_2Cr_2O_7$ is less). A mixture is analyzed by measuring the absorbance at these two wavelengths with the following results: $A_{440} = 0.405$, $A_{545} = 0.712$ in a 1-cm cell (approximate; exact length not known). The absorbances of pure solutions of $K_2Cr_2O_7$ (1.00×10^{-3} M) and $KMnO_4$ (2.00×10^{-4} M) in 1 M H_2SO_4, using the same cell gave the following results: $A_{Cr,440} = 0.374$, $A_{Cr,545} = 0.009$, $A_{Mn,440} = 0.019$, $A_{Mn,545} = 0.475$. Calculate the concentrations of dichromate and permanganate in the sample solution.

SOLUTION

The path length b is not known precisely; but since the same cell is used in all measurements, it is constant. We can calculate the product ϵb from the calibration measurements and use this constant in calculations (call the constant k):

$$0.374 = k_{Cr,440} \times 1.00 \times 10^{-3}; \quad k_{Cr,440} = 374$$
$$0.009 = k_{Cr,545} \times 1.00 \times 10^{-3}; \quad k_{Cr,545} = 9$$
$$0.019 = k_{Mn,440} \times 2.00 \times 10^{-4}; \quad k_{Mn,440} = 95$$
$$0.475 = k_{Mn,545} \times 2.00 \times 10^{-4}; \quad k_{Mn,545} = 2.38 \times 10^3$$
$$A_{440} = k_{Cr,440}[Cr_2O_7^{2-}] + k_{Mn,440}[MnO_4^-]$$
$$A_{545} = k_{Cr,545}[Cr_2O_7^{2-}] + k_{Mn,545}[MnO_4^-]$$
$$0.405 = 374[Cr_2O_7^{2-}] + 95[MnO_4^-]$$
$$0.712 = 9[Cr_2O_7^{2-}] + 2.38 \times 10^3[MnO_4^-]$$

Solving simultaneously,

$$[Cr_2O_7^{2-}] = 1.01 \times 10^{-3}\ M; \quad [MnO_4^-] = 2.95 \times 10^{-4}\ M$$

Note that for Cr at 545 nm, where it overlaps the main Mn peak, the absorbance was measured to only one figure, since it was so small. This is fine. The smaller the necessary correction, the better. Ideally, it should be zero.

If the two spectral curves overlap only at one of the wavelengths, the solution becomes simpler. For example, if the spectrum of x does not overlap with that of y at wavelength 2, the concentration of y can be determined from a single measurement at wavelength 2, just as if it were not in a mixture. The concentration of x can then be calculated from the absorbance at wavelength 1 by subtracting the contribution of y to the absorbance at that wavelength, that is, from Equation 13.16. The molar absorptivity of y must, of course, be determined at wavelength 1. If there is no overlap of either spectrum at the wavelength of measurement (usually at maximum absorbance), then each substance can be determined in the usual manner.

In making these difference measurements, we have assumed that Beer's law holds over the concentration ranges encountered. If one substance is much more concentrated than the other, then its absorbance may be large at both wavelengths compared to that of the other substance, with the result that the determination of this other substance will not be very accurate.

Quantitative Measurements from Infrared Spectra

Infrared instruments usually record the percent transmittance as a function of wavelength. The presence of scattered radiation, especially at higher concentrations in infrared work, makes direct application of Beer's law difficult. Also, due to rather weak sources, it is necessary to use relatively wide slits (which give rise to apparent deviations from Beer's law—see below). Therefore, empirical methods are usually employed in quantitative infrared analysis, keeping experimental conditions constant. The **baseline** or **ratio method** is often used, and this is illustrated in Figure 13.11. A peak is chosen that does not fall too close to others of the test substance or of other substances. A straight line is drawn at the base of the band, and P and P_0 are measured at the absorption peak. (The curve is upside down from the usual absorption spectrum, because transmittance is recorded against wavelength.) Log P_0/P is plotted against concentration in the usual manner. Unknowns are compared against standards run under the same instrumental conditions. This technique minimizes relative errors which are in proportion to the sample size, but it does not elminate simple additive errors, such as those that offset the baseline.

13.6 PRINCIPLES OF INSTRUMENTATION

A *spectrometer* is an instrument that will resolve polychromatic light into different wavelengths. A block diagram of a spectrometer is shown in Figure 13.12. All spectrometers require (1) a **source** of continuous radiation over the wavelengths of interest, (2) a **monochromator** for selecting a narrow band of wavelengths from the source spectrum, (3) a **detector,** or transducer, for converting radiant energy into electrical energy, and (4) a device to read out the response of the detector. The sample may precede or follow the monochromator. Each of these, except the readout device, will vary depending on the wavelength region.

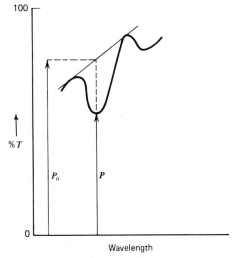

FIGURE 13.11
The baseline method for quantitative determination in the infrared region of the spectrum.

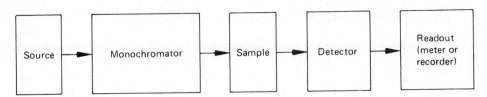

FIGURE 13.12
Block diagram of a spectrometer.

Sources

The source should have a readily detectable output of radiation over the wavelength region for which the instrument is designed to operate. No source, however, has a constant spectral output. The most commonly employed source for the **visible** region is a *tungsten filament incandescent lamp*. The spectral output of a typical filament bulb is illustrated in Figure 13.13. The useful wavelength range is from about 325 or 350 nm to 3 µm, so it can also be used in the near-ultraviolet and near-infrared regions. The wavelength of maximum emission can be shifted to shorter wavelengths by increasing the voltage to the lamp and hence the temperature of the filament, but its lifetime is shortened. For this reason, a stable, regulated power supply is required to power the lamp. This is true for sources for other regions of the spectrum also. Sometimes, a 6-V storage battery is used as the voltage source.

For the **ultraviolet** region, a low-pressure *hydrogen* or *deuterium discharge tube* is generally used as the source. Each of these can be used from 185 to about 375 nm, but the deuterium source has about three times the spectral output of the hydrogen source. Ultraviolet sources must have a quartz window, because glass is not transparent to ultraviolet radiation. They are frequently water-cooled to dissipate the heat generated.

In **infrared** spectrometers, which usually operate from about 2 to 15 µm, a *Nernst glower* is used as the source. This is a rod consisting of a mixture of rare

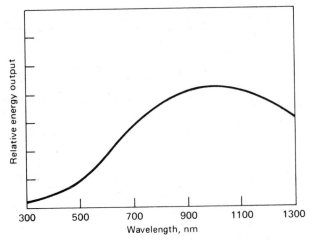

FIGURE 13.13
Intensity of radiation as a function of wavelength for a typical tungsten bulb at 3000 K.

earth oxides. It has a negative temperature coefficient of resistance and is nonconducting at room temperature. Therefore, it must be heated to excite the element to emit radiation, but once in operation it becomes conducting and furnishes maximum radiation at about 1.4 μm, or 7100 cm^{-1} (1500 to 2000°C). Another infrared source is the *Globar*. This is a rod of sintered silicon carbide heated to about 1300 to 1700°C. Its maximum radiation occurs at about 1.9 μm (5200 cm^{-1}), and it must be water-cooled. The Globar is a less intense source than the Nernst glower, but it is more satisfactory for wavelengths longer than 15 μm because its intensity decreases less rapidly. IR sources have no protection from the atmosphere, as no satisfactory envelope material exists.

We shall see below how the instruments can be adjusted to account for the variations in source intensity with wavelength as well as for the variation in detector sensitivity with wavelength.

Monochromators

A monochromator consists chiefly of *lenses* or mirrors to focus the radiation, entrance and exit *slits* to restrict unwanted radiation and help control the spectral purity of the radiation emitted from the monochromator, and a *dispersing medium* to "separate" the wavelengths of the polychromatic radiation from the source. There are two basic types of dispersing elements, the *prism* and the *diffraction grating*.

1. PRISMS. When electromagnetic radiation passes through a prism, it is refracted, because the index of refraction of the prism material is different from that in air. The index of refraction depends on the wavelength and, therefore, so does the degree of refraction. Shorter wavelengths are refracted more than longer wavelengths. The effect of refraction is to "spread" the wavelengths apart into different wavelengths (Figure 13.14). By rotation of the prism, different wavelengths of the spectrum can be made to pass through an exit slit and through the sample. A prism works satisfactorily in the ultraviolet and visible regions and can also be used in the infrared region. However, because of its **nonlinear dispersion,** it works more effectively for the shorter wavelengths. Glass prisms and lenses can be used in the visible region, but quartz or fused silica must be used in the ultraviolet region. The latter can also be used in the visible region.

In the infrared region, glass and fused silica transmit very little, and the prisms and other optics must be made from large crystals of alkali or alkaline earth halides, which are transparent to infrared radiation. Sodium chloride (rock salt) is used in most instruments and is useful for the entire region from 2.5 to 15.4 μm (4000 to 650 cm^{-1}). For longer wavelengths, KBr (10 to 25 μm) or CsI (10 to 38 μm) can be used. These (and the monochromator compartment) must be kept dry.

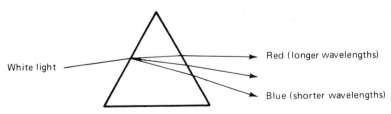

FIGURE 13.14
Dispersion of polychromatic light by a prism.

2. DIFFRACTION GRATINGS. These consist of a large number of parallel lines (grooves) ruled on a highly polished surface such as aluminum, about 15,000 to 30,000 per inch for the ultraviolet and visible regions and 1500 to 2500 per inch for the infrared region. The grooves act as scattering centers for rays impinging on the grating. The result is equal dispersion of all wavelengths of a given order, that is, **linear dispersion** (Figure 13.15). The resolving power depends on the number of ruled grooves, but generally the resolving power of gratings is better than that of prisms, and they can be used in all regions of the spectrum. They are particularly well suited for the infrared region because of their equal dispersion of the long wavelengths. Gratings are difficult to prepare and original gratings are expensive. However, many **replica gratings** can be prepared from an original grating. This is done by coating the grating with a film of an epoxy resin that, after setting, is stripped off to yield the replica. It is made reflective by aluminizing the surface. These replica gratings are much less expensive and are even used in small inexpensive instruments.

The intensity of radiation reflected by a grating varies with the wavelength, the wavelength of maximum intensity being dependent on the angle from which the radiation is reflected from the surface of the groove in the blazed grating. Hence, gratings are blazed at specific angles for specific wavelength regions, and one blazed for the blue region would be poor for an infrared spectrometer. Gratings also will produce radiation at *multiples* of the incident radiation (see Figure 13.15). These multiples are called **higher orders** of the radiation. The primary order is called the first order, twice the wavelength is the second order, three times the wavelength is the third order, and so on. So a grating produces first-order spectra, second-order spectra, and so on. The higher-order spectra are more greatly dispersed and the resolution increased. Because of the occurrence of higher orders, radiation at wavelengths less than the spectral region must be filtered out, or else

FIGURE 13.15
Diffraction of radiation from a grating.

its higher orders will overlap the radiation of interest. This can be accomplished with various types of optical filters that pass radiation only above a certain wavelength. For example, if incident radiation from a radiating sample (replaces the source on a spectrophotometer) in the 400 to 700-nm range is being dispersed and measured (e.g., fluorescence), any radiation by the sample at, for example, 325 nm, would have a second order at 650 nm, which would overlap first order radiation at 650 nm. This can be filtered out by placing a filter between the radiating sample and the grating that blocks radiation of ≤400 nm in the path of the incident beam; then the 325-nm radiation will not reach the grating.

Ruled gratings have a problem of "ghosting" associated with periodic errors in the ruling engine drive screws, particularly if the gratings are used with high-intensity radiation sources (e.g., in fluorescence instruments—see below). This stray light is greatly reduced with **holographic gratings.** These are manufactured by exposing a photoresist layer, on a suitable substrate, to the interference pattern produced by two monochromatic laser beams, followed by photographic development to produce grooves, and then a reflective coating process. The smoother line profile results in reduced light scatter. Also, these gratings can be produced on curved surfaces and used to collimate light, eliminating mirrors or lenses that result in loss of light. The cost of these gratings is significantly higher than that of the more conventional type, but they are finding use in spectrometers used for measurement of radiating samples such as in fluorescence analysis.

Cells

The cell holding the sample (usually a solution) must, of course, be transparent in the wavelength region being measured. The materials described above for the optics are used for the cell material in instruments designed for the various regions of the spectrum.

The cells for use in visible and ultraviolet spectrometers are usually square cuvets 1 cm thick (*internal* distance between parallel walls). These are illustrated in Figure 13.16. For infrared instruments, various assorted cells are used. The most common is a cell of sodium chloride windows. Fixed-thickness cells are available for these purposes and are the most commonly used. The solvent, of course, must not attack the windows of the cell. Sodium chloride cells must be protected from atmospheric moisture (stored in desiccators) and moist solvents. They require periodic polishing to remove "fogging" due to moisture contamination. Silver chloride windows are often used for wet samples or aqueous solutions. These are soft and will gradually darken due to exposure to visible light.

The short path lengths required in infrared spectrometry are difficult to repro-

FIGURE 13.16
Some typical uv and visible absorption cells.

duce, especially when the windows must be repolished, and so quantitative analysis is not as accurate in this region. Use of an internal standard helps. The path length of the empty cell can be measured from the interference fringe patterns. See Reference 5 at the end of the chapter. Variable path length cells are also available in thicknesses from about 0.002 to 3 mm.

When samples exist as pure liquids, they are usually run without dilution ("neat") in the infrared region, as is often the case when an organic chemist is trying to identify or confirm the structure of an unknown or new compound. For this purpose, the cell length must be short in order to keep the absorbance within the optimum region, generally path lengths of 0.01 to 0.05 mm. If a solution of the sample is to be prepared, a fairly high concentration is usually run, because no solvent is completely transparent in the infrared region, and this will keep the solvent absorbance minimal. So again, short path lengths are required, generally 0.1 mm or less.

Solids are often not sufficiently soluble in the available solvents to give a high enough concentration to measure in the infrared region. However, powders may be run as a suspension or thick slurry (mull) in a viscous liquid having about the same index of refraction in order to reduce light scattering. The sample is ground in the liquid, which is often Nujol, a mineral oil (see Figure 13.4). Chlorofluorocarbon greases are useful when the Nujol masks any C—H bands present. The mull technique is useful for qualitative analysis, but it is difficult to reproduce for quantitative work. Samples may also be ground with KBr (which is transparent in the infrared region) and pressed into a pellet for mounting for measurement.

Gases may be analyzed by infrared spectrometry, and for this purpose a long-path cell is used, usually 10 cm in length, although cells as long as 20 m and up have been used in special applications. Some typical infrared cells are shown in Figure 13.17.

(a)   (b)

FIGURE 13.17
Typical infrared cells. (a) Fixed path cell. (Courtesy of Barnes Engineering Co.) (b) Variable path length cell. (Courtesy of Wilks Scientific Corporation.)

Detectors

Again, the detectors will also vary with the wavelength region to be measured. A **phototube** (or photocell) is commonly used in the ultraviolet and visible regions. This consists of a photoemissive cathode and an anode. A high voltage is impressed between the anode and cathode. When a photon enters the window of the tube and strikes the cathode, an electron is emitted and attracted to the anode, causing current to flow that can be amplified and measured. The response of the photoemissive material is wavelength dependent, and different phototubes are available for different regions of the spectrum. For example, one may be used for the blue and ultraviolet portions and a second for the red portion of the spectrum.

A **photomultiplier tube** is more sensitive than a phototube for the visible and ultraviolet regions. This is essentially several successive phototubes built into one envelope. It consists of a series of electrodes (dynodes), each at a more positive potential (50 to 90 V) than the one before it. When a primary electron is emitted from the cathode by a photon, it is accelerated toward the next photoemissive electrode, where it releases many more secondary electrons. These, in turn, are accelerated to the next electrode where each secondary electron releases more electrons, and so on, up to about 10 stages of amplification. The final output of the photomultiplier tube may, in turn, be electronically amplified. The most popular photomultiplier tube is the IP28 since it can be used in both the ultraviolet and visible regions.

Again, different photomultiplier tubes have different response characteristics, depending on the wavelength. Because of the greater sensitivity of photomultiplier tubes, less intense radiation is required and narrower slit widths can be used for better resolution of the wavelengths.

Photomultiplier tubes have also been developed with response limited to the ultraviolet region (160 to 320 nm), the so-called **solar-blind photomultipliers.** They are helpful in reducing stray light effects from visible radiation and are useful as UV detectors in nondispersive systems.

Spectrometers that use phototubes or photomultiplier tubes are detectors are generally called **spectrophotometers,** and the corresponding measurement is called **spectrophotometry.** More strictly speaking, the journal *Analytical Chemistry* defines a spectrophotometer as a spectrometer that measures the *ratio* of the radiant power of two beams, that is, P/P_0, and so it can record absorbance. The two beams may be measured simultaneously or separately, as in a double-beam or a single-beam instrument—see below. Phototube and photomultiplier instruments in practice are almost always used in this manner. An exception is when the radiation source is replaced by a radiating sample whose spectrum and intensity are to be measured, as in fluorescence spectrometry—see below. If the prism or grating monochromator in a spectrophotometer is replaced by an optical filter that passes a narrow band of wavelengths, the instrument may be called a photometer.

As with sources, detectors used in the ultraviolet and visible regions do not work in the infrared region. But infrared radiation possesses the property of heat, and heat detectors that transduce heat into an electrical sign can be used. Thermocouples and bolometers are used as detectors. A **thermocouple** consists of two dissimilar metal wires connected at two points. When a temperature difference exists between the two points, a potential difference is developed, which can be measured. One of the junctions, then, is placed in the path of the light from the mochromator. **Bolometers** and **thermistors** are materials whose *resistance* is temperature dependent. Their change in resistance is measured in a Wheatstone bridge circuit. The advantage of these over thermocouples is the more rapid response time (4 ms, compared with 60 ms), and thus improved resolution and faster scanning rates can be accomplished. The response of thermal detectors is essentially independent of the wavelengths measured.

Slit Width

We mentioned above that it is impossible to obtain spectrally pure wavelengths from a monochromator. Instead, a **band** of wavelengths emanates from the monochromator and the width of this band will depend on both the dispersion of the grating or prism and the exit slit width. The dispersive power of a prism depends on the wavelength and on the material from which it is made, as well as on its geometrical design, while that of a grating depends on the number of grooves per inch. Dispersion is also increased as the distance to the slit is increased.

　　After the radiation has been dispersed, a certain portion of it will fall on the exit slit, and the width of this slit determines how broad a band of wavelengths the sample and detector will see. Figure 13.18 depicts the distribution of wavelengths leaving the slit. The **nominal wavelength** is that set on the instrument and is the wavelength of maximum intensity passed by the slit. The intensity of radiation at wavelengths on each side of this decreases, and the width of the band of wavelengths passed at one-half the intensity of the nominal wavelength is the **spectral bandwidth,** or **bandpass.** The **spectral slit width** is theoretically twice the spectral bandwidth (Figure 13.18 is theoretically an isosceles triangle), and this is a measure of the total wavelength spread that is passed by the slit. Note that the spectral slit width is *not* the same as the mechanical slit width, which may vary from a few micrometers to a millimeter or more (the spectral slit width is the band of radiation passed by the mechanical slit and is measured in units of wavelength). Seventy-

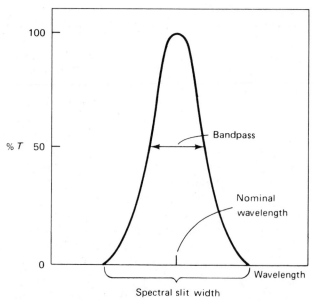

FIGURE 13.18
Distribution of wavelengths leaving the slit of a monochromator.

five percent of the radiation intensity is theoretically contained within the wavelengths of the spectral band width.

If the intensity of the source and the sensitivity of the detector are sufficient, the spectral purity can be improved (the bandpass decreased) by decreasing the slit width. The decrease may not be linear, however, and a limit is reached due to aberrations in the optics and diffraction effects caused by the slit at very narrow widths. The diffraction effectively increases the spectral slit width. In actual practice, the sensitivity limit of the instrument is usually reached before diffraction effects become too serious.

The bandwidth or the spectral slit width is essentially constant with a grating dispersing element for all wavelengths of a given spectral order at a constant slit width setting. This is not so with a prism, because of the variation of dispersion with changing wavelength. The bandwidth will be smaller at shorter wavelengths and larger at longer wavelengths.

13.7 TYPES OF INSTRUMENTS

Although all spectrometric instruments have the basic design presented in Figure 13.12, there are many variations depending on the manufacturer, the wavelength regions for which the instrument is designed, the resolution required, and so on. It is beyond the scope of our discussion to go into these, but we will indicate a few of the important general types of design and the general operation of a spectrometer.

Single-Beam Spectrometers

These are the most common, since they are less expensive than more sophisticated instruments, and excellent results can be obtained with them. A diagram of the popular Bausch and Lomb Spectronic 20 spectrophotometer (phototube instrument) is shown in Figure 13.19. It consists of a tungsten lamp visible-light source and an inexpensive replica grating of 600 grooves per millimeter to disperse the radiation,

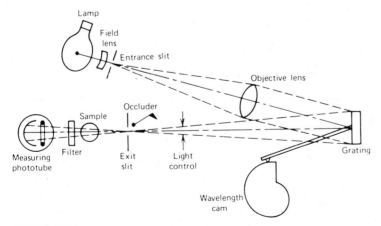

FIGURE 13.19
Optical diagram of the Bausch and Lomb Spectronic 20 spectrophotometer (top view). (Courtesy of Bausch and Lomb, Inc.)

ranging in wavelength from 330 to 950 nm. The exit slit allows a band of 20 nm of radiation to pass. If the wavelength is set at 480 nm, for example, radiation from 470 to 490 nm passes through the exit slit. By turning the wavelength cam, the grating is rotated to change the band of 20 nm of wavelengths passing through the exit slit (the path of only one 20-nm band is shown after reflection from the grating in the figure). The filter removes second-order and higher orders of diffraction from the grating that may pass the slit (stray light). The selection of the filter depends on what radiation must be restricted. For most applications, a cutoff-type filter is used that passes radiation below a certain wavelength where measurements are to be made, but not longer wavelengths where higher orders may appear. Narrower-range filters may be better for some applications, for example, **a red filter to remove any nonred light so the detector sees essentially pure red** (see below).

Any radiation not absorbed by the sample falls on the detector where the intensity is converted to an electrical signal that is amplified and read on a meter. The measuring phototube for the visible region has maximum response at 400 nm, with only 5% of this response at 625 nm. Measurements above 625 nm are best made by substituting a red-sensitive phototube (RCA 6953) along with a red filter to remove second-order diffraction from the grating (it passes the desired red radiation but not undesired higher orders).

We have illustrated that the spectral intensity of the sources and the spectral response of the detectors are dependent on the wavelength. Therefore, some means must be employed to adjust the electrical output of the detector to the same magnitude at all wavelengths. This can be accomplished by one of two ways: by adjusting the slit width to allow more or less light to fall on the detector, or by adjusting the gain on the detector (the amount of amplification of the signal).

A single-beam instrument will have a shutter that can be placed in front of the detector so that no light reaches it. This is the occluder in the Spectronic 20, and it drops into place whenever there is no measuring cell placed in the instrument. With the shutter in position, a "dark current" adjusting knob is used to set the scale reading to zero percent transmittance (infinite absorbance). The **dark current** is a small current that may flow in the absence of light, owing to thermal emission of electrons from the cathode of the phototube. In the above operation, the dark current is set to zero scale reading by effectively changing the voltage on the tube. Now, the cell filled with solvent is placed in the beam path and the shutter is opened. By means of a slit width control to adjust the amount of radiation passed or a "sensitivity" knob (gain control), the output of the detector is adjusted so that the scale reading is 100% transmittance (zero absorbance). The dark current and 100% transmittance adjustments are usually repeated to make certain the adjustment of one has not changed the other. The instrument scale is now calibrated and it is ready to read an unknown absorbance. The above operations must be repeated at each wavelength.

Each time a series of samples is run, the absorbance of one or more blank solutions[4] is read versus pure solvent; and, if appreciable ($\geq$0.01 A with a Spectronic 20), this is subtracted from all analyte solution readings. Actually, if the blank solution is essentially colorless (i.e., its absorbance is small), this solution is often used in place of the solvent for adjusting the 100% transmittance reading. Any blank absorbance is then automatically corrected for (subtracted). This method

---

[4]This contains all reagents used in the sample, but no analyte.

should only be used if the blank reading is small and has been demonstrated to be constant. A large blank reading would be more likely to be variable, and it would require a large gain on the detector, causing an increase in the noise level. An advantage of zeroing the instrument with the blank is that one reading, which always contains some experimental error, is eliminated. If this technique is used, it would be a good practice to check the zero with all the blank solutions to make sure the blank is constant.

Double-Beam Spectrometers

These are in practice rather complex instruments, but they have a number of advantages. They are used largely as recording instruments, that is, instruments that automatically vary the wavelength and record the absorbance as a function of wavelength. The instrument has two light paths, one for the sample and one for the blank or reference. In a typical setup, the beam from the source strikes a vibrating or rotating mirror that alternately passes the beam through the reference cell and the sample cell and, from each, to the detector. In effect, the detector alternately sees the reference and the sample beam and the output of the detector is proportional to the ratio of the intensities of the two beams (P/P_0).

The output is an alternating signal whose frequency is equal to that of the vibrating or rotating mirror. An ac amplifier is used to amplify this signal, and stray dc signals are not recorded. The wavelength is changed by a motor that drives the dispersing element at a constant rate, and the slit is continually adjusted by a servomotor to keep the energy from the reference beam at a constant value; that is, it automatically adjusts to 100% transmittance through the reference cell (which usually contains the blank or the solvent).

This is a simplified discussion of a double-beam instrument. There are variations on this design and operation, but it illustrates the utility of these instruments. They are very useful for qualitative work in which the entire spectrum is required, and they automatically compensate for absorbance by the blank, as well as for drifts in source intensity.

High-Absorbance-Range Spectrophotometers

We shall see below in discussing the spectrophotometric error that, for most accurate measurements, the absorbance reading of spectrometric measurements using conventional spectrophotometers should be between 0.1 and 1.0 or 1.5 absorbance units. Top-line-model spectrophotometers utilize feedback mechanisms and stabilizing electronics that allow measurements of absorbance up to 3 units. For example, an automatic photomultiplier feedback circuit adjusts the voltage applied to the dynodes (electrodes) of a photomultiplier (PM) tube in inverse relation to the amount of light falling on the cathode. Hence, for high light levels, the sensitivity of the PM tube is decreased, while with low light levels, it is increased. An essentially constant-current flow in the tube results. This feedback arrangement permits operation of the phototube at an extremely low anode output current regardless of the amount of light impinging on the photocathode. The voltage required to maintain this anode current constant is the source of the output signal. In effect, the high end of the absorbance scale is expanded over that of conventional instruments. Such expansion is illustrated in Figure 13.20. The chief source of spectrometric errors usually comes in reading the compressed scale above 1 absorbance unit, but in these instruments, this portion of the scale is expanded.

FIGURE 13.20
Illustration of expansion of the high end of the absorbance scale in the Gilford spectrophotometer. (Reproduced by permission of Gilford Instrument Laboratories, Inc.)

13.8 SPECTROMETRIC ERROR

There will always be a certain amount of error or irreproducibility in reading an absorbance or transmittance scale. Uncertainty in the reading will depend on a number of instrumental factors and on the region of the scale being read, and hence on the concentration.

It is probably obvious to you that if the sample absorbs only a very small amount of the light, an appreciable *relative* error may result in reading the small decrease in transmittance. At the other extreme, if the sample absorbs nearly all the light, an extremely stable instrument would be required to read the large decrease in the transmittance accurately. There is, therefore, some optimum transmittance or absorbance where the relative error in making the reading will be minimal.

The transmittance for minimum relative error can be derived from Beer's law by calculus, assuming that the error results essentially from the uncertainty in reading the instrument scale and the absolute error in reading the transmittance is constant, independent of the value of the transmittance. The result is the prediction that the minimum relative error in the concentration theoretically occurs when $T = 0.368$ or $A = 0.434$.

Figure 13.21 illustrates the dependence of the relative error on the transmittance, calculated for a constant error of 0.01 in T in reading the scale. It is evident from the figure that, while the minimum occurs at 36.8% T, a nearly constant minimum error occurs over the range of 20 to 65% T (0.7 to 0.2 A). The percent transmittance should fall within 10 to 80% T ($A = 1$ to 0.1) in order to prevent large errors in spectrophotometric readings.

Figure 13.21 in practice approximates the error only for instruments with "Johnson" or *thermal noise-limited detectors,* such as photoconductive detectors like CdS or PbS detectors (400 to 3500 nm) or thermocouples, bolometers, and Golay detectors in the infrared region. Johnson noise is produced random thermal motion in resistance circuit elements. With phototubes and photomultiplier-type detectors (photoemissive detectors, ultraviolet to visible range), thermal noise becomes insignificant as compared to "shot noise." Shot noise is the random fluctuation of the electron current from an electron-emitting surface (i.e., across a junction from cathode to anode), and in PM tubes this is amplified and becomes the noise-limiting fluctuation. In instruments with these detectors, the absolute error is not constant at all values of T, and the expressions for the spectrophotometric error become more complicated. It has been calculated that, for these cases, the minimal

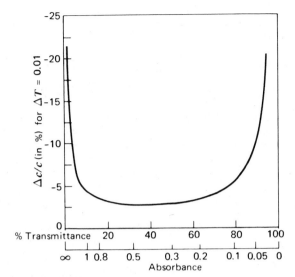

FIGURE 13.21
Relative concentration error as a function of transmittance for a 1% uncertainly in %T.

error should occur at $T = 0.136$ or $A = 0.87$. These instruments have a working range of about 0.1 to 1.5 A.

13.9 DEVIATION FROM BEER'S LAW

It can not always be assumed that Beer's law will apply, that is, that a linear plot of absorbance versus concentration will occur. Deviations from Beer's law occur as the result of chemical and instrumental factors. Most "deviations" from Beer's law are really only "apparent" deviations, because if the factors causing nonlinearity are accounted for, the true or corrected absorbance-versus-concentration curve will be linear. True deviations from Beer's law will occur when the concentration is so high that the index of refraction of the solution is changed from that of the blank. A similar situation would apply for mixtures of organic solvents with water, and so the blank solvent composition should closely match that of the sample. The solvent may also have an effect on the absorptivity of the analyte.

Chemical Deviations

Chemical causes for nonlinearity occur when nonsymmetrical chemical equilibria exist. An example is a weak acid that absorbs at a particular wavelength but has an anion that does not:

$$\underset{\text{(absorbs)}}{HA} \quad \rightleftharpoons H^+ + \underset{\text{(transparent)}}{A^-}$$

The ratio of the acid form to the salt form will, of course, depend on the pH (Chapter 5). So long as the solution is buffered or is very acid, this ratio will remain constant at all concentrations of the acid. However, in unbuffered solution, the degree of ionization will increase as the acid is diluted, that is, the above equilibrium will shift to the right. Thus, a smaller fraction of the species exists in the acid form available for absorption for dilute solutions of the acid. The result will be a positive deviation from linearity at higher concentrations (where the fraction dissociated is smaller). If the anion form were the absorbing species, then the deviation would be negative. Similar arguments apply to colored (absorbing) metal ion complexes or chelates in the absence of a large excess of the complexing agent. That is, in the absence of excess complexing agent, the degree of dissociation of the complex will increase as the complex is diluted. Here, the situation may be extremely complicated, because the complex may dissociate stepwise into successive complexes which may or may not absorb at the wavelength of measurement. pH also becomes a consideration in these equilibria.

Apparent deviations may also occur when the substance can exist as a dimer as well as a monomer. Again, the equilibrium depends on the concentration. An example is the absorbance by methylene blue, which exhibits a negative deviation at higher concentrations due to association of the methylene blue. Most of these difficulties can be minimized by proper adjustment of the solutions (buffering, etc.) or by use of a calibration curve.

If both species of a chemical equilibrium absorb, and if there is some overlap of their absorption curves, the wavelength at which this occurs is called the **isosbestic point,** and the molar absorptivity of both species is the same. Such a point is illustrated in Figure 13.22. The spectra are plotted at different pH values since the pH generally causes the shift in the equilibrium. Obviously, the effect of pH could be eliminated by making measurements at the isosbestic point, but the sensitivity is decreased. By making the solution either very acid or very alkaline, one species predominates and the sensitivity is increased by measuring at this condition.

The existence of an isosbestic point is a necessary (although not sufficient) condition to prove that there are only two absorbing substances in equilibrium with

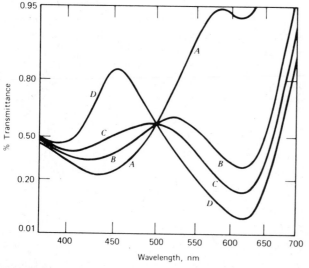

FIGURE 13.22
Illustration of isosbestic point of bromthymol blue (501 nm). (A) pH 5.45. (B) pH 6.95 (C) pH 7.50. (D) pH 11.60.

overlapping absorption bands. If both of the absorbing species follow Beer's law, then the absorption spectra of all equilibrium mixtures will intersect at a fixed wavelength. For example, the different colored forms of indicators in equilibrium (e.g., the red and yellow forms of methyl orange) often exhibit an isosbestic point, supporting evidence that two and only two colored species participate in the equilibrium.

Instrumental Deviations

The basic assumption in applying Beer's law is that monochromatic light is used. We have seen in the discussions above that it is impossible to extract monochromatic radiation from a continuum source. Instead, a band of radiation is passed, the width of which depends on the dispersing element and the slit width. In an absorption spectrum, different wavelengths are absorbed to a different degree; that is, the absorptivity changes with wavelength. At a wavelength corresponding to a fairly broad maximum on the spectrum, the band of wavelengths will all be absorbed to nearly the same extent. However, on a steep portion of the spectrum, they will be absorbed to different degrees. The slope of the spectrum increases as the concentration is increased, with the result that the fractions of the amounts of each wavelength absorbed may change, particularly if the instrument setting should drift over the period of the measurement. So a negative deviation in the absorbance-versus-concentration plot will be observed. The greater the slope of the spectrum, the greater is the deviation.

Obviously, it is advantageous to make the measurement on an absorption peak whenever possible, in order to minimize this curvature, as well as to obtain maximum sensitivity. Because a band of wavelengths is passed, the absorptivity at a given wavelength may vary somewhat from one instrument to another, depending on the resolution, slit width, and sharpness of the absorption maximum. Therefore, you should check the absorptivity and linearity on your instrument rather than rely on reported absorptivities. It is common practice to prepare calibration curves of absorbance versus concentration rather than to rely on direct calculations of concentration from Beer's law.

If there is a second (interfering) absorbing species whose spectrum overlaps with that of the test substance, nonlinearity of the total absorbance as a function of the test substance concentration will result. It may be possible to account for this in preparation of the calibration curve by adding the interfering compound to standards at the same concentration as in the samples. This will obviously work only if the concentration of the interfering compound is essentially constant, and the concentration should be relatively small. Otherwise, simultaneous analysis as described earlier will be required.

Other instrumental factors that may contribute to deviations from Beer's law include stray radiation entering the monochromator and being detected, internal reflections of the radiation within the monochromator, and mismatched cells (in terms of path length) used for different analyte solutions or used in double-beam instruments (when there is appreciable absorbance by the blank or solvent in the reference cell). Nonuniform cell thickness can affect a quantitative analysis. This is potentially a problem, especially in infrared spectrometry, where cell spacers are used. Air bubbles can affect the path length and stray light, and it is important to eliminate these bubbles, again especially in the infrared cells.

13.10 FLUOROMETRY

Fluorometric analysis is extremely sensitive and is used widely by biochemists, clinical chemists, and analytical chemists in general.

Principles of Fluorescence

When a molecule absorbs electromagnetic energy, this energy is usually lost as heat, as the molecule is deactivated via collisional processes. With certain molecules (ca. 5 to 10%), however, particularly when absorbing high-energy radiation such as UV radiation, only part of the energy is lost via collisions, and then the electron drops back to the ground state by emitting a photon of lower energy (longer wavelength) than was absorbed. Refer to Figure 13.23.

A molecule at room temperature normally resides in the ground state. The ground state is usually a **singlet state** (S_0), with all electrons paired. Electrons that occupy the same molecular orbital must be "paired," that is, have opposite spins. In a singlet state, the electrons are paired. If electrons have the same spin, they are "unpaired" and the molecule is in a **triplet state.** Singlet and triplet states refer to the **multiplicity** of the molecule. The process leading to the emission of a fluorescent photon begins with the absorption of a photon (a process that takes 10^{-15} s) by the fluorophore, resulting in an electronic transition to a higher-energy (excited) state. In most organic molecules at room temperature, this absorption corresponds to a transition from the lowest vibrational level of the ground state to one of the vibrational levels of the first or second electronic excited state of the same multiplicity (S_1, S_2). The spacing of the vibrational levels and rotational levels in these higher electronic states gives rise to the absorption spectrum of the molecule.

If the transition is to an electronic state higher than S_1, a process of **internal conversion** rapidly takes place. It is thought that the excited molecule passes from

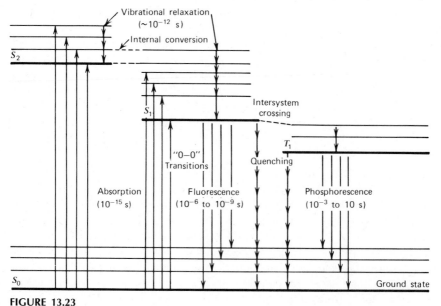

FIGURE 13.23
Energy level diagram showing absorption processes, relaxation processes, and their rates.

the vibrational level of this higher electronic state to a high vibrational level of S_1 that is isoenergetic with the original excited state. Collision with solvent molecules at this point rapidly removes the excess energy from the higher vibrational level of S_1; this process is called **vibrational relaxation.** These energy degration processes (internal conversion and vibrational relaxation) occur rapidly ($\sim 10^{-12}$ s). Because of this rapid energy loss, emission fluorescence from higher states than the first excited state is rare.

Once the molecule reaches the first excited singlet, internal conversion to the ground state is a relatively slow process. Thus, decay of the first excited state by emission of a photon can effectively compete with other decay processes. This emission process is **fluorescence.** Generally, fluorescence emission occurs very rapidly after excitation (10^{-6} to 10^{-9} s). Consequently, it is not possible for the eye to perceive fluorescence emission after removal of the excitation source. Because fluorescence occurs from the lowest excited state, the fluorescence spectrum, that is, the wavelengths of emitted radiation, is independent of the wavelength of excitation. The intensity of emitted radiation, however, will be proportional to the intensity of incident radiation (i.e., the number of photons absorbed).

Another feature of excitation and emission transitions is that the longest wavelength of excitation corresponds to the shortest wavelength of emission. This is the "0–0" band corresponding to the transitions between the 0 vibrational level of S_0 and the 0 vibrational level of S_1 (Figure 13.23).

While the molecule is in the excited state, it is possible for one electron to reverse its spin and the molecule transfers to a lower-energy triplet state by a process called **intersystem crossing.** Through the processes of internal conversion and vibrational relaxation, the molecule then rapidly attains the lowest vibrational level of the first excited triplet (T_1). From here, the molecule can return to the ground state S_0 by emission of a photon. This emission is referred to as **phosphorescence.** Since transitions between states of different multiplicity are "forbidden," T_1 has a much longer lifetime than S_1 and phosphorence is much longer-lived than fluorescence ($> 10^{-4}$ s). Consequently, one can quite often perceive an "afterglow" in phosphorescence when the excitation source is removed. In addition, because of its relatively long life, radiationless processes can complete more effectively with phosphorescence than fluorescence. For this reason, phosphorescence is not normally observed from solutions due to collisions with the solvent or with oxygen. Phosphorescence measurements are made by cooling samples to liquid nitrogen temperature ($-196°C$) to freeze them and minimize collision with other molecules. Solid samples will also phosphoresce, and many inorganic minerals exhibit long-lived phosphorescence. Recent studies have been made in which molecules in solution are adsorbed on a solid support from which they can phosphoresce.

A typical excitation and emission spectrum of a fluorescing molecule is shown in Figure 13.24. The excitation spectrum usually corresponds closely in shape to the absorption spectrum of the molecule. There is frequently (but not necessarily) a close relationship between the structure of the excitation spectrum and the structure of the emission spectrum. In many relatively large molecules, the vibrational spacings of the excited states, especially S_1, are very similar to those in S_0. Thus, the form of the emission spectrum resulting from decay to the various S_0 vibrational levels tends to be a "mirror image" of the excitation spectrum arising from excitation to the various vibrational levels in the excited state, such as S_1. Substructure, of course, results also from different rotational levels at each vibrational level.

FIGURE 13.24
Excitation and emission spectra of a fluorescing molecule.

The longest wavelength of absorption and the shortest wavelength of fluorescence tend to be the same (the "0–0" transition in Figure 13.23). More typically, however, this is not the case due to solvation differences between the excited molecule and the ground-state molecule. The heats of solvation of each are different, which results in a decrease in the energy of the emitted photon by an amount equal to these two heats of solvation.

Only those molecules that will absorb radiation, usually ultraviolet radiation, can fluoresce, and of those that do absorb, only about 5 to 10% fluoresce. This is an advantage when considering possible interference in fluorescence. The emitted radiation may be in the ultraviolet region, especially if the compound absorbs at less than 300 nm, but it is usually in the visible region. It is the emitted radiation that is measured and related to concentration.

Chemical Structure and Fluorescence

In principle, any molecule that absorbs ultraviolet radiation could fluoresce. There are many reasons why they do not; but we will not go into these, other than to point out, in general, what types of substances may be expected to fluoresce.

First of all, the greater the absorption by a molecule, the greater its fluorescence intensity. Many aromatic and heterocyclic compounds fluoresce, particularly if they contain certain substituted groups. Compounds with multiple conjugated double bonds are favorable to fluorescence. One or more electron-donating groups such as —OH, —NH$_2$, and —OCH$_3$ enhances the fluorescence. Polycyclic compounds such as vitamin K, purines, and nucleosides and conjugated polyenes such as vitamin A are fluorescent. Groups such as —NO$_2$, —COOH, —CH$_2$COOH, —Br, —I, and azo groups tend to *inhibit* fluorescence. The nature of other substituents may alter the degree of fluorescence. The fluorescence of many molecules is greatly pH dependent because only the ionized or unionized form may be fluorescent. For example, phenol, C$_6$H$_5$OH, is fluorescent but its anion, C$_6$H$_5$O$^-$, is not.

If a compound is nonfluorescent, it may be converted to a fluorescent derivative. For example, nonfluorescent steroids may be converted to fluorescent compounds by dehydration with concentrated sulfuric acid. These cyclic alcohols are converted to phenols. Similarly, dibasic acids, such as malic acid, may be reacted with β-naphthol in concentrated sulfuric acid to form a fluorescing derivative. C. E. White and co-workers have developed fluorometric methods for many metals by forming chelates with organic compounds (see Reference 19). Antibodies may be made to fluoresce by condensing them with fluorescein isocyanate, which reacts with the free amino groups of the proteins. NADH, the reduced form of nicotinamide adenine dinucleotide, fluoresces. It is a product or reactant (cofactor) in many enzyme reactions (see Chapter 18), and its fluorescence serves as the basis of the sensitive assay of enzymes and their substrates. Most amino acids do not fluoresce, but fluorescent derivatives are formed by reaction with dansyl chloride.

Fluorescence Quenching

One difficulty frequently encountered in fluorescence is that of **fluorescence quenching** by many substances. These are substances that, in effect, compete for the electronic excitation energy and decrease the quantum yield (the efficiency of conversion of absorbed radiation to fluorescent radiation—see below). Iodide ion is an extremely effective quencher. Iodide and bromide substituent groups decrease the quantum yield. Substances such as this may be determined indirectly by measuring the extent of fluorescence quenching. Some molecules do not fluoresce because they may have a bond whose dissociation energy is less than that of the radiation. In other words, a molecular bond may be broken, preventing fluorescence.

A colored species in solution with the fluorescing species may interfere by absorbing the fluorescent radiation. This is the so-called **"inner-filter" effect.** For example, in sodium carbonate solution, potassium dichromate exhibits absorption peaks at 245 and 348 nm. These overlap with the excitation (275 nm) and emission (350 nm) peaks for tryptophan and would interfere. The inner-filter effect can also arise from too high a concentration of the fluorophore itself. Some of the analyte molecules will reabsorb the emitted radiation of others (see the discussion of fluorscence intensity and concentration below).

Practical Considerations in Fluorometry

For reasons more apparent below, fluorometric analysis is extremely sensitive, and determinations at the part-per-billion level are common. In fact, the technique is limited to low concentrations. The high sensitivity presents problems not normally encountered with more concentrated solutions. Dilute solutions are less stable. Similar deterioration may occur in more concentrated solutions, but it is a negligible percentage of the sample. Adsorption onto the surfaces of the containers is a serious problem. Organic substances at less than 1 ppm are especially adsorbed onto glass surfaces from organic solvents. Addition of a small amount of a more polar solvent may help. In analysis of blood samples, a protein-free filtrate is usually prepared. Some trace organics may be adsorbed onto the freshly precipitated protein, and this possibility should always be checked.

Oxidation of trace substances can be a problem. For example, traces of peroxides in ether used for solvent extraction of organic substances may cause oxidation of the test substance. Even dissolved oxygen is sometimes a problem at these concentrations. Appreciable photodecomposition is more likely to occur at low concentrations, and so dilute solutions of labile compounds should be protected from light. Photodecomposition can be a serious problem in the fluorescence measurement because the energy of the exciting radiation may cause the substance to decompose. A high-intensity source is used (see below) which adds to the danger of photodecomposition. The measurement should, therefore, be made rapidly. Another reason for making measurements rapidly is to minimize increased collisional deactivation as the solution is heated by the source.

Relationship Between Concentration and Fluorescence Intensity

It can be readily derived from Beer's law (Problem 44) that the fluorescence intensity F is given by

$$F = \phi P_0(1 - 10^{-abc}) \qquad (13.18)$$

where ϕ is the **quantum yield,** a proportionality constant and a measure of the fraction of absorbed photons that are converted into fluorescent photons. The quantum yield is, therefore, less than or equal to unity. The other terms in the equation are the same as for Beer's law. It is evident from the equation that if the product abc is large, the term 10^{-abc} becomes negligible compared to 1, and F becomes constant:

$$F = \phi P_0 \qquad (13.19$$

On the other hand, if abc is small (≤ 0.01), it can be shown[5] by expanding Equation 13.18 that as a good approximation,

$$F = 2.303 \ \phi P_0 abc \qquad (13.20)$$

Thus, for low concentrations, the fluorescence intensity is directly proportional to the concentration. Also, it is proportional to the intensity of the incident radiation.

This equation generally holds for concentrations up to a few parts per million, depending on the substance. At higher concentrations, the fluorescence intensity may decrease with increasing concentration. The reason can be visualized as follows. In dilute solutions, the absorbed radiation is distributed equally through the entire depth of the solution. But at higher concentrations, the first part of the solution in the path will absorb more of the radiation. So the equation holds only when most of the radiation goes through the solution, when more than about 92% is transmitted.

Fluorescence Instrumentation

For fluorescence measurements, it is necessary to separate the emitted radiation from the incident radiation. This is most easily done by measuring the fluorescence at right angles to the incident radiation. The fluorescence radiation is emitted in all directions, but the incident radiation passes straight through the solution.

---

[5]It is known that $e^{-x} = 1 - x + x^2/2! \ldots$ and that $10^{-x} = e^{-2.303x}$. Therefore, $1 - e^{-2.303abc} = 1 - (1 - 2.303abc + (2.303abc)^2/2! \ldots)$. The squared term and higher-order terms can be neglected if $abc \leq 0.01$, and so the expanded term reduces to $2.303abc$. This is a Taylor expansion series.

A simple fluorometer design is illustrated in Figure 13.25. An ultraviolet source is required. Most fluorescing molecules absorb ultraviolet radiation over a band of wavelengths, and so a simple line source is sufficient for many applications. Such a source is a mercury vapor lamp. A spark is passed through mercury vapor at low pressure, and principal lines are emitted at 2537, 3650, 5200 (green), 5800 (yellow), and 7800 (red) Å. **Wavelengths shorter than 3000 Å are harmful to the eyes,** and one must never look directly at a short-wavelength UV source. The mercury vapor itself absorbs most of the 2537-Å radiation (self-absorption), and a blue filter in the envelope of the lamp may be added to remove most of the visible light. The 3650-Å line is thus the one used primarily for the activation. A high-pressure xenon arc (a continuum source) is usually used as the source in more sophisticated instruments that will scan the spectrum (spectrofluorometers). The lamp pressure is 7 atm at 25°C and 35 atm at operating temperatures. Take care!

In the simple instrument in Figure 13.25, a primary filter (Filter 1) is used to filter out the wavelengths close to the wavelength of the emission because, in practice, some radiation is scattered. The primary filter allows the passage of only the wavelength of excitation. The secondary filter (Filter 2) passes the wavelength of emission but not the wavelength of excitation (which may be scattered). Glass will pass appreciable amounts of the 3650-Å line, and so some instruments employ glass cuvets and filters. However, it is better to use quartz (special nonfluorescing grades are available). This simple setup is satisfactory for many purposes.

We can see why fluorometric methods are so sensitive if we compare them with absorption spectrometry. In absorption methods, the difference between two finite signals, P_0 and P, is measured. The sensitivity is then governed by the ability to distinguish between these two, which is dependent on the stability of the instrument, among other factors. In fluorescence, however, we measure the difference between zero and a finite number, and so, in principle, the limit of detection is governed by the intensity of the source and the sensitivity and stability of the detector (the "shot noise"). Also, in fluorescence, the signal depends linearly on concentration, and a much wider dynamic range of concentration can be measured; a dynamic range of 10^3 to 10^4 is not uncommon. In addition to the enhanced sen-

FIGURE 13.25
A simple fluorometer design.

sitivity, much wider ranges of concentrations can be measured; a 1000-fold or greater range is not uncommon.

In a **spectrofluorometer,** the measurement is also made at right angles to the direction of the incident radiation. But instead of using filters, the instrument incorporates two monochromators, one to select the wavelength of excitation and one to select the wavelength of fluorescence. The wavelength of excitation from a continuum source can be scanned and the fluorescence measured at a set wavelength to give a spectrum of the excitation wavelengths. This allows the establishment of the wavelength of maximum excitation. Then, by setting the excitation wavelength for maximum excitation, the emission wavelength can be scanned to establish the wavelength of maximum emission. When this spectrum is scanned, there is usually a "scatter peak" corresponding to the wavelength of excitation.

In spectrofluorometers, it is difficult to correct for variations in intensity from the source or response of the detector at different wavelengths, and calibration curves are generally prepared under a given set of conditions. Since the source intensity or detector response may vary from day to day, the instrument is usually calibrated by measuring the fluorescence of a standard solution and adjusting the gain to bring the instrument reading to the same value. A dilute solution of quinine in dilute sulfuric acid is usually used as the calibrating standard.

Sometimes it is desirable to obtain "absolute" spectra of a fluorescing compound to calculate quantum efficiencies for different transitions. This would require point-by-point correction for variations in the recorded signal due to variations in the instrumental parameters. Commercial instruments are available that will provide "corrected spectra." These adjust for variations in the source intensity with wavelength, so the sample is irradiated with constant energy, and they also correct for variations of the detector response. The recorded emission spectrum is presented directly in quanta of photons emitted per unit bandwidth.

QUESTIONS

Absorption of Radiation

1. Describe the absorption phenomena taking place in the far-infrared, near-infrared, and visible-ultraviolet regions of the spectrum.
2. What types of electrons in a molecule are generally involved in the absorption of UV or visible radiation?
3. What are the most frequent electronic transitions during absorption of electromagnetic radiation? Which results in more intense absorption? Give examples of compounds that exhibit each.
4. What is a necessary criterion for absorption to occur in the infrared region?
5. What types of molecular vibration are associated with infrared absorption?
6. Define the following terms: chromophore, auxochrome, bathocromic shift, hypsochromic shift, hyperchromism, and hypochromism.
7. Which of the following pairs of compounds is likely to absorb radiation at the longer wavelength and with greater intensity?

(a)

$$CH_3CH_2CO_2H \quad \text{or} \quad CH_2\!\!=\!\!CHCO_2H$$

(b)

$$CH_3CH\!\!=\!\!CHCH\!\!=\!\!CHCH_3 \quad \text{or} \quad CH_3C \equiv C - C \equiv CCH_3$$

(c)

or

8. In the following pairs of compounds, describe whether there should be an increase in the wavelength of maximum absorption and whether there should be an increase in absorption intensity in going from the first compound to the second:

(a)

(b)

(c)

9. Why do acid–base indicators change color in going from acid to alkaline solution?

10. What are the mechanisms by which a metal complex can absorb radiation?

Quantitative Relationships

11. Define absorption, absorbance, percent transmittance, and transmittance.

12. Define absorptivity and molar absorptivity.

13. Why is a calibration curve likely to be linear over a wider range of concentrations at the wavelength of maximum absorption compared to a wavelength on a shoulder of the absorption curve?

14. List some solvents that can be used in the ultraviolet, visible, and infrared regions, respectively. Give any wavelength restrictions.

15. What is an isosbestic point?

16. Describe and compare different causes for deviations from Beer's law. Distinguish between real and apparent deviations.

Instrumentation

17. Describe radiation sources and detectors for the ultraviolet, visible, and infrared regions of the spectrum.

18. Distinguish between the two types of monochromators (light dispersers) used in spectrophotometers and list the advantages and disadvantages of each.

19. Discuss the effect of the slit width on the resolution of a spectrophotometer and the adherence to Beer's law. Compare it with the spectral slit width.

20. Compare the operations of a single-beam spectrophotometer, a double-beam spectrophotometer, and a high-absorbance spectrophotometer.

21. Referring to Figure 13.22, what would be the color of an acid solution and an alkaline solution at maximum absorption? What color filter would be most applicable for the analysis of each in a filter colorimeter? (A filter replaces the prism and slit arrangement).

Fluorescence

22. Describe the principles of fluorescence. Why is fluorescence generally more sensitive than absorption measurements?
23. Under what conditions is fluorescence intensity proportional to concentration?
24. Describe the instrumentation required for fluorescence analysis. What is a primary filter? A secondary filter?
25. Suggest an experiment by which you could determine iodide ion by fluorescence.

PROBLEMS

Wavelength/Frequency/Energy

26. Express the wavelength 2500 Å in micrometers and nanometers.
27. Convert the wavelength 4000 Å into frequency (Hz) and into wavenumbers (cm^{-1}).
28. The most widely used wavelength region for infrared analysis is about 2 to 15 μm. Express this range in angstroms and in wavenumbers.
29. One mole of photons (Avogadro's number of photons) is called an *einstein* of radiation. Calculate the energy, in calories, of one einstein of radiation at 3000 Å.

Beer's Law

30. Several spectrophotometers have scales that are read either in absorbance or in percent transmittance. What would be the absorbance reading at 20% T? At 80% T? What would the transmittance reading be at 0.25 absorbance? At 1.00 absorbance?
31. A 20-ppm solution of a DNA molecule (unknown molecular weight) isolated from *Escherichia coli* was found to give an absorbance of 0.80 in a 2-cm cell. Calculate the absorptivity of the molecule.
32. A compound of formula weight 280 absorbed 65.0% of the radiation at a certain wavelength in a 2-cm cell at a concentration of 15.0 μg/mL. Calculate its molar absorptivity at that wavelength.
33. Titanium is reacted with hydrogen peroxide in 1 M sulfuric acid to form a colored complex. If a 2.00×10^{-5} M solution absorbs 31.5% of the radiation at 415 nm, what would be (a) the absorbance and (b) the transmittance and percent absorption for a 6.00×10^{-5} M solution?
34. A compound of formula weight 180 has an absorptivity of 286 cm^{-1} g^{-1} L. What is its molar absorptivity?
35. Aniline, $C_6H_5NH_2$, when reacted with picric acid gives a derivative with a molar absorptivity of 134 cm^{-1} g^{-1} L at 359 nm. What would be the absorbance of a 1.00×10^{-4} M solution of reacted aniline in a 1.00-cm cell?

Quantitative Measurements

36. The drug tolbutamine (f. wt. = 270) has a molar absorptivity of 703 at 262 nm. One tablet is dissolved in water and diluted to a volume of 2 L. If the solution exhibits an absorbance in the UV region at 262 nm equal to 0.687 in a 1-cm cell, how many grams tolbutamine are contained in the tablet?
37. Amines (weak bases) form salts with picric acid (trinitrophenol), and all amine picrates exhibit an absorption maximum at 359 nm with a molar absorptivity of 1.25×10^4. A 0.200-g sample of aniline, $C_6H_5NH_2$, is dissolved in 500 mL water. A 25.0-mL aliquot

is reacted with picric acid in a 250-mL volumetric flask and diluted to volume. A 10.0-mL aliquot of this is diluted to 100 mL and the absorbance read at 359 nm in a 1-cm cell. If the absorbance is 0.425, what is the percent purity of the aniline?

38. Phosphorus in urine can be determined by treating with molybdenum(VI) and then reducing the phosphomolybdo complex with aminonaphtholsulfonic acid to give the characteristic molybdenum blue color. This absorbs at 690 nm. A patient excreted 1270 mL urine in 24 hours, and the pH of the urine was 6.5. A 1.00-mL aliquot of the urine was treated with molybdate reagent and aminonaphtholsulfonic acid and was diluted to a volume of 50.0 mL. A series of phosphate standards was similarly treated. The absorbance of the solutions at 690 nm, measured against a blank, were as follows:

Solution	Absorbance
1.00 ppm P	0.205
2.00 ppm P	0.410
3.00 ppm P	0.615
4.00 ppm P	0.820
Urine sample	0.625

(a) Calculate the number of grams phosphorus excreted per day.
(b) Calculate the phosphate concentration in the urine as millimoles per liter.
(c) Calculate the ratio of HPO_4^{2-} to $H_2PO_4^-$ in the sample:

$$K_1 = 1.1 \times 10^{-2}; \quad K_2 = 7.5 \times 10^{-8}; \quad K_3 = 4.8 \times 10^{-13}$$

39. Iron(II) is determined spectrophotometrically by reacting with 1,10-phenanthroline to produce a complex that absorbs strongly at 510 nm. A stock standard iron(II) solution is prepared by dissolving 0.0702 g ferrous ammonium sulfate, $Fe(NH_4)_2SO_4 \cdot 6H_2O$, in water in a 1-L volumetric flask, adding 2.5 mL H_2SO_4, and diluting to volume. A series of working standards is prepared by transferring 1.00-, 2.00-, 5.00-, and 10.00-mL aliquots of the stock solution to separate 100-mL volumetric flasks and adding hydroxylammonium chloride solution to reduce any iron(III) to iron(II), followed by phenanthroline solution and then dilution to volume with water. A sample is added to a 100-mL volumetric flask and treated in the same way. A blank is prepared by adding the same amount of reagents to a 100-mL volumetric flask and diluting to volume. If the following absorbance readings measured against the blank are obtained at 510 nm, how many milligrams iron are in the sample?

Solution	A
Standard 1	0.081
Standard 2	0.171
Standard 3	0.432
Standard 4	0.857
Sample	0.463

40. Nitrate nitrogen in water is determined by reacting with phenoldisulfonic acid to give a yellow color with an absorption maximum at 410 nm. A 100-mL sample that has been stabilized by adding 0.8 mL H_2SO_4/L is treated with silver sulfate to precipitate chloride ion, which interferes. The precipitate is filtered and washed (washings added to filtered sample). The sample solution is adjusted to pH 7 with dilute NaOH and evaporated just to dryness. The residue is treated with 2.0 mL phenoldisulfonic acid solution and heated in a hot water bath to aid dissolution. Twenty milliliters distilled water and 6 mL ammonia are added to develop the maximum color, and the clear solution is transferred to a 50-mL volumetric flask and diluted to volume with distilled water. A blank is

prepared using the same volume of reagents, starting with the disulfonic acid step. A standard nitrate solution is prepared by dissolving 0.722 g anhydrous KNO_3 and diluting to 1 L. A standard addition calibration is performed by spiking a separate 100-mL portion of sample with 1.00 mL of the standard solution and carrying through the entire procedure. The following absorbance readings were obtained: blank, 0.032; sample, 0.270; sample plus standard, 0.854. What is the concentration of nitrate nitrogen in the sample in parts per million?

41. Two colorless species, A and B, react to form a colored complex AB that absorbs at 550 nm with a molar absorptivity of 450. The dissociation constant for the complex is 6.00×10^{-4}. What would the absorbance of a solution, prepared by mixing equal volumes of 0.0100 M solutions of A and B in a 1.00-cm cell, be at 550 nm?

Mixtures

42. Compounds A and B absorb in the ultraviolet region. A exhibits an absorption maximum at 267 nm ($a = 157$) and a tailing shoulder at 312 nm ($a = 12.6$). B has an absorption maximum at 312 nm ($a = 186$) and does not absorb at 267 nm. A solution containing the two compounds exhibits absorbances (using a 1-cm cell) of 0.726 and 0.544 at 267 and 312 nm, respectively. What are the concentrations of A and B in mg/L?

43. Titanium(IV) and vanadium(V) form colored complexes when treated with hydrogen peroxide in 1 M sulfuric acid. The titanium complex has an absorption maximum at 415 nm, and the vanadium complex has an absorption maximum at 455 nm. A 1.00×10^{-3} M solution of the titanium complex exhibits an absorbance of 0.805 at 415 nm and of 0.465 at 455 nm, while a 1.00×10^{-2} M solution of the vanadium complex exhibits absorbances of 0.400 and 0.600 at 415 and 455 nm, respectively. A 1.000-g sample of an alloy containing titanium and vanadium was dissolved, treated with excess hydrogen peroxide, and diluted to a final volume of 100 mL. The absorbance of the solution was 0.685 at 415 nm and 0.513 at 455 nm. What were the percentages of titanium and vanadium in the alloy?

Fluorescence

44. Derive Equation 13.18 relating fluorescence intensity to concentration.

RECOMMENDED REFERENCES

General

1. E. Olsen, *Modern Optical Methods of Analysis*. New York: McGraw-Hill, 1975.

2. D. F. Swinehart, "The Beer-Lambert Law," *J. Chem. Educ.*, **39,** 333 (1962).

3. J. D. Winfordner, *Spectrochemical Methods of Analysis*. New York: Wiley-Interscience, 1971.

Colorimetry/Spectrophotometry

4. D. F. Boltz and J. A. Howell, *Colorimetric Determination of Nonmetals*. New York: Wiley-Interscience, 1978.

5. R. T. Conley, *Infrared Spectroscopy*, 2nd ed. Boston: Allyn and Bacon, 1972.

6. C. H. Morrison and H. Freiser, *Solvent Extraction in Analytical Chemistry*. New York: John Wiley & Sons, 1975, pp. 189–247.

7. E. B. Sandell and H. Onishi, *Photometric Determination of Traces of Metals. General Aspects*. New York: Wiley-Interscience, 1978.

8. R. M. Silverstein, G. C. Bassler, and T. C. Morrillo, *Spectrometric Identification of Organic Compounds*, 3rd ed. New York: John Wiley & Sons, 1974.

Catalogued Spectra

9. *Catalogue of Infrared Spectral Data.* Washington, D.C.: American Petroleum Institute, Research Project 44. Multivolume series started in 1943 and continuing to the present date.

10. *Catalogue of Ultraviolet Spectral Data.* Washington, D.C.: American Petroleum Institute, Research Project 44. Multivolume series started in 1945 and continuing to the present date.

11. "Infrared Prism Spectra," in *The Sadtler Standard Spectra,* Vols. **1–36;** "Standard Infrared Grating Spectra," in *The Sadtler Standard Spectra,* Vols. **1–16.** Philadelphia: Sadtler Research Laboratories.

12. L. Lang, ed., *Absorption Spectra in the Ultraviolet and Visible Regions,* Vols. **1–23.** New York: Academic Press, 1961–1979.

13. *U. V. Atlas of Organic Compounds,* Vols. **1–4.** New York: Plenum Press.

14. "Ultraviolet Spectra," in *The Sadtler Standard Spectra,* Vol. **1–62.** Philadelphia: Sadtler Research Laboratories. A comprehensive catalog of ultraviolet spectra of organic compounds.

Fluorometry

15. "Fluorometric Analysis," *Anal. Chem.,* Biannual Reviews, alternate years.

16. G. G. Guilbault, *Practical Fluorescence.* New York: Marcel Dekker, 1973.

17. S. Udenfriend, *Fluorescence Assay in Biology and Medicine,* 2 volumes. New York: Academic Press, 1962, 1969.

18. E. L. Wehry, ed., *Modern Fluorescence Spectroscopy,* Vols. **1–4.** New York: Plenum Publishing Corp., 1976–1981.

19. C. E. White and R. J. Argauer, *Fluorescence Analysis: A Practical Approach.* New York: Marcel Dekker, 1970.

chapter 14

ATOMIC SPECTROMETRIC METHODS

Chapter 13 dealt with the spectrometric determination of substances in solution, that is, the absorption of energy by molecules. This chapter deals with the spectroscopy of atoms. Since atoms cannot rotate or vibrate as a molecule does, only electronic transitions can take place when energy is absorbed. Because the transitions are discrete (quantized), line spectra are observed. There are various ways to obtain free atoms and to measure the absorption or emission of radiation by these. The principal techniques described in this chapter are emission spectroscopy, in which atoms are excited in an electron arc or spark; flame emission spectrometry, in which they are excited by a flame; and atomic absorption spectrophotometry, in which the amount of radiation absorbed by ground-state atoms in a flame is measured. Included is a discussion of types of flames, interferences in flames, and nonflame, or electrothermal, atomizers for extremely sensitive atomic absorption measurements.

14.1 EMISSION SPECTROSCOPY

This is a well-established and extremely useful technique for determining inorganic constituents in various types of samples. The most common use is the direct analysis of solid samples. A diagram of a typical **spectrograph** is shown in Figure 14.1. A spectrograph utilizes photographic detection.

Basically, the powdered sample is placed in a graphite electrode (shaped in the form of a cup) and a high-voltage arc or spark is struck between this and a counterelectrode. This causes the sample to vaporize, forming atomic vapor of the elements present, and the individual elements are raised to excite electronic energy

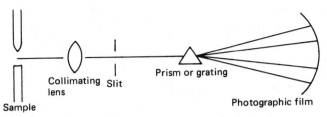

FIGURE 14.1
Schematic diagram of a spectrograph.

levels. The excited levels have a short lifetime, and they drop back to the ground state, emitting a photon[1] with energy equal to $h\nu$. Since there are many possible electronic transitions, several discrete wavelengths of light are emitted by each element. The emitted light is passed through a slit into a dispersing element, either a prism or a grating, where it is separated into the individual wavelengths. A grating is usually used because its linear dispersion allows much easier interpretation of the spectra. Since there are so many wavelengths, it is most convenient to use as the detector a photographic film or plate. Each emitted wavelength registers as a dark line on the film when it is developed (the equivalent of a photographic "negative").

A typical emission spectrogram (the "positive" of it) is shown in Figure 14.2. You can see here the large number of lines emitted by each element. The intensity of the lines will depend on the time of exposure, the concentration of the element in the matrix, the sensitivity of the photographic emulsion, the grain size of the emulsion, conditions of developing the film, and so forth. The emulsion must be sensitive to the wavelengths emitted, and most of these occur in the ultraviolet region. Because of difficulties in standardizing the matrix (matrix effects can alter the intensity) and in preparing standards, and because of selective volatalization of one element over another and several other factors, it is relatively time consuming to calibrate this technique for quantitative analysis. However, once the procedure

FIGURE 14.2
Typical emission spectrogram. The bottom spectrum is that of copper and the second one up is an iron spectrum. The remainder are spectra from ore samples. This figure is printed as a "positive" and so the lines appear white on a dark background.

---

[1]This is in opposition to excited molecules in solution, where there is much greater probability for collisions with solvent and other molecules. In the gaseous state, there is less probability for collision and, therefore, many of the atoms lose their energy of excitation as electromagnetic radiation rather than as heat.

is calibrated, many samples can be run routinely for a large number of elements simultaneously. Consequently, emission spectroscopy is used widely in industry for quantitative analysis. Because of the importance of careful attention to the details mentioned above, you are referred to a book devoted specifically to emission spectroscopy for a thorough discussion of quantitative analysis (see Reference 5 at the end of the chapter).

Once a particular method is standardized, the density of the lines, using one line for each element, is measured with a **densitometer.** This is nothing more than a spectrophotometer that measures the amount of light "absorbed" by the line. The density of the line is then related to the concentration of the metal in the sample.

For nonroutine analysis, emission spectroscopy finds its widest use in qualitative analysis. In this mode, it is relatively simple to operate. Several samples can be run on one photographic film or plate. Following each exposure, the film is moved up and then the next sample is run, being exposed just below the previous one. In order that the photographic film or plate may be calibrated with respect to wavelength and resolution, a spectrum of a standard material, usually iron, is always run with each set of samples. The analyst, by knowing the wavelength of certain iron lines that are easily picked out, can then line up the sample spectra with a known spectrum of the element being tested for. This is usually done in a densitometer, which projects and magnifies the film image on a screen. In this manner, the overlapping of certain lines with those of the suspected element (or the lack of the lines) can be noted and thus confirm the presence or absence of the element. Since emission spectra are complicated, it is likely that lines of certain elements will overlap, and so it is best to confirm at least three of the major (darkest) lines of the suspected element.

When using emission spectroscopy, the eyes should always be protected, because of harmful ultraviolet radiation emitted. Avoid looking directly at the electrodes. Also, take precaution not to touch the electrodes.

14.2 FLAME EMISSION SPECTROMETRY

A technique familiar to clinical chemists is flame emission spectrometry (formerly called **flame photometry**).[2] It is in principle similar to emission spectroscopy. However, the source of excitation energy is a flame. This is a much lower-energy source, and so the emission spectrum is much simpler, as there are fewer lines. Also, the sample is introduced into the flame in the form of a solution, and so the technique is very easy to quantify.

There are numerous types of aspirator burners used. Basically, the solution is introduced into the flame as a fine spray. The mechanism of obtaining atomic vapor is complex, but an attempt at explaining the basic processes is illustrated in Figure 14.3. The solvent evaporates, leaving the dehydrated salt. The salt is dissociated into free gaseous atoms in the ground state. A certain fraction of these atoms can absorb energy from the flame and be raised to an excited electronic state. These excited atoms, on returning to the ground state, emit photons of characteristic wavelength. These can be detected with a conventional monochromator–detector setup. Because relatively few lines are emitted, it is not always necessary to

---

[2]See Chapter 13 for the distinction between spectrometry and spectrophotometry.

FIGURE 14.3
Processes occurring in the flame.

have a high-resolution monochromator, and even simple interference filters may suffice (e.g., for the alkali metals). But because the line must be resolved from the flame background, which may be significant, it is desirable to have a fairly good monochromator.

The intensity of emission is directly proportional to the concentration of the analyte in the solution being aspirated. So a calibration curve of emission intensity as a function of concentration is prepared.

As indicated in the figure, side reactions in the flame may decrease the population of free atoms and hence the emission signal. These will be discussed under atomic absorption below.

In the early years of flame photometry, only relatively cool flames were used. We shall see below that only a small fraction of most elements is excited by flames and that the fraction excited increases as the temperature is increased. Consequently, relatively few elements have been determined routinely by flame emission spectrometry, especially few of those that emit line spectra (several can exist in flames as molecular species, particularly as oxides, which emit molecular band spectra). Only the alkali metals sodium and potassium are routinely determined by flame emission spectrometry in the clinical laboratory. However, with flames such as oxyacetylene and nitrous oxide–acetylene, over 60 elements can now be determined by flame emission spectrometry. The is in spite of the fact that a small fraction of excited atoms is available for emission. Good sensitivity is achieved because, as with fluorescence (Chapter 13), we are, in principle, measuring the difference between zero and a small but finite signal, and so the sensitivity is limited by the sensitivity and stability of the detector and the stability (noise level) of the flame aspiration system.

14.3   PLASMA EMISSION SPECTROMETRY

The use of plasma as excitation sources for atomic emission has become very important in recent years. Inductively coupled plasma (ICP) became commercially available in 1974.

The ICP discharge is caused by the effect of a radio-frequency field on a flowing gas. Argon gas flows upward through a quartz tube, around which is wrapped a copper coil or solenoid. The coil is energized by a radio-frequency generator operating at 5 to 75 MHz and 1 to 2 kW power, creating a changing magnetic field in the flowing gas inside the coil. This induces a circulating eddy current in a conductor (the gas), which, in turn, heats it. Argon is not a conductor at room temperature, but it can be made electrically conducting by heating it. To initiate the ICP discharge, a discharge from a Telsa coil or a pilot spark is applied to the flowing argon. The argon is quickly heated, with a stable plasma being produced having a core temperature of about 9000 to 10,000 K.

This type of excitation source has a number of advantages. First, samples can be introduced in solution form through a spray chamber nebulizer, as in flame emission spectrometry. This makes quantitative analysis and sample handling much easier than in conventional emission spectroscopy. Modern-day instruments replace the photographic film detector in an emission spectrograph with a series of photomultiplier tubes set at the wavelengths for specific elements (e.g., 40 elements), and so quantitative multielement analysis of solutions becomes routine. The high temperature of the plasma eliminates many chemical interferences present in a flame (see below), and most elements are readily excited. The plasma is well suited for the refractory (oxide-forming) elements such as boron, phosphorus, uranium, and tungsten and for difficult-to-excite elements such as zinc and cadmium. Detection limits are very competitive with those using atomic absorption or flame emission spectrometry.

14.4 DISTRIBUTION BETWEEN GROUND AND EXCITED STATES

The relative populations of ground-state (N_0) and excited-state (N_e) populations at a given flame temperature can be estimated from the **Maxwell–Boltzmann expression:**

$$\frac{N_e}{N_0} = \frac{g_e}{g_0} e^{-(E_e - E_0)/kT} \tag{14.1}$$

where g_e and g_0 are the *statistical weights* of the excited and ground states, respectively; E_e and E_0 are the energies of the two states ($= h\nu$; E_0 is usually zero); k is the Boltzmann constant (1.3805×10^{-16} erg K^{-1}); and T is the absolute temperature. The statistical weights represent the probability that an electron will reside in a given energy level, and they are available from quantum mechanical calculations.[3] See Problem 21 for an example calculation.

Table 14.1 summarizes the relative population ratios for a few elements at 2000 and 3000 K. We see that even for a relatively easily excited element such as sodium, the excited-state population is small. Short-wavelength elements (higher energy, $h\nu$) require much more energy for excitation and exhibit poor sensitivity by flame emission spectrometry where temperatures rarely exceed 3000 K. Those with long-wavelength emissions will exhibit better sensitivity. Measurement of ground-state atoms, as in atomic absorption below, will be less dependent on the wavelength or element. We see also from Table 14.1 that the fraction of excited-state

---

[3]The statistical weight is given by $2J + 1$, where J is the Russel–Saunders coupling and is equal to $L + S$ or $L - S$; L is the total orbital angular momentum quantum number, represented by the sharp (S), principal (P), diffuse (D), and fundamental (F) series ($L = 0, 1, 2,$ and 3, respectively); and S is spin ($\pm\frac{1}{2}$). The information is supplied in the form of term symbols, $N^M L_J$, where N is the principal quantum number and M is the multiplicity. For example, the transition for the sodium 589.0-nm line omitting the principal quantum number N, is $^2S_{1/2} - {}^2P_{1/2}$, and $g_e/g_0 = [2(\frac{1}{2}) + 1]/[2(\frac{1}{2}) + 1] = 2/2 = 1$.

TABLE 14.1
Values of N_e/N_0 for Different Resonance Lines

Line, nm	N_e/N_0	
	2000 K	3000 K
Na 589.0	9.9×10^{-6}	5.9×10^{-4}
Ca 422.7	1.2×10^{-7}	3.7×10^{-5}
Zn 213.8	7.3×10^{-15}	5.4×10^{-10}

atoms is very temperature dependent, whereas the fraction in the ground state is virtually constant (since nearly 100% reside there at all temperatures).

In flame emission methods, we measure the excited-state population; and in atomic absorption methods (below), we measure the ground-state population. Because of chemical reactions that occur in the flame, differences in flame emission and atomic absorption sensitivities above 300 nm are, in practice, not as great as one would predict from the Boltzmann distribution. For example, many elements react partially with flame gases to form metal oxide or hydroxide species, and this reaction detracts from the atomic population equally in either method and is equally temperature dependent in either.

14.5   ATOMIC ABSORPTION SPECTROPHOTOMETRY

A technique closely related to flame emission spectrometry is atomic absorption spectrophotometry, because they each use a flame as the atomizer. We discuss here the factors affecting absorption; and because of the close relationship of atomic absorption and flame photometry, we shall make comparisons between the two techniques where appropriate.

Principles

The sample solution is aspirated into a flame as in flame emission spectrometry, and the sample element is converted to atomic vapor. The flame then contains atoms of that element. Some are thermally excited by the flame, but most remain in the ground state. These ground-state atoms can absorb radiation given off by a special source made from that element (see "Sources"). The wavelengths of radiation given off by the source are the same as those absorbed by the atoms in the flame.

Atomic absorption spectrophotometry is identical in principle to absorption spectrophotometry described in the previous chapter. The absorption follows Beer's law. That is, the *absorbance* is directly proportional to the path length in the flame and to the concentration of atomic vapor in the flame. Both of these variables are difficult to determine, but the path length can be held constant and the concentration of atomic vapor is directly proportional to the concentration of the analyte in the solution being aspirated. The procedure used is to prepare a calibration curve of concentration in the solution versus absorbance.

The major disadvantage of making measurements by atomic absorption, as we shall see below, is that a different source is required for each element.

Instrumentation

As in regular absorption spectrophotometry, the requirements for atomic absorption spectrophotometry are a light source, a cell (the flame), a monochromator, and a detector. The flame is placed between the source and the monochromator. A schematic diagram of an atomic absorption spectrophotometer is shown in Figure 14.4. This is for a double-beam instrument that measures the ratio of P_0/P. The source beam is alternately sent through the flame and around the flame by the chopper. The detector measures these alternately and the logarithm of the ratio is displayed. The detector amplifier is tuned to receive only radiation modulated at the frequency of the chopper, and so dc radiation emitted by the flame is discriminated against.

The various components of an atomic absorption spectrophotometer are described as follows.

1. SOURCES.   A sharp-line source is required in atomic absorption because the width of the *absorption line* is very narrow, a few hundredths to one-tenth of an angstrom, at most. Because the absorption line is so narrow, only a small fraction of the radiation passed by the slit and reaching the detector from a continuum source would be absorbed.

The source used almost exclusively is a **hollow-cathode lamp.** This is a sharp-line source that emits specific (essentially monochromatic) wavelengths, and the basic construction is illustrated in Figure 14.5. It consists of a cylindrical hollow cathode made of the element to be determined or an alloy of it, and a tungsten anode. These are enclosed in a glass tube usually with a quartz window, since the lines of interest are often in the ultraviolet region. The tube is under reduced pressure and filled with an inert gas such as argon or neon. A high voltage is impressed across the electrodes, causing the gas atoms to be ionized at the anode. These positive ions are accelerated toward the negative cathode. When they bombard the cathode, they cause some of the metal to "sputter" and become vaporized. The vaporized metal is excited to higher electronic levels by continued collision with the high-energy gas ions. When the electrons return to the ground state, the characteristic lines of the element are emitted. Also emitted are lines of the filler gas, but these are not usually close enough to the element lines to interfere.

These lines are passed through the flame, and certain ones become absorbed by the test element because they possess just the right energy (the right wavelength) to result in the discrete electronic transitions. The most strongly absorbed line is often, but not always, the one corresponding to the electronic transition from the ground state to the lowest excited state. This is called the **resonance line.** The lines

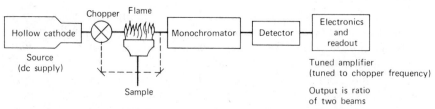

FIGURE 14.4
Schematic diagram of an atomic absorption instrument. From G. D. Christian and F. J. Feldman. *Atomic Absorption Spectroscopy. Applications in Agriculture, Biology, and Medicine.* New York: Interscience, 1970 (Reproduced by permission of John Wiley & Sons, Inc.)

FIGURE 14.5
Design of a hollow-cathode lamp.

from a hollow-cathode lamp are narrower than the absorption line of the element in the flame because of broadening of the absorption line at the higher temperature and pressure of the flame. So the entire source line width is absorbed. Greater specificity also results for the reason that, with a continuum source, an element with an absorption line falling anywhere within the spectral slit width would absorb part of the source radiation. However, a line source will not be absorbed so long as an element's absorption lines does not overlap with it. There are very few instances where line overlap does occur among lines of different elements.

It is sometimes possible to use an alloy of several elements for the hollow cathode, and with such lamps, the lines of all the elements are emitted. These are the so-called multielement hollow-cathode lamps and can be used as a source for usually two or three elements. They may exhibit shorter lifetimes than do single-element lamps due to selective volatilization ("distillation") of one of the elements from the cathode with condensation on the walls of the lamp.

2. BURNERS. There are two basic types of aspirator–burners that are used in atomic absorption. The first is the **total-consumption burner,** illustrated in Figure 14.6. The fuel and oxidant (support) gases are mixed and combust at the tip of the burner. The sample is drawn up into the flame by the "Venturi effect," by the support gas. The gas creates a partial vacuum above the capillary barrel, causing the sample to be forced up the capillary. It is broken into a fine spray at the tip where the gases are turbulently mixed and burned. This is the usual process of "nebulization."[4] The burner is called total consumption because the entire aspirated sample enters the flame.

The second type of burner, used in most commercial instruments, is the **premix chamber burner,** sometimes called the **laminar-flow burner.** This is illustrated in Figure 14.7. The fuel and support gases are mixed in a chamber before they enter the burner head (through a slot) where they combust. The sample solution is again aspirated through a capillary by the "Venturi effect" using the support gas for the aspiration. Large droplets of the sample condense and drain out of the chamber. The remaining fine droplets mix with the gases and enter the flame. As much as 90% of the droplets condense out, leaving only 10% to enter the flame.

There are advantages and disadvantages to each type of burner. The total-con-

---

[4]In atomic absorption spectrophotometry, we often speak of "atomization" in referring to the process of obtaining *atomic vapor.* This is not to be confused with the above process.

FIGURE 14.6
Total-consumption burner: *A*, oxidant flow; *B*, surface mixing; *C*, fuel flow. From G. D. Christian and
F. J. Feldman. *Atomic Absorption Spectroscopy. Applications in Agriculture, Biology, and Medicine.*
New York: Interscience, 1970 (Reproduced by permission of John Wiley & Sons, Inc.)

FIGURE 14.7
Premix burner. (a) Nebulizer, chamber, and burner. (b) Burner head. From G. D. Christian and F. J.
Feldman. *Atomic Absorption Spectroscopy. Applications in Agriculture, Biology, and Medicine.* New
York: Interscience, 1970. (Reproduced by permission of John Wiley & Sons, Inc.)

sumption burner obviously uses the entire aspirated sample, but it has a shorter path length and many larger droplets are not vaporized in the flame. The larger droplets may vaporize partially, leaving solid particles in the light path. This may result in light scattering, which is registered as an absorbance. The absorbance by the sample, that is, the atomic vapor population, is generally more dependent on the gas flow rates and the height of observation in the flame than with the premix burners. The viscosity of the sample will more greatly affect the atomization efficiency (production of atomic vapor) in the total-consumption burner. However, the total-consumption burner can be used to aspirate viscous and "high solids" samples with more ease, such as undiluted serum and urine. Also, this burner can be used for most types of flames, both low- and high-burning velocity flames (see below).

The premix burners are generally limited to relatively low-burning velocity flames. Although a large portion of the aspirated sample is lost in the premix burner, the "atomization efficiency" (efficiency of producing atomic vapor) of that portion of the sample that enters the flame is greater, because the droplets are finer. Also, the path length is longer. So sensitivities are comparable with either burner in most cases. Combustion with the premix burners is very quiet, while with total-consumption burners it is audibly noisy. Most commercial instruments use premix burners with the option of using total-consumption burners. A popular version of the premix burner is the "Boling" burner. This is a three-slot burner head that results in a broader flame and less distortion of the radiation passing through at the edges of the flame (see Figure 14.7). This burner warps more easily than others, though, and care must be taken not to overheat it when using organic solvents.

Varieties of these burners have been used for flame photometers. Advances in flame photometry with different flames have been achieved with the same burners and instrumentation as with atomic absorption.

3. FLAMES. The chief flames that are used for atomic absorption and emission spectrometry are listed in Table 14.2 together with their maximum burning temperatures. The most widely used flames for atomic absorption are the air–acetylene flame and the nitrous oxide–acetylene flame with premix burners. The latter high-temperature flame is not required and may even be detrimental for many cases in atomic absorption because it will cause ionization of the gaseous atoms (see be-

TABLE 14.2
Burning Temperatures and Velocities of Commonly Used Flames[a]

Flame Mixture	Maximum Flame Speed, cm/s	Maximum Temperature, °C
Hydrogen–oxygen	—	2677
Hydrogen–air	—	2045
Propane–air	—	1725
Propane–oxygen	—	2900
Acetylene–air	160	2250
Acetylene–oxygen	1130	3060
Acetylene–nitrous oxide	180	2955
Hydrogen–argon-entrained air	—	1577

[a]From G. D. Christian and F. J. Feldman, *Atomic Absorption Spectroscopy. Applications in Agriculture, Biology, and Medicine.* New York: Wiley-Interscience, 1970. Reproduced by permission of John Wiley & Sons, Inc.

low). However, it is very useful for those elements that tend to form heat-stable oxides in the air–acetylene flame (the "refractory elements"). The air–acetylene and other hydrocarbon flames absorb a large fraction of the radiation at wavelengths below 200 nm, and an argon–hydrogen-entrained air flame is preferred for this region of the spectrum for maximum detectability. This is a colorless flame, and entrained air is the actual oxidant gas. It is used for elements such as arsenic (193.5 nm) and selenium (197.0 nm), when they are separated from the sample solution by volatilization as their hydrides (AsH_3, H_2Se) and passage of these gases into the flame. This is necessary because this cool flame is more subject to chemical interferences than other flames (see the following paragraphs). A nitrous oxide–acetylene flame offers an advantage in this region of the spectrum when danger of interference exists; the flame absorption is relatively small at short wavelengths.

In flame emission spectrometry, a hot flame is required for the analysis of a large number of elements, and either the oxygen–acetylene ("oxyacetylene") flame or the nitrous oxide–acetylene flame is used. The oxyacetylene flame has a high burning velocity and cannot be used with a conventional premix burner. The nitrous oxide–acetylene flame can, however, be used with a premix burner. Because of its high temperatures, a special, thick, stainless steel burner head must be used to prevent it from melting. A "cool" air–propane or similar flame is preferred for the flame emission spectrometry of the easily excited elements sodium and potassium because of decreased ionization of these elements.

Interferences

These fall under three classes, spectral, chemical, and physical. We will discuss these briefly and point out their relative effects in emission and absorption measurements.

1. SPECTRAL INTERFERENCES. In emission analyses when either another emission line or a molecular emission band is close to the emitted line of the test element and is not resolved from it by the monochromator, spectral interference occurs. The most probable danger is from molecular emission, such as from oxides of other elements in the sample. Similar interferences would occur in atomic absorption if a dc instrument were used, but it is eliminated if an ac instrument is used. If, on the other hand, an element or molecule is capable of absorbing the source radiation, a positive interference would occur in atomic absorption. With line sources, this danger is minimized but not eliminated.

Light scatter or absorption by solid particles, unvaporized solvent droplets, or molecular species in the flame will cause a positive interference in atomic absorption spectrophotometry. This is especially a problem for wavelengths less than 300 nm, when solutions of high salt content are aspirated because the salt may not be completely desolvated or its molecules dissociated into atoms. Such **background absorption** can be corrected for by measuring the absorbance of a line that is close to the absorption line of the test element but that is not absorbed by the element itself. Since the interfering absorption occurs over a band of wavelengths, the absorbance will be essentially the same at several angstroms removed from the resonance line.

The measurement should be made at least two band passes (Chapter 13) away from the absorption line. The line used for correction can be a filler gas line from the hollow-cathode lamp or a "nonresonance" line of the element that is not absorbed, or a nearby line from a second hollow-cathode lamp can be used. A solution of the test element should always be aspirated to check that it does not absorb the background correction line. This technique requires two separate measurements on the sample.

A **background correction** for broadband absorption can also be made in the UV region (where most elements absorb and background absorption is most serious) with a hydrogen or deuterium continuum source. In the visible region, a tungsten continuum source may be used. The monochromator is set at the same wavelength as the resonance line. Sharp-line absorption of the continuum source by the test element is assumed negligible compared to that by the background over the bandwidth of the monochromator, so the absorbance of the continuum source can simply be subtracted from the absorbance of the resonance line from the hollow-cathode lamp. This is the basis of commercially available automatic background correctors. A mirror alternately passes the hollow-cathode radiation and the continuum radiation, the absorption of each is measured, and the continuum source absorbance is automatically subtracted from the hollow-cathode absorbance to obtain the net sharp-line absorbance by the test element. So only a single measurement is required.

2. IONIZATION INTERFERENCE. An appreciable fraction of alkali and alkaline earth elements and several other elements in hot flames may be ionized in the flame. Since we are measuring the un-ionized atoms, either emission or absorption signals will be decreased. This in itself is not necessarily serious, except that the sensitivity and linearity may be decreased. However, the presence of other easily ionized elements in the sample will add free electrons to the flame and suppress ionization of the test element. This will result in enhanced emission or absorption and a positive interference. Ionization interference can usually be overcome either by adding the same amount of the interfering element to the standard solutions, or more simply by adding large amounts to both the samples and the standards, to make the enhancement constant. Ionization can usually be detected by noting that the calibration curve has a positive deviation or curvature upward at higher concentrations, because a larger fraction of the atoms are ionized at lower concentrations.

3. REFRACTORY COMPOUND FORMATION. The sample solution may contain a chemical, usually an anion, that will form a refractory (heat-stable) compound with the test element in the flame. For example, phosphate will react with calcium ions and in the flame produce calcium pyrophosphate, $Ca_2P_2O_7$. This causes a reduction in the absorbance, since the calcium must be in the atomic form to absorb its resonance line. Generally, this type of solution interference can be reduced or eliminated chemically. In the above example, a high concentration (ca. 1%) of strontium chloride or lanthanum nitrate can be added to the solution. The strontium or lanthanum will preferentially combine with the phosphate and prevent its reaction with the calcium. Alternatively, a high concentration of EDTA can be added to the solution to form a chelate with the calcium. This prevents its reaction with phosphate, and the calcium–EDTA chelate is dissociated in the flame to give free calcium vapor. These types of interferences will occur with both atomic absorption and flame emission spectrometry. They may be eliminated also by using a high-temperature flame such as the nitrous oxide–acetylene flame.

A serious situation occurs when the analyte metal reacts with gases present in the flame. The refractory elements, such as aluminum, titanium, molybdenum, vanadium, and so on, will react with O and OH species in the flame to produce thermally stable metal oxides and hydroxides. These can be decomposed only by using high-temperature flames. Several of these elements exhibit no appreciable absorption or emission in the conventional air–acetylene flame. A more useful flame for these elements is the nitrous oxide–acetylene flame. It is usually used in the reducing (fuel-rich) condition, in which a large red-feather secondary-reaction zone is present. This red zone arises from the presence of CN, NH, and other highly reducing radicals. These (or the lack of oxygen-containing species), combined with the high temperature of the flame, decompose and/or prevent the formation of refractory oxides so that atomic vapor of the metal can be produced. The fuel-rich oxyacetylene flame is also useful for the flame emission spectrometry of monoxide-forming metals.

4. PHYSICAL INTERFERENCES.   Most parameters that affect the rate of sample uptake in the burner and the atomization efficiency can be considered physical interferences. This includes such things as variations in the gas flow rates, variation in sample viscosity due to temperature, or solvent variation, high solids content, changes in the flame temperature, and so forth. These can generally be accounted for by frequent calibration. Some instruments offer the capability of using internal standards that can partially compensate for changes in physical parameters. See below.

Use of Organic Solvents

The overall atomization efficiency for producing atomic vapor in the flame is increased by use of organic solvents. Such increase is due to a complex variety of causes, including increased rate of aspiration, finer droplets, more efficient evaporation or combustion of the solvent, and so forth. Thus, increased sensitivity is generally obtained by adding a miscible organic solvent such as acetone to the solution. A threefold increase is common. The problem is that by adding the miscible solvent, the sample solution is diluted, which more or less defeats the purpose. Therefore, the technique of solvent extraction is usually employed to obtain increased sensitivity. Solvent extraction is discussed in detail in Chapter 15. It involves extracting the metal from the aqueous solution, usually in the form of an uncharged chelate, into an immiscible organic solvent. The organic phase containing the metal is then aspirated into the flame. A number of advantages accrue. The test element is separated from the bulk matrix of the sample, thereby frequently eliminating possible interferences. It is obtained in a pure organic solvent, which results in maximum atomization efficiency. A 10-fold enhancement in the signal for a given concentration can be obtained. Finally, the test element can be concentrated by extracting it into a smaller volume of organic solvent. A 10- to 100-fold concentration can be obtained in many cases. Methylisobutyl ketone (MIBK) is one of the best solvents for extraction and aspiration into a flame. When organic solvents are aspirated into a flame, an oxidizing (fuel-lean) flame must be used, because the solvent must be burned.

Sample Preparation

Sample preparation with flame or plasma methods can often be kept to a minimum. As long as chemical or spectral interferences are absent, essentially all that is required is to obtain the sample in the form of a solution. It often makes no difference what the chemical form of the analyte is, because it will be dissociated to the free elemental vapor in the flame. Thus, several elements can be determined in blood, urine, cerebral spinal fluid, and other biological fluids by direct aspiration of the sample. Usually, dilution with water will be required to prevent clogging of the burner.

Note that in the preparation of standards, the matrix of the analyte must always be matched. Thus, if lead in gasoline is to be determined, a simulated solvent matrix must be used for standards, not water.

Chemical interferences can often be overcome by simple addition of (dilution with) a suitable reagent solution. Thus, serum is diluted 1:20 with a solution containing EDTA for the determination of calcium in order to prevent interference from phosphate. Sodium and potassium, in concentrations equal to those in serum, are added to calcium standards to prevent ionization interference.

If solvent extraction is performed on the sample, care must be taken that an interfering anion is not present that will prevent the extraction. For example, calcium–EDTA is often administered in the therapy for heavy metal poisoning. The presence of this in the urine may prevent solvent extraction of some metal chelates, such as those of lead ion, by forming a stable EDTA complex.[5] In cases such as this, it will be necessary to perform an acid digestion or dry ashing of the sample to destroy the interfering organic constituents. If the aqueous sample can be aspirated directly, the EDTA does not interfere.

Reference 6 at the end of the chapter gives a review of applications of atomic absorption spectrophotometry to biological samples. This technique is widely used for metal analysis in biological fluids and tissues, in environmental samples such as air and water, and in occupational health and safety areas. Very few routine applications of flame emission spectrometry to biological samples have been described, except for the alkali and alkaline earth metals, but in many cases, the sample preparation would be similar to that used for atomic absorption spectrophotometry.

Relative Detectabilities of Atomic Absorption and Flame Emission Spectrometry

Table 14.3 lists some representative detection limits of various elements by atomic absorption and flame emission spectrometry. We should distinguish here between the sensitivity and detection limits in atomic absorption. The former term is frequently used in the atomic absorption literature. **Sensitivity** is defined as the concentration required to give 1% absorption (0.0044 A). It is a measure of the slope of the analytical calibration curve and says nothing of the signal-to-noise ratio (S/N). **Detection limit** is generally defined as the concentration required to give a signal equal to three times the standard deviation of the baseline (blank).

Generally speaking, atomic absorption shows superior detectability for those elements that emit below 300 nm because of the high thermal energy required to excite the atoms for emission at these wavelengths. But at wavelengths between 300 and 400 nm, either method may exhibit comparable detectability, while flame emission is generally superior in the visible region.

---

[5]The EDTA chelates are charged and, as discussed in Chapter 15, the complex must be neutral before it will dissolve in the organic solvent.

TABLE 14.3
**Representative Detection Limits by Atomic Absorption (AAS) and
Flame Emission (FES) Spectrometry**

		Detection Limit, ppm	
Element	Wavelength, nm	AAS[a]	FES[b]
Ag	328.1	0.001(A)	0.01
Al	309.3	0.1(N)	
	396.2		0.08
Au	242.8	0.03(N)	
	267.6		3
Ca	422.7	0.003(A)	0.0003
Cu	324.8	0.006(A)	0.01
Eu	459.4	0.06(N)	0.0008
Hg	253.6	0.8(A)	15
K	766.5	0.004(A)	0.00008
Mg	285.2	0.004(A)	0.1
Na	589.0	0.001(A)	0.0008
Tl	276.8	0.03(A)	
	535.0		0.03
Zn	213.9	0.001(A)	15

[a]Fuel is acetylene. Letter in parentheses indicates oxidant: A = air, N = nitrous oxide.

[b]Nitrous oxide–acetylene flame.

Electrothermal Atomizers

Although aspiration into a flame is the most convenient and reproducible means of obtaining atomic vapor, it is one of the least efficient in terms of converting all the sample elements to atomic vapor and presenting this to the optical path. The overall efficiency of atomic conversion and measurement of ions present in aspirated solutions has been estimated to be as little as 0.1%. Also, aspiration methods usually require several milliliters of solution for analysis.

Electrothermal atomizers are generally a type of minifurnace in which a drop of the sample is dried and then decomposed at a high termperature to produce an atomic vapor cloud.

Electrothermal atomizers have conversion efficiencies approaching 100%, so absolute detection limits are often 100 to 1000 times improved over those of flame aspiration methods. Our discussion will center on resistively heated atomizers. Although these are not generally useful for emission measurements, they are well suited for atomic absorption measurements. A schematic of a typical electrothermal atomizer is shown in Figure 14.8.

In most of the electrothermal techniques, a few microliters of sample is placed in a horizontal graphite tube or on a carbon rod or tantalum ribbon. The tube or rod is heated resistively by passing a current through it. The sample is first dried at a low temperature for a few seconds (~100 to 200°C), followed by pyrolysis at 500 to 1400°C to destroy organic matter that produces smoke and scatters the light source during measurement; the smoke from pyrolysis is flushed out by flowing argon gas. Finally, the sample is rapidly thermally atomized at a high temperature, up to 3000°C.

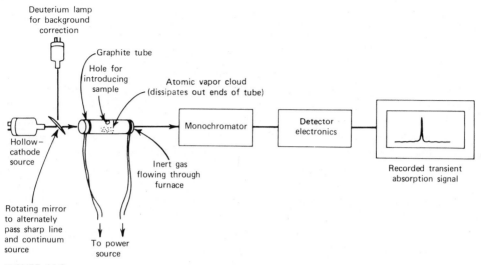

FIGURE 14.8
Electrothermal atomization.

The light path passes over the atomizer (or through the tube). A sharp peak of absorbance versus time is recorded as the atomic cloud passes through the light path (Figure 14.8). Either the height of the observed peak or its area is directly related to the quantity of metal vaporized. The heating is done in an inert atmosphere (e.g., argon gas) to prevent oxidation of the graphite or carbon at the high temperatures involved, and also to prevent formation of refractory metal oxides.

A major difficulty with electrothermal atomization methods is that interelement effects are generally much more pronounced than in flames. The interferences can sometimes be compensated for by using a standard additions method for calibration in which the standard is added to a separate aliquot of the sample and the increase in the measured signal is proportional to the concentration added. In this manner, the standard is subjected to the same matrix as the sample (see below).

Often when the matrix concentration is changed, there is a change in the peak height and in the *shape* of the analytical peak. In such instances, less influence of the matrix and better accuracy can sometimes be achieved by integrating the signal or measuring its area, rather than measuring its maximum intensity. Of course, this requires more sophisticated instrumentation.

Even with pyrolysis before the atomization step, background absorption in electrothermal methods tends to be more prominent than in flame methods, especially with biological and environmental samples. This is due to residual organic material or vaporized matrix salts. Hence, automatic background correction using a continuum source (see above) is generally required.

Detection limits quoted by the manufacturers of electrothermal atomizers are typically in the range of 10^{-10} to 10^{-12} g or even less! The concentrational detection limit will depend on the sample volume. This will depend on the sample matrix composition and the concentration of the test element. Assume that a 10-μL sample is analyzed for an element with a detection limit of 10^{-11} g. Then the concentration detection limit would be 10^{-11} g/0.01 mL or 10^{-9} g/mL. This is equal to 1 ng/mL or 1 part per billion. The extreme sensitivity of these techniques is, therefore, quite apparent, even when dealing with very small sample volumes.

Electrothermal methods are complementary to flame methods. The latter are better suited when the analyte element is at a sufficiently high concentration to measure and adequate solution volume is available. They provide excellent reproducibility, and interferences are usually readily dealt with. The electrothermal techniques, on the other hand, are required when the concentrations are very small or the sample size is limited. It is possible to analyze solid samples directly without preparing a solution. The calibration and use of electrothermal methods generally requires more care. These are among the most sensitive of all analytical methods. A comparable method in sensitivity is anodic-stripping voltammetry (Chapter 12). Anodic-stripping voltammetry generally requires larger solution volumes, but may be less subject to matrix difficulties. It is prone to impurity contamination from the supporting electrolyte, which may limit the sensitivity. It is capable of measuring several elements simultaneously. Deaeration of the solution is usually required and relatively long electrolysis times may be needed, and so analyses generally are more time consuming than with atomic absorption methods. Calibration is required less frequently with anodic stripping than with atomic absorption.

Flameless Mercury Determination

An extremely sensitive determination of mercury is accomplished by a nonflame atomization at room temperature. In this technique, mercury (e.g., in air, water, or occupational samples) is collected in a potassium permanganate solution. The permanganate oxidizes organomercury compounds, as well as elemental mercury, to produce mercuric ions. For analysis, the excess permanganate is destroyed with hydroxylamine, and a reducing agent such as $SnCl_2$ is added to reduce the mercury ions to elemental mercury. This elemental mercury has an appreciable vapor pressure at room temperature; and by bubbling a gas such as nitrogen through the solution in an Erlenmeyer flask, the mercury vapor can be swept into a glass tube (with quartz end windows to pass the UV radiation), aligned in the right path of the atomic absorption spectrophotometer. The mercury vapor absorbs the lines of mercury just as atomic mercury in a flame does. Again, the atomization efficiency is greatly improved over flame atomization methods, with typical detection limits of a few nanograms or less. This method is preferred over electrothermal ones because of volatility losses of mercury during drying and charring cycles before the atomization.

Care must be taken to avoid the presence of UV-absorbing organic solvent vapors in this method, many of which absorb strongly at the mercury 253.7-nm line usually used for measurements.

14.6 INTERNAL STANDARD AND STANDARD ADDITION CALIBRATION

In atomic spectrometric methods, signals can frequently vary with time due to fluctuations in gas flow rates, aspiration rates, and so forth. Precision can be improved by the technique of **internal standards.** As an example, a simple flame emission spectrometer designed for simultaneous measurement of sodium and potassium in serum, using fixed wavelengths and two separate detectors, will usually contain a third fixed-wavelength channel-detector channel for lithium. A fixed concentration of lithium is added to all standards and samples. The instrument records and reads out the *ratios* of the K/Li and Na/Li signals. If the aspiration rate, for example, fluctuates, each signal is affected to the same extent and the ratio, at a given K or Na concentration, remains constant.

The internal standard element should be chemically similar to the analyte element, and their wavelengths should not be too different. See Reference 10 at the end of the chapter for a discussion of the selection of the internal standard.

A second difficulty often encountered in flame-spectrometric methods is a suppression (or sometimes enhancement) of the signal by the sample matrix, for example, due to high viscosity, chemical reaction with the analyte, and so forth. The technique of **standard addition calibration** can be utilized to minimize errors of this type. The sample is measured in the usual way to produce a given signal. A separate portion of the sample is taken and spiked with a known amount of standard and followed through the analytical procedure, and a new signal is recorded. The standard, then, is subjected to the same matrix as the unknown analyte. The increase in signal is due to the standard, and the original signal is due to the analyte. It is important to perform blank corrections. A simple proportionality applies, assuming you are in a linear portion of the calibration curve. Two additions of standard are recommended to assure linearity.

EXAMPLE 14.1   A serum sample is analyzed for potassium by flame emission spectrometry using the method of standard additions. Two 0.500-mL aliquots are added to 5.00-mL portions of water. To one portion is added 10.0 μL of 0.0500 M KCl solution. The net emission signals in arbitrary units are 32.1 and 58.6. What is the concentration of potassium in the serum?

SOLUTION
The amount of standard added is

$$0.0100 \text{ mL} \times 0.0500 \text{ } M = 5.00 \times 10^{-4} \text{ mmol}$$

This produces a signal of

$$58.6 - 32.1 = 26.5 \text{ arbitrary units}$$

The millimoles potassium in the sample, then, is

$$5.00 \times 10^{-4} \text{ mmol} \times \frac{32.1 \text{ units}}{26.5 \text{ units}} = 6.06 \times 10^{-4} \text{ mmol}$$

This is contained in 0.500 mL serum, so the concentration is

$$\frac{6.06 \times 10^{-4} \text{ mmol}}{0.500 \text{ mL serum}} = 1.21 \times 10^{-3} \text{ mmol/mL serum}$$

QUESTIONS

Principles

1. Describe the principles of emission spectroscopy. What are the requirements for the instrumentation?
2. Describe the principles of flame emission spectrometry and of atomic absorption spectrophotometry.

3. Compare flame emission and atomic absorption spectrophotometry with respect to instrumentation, sensitivity, and interferences.

4. Why is a sharp-line source desirable for atomic absorption spectroscopy?

5. Explain why flame emission spectrometry is often as sensitive as atomic absorption spectrophotometry, even though only a small fraction of the atoms may be thermally excited in the flame.

6. The Maxwell–Boltzmann expression predicts that the fraction of excited-state atoms in a flame is both highly temperature dependent and wavelength dependent, while the fraction of ground-state atoms remains large in all cases. Yet flame emission and atomic absorption spectrometry in practice do not exhibit large differences in dependence for many elements if the wavelength is greater than about 300 nm. Why is this?

7. Explain why absorption spectra for atomic species consist of discrete lines at specific wavelengths rather than broad bands, for molecular species.

8. Explain why solution in an organic solvent generally results in enhanced sensitivity in flame-spectrometric methods.

9. Explain why electrothermal atomizers result in greatly enhanced sensitivity in atomic absorption spectrophotometry.

10. Explain why an internal-standard element can improve the precision of atomic spectrometry measurements.

Instrumentation

11. Explain the mechanism of operation of a hollow-cathode lamp.

12. Describe the premix chamber burner and the total-consumption burner. Compare them with respect to efficiency and sensitivity.

13. Explain why the radiation source in atomic absorption instruments is usually modulated.

Interferences

14. Lead in seawater was determined by atomic absorption spectrophotometry. The APCD chelate was extracted into methylisobutyl ketone and the organic solvent was aspirated. A standard and reagent blank were treated in a similar manner. The blank reading was essentially zero. Measurements were made at the 283.3-nm line. An independent determination using anodic-stripping voltammetry revealed the atomic absorption results to be high by nearly 100%. Assuming the anodic-stripping voltammetry results are correct, suggest a reason for the erroneous results and how they might be avoided in future analyses.

15. Why is a high-temperature nitrous oxide–acetylene flame sometimes required in atomic absorption spectrophotometry?

16. Why is a high concentration of a potassium salt sometimes added to standards and samples in flame absorption or emission methods?

17. Chemical interferences are more prevalent in "cool" flames such as air–propane, but this flame is preferred for the determination of the alkali metals. Suggest why.

18. An analyst notes that a 1-ppm solution of sodium gives a flame emission signal of 110, while the same solution containing also 20 ppm potassium gives a reading of 125. It was determined that a 20-ppm solution of potassium exhibited no blank reading. Explain the results.

PROBLEMS

Sensitivity

19. A 12-ppm solution of lead gives an atomic absorption signal of 8.0% absorption. What is the atomic absorption sensitivity?

20. Silver exhibits an atomic absorption sensitivity of 0.050 ppm under a given set of conditions. What would be the expected percent absorption for a 0.70-ppm solution?

Boltzmann Distribution

21. The transition for the cadmium 228.8-nm line is a $^1S_0-^1S_1$ transition. Calculate the ratio of N_e/N_0 in an air–acetylene flame. What percent of the atoms is in the excited state? The velocity of light is 3.00×10^{10} cm/s, Planck's constant is 6.62×10^{-27} erg-s, and the Boltzmann constant is 1.380×10^{-16} erg K^{-1}.

Quantitative Calculations

22. Calcium in a sample solution is determined by atomic absorption spectrophotometry. A stock solution of calcium is prepared by dissolving 1.834 g $CaCl_2 \cdot 2H_2O$ in water and diluting to 1 L. This is diluted 1:10. Working standards are prepared by diluting the second solution, respectively, 1:20, 1:10, and 1:5. The sample is diluted 1:25. Strontium chloride is added to all solutions before dilution, sufficient to give 1% (wt/vol) to avoid phosphate interference. A blank is prepared to give 1% $SrCl_2$. Absorbance signals on the stripchart recorder, when the solutions are aspirated into an air–acetylene flame, are as follows: blank, 1.5 cm; standards, 10.6, 20.1, and 38.5 cm; sample, 29.6 cm. What is the concentration of calcium in the sample in parts per million?

23. Lithium in the blood serum of a manic-depressive patient treated with lithium carbonate is determined by flame emission spectrophotometry, using a standard additions calibration. One hundred microliters of serum diluted to 1 mL gives an emission signal of 6.7 cm on the recorder chart. A similar solution to which 10μL of a 0.010 M solution of $LiNO_3$ has been added gives a signal of 14.6 cm. Assuming linearity between the emission signal and the lithium concentration, what is the concentration of lithium in the serum, in parts per million?

24. Chloride in a water sample is determined indirectly by atomic absorption spectrophotometry by precipitating it as AgCl with a measured amount of $AgNO_3$ in excess, filtering, and measuring the concentration of silver remaining in the filtrate. Ten-milliliter aliquots each of the sample and a 100-ppm chloride standard are added to separate dry 100-mL Erlenmeyer flasks. Twenty-five milliliters of a silver nitrate solution is added to each with a pipet. After allowing time for the precipitate to form, the mixtures are transferred partially to dry centrifuge tubes and are centrifuged. Each filtrate is aspirated for atomic absorption measurement of silver concentration. A blank is similarly treated in which 10 mL deionized distilled water is substituted for the sample. If the following absorbance signals are recorded for each solution, what is the concentration of chloride in the water sample?

Blank:	12.8 cm
Standard:	5.7 cm
Sample:	6.8 cm

RECOMMENDED REFERENCES

General
1. E. Olsen, *Modern Optical Methods of Analysis*. New York: McGraw-Hill, 1975.
2. J. D. Winefordner, *Spectrochemical Methods of Analysis*. New York: Wiley-Interscience, 1971.

Emission Spectroscopy
3. R. M. Barnes, *Emission Spectroscopy*. Stroudsburg, Pa.: Dowden, Hutchinson & Ross, 1975.

4. G. R. Harrison, *M.I.T. Wavelength Tables,* 2nd ed. Cambridge, Mass.: MIT Press, 1969.

5. N. H. Nachtrieb, *Spectrochemical Analysis.* New York: McGraw-Hill, 1950.

Flame Emission and Atomic Absorption Spectrometry

6. G. D. Christian, "Medicine, Trace Elements, and Atomic Absorption Spectroscopy," *Anal. Chem.,* **41** (1) (1969) 24A.

7. G. D. Christian and F. J. Feldman, *Atomic Absorption Spectroscopy. Applications in Agriculture, Biology, and Medicine.* New York: Wiley-Interscience, 1970.

8. J. A. Dean, *Flame Photometry.* New York: McGraw-Hill, 1960.

9. J. A. Dean and T. C. Rains, eds., *Flame Emission and Atomic Absorption Spectrometry,* Vol. I, *Theory* (1969); Vol. 2, *Components and Techniques* (1971); Vol. 3, *Applications* (1975). New York: Marcel Dekker.

10. F. J. Feldman, "Internal Standardization in Atomic Emission and Absorption Spectrometry," *Anal. Chem.,* **42** (1970) 719.

11. P. M. Hald, "Sodium and Potassium by Flame Photometry," in *Standard Methods of Clinical Chemistry,* Vol. II, D. Seligson, ed. New York: Academic Press, 1958, pp. 165–185. Describes determination in serum.

12. J. Hosking, K. Oliver, and B. Sturman, "Errors in Atomic Absorption Spectroscopy Determination of Calcium by the Standard Addition Method in the Presence of Silicon, Aluminum, and Phosphate," *Anal. Chem.,* **51** (1979) 307.

13. G. F. Kirkbright and M. Sargent, *Atomic Absorption and Fluorescence Spectroscopy.* London: Academic Press, 1974.

14. R. Mavrodineanu, ed., *Analytical Flame Spectroscopy: Selected Topics.* New York: Springer-Verlag, 1970.

chapter 15

SOLVENT EXTRACTION

Solvent extraction involves the distribution of a solute between two immiscible liquid phases. This technique is extremely useful for very rapid and "clean" separations of both organic and inorganic substances. In this chapter, we discuss the distribution of substances between two phases and how this can be used to form analytical separations. The solvent extraction of metal ions into organic solvents is described. Finally, multiple extractions are described for performing difficult separations, including countercurrent distribution. The latter technique, which involves sequential multistep separations, serves as an introduction to chromatographic processes described in the following chapter.

15.1 THE DISTRIBUTION COEFFICIENT

A solute S will distribute itself between two phases (after shaking and allowing the phases to separate) and, within limits, the ratio of the concentrations of the solute in the two phases will be a constant:

$$K_D = \frac{[S]_1}{[S]_2}$$

(15.1)

where K_D is the **distribution coefficient** and the subscripts represent solvent 1 (e.g., an organic solvent) and solvent 2 (e.g., water). If the distribution coefficient is large, the solute will tend toward quantitative distribution in solvent 1.

The apparatus used for solvent extraction is the **separatory funnel,** illustrated in Figure 15.1. Most often, a solute is extracted from an aqueous solution into an

FIGURE 15.1
A separatory funnel.

immiscible organic solvent. After the mixture is shaken for about a minute, the phases are allowed to separate and the bottom layer (the denser solvent) is drawn off in a completion of the separation.

Many substances are partially ionized in the aqueous layer as weak acids. This introduces a pH effect on the extraction. Consider, for example, the extraction of benzoic acid from an aqueous solution. Benzoic acid (HBz) is a weak acid in water with a particular ionization constant K_a (given by Equation 15.4 below). The distribution coefficient is given by

$$K_D = \frac{[HBz]_e}{[HBz]_a} \qquad (15.2)$$

where e represents the ether solvent and a represents the aqueous solvent. However, part of the benzoic acid in the aqueous layer will exist as Bz^-, depending on the magnitude of K_a and on the pH of the aqueous layer; hence, quantitative separation may not be achieved.

15.2   THE DISTRIBUTION RATIO

It is more meaningful to describe a different term, the **distribution ratio,** which is the ratio of the concentrations of *all* the species of the solute in each phase. In this example, it is given by

$$D = \frac{[HBz]_e}{[HBz]_a + [Bz^-]_a} \qquad (15.3)$$

We can readily derive the relationship between D and K_D from the equilibria involved. The acidity constant K_a for the ionization of the acid in the aqueous phase is given by

$$K_a = \frac{[H^+]_a[Bz^-]_a}{[HBz]_a} \qquad (15.4)$$

Hence,

$$[Bz^-]_a = \frac{K_a[HBz]_a}{[H^+]_a} \qquad (15.5)$$

From Equation 15.2,

$$[HBz]_e = K_D[HBz]_a \qquad (15.6)$$

Substitution of Equations 15.5 and 15.6 into Equation 15.3 gives

$$D = \frac{K_D[HBz]_a}{[HBz]_a + K_a[HBz]_a/[H^+]_a} \qquad (15.7)$$

$$\boxed{D = \frac{K_D}{1 + K_a/[H^+]_a}} \qquad (15.8)$$

This equation predicts that when $[H^+]_a \gg K_a$, D is nearly equal to K_D, and if K_D is large, the benzoic acid will be extracted into the ether layer; D is maximum under these conditions. If, on the other hand, $[H^+] \ll K_a$, then D reduces to $K_D[H^+]_a/K_a$, which will be small, and the benzoic acid will remain in the aqueous layer. That is, in alkaline solution, the benzoic acid is ionized and cannot be extracted, while in acid solution, it is largely undissociated. These conclusions are what we would intuitively expect from inspection of the chemical equilibria.

Equation 15.8, like Equation 15.1, predicts that the *extraction efficiency will be independent of the original concentration of the solute*. This is one of the attractive features of solvent extraction; it is applicable to tracer (e.g., radioactive) levels and to macro levels alike, a condition that applies only so long as the solubility of the solute in one of the phases is not exceeded and there are no side reactions such as dimerization of the extracted solute.

Of course, if the hydrogen ion concentration changes, the extraction efficiency (D) will change. In this example, the hydrogen ion concentration would increase with increasing benzoic acid concentration, unless an acid–base buffer is added to maintain the hydrogen ion concentration constant (see Chapter 5 for a discussion of buffers).

In deriving Equation 15.8, we actually neglected to include in the numerator of Equation 15.3 a term for a portion of the benzoic acid that exists as the dimer in the organic phase. The extent of dimerization tends to increase with increased concentration, and by Le Châtelier's principle, this will cause the equilibrium to shift in favor of the organic phase with increased concentration. So, in cases such as this, the efficiency of extraction will actually increase at higher concentrations. As an exercise, derivation of the more complete equation is presented in Problem 10 at the end of the chapter.

15.3 THE PERCENT EXTRACTED

The distribution ratio D is a constant independent of the volume ratio. However, the fraction of the solute extracted will depend on the volume ratio of the two solvents. If a larger volume of organic solvent is used, more solute must dissolve in this layer to keep the concentration ratio constant and to satisfy the distribution ratio.

The fraction of solute extracted is equal to the millimoles of solute in the organic layer divided by the total number of millimoles of solute. The millimoles are given by the molarity times the milliliters. Thus, the percent extracted is given by

$$\% \ E = \frac{[S]_o V_o}{[S]_o V_o + [S]_a V_a} \times 100\% \tag{15.9}$$

where V_o and V_a are the volumes of the organic and aqueous phases, respectively. It can be shown from this equation (see Problem 9) that the percent extracted is related to the distribution ratio by

$$\boxed{\% \ E = \frac{100D}{D + (V_a/V_o)}} \tag{15.10}$$

If $V_a = V_o$, then

$$\boxed{\% \ E = \frac{100D}{D + 1}} \tag{15.11}$$

In the case of equal volumes, the solute can be considered quantitatively retained if D is less than 0.001. It is essentially quantitatively extracted if D is greater than 1000. The percent extracted changes only from 99.5% to 99.9% when D is doubled from 500 to 1000.

EXAMPLE 15.1 Twenty milliliters of an aqueous solution of 0.10 M butyric acid is shaken with 10 mL ether. After the layers are separated, it is determined by titration that 0.5 mmol butyric acid remains in the aqueous layer. What is the distribution ratio, and what is the percent extracted?

SOLUTION
We started with 2.0 mmol butyric acid, and so 1.5 mmol was extracted. The concentration in the ether layer is 1.5 mmol/10 mL = 0.15 M. The concentration in the aqueous layer is 0.5 mmol/20 mL = 0.025 M. Therefore,

$$D = \frac{0.15}{0.025} = 6.0$$

Since 1.5 mmol was extracted, the percent extracted is (1.5/2.0) $\times$ 100% = 75%. Or

$$\% \ E = \frac{100 \times 6.0}{6.0 + (20/10)} = 75\%$$

Equation 15.10 shows that the fraction extracted can be increased by decreasing the ratio of V_a/V_o, for example, by increasing the organic phase volume. However, a more efficient way of increasing the amount extracted using the same volume of organic solvent is to perform successive extractions with smaller individual volumes of organic solvent. For example, with a D of 10 and $V_a/V_o = 1$, the percent extracted is about 91%. Decreasing V_a/V_o to 0.5 (doubling V_o) would result in an increase of % E to 95%. But by performing two successive extractions with $V_a/V_o = 1$ would give an overall extraction of 99%. See the section below on multiple extractions for further discussions.

15.4   SOLVENT EXTRACTION OF METALS

Solvent extraction has one of its most important applications in the separation of metal cations. This separation can be accomplished in several ways. You have noted above that the uncharged organic molecules tend to dissolve in the organic layer while the charged anion from the ionized molecules remains in the polar aqueous layer. This is an example of "like dissolves like." Metal ions do not tend to dissolve appreciably in the organic layer. For them to become soluble, their charge must be neutralized and something must be added to make them "organic-like." There are two principal ways of doing this.

Ion Association Complexes

In one method, the metal ion is incorporated into a bulky molecule and then associates with another ion of the opposite charge to form an **ion pair,** or the metal ion associates with another ion of great size (organic-like). For example, it is well known that iron(III) can be quantitatively extracted from hydrochloric acid medium into diethyl ether. The mechanism is not completely understood, but evidence exists that the chloro complex of the iron is coordinated with the oxygen atom of the solvent (the solvent displaces the coordinated water), and this ion associates with a solvent molecule that is coordinated with a proton:

$$\{(C_2H_5)_2O:H^+, FeCl_4[(C_2H_5)_2O]_2^-\}$$

Similarly, the uranyl ion UO_2^{2+} is extracted from aqueous nitrate solution into iso-butanol by associating with two nitrate ions (UO_2^{2+}, $2NO_3^-$), with the uranium probably being solvated by the solvent to make it "solvent-like." Permanganate forms an ion pair with tetraphenylarsonium ion ($(C_6H_5)_4As^+$, MnO_4^-), which makes it organic-like, and it is extracted into methylene chloride. There are numerous other examples of ion-association extractions.

Metal Chelates

The most widely used method of extracting metal ions is formation of a chelate molecule with an organic chelating agent.

As mentioned in Chapter 9, a chelating agent contains two or more complexing groups. Many of these reagents form colored chelates with metal ions and form the basis of spectrophotometric methods for determining the metals. The chelates are often insoluble in water and will precipitate. They are, however, usually soluble in organic solvents such as chloroform, carbon tetrachloride, or methylene chloride.[1] Many of the organic precipitating agents listed in Chapter 6 are used as extracting agents.

The Extraction Process

Most chelating agents are weak acids that ionize in water; the ionizable proton is displaced by the metal ion when the chelate is formed, and the charge on the

---

[1]These solvents are widely used in solvent extraction work. Care should be taken in handling them because of their toxic nature. They should not be inhaled any more than necessary. Whenever possible, methylene chloride should be used because of its lower toxicity. Chloroform is about one-tenth as toxic as carbon tetrachloride.

organic compound neutralizes the charge on the metal ion. An example is **diphen-ylthiocarbazone (dithizone),** which forms a chelate with lead ion:

The usual practice is to add the chelating agent to the organic phase. The extraction process can be thought to consist of four equilibrium steps, each with an equilibrium constant. These steps are illustrated in Figure 15.2. First, the chelating agent HR distributes between the aqueous and the organic phases:

$$(HR)_o \rightleftharpoons (HR)_a \quad \text{and} \quad K_{DHR} = \frac{[HR]_o}{[HR]_a} \tag{15.12}$$

Second, the reagent in the aqueous phase ionizes:

$$HR \rightleftharpoons H^+ + R^- \quad \text{and} \quad K_a = \frac{[H^+][R^-]}{[HR]} \tag{15.13}$$

Third, the metal ion chelates with the reagent anion to form an uncharged molecule:

$$M^{n+} + nR^- \rightleftharpoons MR_n \quad \text{and} \quad K_f = \frac{[MR_n]}{[M^{n+}][R^-]^n} \tag{15.14}$$

Finally, the chelate distributes between the organic and aqueous phases:

$$(MR_n)_a \rightleftharpoons (MR_n)_o \quad \text{and} \quad K_{DMRn} = \frac{[MR_n]_o}{[MR_n]_a} \tag{15.15}$$

K_{DHR} and K_{DMRn} are the distribution coefficients of the reagent and the chelate, respectively; K_a is the ionization constant of the reagent; and K_f is the formation constant of the chelate.

Assuming that the chelated portion of the metal distributes largely into the organic phase and that the chelate is essentially undissociated in the nonpolar organic solvent, the distribution ratio is given by

$$D \approx \frac{[MR_n]_o}{[M^{n+}]_a} \tag{15.16}$$

By substituting the appropriate equilibrium concentrations from Equations 15.12 through 15.15 into Equation 15.16 (similar to the procedure used in deriving Equation 15.8 above), so that all four equilibrium constants are included, the following equation can be derived:

$$D = \frac{K_{DMRn} K_f K_a^n}{K_{DHR}^n} \cdot \frac{[HR]_o^n}{[H^+]_a^n} = K \cdot \frac{[HR]_o^n}{[H^+]_a^n} \tag{15.17}$$

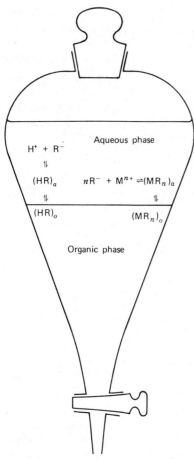

$$H^+ + R^-$$
$$\updownarrow$$
$$(HR)_a \qquad nR^- + M^{n+} \rightleftharpoons (MR_n)_a$$
$$\updownarrow \qquad\qquad \updownarrow$$
$$(HR)_o \qquad\qquad (MR_n)_o$$

Aqueous phase

Organic phase

FIGURE 15.2
Equilibria involved in the solvent extraction of metal chelates.

On examining the terms in Equation 15.17, we see that *the distribution ratio is independent of the concentration of the solute* (the metal ion, in this case), provided that the solubility of the metal chelate in the organic layer is not exceeded. The extraction efficiency can be affected only by changing the reagent concentration or by changing the pH. A 10-fold increase in the reagent concentration will increase the extraction efficiency by the same amount as an increase in the pH of one unit (10-fold decrease in the hydrogen ion concentration). Each affect is greater as n becomes larger. By using a high reagent concentration, extractions can be performed in more acid solutions.

The more stable the chelate (the larger the K_f), the greater the extraction efficiency, and this principle serves as the basis for the separation of many metals. An acidic reagent (large K_a) that is relatively soluble in water (small $K_{D_{HR}}$) favors good extraction. But chelate stability generally decreases as the reagent acidity increases (when it prefers to lose protons, it also prefers to lose metal ions), and the effects of K_a and K_f must be considered together for a series of different chelating agents.

For the most part, the nature of the organic solvent is not too critical a factor in the success of an extraction. Thus, the choice of toluene or methylene chloride is determined more by the requirement of whether the solvent should be more dense (CH_2Cl_2) or less dense (toluene) than the aqueous phase. For multivalent metals ($n > 1$), the solvent may affect the distribution ratio. This occurs because it affects the distribution coefficients of both the reagent and chelate; that is, their solubilities vary with the solvent. The relative change in the two solubilities is usually similar, but since $K_{D_{HR}}$ is raised to the n power, *a solvent in which the reagent is more soluble would result in a lower D value*. Dithizone and its chelates, for example, are more soluble in chloroform than in carbon tetrachloride, and so extractions with the former solvent require a higher pH than those with the latter solvent.

The Separation Efficiency of Metal Chelates

The separation efficiency of two metals at a given pH and reagent concentration can be predicted from Equation 15.17. The **separate factor β** is equal to the ratio of the distribution ratios of the two metal chelates formed with a given reagent. Since only K_f and $K_{D_{MRn}}$ will be a function of the metal, then

$$\beta = \frac{D_1}{D_2} = \frac{K_{f(1)}K_{D_{MRn(1)}}}{K_{f(2)}K_{D_{MRn(2)}}} \tag{15.18}$$

So, the separation efficiency depends on the relative formation constants and on the relative solubilities of the chelates.

The order of stability of complexes of a limited number of divalent metals has been shown, other things being equal, to be fairly independent of the nature of the chelating agents, with the following stability sequence:

$$Pd > Cu > Ni > Pb > Co > Zn > Cd > Fe > Mn > Mg$$

The order of extraction may be altered from this by differences in solubilities of the chelates. Also, steric hindrance (in which a functional group on the reagent molecule may hinder its reaction with a given metal ion) may affect the specificity of extraction. For example, 8-hydroxyquinoline (oxine) forms a chelate with many metals, including aluminum:

The oxine derivative 2-methyl-8-hydroxyquinoline,

will not form a complex with aluminum because the added methyl group does not allow room for three molecules to group around this small ion. Therefore, other metals can be extracted in the presence of aluminum with this reagent.

A more general approach to increasing the selectivity of extractions is the adding of **masking agents.** These are competing complexing agents that form *charged* complexes that are more stable for certain metals. EDTA (see Chapter 9) and cyanide ion are commonly used masking agents. Thus, although Cu^{2+} forms a more stable complex with oxine than VO_2^{2+} does, the vanadium may be extracted in the presence of copper by adding EDTA, which forms an even more stable complex with the copper ($Cu–EDTA^{2-}$).

The selectivity of an extraction can often be controlled by proper pH adjustment. Figure 15.3 illustrates the effect of pH on the percent extracted for a series of metals using the very useful chelating agent dithizone. We see that mercury(II) could be separated from all the metals except silver and copper by extraction at pH 1. Conversely, at pH 10, all the metals could be extracted and separated from cadmium. Additional separations may be executed by means of **back-extraction.** Consider a mixture of mercury, bismuth, tin, lead, and cadmium in which it is desired to separate tin from the other elements. At pH 6, the first three elements are extracted and separated from the last two. After the extraction, the tin can be back-extracted into an aqueous phase at pH 3.5, leaving the mercury and most of the bismuth in the organic phase. Remember, however, that the positions of the curves in Figure 15.3 are dependent on the dithizone concentration.

15.5 ANALYTICAL SEPARATIONS

One of the most important applications of solvent extraction is the spectrophotometric determination of metals in the visible region. Many organic reagents form colored chelates with metals, but most of the chelates are insoluble in water. They are, however, soluble in organic solvents and can easily be extracted. Since the chelating agent is often itself colored, its absorption spectrum may overlap that of the metal chelate, and a blank correction must be made or the reagent washed from the organic phase. The selectivity of the determination can be adjusted by the factors discussed above.

Solvent extraction is widely employed in drug analysis. See Reference 6 at the end of the chapter, and Chapter 20.

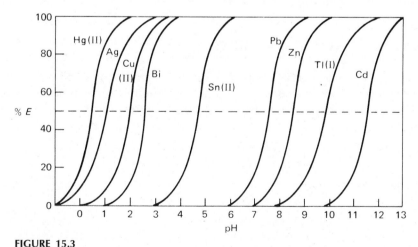

FIGURE 15.3
Qualitative extraction curves for metal dithizonates in carbon tetrachloride. (From G. H. Morrison and H. Freiser in *Comprehensive Analytical Chemistry*, Vol. **IA,** C. L. Wilson and D. Wilson, Eds. New York: Elsevier Publishing Company, 1959. Reproduced by permission of Elsevier Publishing Company.)

The test substance can be easily concentrated by employing a small volume of the organic solvent. Enhanced sensitivity results when the concentration of the substance is determined in the organic layer (e.g., in spectrophotometry or atomic absorption).

An example of the use of solvent extraction in the clinical laboratory is for the separation and determination of blood lead. After the organic matter is destroyed (see Chapter 3), the dithizone chelate of lead is formed and the chelate is extracted at pH 8 to 10 into methylene chloride where it is determined spectrophotometrically. This determination requires some care, however, to achieve selectivity. By judicious use of masking agents and proper adjustment of pH prior to solvent extraction of the lead complex, many major interferences can be eliminated. Figure 15.4 outlines one possible scheme.

Referring again to Figure 15.3, we see that if the extraction is performed at pH 8, Hg, Ag, Cu, Bi, Sn(II), and (to a lesser extent) Zn will extract with the Pb (the Pb is not quantitatively extracted at this pH, but this is all right so long as standards are treated similarly). Hg, Ag, and Sn are not likely to be present in blood, but Cu and Zn are, and Bi is chemically similar to Pb and may be mistaken for it. The degree of interference from other metals will depend on the extent of overlap of the absorption spectra of their dithizone chelates with the spectrum of the lead–dithizone chelate. By adding a masking agent, such as cyanide in base, the zinc group of metals (Zn, Cd, Hg) and Cu are complexed and will not extract. The Bi can be removed by preextraction at pH 4. Citrate is generally added also to complex Fe and prevent precipitation of its hydroxide, and hexametaphosphate is added to prevent precipitation of calcium phosphate at the alkaline pH involved.

It is important that Tl not extract, because it behaves toxicologically somewhat similar to Pb. However, this requires careful pH control (pH 8). Extractions are frequently carried out at about pH 10 where careful pH control is not necessary for efficient extraction of the Pb. In this case, a back-extraction of the Tl could be performed at pH 8, only if a positive test is obtained at pH 10; Pb would remain in the organic layer, but Tl would distribute into the aqueous layer.

Most of the metals listed in Table 6.2 in Chapter 6 can be extracted into methylene chloride or other organic solvents using the reagents listed.

15.6 MULTIPLE BATCH EXTRACTIONS

It may be desired in some cases to obtain complete separation of a solute when quantitative extraction is not accomplished in a single step. We have mentioned that the percent extracted can be increased by increasing the volume. However,

FIGURE 15.4
Selective solvent extraction of lead from blood into methylene chloride using dithizone reagent.

quantitative extraction is most efficiently carried out by performing multiple extractions with smaller portions of the same volume of solvent.

It is useful for calculations with multiple extractions to compute the fraction of analyte remaining unextracted after a given number of extractions. This is readily obtained from Equation 15.10. Note that division of this equation by 100 gives the *fraction* extracted $[D/(D + V_a/V_o)]$. Subtraction of this from 1 gives the *fraction F remaining unextracted*:

$$F = 1 - \frac{D}{D + (V_a/V_o)} \qquad (15.19)$$

which can be rearranged to

$$F = \frac{V_a}{DV_o + V_a} \qquad (15.20)$$

V_a and V_o can be any two immiscible phases, but are usually aqueous and organic; we have let V_a represent the volume containing the initial solute. Multiplication of W_a, the initial quantity (grams or moles) or concentration of analyte in the initial (aqueous) phase, by this fraction gives the quantity or concentration W remaining in the phase after one extraction:

$$W = W_a\left(\frac{V_a}{DV_o + V_a}\right) \qquad (15.21)$$

If n separate extractions are performed with equal volumes of V_o, then W_a should be multiplied by F a total of n times:

$$W = W_a\left(\frac{V_a}{DV_o + V_a}\right)^n \qquad (15.22)$$

EXAMPLE 15.2   One gram of a solute is contained in 100 mL of an aqueous solution. Calculate the amount remaining in the aqueous phase after (a) a single extraction with 90 mL of an organic solvent with appropriate reagents, (b) a single extraction with 30 mL solvent, and (c) three successive extractions with 30 mL solvent, assuming that the distribution ratio for the extraction is 10.

SOLUTION
(a)

$$W = 1.00\left(\frac{100}{10 \times 90 + 100}\right) = 0.100 \text{ g}$$

(b)

$$W = 1.00\left(\frac{100}{10 \times 30 + 100}\right) = 0.250 \text{ g}$$

(c)

$$W = 1.00\left(\frac{100}{10 \times 30 + 100}\right)^3 = 0.0156 \text{ g}$$

Thus, 90% of the solute is extracted with 90 mL solvent, 75% is extracted with 30 mL solvent, but 98.4% is extracted by three 30-mL portions of the solvent.

We see from this example that quantitative extraction is best carried out by multiple extraction with small volumes of solvent. (This is also the reason analytical glassware should be washed with multiple portions of solvent or reagent.) The individual solvent portions are usually combined for completion of the analysis.

15.7 COUNTERCURRENT DISTRIBUTION

When two solutes to be separated have distribution ratios of the same order of magnitude, their separation efficiency in a single-batch extraction is very poor. By a technique of successive extractions with fresh solvent, complex mixtures with similar distribution may be separated. Such a system is **countercurrent distribution.**
    A series of contacting tubes called Craig countercurrent distribution tubes are used, and these are illustrated in Figures 15.5 and 15.6. There are several of these tubes in a series, each containing two separate chambers. Chamber A in all the tubes is filled with solvent 1 that is more dense than the extracting solvent 2, as in Figure 15.5. This may be an aqueous phase containing appropriate reagents and buffers. The volume of solvent 1 is small enough that when the tube is tilted 90° as in Figure 15.5, it will not flow through tube C. Tube 1 contains the sample. Extracting solvent is introduced into the first tube through inlet B. After shaking by rocking back and forth and after allowing the phases to separate, the tube is tilted (rotated) 90° as in Figure 15.5. The less dense solvent 2 flows through tube C into chamber D. Now the assembly is rotated back to its original position, and solvent 2 flows out through tube E into the next tube (via its tube B), leaving the original solvent 1 in chamber A of the first tube. Fresh solvent 2 is added to the first tube, and the process is repeated, this time the solutes being distributed between the two phrases in two tubes: the solute originally extracted into solvent 2 has been transferred to the second tube containing solvent 1 (no sample), where it is redistributed. This procedure is repeated several times, depending on the separation efficiency or the number of tubes.
    The fraction, $F_{r,n}$, of solute contained in the rth tube after n transfers using equal volumes of the two solvents is given by the following binomial expansion[2]:

FIGURE 15.5
Craig countercurrent distribution tubes.

---

[2]The symbol ! is the factorial function. As an example, $4! = 4 \times 3 \times 2 \times 1 = 24$.

FIGURE 15.6
Typical countercurrent apparatus. (Courtesy of Spectrum Medical Industries Inc.)

$$F_{r,n} = \frac{n!}{r!(n-r)!}\left(\frac{1}{D+1}\right)^n D^r$$

(15.23)

If the volumes are unequal, D is multiplied by the ratio of the volumes of the upper and lower phases, V_{upper}/V_{lower}. The distribution of two solutes (with different distribution ratios) in the tubes after different numbers of transfers is shown in Figure 15.7. Note that the peaks broaden out as the number of tubes increases; that is, as the number of transfers increases, the solute is spread through a larger number of tubes. The separating ability, however, increases with an increased number of transfers. After the run is complete, each tube is analyzed for the solute(s). In a preparative run, the contents of the tubes containing a particular solute are combined.

Craig countercurrent instruments are available for performing the above operations automatically, containing from a few dozen tubes up to 1000 or more. The technique has been used widely by biochemists for separating mixtures of vitamins, polypeptides, nucleic acids, hormones, and even proteins. It is also important in the pharmaceutical industry for purification of pharmaceutical preparations. Some of the separations are more easily accomplished now by chromatographic techniques, but countercurrent distribution is still useful for specific applications, especially on a preparative scale.

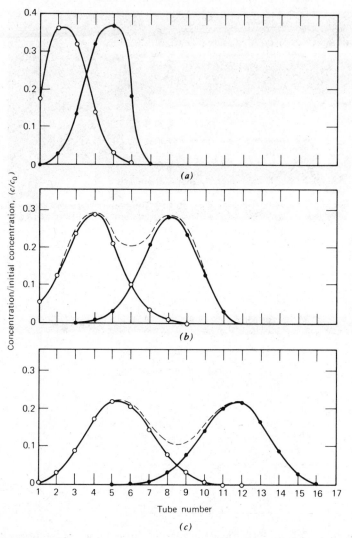

FIGURE 15.7
Distribution of two solutes in the tubes after different numbers of transfers. Distribution ratio for left peak is 7/3 and that for right peak is 3/7. (a) After 5 transfers ($n = 5$). (b) After 10 transfers ($n = 10$). (c) After 15 transfers ($n = 15$). From H. Purnell, *Gas Chromatography*. New York: Wiley, 1962, p. 95.

QUESTIONS

1. What is the distribution coefficient? The distribution ratio?
2. Suggest a method for the separation of aniline, $C_6H_5NH_2$, an organic base, from nitrobenzene, $C_6H_5NO_2$ (extremely toxic!).
3. Describe two principle solvent extraction systems for metal ions. Give examples of each.
4. Describe the equilibrium processes involved in the solvent extraction of metal chelates.
5. What is the largest concentration of a metal chelate that can be extracted into an organic solvent? The smallest concentration?

6. By referring to Figure 15.3, suggest a scheme for the separation of (a) silver, lead, and cadmium; (b) silver, bismuth, and tin.

7. Discuss the effect of the pH and of the reagent concentration on the solvent extraction of metal chelates.

8. Discuss the principles of countercurrent distribution.

PROBLEMS

Extraction Efficiencies

9. Derive Equation 15.10 from Equation 15.9.

10. In deriving Equation 15.8, we neglected the fact that benzoic acid partially forms a dimer in the organic phase $(2HBz \rightleftharpoons (HBz)_2$; $K_p = [(HBz)_2]/[HBz]^2$, where K_p is the dimerization constant). Derive an expression for the distribution ratio taking this into account.

11. Ninety-six percent of a solute is removed from 100 mL of an aqueous solution by extraction with two 50-mL portions of an organic solvent. What is the distribution ratio of the solute?

12. The distribution ratio between 3 M HCl and tri-n-butylphosphate for $PdCl_2$ is 2.3. What percent $PdCl_2$ will be extracted from 25.0 mL of a 7.0×10^{-4} M solution into 10.0 mL tri-n-butylphosphate?

13. Derive Equation 15.17 from Equation 15.16, using Equations 15.12 through 15.15.

14. Ninety percent of a metal chelate is extracted when equal volumes of aqueous and organic phases are used. What will be the percent extracted if the volume of the organic phase is doubled?

Separation Efficiencies

15. Metal A was determined to be 95% extracted into methylene chloride with dithizone at pH 6 when equal volumes of aqueous and nonaqueous solvents were used. Metal B was 5% extracted under the same conditions. What is the separation efficiency for these two metals at pH 6?

16. A divalent metal forms chelates with two different chelating agents, HA and HB. K_a of HA is 10^{-5} and K_a of HB is 10^{-10}. The two chelates have equal stability and are equally soluble in the two phases $(K_{D(MA_2)} = K_{D(MB_2)})$, as are the chelating agents $(K_{D(HA)} = K_{D(HB)})$. Calculate the relative percent of each chelate extracted at a given pH and reagent concentration and with equal volumes of the two phases.

Multiple Extractions

17. For a solute with a distribution ratio of 25.0, show by calculation which is more effective, extraction of 10 mL of an aqueous solution with 10 mL organic solvent or extraction with two separate 5.0-mL portions of organic solvent.

18. Arsenic(III) is 70% extracted from 7 M HCl into an equal volume of toluene. What percentage will remain unextracted after three individual extractions with toluene?

19. If the distribution ratio for the solvent extraction of iodine from water into carbon tetrachloride is 65.7, calculate (to the nearest 0.01%) the percent extracted from the 50.0 mL of water after three extractions with 10.0-mL portions of carbon tetrachloride.

20. A metal–APCD chelate has a distribution ratio of 5.96 for extraction from aqueous solution at pH 3 into methylisobutyl ketone (MIBK). Calculate the number of extractions necessary using 25.0-mL portions of MIBK to extract 99.9% of the metal from 50.0 mL urine at pH 3.

21. For the $PdCl_2$ solution in Problem 12, how many extractions must be performed with separate 10.0-mL portions of tri-*n*-butylphosphate to remove 99% of the $PdCl_2$?

Countercurrent Distribution

22. A solute with a distribution ratio between two solvents of 2.6 is to be purified by countercurrent distribution. (a) What fraction of the solute will be contained in the first tube after 20 transfers? (b) What fraction in the tenth tube? (c) The nineteenth tube?

RECOMMENDED REFERENCES

1. D. Dyrssen, J. O. Liljenzin, and J. Rydberg, eds., *Solvent Extraction Chemistry*. New York: Wiley-Interscience, 1967.

2. H. Irving and R. J. P. Williams, "Liquid–Liquid Extraction," in *Treatise on Analytical Chemistry*, Part I, Vol. **3**, Kolthoff, I. M. and P. J. Elving, eds. New York: Interscience, 1961, Chapter 31.

3. C. J. O. R. Morris and P. Morris, *Separation Methods in Biochemistry*. New York: Interscience, 1963, p. 559.

4. G. H. Morrison and H. Freiser, *Solvent Extraction in Analytical Chemistry*. New York: John Wiley & Sons, 1957.

5. J. Stary, *The Solvent Extraction of Metal Chelates*. New York: The Macmillan Company, 1964.

6. *The United States Pharmacopeia*. Rockville, Md.: United States Pharmacopeial Convention, Inc., 1980.

chapter 16

CHROMATOGRAPHIC METHODS

In 1906, the Russian scientist Tswett reported separating different colored constituents of leaves by passing an extract of the leaves through a column of calcium carbonate, alumina, and sucrose. He coined the term **chromatography,** from the Greek words meaning "color" and "to write." Tswett's original experiments went virtually unnoticed in the literature for decades, but eventually other methods were developed and today there are several different types of chromatography. Chromatography is taken now to refer generally to the separation of components in a sample by distribution of the components between two phases—one that is stationary and one that moves, usually but not necessarily in a column. Probably no other technique has been more valuable in the separation and analysis of highly complex mixtures. We briefly discuss the principles of chromatography and then separately describe some of the different types of chromatography.

Types of chromatography discussed include size exclusion chromatography, in which molecules are separated based on their size by passage through a porous structure stationary phase; ion exchange chromatography, in which ions are separated based on their charge; and gas chromatography, in which gaseous substances are separated based on their adsorption on or solubility in the stationary phase. High-performance liquid chromatography is described. This is a modern development based on the above principles but using micrometer-sized particles for the stationary phase so that equilibrium is achieved rapidly and separations are performed rapidly. Finally, various types of plane chromatography are described in which the stationary phase is in the form of a sheet or other flat surface. This includes paper chromatography, thin-layer chromatography, and electrophoresis, in which an electrical gradient is applied across the sheet to cause molecules to migrate according to the sign and magnitude of their charge.

Before beginning this chapter, it would be well to review solvent extraction in the previous chapter, particularly the multistep countercurrent distribution technique.

16.1 PRINCIPLES

We can analyze the principles of one type of chromatography (liquid–liquid chromatography) in terms of solvent extraction, but all types of chromatography will be based on similar equilibria. The physical design of chromatographic apparatus differs according to the technique, but the most common and typical form is a column. In liquid–liquid chromatography (elution analysis), a liquid, usually immiscible with water, is supported (adsorbed) on an inert solid, and the column is filled with a second solvent system (may be a mixed solvent or containing buffers, etc.) immiscible with the first. A solution of the sample is placed on top of the column and is washed down the column by means of adding the second solvent and allowing it to flow out of the bottom end of the column. The solutes establish an equilibrium between the two phases (the stationary and the mobile or eluting phase), with those favoring the stationary phase moving down the column more slowly than those favoring the mobile phase. This serves as the basis of separations.

Consider the "matchbox" model in Figure 16.1. Column 1 of the boxes represents the stationary liquid (the one supported on the inert solid), and column 2 of the boxes represents the mobile solvent that is immiscible with the first. Now suppose we place a mixture of substances A and B in equal concentrations in the first box in column 2, as in Figure 16.1a. If neither substance were soluble in solvent 1, then when additional solvent 2 is added, they both would move down the column and emerge together. Assume, however, that substance A is equally soluble in the two solvents but that substance B preferentially dissolves 75% in solvent 2. The two substances will then equilibrate between the two solvents as in Figure 16.1b. Now suppose we place a box of pure solvent 2 on top of column 2 and push all the boxes of column 2 down by one. The A and B in the first box of column 1 will ideally reequilibrate between the two solvents as before, as the A and B in the now second box in column 2 will. At equilibrium, the distribution in the four boxes will be as in Figure 16.1c. Now place another box of pure solvent 2 on column 2 and push all boxes down one. All species will redistribute so that the ratio will be the same at each level of boxes as in Figure 16.1d. For example, in the second box down, there will be a total fraction of $\frac{2}{4}$A $(\frac{1}{4} + \frac{1}{4})$ and $\frac{12}{32}$B $(\frac{3}{32} + \frac{9}{32})$ to distribute between the two solvents, with A equally distributed, and B 75% in solvent 2 $(\frac{9}{32})$. Figures 16.1e and 16.1f show the fraction distribution after two more successive additions of solvent.

It is obvious that both substances are gradually moving down the column; B moves at a faster rate than A, since it is more soluble in solvent 2. Thus, a gradual separation of the two species is occurring. In the extreme case where A dissolves completely in solvent 1 and B in solvent 2, substance A would not move at all and B would move down the column, remaining all in the original box at the front of the added solvent. This would represent maximum separation efficiency and minimum "spread" along the column. In principle, only one box of solvent would have to be added in this case to achieve complete separation.

Returning to our example, Figure 16.2 shows the distribution of the two substances in solvent 2 along the column after five additions of solvent. Substance B

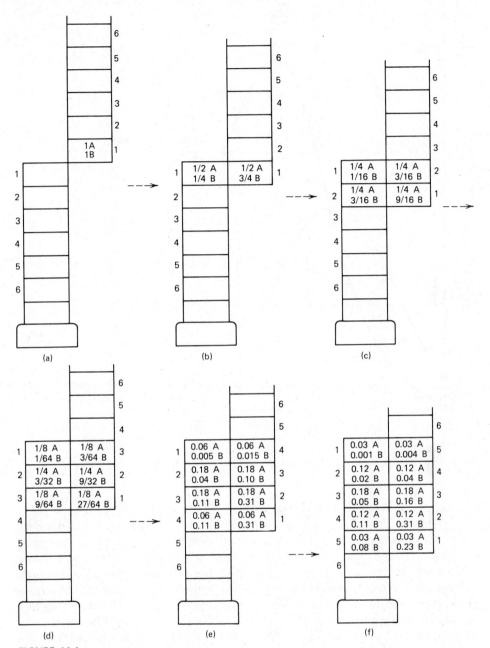

FIGURE 16.1
Matchbox model of chromatographic separation. Substance A dissolves equally in phase 1 and phase 2, while substance B dissolves 75% in phase 2.

is moving ahead of substance A, and since B is more soluble in this solvent, it is less spread out. Theoretically, there will always be a very small fraction of A and B in the first box because there will always be a small amount in the first box of solvent 1 to distribute between the two solvents, but this soon becomes a very negligible portion as the number of boxes increases.

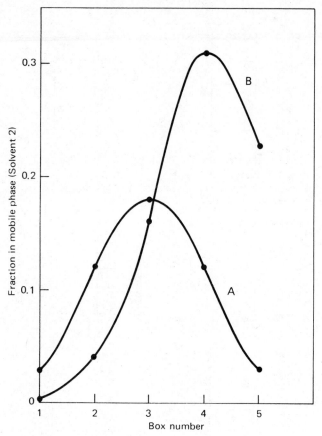

FIGURE 16.2
Plot of distribution of A and B along the column of Figure 16.1 after five additions of solvent.

A fraction distribution plot similar to Figure 16.2 is obtained if the total amounts of A and B in both phases at each box number are instead plotted, but the peak for A would be higher relative to B, since half of it is in the first phase compared to one-fourth for B. This would be a better representation of the situation required to achieve complete separation *on the column*. But in practice, we collect fractions of the mobile phase and measure the analyte in these, or we monitor the mobile phase continuously as it emerges from the column, so the plot in Figure 16.2 represents the situation for substances emerging from the column. The fractions of each in phase 2 of the last box on the column represent what emerges from the column. When there is none left in phase 2 of the last box, there will necessarily be none left on the column—phase 1.

These boxes can be considered representative of the **height equivalent to a theoretical plate** (HETP), which is a concept originating with distillation (to describe the ability of a distillation column to separate substances having different boiling points). The efficiency of a chromatographic column is similarly described in terms of theoretical plates. As we have indicated above, a chromatographic separation involves a series of equilibrations of solute between a stationary phase and a mobile phase. The theoretical plate operationally can be thought of as representing a single equilibrium step. We shall discuss this later in more detail, but suffice it to say here that, in practice, a column will have several hundred or several thousand theoretical plates (the HETP is small). The larger the number of theoretical plates, the more efficient the separation.

Figure 16.3 illustrates the distribution of A and B along the mobile phase of the column in a typical separation as increasing amounts of solvent are added. As more solvent is added and the peaks move down the column, they separate and spread out. The total amounts of A and B remain constant throughout, and if they are equal, the area under each curve remains the same.

Although there are several different forms of chromatography, this simplified model typifies the mechanism of each. That is, *there is nominally an equilibrium between two phases, one mobile and one stationary.* (True equilibrium is never really achieved.) As pure solvent is added in the mobile phase, the substances will distribute between the two phases and eventually be eluted, and if the distribution is sufficiently different for the different substances, they will be separated.

16.2   CLASSIFICATIONS OF CHROMATOGRAPHIC TECHNIQUES

Chromatographic processes can be classified according to the type of equilibration process involved, which is governed by the type of stationary phase. Various bases of equilibration are: (1) sorption, (2) solubility, (3) ion exchange, and (4) pore penetration.

Sorption Chromatography

The stationary phase is a solid on which the sample components are adsorbed. The mobile phase may be a liquid (liquid–solid chromatography) or a gas (gas–solid chromatography); the components distribute between the two phases through a combination of sorption and desorption processes. Thin-layer chromatography (TLC) is a special example of sorption chromatography in which the stationary phase is a *plane,* in the form of a solid supported on an inert plate.

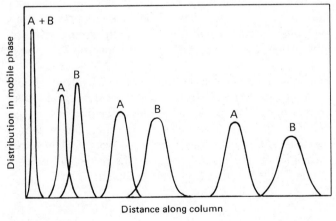

FIGURE 16.3
Distribution of two substances, A and B, along a chromatographic column in a typical chromatographic separation.

Partition Chromatography

The stationary phase of partition chromatography is a liquid supported on an inert solid. Again, the mobile phase may be a liquid (liquid–liquid partition chromatography) or a gas (gas–liquid chromatography, GLC). Paper chromatography is a type of partition chromatography in which the stationary phase is a layer of water adsorbed on a sheet of paper.

Ion Exchange and Size Exclusion Chromatography

Ion exchange chromatography uses an ion exchange resin as the stationary phase. The mechanism of separation is based on ion exchange equilibria. In size exclusion chromatography, solvated molecules are separated according to their size by their ability to penetrate a sievelike structure (the stationary phase).

These are arbitrary classifications of chromatographic techniques, and some types of chromatography are considered together as a separate technique, such as "gas chromatography" for gas–solid chromatography. In every case, successive equilibria are at work that determine to what extent the analyte stays behind or moves along with the eluent (mobile phrase). The individual types of chromatographic techniques mentioned under the above classifications will be described in more detail below.

16.3   TECHNIQUES OF COLUMN CHROMATOGRAPHY

A column is prepared by carefully packing the solid material (small particles) in a column, usually by adding it to the column filled with solvent or by pouring a slurry of it into the column and allowing this to settle. The column can be mechanically vibrated or the solid material tamped with a long plunger during packing. Care must be taken to keep out air bubbles or channeling will result in the column, rendering it less effective.

A typical column is shown in Figure 16.4. A sintered-glass frit or glass wool is placed in the bottom of the column to support the solid. A buret can be used as the column. The dimensions will depend on the separation efficiency required, the size of the sample, the type of chromatography, and so on. A typical column may range from a few millimeters in diameter and a few centimeters in length to a few centimeters in diameter and several dozen centimeters in length. **Preparative columns** may have dimensions in feet and yards, but we are concerned only with **analytical columns.**[1]

The sample is placed in a small volume on top of the column. A solvent should be used in which the sample is readily soluble, so that the volume can be kept small. But there should be a compromise for desirable eluting characteristics. An eluting (mobile) solvent that elutes (moves) the solutes too fast (i.e., one in which the solutes may be very soluble) will not separate the constituents. On the other hand, the solvent should not elute the solutes too slowly or the peaks will be very broad and difficult to detect and the solutes will be greatly diluted.

---

[1]Preparative columns are used for the separation and purification of large quantities of materials, which may range from a few grams or less to several pounds. Analytical separations, on the other hand, deal with much smaller quantities, usually in the milligram and submilligram range.

Eluting solvent added

— Sample

— Solid material

— Glass wool plug

— Eluent collected

FIGURE 16.4
Typical chromatographic column.

The solvent is allowed to flow slowly from the column until all of the sample just reaches the top of the solid (*not* below the surface). Then, solvent is added at the top of the column at the same rate that it flows out, or else a given volume is added at once. The solvent level must not drop to below the top of the solid phase or air channels in the column will result. The solvent must be eluted slowly, so that there is adequate time to establish continuous equilibrium. This will decrease "tailing" of the chromatographic peaks in which that part of the solute in the stationary phase does not have time to equilibrate with the mobile phase and hence lags behind the solute in the mobile phase.

The elution must not be so slow, however, that diffusion processes will set in; the sample components (solutes) will diffuse along the column even in the absence of solvent flow and cause the peak to spread. Again, the flow rate will depend on the type of chromatography, the column size, and so forth. With a typical column, about 1.5 cm diameter × 5 cm length, a flow rate of about 1 to 2 mL/min is common. More generally, a flow rate of about 1 cm/min for different-sized columns can be used.

The solvent **(eluent)** is collected as it emerges from the column. Usually, equal-volume aliquots of eluent are collected in individual test tubes (for example, 1-mL increments). If the flow rate is constant, then samples may be collected at equal time intervals. There are automatic **fraction collectors** that will perform either type of collection. A typical fraction collector is shown in Figure 16.5. After a certain volume of sample has been collected in the intermediate sample collector, it is emptied into a test tube (this may be signaled by completion of an electrical circuit by the sample solution in the sampler). The rack holding the test tubes rotates to the next tube. After various fractions have been collected, they are analyzed for the sample components, and the amount or concentration in each tube is plotted as a function of the tube number or volume of solvent collected. A series of peaks is obtained, each peak representing a different component, as shown in Figure 16.6. The left peak is the first to emerge and is less likely to spread out.

Once the elution order has been established under a given set of conditions, all the tubes containing a certain substance may be combined and analyzed singularly for that substance. Or, if the contents of each tube are analyzed separately, the area under the peak is proportional to the amount of the solute. If two peaks overlap, as the last two of the figure do, it may be possible to extrapolate their bases to obtain the separate areas.

Rather than perform specific analyses for individual components, it is more common to use a nonspecific method such as absorption of electromagnetic radiation at a certain wavelength and to plot the signal against tube number. Each peak is identified by running pure known substances through the column under the same conditions to determine its elution volume. Sometimes, it is possible to record the detection signal (e.g., light absorption) continually as the eluent flows from the column, thus eliminating the need to collect individual fractions. This is the com-

FIGURE 16.5
A fraction collector. (Courtesy of Arther H. Thomas Company.)

FIGURE 16.6
A typical chromatogram.

mon practice, for example, in high-performance liquid chromatography (below) and gas chromatography.

Often, all substances are not eluted from the column with a single solvent. After elution of part of the solutes, a different solvent or solvent mixture is added that shows a greater eluting power for the remaining solutes. In **solvent gradient analysis,** mixed solvents are used, but their volume ratios are varied continuously throughout the elution, usually in a linear fashion. Or a concentrated salt or other reagent in the eluting solvent may be continuously diluted, as in **salt gradient analysis.** The salt may act as a salting-out agent (precipitating) for the solutes, and the salting-out ability decreases continuously. So, a combination of salting-out and chromatographic effects is achieved. **Temperature gradient analysis** is common in gas chromatography; the gas phase equilibria are greatly affected by the temperature.

16.4   SIZE EXCLUSION CHROMATOGRAPHY

Size exclusion chromatography (also called molecular or gel chromatography) is a type of chromatography in which the stationary phase is a *molecular sieve*. These are polymeric carbohydrates and acrylamides that have an open network formed by the crosslinking of the polymeric chains. They are hydrophilic and, therefore, capable of absorbing water, whereupon swelling causes an opening of this structure. The degree of crosslinking will determine the size of the "holes."

Solvated molecules larger than the largest pores of the swollen gel cannot penetrate the gel particles and, therefore, will pass straight through the column through the spaces between the individual particles. Smaller molecules, however, will penetrate the open network in the particles to a varying degree, depending on their size and shape. They are, therefore, retarded to varying degrees and will be eluted in order of decreasing molecular size. Gels with a high degree of swelling are used to fractionate large molecules (generally high-molecular-weight substances), whereas the denser (lower swelling) gels are used for separation of low-molecular-weight compounds.

Names such as **gel filtration chromatography** (mobile phase is water), used by biochemists, and **gel permeation chromatography,** (mobile phase is an organic solvent) used by polymer chemists, describe this technique. **Size exclusion chromatography,** however, is the recommended term. Molecular weight distribution of polymers can be obtained by this technique, and proteins, enzymes, peptides, nucleic acids, hormones, polysaccharides, and so on, can be separated.

The **exclusion limit** is the molecular weight of that molecule that will just permeate the gel and be retarded. This can range from a molecular weight of 1000 to several million, depending on the gel. It should be emphasized that separations are based on a molecule's size and configuration rather than just its molecular weight, but there is generally a correlation. Generally, *molecules smaller than the exclusion limit can be fractionated down to a certain limiting size* (see Table 16.1).

The gels must be equilibrated for a few hours to a day or more with the solvent to be used, depending on the solvent uptake. Those with loose crosslinking designed for high-molecular-weight substances require the longer periods of soaking.

Sephadex is a popular molecular-sieve material for the separation of proteins. It is a polymeric carbohydrate material that, because of hydroxyl groups along the polymer chain, is fairly polar and so will adsorb water. The amount of crosslinking in the preparation can be carefully controlled to give different pore sizes and exclusion limits. Gels are characterized with respect to their swelling ability by their "water regain." This represents the amount of water imbibed by the gels on swelling. The type numbers of the Sephadex gels refer to the water-regaining values of the gels. Sephadex G-10, thus, has a water-regaining value of about 1 mL/g dry gel, and Sephadex G-200 has a value of about 20 mL/g. Several types of Sephadex gels and the fractionation range of molecules are listed in Table 16.1. These gels are insoluble in water and are stable to mild redox agents as well as to bases and weak acids.

Bio-Gel is a more chemically inert series of molecular-sieve gels, consisting of polyacrylamides. These are insoluble in water and common organic solvents and can be used in the pH range of 2 to 11. The inert gel decreases the possibility of

TABLE 16.1
Sephadex Gels[a] and Bio-Gels[b]

Sephadex Type	Fractionation Range[c] for Peptides and Globular Proteins, M.W.	Bio-Gel Type	Fractionation Range, M.W.
G-10	Up to 700	P-2	100–1800
G-15	Up to 1500	P-4	800–4000
G-25	1000 to 5000	P-6	1000–6000
G-50	1500 to 30,000	P-10	1500–20,000
G-75	3000 to 70,000	P-30	2500–40,000
G-100	4000 to 150,000	P-60	3000–60,000
G-150	5000 to 400,000	P-100	5000–100,000
G-200	5000 to 800,000	P-150	15,000–150,000
		P-200	30,000–200,000
		P-300	60,000–400,000

[a]Courtesy of Pharmacia Fine Chemicals Inc.

[b]Courtesy of Bio · Rad Laboratories.

[c]Upper limit is the exclusion limit.

adsorption of polar substances; adsorption can be a variable with Sephadex, causing changes in the chromatographic behavior of these substances. Table 16.1 lists the different Bio-Gel preparations and their separation properties.

Styragel is a polystyrene gel that is useful for purely nonaqueous separations in methylene chloride, toluene, trichlorobenzene, tetrahydrofuran, cresol, dimethylsulfoxide, and so on. It cannot be used with water, acetone, or alcohols. Gels of this can be prepared with exclusion limits for molecular weights of from 1600 to 40,000,000.

Molecular sieves are useful for the desalting of proteins that have been partially fractionated by salting out with a high concentration of some salt. A gel with a low exclusion limit, such as Sephadex 25, will allow the proteins to pass right through the column while the salts are retained. The protein dilution is limited to the elution volume of the column (the volume external of the swelled gel to fill the column).

16.5   ION EXCHANGE CHROMATOGRAPHY

While most other types of chromatography are used principally for separations of complex organic substances, ion exchange chromatography is particularly well suited for the separation of inorganic ions, both cations and anions, because the separation is based on exchange of ions in the stationary phase. It has also proved to be extremely useful for the separation of amino acids.

The stationary phase in ion exchange chromatography consists of beads made of a polystyrene polymer crosslinked with divinylbenzene. The crosslinked polymer (resin) has free phenyl groups attached to the chain, which can easily be treated to add ionic functional groups. There are basically four types of ion exchange resins used in analytical chemistry, and these are summarized in Table 16.2.

TABLE 16.2
Types of Ion Exchange Resins

Type of Exchanger	Functional Exchanger Group	Trade Name
Cation Strong acid	Sulfonic acid	Dowex[a] 50; Amberlite[b] IR120; Ionac[c] CGC-240; Rexyn[d] 101; Permutit[e] Q
Weak acid	Carboxylic acid	Amberlite IRC 50; Ionac CGC-270; Rexyn 102; Permutit H-70
Anion Strong base	Quaternary ammonium ion	Dowex 1; Amberlite IRA 400; Ionac AGA-542; Rexyn 201; Permutit S-1
Weak base	Amine group	Dowex 3; Amberlite IR 45; Ionac AGA-316; Rexyn 203; Permutit W

[a]Dow Chemical Company.
[b]Mallinckrodt Chemical Works.
[c]J. T. Baker Chemical Company.
[d]Fisher Scientific Company.
[e]Matheson Coleman & Bell.

Cation Exchange Resins

These resins contain acidic functional groups added to the aromatic ring of the resin. The strong-acid cation exchangers have sulfonic acid groups, $-SO_3H$, which are strong acids much like sulfuric acid. The weak-acid cation exchangers have carboxylic acid groups, $-CO_2H$, which are only partially ionized. The protons on these groups can exchange with other cations:

$$nRzSO_3^-H^+ + M^{n+} \rightleftharpoons (RzSO_3)_nM + nH^+ \tag{16.1}$$

and

$$nRzCO_2^-H^+ + M^{n+} \rightleftharpoons (RzCO_2)_nM + nH^+ \tag{16.2}$$

where Rz represents the resin. The equilibrium can be shifted to the left or right by increasing $[H^+]$ or $[M^{n+}]$ or decreasing one with respect to the amount of resin present.

Cation exchange resins are usually supplied in the hydrogen ion form, but they can easily be converted to the sodium ion form by treating with a sodium salt. The sodium ions then undergo exchange with other cations. The **exchange capacity** of a resin is the total number of equivalents of replaceable hydrogen per unit volume or per unit weight of resin, and is determined by the number and strength of fixed ionic groups on the resin. The ion exchange capacity affects solute retention, and exchangers of high capacity are most often used for separating complex mixtures, where increased retention improves resolution.

Weak-acid cation exchange resins are more restricted in the pH range in which they can be used, from 5 to 14, while the strong-acid resins can be used from pH 1 to 14. At low pH values, the weak-acid exchangers will "hold on" to the protons too strongly for exchange to occur. Also, the weak-acid cation exchangers will not completely remove the cations of very weak bases, while strong-acid resins will. This is analogous to the incompleteness of a weak acid–weak base reaction. Weak-acid resins are generally used for separating strongly basic or multifunctional ionic substances such as proteins or peptides that are often firmly retained on strong-acid exchangers, while strong-acid resins are more generally preferred, especially for complex mixtures.

Anion Exchange Resins

Basic groups on the resin in which the hydroxyl anion can be exchanged with other anions make up the anion exchange resins. There are strong-base groups (quaternary ammonium groups) and weak-base groups (amine groups). The exchange reactions can be represented by

$$nRzNR_3^+OH^- + A^{n-} \rightleftharpoons (RzNR_3)_nA + nOH^- \tag{16.3}$$

and

$$nRzNH_3^+OH^- + A^{n-} \rightleftharpoons (RzNH_3)_nA + nOH^- \tag{16.4}$$

where R represents organic groups, usually methyl.

The strong-base exchangers can be used over the pH range 0 to 12, but the weak-base exchangers only over the range of 0 to 9. The latter exchangers will not remove very weak acids, but they are preferred for strong acids that may be retained by strong-base resins, such as sulfonates.

Crosslinkage

The greater the crosslinkage of the resin, the greater the difference in selectivities. Increasing the crosslinkage is expressed by manufacturers as percent of divinyl-benzene. Generally, crosslinkage also increases the rigidity of the resin, reduces swelling, reduces porosity, and reduces the solubility of the resin. In general, medium-porosity materials are used for low-molecular-weight ionic species and high-porosity materials are used for high-molecular-weight ionic species. The degrees of crosslinkage is expressed by the manufacturers as percent of divinylbenzene. Generally, crosslinkage of 8 to 10% is used.

Effect of pH—Separation of Amino Acids

The ionic forms of many substances will be affected by the pH of the effluent solution. Hydrolysis of metal ions and of salt of weak acids and bases is controlled by adjusting the pH. Weak acids will not dissociate in high acid concentrations and will not exchange, and the same is true for weak bases in high alkaline concentrations. Control of pH is especially important in the separation of amino acids, which are **amphoteric** (can act as acids or bases). There are three possible forms:

$$
\underset{\underset{\text{A}}{NH_3^+}}{R{-}CH{-}CO_2H} \overset{+H^+}{\underset{}{\rightleftharpoons}} \underset{\underset{\text{B}}{NH_3^+}}{R{-}CH{-}CO_2^-} \overset{-H^+}{\underset{}{\rightleftharpoons}} \underset{\underset{\text{C}}{NH_2}}{R{-}CH{-}CO_2^-}
$$

Form B, called a **zwitterion,** is the dominant form at the pH corresponding to the **isoelectric point** of the amino acid (see Chapter 5). The isoelectric point is the pH at which the net charge on the molecule is zero. In more acid solutions than this, the $-CO_2^-$ group is protonated to form a cation (form A), while in more alkaline solutions, the $-NH_3^+$ group loses a proton to form an anion (form C). The isoelectric point will vary from one amino acid to another, depending on the relative acidity and basicity of the carboxylic acid and amino groups. Thus, group separations based on the isoelectric points are possible by pH control.

At a given pH, the amino acids can be separated into three groups by being passed successively through an anion and a cation exchange column. The uncharged zwitterions (isoelectric point) will pass through both columns, while the positively and negatively charged amino acids will each be retained by one of the columns. The groups can be further subdivided by changing the pH.

Moore and Stein [*J. Biol. Chem.,* **192** (1951), 663] have successfully separated up to 50 amino acids and related compounds on a single Dowex-50 cation exchange column by a combination of pH and temperature control. (The temperature affects the equilibria involved.)

Automatic amino acid analyzers based on ion exchange separation are commercially available. The elution of each amino acid is automatically recorded by measuring the color formed between the amino acid and ninhydrin as it is eluted. By operating at pressures of several hundred psi, these perform separations in about 200 minutes. They have proved valuable to the biochemist as an aid in the elucidation of protein structure. The protein fragments are degraded to amino acids, which must be determined.

Effect of Complexing Agents—Separation of Metal Ions on Anion Exchange Columns

Many metals can be separated on anion exchange columns by being converted to anions by complexation. The complexing agent is an anion such as chloride, bromide, or fluoride. Uncharged complexing agents also affect the equilibrium if they form a complex that must dissociate before the metal is exchanged or if they change its size. Many complexing agents are either weak acids or bases or are salts of these, and so a complicated interdependence of pH and complexation often results.

Some of the most successful separations of metals have been on anion exchange columns. A complexing acid is added in high concentration to form anionic complexes of the metals. Concentrated hydrochloric acid forms anionic chlorocomplexes with all the common metals, with the exception of the alkali and alkaline earth metals and Al(III), Ni(II), and Cr(III), and so all of these can be adsorbed on a quaternary ammonium anion exchange column.

Distribution coefficients of metals on Dowex-1 anion exchange resin as a function of hydrochloric acid concentration are summarized in Figure 16.7 for different valences of the metals. Separations can be achieved by choosing a hydrochloric acid concentration at which one element has a high distribution coefficient and the other a low distribution coefficient. The former will then be strongly retained by the resin, while the other will be eluted. Then, by decreasing the hydrochloric acid concentration to a level at which the distribution coefficient of the second metal is low, the metal can be eluted. Several metals can be successively eluted in this manner. The sample is usually placed on the column in 10 M or 11 M HCl.

As an example, a mixture of Ni^{2+}, Mn^{2+}, Co^{2+}, Cu^{2+}, Fe^{3+}, and Zn^{2+} can be separated as follows. The Ni(II) is not anionic and is eluted in the first column volume of 12 M HCl. Manganese(II) is essentially not adsorbed in 6 M HCl (Figure

FIGURE 16.7
Distribution coefficients of the elements on Dowex-1 anion exchange resin as a function of the hydrochloric acid concentration. From K. A. Kraus and F. Nelson, *Proceedings of the First United Nations International Conference on the Peaceful Uses of Atomic Energy*, **7** (1956) 113.

16.7, while the other metals have rather large distribution coefficients (remember that the values in Figure 16.7 are log D, and so a value of 2 means D is equal to 100). Thus, manganese is eluted with 6 M HCl. At 4 M HCl, D for cobalt(II) is 1 (log D = 0), and it is eluted. The copper is only partially eluted in the volume required to elute the manganese, but it is completely eluted in the 2.5 M HCl. Note that iron and zinc still have large D values at this concentration. At 0.5 M HCl, the iron(III) D value is near unity, and it becomes eluted. The zinc is so strongly adsorbed that the acid concentration must be decreased to 0.005 M HCl, and even then it requires several column volumes for complete elution. It can be more easily eluted with 2 M HNO$_3$ (after the others are eluted).

We attempt to chose conditions so that each given metal is eluted quickly in essentially one or two column volumes ($D \leq 1$).

Some Applications of Ion Exchange Chromatography

1. PURIFICATION. One of the most important applications of ion exchange is the deionization of water, which, although nonanlytical, offers great advantage for the analytical chemist. The water is passed through a **mixed-bed ion exchange resin** (commercially available) that contains both a strong-acid cation and a strong-base anion exchange resin. When water containing a salt such as CaCl$_2$ is passed through the column, the Ca^{2+} ion is exchanged for two H$^+$ ions and the two Cl$^-$ ions are exchanged for two OH$^-$ ions. The net result is that the salt is exchanged for H$_2$O.

Water with resistance of several megohms can be obtained in this way. Organic constituents, however, are not removed, and sometimes the water is passed through a column of activated charcoal for removal of organic matter. Traces of organic substances in the water are always a danger when it is passed through a mixed-bed ion exchange column. If these interfere with a particular analysis, it might be preferable to distill the water.

2. CONCENTRATION. Ionic materials that exist at very low concentrations can often be concentrated by collecting them on an ion exchange column and then eluting them with a much smaller volume of an appropriate eluting agent. In this manner, the ions are also often removed from the bulk matrix so they may be obtained in purer form. The concentration of trace elements in seawater by ion exchange is commonly employed.

3. ANALYTICAL SEPARATIONS. The most important application of ion exchange chromatography is seen in analytical separations.

We have mentioned some representative analytical separations earlier—of metal ions and amino acids. Halide ions can be separated on a Dowex 2 column in the order F$^-$, Cl$^-$, Br$^-$, I$^-$, with 1 M NaNO$_3$ at pH 10.4 (adjusted with NaOH). The alkali metals can be separated on a Dowex 50 or Amberlite IR 120 sulfonated cation exchange resin in the order Li$^+$, Na$^+$, K$^+$, by eluting with 0.7 M HCl. The H$^+$ acts as the eluent via mass action. The alkaline earth metals Ca^{2+}, Sr^{2+}, and Ba^{2+} can be eluted from a Dowex 50 column in that order by 1.2 M ammonium lactate.

Quantitative ion exchange analysis of parts-per-million concentrations of ions can be done automatically using the technique of **ion chromatography,** discussed under high-performance liquid chromatography below.

16.6   GAS CHROMATOGRAPHY

There are two types of gas chromatography: gas–solid (adsorption) chromatography and gas–liquid (partition) chromatography. The most important of the two is gas–liquid chromatography (GLC). Gas chromatography is the most widely used form of chromatography, particularly by organic chemists, and it undoubtedly ranks as one of the most important new analytical techniques since its development in 1952. The separation of benzene and cyclohexane (b.p. 80.1 and 80.8°C) is extremely simple by gas chromatography, but it is virtually impossible by conventional distillation. Very complex mixtures can be separated by this technique.

Principles

In gas chromatography, the sample is converted to the vapor state (if it is not already a gas) and the eluent is a gas (the *carrier gas*). The stationary phase is generally a nonvolatile liquid supported on an inert solid such as firebrick (Chromosorb-P or W, etc.) or diatomaceous earth. There are a large number of liquid phases available, and it is by changing this phase rather than the mobile gas phase that different separations are accomplished.

A schematic diagram of a gas chromatograph is given in Figure 16.8. The sample is rapidly injected be means of a hypodermic syringe through a rubber septum into the column. The sample injection port, column, and detector are heated to temperatures at which the sample has a vapor pressure of at least 10 torr. The injection port and detector are usually kept somewhat warmer than the column to promote rapid vaporization of the injected sample and prevent sample condensation in the detector. Liquid samples of 0.1 to 10 μL are injected, while gas samples of 1 to 10 mL are injected. Gases may be injected by means of a gas-tight syringe or through a special gas inlet chamber of constant volume (gas sampling valve).

Separation occurs as the vapor constituents partition between the gas and the liquid phases in the same manner as in other chromatographic processes. The carrier gas is a chemically inert gas available in pure form, such as argon, helium, nitrogen, or carbon dioxide. A high-density gas gives best efficiency, but a low-

FIGURE 16.8
Schematic diagram of a gas chromatograph.

density gas gives faster speed. The choice of gas is often dictated by the type of detector.

The sample is automatically detected as it emerges from the column (at a constant flow rate), using a variety of detectors whose response is dependent upon the composition of the vapor (see below). Usually, the detector contains a *reference side* and a *sampling side*. The carrier gas is passed through the reference side before entering the column and emerges from the column through the sampling side. The response of the sampling side relative to the reference side signal is fed to a recorder where the chromatographic peaks are recorded as a function of time. By measuring the **retention time** (the minutes between the time the sample is injected and the time the chromatographic peak is recorded) and comparing this time with that of a standard of the pure substance, it may be possible to identify the peak (agreement of retention times of two compounds does not guarantee the compounds are identical). The area under the peak is proportional to the concentration, and so the amount of substance can be quantitatively determined. If the peak is symmetrical, then the area is given by the product of the peak height and the width at half the peak height. The peaks are often very sharp and, if so, the peak height is taken and compared with a calibration curve prepared in the same manner. Problem 27 shows use of *relative* retention times.

The separation ability of this technique is illustrated in a chromatogram in Figure 16.9. A gas can move much more rapidly through a packed column than a liquid can, and so chromatographic separations require only minutes, as compared with much longer times for other chromatographic techniques (an exception is high-performance liquid chromatography—see below). In addition, since the peaks are

FIGURE 16.9
Typical gas chromatogram. From T. Johns, *Gas Chromatography Applications Manual*, Beckman Instruments, Inc., 1964, p. 29. (Reproduced by permission of Beckman Instruments, Inc.)

automatically recorded, the entire analysis time is amazingly short. This, coupled with the very small sample required, explains the popularity of the technique. This is not to exclude the more important reason that many of the analyses performed simply cannot be done by other methods.

With complex mixtures, it is not a simple task to identify the many peaks. The individual fractions (each peak may be a mixture of components) can be collected by cooling the collecting vessel in a dry ice–acetone bath to condense the constituents, and then they can be analyzed by infrared spectrophotometry or mass spectrometry to aid in their identifications. Instruments are commercially available in which the gas effluent is automatically fed into a mass spectrometer where they are positively identified according to mass (formula weight and fragmentation pattern). This important analytical technique is called **gas chromatography–mass spectrometry** (GC–MS). The mass spectrometer is a very sensitive and selective detector, and when a capillary GC column (very high resolution—see columns below) is used (capillary GC–MS), this technique is capable of identifying and quantifying unbelievably complex mixtures of trace substances. For example, hundreds of compounds may be identified in sewage effluents, and traces of complex drugs in urine or blood or pollutants in water can be determined.

Gas Chromatography Columns

Chromatographic columns can be in any shape that will fit the heating oven. Common forms are a U-shaped tube, a W-shaped tube, and a coiled tube (Figure 16.8). Typical packed columns are 1 to 2 m long and 0.3 to 0.6 cm in diameter. Short columns can be made of glass, but longer columns may be made of stainless steel so they can be straightened out for filling and packing. For inertness, glass is still preferred for long columns. Besides packed columns, there are **capillary,** or **Golay, columns.** These are open tubular columns, typically about 0.01 cm i.d. and 20 or more m long, in which the liquid stationary phase is coated on and supported by the walls of the tube. The original capillary columns had a very thin liquid film coated on a smooth metal or glass surface. The small diameter necessitated a slow flow rate (1 mL/min). The columns could be made very long (15 to 35 m) since there is a low pressure drop in the open tube, and they possessed very high separation efficiency (large number of theoretical plates—see Column Efficiency below); but they could handle only very small samples, about 10^{-2} or 10^{-3} μL of liquid. This necessitated the use of special gas-splitting devices so that only a fraction of the injected sample was introduced into the column. Only the most sensitive detectors could be used, and analysis of trace constituents was difficult.

The second-generation capillary columns overcome some of the restrictions of the earlier columns. These are the **support-coated open tubular** columns (S.C.O.T.). In these, a layer of celite or other solid support is adsorbed onto the tubing wall and a liquid phase is adsorbed onto the celite. Tubing diameter is 0.05 to 0.15 cm. The advantages of low pressure drop and long columns is maintained, but capacity of the columns approaches that of packed columns. Flow rates are faster (4 to 10 mL/min) and dead volume connections at the inlet and detector are less critical. Analysis of trace constituents is possible because larger samples can be injected. Sample splitting is not required in many cases so long as the sample volume is 0.5 μL or less. If a separation requires more than 10,000 theoretical plates (see Column Efficiency), then a S.C.O.T. column should be considered.

Gas Chromatography Detectors

Since the initial experiments with gas chromatography were begun, over 40 detectors have been developed. Some are designed to respond to most compounds in general, while others are designed to be selective for particular types of substances. We describe some of the more widely used detectors. Table 16.3 lists and compares some commonly used detectors with respect to application, sensitivity, and linearity.

The most widely used is the **thermal conductivity, or hot wire, detector.** As a gas is passed over a heated filament wire, the temperature and thus the resistance of the wire will vary according to the thermal conductivity of the gas. The pure carrier is passed over one filament, and the effluent gas containing the sample constituents is passed over another. These filaments are in opposite arms of a Wheatstone bridge circuit that measures the *difference* in their resistance. So long as there is no sample gas in the effluent, the resistance of the wires will be the same. But whenever a sample component is eluted with the carrier gas, a small resistance change will occur in the effluent arm. The change in the resistance, which is proportional to the concentration of the sample component in the carrier gas, is registered on the recorder.

Hydrogen and helium carrier gases are preferred with thermal conductivity detectors because they have a very high thermal conductivity compared with most other gases, and so the largest change in the resistance occurs in the presence of sample component gases (helium is preferred for safety reasons). The thermal conductivity of hydrogen is 53.4×10^{-5} and that of helium is 41.6×10^{-5} cal/°C-mol at 100°C, while those of argon, nitrogen, carbon dioxide, and most organic vapors are typically one-tenth of these values. The advantages of thermal conductivity detectors are their simplicity and approximately equal response for most substances. Also, their response is very reproducible. They are not the most sensitive detectors, however.

Most organic compounds form ions in a flame. This forms the basis of an extremely sensitive detector, the **flame ionization detector.** The ions are measured (collected) by a pair of oppositely charged electrodes. The response (number of ions collected) depends on the number of carbon atoms in the sample and on the oxidation state of the carbon. Those atoms that are completely oxidized do not ionize. This detector gives excellent sensitivity, permitting measurement of components in the ppb concentration range. This is about 1000 times more sensitive than the thermal conductivity detector. However, the dynamic range is more limited, and samples of pure liquids are generally restricted to 0.1 μL or less. The carrier gas is relatively unimportant. Helium, nitrogen, and argon are most frequently employed. The flame ionization detector is insensitive to most inorganic compounds, including water, and so aqueous solutions can be injected. If oxygen is used as the flame support gas in place of air, then many inorganic compounds can be detected because a hotter flame is produced which can ionize them. Since the flame ionization detector is so sensitive, a portion of the sample can be diverted by an appropriate stream splitter so that it can be collected and analyzed further if necessary.

The **flame thermionic detector** is essentially a two-stage flame ionization detector, which is designed to give an increased specific response for halogen- and phosphorus-containing substances. A second flame ionization detector is mounted above the first, with the flame gases from the first passing into the second flame.

TABLE 16.3
Comparison of Gas Chromatographic Detectors

Detector	Application	Sensitivity	Linearity	Remarks
Thermal conductivity	General, responds to all substances	Fair	Good, except thermistors at higher temperatures	Sensitive to temperature and flow changes; concentration sensitive
Flame ionization	All organic substances; some oxygenated products respond poorly	Very good; 10^{-11} g/mL in carrier gas	Excellent, up to 10^6	Requires very stable gas flow; response for water is 10^4–10^6 times weaker than for hydrocarbons; concentration-sensitive (organics)
Flame thermionic	All halogen and phosphorous-containing substances	Excellent		Need recoating of sodium salts on screen; mass-sensitive
Rubidium silicate bead	Specific for nitrogen- and phosphorous-containing substances	Excellent		Mass-sensitive
Argon ionization (β-ray)	All organic substances; with ultrapure He carrier gas, also for inorganic and permanent gases	Very good; 10^{-10}–10^{-13} g/mL in carrier gas	Good	Very sensitive to impurities and water; needs very pure carrier gas; concentration sensitive
Electron capture	All substances that have affinity to capture electrons; no response for aliphatic and naphthenic hydrocarbons	Excellent for halogen-containing substances	Poor	Very sensitive to impurities and temperature changes; quantitative analysis complicated; concentration-sensitive

The two stages are divided by a wire mesh screen coated with an alkali salt or base such as sodium hydroxide.

The column effluent enters the lower flame, which acts as a conventional FID (flame ionization detector) whose response may be recorded. A small current normally flows in the second flame due to evaporation and ionization of sodium from the screen. However, if a substance containing a halogen or phosphorus is burned in the lower flame, the ions resulting from these greatly increases the volatilization of the alkali metal from the screen. This results in a response that is much greater (at least 100 times) than the response of the lower flame to the halogen or phosphorus. By recording the signals from both flames, one can obtain the usual chromatogram of a FID; a second chromatogram is obtained where the peaks corresponding to the halogen- and phosphorus-containing compounds are amplified over the others, which will be practically missing.

In the **β-ray,** or **argon ionization, detector,** the sample is ionized by bombardment with β rays from a radioactive source (for example, strontium 90). The carrier gas is argon, and the argon is excited to a metastable state by the β particles. Argon has an excitation energy of 11.5 eV, which is greater than the ionization potential of most organic compounds, and the sample molecules are ionized when they collide with the excited argon atoms. The ions are detected in the same manner as in the flame ionization detector. This detector is very sensitive but less accurate than others, and the β-ray source is a potential hazard, although with proper shielding, no danger exists. The sensitivity is about 300 times greater than that of the conventional thermal conductivity cell.

The **electron capture detector** is extremely sensitive for compounds that contain electronegative atoms and is selective for these. It is similar in design to the β-ray detector, except that hydrogen or nitrogen is used as the carrier gas. These gases have low excitation energies compared to argon and only compounds that have high electron affinity are ionized, by capturing electrons.

The detector cathode consists of a metal foil impregnated with a β-emitting element, usually tritium or nickel 63. The former isotope gives greater sensitivity than the latter, but it has an upper temperature limit of 220°C because of losses of tritium at high temperatures; nickel 63 can be used routinely at temperatures up to 350°C. Also, nickel is easier to clean than the tritium source; these radioactive sources inevitably acquire a surface film that decreases the beta-emission intensity and hence the sensitivity. A 30% KOH solution is usually used to clean the sources.

The cell is normally polarized with an applied potential, and electrons (β rays) emitted from the source at the cathode migrate to the anode, establishing a current. When a compound possessing electron affinity is introduced into the cell, it captures electrons to create a large negative ion. The negative ion has a mobility in an electric field about 100,000 less than electrons, and so a decrease in current results.

Relatively few compounds show significant electronegativities, and so electron capture is quite selective, allowing the determination of trace constituents in the presence of noncapturing substances. High-electron-affinity atoms or groups include halogens, carbonyls, nitro groups, certain condensed ring aromatics, and certain metals. Electron capture has very low sensitivity for hydrocarbons other than aromatics.

Compounds with low electron affinities may be determined by preparing appropriate derivatives. Most important biological compounds, for example, possess low electron affinities. Steroids such as cholesterol can be determined by preparing their chloroacetate derivatives. Trace elements have been determined at nanogram and picogram levels by preparing volatile trifluoroacetylacetone chelates. Examples are chromium, aluminum, copper, and beryllium. Methylmercuric chloride, present in contaminated fish, can be determined at the nanogram level.

The **coulometric detector** was developed in 1960 by Coulson and co-workers. It was originally designed for the selective detection of chlorine containing substances and is widely used for pesticide analysis. The column effluent is automatically combusted by passing through an oven at 800°C to give HCl, which is absorbed in a cell containing a solution of silver nitrate. The cell contains a potentiometric silver-indicating electrode that detects changes in the silver concentration as AgCl is precipitated. The variation in potential relative to a reference electrode is amplified and triggers a generating current circuit; the circuit contains a pair of generating electrodes, the anode being a silver wire. Silver ion is produced from the silver anode to compensate exactly for the amount that has been removed from the solution. The generating current is varied as needed to maintain a constant silver concentration in the solution and this current is recorded as a function of time to provide the chromatogram. The principles of coulometric titration are described in more detail in Chapter 12.

Sulfur and phosphorus are determined in a similar fashion when burned in a reducing gas (hydrogen). Sulfur is converted to H_2S, which precipitates Ag_2S, while phosphorus gives PH_3, which reacts with silver to give Ag_2PH. Nitrogen-containing compounds are determined by passing over a nickel catalyst to convert the nitrogen to NH_3, which is absorbed in an acid solution and titrated with H^+ generated coulometrically at a platinum anode by electrolysis of water. Sulfur can be determined in the presence of halides by combustion in O_2 to give SO_2 which is absorbed in an iodine solution and titrated with coulometrically generated iodine by oxidation of iodide at a platinum anode.

Column Efficiency in Gas Chromatography

1. THEORETICAL PLATES.   The separation efficiency of a column can be expressed in terms of the number of theoretical plates in the column. A theoretical plate is a concept that can be visualized from the "matchbox" model in Figure 16.1, and is defined in such a way that it is a useful empirically defined parameter. As mentioned before, it can be thought of as representing a single equilibrium step. For high efficiency, a large number of theoretical plates is necessary. The **height equivalent to a theoretical plate** (HETP) is the length of a column divided by the number of theoretical plates. To avoid a long column, then, the HEPT should be as short (thin or small) as possible. These concepts apply to all forms of column chromatography and also to distillation separations, but the parameters are easier to determine in gas chromatography.

The **number of theoretical plates** can be obtained from a chromatogram from the expression

$$n = 16\left(\frac{t_R}{w}\right)^2 \tag{16.5}$$

where n is the number of theoretical plates of a column *toward a particular compound*, t_R is the retention time, and w is the peak width measured in the same units as t_R. These are illustrated in Figure 16.10. Retention volume V_R may be used in place of t_R. It should be noted that w is not the base width of the peak, but is the width obtained from the intersection of the baseline with tangents drawn through the inflection points at each side of the peak.

An alternative way to estimate the number of theoretical plates is from the width of the peak measured at a height of one-half the peak height, $w_{1/2}$:

$$n = \frac{5.55 t_R^2}{w_{1/2}^2} \qquad (16.6)$$

Once the number of theoretical plates is known, the HETP can be obtained by dividing the length of the column by n. The width of a peak, then, is related to the HETP, being narrower with a smaller HETP. The units for HETP are cm/plate.

2. THE VAN DEEMTER EQUATION.   Van Deemter has shown that the broadening of a peak is the summation of somewhat interdependent effects from several sources. The **van Deemter equation** expresses these in terms of the HETP, the height equivalent to a theoretical plate:

$$\boxed{\text{HETP} = a + \frac{B}{\mu} + C\overline{\mu}} \qquad (16.7)$$

where A, B, and C are constants for a given system and are related to the three major factors affecting the HETP, and $\overline{\mu}$ is the average linear velocity of the carrier gas in cm/s. The significance of the three terms A, B, and C is illustrated in Figure 16.11, which is a plot of the HETP determined as a function of the carrier gas velocity. Here, A represents **eddy diffusion** and is due to the variety of tortuous (variable-length) pathways available betweeen the particles in the column and is independent of the gas velocity; B represents **longitudinal** or **molecular diffusion**

FIGURE 16.10
Determination of number of theoretical plates.

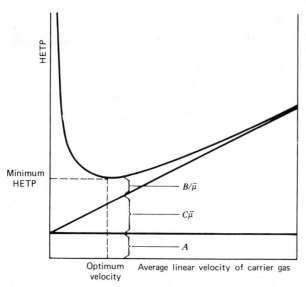

FIGURE 16.11
Illustration of the van Deemter equation.

of the sample components in the carrier gas, due to concentration gradients within the column; and C represents the **rate of mass transfer** and is due to the finite time required for solute equilibrium to be established between the two phases. The last two terms in the van Deemter equation are dependent on the gas velocity, but in opposite manners. A is a characteristic of the column packing and can be decreased with smaller and more uniform particles and tighter packing, which reduces the effective HETP and thus increases efficiency. However, an extremely fine solid support is difficult to pack uniformly, and the uniformity of packing also affects the eddy diffusion. The molecular diffusion (B) is a function of both the sample and the carrier gas. Since the sample components are fixed in a given analysis, the only way to change B or $B/\bar{\mu}$ is by varying the type, pressure, and flow rate of the carrier gas. High flow rates reduce molecular diffusion, as do denser gases, such as nitrogen or carbon dioxide versus helium or hydrogen. The mass transfer term (C) is influenced by the partition coefficient and, therefore, by the relative solubility of the sample in the liquid phase (i.e., by the type and amount of liquid phase as well as the temperature). Or, in the case of gas–solid chromatography, it is influenced by the adsorbability of the solute on the solid phase. Increasing the solubility of the vapor components of the sample in the liquid phase by decreasing the temperature may decrease C, provided the viscosity of the liquid phase is not increased so much that the exchange equilibrium becomes slower. The term $C\bar{\mu}$ is also decreased by decreasing the flow rate, allowing more time for equilibrium. In addition, it is minimized by keeping the liquid-phase film as thin as possible to minimize diffusion within this phase.

Since the flow rate of the carrier gas (e.g., L/min) is proportional to the linear velocity, a qualitatively similar curve to Figure 16.11 will be obtained by plotting HETP as a function of flow rate. The constants A, B, and C would have different numerical values (using flow rate in place of $\bar{\mu}$ in Equation 16.7).

The conditions (e.g., flow rate) must be adjusted to obtain a balance between molecular diffusion and mass transfer. The three terms A, $B/\bar{\mu}$, and $C\bar{\mu}$ are kept as small as possible to provide the minimum HETP for the sample solute that is the most difficult to elute (last to be eluted from the column). A van Deemter plot can aid in optimizing conditions. A, B, and C can be determined from three points and a solution of the three simultaneous van Deemter equations. Theoretically, a plot of Equation 16.7 results in a minimum (HETP)$_{min}$, of $A + 2\sqrt{BC}$ at $\overline{\mu}_{opt} = \sqrt{B/C}$. Note the importance of the slope beyond $\overline{\mu}_{opt}$. The smaller the slope, the better since the efficiency will then suffer little at velocities in excess of $\overline{\mu}_{opt}$.

An efficient chromatography column will have several thousand theoretical plates, and capillary columns will have in excess of 10,000 theoretical plates. The HETP for a 1-m column with 10,000 theoretical plates would be 100 cm/10,000 plates = 0.01 cm/plate.

3. CAPACITY FACTOR. The **capacity factor** k' is defined by

$$k' = \frac{t_r - t_m}{t_m} = \frac{t_r'}{t_m} \tag{16.8}$$

where t_r is the **retention time** (time required for the analyte peak to appear) and t_m is the time required for the mobile phase to traverse the column and is the time it would take for an unretained solute to appear. The difference $t_r - t_m$ is called the **adjusted retention time** t_r'. An air peak often appears in a chromatogram from unretained air injected with the sample, and the time for this to appear is taken as t_m. A large capacity factor favors good separation. However, large capacity factors mean increased elution time, so there is a compromise between separation efficiency and separation time. The capacity factor can be increased by increasing the stationary phase volume. A change in the capacity factor is an indication of degradation of the stationary phase.

The volume of a chromatographic column consists of the stationary phase volume and the **void volume,** the volume occupied by the mobile phase. The latter can be determined from t_m and the flow rate. One void volume of the mobile phase is required to flush the column once.

4. RESOLUTION. The resolution of two chromatographic peaks is defined by

$$\text{resolution} = \frac{\Delta t_r}{w} \tag{16.9}$$

where t_r is the difference in the retention times of the two peaks and w is the width of the peaks and is assumed to be the same for each. A resolution of 1.0 results in 2.3% overlap of two peaks, and a resolution of 1.5 in an overlap of only 0.1%. The resolution is proportional to the square root of the number of theoretical plates, $\sqrt{n}$, and hence to the square root of the length of the column, $\sqrt{l}$. So, doubling the column length increases the resolution by $\sqrt{2}$, or 1.4. A fourfold increase would double the resolution. Retention times, of course, would be increased in direct proportion to the length of the column.

Some Practical Aspects of Gas Chromatography

There are innumerable examples of gas-chromatographic separations, and it would be impossible to consider all the different types and variations of the technique. The most important factor in gas chromatography is the selection of the proper column (stationary phase) for the particular separation to be attempted. The nature of the liquid or solid phase will determine the exchange equilibrium with the sample components and this will depend on the solubility or adsorbability of the sample, the polarity of the stationary phase and sample molecules, the degree of hydrogen bonding, specific chemical interactions, and so on. Most of the useful separations have been determined empirically, although more quantitative information is now available (see Selecting the Liquid Phase below). Table 16.4 lists some common substrates (stationary phases) used along with some types of compounds for which they are useful.

1. SOLID SUPPORTS. The solid support for a liquid phase should have a high specific surface area that is chemically inert but wettable by the liquid phase. It must be thermally stable and available in uniform sizes. The most commonly used supports are prepared from diatomaceous earth, a spongy siliceous material. They are sold under many different trade names. Chromosorb W is diatomaceous earth that has been heated with an alkaline flux to decrease its acidity; it is light in color. Chromosorb P is crushed firebrick that is much more acidic than Chromosorb W, and it tends to react with polar solutes, especially those with basic functional groups.

The polarity of Chromosorb P can be greatly decreased by silanizing the surface with hexamethyldisilazane, $[(CH_3)_3Si]_2NH$. Ottenstein (Reference 10 at the end of the chapter) has reviewed the selection of solid supports, both diatomaceous earth and porous polymer types.

2. COATING THE SOLID SUPPORT. Column packing support material is coated by mixing with the correct amount of liquid phase dissolved in a low-boiling solvent such as acetone or pentane. About a 5 to 10% coating (wt/wt) will give a thin layer. After coating, the solvent is evaporated by heating and stirring;

TABLE 16.4
Some Typical Substrates for Gas Chromatography

Substrate	Type	Solutes
Solids		
Poropak	Polystyrene crosslinked with divinylbenzene	Polar compounds: H_2O, acids, amines, alcohols; useful as solid support for liquids
Molecular sieves 5A and 13X		Gases: H_2, Ne, O_2, Ar, N_2, CH_4, CO; retains polar gases CO_2, H_2S
Silica gel		Separation of H_2, air, CH_4, C_2H_6, CO_2
Liquids		
Apiezons J, K, L, and M ($\leq$ 300°C)	High-vacuum greases, useful at high temperatures	High-boiling hydrocarbons, ethers, esters, boranes, polar compounds
Silicone oils ($\leq$200°C)	Low to moderate polarity	Many types
Squalene ($\leq$200°C)	High-boiling, low-polarity, saturated hydrocarbon oil	Saturated hydrocarbons
Polyethylene glycols, 200, 400, 600, 1000 (carbowaxes) ($\leq$250°C)		n-Paraffins, aromatics, alcohols, ketones, amines
Silicone rubber gum (SE-30) ($\leq$350°C)		Steroids, alkaloids, pesticides, acids, methyl esters

last traces may be removed in vacuum. A newly prepared column should be conditioned at elevated temperature by passing carrier gas through it for several hours.

Open tubular columns are coated by slowly passing a dilute solution of the liquid phase through the columns. The solvent is evaporated by passing carrier gas through the column.

3. SELECTING THE LIQUID PHASE.   There are thousands of commercially available column materials, and several attempts have been made to predict the proper selection of liquid immobile phase without resorting exclusively to trial-and-error techniques. These methods attempt to group phases according to their retention properties, for example, according to polarity. The **Kovats indeces** and **Rohrschneider constant** are two approaches used to group different materials. Supina and Rose (Reference 11, at the end of the chapter) have tabulated the Rohrschneider constants for 80 common liquid phases, which enables one to decide, almost by inspection, if it is worth trying a particular liquid phase. Equally important, it is easy to identify phases that are very similar and differ only in trade name. McReynolds described a similar approach, defining phases by their **McReynolds constants** (Reference 8).

Another useful literature reference for the selection of stationary phases is a booklet entitled *Guide to Stationary Phases for Gas Chromatography,* compiled by Analabs, Inc., North Haven, Conn., 1977.

The **Kovats retention index** is useful also for identifying a compound from its retention time relative to those of similar compounds in a homologous series (those that differ in the number of carbon atoms in a similar structure, as in alkane chains). The index *I* is defined as

$$I = 100\left[n_s + \frac{\log t'_{r(unk)} - \log t'_{r(ns)}}{\log t'_{r(nl)} - \log t'_{r(ns)}}\right] \tag{16.10}$$

where n_s is the number of carbon atoms in the smaller alkane, and n_l refers to the larger alkane. The Kovats index for an unknown compound can be compared with catalogued indices on various columns to aid in its identification. The logarithm of the retention time, $\log t'_r$, is generally a linear function of the number of carbon atoms in a homologous series of compounds.

4. POLAR VERSUS NONPOLAR LIQUIDS.   Nonpolar liquid phases are generally nonselective. The reason for this is that there are few forces between the solute and solvent, and so the volatility of the solute is determined primarily by its vapor pressure. Hence, separations will tend to follow the order of the boiling points of the solutes, the lower boiling being eluted first. With polar liquid phases, however, volatility is not so simply determined because of interactions between the solute and the solvent, such as dipole–dipole interactions, hydrogen bonds, induction forces, and so on.

5. TEMPERATURE SELECTION.   The proper temperature selection in gas chromatography is a compromise between several factors. The *injection temperature* should be relatively high, consistent with thermal stability of the sample, to give the fastest rate of vaporization to get the sample into the column in a small volume; decreased spreading and increased resolution result. Too high an injection temperature, though, will tend to degrade the rubber septum and cause dirtying of the injection port. The *column temperature* is a compromise between *speed, sensitivity,* and *resolution.* At high column temperatures, the sample components spend most of their time in the gas phase and so they are eluted quickly, but resolution is poor. At low temperatures, they spend more time in the stationary phase and elute slowly; resolution is increased but sensitivity is decreased due to increased spreading of the peaks. The *detector temperature* must be high enough to prevent condensation of the sample components. The sensitivity of the thermal conductivity detector decreases as the temperature is increased and so its temperature is kept at the minimum required.

Some separations can be facilitated by **temperature programming.** The temperature is automatically increased at a preselected rate during the running of the chromatogram; this may be linear, exponential, steplike, and so on. In this way, the compounds eluted with more difficulty can be eluted in a reasonable time without forcing the others from the column too quickly.

If the constituent to be determined is not volatile at the accessible temperatures, it may be converted to a *volatile derivative*. For example, nonvolatile fatty acids are converted to their volatile methyl esters. Some inorganic halides are sufficiently volatile that at high temperatures they can be determined by gas chromatography. Metals may be made volatile by complexation, for example with trifluoroacetylacetone.

Column lengths of 2 to 3 m are commonly used with packed columns. The resolution generally increases only with the square root of the length of the column. Long columns require high pressure and longer analysis times and are used only when necessary.

6. COLUMN LENGTH. Separations are generally attempted by selecting columns in lengths of multiples of three, such as 1 m, 3 m, and so on. If a separation isn't complete in the shorter column, the next longer one is tried.

Quantitative Measurement of Chromatographic Peaks

We mentioned that the concentrations of eluted solutes are proportional to the areas under the recorded peaks. A variety of methods can be used to estimate the areas, including mechanical or electronic integrators (as a part of the recorder) or cutting out and weighing the peaks. And we have mentioned the possibility of measuring the peak height to construct a calibration curve or taking the product of the peak height and its width at half-height as a measure of the area. The linearity of a calibration curve should always be established.

The method of **standard additions** is a useful technique for calibrating, especially for occasional samples. One or more aliquots of the sample are spiked with a known concentration of standard, and the increase in peak area is proportional to the added standard. This method has the advantage of verifying that the retention time of the unknown analyte is the same as that of the standard.

A more important method of quantitative analysis is the use of **internal standards.** Here, the sample and standards are spiked with an equal amount of a solute whose retention time is near that of the analyte. The ratio of the area of the standard or analyte to that of the internal standard is used to prepare the calibration curve and determine the unknown concentration. This method compensates for variations in physical parameters, especially inaccuracies in pipetting microliter volumes of samples. Also, the *relative* retention should remain constant, even if the flow rate should vary somewhat.

16.7 HIGH-PERFORMANCE LIQUID CHROMATOGRAPHY

Gas chromatography, because of its speed and sensitivity, has been used much more widely since its development than the liquid column-chromatographic techniques described in the previous sections. The latter techniques, however, have potentially broader use because approximately 85% of all known compounds are not sufficiently volatile or stable to be separated by gas chromatography. The traditional approach in liquid chromatography has been to use relatively large-diameter columns and small flow rates under gravity feed or low pressure pumping. Separation times are frequently on the order of hours, and the collected fractions must usually be analyzed separately, adding more hours to the analysis. The wealth of chromatographic theory accumulated, primarily from gas chromatography, has led to the development of techniques and equipment for **high-performance liquid chromatography,** that allows separations and measurements to be made in a matter of minutes.

Principles

The rate of distribution of solutes between the stationary and the mobile phase in traditional liquid chromatography is largely diffusion-controlled. Diffusion in liquids is extremely slow compared to that in gases. In order to minimize diffusion and the time required for the movement of sample components to and from the interaction sites in the column, two criteria should be met. First, the packing should be finely divided and have high spherical regularity to allow for optimum homogeneity and packing density; and second, the stationary liquid phase should be in the form of a thin uniform film with no stagnant pools. The former results in a smaller A value in Equation 16.7 (smaller eddy diffusion) and the latter results in a small C value (more rapid mass transport between the phases—necessary for high flow rates). (Molecular diffusion in liquids is much smaller than in gases, and the B term in Equation 16.7 is small. Hence, the detrimental increase in HETP at slow flow rates does not occur as in Figure 16.11).

Several column supports or packings tailored to these requirements are commercially available. An early example is du Pont's Zipax, a controlled surface porosity support consisting of a solid core with a thin porous layer of silica, shown in Figure 16.12. The porous layer is relatively open, and thin films of sorbent (stationary liquid phase) can be uniformly dispersed on it. The solid center eliminates the possibility of deep pools of stationary liquid phase.

Including the Zipax-type particles, there are, in general, three types of particles used in high-performance liquid chromatography. These are illustrated in Figure 16.13 for ion exchange resins. **Microporous particles** (called *microreticular resins* for ion exchange particles) contain crosslinked structural networks. They have relatively small-molecular-size openings and particle diameters of 5 to 10 μm. Only small solute molecules are accessible to the pores where they can interact with the stationary phase. **Macroporous particles** (called *macroreticular resins* for ion exchange particles) contain macro pores in addition to micro pores, the former several hundred angstroms in diameter. The particle diameter is greater than 60 μm. These particles have high porosities and render the particle accessible to large molecules as well as small. **Pellicular particles** (diameter 35 to 45 μm) have an inert core surrounded by a stationary-phase film. Zipax mentioned above is this type. Other commercial pellicular particles include Corasil I, Corasil II, Vydac, Perisorb,

FIGURE 16.12
(a) Diagram of Zipax controlled surface-porosity support. The solid core of the packing material and the thin porous shell surrounding the core are illustrated. Average diameter of a support particle is 30 μm and the average thickness of the porous shell is 1 μm. (b) Photomicrograph cross section. This shows the uniform, porous structure of the particle's thin outer shell. From R. E. Leitch and J. J. Kirkland, *Industrial Research* (August 1970) 36. Reproduced by permission of Industrial Research.

and Pellidon. Pellicular particles offer very high resolution, but their capacity is very low compared to the other types, and so very small sample sizes must be used, with very sensitive detectors.

As indicated above, a modified van Deemter equation can be written to describe the efficiency of HPLC. The *B* (longitudinal diffusion) term in Equation 16.7, except at very low mobile-phase velocities, is nearly zero and can be neglected. Empirically, the equation for the HETP in a high-performance liquid chromatograph is given by

$$HEPT = A + C\bar{\mu}^n \qquad (16.19)$$

where *n* is an empirical constant in the range 0.3 to 0.6. The *A* (eddy diffusion) term turns out to be small and almost a constant value and is, therefore, usually

   (a) Microreticular resin   **(b)** Macroreticular resin   **(c)** Pellicular resin

FIGURE 16.13
Structural types of ion exchange resins used in high-speed liquid chromatography. (From C. D. Scott in *Modern Practice of Liquid Chromatography*, J. J. Kirkland, Ed. New York: Interscience, 1971, p. 292. With permission.)

neglected. So HETP $\approx C\bar{\mu}^n$. The $C\bar{\mu}^n$ term approaches a limiting value as $\bar{\mu}$ increases. A representative HETP-versus-$\bar{\mu}$ plot for HPLC is shown in Figure 16.14. At very slow velocities, molecular diffusion does become appreciable and HETP increases slightly. The approximate contributions of molecular diffusion (H_B) and mass transport phenomenon (H_C) to HETP are shown in the figure.

While in gas chromatography it is often common practice to adjust the carrier gas flow to obtain a minimum HETP, in HPLC this is not the practice, because separations would take inordinately long times at the very slow velocities required.

Equipment for HPLC

The price that must be paid for faster, more efficient separations using finer column packings is pressure and the special hardware to handle this. Pressures of 1000 to 3000 psi are required to provide flow rates of 1 to 2 mL/min in columns of 2- to 4-mm diameter and 10 to 50 cm long, although in certain instances pressures up to 6000 psi may be required. Probably 80 to 90% of HPLC separations are performed with pressures of less than 1200 psi, and even some polyurethane column materials require very low pressures near atmospheric pressures.

High-performance liquid chromatography apparatus consists of four principal parts:

1. MOBILE-PHASE SUPPLY SYSTEM. The basic components are illustrated in Figure 16.15. The mobile supply contains a pump to provide the high pressures required and usually contains some means of providing gradient elution (i.e., changing concentrations of the eluent, such as salt or H^+). The solvent reservoirs can be filled with a range of solvents of different polarities, providing they are miscible, or they can be filled with solutions of different pH and are mixed in the buffer volume. The solvents must be pure and be degassed.

2. SAMPLE INJECTION SYSTEM. A typical injection system is shown in Figure 16.16. This consists of a stainless steel ring with six different parts, one to the column. A movable Teflon cone within the ring has three open segments, each of

FIGURE 16.14
Van Deempter plot for HPLC.

FIGURE 16.15
Basic components of a high-performance liquid chromatograph. (Reproduced by permission from An-alabs, Inc. *Research Notes*. Copyright © 1971.)

FIGURE 16.16
Sample loop injector. (From C. D. Scott, in *Modern Practice of Liquid Chromatography*, J. J. Kirkland, ed. New York: Interscience, 1971. With permission.)

which connects a pair of external parts. Two of the parts are connected by an external sample loop of known or fixed volume. In one configuration, the cone permits direct flow of effluent into the column, and the loop can be filled with the sample. The cone can then be rotated 30° to make the sample loop part of the moving stream, which sweeps the sample into the column. Samples of a few microliters can be injected at pressures up to 6000 psi.

3. COLUMN. Straight lengths of stainless steel tubing make excellent columns. Some rules for selecting column stationary and mobile phases are given below. Temperature control of the column is usually not necessary for liquid–solid chromatography, unless it has to be operated at elevated temperatures, but is generally required for other forms of liquid chromatography (liquid–liquid, size exclusion, ion exchange). Some detectors, especially refractometers, are very sensitive to temperature changes; and so, if the column is operated at greater than ambient temperature, a cooling jacket should be placed between the end of the column and the detector to bring the mobile phase back to ambient temperature.

4. DETECTOR. Detectors with high sensitivity are required in high-performance liquid chromatography, usually with sensitivities in the microgram to nanogram range. Widely used detectors are refractometer detectors and ultraviolet (UV) detectors. The **differential refractometer detector** is often called a "universal detector." This measures changes in refractive index of the eluent that result from the presence of solutes as they emerge from the column. It cannot be used effectively with gradient elution due to a change in the baseline (a change in the solvent index of refraction as the gradient is changed) nor when the solvent has an index of refraction close to that of the solutes. As mentioned, it is very sensitive to temperature changes. This detector is rugged and will detect concentrations of about 10^{-5} to 10^{-6} g/mL (10 to 1 ppm). The **ultraviolet detector** has much better sensitivity, about 10^{-8} g/mL (0.01 ppm). It is not temperature sensitive, it is relatively inexpensive, and it can be used with gradient elution. It is sensitive to a large number of organic compounds. Because of its advantages, the UV detector represents about 80% of the measurements made. Of course, it cannot be used with solvents that have significant absorption in the UV or with sample components that do not absorb in the UV.

Most UV detectors are simple interference filter devices that can measure the absorbance at only a few selected wavelengths. The more expensive detectors have a monochromator that allows selection of a particular wavelength. Scanning of the spectrum can even be achieved for qualitative identification by momentarily stopping the flow of the mobile phase.

Fluorescence detectors can give improved selectivity over ultraviolet absorption detectors because fewer compounds fluoresce than absorb (Chapter 13). Sensitivities at least as good as and perhaps better than with the UV detector are achieved, depending on the geometry of the excitation source–detector arrangement, the intensity of the source, and the quantum efficiency of the fluorophore. The **amperometric detector** (see Chapter 12) is useful for detecting electroactive substances and has found considerable use in biological applications, for example, in the HPLC separation and detection of trace quantities of catecholamines from the brain.

In arranging the apparatus, there must be a minimum of "dead volume" between the injection port and column and between the column and detector, to minimize spreading of the peaks and to obtain maximum separation efficiency. A 20-cm length of stainless steel capillary tubing can generally be used to connect the column to the detector without significantly affecting column performance. The detector volume must also be small, and typical volumes are 20 μL or less. Because flammable organic solvents are frequently used under high pressure in closed spaces, a safety feature should be provided in which nitrogen can be used to flush column ovens (used for temperature control) and to provide an inert atmosphere over the solvent reservoir.

Choice of Column Materials for HPLC

High-performance liquid chromatography is used for liquid–solid adsorption chromatography, liquid–liquid partition chromatography, size exclusion chromatography, and ion exchange chromatography. Below are described some general rules for selecting the column materials and preparing the column.

1. PREPARATION OF THE COLUMN. Supports are generally dried in an oven before coating them, and solid adsorbents are activated by heating. Generally, the smaller the particle diameter of the support or adsorbent, the greater the separation efficiency but the higher the required inlet pressure. To minimize stagnant pools in the column requires optimum loading of the support; that is, the pores of the support must be completely filled with stationary phase. For Celite, a commonly used liquid support, this represents a loading of 35 or 40% (wt/wt). Solid adsorbents should be packed dry, as in gas chromatography. When the packing is complete, the mobile phase should be passed through the column for about 2 hours at $\frac{1}{2}$ mL/min to remove entrained air and allow the adsorbent to equilibrate with the mobile phase. Because there is usually some solubility of the stationary phase in the mobile phase, the mobile phase must be presaturated with stationary liquid in order to prevent gradual stripping of this phase from the column. This is accomplished by placing a **precolumn** between the pump and the sample inlet.

The high pressure and speeds involved also sometimes tend to mechanically remove some of the stationary phase from the support. Column packings with **chemically bonded stationary phases** have been developed to overcome these disadvantages. For example, the Permaphase chromatographic packings produced by du Pont consist of Zipax support that has been reacted with organic modifiers to produce nonextractable, stable polymeric coatings having a variety of functional groups with which interactions can take place. These materials are unaffected by highly polar solvents, such as alcohol–water mixtures that are used to rapidly elute strongly retained sample components.

A properly packed liquid chromatographic column should provide an efficiency of about 400 theoretical plates per centimeter with 5- to 10-μm-sized particles.

2. SELECTION OF THE SUPPORT. A number of support materials are available for liquid chromatography. For general application, Corasil, Zipax, or other similar materials are widely used supports for liquid–liquid systems; a polar liquid is normally used as the stationary phase, with a nonpolar mobile phase. If a nonpolar stationary phase is to be employed (called **reversed-phase** liquid–liquid chromatography), then the support has to be silanized by exposing to chlorosilane vapor in a desiccator for a few days, followed by washing with alcohol. Then it is ready for coating with the stationary phase. Silanized particles are commercially available, and reversed-phase HPLC is the most commonly used for separating organic compounds.

3. CHOICE OF TECHNIQUE—SELECTION OF THE MOBILE PHASE. A major difficulty with liquid chromatography is predicting the proper phase systems to use, and in most instances these must be determined experimentally. Both liquid–liquid (partition) and liquid–solid (adsorption) processes operate on differences in solute polarities since polarity is important in determining both adsorption and solubility. Liquid–liquid partition processes are quite sensitive to small molecular-weight differences in solutes and so are preferred for the separation of members of a homologous series. Adsorption processes, on the other hand, are sensitive to steric effects and are preferred for the separation of similar compounds having different steric configurations.

As a general rule, highly polar materials are best separated using partition chromatography, while very nonpolar materials are separated using adsorption chromatography. Between extremes, either process might be applicable. Compound polarity follows the approximate order of: hydrocarbons and derivatives $<$ oxygenated hydrocarbons $<$ proton donors $<$ ionic compounds; that is, $RH < RX <$ $RNO_2 <$ ROR (ethers) $=$ RCO_2R (esters) $=$ RCOR (ketones) $=$ RCHO (aldehydes) $=$ RCONHR (amides) $<$ RNH_2, R_2NH, R_3N (amines) $<$ ROH (alcohols) $<$ $H_2O < ArOH$ (phenols) $< RCO_2H$ (acids) $<$ nucleotides $<$ $^+NH_3RCO_2^-$ (amino acids). The adsorbent is usually kept constant and the eluting solvent polarity is increased until elution is achieved. Some commonly used solvents in order of increasing polarity are: light petroleum solvents (hexane, heptane, petroleum ether) $<$ cyclohexane $<$ trichloroethane $<$ toluene $<$ dichloromethane $<$ chloroform $<$ ethyl ether $<$ ethyl acetate $<$ acetone $<$ n-propanol $<$ ethanol $<$ water. In some circumstances, such a highly polar solvent may be required for elution that many solutes are eluted together rather than being separated. A less polar adsorbent should then be used.

In partition chromatography, a polar stationary phase and a nonpolar mobile phase are normally used, that is, reversed-phase chromatography. By changing the polarity of the stationary phase, the liquid–liquid systems can be made selective to either polar or nonpolar materials.

Figure 16.17 illustrates the separation efficiency in a typical HPLC application and compares it with conventional liquid chromatography. Note the differences in the time axes. Only microgram quantities of sample components are usually required. Liquid chromatography generally provides a 3- to 10-fold increase in resolution over thin-layer chromatography (TLC), a widely used technique (see below). In some instances, individual TLC spots have been separated into as many as 10 components using liquid chromatography. HPLC is useful for multicomponent analysis of serum (see, e.g., Miller and Tucker, *Am. Lab.*, January 1979, p. 18).

Ion Chromatography

The application of HPLC techniques to ion exchange chromatography has become known as ion chromatography. This technique combines the separating power of ion exchange with the universality of the conductivity detector. In ordinary ion exchange chromatography, a conductivity detector is limited in use because of the high background conductance (millimhos) of the eluting agent. In 1970, William Bauman at Dow Chemical Company suggested a way to remove the background eluent using a second ion exchange column, and thus permit detection of analyte ions with a highly sensitive conductivity detector (micromhos). This second column is called the **suppressor column.** For anion analysis, this is a cation exchange col-

FIGURE 16.17
Separation of nucleic acids. (a) on Zipax cation-exchange packing using high-performance liquid chromatography. (b) Using conventional ion-exchange chromatography. [From R. E. Leitch and J. J. Kirkland, *Industrial Research* (August 1970) 36. Reproduced by permission of Industrial Research.]

umn in the acid form; and for cation analysis, it is an anion exchange column in the base form. The principles are illustrated in Figure 16.18.

Suppose the salt MA of an anion A$^-$ is separated on an anion exchange resin and eluted with NaOH. Eluting from the column will be a mixture of MA and NaOH. Upon passing through the cation exchange suppressor column, the following reactions take place:

$$RzSO_3^- H^+ + M^+ + A^- \rightarrow RzSO_3^- M^+ + H^+ + A^-$$
$$RzSO_3^- H^+ + Na^+ + OH^- \rightarrow RzSO_3^- Na^+ + H_2O$$

So the NAOH is converted to H$_2$O, and the analyte ion is converted to the corresponding acid HA. The acid HA is detected by the conductivity detector. For cation analytes, a mixture of MA and HCl emerges from the first (analytical) column. These react in the suppressor anion exchange column as follows:

$$RzNR_3^+ OH^- + M^+ + A^- \rightarrow RzNR_3^+ A^- + M^+ + OH^-$$
$$RzNR_3^+ OH^- + H^+ + Cl^- \rightarrow RzNR_3^+ Cl^- + H_2O$$

The HCl is converted to H$_2$O and the analyte cation is converted to the corresponding base MOH, which is detected by the conductivity detector.

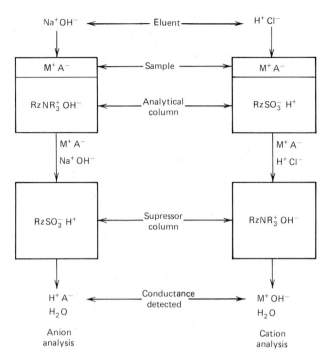

FIGURE 16.18
Principles of ion chromatography.

The suppressor column obviously will eventually become depleted and will have to be regenerated (with HCl for the cation exchanger and with NaOH for the anion exchanger). The suppressor column is usually a small-volume bed of a high-capacity resin to minimize band spreading in the column. Since microgram or smaller quantities of analytes are usually separated, a low-capacity analytical column is used coupled with relatively low eluent concentration (1 to 10 mM).

Ion chromatography is particularly useful for the determination of anions. A typical eluting agent consists of a mixture of $NaHCO_3$ and Na_2CO_3, and these are converted to low-conductivity carbonic acid. Anions such as F^-, Cl^-, Br^-, I^-, NO_2^-, NO_3^-, SO_4^{2-}, PO_4^{3-}, SCN^-, IO_3^-, and ClO_4^-, as well as organic acids or their salts, can be readily determined in a matter of minutes, down to parts-per-million or lower levels.

If solutions are too dilute for direct analysis, analytes can be concentrated first on an ion exchange concentrator column. Low parts-per-billion concentrations can be measured in this way.

Commercial ion chromatography systems utilizing a suppressor column became available in 1975. Systems are also available in which a low-capacity analytical column combined with low-concentration eluent allow conductivity detection with reasonable sensitivity without the use of a suppressor column.

16.8   PAPER CHROMATOGRAPHY

Paper chromatography is a very simple form that is used widely for qualitative identification, although quantitative analysis can be done. A sample is spotted onto a strip of filter paper with a micropipet, and the chromatogram is "developed" by placing the bottom of the paper (but not the sample spot) in a suitable solvent (see Figure 16.19). The solvent is drawn up the paper by capillary action, and the sample components move up the paper at different rates, depending on their solubility and their degree of retention by the paper. Following development, the individual solute spots are noted or are made visible by treatment with a reagent that forms a colored derivative. The spots will generally move at a certain fraction of the rate at which the solvent moves, and they are characterized by the **R_f value:**

$$R_f = \frac{\text{distance solute moves}}{\text{distance solvent front moves}} \qquad (16.12)$$

where the distances are measured from the center of where the sample was spotted at the bottom of the paper. The solvent front will be a line across the paper. The distance the solute moves is measured at the center of the solute spot or at its maximum density, if tailing occurs. The R_f value, then, is characteristic for a given paper and solvent combination. Because of slight variation in papers, it is always a good idea to determine the R_f value on each set of papers.

The cellulose filter paper used is very hydrophilic and will normally have a thin coat of water adsorbed from the air. So the mechanism of separation is a form of liquid–liquid partition chromatography in which the sample distributes between the stationary water phase and the developing solvent. The developing solvent is usually a mixture of an organic solvent with water that may be buffered at a certain pH. The water, as it moves up the paper, is adsorbed and causes the paper to swell. Other polar solvents will also be adsorbed by the paper. Sometimes a water-immiscible solvent is used to develop the chromatogram.

Developing the Chromatogram

A typical setup for paper chromatography is shown in Figure 16.19. A thin pencil line is drawn across the paper, a few centimeters from the bottom and the sample

FIGURE 16.19
Paper chromatography setup.

is spotted on this for future reference in R_f measurements. The spot must be made as small as possible for maximum separation and minimum tailing. It is best done dropwise with a warm air blower (e.g., hair dryer) to evaporate the solvent after each drop. The paper is placed in a chamber with its end dipping in the developing solvent. A closed (presaturated) chamber must be used to saturate the atmosphere with the solvent and prevent it from evaporating from the surface of the paper as it moves up. The developing may take 1 hour or longer, but it requires no operator time. The amount of development will depend on the complexity of the mixture of solutes being separated. If a wide paper is used, several samples and standards can be spotted along the bottom and developed simultaneously.

A principal advantage of this technique is that greater separating power can be achieved by using **two-dimensional paper chromatography;** a large square piece of paper is used, and the sample is spotted at a bottom corner of the paper. After development with a given solvent system, the paper is turned 90° and further development is obtained with a second solvent system. Thus, if two or more solutes are not completely resolved with the first solvent, it may be possible to resolve them with a second solvent. Proper control of the pH is often important in achieving efficient separations.

Detection of the Spots

If the solutes fluoresce (aromatic compounds), they can be detected by shining an ultraviolet light on the paper. A pencil line is drawn around the spots for permanent identification. Color-developing reagents are often used. For example, amino acids and amines are detected by spraying the paper with a solution of ninhydrin, which is converted to a blue or purple color. Exposure to iodine vapor often produces a color with colorless solutes. After the spots are identified, they may be cut out and the solutes washed off (eluted) and determined quantitatively by a micromethod.

Paper chromatography is useful for the separation of very small amounts of substances and has proved extremely valuable in biochemistry, where small and complex samples are often found. Simple mixtures of amino acids can be separated by developing with water-saturated phenol. Two-dimensional chromatography is required for complex mixtures, such as protein hydrolysates. Almost any mixture of organic components can be separated. Inorganic substances are also easily resolved. Modified forms of paper are available impregnated with ion exchange resins, silica gel, alumina, and other substances, but the principle of the operation is the same.

16.9 THIN-LAYER CHROMATOGRAPHY

Thin-layer chromatography (TLC) is very similar to paper chromatography, except that the stationary phase is a thin layer of finely divided adsorbent supported on a glass or aluminum plate, plastic strip, and so on. Any of the solids used in column chromatography can be used, provided a suitable binder can be found for good adherence to the plate.

Stationary Phases for TLC

The stationary phase consists of a finely divided powder (particle size 5 to 50 μm). It can be an adsorbent, an ion exchanger, or a molecular sieve, or it can serve as the support for a liquid film. An aqueous slurry of the powder is prepared, usually with a binder such as plaster of paris, gypsum, or poly(vinyl alcohol) to help it adhere to the backing material. The slurry is spread on the plate in a thin film, typically 0.1 to 0.3 mm, using a spreading adapter to assure uniform thickness. Adapters are commercially available. The solvent is evaporated off and adsorbents are activated by placing in an oven at 110°C for several hours. Commercially prepared plates and strips on plastic are available.

The most commonly used stationary phases are **adsorbents.** Silica gel, alumina, and powdered cellulose are the most popular. Silica gel particles contain hydroxyl groups on their surface which will hydrogen bond with polar molecules. Adsorbed water prevents other polar molecules from reaching the surface, so the gel is activated by heating to remove the adsorbed water. Alumina also contains hydroxyl groups or oxygen atoms. Alumina is preferred for the separation of weakly polar compounds, but silica gel is preferred for polar compounds such as amino acids and sugars. Magnesium silicate, calcium silicate, and activated charcoal may also be used as adsorbents. Adsorbents are sometimes not activated by heating, in which case the residual water acts as the stationary phase.

Thin-film **liquid stationary phases** can be prepared for separation by liquid–liquid partition chromatography. The film, commonly water, is supported on materials such as silica gel or diatomaceous earth, as in column chromatography. Either silica gel or diatomaceous earth may be silanized to convert the surfaces to nonpolar methyl groups for reversed-phase thin-layer chromatography.

Ion exchange resins are available in particle sizes of 40 to 80 μm, suitable for preparing thin-layer plates. Examples are Dowex 50W strong-acid cation exchange and Dowex 1 strong-base anion exchange resins, usually in the sodium or hydrogen or the chloride forms, respectively. An aqueous slurry of six parts resin to one part cellulose powder is suitable for spreading into a 0.2 to 0.3-mm layer.

Size exclusion thin layers can be prepared from Sephadex Superfine. The gel is soaked in water for about three days to complete the swelling, and then spread on the plate. The plates are not dried, but stored wet. The capillary action through these molecular sieves is much slower than with most other thin layers, typically only 1 to 2 cm/hour, and so development takes 8 to 10 hours, compared to about 30 minutes for other stationary phases.

Mobile Phases for TLC

Considerations in selection of the developing solvent generally parallel those we discussed earlier for column chromatography. In adsorption chromatography, the eluting power of solvents increases in the order of their polarities (e.g., from hexane to acetone to alcohol to water). See the list described under choice of column materials. A single solvent, or at most two or three solvents, should be used whenever possible, because mixed solvents tend to chromatograph as they move up the thin layer, causing a continual change in the solvent composition with distance on the plate. This may result in varying R_f values depending on how far the spots are allowed to move up the plate.

The developing solvents must be of high purity. The presence of small amounts of water or other impurities can produce irreproducible chromatograms.

Developing the Chromatogram

The technique of development in thin-layer chromatography is the same as for paper chromatography. The development is generally more rapid, taking from 10 minutes to an hour, and is more reproducible. Development times of 5 minutes can be accomplished using small microscope slides for TLC plates, and preliminary separations with these can be conveniently used to determine the optimum developing conditions. There is less tendency for tailing, with the result of sharper separations. Typically, sample sizes range from 10 to 100 μg per spot (e.g., 1 to 10 μL of a 1% solution). Sample spots should be 2 to 5 mm in diameter.

Detection of the Spots

The detection of spots is sometimes easier than with paper chromatography because more universal techniques can be used. Frequently, colorless or nonfluorescent spots can be visualized by exposing the developed plate to iodine vapor. The iodine vapor interacts with the sample components, either chemically or by solubility, to produce a color. Thin-layer plates and sheets are commercially available which incorporate a fluorescent dye in the powdered adsorbent. When held under ultraviolet light, dark spots appear where sample spots occur due to quenching of the plate fluorescence.

A common technique for organic compounds is spraying the plate with a sulfuric acid solution and then heating it to char the compounds and develop black spots. This precludes quantitative analysis by scraping the spots off the plate and eluting for measurement.

A representative developed thin-layer plate is shown in Figure 20.3 in Chapter 20.

16.10 ELECTROPHORESIS

Electrophoretic methods are used to separate substances based on their charge-to-mass ratios, using the effect of an electric field on the charges of these substances. These techniques are widely used for charged colloidal particles or macromolecular ions such as those of proteins, nucleic acids, and polysaccharides. There are several types of electrophoresis, **zone electrophoresis** being one of the most common.

In zone electrophoresis, proteins are supported on a solid so that, in addition to the electric migration forces, conventional chromatographic forces may enter into the separation efficiency. There are several types of zone electrophoresis according to the different supports. The common supports include starch gels, polyacrylamide gels, polyurethane foam, and paper. Starch gel electrophoresis has been a popular technique, although it is now somewhat superseded by the use of polyacrylamide gels, which minimize convection and diffusion effects. A block or "plate" of starch gel is prepared, and the sample is applied in a narrow band (line) across the block about midway between the ends, which are contacted with electrodes through a connecting bridge. When current is passed through the cell, the different components of a mixture move with velocities that depend on their electric charges, their sizes, and their shapes. As electrophoresis proceeds, the negatively charged components migrate toward the anode and the positively charged components migrate toward the cathode. The result is a series of separated bands or lines of sample constituents, such as visualized by a stain.

Very complex mixtures can be resolved with zone electrophoresis. For example, the starch gel-electrophoretic separation of plasma proteins reveals 18 components. A densitometer can be used to measure the intensity of the colored zones and thereby obtain quantitative information.

QUESTIONS

1. Describe the basic principle underlying all chromatographic processes.
2. Classify six different types of chromatography, describing the stationary and the mobile phases.
3. Describe the principle of size exclusion chromatography. What is the exclusion limit?
4. What is a molecular sieve?
5. Explain the difference between a cation exchange resin and an anion exchange resin.
6. Describe some of the factors that affect the selectivity of ion exchange resins.
7. Describe the principles of ion chromatography.
8. Describe the principles of gas chromatography.
9. What is the van Deemter equation? Define terms.
10. Describe the principle of the following gas-chromatographic detectors: (a) thermal conductivity; (b) flame ionization; (c) electron capture.
11. For the detectors in Question 10, compare them with respect to sensitivity and types of compounds that can be detected.
12. Describe the principal differences between conventional liquid chromatography systems and high-performance liquid chromatography.
13. Describe some commonly used detectors in liquid chromatography and their basis of operation.
14. In what order would the following compounds be eluted from an alumina column using n-hexane as the eluting solvent? CH_2CH_2OH; CH_3CHO; CH_3CO_2H.
15. What solvent would you choose to separate a group of hydrocarbons, $CH_3(CH_2)_xCH_3$, on an alumina column?
16. What is the retention time? The R_f value?
17. What are the largest and smallest R_f values possible?
18. Compare the advantages and disadvantages of paper chromatography and thin-layer chromatography.
19. Describe the basis of separation in electrophoresis.

PROBLEMS

Ion Exchange Chromatography

20. Alkali metal ions can be determined volumetrically by passing a solution of them through a cation exchange column in the hydrogen form. They displace an equivalent amount of hydrogen ions that appear in the effluent and can be titrated. How many millimoles of potassium ion are contained in a liter of solution if the effluent obtained from a 5.00-mL aliquot run through a cation exchange column requires 26.7 mL of 0.0506 M NaOH for titration?
21. By referring to Figure 16.7, suggest ion exchange separation schemes for the separation of (a) Sn(IV), Ge(IV), and Pb(II); (b) As(III), Sb(III), and Na^+; (c) Fe(III) and Fe(II); (d) Mo(VI), Ag(I), and Zr(IV).
22. The sodium ion in 200 mL of a solution containing 10 g/L NaCl is to be removed by passing through a cation exchange column in the hydrogen form. If the exchange capacity of the resin is 5.1 meq/g of dry resin, what is the minimum weight of dry resin required?
23. What will be the composition of the effluent when a dilute solution of each of the following is passed through a cation exchange column in the hydrogen form? (a) NaCl; (b) Na_2SO_4; (c) $HClO_4$; (d) $FeSO_4 \cdot (NH_4)_2SO_4$.

Gas Chromatography

24. A gas-chromatographic peak had a retention time of 65 seconds. The base width obtained from intersection of the baseline with the extrapolated sides of the peak was 5.5 seconds. If the column was 3 feet in length, what was the HETP in cm/plate?

25. It is desired to just resolve two gas-chromatographic peaks with retention times of 85 and 100 seconds, respectively, using a column that has a HETP of 1.5 cm/plate under the operating conditions. What length column is required? Assume the two peaks have the same base width.

26. Gas reduction valves used on gas tanks in gas chromatography usually give the pressure in psig (pounds per square inch above atomspheric pressure). Given that normal atmospheric pressure (760 torr) is 14.7 psi, calculate the inlet pressure to the gas chromatograph in torr, for 40.0 psig, if the ambient pressure is 745 torr.

27. The following gas-chromatographic data were obtained for individual 2-μL injections of n-hexane in a gas chromatograph with a 3-m column. Calculate the number of theoretical plates and HETP at each flow rate, and plot HETP versus the flow rate to determine the optimum flow rate. Use the *relative* retention time t_R' ($t_R - t_0$).

Flow Rate, mL/min	t_0("Air Peak"), min	t_R', min	Peak Width, min
120.2	1.18	5.49	0.35
90.3	1.49	6.37	0.39
71.8	1.74	7.17	0.43
62.7	1.89	7.62	0.47
50.2	2.24	8.62	0.54
39.9	2.58	9.83	0.68
31.7	3.10	11.31	0.81
26.4	3.54	12.69	0.95

Thin-Layer Chromatography

28. Refer to Figure 20.3 in Chapter 20. Taking the top horizontal line of the figure as the solvent front, calculate the R_f values for methadone, amphetamine, quinine, and morphine.

RECOMMENDED REFERENCES

General

1. J. A. Dean, *Chemical Separation Methods*. New York: Van Nostrand Reinhold Co., 1969.

Size Exclusion Chromatography

2. *Sephadex-Gel Filtration in Theory and Practice,* Pharmacia Fine Chemicals, Uppsala, Sweden. This company also supplies yearly booklets of literature references on gel permeation chromatography.

Ion Exchange Chromatography

3. J. S. Fritz and B. B. Garralda, "Cation Exchange Separation of Metal Ions with Hydrobromic Acid," *Anal. Chem.,* **34** (1962) 102.

4. J. Inczedy, *Analytical Applications of Ion Exchangers*. Oxford: Pergamon Press, 1966.

5. O. Samuelson, *Ion Exchange Separation in Analytical Chemistry*. New York: John Wiley & Sons, 1963.

Gas Chromatography

6. R. Grob, *Modern Practice of Gas Chromatography*. New York: John Wiley & Sons, 1978.
7. W. Jennings, *Gas Chromatography with Glass Capillary Columns,* 2nd Ed. New York: Academic Press, 1980.
8. W. O. McReynolds, "Characterization of Some Liquid Phases," *J. Chromatogr. Sci.,* **8** (1970) 685.
9. R. W. Moshier and R. E. Sievers, *Gas Chromatography of Metal Chelates*. Oxford: Pergamon Press, 1965.
10. D. M. Ottenstein, "Column Support Materials for Use in Gas Chromatography," *J. Gas Chromatogr.,* **1**(4), (1963) 11.
11. W. R. Supina and L. P. Rose, "The Use of Rohrschneider Constants for Classification of GLC Columns," *J. Chromatogr. Sci.,* **8** (1970) 214.
12. J. A. Yancy, *Analabs Guide to Stationary Phases for Gas Chromatography*. New Haven, Conn.: Analabs, 1977.
13. A. Zlatkis and C. F. Poole, eds., *Electron Capture-Theory and Practice in Chromatography*. New York: Elsevier, 1981.

High-Performance Liquid Chromatography

14. P. R. Brown, *High Pressure Liquid Chromatography: Biochemical and Biomedical Applications*. New York: Academic Press, 1973.
15. P. H. Dixon, C. H. Gray, C. K. Lim, and M. S. Stoll, eds., *High Pressure Liquid Chromatography in Clinical Chemistry*. New York: Academic Press, 1976.
16. P. K. Kabra and L. J. Marton, eds., *Liquid Chromatography in Clinical Analysis*. Clifton, N.J.: Human Press, 1981.
17. J. J. Kirkland and L. R. Snyder, *Introduction to Modern Liquid Chromatography*. New York: Wiley-Interscience, 1974.
18. R. R. Ryall and H. D. Kessler, Jr., "Microbore Columns for HPLC," *Anal. Lab.,* May 1982, p. 49.

Paper and Thin-Layer Chromatography

19. E. J. Shellard, ed., *Quantitative Paper and Thin-Layer Chromatography*. New York: Academic Press, 1968.
20. E. Stahl, ed., *Thin Layer Chromatography*. Berlin: Springer-Verlag, 1965.
21. G. Zweig and J. R. Whitaker, *Paper Chromatography and Electrophoresis*. New York: Academic Press, 1967.

chapter 17

AUTOMATION IN
THE LABORATORY

The services of the analytical chemist are constantly increasing as more and better analytical tests are developed, particularly in the environmental and clinical laboratories. The analyst often must handle a large number of samples and/or process vast amounts of data. Instruments are available that will automatically perform many or all of the steps of an analysis, greatly increasing the load capacity of the laboratory. The data generated can often be processed best by computer techniques; computers may even be interfaced to the analytical instruments. An important type of automation is in process control whereby the progress of an industrial plant process is monitored in real time (i.e., on line), and continuous analytical information is fed to control systems that maintain the process at preset conditions.

In this chapter, we briefly consider the types of automated instruments and devices commonly used and the principles behind their operation. Their application to process control is discussed. Two widely used automatic clinical instruments are described. These are the multichannel AutoAnalyzer for simultaneously determining several constituents in a sample, and the centrifugal (parallel fast) analyzer for simultaneously analyzing several samples for a single analyte. The continuous-flow technique of flow injection analysis is described for the rapid analysis of microliter volume samples. The newer "smart" instruments which contain dedicated microprocessor computers for performing a variety of features from instrument operation to data handling are also discussed. Finally, the interfacing of computers to instruments for on-line data acquisition will be described.

17.1 PRINCIPLES OF AUTOMATION

There are two basic types of automation equipment. **Automatic devices** perform specific operations at given points in an analysis, frequently the measurement step. Thus, an *automatic titrator* will stop a titration at the end point, either mechanically or electrically, upon sensing a change in the property of the solution. An example given in Chapter 12 was the automatic coulometric chloride titrator in which a surge in the amperometric indicating electrode current, due to excess silver ion titrant, triggered a relay that stopped the titration. **Automated devices,** on the other hand, control and regulate a process without human intervention. They do so through mechanical and electronic devices that are regulated by means of *feedback information* from a sensor or sensors. Hence, an *automated titrator* may maintain a sample pH at a preset level by addition of acid or base as drift from the set pH is sensed by a pH electrode. Such an instrument is called a *pH-stat* and may be used, for example, in maintaining the pH during an enzyme reaction which releases or consumes protons. The rate of the reaction can actually be determined from a recording of the rate of addition of acid or base to keep the solution pH constant.

Automated devices are widely used in process control systems, whereas automatic instruments of various sophistication are used in the analytical laboratory for performing analyses. The latter may perform all steps of an analysis, from sample pickup and measurement through data reduction and display.

17.2 AUTOMATED INSTRUMENTS: PROCESS CONTROL

Automated devices with control loops are largely used in industrial plants where a production process can be critically dependent on rapid analysis of certain constituents at different stages of the process. Other applications are, of course, possible (e.g., the pH-stat mentioned above). These devices may be classed as continuous or discrete (batch) instruments. The **continuous instrument** constantly measures some physical or chemical property of the sample and yields an output that is a continuous (smooth) function of time. A **discrete,** or **batch, instrument** analyzes a discrete or batch-loaded sample, and information is supplied only in discrete steps. In either case, information on the measured variable is fed back to monitoring or control equipment. Each technique utilizes conventional analytical measurement procedures and must be capable of continuous unattended operation.

Continuous Analyzers

Continuous-process control instruments may make measurements directly in a flowing stream. This generally precludes any analytical operation on the sample and direct sensing devices such as electrodes must be used. If a sample dilution, temperature control, or reagent addition is required, or measurements are made with nonprobe-type instruments, then a small fraction of the stream is diverted into a test stream where reagents may be mixed continuously and automatically with the sample, and the test measurement is made.

Process control instruments operate by means of a **control loop,** which consists of three primary parts:

1. A **sensor** or measuring device that monitors the variable being controlled.
2. A **controller** that compares the measured variable against a reference value (set point) and feeds the information to an operator.
3. An **operator** that activates some device such as a valve to bring the variable back to the set point.

The control loop operates by means of a **feedback mechanism,** illustrated in Figure 17.1. The process may be any industrial process that produces some desired product. It has one or more inputs that can be controlled to provide the desired product (output). The sensor measures the variable to be controlled (e.g., pH, temperature, etc.). The information is fed to the controller, which compares the measured variable against a reference set point. The difference is fed to the opeator that actuates the valve (opens or closes it) or some other appropriate device to adjust the variable back to the set point.

These devices are characterized by their **dead time.** This is equal to the time interval, after alteration of the variable at the input, during which no change in the variable is sensed by the detector at the output. It includes the analytical dead time. It can be minimized by keeping the detector as near the input as possible and by high flow rate. The sensor may actually be placed at the input, before the process. In this manner, corrective action on the manipulated variable (if it changes erroneously) is taken before the process occurs, and an error need not occur in the output prior to corrective action. However, the result is not detected at the output. Such systems may be called **feedforward systems,** and are **open-loop controls,** as opposed to **closed-loop controls** in feedback systems.

Discrete Analyzers

In discrete analyzers, a batch sample is taken at selected intervals and then analyzed, with the information being fed to the controller and operator in the usual fashion. Obviously, the sampling and analytical dead times are increased over continuous analyzers, and the manipulated variable is held at a fixed value between measurements. If a transient error occurs between measurements, it may not be detected and corrected for. On the other hand, a short transient error may be detected during the measurement interval and a correction for this applied during the entire interval between measurements.

Discrete measurements must be made when the sensing instrument requires discrete samples, as in a chromatograph.

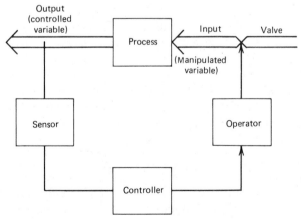

FIGURE 17.1
Feedback control loop.

Instruments Used in Automated Process Control

In principle, any conventional measuring device or technique can be used in process control. The choice is dictated by cost and applicability to the problem. The most widely used methods include spectrophotometry to measure color, ultraviolet or infrared absorption, turbidity, film thickness, and the like; electrochemistry, primarily potentiometry, for the measurement of pH and cation or anion activity; and gas and liquid chromatography, especially in the petrochemical industry where complex mixtures from distillation towers are monitored.

17.3 AUTOMATIC INSTRUMENTS

Automatic instruments, as mentioned before, are not feedback control devices but rather are designed to automate one or more steps in an analysis. They are generally intended to analyze multiple samples, either for a single analyte or for several analytes.

Automatic instruments will perform one or more of the following operations:

1. Sample pickup (e.g., from a small cup on a turntable or assembly line).
2. Sample dispensing.
3. Dilution and reagent addition.
4. Incubation.
5. Placing of the reacted sample in the detection system.
6. Reading and recording the data.
7. Processing of the data (e.g., correct for blanks, correct for nonlinear calibration curves, calculate averages or precisions, correlate with the sample number, etc).

Clinical instruments may also contain a feature for deproteinizing the sample. Instruments that provide only a few of these steps, primarily electronic data processing, are called **semiautomatic instruments.**

While all automatic instruments are discrete in the terminology of automated process analyzers, in that they analyze individual discrete samples, they may be classified as follows:

1. **Discrete sampling instruments.** In discrete sampling, each sample undergoes reaction (and usually measurement) in a separate cuvet or chamber. These samples may be analyzed sequentially or in parallel (see below).
2. **Continuous-flow sampling instruments.** In continuous-flow sampling, the samples flow sequentially and continuously in a tube, perhaps being separated by air bubbles. They are each sequentially mixed with reagents in the same tube at the same point downstream and then flow sequentially into a detector.

Discrete samplers have the advantage of minimizing or avoiding cross contamination between samples. But continuous-flow instruments require fewer mechanical manipulations and can provide very precise measurements, and probably the most successful automatic instrument has been a continuous-flow instrument (see Section 17.5, The AutoAnalyzer).

Discrete instruments may be designed to analyze samples for one analyte at a time. These are the so-called **batch instruments,** or **single-channel analyzers.** See

FIGURE 17.2
Discrete analyzers. (a) Single-channel (batch). (b) Multichannel.

Figure 17.2. However, those that analyze the samples in parallel, that is, simultaneously rather than sequentially one at a time, can analyze a large number very quickly; and they can readily be changed to perform different analyses (see, for example, The Centrifugal Analyzer). Discrete analyzers may also analyze separate aliquots of the same sample (one in each cup) in parallel for several different analytes. These are **multichannel analyzers.**

Continuous-flow instruments may also be single-channel (batch) instruments that analyze a continuous series of samples sequentially for a single analyte (Figure 17.3). Or they may be multichannel instruments, in which the samples are split at one or more points downstream into separate streams for different analyte analyses, or separate aliquots of samples may be taken with separate streams in parallel.

Modern-day instruments are so sophisticated that they actually possess *automated* features whether they perform an analysis automatically or not. For example, they may monitor sample chamber temperature and by feedback to a regulator maintain it constant (important in enzyme reactions).

17.4 SEMIAUTOMATIC INSTRUMENTS

Many instruments less sophisticated than the above fully automatic ones will perform the measurement and readout steps automatically. Automatic titrators, for example, will stop addition of a titrant by triggering a solenoid switch signaled by a change in the potential of an indicating electrode, a change in indicator color, and so forth.

In the clinical laboratory, many semiautomatic instruments are **dedicated instru-**

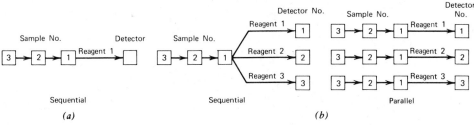

FIGURE 17.3
Continuous-flow analyzers. (a) Single channel (batch.) (b) Multichannel.

ments designed to analyze for a specific component or a few specific diagnostically related components. They are not readily adapted to other analyses.

An automatic glucose analyzer is shown in Figure 17.4. The instrument performs measurements in a matter of seconds and reads out the glucose concentration digitally. The determination is based on the amperometric measurement of oxygen consumed in the enzymatic reaction of glucose with oxygen. A 10 μL sample is rapidly injected into the instrument which contains a solution of the enzyme glucose oxidase that is saturated with air. The initial rate of consumption of oxygen ($-d[O_2]/dt$, the derivative) is rapid and quickly goes through a sharp maximum. The height of the maximum, which is proportional to the glucose concentration, is recorded and read out on the digital counter. Even though the sample is added manually, the rapidity of the measurement allows the analysis of several hundred samples in a day. Instruments of this type, which are considerably less expensive and will compete favorably with completely automatic instruments on an individual analyte basis, fill a real need in many laboratories.

Many of these dedicated instruments are also fully automatic for unattended sampling and operation.

17.5 THE AUTOANALYZER

The first widely used automatic analyzer in the clinical laboratory was the Auto-Analyzer, manufactured by Technicon Corporation. It is a continuous segmented-flow sequential analyzer. The first AutoAnalyzer was built to perform urea nitrogen (BUN) analysis, because this is one of the most frequently requested analyses in the clinical laboratory (see Chapter 19). It was a *single-channel* instrument, capable of performing one analysis at a time. However, the instrument could be adapted to other analyses by changing the reagents. It is widely used for the determination

FIGURE 17.4
Automatic glucose analyzer. (Courtesy Beckman Instruments, Inc.)

of air and water pollutants, for example, sulfate in air or phosphate and nitrate in water.

Figure 17.5 shows a diagram of a single-channel AutoAnalyzer. The basic components of the instrument are a turntable, a proportioning pump, mixing coils, a dialyzer, and a photoelectric colorimeter adapted for a flowing stream. The principle of operation is as follows. The serum samples are placed in individual sample cups on the turntable in numbered sequence. Each sample is successively drawn into a tube at timed intervals; the turntable automatically turns after each sample has been drawn. Reagents are also drawn into individual tubes to be mixed with the sample. For BUN analysis, the sample is pumped and mixed with a buffered urease solution and sent through a preheating coil where the two are incubated at 55°C in a flowing stream and then passed continuously through the dialyzer. Ammonia is produced from the action of the enzyme urease on urea, and this ammonia diffuses through the dialysis membrane and is picked up in a stream of water and air. A solution of Nessler's reagent is added to this stream, producing a yellow color whose intensity is proportional to the amount of ammonia present and hence to the amount of urea in the original sample. The colored solution flows through the colorimeter, where a filter for the proper wavelength allows spectrophotometric measurement of the color intensity, which is read on a recorder.

The instrument possesses several essential features:

1. The sample and the reagents are both pumped at constant rates. This **proportional pumping** eliminates the need for measuring the volumes, since the solutions are mixed at a set proportion.
2. The flowing streams are segmented by air bubbles, and so successive samples are completely separated from each other, as are successive additions of each reagent. This prevents mixing of the samples and allows successive samples to be drawn and run through the system at frequent intervals.

FIGURE 17.5
Single-channel AutoAnalyzer. (Courtesy of the Technicon Instruments Corporation.)

3. The dialyzer produces a continuous protein-free stream that contains all of the diffusable constituents of the sample, and so a number of different analyses are possible.

4. The components of the flow circuit can be rearranged in different ways to provide flexibility and allow different analyses to be performed or variations in the procedure employed for a particular analysis. Urea, for example, can be analyzed by nonenzymatic colorimetric reactions, rather than the one described here.

An instrument of this type will analyze several hundred samples a day (about 40 per hour), and by a change of reagents, it can be used to perform different analyses. As will be pointed out in Chapter 19, the physician can often better diagnose a disease from a knowledge of a battery of clinical tests. However, the requirement of performing all the analyses on a single instrument presents certain difficulties. Suppose BUN and glucose analyses are to be performed by the instrument. First, after all the samples are collected, they must be separated into groups for the two analyses. If some samples are to be analyzed for both constituents, then two separate aliquots must be taken. The glucose analyses must wait for the completion of the BUN analyses if these are run first, and if both results are to be correlated, then the reporting of the BUN results will have to wait until the glucose results are also ready to be reported.

These problems have led to the development of Sequential Multiple Auto-Analyzers (SMAs) that analyze for several constituents in each sample simulta-

FIGURE 17.6
Flow diagram of the 8-channel AutoAnalyzer. Pump tubes are shown as small open circles. Rates of pumping (in mL per minute) are given in Arabic numerals. Roman numerals indicate pump number. A1, C3, D2, G2, HO, K4 are Technicon designations for fittings used to join lines; SMC = single-mix coil; DMC = double-mix coil; HABA = 2-(4-hydroxybenzeneazo)benzoic acid; DAM = diacetyl monoxime. From L. T. Skeggs, Jr., *CLin. Chem.*, **10** (1964) 918. (Reproduced by permission of the American Association of Clinical Chemists.)

neously. The principle can be described as follows for the original eight-channel instrument. This instrument, shown in Figure 17.6, was designed to determine the electrolytes sodium, potassium, chloride, and carbon dioxide (bicarbonate), along with total protein, albumin, urea, and glucose. Sodium and potassium are measured by flame photometry after dilution with a lithium internal standard, while the other constituents are measured colorimetrically. Carbon dioxide and proteins are determined in the nondiffusable portion of the stream.

A flow diagram is shown in Figure 17.6. It contains four pumps with 32 tubes, one heating bath with two coils, a multiple colorimeter, and a flame photometer. Only a single sample is required for all eight tests. The samples are changed every 3 minutes, so that 20 per hour are analyzed for eight constituents, or a total of 160 analyses per hour. The multiple colorimeter contains a single light source and detector with six separate flow cells contained in a single cell block. Each flow cell contains the proper light filter for the particular measurement. Each successive aliquot from the sample (eight per sample) is staggered to arrive at either the colorimeter or the flame photometer 22.5 seconds apart ($8 \times 22.5 = 180$ s, or 3 min, the staggering time between successive samples). The absorbance of a given solution is recorded for 22.5 seconds and then the cell block is motor driven to place the next flow cell in the light path, and then its absorbance is recorded, and so on. All the data for a single sample are recorded on a single recorder. Different analyses can be performed besides the ones illustrated here.

A 12-channel analyzer was introduced in 1967 (the SMA 12/60). The sample is split into 12 channels, and 12 results on each sample are generated sequentially at the rate of 60 samples, standards, and controls per hour (720 tests). The next generation of this instrument was the computer-controlled SMAC, a 20-channel instrument introduced in 1974. It provides 20 results on 150 samples, standards, and controls per hour (3000 tests!). About 2000 useful sample tests per hour are effectively achieved, after accounting for standards, controls, and out-of-range samples. Less than 1 mL serum is required. The instrument utilizes a computer and keyboard, with an oscilloscope display for communication, to enter control variables, information on standards, and so forth. Digital printout of results is provided. Two operators are required to maintain and run the system, including loading of samples.

These instruments are characterized by a significant dead time for the reading of the first sample. For example, a 90-minute start-up time is required for the SMAC.

17.6 THE CENTRIFUGAL ANALYZER

The centrifugal analyzer was developed by scientists at the Oak Ridge National Laboratory. Its development was sponsored by the Institute of General Medical Sciences of the National Institutes of Health and the Atomic Energy Commission. Hence, it has been dubbed the GeMSAEC Analyzer. It is also called a **parallel fast analyzer,** a type of parallel discrete single-channel analyzer described above. This is a photometric analyzer in which samples are mixed with reagents in a centrifuge, and then the color developed upon reaction is automatically read with a spectrophotometer.

Figure 17.7 illustrates the centrifuge rotor and photometric analyzer. The sample and reagent are pipetted into separate cups in the rotor. There are 15 to 42 or more sets of two cups for as many samples, arranged as two concentric circles. When

FIGURE 17.7
Cross section of sample disk and centrifugal analyzer section of GeMSAEC Analyzer. (From R. L. Coleman, W. D. Schults, M. T. Kelly, and J. A. Dean, *Am. Lab.*, (July 1971, p. 26. (Reproduced by permission of International Scientific Communications, Inc.)

the rotor attains a speed of about 350 rpm, the sample and reagent are thrown
through a transfer cavity into a measuring cell where the transmittance of the mixture is measured. Complete mixing is facilitated by pulling air through siphonlike
exit tubes in each cell by means of a vacuum line. The total volume of the analysis
mixture is 450 to 800 μL.

The transmittance of each cell is read sequentially using an oscilloscope to monitor the traces, and the data are tabulated with a dedicated computer. The oscilloscope records the transmittance as a function of time as the different cells pass
through the space between the light source and the photomultiplier detector, similar to a strip chart recorder but capable of expanding the time scale so that transmittance changes on the order of milliseconds or less can be seen. Hence, the
transmittance is zero between cells and then increases as each cell passes through
the measuring zone. The transmittance is usually read when the rotor speed is
about 600 rpm. The sweep rate of the oscilloscope is keyed to the rotation rate to
synchronize the readout to each cell. Figure 17.8 shows how the transmittance
changes as a function of time. The transmittance due to the sample (lower portion
of the signal) is flat with a width of about 650 μs at 600 rpm.

The computer, which stores the transmittance data, reads and records the observed detector signal 16 times during the 650 μs flat portion of the curve and
averages the readings. This step can be repeated on up to 16 successive passes of
the cell to give a final average of 256 observations. Similarly, the computer records
the transmittance data on each of the other cells as they pass the detector.

In operation, one cell contains blank reagents, and several other cells contain
standards at increasing concentrations. The computer constructs a calibration curve
from the blank corrected standard readings, and from this, plus the transmittance
values of the samples, calculates and prints out the concentrations of the samples
contained in the remaining cells.

As opposed to the AutoAnalyzer, only a single determination can be made for
each sample during a given run. However, determinations are made in a matter of

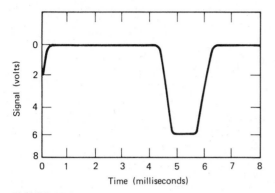

FIGURE 17.8
Drawing of oscilloscopic readout showing signal from a single cell in GeMSAEC Analyzer. (From R. L. Coleman, W. D. Schults, M. T. Kelley, and J. A. Dean, *Am. Lab.*, July 1971, p. 26. Reproduced by permission of International Scientific Communications, Inc.)

seconds after samples and reagents are loaded, and only microliter quantities of sample and reagents are required for each determination. Automatic high-speed micropipets are used to load the samples and reagents (about 4 minutes for a 15-place rotor disc). A prime advantage of the system is that samples and standards are observed virtually simultaneously and under the same conditions. For systems in which the rate of reaction is slow, the analysis time can be optimized without waiting for maximum or equilibrium color development; this is particularly advantageous for enzyme measurements. It is possible to program the computer to take a series of readings over 1 minute, for example, to obtain a kinetic curve for rate-limited reactions.

Procedures can be changed simply by changing the reagents in the rotor and changing the wavelength setting. Ultraviolet as well as visible measurements can be made. Operation is simplified with the availability of disposable discs, so there is no cleanup required.

A number of other automated instruments are marketed, particularly for the clinical laboratory. Most, like the GeMSAEC analyzer, are "discrete" analyzers, in which each sample is pipetted or sampled and analyzed in separate instrument channels, thus avoiding danger of cross contamination between samples. Although only one test can generally be made on a given run, several hundred samples may be run in 1 hour. An important advantage of most of these discrete analyzers is that "stat" or nonroutine analysis of a small number of samples for a given test can be done quickly, since the instrument dead time is generally short.

17.7 FLOW INJECTION ANALYSIS

In the early 1970s, it was demonstrated that air segmentation in continuous-flow analysis was not essential and that a sample can be injected directly into a flowing reagent stream. This technique of unsegmented continuous-flow analysis has become known as flow injection analysis (FIA), and was developed in large part by Ruzicka and co-workers in Denmark (see Reference 8 at end of chapter), as well as by Stewart and co-workers in the U.S. The basic components for FIA are illustrated in Figure 17.9. A few microliters of sample is injected as a plug into a flowing reagent stream via a syringe or valve (typically a loop injector valve as

FIGURE 17.9
Basic components for flow injection analysis

used in high-performance liquid chromatography), and after a few seconds, it passes through a detector where a transient peak signal is recorded in proportion to the analyte concentration. A peristaltic pump may be used to propel the stream. Mixing with the reagent in the flowing stream occurs mainly by diffusion-controlled processes and is sufficiently rapid that for ionic reactions (instantaneous), a measureable product is usually achieved within 15 seconds. Any appropriate flow cell detector may be used, such as spectrophotometric, fluorometric, potentiometric, atomic absorption, and so forth.

The absence of air segmentation allows high sample throughout. Sampling rates of 700 samples per hour or more have been reported. In addition, the dead time is very short, $\frac{1}{4}$ to 1 minute, and FIA is well suited for stat-type analyses that must be done quickly but infrequently. Typical dimensions are 0.5-mm-i.d. tubing, 100 cm long, with flow rates of 1 to 3 mL per minute. Sample volumes are on the order of 10 to 50 μL, with 20 μL being typical.

FIA is capable of very high precision, better than 1%, as a result of very controlled sample dispersion in a continuous flowing stream. Kinetic reactions may be measured by precisely stopping the flow at a fixed point on the peak (in the detector cell) and allowing the reaction to proceed several seconds to follow the rate of product formation. This is readily accomplished with automatic timing circuits.

You are referred to the book by Ruzicka and Hansen (Reference 8) for a complete description of the many variations and applications of FIA.

17.8 MICROPROCESSOR-CONTROLLED INSTRUMENTS: "SMART" INSTRUMENTS

We have indicated earlier the use of computers in complex automatic instruments. These are generally sophisticated minicomputers that can be programmed to provide a variety of functions. We describe in the next section the interfacing of computers to one or more instruments for automating specific functions and processing data.

Many modern instruments, however, do not require a separate computer for automating their control and operation. Instead, they have their own dedicated **microcomputer** built in. These are the so-called "smart" instruments. The computer is a **microprocessor** and peripherals constructed from spage age microelectronic circuits. Many instruments are preprogrammed in "machine language," but others have basic programs that can be modified by the operator to perform specific dedicated functions. The design and engineering of a microprocessor may require tens or hundreds of thousands of dollars, but the electronic circuits are etched in microchips and can be mass-produced inexpensively. An example is the popular digital watch that can be purchased inexpensively and has a variety of stored information such as dates, and has programs in which a specific time to buzz an alarm can be entered, and so forth.

A representative microprocessor-controlled instrument is shown in Figure 17.10. This is an IR computing spectrophotometer, which is capable of giving both qualitative and quantitative results. The heart of the instrument is a computing microprocessor which allows the analyst to set and control operating parameters in routine scanning modes, such as the starting wavelength, the scan rate, and so forth. It also allows the operator to initiate several nonroutine modes of operation such as repetitive scanning and the time interval between each. But most importantly, it also provides the ability to solve a series of simultaneous equations for quantitative analysis of unknown mixtures containing up to a maximum of six components.

Communication between the analyst and the instrument is established through a keyboard array which accepts command inputs (e.g., a repetitive scan button) and numerical data inputs (e.g., to enter the number of scans desired). The instrument provides digital display and printout in addition to the recorded spectrum.

Microprocessor-controlled instruments can provide a number of other functions, depending on the type of instrument and the desired capabilities. These include such parameters as automatic background correction, first- or second-derivative spectra, integration or averaging of signals for a preselected interval for improved signal-to-noise ratio, and statistical evaluation of data, for example, the standard deviation. Calibration to read out in any desired concentration units is simply a matter of putting in a standard, keying in its concentration and pressing an appropriate calibrate button. If nonlinear analytical curves are obtained, the microprocessor may determine the algorithm of the curve and automatically correct the nonlinearity. This is particularly valuable in atomic absorption instruments. Periodic rereading of one or more standards may be used to correct automatically for calibration drift. Least-squares fitting of calibration curves may be utilized.

Instruments designed for enzyme and other kinetic assays will be programmed to measure the signal continuously over a given time interval (e.g., absorbance)

FIGURE 17.10
A microprocessor-controlled infrared spectrophotometer (Courtesy of Beckman Instruments, Inc.)

and calculate the rate of the reaction: if properly calibrated, the activity of the enzyme or the concentration of the substrate will be presented.

In addition to data processing, the instruments may be programmed to provide data evaluation. We have already mentioned standard deviation. For example, the linearity of an enzyme rate curve may be monitored, and if incorrect, indicated on the readout. Results of measurements that fall outside the measurement range of the instrument may be rejected. Abnormal clinical analyses results may be flagged for ready identification.

The instrument may contain closed-loop feedback control (automated) features. We have mentioned the possibility of temperature control of the reaction chamber for kinetic measurements. An automatic titrator may sense the approach of an end point and slow down the addition of titrant near the end point for improved precision in locating the end of the titration, just as a human operator would, but probably in a more reproducible fashion. The instrument may monitor its various functions and shut itself down if it malfunctions.

17.9   COMPUTERS IN ANALYTICAL CHEMISTRY

Rather than have individually microcomputer-controlled instruments, or in order to improve the versatility and operator control of the computer functions, it may be desirable to utilize larger computers that can be programmed. One of the chief developments in laboratory automation has come through the use of programmable digital computers, as described below. These systems can be programmed to acquire data, perform analyses (calculations), and control processes and instrumentation. A small computer can analyze and report results from several analytical instruments simultaneously.

The main difficulty in transmitting instrument data output to a computer is that the data are generally *analog* in nature (electrical signals such as voltage) while the computer is *digital* in nature and accepts digital data. Therefore, an **analog-to-digital converter** (A/D converter) accepts the data from the instrument and then converts it into language compatible with the computer to complete the interfacing. The computer may present the calculated results in digital form or it may transfer them to a **digital-to-analog** (D/A) **converter** to present them in analog form, as on a recorder.

We will not go into the actual programming of computers. This is available through standard computer programming courses. Computer languages such as FORTRAN and BASIC are used to enter a program of the desired operations (e.g., based on mathematical manipulations). Complex computations can be performed rapidly. Although data processing can be accomplished by manual entry of the data into a computer, we are concerned here with automating the process by interfacing the instrument to the computer.

The Small Laboratory Computer

A number of small and medium-sized general-purpose laboratory or personal computers are available. These can be equipped to process on-line data from one or several instruments.

A block diagram of a system for handling several instruments is shown in Figure 17.11. The system consists of a central processing station plus sufficient data ac-

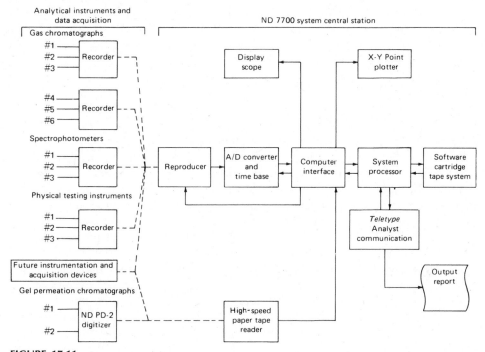

FIGURE 17.11

Typical laboratory installation and Nuclear Data 7700 system block diagram. (From R. C. Master, *Am. Lab.*, February 1970, p. 47. Reproduced by permission of American Laboratory, Inc.)

quisition stations (recorders in the figure) to provide the instrument interfacing required in the laboratory.

The recorder is an analog magnetic tape recorder. A printing digitizer is used as the method of data acquisition for long experiments with slow generation of data, such as size exclusion chromatography. The machine digitizes the analog data and generates a punched paper-tape record. The reproducer has the necessary electronics to demodulate the recorded signal. It is operated under computer control wherein the data cassette (the magnetic tape reel of the recorder) is rewound and aligned at the beginning of the tape and commanded to begin the run phase automatically. At this phase of data reduction, the entire computer system is dedicated to this one analysis. The data from the reproducer are fed to the A/D converter.

The **computer interface** is the primary **input/output** (I/O) **device** for the system. It provides the I/O decoding for data transfer in a format compatible to the computer. It also includes a D/A converter so that analog results can be displayed on an X–Y recorder or an oscilloscope. The system processor is the computer, and it is directed to use the appropriate program from the software system, where a library of programs is stored on tape. For example, with gas chromatography, the computer is directed to determine the areas of all peaks, including overlapping peaks, corrected for the baseline. The gas chromatography program also has a storage of compound identification dictionaries containing compound names, peak weighting factors, and relative retention time windows, as well as standard-area correlation for quantitation. Once the areas are computed, a search for the chosen dictionary is performed and the compound identification sequence begins. The generated output report gives compound names, peak retention times, and percent concentration for the analysis.

Similar software packages (programs) can be written or introduced into the computer memory for other instruments.

The Hierarchical System

A small or medium-sized computer is sometimes faced with a problem more readily handled by a larger computer. However, the smaller computer is capable of manipulating the data partially so they will be easier for the other computer to handle. This is known as **"preprocessing."** Through preprocessing, the smaller computer sends those problems it cannot solve to the larger computer via some system of data transmission. The larger machine solves the problem, sends the solution back to the smaller computer, and the smaller computer presents the answer and/or uses it is in subsequent problems.

This small-to-large computer arrangement is known as the **hierarchical computer system.** An example of its use might involve an extensive "fingerprinting" problem in spectrophotometry. The small computer could acquire the spectrum, smooth it, and normalize it. The spectrum would then be sent to the larger computer to be matched with one of perhaps 100,000 filed spectra, and then the compound name would be sent back to the small computer for local reporting. The smaller computer does most of the work so that the large computer is used only infrequently, which is more economical than using only the single giant computer.

Hospitals and clinics may couple computers to clinical analyzers to obtain direct printout of results; in some cases, the data are transmitted directly to the physician's office, together with other information about the patient. The analytical results may be forwarded to a centralized computer that will handle large files and other bookkeeping functions. Thus, patient histories can be forwarded with results to the physician.

QUESTIONS

1. Distinguish between an automatic instrument and an automated instrument.
2. Distinguish between discrete and continuous automated devices.
3. Distinguish between discrete and continuous-flow sampling automatic instruments.
4. What is a feedback control loop?
5. Describe the principles of the AutoAnalyzer. How does a multichannel analyzer differ from a single-channel analyzer?
6. Describe the principles of the centrifugal analyzer.
7. List an advantage for each, the AutoAnalyzer and the centrifugal analyzer.
8. Consult some recent issues of the journal *Clinical Chemistry* and report two examples each of methods performed using the AutoAnalyzer and a centrifugal analyzer.
9. Describe the principles of flow injection analysis.
10. Summarize the principal uses of computers in the analytical laboratory.
11. What is the hierarchical system of computing?
12. What is an A/D converter?

RECOMMENDED REFERENCES

General
1. H. H. Bauer, G. D. Christian, and J. E. O'Reilly, *Instrumental Analysis*. Boston: Allyn and Bacon, 1978. Chapter 24, "Automation in the Laboratory," by K. S. Fletcher and N. C. Alpert. Provides an excellent detailed but brief description of various types and operations of automated and automatic instruments.
2. N. Gochman and L. J. Bowie, "Automated Systems for Radioimmunoassay," *Anal. Chem.,* **49** (1977) 1183A.

AutoAnalyzer
3. L. T. Skeggs, "An Automatic Method for Colorimetric Analysis," *Am. J. Clin. Pathol.,* **28** (1957) 311.
4. L. T. Skeggs, "New Dimensions in Medical Diagnosis," *Anal. Chem.,* **38** (6) (1966) 31A. Describes development and use of AutoAnalyzer.
5. L. T. Skeggs and H. Hochstrasser, "Multiple Automatic Sequential Analysis," *Clin. Chem.,* **10** (1964) 918.

Centrifugal Analyzer
6. C. A. Burtis, W. F. Johnson, J. E. Attrill, C. D. Scott, and N. Cho, "Increased Rate of Analysis by Use of a 42-Cuvet GeMSAEC Fast Analyzer," *ibid.,* **17** (1971) 686.
7. M. A. Pesce, "Centrifugal Analyzers—A New Concept in Automation for the Clinical Chemistry Laboratory," *J. Chem. Educ.,* **51** (1974) A521.

Flow Injection Analysis
8. J. Ruzicka and E. H. Hansen, *Flow Injection Analysis*. Wiley Chemical Analysis Series, Vol. 62. New York: John Wiley and Sons, 1981.

Computers
9. C. E. Klopfenstein and C. L. Wilkins, eds., *Computers in Chemical and Biochemical Research,* Vols. 1 and 2. New York: Academic Press, 1972 and 1974.
10. S. P. Perone and D. O. Jones, *Digital Computers in Scientific Instrumentation: Applications to Chemistry*. New York: McGraw-Hill, 1973.
11. M. H. Shames, *Automation in Clinical Chemistry*. An audio short course on tape. Washington, D.C.: American Chemical Society, 1973.

chapter 18

KINETIC METHODS OF ANALYSIS

We mentioned in Chapter 10 the use of catalysts to alter the reaction rate in certain titrations. The catalyst was added in sufficient concentration to make the reaction occur immediately. Here, we describe the use of rate-limited reactions for the determination of substances that speed up or catalyze the reactions, where the rate of the reaction is proportional to the concentration of small amounts of the catalysts. Second, we describe the use of specific catalysts, called enzymes, for the highly selective determination of their substrates.

18.1 KINETICS

Kinetics is the description of *reaction rates*. The *order* of a reaction defines the dependence of reaction rates on the concentrations of reacting species. Order is determined empirically and is not necessarily related to the stoichiometry of the reaction. Rather, it is governed by the *mechanism* of the reaction, that is, by the number of species that must collide for the reaction to occur.

First-Order Reactions

Reactions in which the rate of the reaction is directly proportional to the concentration of a single substance are known as first-order reactions. Consider the reaction

$$A \rightarrow P \tag{18.1}$$

A might be a compound that is decomposing to one or more products. The rate of the reaction is equal to the rate of disappearance of A, and it is proportional to the concentration of A:

$$-\frac{dA}{dt} = k[A] \qquad (18.2)$$

This is a **rate expression,** or **rate law.** The minus sign is placed in front of the term on the left side of the equation to indicate that A is disappearing as a function of time. The constant k is the **specific rate constant** at the specified temperature and has the dimensions of reciprocal time for example, s^{-1}. The **order of a reaction** is the sum of the exponents to which the concentration terms in its rate expression are raised. Thus, this is a first-order reaction and its rate depends only on the concentration of A.

Equation 18.2 is known as the **differential form** of the first-order rate law. The **integrated form** of the equation is

$$\log [A] = \log [A]_0 - \frac{kt}{2.303} \qquad (18.3)$$

where $[A]_0$ is the initial concentration of A ($t = 0$) and $[A]$ is its concentration at time t after the reaction is started. This equation gives the amount of A that has reacted after a given time interval. It is a straight-line equation, and if t is plotted versus $\log [A]$ (which can be measured at different times), a straight line with slope $-k/2.303$ and intercept $\log [A]_0$ is obtained. Thus, the rate constant can be determined.

Note that, from Equation 18.2, the rate of the reaction (*not* the rate constant) will decrease as the reaction proceeds, because the concentration of A decreases. Since $[A]$ decreases logarithmically with time (see Equation 18.3), it follows that the rate of the reaction will decrease exponentially with time. The time for one-half of a substance to react is called the **half-life** of the reaction, $t_{1/2}$. The ratio of $[A]/[A]_0$ at this time is $\frac{1}{2}$. By inserting this in Equation 18.3 and solving for $t_{1/2}$, we see that for a first-order reaction

$$t_{1/2} = \frac{0.693}{k} \qquad (18.4)$$

After the reaction is half complete, then one-half of the remaining reacting substances will react in the same time $t_{1/2}$, and so on. This is the exponential decrease we mentioned. Theoretically, it would take an infinitely long time for the reaction to go to completion, but for all practical purposes, it is complete (99.9%) after 10 half-lives. It is important to note that the half-life, and hence the time for the reaction to go to completion, is independent of the concentration for first-order reactions.

Radioactive decay is an important example of a first-order reaction.

Second-Order Reactions

Suppose we have the following reaction:

$$A + B \rightarrow P \qquad (18.5)$$

The rate of the reaction is equal to the rate of disappearance of either A or B. If it is empirically found to be

$$-\frac{dA}{dt} = -\frac{dB}{dt} = k[A][B] \tag{18.6}$$

then the reaction is first order with respect to [A] and to [B] and second order overall (the sum of the exponents of the concentration terms is 2). The specific reaction rate constant has the dimensions of reciprocal time and molarity, for example, $s^{-1} M^{-1}$.

The integrated form of Equation 18.6 depends on whether the initial concentrations of A and B ($[A]_0$ and $[B]_0$) are equal. *If they are equal*, the equation is

$$kt = \frac{[A]_0 - [A]}{[A]_0[A]} \tag{18.7}$$

If $[A]_0$ and $[B]_0$ *are not equal*, then

$$kt = \frac{2.303}{[B]_0 - [A]_0} \log \frac{[A]_0[B]}{[B]_0[A]} \tag{18.8}$$

If the concentration of one species, say B, is very large compared with the other and its concentration remains essentially constant during the reaction, then Equation 18.6 reduces to that of a first-order rate law:

$$\frac{-dA}{dt} = k'[A] \tag{18.9}$$

where k' is equal to $k[B]$; the integrated form becomes

$$kt = \frac{2.303}{[B]_0} \log \frac{[A]_0}{[A]} \tag{18.10}$$

Since $[B]_0$ is constant, Equation 18.10 is identical in form to Equation 18.3. This is a **pseudo first-order reaction.**

The half-life of a second-order reaction in which $[A]_0 = [B]_0$ is given by

$$t_{1/2} = \frac{1}{k[A]_0} \tag{18.11}$$

Thus, unlike the half-life of a first-order reaction, the half-life here depends on the initial concentration.

A reaction between A and B need not necessarily be second-order. Reactions of a fraction rate order are common. A reaction such as $2A + B \rightarrow P$ may be third order (rate $\propto [A]^2[B]$), or it may be second order (rate $\propto [A][B]$) or a more complicated order (even a fractional order).

Reaction Time

The time for a reaction to go to completion will depend on the rate constant k and, in the case of second-order reactions, the initial concentration. A first-order reaction is essentially instantaneous if k is greater than 10 s^{-1} (99.9% complete in less than 1 second). When k is less than 10^{-3} s^{-1}, the time for 99.9% reaction exceeds 100 minutes. Although it is more difficult to predict the time for second-order reactions, they can generally be considered to be instantaneous if k is greater than about 10^3 or 10^4 s^{-1} M^{-1}. If k is less than 10^{-1} s^{-1} M^{-1}, the reaction requires hours for completion.

18.2 CATALYSIS

The rates of some reactions can be speeded up by the presence of a **catalyst.** A catalyst can be defined as a substance that alters the rate of reaction without shifting the equilibrium and without itself undergoing any net change. We are interested in catalysis in which the rate of the reaction is proportional to the concentration of the catalyst. This forms the basis for the quantitative analysis of a catalyst. The rate of decrease of a reactant or the rate of increase of a product is measured and related to the catalyst concentration. These techniques are extremely sensitive for many catalysts.

Most analytical reactions used for catalytic analyses involve the slow reaction between two substances. If these substances are A and B and the catalyst is C, then a general rate expression can be written:

$$\text{rate} = k[A]^a[B]^b[C]^c \tag{18.12}$$

where a, b, and c are the reaction orders for each species. Characteristically, the reaction will be first order in A and B and perhaps in C. The reaction can be followed by a measuring of the decrease in the concentration of A or B or the increase in the concentration of one of the products. The reaction is simplified if the concentration of either A or B (the one not being measured) is made large so that it remains essentially constant (assuming the rate is not increased so much that it becomes difficult to make accurate measurements).

Measurement of Catalytic Reactions

The two basic ways of measuring the rates of catalyzed reactions and relating these to the catalyst concentration are illustrated in Figures 18.1 and 18.2. The usual

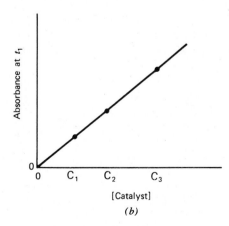

FIGURE 18.1
Determination of catalyst concentration from measurement of absorbance at a fixed time, t_1. (a) Reaction rate curves. (b) Calibration curve.

practice is to measure the concentration of one species of the reaction (spectrophotometrically, for example) at a certain time in the reaction, preferably a product (Figure 18.1). A calibration curve is prepared by plotting the absorbance at this time (corrected for a blank) against concentration of the catalyst. Such a measurement gives, essentially, the rate of change of absorbance per unit time.

These relationships apply for the initial rates of reactions.

There are automated instruments that will continuously record the change in absorbance as a function of time, and the initial slope of the plot is related to the catalyst concentration, being steeper the greater the catalyst concentration.

The second manner in which the rate may be measured is shown in Figure 18.2. Here, the time required to reach a certain absorbance (i.e., concentration) is measured, and this is related to the catalyst concentration. It should be emphasized that reaction rates are very temperature dependent (approximately doubling for each 10°C rise in temperature, i.e., about 10% per degree), and so the temperature must generally be regulated or held constant within a narrow range (a few tenths of a degree or better).

The more effective the catalyst is in enhancing the reaction rate, the lower the concentration that can be detected. Concentrations on the order of 10^{-6} M are not uncommon, and many catalysts can be determined at concentrations of a few hundredths of a nanogram (10^{-9} g) in several milliliters of solution, or about 10^{-10} M!

Nonspecificity of Catalysts

The problem, analytically, with catalytic reactions is that the catalyst being determined is frequently not specific and other substances may accelerate or decelerate the reaction. A preseparation procedure, such as ion exchange chromatography (Chapter 16), may satisfactorily eliminate interferences.

A substance may *inhibit* the catalytic effect of another substance. An example is the complexation of the catalyst iodide by mercury. The retarding action of the mercury can form the basis for the indirect determination of mercury (in the presence of a constant amount of added iodide).

FIGURE 18.2
Determination of concentration from measurement of time to reach a fixed absorbance, A_1. *(a)* Reaction rate curves. *(b)* Calibration curve.

Types of Reactions Catalyzed

Many analyses of catalysts utilize redox reactions. These reactions are often slow when unequal numbers of electrons are involved in the two half-reactions. A list of some catalyzed reactions and the catalysts are given in Table 18.1. The most common application of catalytic analysis in the clinical laboratory is determination of protein-bound iodine (PBI) in serum by its catalytic action on the Ce(IV)–As(III) reaction. There are no common substances in ashed serum that will interfere with the catalysis or that will react with the cerium(IV) or arsenic(III). The presence of the chloride in serum is required for the catalysis if only small amounts of iodide are present. This apparently prevents a side reaction with the production of iodate by forming instead a different intermediate compound with the iodine (ICl).

The normal concentration of protein-bound iodine in serum is only about 0.05 ppm, and so a highly sensitive method such as this is required. In the procedure, the proteins are precipitated, filtered, and dry-ashed in a furnace. The ash is taken up in sulfuric acid plus hydrochloric acid, and at time zero, the cerium(IV) and arsenic(III) are added. After a preselected time (for example, 15 minutes), the decrease in absorbance of the cerium(IV) is read at 420 nm.

As little as 0.01 ng cobalt can be determined in 5 mL solution by its catalytic effect on the reaction of alizarin with potassium perborate. The concentration of cobalt in blood is in the range of 0.3 to 15 ppb; the level is so low that it is difficult to analyze accurately, and the concentration is not known accurately. However, with a method as sensitive as this one, the analysis should be performed easily, provided the cobalt can be isolated to obtain the required selectivity.

Obviously, great care must be taken to prevent contamination in these methods.

TABLE 18.1
Determination of Catalysts by Reaction Rate[a]

Oxidant	Reductant	Catalysts
Cerium(IV)	Arsenic(III)	I^-, Os
Permanganate	Arsenic(III)	Os
Cerium(IV)	Chloride	Ag
Peroxydisulfate	Manganese(II)	Ag
Iron(III)	Thiosulfate	Cu
Iodine	Azide	S^{2-}, $S_2O_3^{2-}$, SCN^-
Chlorate	Iodide	Os, Ru, V
Malachite green	Titanium(III)	Mo, W
Methylene blue	Sulfide	Se
Oxygen	Ascorbic acid	Cu
Oxygen	Resorcinol	Cu
Hydrogen peroxide	Hydrogen peroxide	Mn, Pd, Cu
Hydrogen peroxide	p-Phenylenediamine	Cu, Fe, Os
Hydrogen peroxide	Iodide	Mo, W, V
Iron(III)	p-Phenylenediamine	$S_2O_3^-$

[a]From H. A. Laitinen, *Chemical Analysis*, p. 463. Copyright McGraw-Hill, Inc., New York, 1960. Used with permission of McGraw-Hill Book Company.

18.3 ENZYME CATALYSIS

Enzymes are remarkable naturally occurring proteins that catalyze *specific* reactions with a high degree of efficiency. Enzymes range in formula weight from 10,000 to 2,000,000. They are, of course, intimately involved in biochemical reactions in the body, that is, the life process itself. The determination of certain enzymes in the body is, therefore, important in the diagnosis of diseases. But aside from this, enzymes have proved extremely useful for the determination of **substrates,** the substances whose reaction the enzymes catalyze.

Enzyme Kinetics

We can describe the rate equation for enzyme reactions from a simple reaction model. The typical enzyme-catalyzed reaction can be represented as follows:

$$E + S \underset{k_2}{\overset{k_1}{\rightleftharpoons}} ES \overset{k_3}{\rightarrow} P + E \qquad (18.13)$$

where E is the enzyme, S is the substrate, ES is the **activated complex** that imparts a lower energy barrier to the reaction, P is the product(s), and the k's are the rate constants for each step. That is, the enzyme forms a complex with the substrate which then dissociates to form product. The rate of Reaction, R, is proportional to the complex concentration and, therefore, to the substrate and enzyme concentrations:

$$R = k_3[ES] = k[S][E] \qquad (18.14)$$

Assuming that k_1 and k_2 are much larger than k_3, we find the rate of the reaction to be limited by the rate of dissociation of the activated complex.

The dependence of an enzyme-catalyzed reaction rate on substrate concentration is illustrated in Figure 18.3. An enzyme is characterized by the number of molecules of substrate it can complex per unit time and convert to product, that is, the **turnover number.** As long as the substrate concentration is small enough with respect to the enzyme concentration that the turnover number is not exceeded, the reaction rate is directly proportional to substrate concentration, that is,

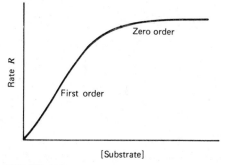

FIGURE 18.3

Dependence of enzyme catalyzed reaction rate on substrate concentration. At high concentrations, the enzyme becomes saturated with substrate and the reaction rate becomes maximum and constant since [ES] becomes constant (Equation 18.14).

it is first order with respect to substrate (Equation 18.14). If the enzyme concentration is held constant, then the overall reaction is first order and directly proportional to substrate concentration ($k[E]$ = constant in Equation 18.14). This serves as the basis for substrate determination.[1] However, if the amount of substrate exceeds the turnover number for the amount of enzyme present, the enzyme becomes *saturated* with respect to the number of molecules it can complex (saturated with respect to substrate) and the reaction rate reaches a maximum value. At this point, the reaction becomes independent of further substrate concentration increases, that is, becomes **pseudo zero order** if the enzyme concentration is constant (Figure 18.3); in Equation 18.14, [ES] becomes constant and R = constant.

When the enzyme is saturated with respect to substrate, then the overall reaction is first order with respect to enzyme concentration ($k[S]$ = constant in Equation 18.14). This becomes the basis for enzyme determination since a linear relationship between reaction rate and enzyme concentration will exist. Since substrate is consumed in the reaction, however, it must be kept at a high enough concentration that the reaction remains zero order with respect to substrate during the time of the reaction (i.e., the enzyme remains saturated). Eventually, at high enzyme concentrations, insufficient substrate will be available for saturation, and a plot similar to Figure 18.3 will result.

Properties of Enzymes

The rate of an enzymatic reaction depends on a number of factors, including the temperature, pH, ionic strength, and so forth. The rate of the reaction will increase as the temperature is increased, up to a point. Above a certain temperature, the activity of the enzyme is decreased, because, being a protein, it becomes *denatured*, that is, the tertiary structure of the enzyme is destroyed as hydrogen bonds are broken. The steric nature of an enzyme is critical in its catalytic mechanism. Most animal enzymes become denatured at temperatures above about 40°C.

As with other catalytic reactions, temperature changes as small as 1 or 2°C may result in changes as high as 10 to 20% in the reaction rate under analytical conditions. So it is important that the temperature be controlled during the measurements of enzyme reactions.

Enzymes should be stored at 5°C or less since they are eventually deactivated over a period of time at even moderate temperatures. Some enzymes lose activity when frozen.

The reaction rate will be at a maximum at a certain pH, owing to complex acid–base equilibria such as acid dissociation between the substrate, the activated complex, and the products. Also, the maximum rate may depend on the ionic strength and on the type of buffer used. For example, the rate of aerobic oxidation of glucose in the presence of the enzyme glucose oxidase is maximum in an acetate buffer at pH 5.1, but in a phosphate buffer of the same pH, it is decreased.

The **activity** of an enzyme preparation will vary from one source to another, because the enzymes are usually not purified to 100% enzyme. That is, the percent enzyme will vary from one preparation to another. The activity of a given prepa-

---

[1]Substrates need not be determined from the reaction rate. Instead, the reaction may be allowed to proceed until the substrate is completely converted to product. The concentration of the product is measured before (blank) and after the reaction. Each of these techniques is discussed in more detail below under the determination of enzymes and of enzyme substrates.

ration is expressed in **international units** (I.U.). An international unit has been defined by the International Union of Biochemistry as "the amount that will catalyze the transformation of one micromole of substrate per minute under defined conditions." The defined conditions will include temperature and pH. For example, a certain commercial preparation of glucose oxidase may have an activity of 30 units per milligram. Thus, for the determination of a substrate, a certain number of units of enzyme is taken. The **specific activity** is the units of enzyme per milligram of *protein*. **Molecular activity** is defined as units per molecule of enzyme, that is, it is the number of molecules of substrate transformed per minute per molecule of enzyme. The **concentration** of an enzyme in solution should be expressed as international units per milliliter or liter.

Enzyme Inhibitors and Activators

Although enzymes catalyze only certain reactions or certain types of reaction, they are still subject to interference. When the activated complex is formed, the substrate is adsorbed at an *active site* on the enzyme. Other substances of similar size and shape may be adsorbed at the active site. Although adsorbed, they will not undergo any transformation. However, they do compete with the substrate for the active sites and slow down the rate of the catalyzed reaction. This is called **competitive inhibition.** For example, the enzyme succinic dehydrogenase will specifically catalyze the dehydrogenation of succinic acid to form fumaric acid. But other compounds similar to succinic acid will competitively inhibit the reaction. Examples are other diprotic acids such as malonic and oxalic acids. Competitive inhibition can be reduced by increasing the concentration of the substrate relative to that of the interferent so that the majority of enzyme molecules combine with the substrate.

    Noncompetitive inhibition occurs when the inhibition depends only on the concentration of the inhibitor. This is usually caused by adsorption of the inhibitor at a site other than the active site but one which is necessary for activation. In other words, an inactive derivative of the enzyme is formed. Examples are the reaction of the heavy metals mercury, silver, and lead with sulfhydryl groups ($-SH$) on the enzyme. The sulfhydryl group is tied up by the heavy metal ($ESH + Ag^+ \rightarrow ESAg + H^+$), and this reaction is irreversible. This is why heavy metals are poisons; they inactivate enzymes in the body.

    Some enzymes require the presence of a certain metal for activation, perhaps to form a complex of the proper stereochemistry. Any substance that will complex with the metal ion may then become an inhibitor. For example, magnesium ion is required as an activator for a number of enzymes. Oxalate and fluoride will complex the magnesium, and they are inhibitors. Activators of enzymes are sometimes called **coenzymes.**

    Substrate inhibition sometimes occurs when excessive amounts of substrate are present. In cases such as this, the reaction rate actually decreases after the maximum velocity has been reached. This is believed to be due to the fact that there are so many substrate molecules competing for the active sites on the enzyme surfaces that they block the sites, preventing other substrate molecules from occupying them.

The Michaelis Constant

As explained previously, an enzyme at a given concentration eventually becomes saturated with respect to substrate as the substrate concentration is increased, and the reaction rate becomes maximum, R_{max}. The **Lineweaver–Burk equation** describes the relationship between the enzyme effectiveness as a catalyst and the maximum rate:

$$\frac{1}{R} + \frac{1}{R_{max}} + \frac{K_m}{R_{max}[S]}$$ (18.15)

where K_m is the **Michaelis constant.** The Michaelis constant is a measure of the enzyme activity and can be shown to be equal to $(k_2 + k_3)/k_1$ in Equation 18.13. It is also equal to the substrate concentration at one-half the maximum rate, $R_{max}/2$, as derived from Equation 18.15 by setting $R = R_{max}/2$. A plot of $1/[S]$ versus $1/R$ gives a straight line whose **intercept** is $1/R_{max}$ and whose **slope** is K_m/R_{max}. Thus, the Michaelis constant, which is characteristic for an enzyme with a substrate, can be determined.

The significance of K_m is illustrated in Figure 18.4 (similar to Figure 18.3). When the reaction rate increases rapidly with substrate concentration, K_m is small (curve 1). The substrate that gives the lowest K_m for a given enzyme is often (but not necessarily) the enzyme's natural substrate, hence the reason for the rapid increase in rate with increased substrate concentration. A small K_m indicates that the enzyme becomes saturated at small concentrations of substrate. Conversely, a large K_m indicates that high concentrations of substrate are required to achieve maximum reaction velocity. In such as case, it would be difficult to achieve zero-order rate with respect to substrate, and the substrate would not be a good one for determining the enzyme.

Enzyme Specificity

In general, there are four types of enzyme specificity: (1) **absolute specificity,** in which the enzyme will catalyze only one reaction; (2) **group specificity,** in which the enzyme will act on molecules with certain functional groups, such as amino, phosphate, or methyl groups; (3) **linkage specificity,** in which the enzyme will act on a particular type of chemical bond; and (4) **stereochemical specificity,** in which the enzyme will act on a particular steric or optical isomer.

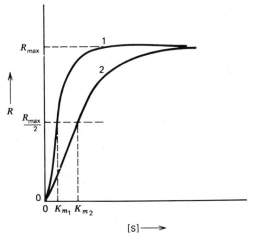

FIGURE 18.4
Illustration of relationship of Michaelis constant, K_m ($= [S]$ at $R_{max}/2$), to reaction rate. Curve 1. Small K_m. Curve 2. Large K_m.

In addition to the substrate being acted on, many enzymes require a second cosubstrate. Such a cosubstrate may activate many enzymes and is an example of a **cofactor** or **coenzyme** (described previously). An example is nicotinamide adenine dinucleotide (NAD^+), which is a cofactor for many dehydrogenase reactions by acting as a hydrogen acceptor:

$$SH_2 + NAD^+ \stackrel{enzyme}{\rightleftharpoons} S + NADH + H^+ \qquad (18.16)$$

where SH_2 is the reduced form of the substrate, S is its oxidized (dehydrogenated) form, and NADH is the reduced form of NAD^+.

Enzyme Nomenclature

Enzymes are classified according to the type of reaction and the substrate, that is, according to their reaction specificity and their substrate specificity. Most enzyme names end in "ase." Enzymes can be divided into four groups based on the kind of chemical reaction catalyzed: (1) Those that catalyze addition *(hydrolases)* or removal *(hydrases)* of water. Hydrolases include esterases, carbohydrases, nucleases, and deaminases, while hydrases include enzymes such as carbonic anhydrase and fumarase. (2) Those that catalyze the transfer of electrons: *oxidases* and *dehydrogenases.* (3) Those that catalyze transfer of a radical, such as *transaminases* (amino groups), *transmethylases* (methyl groups), *transphosphorylases* (phosphate groups), and so on. (4) Those that catalyze splitting or forming of a C—C bond: *desmolases.* For example, α-glucosidase acts on any α-glucoside. The rate of reaction may be different for different glucosides. More generally, however, enzymes show absolute specificity to one particular substrate. Thus, glucose oxidase catalyzes the aerobic (oxygen) oxidation of glucose to gluconic acid plus hydrogen peroxide:

$$C_6H_{12}O_6 + O_2 + H_2O \xrightarrow{\text{glucose oxidase}} C_6H_{12}O_7 + H_2O_2 \qquad (18.17)$$

Actually, this enzyme shows almost complete specificity for β-D-glucose. α-D-Glucose reacts at a rate of 0.64 relative to 100 for the β form. In the latter form, all the hydrogens are axial and the hydroxyl groups are equatorial, allowing the molecule to lie down flat on the enzyme active site and form the enzyme–substrate complex. The α form does not have the same arrangement of hydrogens and hydroxyls and cannot lie flat on the enzyme. Thus, the aerobic conversion of glucose (usually 36% α and 64% β) depends on the mutarotation of the α form to the β form. The mutarotation (equilibrium) is shifted as the β form is removed. Another enzyme, mutarotase, will affect the mutarotation, but this is usually not necessary. There is one other substance, 2-deoxy-D-glucose, that is affected by glucose oxidase. The relative rate of its reaction is about 10% of that of β-D-glucose, and it is usually not present in blood samples being analyzed for glucose.

There are thousands of enzymes in nature, and most of these exhibit absolute specificity.

Determination of Enzymes

Enzymes themselves can be analyzed by measuring the amount of substrate transformed in a given time or the product that is produced in a given time. The substrate should be in excess so that the reaction rate depends only on the enzyme concentration. The results are expressed as international units of enzyme. For example, the activity of a glucose oxidase preparation can be determined by measuring manometrically or amperometrically the number of micromoles of oxygen consumed per minute. On the other hand, the use of enzymes to develop specific procedures for the determination of substrates, particularly in clinical chemistry, has proved to be extremely useful. In this case, the enzyme concentration is in excess so the reaction rate is dependent on the substrate concentration.

Determination of Enzyme Substrates

Two general techniques may be used for measuring enzyme substrates. First, **complete conversion** of the substrate may be utilized. Before and following completion of the enzymatically catalyzed reaction, a product is analyzed or the depletion of a reactant that was originally in excess is measured. The analyzed substance (net change) is then related to the original substrate concentration. These reactions are often not stoichiometric with respect to the substrate concentration because of side reactions, instability of products or reactants, and so forth. Also, the reaction may require extraordinarily long times for completion. For these reasons, the analytical procedure is usually standardized by preparing a calibration curve of some type in which the measured quantity is related to known concentrations or quantities of the substrate.

The second technique employed for substrate determination is the measurement of the **rate** of an enzymatically catalyzed reaction, as is used to determine enzyme activity. This may take one of three forms, identical to the procedures described for catalytic reactions above. First, the time required for the reaction to produce a preset amount of product or to consume a preset amount of substrate may be measured. Second, the amount of product formed or substrate consumed in a given time may be measured. These are single point measurements (called *end-point measurements*) and require well-defined reaction conditions. They are easy to automate or may be performed manually. A third procedure is continuous measurement of a product or substrate concentration as a function of time to give the slope of the reaction rate curve, $\Delta c/\Delta t$. These are the so-called "true" rate measurements. The measurements must generally be made during the early portion of the reaction where the rate is pseudo first order.

Rate methods are generally more rapid than end point methods or complete conversion reactions. Complete conversion reactions, on the other hand, are less subject to interference from enzyme inhibitors or activators as long as sufficient time is allowed for complete conversion. See also the discussion of enzyme electrodes in Chapter 11 for a different approach to measuring substrates.

Example Enzymatic Analyses

Spectrophotometric methods are widely used to measure enzyme reactions. The reaction product may have an absorption spectrum quite different from the substrate, allowing simple measurement of the product or substrate. In other cases, a dye-forming reagent is employed that will react with the product or the substrate and the increase or decrease in color is measured. Frequently, the chromogen is enzymatically coupled with the product using a second enzyme.

1. DEHYDROGENASE REACTIONS. The reduced (NADH) and oxidized (NAD^+) forms of nicotinamide adenine dinucleotide exhibit marked differences in their ultraviolet absorption spectra and are, therefore, widely used for following the course of dehydrogenase reactions. The ultraviolet absorption spectra for NAD^+ and NADH are given in Figure 18.5. NAD has negligible absorption at 340 nm while NADH has an absorption maximum, and so it is a simple matter to monitor the increase or decrease in NADH concentration.

An example using NADH for measurement is the determination of the enzyme **lactic acid dehydrogenase** (LDH), which is important in confirming myocardial infarction (heart attack). NAD^+ is required in the LDH-catalyzed oxidation of lactic acid to pyruvic acid:

$$
\begin{array}{lc}
CH_3 & CH_3 \\
| & | \\
CHOH + NAD^+ \overset{LDH}{\rightleftharpoons} C{=}O + NADH + H^+ & \text{(18.18)} \\
| & | \\
CO_2H & CO_2H \\
\text{lactic acid} & \text{pyruvic acid}
\end{array}
$$

The reaction is reversible and can be employed in either direction. In the forward reaction, serum containing an unknown amount of LDH would be added to a solution containing enzyme saturating concentrations of lactic acid and NAD, and the increase in absorbance at 340 nm would be measured as a function of time.

NADH can sometimes be used to follow enzyme reactions in which it is not directly involved by using it as a coupling agent in a secondary reaction with the product. For example, serum **glutamic–oxaloacetic transaminase** (GOT) catalyzes the reaction of α-ketoglutarate and asparate and the product is reduced by NADH in the presence of another enzyme, malic acid dehydrogenase (MDH):

$$\text{α-ketoglutarate} + \text{asparate} \overset{GOT}{\rightleftharpoons} \text{glutamate} + \text{oxaloacetate} \qquad \text{(18.19)}$$

$$\text{oxaloacetate} + \text{NADH} \overset{MDH}{\rightleftharpoons} \text{malate} + NAD^+ \qquad \text{(18.20)}$$

The second reaction is fast compared to the first in the presence of an excess of MDH, and so the rate of decrease of NADH concentration is directly proportional to the GOT activity.

2. COMMONLY DETERMINED SUBSTRATES. A list of some substrates determined in blood or urine is given in Table 18.2. They are discussed in order in the paragraphs following.

Urease was the first enzyme to be isolated and crystallized. It quantitatively converts **urea** to ammonia and carbon dioxide. The amount of urea is calculated

FIGURE 18.5
Ultraviolet absorption spectra of NAD and NADH. (Courtesy of Worthington Biochemical Corporation.)

TABLE 18.2
Examples of Commonly Used Enzyme Reactions for Determining Substrates in Clinical Chemistry

Substrate Determined	Enzyme	Reaction
Urea	Urease	$NH_2-\overset{\overset{\displaystyle O}{\|\|}}{C}-NH_2 + H_2O \rightarrow 2NH_3 + CO_2$
Glucose	Glucose oxidase	$C_6H_{12}O_6 + H_2O + O_2 \rightarrow C_6H_{12}O_7 + H_2O_2$ (glucose) (gluconic acid)
Uric acid	Uricase	$C_5H_4O_3 + 2H_2O + O_2 \rightarrow C_4H_6O_3N_4 + CO_2 + H_2O_2$ (uric acid) (allantoin)
Galactose	Galactose oxidase	D-Galactose $+ O_2 \rightarrow$ D-galactohexodialdose $+ H_2O_2$
Blood alcohol	Alcohol dehydrogenase	Ethanol $+ NAD^+ \rightarrow$ acetaldehyde $+ NADH + H^+$

from a determination of either the ammonia or the carbon dioxide produced, usually the former. This can be done spectrophotometrically by reacting the ammonia with a color reagent.

Glucose is usually determined by measuring the hydrogen peroxide produced upon addition of glucose oxidase. This is done spectrophotometrically by coupling the hydrogen peroxide as it is produced with a reagent such as o-toluidine. This coupling occurs in the presence of a second enzyme, *horseradish peroxidase,* and a colored product results. Commercial preparations of glucose oxidase usually contain impurities that react with and consume part of the hydrogen peroxide and so the conversion is not stoichiometric. *Catalase* is an enzyme impurity, for example, that is specific for the decomposition of hydrogen peroxide. Nevertheless, the fraction of hydrogen peroxide converted to the dye product is constant and a calibration curve can be prepared using different concentrations of glucose.

There are a number of possibile inhibitors in the glucose determination. Most of them, however, occur in the second enzymatic reaction. The glucose oxidase method should be more specific, then, if the hydrogen peroxide were measured directly without the need for a second enzyme. For example, added iodide ion, in the presence of a molybdenum(VI) catalyst, is rapidly oxidized to iodine (Table 18.1). The iodine concentration can be followed amperometrically (Chapter 12). An alternative is to measure the depletion of oxygen amperometrically (see Figure 17.4, Chapter 17).

Uric acid is usually determined by measuring its ultraviolet absorption at 292 nm. However, the amount of uric acid in blood is small and the absorption is not specific. So, after the measurement, the uric acid is destroyed by adding the enzyme uricase. The absorbance is measured again. The *difference* in absorbance is due to the uric acid present. Since only uric acid will be decomposed by uricase, the method becomes specific. A similar procedure for the colorimetric determination of uric acid involves the oxidation of uric acid with molybdate to form molybdenum blue, a molybdenum(V) compound. In principle, uric acid could be determined as glucose was, but impurities in uricase preparations usually rapidly destroy the very small amount of hydrogen peroxide produced.

Galactose is determined in the same manner as glucose, by oxidizing the chromogen by H_2O_2 in the presence of peroxidase. **Blood alcohol** can be determined by UV measurement of the NADH produced.

3. COMMONLY DETERMINED ENZYMES. Table 18.3 summarizes reactions used to determine the activity of some enzymes frequently determined in the clinical laboratory. The pyruvate formed in the **GPT** reaction is coupled with NADH in the presence of added LDH for measurement. **CK** catalyzes the transfer of a phosphate group from creatinine phosphate to the nucleotide adenosine diphosphate (ADP) to produce adenosine triphosphate (ATP). The ATP is reacted with glucose in the presence of the enzyme hexokinase to form glucose-6-phosphate, which can then be reacted with NAD in the presence of glucose-6-phosphodehydrogenase. CK is also determined now by high-performance liquid chromatography, size exclusion chromatography, ion exchange chromatography, or electrophoresis.

Natural **LDH** consists of five components called **isozymes,** or **isoenzymes,** the ratio of which varies with the tissue source. The two electrophoretically fastest components occur in high percentage in LDH from heart muscle, and the level of these is preferentially increased in blood after heart muscle damage. The LDH method measures total LDH isozymes, which will usually indicate the heart damage. The two heart muscle isozymes mentioned, however, more readily catalyze the reduction of α-ketobutyrate than the slower-moving components of hepatic origin and are referred to as α-hydroxybutyrate dehydrogenase (HBD) active. Elevated HBD can be determined by the reaction given and is more specific for myocardial infarction than LDH; it also remains elevated for longer periods after infarct.

The **phosphatases** are determined by measuring the blue color of thymolphthalein in highly alkaline solution after a specific time; the high alkalinity stops the enzyme reactions.

Enzyme reactions can, of course, be measured by other techniques than spectrophotometry. Techniques that have been used include amperometry, conductivity, coulometry, and ion-selective electrodes. Certain enzyme reactions have been

TABLE 18.3
Examples of Commonly Determined Enzymes in Clinical Chemistry

Enzyme	Abbreviation	Reaction
Glutamic–pyruvic transaminase	GPT	α-Ketoglutarate + L-alanine $\overset{GPT}{\rightleftharpoons}$ glutamate + pyruvate Pyruvate + NADH + H$^+$ $\overset{LDH}{\rightleftharpoons}$ lactate + NAD$^+$
Glutamic–oxaloacetic transaminase	GOT	α-Ketoglutarate + aspartate $\overset{GOT}{\rightleftharpoons}$ glutamate + oxaloacetate Oxaloacetate + NADH + H$^+$ $\overset{MDH}{\rightleftharpoons}$ malate + NAD$^+$
Creatinine phosphokinase	CK	Creatinine phosphate + ADP $\overset{CPK}{\rightleftharpoons}$ creatine + ATP ATP + glucose $\overset{hexokinase}{\longrightarrow}$ ADP + glucose-6-phosphate Glucose-6-phosphate + NAD$^+$ $\overset{G\text{-}6PDH}{\longrightarrow}$ 6-phosphogluconate + NADH + H$^+$
Lactate dehydrogenase	LDH	L-Lactate + NAD$^+$ $\overset{LDH}{\rightleftharpoons}$ pyruvate + NADH + H$^+$
α-Hydroxybutyrate dehydrogenase	HBD	α-Ketobutyrate + NADH + H$^+$ $\overset{HBD}{\rightleftharpoons}$ α-hydroxybutyrate + NAD$^+$
Alkaline phosphatase		Na thymolphthalein monophosphate $\overset{pH\ 10.1}{\longrightarrow}$ Na thymolphthalein + phosphate
Acid phosphatase		Same as alkaline phosphatase, except pH 6.0

measured using **enzyme electrodes** whose response is specific for the particular reaction. These are described in Chapters 11 and 12.

Enzyme inhibitors and **activators** may be analyzed by employing enzyme reactions. The easiest technique is to measure the decrease or increase in the rate of the enzymatic reaction. Or the enzyme may be "titrated" with an inhibitor (or vice versa), and the amount of inhibitor required to completely inhibit the reaction measured. Trace elements, for example, have been determined by their inhibition or activation of enzyme reactions.

Many enzymes ideally represent the analytical chemist's dream of an absolutely specific reactant, but because of inhibitor effects, as well as problems associated with pH and ionic strength control, they must be used with some caution.

Most of the above discussed procedures can be adapted to automatic rate monitoring systems for enzymatic analysis. See, for example, Reference 8.

QUESTIONS

General

1. Distinguish between a first-order and a second-order reaction.
2. What is the half-life of a reaction? How many half-lives does it take for a reaction to go to completion?
3. What is a pseudo first-order reaction?
4. Suggest a way to determine whether a particular reaction between two substances A and B is first order or second order.

Enzymes

5. What is an international unit?
6. What is the difference between competitive inhibition and noncompetitive inhibition of an enzyme?
7. Why are heavy metals often poisons in the body?
8. What are coenzymes?
9. Suggest a way to test whether an enzyme inhibitor is competitive or noncompetitive.
10. Suggest how a Lineweaver–Burk plot can be used to determine whether an inhibitor is competitive or noncompetitive.

PROBLEMS

Kinetics

11. A first-order reaction requires 10.0 minutes for 50% conversion to products. How much time is required for 90% conversion? For 99% conversion?
12. A first-order reaction required 25.0 seconds for 30% conversion to products. What is the half-life of the reaction?
13. A solution is 0.100 M in substances A and B, which react by a second-order reaction. If the reaction is 15.0% in 6.75 minutes, what is its half-life under these conditions? What would be the half-life if A and B were each at 0.200 M, and how long would it take for 15.0% completion of the reaction?
14. Sucrose is hydrolyzed to glucose and fructose:

$$C_{12}H_{22}O_{11} + H_2O \rightarrow C_6H_{12}O_6 + C_6H_{12}O_6$$

In dilute aqueous solution, the water concentration remains essentially constant, and so the reaction is pseudo first order and follows first-order kinetics. If 25.0% of a 0.500 M sucrose solution is hydrolyzed in 9.00 hours, in how much time will the glucose and fructose concentration be equal to one-half the concentration of the remaining sucrose?

15. Hydrogen peroxide decomposes by a second-order reaction,

$$2H_2O_2 \xrightarrow{\text{catalyst}} 2H_2O + O_2$$

If 35.0% of a 0.1000 M solution decomposes in 8.60 minutes, how much time is required for the evolution of 100 mL O_2 from 100.0 mL of a 0.1000 M solution of H_2O_2 at standard temperature and pressure?

Catalysis

16. Traces of manganese are determined by measuring its catalytic effect on the oxidation of diethylamine by sodium periodate. The rate of formation of the oxidation product is followed spectrophotometrically. One milliliter aliquots of manganese standard solutions containing 0, 0.010, 0.020, 0.030, and 0.050 µg manganese are added to individual tubes containing 5.0 mL diethylamine solution. Five milliliters sodium periodate solution is added at time zero, and the absorbance of the solution is measured after 10 minutes. The absorbances were 0.020, 0.050, 0.081, 0.109, and 0.169, respectively. A 5.0-mL blood sample was dry-ashed and the manganese was separated by ion exchange chromatography. The final volume was 3.0 mL. A 1.0-mL aliquot of this was treated the same way as the standards, and after 10 minutes the absorbance was 0.098. What is the concentration of the manganese in the blood in ppm?

Enzyme

17. The activity of a glucose oxidase preparation is determined by measuring the volume of oxygen gas consumed as a function of time. A 10.0-mg fraction of the preparation is added to a solution containing 0.01 M glucose and saturated in oxygen. After 20.0 minutes, it is determined that 10.5 mL oxygen is consumed (STP). What is the activity of the enzyme preparation expressed in enzyme units per milligram? If the purified enzyme has an activity of 61.3 units/mg, what is the percent purity of this enzyme preparation?

RECOMMENDED REFERENCES

Kinetics

1. H. H. Bauer, G. D. Christian, and J. E. O'Reilly, eds., *Instrumental Analysis*. Boston: Allyn and Bacon, 1978, Chapter 18, "Kinetic Methods," by H. B. Mark, Jr.

2. R. A. Greinke and H. B. Mark, Jr., "Kinetic Aspects of Analytical Chemistry," *Anal. Chem.*, **46** (1974) 413R.

3. H. B. Mark, Jr., and G. A. Rechnitz, *Kinetics in Analytical Chemistry*. New York: Interscience Publishers, 1968.

4. K. B. Yatsimirsskii, *Kinetic Methods of Analysis*. New York: Pergamon Press, 1966.

Catalysis

5. H. A. Laitinen and W. E. Harris, "Reaction Rates in Chemical Analysis," in *Chemical Analysis*, 2nd ed. New York: McGraw-Hill, 1975, Chapter 21. Discusses catalytic methods.

6. P. W. West and T. V. Ramakrishna, "A Catalytic Method for Determining Traces of Selenium," *Anal. Chem.*, **40** (1968) 966.

Enzymes

7. H. U. Bergmeyer, *Methods of Enzymatic Analysis,* 2nd English ed. New York: Academic Press, 1974.

8. N. Gochman and L. J. Bowie, "Automated Rate-Monitoring Systems for Enzyme Analyses," *Anal. Chem.,* **48** (1976) 1189A.

9. G. G. Guilbault, *Enzymatic Methods of Analysis.* New York: Pergamon Press, 1969.

10. *International Union of Biochemistry, Standing Committee on Enzymes.* New York: The Macmillan Company, 1964. Gives recommended enzyme nomenclature.

11. *Manual of Enzyme Measurements,* Worthington Biochemical Corporation, Freehold, N.J., 1972.

chapter 19

CLINICAL CHEMISTRY

The previous chapters have dealt with the methodology and tools that are basic to all types of analyses. In the following chapters, we shall discuss specifically the practical aspects of analyzing some different types of materials. We will draw on your knowledge gained from the previous chapters, and these subsequent chapters will be more of a reference nature for specified analyses, suggesting approaches that may be taken for solving real analytical problems.

In this chapter, we discuss specifically the practical aspects of clinical analyses. The clinically significant constituents of blood and urine are described, including major electrolytes, proteins, and organic substances. Trace elements are also described; and some of the commonly used analytical procedures for important clinical determinations, that is, the normal physiological ranges of the constituents and the conditions under which they may fall outside this range, are given. Also, the rapidly growing and extremely sensitive technique of radioimmunoassay (RIA) is described.

19.1 THE COMPOSITION OF BLOOD

Whole blood can be broken down into its general components as follows: **plasma,** which contains the *serum* and *fibrinogen;* and the **cellular elements,** which contain the *erythrocytes, leukocytes,* and *platelets.* These are described in Chapter 1. Plasma is the liquid portion of circulating blood. The cells are separated from the plasma by centrifuging whole blood. If blood is allowed to clot, the fibrinogen is removed with the cells, leaving **serum.** The majority of clinical analyses are performed on whole blood, plasma, or serum, and most of these use serum. Urine is also frequently analyzed.

Table 19.1 summarizes the normal range of concentrations of some clinically important constituents in human blood. We should emphasize that these normal ranges are approximations. Table 1.4 in Chapter 1 summarizes the major electrolyte composition (cations and anions) in blood. The analysis of some of the more commonly determined constituents will be discussed below. The physiological significance of the results is also discussed.

TABLE 19.1
Information Pertinent to Blood Samples for Chemical Examinations[a]

Determination	Sample[b]	Normal Range
Albumin	S	4 to 5 g/dL
Amino acid nitrogen	B	4 to 6 mg/dL
Ammonia	B	40 to 125 μg/dL
Amylase	S	Up to 150 mg/dL
Bilirubin	S	Direct up to 0.4 mg/dL
		Total up to 1.0 mg/dL
Calcium	S	4.5 to 5.5 meq/L
Carbon dioxide content	S	25 to 32 meq/L
Chloride, serum	S	100 to 108 meq/L
Cholesterol, total	S	140 to 250 mg/dL[c]
Cholesterol, esters	S	50 to 65% of total
Creatinine	S	0.7 to 1.7 mg/dL
Creatinine clearance		100 to 180 mL/min
Lipase	S	Up to 1.5 units
Lipids, total	S	350 to 800 mg/dL
Fatty acids	S	200 to 400 mg/dL
Globulins, total	S	2.5 to 3.5 g/dL
alpha-1		0.1 to 0.4 g/dL
alpha-2		0.3 to 0.7 g/dL
beta		0.4 to 0.9 g/dL
gamma		0.6 to 1.3 g/dL
Iron, serum	S	50 t0 180 μg/dL
Magnesium	S	1.5 to 2.5 meq/L
Nonprotein nitrogen (NPN)	B	25 to 40 mg/dL
Phosphatase, acid	S	Up to 4 Gutman units
Phosphatase, alkaline	S	Up to 4 Bodansky units
Phospholipids	S	100 to 250 mg/dL
as phosphorus		4 to 10 mg/dL
Phosphorus, inorganic	S	3 to 4.5 mg/dL
Potassium	S	3.8 to 5.6 meq/L
Protein, total serum	S	6.5 to 8 g/dL
Protein-bound iodine (PBI)	S	3.5 to 8 μg/dL
Sodium	S	138 to 146 meq/L
Sugar, blood (glucose)	B	65 to 90 mg/dL
Transaminase (SGO)	S	Up to 40 units
Uric acid	S	3 to 6 mg/dL
Urea nitrogen (BUN)	B	Up to 20 mg/dL

[a]Adapted from J. S. Annino, *Clinical Chemistry*. 3rd ed. Boston: Little, Brown and Company 1960.

[b]S, serum; B, blood.

[c]This range varies with age.

19.2   COLLECTION AND PRESERVATION OF SAMPLES

Blood and urine samples are often collected after the patient has fasted for a period of time (e.g., overnight). Few studies have been performed on the effects of eating on the chemical composition of blood, but one study indicates that an average breakfast has no significant effect on the concentration of the blood content of carbon dioxide, chloride, sodium, potassium, calcium, urea nitrogen, uric acid, creatinine, total protein, albumin, cholesterol, and cholesterol esters. Serum phosphorus is slightly depressed at 45 minutes after the meal, but not after 2 hours. The variations in glucose levels are large, and so fasting should precede collection of samples for analysis of this constituent.

When serum is required for the analysis, the blood is collected in a clean and dry tube to prevent contamination and **hemolysis.** Hemolysis is the destruction of red cells, with the liberation of hemoglobin and other cell constituents into the surrounding fluid (serum or plasma). When hemolysis occurs, the serum is noticeably red instead of its normal straw color. A substance may occur at a much higher concentration in the cellular portion of the blood than in the serum or plasma. If it is the concentration in the serum or plasma that is clinically significant, then hemolysis would give erroneously high serum or plasma results, as in the case of potassium, iron, magnesium, zinc, urea, protein (from heomoglobin), and others. For this reason, blood samples should be centrifuged as soon as possible to separate serum or plasma from the cells.

If plasma or whole blood is required for analysis, then the blood is collected in a tube containing an **anticoagulant.** *Heparin* (sodium salt) is frequently used. However, its effect is temporary and heparin is expensive. Therefore, a more widely used anticoagulant is *potassium oxalate,* about 1 mg per mL blood. Oxalate precipitates blood calcium, and the calcium is required in the clotting process. Obviously, plasma prepared in this way cannot be analyzed for calcium or potassium; many other metals are precipitated by oxalate, and so serum is usually analyzed for these. The usual practice in preparing whole blood or plasma samples is to add the required amount of oxalate in solution form to the collection tube and then to dry the tube in an oven at 110°C. By this procedure, the collected blood is not diluted. For example, 0.5 mL of a 2% potassium oxalate solution would be taken and dried for a 10-mL blood sample. Potassium oxalate causes red cells to shrivel, with the result that the intracellular water diffuses into the plasma. Thus, the plasma should be separated as soon as possible.

If a sample must be kept anaerobically, as in the case of CO_2 analysis, mineral oil is added to the collection tube. This is lighter than blood and will cover it. A cork stopper should be used in these cases because the oil causes rubber stoppers to swell.

Sometimes a preservative is added to the sample, usually along with an anticoagulant. Sodium fluoride is widely used as a preservative for samples to be analyzed for glucose. This is an enzyme inhibitor that prevents the enzymatic breakdown of glucose, or **glycolysis.** One milligram sodium fluoride per milliliter blood is adequate. Since it also inhibits other enzymes, including urease, sodium fluoride should not be added to samples to be analyzed for enzymes or for urea based on urease catalysis.

Samples can generally be stored for one or two days by refrigerating them. This slows down enzymatic and bacteriological processes but does not eliminate them, and so it is best to analyze fresh samples when possible. Samples are always brought to room temperature before analyzing. Freezing samples will preserve them for long periods of time, at the same time suspending the activity of enzymes in the blood. Whole blood should not be frozen because the red cells will become ruptured thereby. Serum and plasma samples fractionate into layers of different composition when frozen and so should be shaken gently after thawing.

Samples are more stable if a **protein-free filtrate** (PFF) is prepared. There are a number of procedures for preparing them. In the *Folin-Wu* or *tungstic acid method,* 1 volume of blood, serum, or plasma is mixed with 7 volumes of water and 1 volume of 0.33 *M* sulfuric acid and allowed to turn brown. Then, 1 volume of 10% sodium tungstate (wt/vol, $Na_2WO_4 \cdot 2H_2O$) is added, and after 2 minutes, the precipitated protein is filtered or centrifuged. In the *trichloroacetic acid* (TCA) *method,* 1 volume of blood, serum, or plasma is mixed with 9 volumes of 5% TCA (wt/vol), and after proteins are precipitated, the mixture is filtered or centrifuged. Both of these yield acidic filtrates but are useful for the analysis of many substances.

When glucose is analyzed by being oxidized (for example, with alkaline copper tartrate; see below), high results are obtained if the above filtrates are used. A PFF prepared with *barium hydroxide* and *zinc sulfate* removes most interfering nonglucose-reducing substances (see Experiment 44). This provides essentially a neutral PFF because the $Zn(OH)_2$ product is precipitated:

$$Ba(OH)_2 + ZnSO_4 \rightarrow \underline{BaSO_4} + \underline{Zn(OH)_2}$$

Glucose is stable in a protein-free filtrate because the glycolytic enzymes are removed.

19.3 CLINICAL ANALYSIS

Table 19.2 summarizes some of the more frequently determined constituents in blood and the principles of some of the procedures employed. These are only representative procedures, and in many cases, there are numerous variations or different procedures offering varying degrees of convenience, speed, sensitivity, accuracy, precision and so on.

The major **serum electrolytes,** sodium, potassium, calcium, magnesium, chloride, and bicarbonate (CO_2) are fairly easy to determine. The metals are most readily determined by the use of flame-spectrophotometric methods, although colorimetric methods exist for calcium and magnesium. Calcium and, less frequently, magnesium are also titrated with EDTA. Ion-selective electrodes are used for the routine analysis of sodium, potassium, and calcium. Bicarbonate is analyzed also by titration against standard acid (see Experiment 8) in addition to a manometric method. Chloride is most widely determined by automatic coulometric titration with electrogenerated silver ion.

Blood glucose and **blood urea nitrogen** (BUN) are probably the two most frequently performed clinical tests. In the procedures described in Table 19.2, the total of all reducing sugars is measured, and so results tend to be high. But these methods have been adopted as standard ones for many years. The enzymatic determination of glucose (Chapter 18) is gaining in popularity, and the dedicated enzymatic glucose analyzer described in Chapter 17 is now widely used.

Uric acid is more specifically determined enzymatically than by the described method. The uric acid in a separate aliquot of the sample is destroyed with the enzyme uricase and the decrease in the blue phosphotungstate reduction product is measured. This second aliquot effectively serves as a blank. The uric acid may

TABLE 19.2
Some Procedures Used in Clinical Analysis

Substance Determined	Sample Analyzed[a]	Measurement Method	Principle
Barbiturates	S	Ultraviolet spectrophotometry	Extract barbiturate into $CHCl_3$, back-extract into NaOH solution, and measure UV absorption at alkaline pH at 252 nm
Calcium	S	Atomic absorption spectrophotometry	Dilute 1:20 with 10,000 ppm Na_2EDTA and measure absorbance at 422.7 nm
Carbon dioxide	S	Manometric	Van Slyke manometer method
Chloride	S	Volumetric; coulometric	Titrate with Ag^+ or Hg(II)
Creatinine	S	Spectrophotometric	PFF (TCA) reacted with alkaline picrate and absorbance measured at 490 nm
Iron	S	Spectrophotometric	Iron in PFF is reduced to Fe(II) and complexed with bathophenanthroline; absorbance measured at 535 nm
Magnesium	S	Atomic absorption spectrophotometry	Dilute 1:20 with 1% Na_2EDTA and measure absorbance at 285.2 nm
Nonprotein nitrogen (NPN)	B	Spectrophotometry	Digest PFF to form NH_4HSO_4, add alkaline mercuric iodide (Nessler's reagent) to complex ammonia, and measure absorbance at 480 nm
Phosphatase, acid	S	Spectrophotometry	Incubate with sodium glycerophosphate for 1 hr at pH 4.9 to liberate phosphate; determine phosphate as with inorganic phosphorus
Phosphatase, alkaline	S	Spectrophotometry	Same, but incubate at pH 9.6
Phosphorus, inorganic	S	Spectrophotometry	PFF (TCA); react with Mo(VI) to form phosphomolybdic acid and reduce this to molybdenum blue; measure absorbance at 660 nm
Potassium	S	Flame photometry	Dilute 1:50 with 0.15 M NaCl and measure emission at 766 nm
Protein, total	S	Spectrophotometry	Biuret reagent (alkaline copper tartrate); forms complex with proteins; measure absorbance at 550 nm after 30 min
Protein-bound iodine (PBI)	S	Spectrophotometry (catalytic)	Precipiate proteins with TCA, ash or digest the protein, and measure the catalytic effect of I^- on the Ce(IV)–As(III) reaction rate by measuring absorbance of Ce(IV) at 420 nm after 20 min

Salicylate	S	Spectrophotometry	Form complex with ferric nitrate in acid solution and measure absorbance at 535 nm
Sodium	S	Flame photometry	Dilute 1:50 and measure emission at 589 nm
Sugar (glucose)	B or S	Spectrophotometry	$Ba(OH)_2$–$ZnSO_4$ PFF; oxidize sugar with alkaline Cu(II) to form Cu_2O; the Cu_2O is determined by allowing it to reduce phosphomolybdic acid to form molybdenum blue, which is measured at 420 nm or react sugar with o-toluidine and measure at 635 nm
Uric acid	S	Spectrophotometry	Tungstic acid PFF; oxidize with alkaline phosphotungstate and measure the blue reduction product of phosphotungstate at 680 nm
Urea nitrogen (BUN)	B	Spectrophotometry (enzymatic)	Tungstic acid PFF; incubate with urease enzyme at pH 6.8 to produce NH_3; determine the NH_3 with Nessler's reagent

[a]S, serum; B, blood.

also be measured directly by its absorbance in the ultraviolet region at 290 nm, before and after incubation with uricase.

Albumin and **globulins** can be analyzed by the same procedure as for total proteins after fractionation by salting out with either sodium sulfate or sodium sulfite. More detailed information may be required about the protein fractions (α, β, γ globulins), in which case starch gel electrophoresis can be used to separate the proteins. Micro-Kjeldahl analysis of proteins is used when highly accurate data are required; the biuret method is accurate to about 4%.

Acid and alkaline phosphatases are enzymes contained in the blood and are active in acid and alkaline solution, respectively. The substrate used for their analysis is sodium glycerophosphate, at the appropriate pH. These enzymes hydrolyze the substrate to release phosphate. Other phosphate substrates can be used.

Barbiturates, which are drugs not normally present in the blood, are extractable (un-ionized) from blood into methylene chloride. They can then be back-extracted into 0.45 M sodium hydroxide as the ionized form. The ionized form absorbs in the ultraviolet region, whereas the un-ionized form does not. Plotting an absorption spectrum can qualitatively confirm the presence of barbiturates. The ionized form at pH 13 to 14 exhibits an absorption maximum between 252 and 255 nm, with a minimum between 234 and 237 nm; and at ph 9.8 to 10.5, a different ionized form exhibits a maximum at 240 nm.

A number of the above tests are also performed with urine samples, and frequently the same procedures can be used. Because of different concentrations and interferences, modifications in the procedures may be introduced. Analyses are usually performed with a 24 -hour sample, since the total daily excretion of most substances has much more significance than the concentration in a random sample. The composition of urine compared with that of serum is illustrated in Figure 19.1.

19.4   IMMUNOASSAY

Immunoassay techniques are important for the specific determination of hormones, drugs, vitamins, and other compounds at nanogram and smaller levels. In these techniques, an antigen and a (specific) antibody react to form a complex or precipitate. The first analytical application was in the form of radioimmunoassay (RIA) in which Berson and Yalow demonstrated the ability to selectively measure small quantities of insulin (see Reference 5). It was not until the late 1960s and early 1970s that RIA became widely available for routine analyses. At this time, the methods moved from the research laboratory to the clinical laboratory in what must be record time, demonstrating the real need that existed. It has been predicted that in the future as many as 90% of the tests performed in the clinical laboratory will be radioimmunoassays and related competitive binding assays. While these figures may be overly optimistic, they do point out the importance of immunoassay techniques. The importance attached to the technique is further evidenced by the fact that the Nobel prize in physiology was awarded to Rose Yalow in 1977, following Berson's death, for its development.

Immunoassay techniques generally involve a competitive reaction between an analyte antigen and a standard antigen that has been tagged, for limited binding sites on the antibody to that antigen. The tag may be a radioactive tracer, an enzyme, a fluorophore, and so forth. Radioimmunoassay is one of the most common

FIGURE 19.1

Composition of urine compared with that of blood plasma. Nonelectrolytes are expressed as millimoles on the milliequivalent scale. Values on the scale inclusive for all constituents (the sum of all of them). (From J. A. Gamble, *Chemical Anatomy, Physiology, and Pathology of Extracellular Fluid.* Cambridge, Mass.: Harvard University Press, 1950. Reproduced by permission of Harvard University Press.)

techniques, and we describe below this technique as well as the principles that are common to all the techniques. Fluorescence and enzyme immunoassay are also described.

FIGURE 19.2
Principle of immunological reactions.

Principles of RIA

Radioimmunoassay combines the sensitivity of radiochemistry with the specificity of immunology. Immunology is the study of antigens and their reactions with antibodies, that is, an organism's defense mechanism to foreign bodies through antibodies (Figure 19.2). An **antigen** (e.g., a hormone) is a (foreign) substance capable of producing antibody formation in the body and is able to react with (bind to) that antibody. An **antibody** is a protein endowed with the capacity to recognize, by stereospecific association, a substance foreign to the organism it has invaded, for example, bacteria and viruses. It is usually a *gamma globulin* or *immunoglobulin* that will specifically react with an antigen to form an **antigen–antibody complex.** It is produced in the organism (where it will remain present for some time) only after the organism has had at least one exposure to the intruder (through vaccination, either spontaneous or artificial). An antibody is produced for use in RIA by injecting the antigen into an animal species to which it is foreign and recovering the serum which contains the resultant antibody **(antiserum).**

All RIA procedures are based on the original discovery by Berson and Yalow that low concentrations of antibodies to the antigenic hormone insulin could be detected radiochemically by their ability to bind radiolabeled (^{131}I) insulin. The determination of unknown concentrations of antigen, then, is based on the fact that radiolabeled antigen and unlabeled antigen (from the sample or a standard) compete physiochemically for the binding sites on the antibodies. This is illustrated in Figure 19.3.

The initial reaction vessel contains antibody solution (antiserum), radiolabeled antigen, and the serum sample that may contain unlabeled (natural) **antigen (the substance to be determined).** Upon incubation, the antibody (Ab) will form an antigen–antibody immuno complex (Ag–Ab). In the absence of unlabeled antigen, a certain fraction of the radiolabeled antigen (Ag*) is bound (as Ag*–Ab). But when increasing amounts of unlabeled antigen (Ag) are added, the limited binding sites of the antibody are progressively saturated and the antibody can bind less of the radiolabeled antigen. Following incubation, the bound antigens are separated from

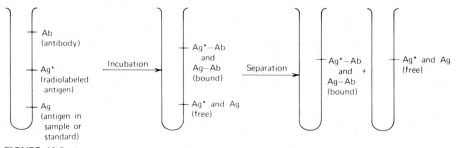

FIGURE 19.3
Principles of radioimmunoassay.

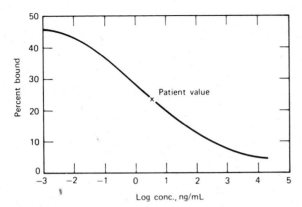

FIGURE 19.4
Typical radioimmunoassay calibration curve.

the unbound (free) antigens, and the radioactivity of either or both phases is measured to determine the percent bound of the labeled antigen.

The antibody solution is initially diluted so that, in the absence of unlabeled standard or unknown antigen, about 50% of the tracer dose of Ag* is bound. A diminished binding of labeled antigen when sample is added indicates the presence of unlabeled antigen. A calibration curve is prepared using antigen standards of known concentration by plotting either the percent bound labeled antigen or the ratio of percent bound to free (B/F) as a function of the unlabeled antigen concentration. From this, the unknown antigen concentration can be ascertained.

A typical calibration curve is shown in Figure 19.4. Note the wide range of concentrations (10^{-2} to 10^3 ng/mL in the most linear region) as well as the high sensitivity. The slope of the calibration curve, and hence the range and sensitivity, is dependent on the initial dilution of the antibody (see Figure 19.5). Sensitivity is greatest with high dilutions of the antibody (curve A), but a wider range of antigen concentrations is covered when a more concentrated antibody is used (curve C). Also, as mentioned above, to obtain a very high sensitivity, it is advisable to use a dilution of antiserum that will bind about 50% of the labeled antigen (using a very small amount) in the absence of unlabeled antigen. This is because a decrease of 50% of the bound/free (B/F) ratio from an initial value of 1.0 (i.e., 50% initially bound) can be more accurately determined than can a similar percentage decrease from an initial value of, say, 10 (91% bound). In the former case, there would be a decrease from 50% bound (B/F = 1.0) to 33% bound (B/F = 0.5), while in the latter case there would be a decrease from 91% (B/F = 10) to 83% bound (B/F = 5). The sensitivity of the initial part of the curve may be increased by using minimal amounts of tracer antigen.

The competitive principle of RIA can be applied to nonimmunologic systems. Any substance that can bind specifically to a macromolecule can be measured using the principle of **competitive protein binding.** The specific binding macromolecule can be a serum protein such as thyroxine binding globulin, a specific receptor, or an antibody.

Specificity of RIA

No antiserum used in RIA is completely specific for a particular antigen. The specificity is influenced by (a) heterogeneity of the antibody, (b) cross reaction with other antigens, and (c) possible interferences of the antigen–antibody reaction from

FIGURE 19.5
Standard curves for the assay of hormone (ng/mL) at three dilutions of the same antiserum. (From J. T. Potts, Jr. L. M. Sherwood, J. L. H. O'Riordan, and G. D. Aurbach, *Advan. Inter. Med.*, **13** (1967) 183. Reproduced by permission of Yearbook Medical Publishers, Inc.)

low-molecular-weight substances that may alter the environment of the reaction. A given antigen induces the formation of multiple antibodies. It will combine with the multiple antibodies to various degrees, depending on the respective equilibrium constants. In addition, a single type of antibody may have a variable number and location of binding sites. The problem of heterogeneity is being diminished with the development of improved techniques for antisera purification. Also, the synthetic production of *monoclonal antibodies,* that is, single antibodies, provides high specificity.

The problem of cross reaction with other antigens is a complex one and must be considered separately for each type of antigen being measured. Special purification techniques may be required. You are referred to Reference 9 at the end of the chapter for details.

Nonspecific factors that may modify the rate of the antigen–antibody reaction include pH, ionic strength, high temperatures, the composition of the incubating medium buffer, heparin, urea, and high bilirubin concentrations. Antigen standards and unknowns should be prepared (diluted) in antigen-free plasma to swamp out differences in composition. The same buffer or other dilutant should be used. If antigen-free plasma of the same species is not available, then plasma from a non-cross-reacting species can be used.

Radiolabeling of the Antigen

A highly purified antigen is required that can be radiolabeled without loss in immunoreactivity. Most polypeptide hormones contain one or more tyrosine residues,

$$\langle \bigcirc \rangle - H_2C - \overset{\overset{\displaystyle NH_2}{\displaystyle |}}{CH} - CO_2H$$

which can be labeled with radioactive iodine (^{131}I or ^{125}I) by iodination of the phenyl ring. Other compounds such as drugs, steroid hormones, and vitamins may be labeled with ^3H or ^{14}C. The iodine isotopes have the advantage of being gamma emitters and hence easy to count. ^3H and ^{14}C are low-energy beta emitters that must be counted via liquid scintillation methods. Their advantage is that they have long half-lives (12 and 5700 years, respectively) compared to 8 days for ^{131}I and 60 days for ^{125}I. Hence, preparations with ^3H and ^{14}C are readily commercially available while iodine preparations are frequently made at or near the time of the analysis. They will generally require purification.

Preparation of the Antibody

A suitable antiserum is, of course, required for the radioimmunoassay. The concentration of the antibody (called its **"titer"**) is important, but the main criterion for establishing a suitable antiserum is its specificity and affinity for the antigen being assayed.

The general method of inducing antibody formation is to inject 0.2 to 2 mg of the pure antigen mixed with "Freund's adjuvant" (a mixture of mineral oil, waxes, and killed bacilli, which enhances and prolongs the antigenic response). Animals used include rabbits, sheep, guinea pigs, goats, chickens, or monkeys—depending on the volume of antiserum desired and the degree of foreign activity of the antigen.

Molecules that are too small to induce antibody formation (formula weight 1000 to 5000, called **haptens** as opposed to antigens, e.g., smaller polypeptides, steroids) are linked to carriers such as proteins or synthetic compounds before they are injected. Common carriers are human gamma globulin and albumin, synthetic peptides, and polymers.

Once an animal has been immunized, it can be injected several times to obtain different lots of antisera. Antisera are commercially available for most assays for which labeled antigens are available. Diluted antisera can be stored for long periods of time when frozen. Once thawed, they should be stored at 4°C.

Incubation Period for the Assay

The time required to reach equilibrium in the antigen–antibody reaction during the assay varies from a few hours to several days, depending on the specific antigen being measured. Lengthy incubations are geneally not desirable (the antigen may be damaged by prolonged exposure to the high concentrations of plasma proteins, oxidants, etc.), and so some procedures may not be carried to the point of equilibrium.

Separation of the Bound and Free Antigen

Various techniques are used to separate the bound antigen from the free antigen after incubation so that the percent bound can be determined from the radioactivity. The immune complex is a protein-containing substance that can be **precipitated** (denatured) with high concentrations of various salts [e.g., $(NH_4)_2SO_4$, Na_2SO_4] or organic solvents (e.g., acetone, ethanol). Precipitation is one of the most widely used separation techniques. The precipitated complex is generally separated by centrifugation (although filtration can be used), and then the radioactivity in either of the phases can be measured.

The **double-antibody technique** is also widely used. Here, a second antibody is employed to precipitate the primary antigen–antibody complex. This second antibody is produced by injecting a second animal with the gamma globulin produced in the first animal used to prepare the first antibody. Although the use of a second antibody introduces another source of error, the double-antibody technique has the advantage of being applicable for almost any RIA, and the separation is complete.

Other separation techniques used include electrophoresis (Chapter 16) and bonding of the antigen or antibody to a solid phase for use as reagent.

Applications of RIA

Radioimmunoassays were originally developed for hormones but subsequently became available for other substances. The organism is capable of forming antibodies to cope with practically any natural or synthetic steric pattern, and so the possibilities seem unlimited.

Substances that are routinely determined using radioiodine-labeled antigens include thyroid-related hormones[1] (T_3, T_4, TSH), digestive hormones (gastrin), sex hormones (FSH, LH, HCG), other hormones (insulin, ACTH, parathormone, cortisol), vasoactive substances (renin, norepinephrine, angiotensin), and other compounds (morphine, digoxin, digitoxin, immunoglobulins, hepatitis-associated antigen). Other compounds are also routinely determined using ^{3}H- or ^{14}C-labeled antigens: drugs (morphine, digoxin, digitoxin, LSD, barbiturates, marijuana), hormones (serotonin, progesterone, testosterone, aldosterone, cortisol, prostaglandins), and others (DNA, vitamin D_3, folate). These are just some representative substances.

A number of systems are available for automatic radioimmunoassays (see, for example, Reference 6).

Fluorescence Immunoassay

Antibodies or antigens can be labeled with fluorescent dyes and the fluorescence used for measurements instead of radioactivity (see Reference 8). This technique is popular because of the greater ease of handling the chemicals than radiochemicals and the problem of radioactivity decreases with time. Fluorescein isothiocyanate (FITC) and lissamine rhodamine B (RB 200) are commonly used dyes. The protein to be conjugated (reacted) with the dye is usually a mixture of immunoglobulins and should be free of other serum protein fractions because albumin and other globulins label more readily than gamma globulin and would render the method nonspecific. After labeling, unreacted dye is removed and the labeled protein purified using size exclusion chromatography.

The fluorescence of a tagged antigen is often quenched upon immunochemical reaction, and the decrease in fluorescence can be measured without going through a physical separation. This serves as the basis for **homogeneous immunoassays,** in contrast to **heterogeneous immunoassays,** which require a separation.

Enzyme Immunoassay

Here, the tag employed is an enzyme, for example, peroxidase, and the activity of the unreacted enzyme–tagged antigen is measured using an appropriate enzyme reaction. Again, in these techniques, the activity of the enzyme is often inhibited

---

[1]The hormone symbols are as follows: T_3, T_4 = iodinated thyronine ($T_4 \equiv$ thyroxine), TSH = thyroid-stimulating hormone, FSH = follicle-stimulating hormone, LH = luteinizing hormone, HCG = human chorionic gonadotropin, ACTH = adrenocorticotropic hormone.

upon immunochemical reaction and the decrease in activity can be measured in a homogeneous immunoassay system.

If an enzyme tag is placed on the antibody, it is inhibited upon reacting with the analyte antigen; in this case, we have a *noncompetitive* protein binding assay (see, for example, Reference 7).

19.5 THE BLOOD GAS ANALYZER

Blood gas analyzers are commercially available, and they automatically or manually measure the pH, p_{CO_2}, and p_{O_2} of blood. The oxygen is measured with a conventional polarographic oxygen electrode. The carbon dioxide is measured by a pH glass electrode covered with a plastic membrane that allows diffusion of only gases. The instrument shown in Figure 19.6 will automatically make these measurements and read out the three results.

19.6 TRACE ELEMENTS IN THE BODY

A number of trace elements in the body are essential to the life process. Little is known about most of these, but many are involved in the activation (or deactivation) of enzymes. Others are involved in vitamin, hormone, skeletal, and other controls. Some evidence has been obtained indicating that trace metals may be associated with RNA. Some elements are essential in trace quantities but become toxic in large amounts. Examples are molybdenum and selenium in animals.

Table 19.3 summarizes the elemental content of the human body. The levels of some metals such as cobalt are not accurately known because of the difficulty in analyzing for these small amounts. Because insufficient knowledge is available to

FIGURE 19.6

Instrumentation Laboratory Blood Gas Analyzer, Model 113-S1, for measurement of pH, P_{O_2}, and P_{CO_2}. (Courtesy of Instrumentation Laboratory, Inc.)

TABLE 19.3
Elemental Content in the Human Body[a]

Element	Serum, ppm	Urine	Tissues, ppm[b]
Li	0.01		
Na	3200	1000–5000 mg/day	0.07 g/gN
K	120–214		
Rb			20–200 (dry)[c]
Be			0.00012 (liver)
Mg	18–29 (ave. 22)	60–120 mg/day	300–500
Ca	90–100	96–800 mg/day	60–90
Sr		0.4 mg/day	0.1–0.5 (dry)
Ba			0.02–0.10 (dry), 1.0 in lung
Ti			0.3–0.6 (dry), 10 in lung
V	0.005 ± 0.008		<0.02–0.03 (dry),[d] 0.6 in lung
Cr	0.03		0.01–0.13,[c] 0.36 in spleen
Mo	0.01–0.16[e]	0.01–0.03 mg/day	0.1–0.2 (dry)
Mn	0.01–0.02	0.04–0.07 mg/day	0.2–1.7
Fe	0.80–1.6 (ave. 1.25), men 0.65–1.3 (ave 0.90), women	0.1–0.3 mg/day	
Co	0.00007–0.017 (whole blood)	0.001–0.007 ppm	0.1–0.2 (dry)
Ni	0.025 (range 0.001–0.08)	0.025 mg/day (range 0.007–0.04)	
Cu	1.05–1.10		5–20 (dry), 10–50 in liver
Ag			<0.01–0.2 (dry)
Au			<0.1–1 (dry) (occurrence <20%)
Zn	1.2	0.3–0.6 mg/day	12–100
Cd	0.0033 ± 0.0024	0.002–0.02 ppm	0.2–0.8, 2.0–60 in kidney
Hg		<0.03 ppm	
B			0.5–1 (dry)
Al	0.13–0.17	0.05 mg/day	0.2–0.6, 20–60 in lung
Tl[f]	3–17	>0.35 ppm	
Si			20–40 (dry), 400 in lung
Sn			5–30 (ash)
Pb	0.3–0.4[g]	0.01–0.07 mg/day	1–10 (dry)
As	0.04–0.2	≤0.1 ppm	0.2–0.3
Bi	0.02		<1 (ash)
Se	0.14		0.2–0.6

[a]Reprinted from G. D. Christian, *Anal. Chem.*, **41** (1) (1969) 24A. Copyright 1969 by the American Chemical Society. Reprinted by permission of the copyright owner.

[b]Fresh weight.

[c]Rats.

[d]Organs.

[e]Sheep and cattle.

[f]Concentrations during thallium poisoning.

[g]Whole blood.

relate trace element levels to pathological conditions, they are not routinely analyzed in the clinical laboratory. Most of those listed in the table can be analyzed by atomic absorption spectrophotometry (Chapter 14). A summary of procedures is given in the reference listed in Table 19.3. About half the elements in blood can be determined at physiological levels using flame atomization. An increased number of elements can be determined using electrothermal atomization techniques (Chapter 14).

The analysis of elevated levels of many elements is important for indicating exposure to toxic quantities. Mercury, lead, thallium, and arsenic are common **poisons.** It has been found that zinc protoporphyrin concentrations can be correlated with lead toxicity, and inexpensive lead screening procedures based on measurement of the fluorescence of this substance have been developed. See *J. Lab. Clin. Med.,* **89** (1977) 712; *Science,* **186** (1976) 936.

Some elements are elevated or depressed in certain diseases. Chronic copper toxicity is found, for example, in **Wilson's disease.** This is not due to increased copper intake. Wilson's disease is a genetically determined disease in which copper accumulates in the tissues, probably as the result of a deficiency or absence of a specific copper-concentrating enzyme system in the liver. This leaves the metal more readily available to diffuse into other tissues, where it attacks certain enzymes. The effect is a decrease in plasma copper to about one-half the normal level and increased urinary excretion of copper. Blood copper is elevated in other diseases, such as **Hodgkin's disease** and leukemia. **Zincurea** (elevated serum zinc) accompanies albuminuria and postalcoholic cirrhosis. **Hypozincemia** (lowered zinc levels) accompanies hepatitis, cardiac infarction, pregnancy, oral contraception, and stagnant skeletal growth.

QUESTIONS

1. What are the principal components of blood?
2. What is hemolysis and why is it important?
3. Why is sodium fluoride often added to blood samples collected for glucose analysis?
4. Why should whole blood samples not be frozen?
5. What are two of the most frequently performed clinical analyses?
6. What main roles do trace elements play in the body?
7. Explain the principles of radioimmunoassay. What are some typical applications?

RECOMMENDED REFERENCES

General

1. D. Glick, ed., *Methods of Biochemical Analysis.* New York: Interscience Publishers. Annual volumes since 1954.
2. R. J. Henry, D. C. Cannon, and J. W. Winkelman, eds., *Clinical Chemistry. Principles and Techniques,* 2nd ed. Hagerstown, Md.: Harper & Row, Publishers, 1974.
3. M. Reiner and D. Seligson, eds., *Standard Methods in Clinical Chemistry.* New York: Academic Press. Multivolume series infrequently since 1953.
4. N. W. Tietz, ed., *Fundamentals of Clinical Chemistry,* 2nd ed. Philadelphia: W. B. Saunders Co., 1976.

Immunoassay

5. S. A. Berson and R. S. Yalow, "Immunoassay of Plasma Insulin," *Immunoassay of Hormones, Ciba Found. Colloq. Endocronol.,* **14** (1962) 182.

6. N. Gochman and L. J. Bowie, "Automated Systems for Radioimmunoassay," *Anal. Chem.,* **49** (1977) 1183A.

7. T. A. Kelly and G. D. Christian, "Homogeneous Enzymatic Fluorescence Immunoassay of Serum IgG by Continuous-Flow Injection Analysis," *Talanta,* **29** (1982) 1109.

8. C. M. O'Donnell and S. C. Suffin, "Fluorescence Immunoassays," *Anal. Chem.,* **51** (1979) 33A.

9. D. S. Skelley, L. P. Brown, and P. K. Besch, "Radioimmunoassay," *Clin. Chem.,* **19** (1973) 146.

Trace Elements

10. E. J. Underwood, *Trace Elements in Human and Animal Nutrition,* 2nd ed. New York: Academic Press, 1962.

chapter 20

DRUG ANALYSIS: NARCOTICS AND DANGEROUS DRUGS

Qualitative and quantitative drug screening tests have been developed that are useful for diagnosing emergency drug overdoses, detection of doping in athletes and race horses, toxicology, identification of samples illegally possessed, and other applications. The determinations include two steps: an isolation and cleanup procedure and the analysis procedure. Solvent extraction and ion exchange chromatography are the two most frequently used isolation techniques. After the preliminary isolation, gas chromatography, high-performance liquid chromatography, and thin-layer chromatography are widely used as analytical techniques for screening purposes; the first two provide quantitative as well as qualitative data. Spectrophotometric measurement technqiues are useful particularly for selected drugs. Each of these will be discussed below. For the identification and assay of specific drugs, see Reference 1.

20.1 CLASSIFICATION OF DRUGS

A proliferation of drugs, both legal and illegal, has occurred in recent years. It would be a monumental task to classify, identify, or screen all of them. The most common drugs of abuse, however, can be classed under the headings **barbiturates, alkaloids, amphetamines,** and **hallucinogens.** Table 20.1 lists some of the more common ones. Trade names of the drugs are given in parentheses.

Barbiturates

Barbiturates are one of the major types of drugs in current use and are among the most frequently abused drugs. Known as "downers," they act as **sedatives** and are the most commonly used sleeping pills. The general structure of barbiturates is

TABLE 20.1
Classifications of Some Common Drugs and Narcotics

Barbiturates	Amphetamines	Alkaloids	Hallucinogens
Amobarbital (Dexamyl)	Methamphetamine · HCl	Methadone	Marijuana
Barbital	(Dexosyn)	Cocaine	Lysergic acid
Butabarbital	Amphetamine sulfate	Morphine	diethylamide
Phenobarbital	(Benzedrine)	Quinine	(LSD)
Pentobarbital (Nembutal)	Dextroamphetamine sulfate	Codeine	Mescaline
Secobarbital (Seconal)	(Dexedrine)	Procaine	
Methyprylon (Noludar)		Heroin	
Diphenylhydantoin (Dilantin)		Phenacetin	
Glutethimide (Doriden)		(Empirin)	

shown in Figure 20.1. The parent compound, barbituric acid, is inactive but derivatives usually possess pharmacological properties. Thousands of derivatives have been synthesized, but the most common ones are those with substituents on carbon number 5 in the ring. Other derivatives include methylation of the ring nitrogens and substitution of sulfur for oxygen on carbon number 2.

The acid (keto) forms of barbiturates are water insoluble but soluble in organic solvents, while the salt (enol) forms are water soluble. Once in the body, they revert to the acid form and are adsorbed on plasma proteins.

The related compound glutethimide (Doriden) is also included in most analytical procedures for barbiturates.

Amphetamines

Amphetamines, in contrast to barbiturates, act as **stimulants** and are known as "pep pills," or "uppers." Structurally, they are phenylisopropylamines (Figure 20.1). While numerous derivatives can be found, only **amphetamine** and **methamphetamine** are commonly available. The d-form of amphetamine (dexedrine) is more active than the racemic form (benzedrine).

Alkaloids

Narcotics have the property of being addictive, in addition to their analgesic and euphoric properties. The **opium alkaloids** are the most widely used narcotics. The major active ingredient of opium is **morphine** (Figure 20.1), although other alkaloids are present. Acetylation of the two hydroxy groups (replacement with acetyl groups, $-O-C-CH_3$) on morphine gives the more water-soluble diacetyl deriv-

$$\underset{O}{\overset{\|}{}}$$

ative **heroin.**

Morphinelike substances have been synthesized as possible analgesic substitutes for morphine without addictive properties. None is completely satisfactory, but methadone (Figure 20.1) is commonly used in programs for the rehabilitation of addicts.

Cocaine is an alkaloid derived from cocoa leaves.

FIGURE 20.1
Structures of some common drugs and narcotics.

Hallucinogens

Hallucinogens are like amphetamines in that they act as **stimulants,** but they often act with unpredictable and unusual effects. **Marijuana** is one of the oldest and most widespread hallucinogens and is a natural extract of the hemp plant, *cannabis sativa.* It is a complex mixture of compounds that varies considerably from one sample to another. This greatly complicates its analytical chemistry, and detection is often based on morphological examination rather than chemical means. Recently, however, interest has centered on a group of compounds known as **cannabinoids,** especially the tetrahydrocannabinols (THC) (Figure 20.1). The Δ^9-3,4 -trans-THC (Δ^9-THC) and Δ^8-3,4 -trans-THC (Δ^8-THC) are of particular interest, with the former being more abundant in marijuana. Analytical methods are being based more and more on the determination of these compounds.

One of the most widely publicized abused synthetic hallucinogens is **lysergic acid diethylamide** (LSD). It is impractical to detect once ingested because only microgram quantities are required for physiological action. It degrades rapidly in the presence of light and heat and so care must be taken in sample handling. Only the D-form of LSD is active, and the analytical procedure should ideally be able to measure this isomer. LSD is usually present as a tartrate or other salt rather than the pure compound.

Other miscellaneous drugs include tranquilizers, such as meprobamate and phenothiazines. The alcohols ethanol, methanol, and isopropanol in addition to acetone are also frequently of diagnostic interest.

Generally, sedatives (barbiturates), tranquilizers, and neutral compounds tend to concentrate in the blood, while amphetamines and narcotics (alkaloids) will be in higher concentration in the urine. Hence, blood serum is recommended for analysis of the former and urine for the latter. For many applications, however, it may be preferable or necessary to use urine for complete screening purposes.

20.2   GAS CHROMATOGRAPHY SCREENING

Prior Drug Isolation Using Solvent Extraction

While ion exchange and other isolation procedures can be used for cleanup prior to gas-chromatographic analysis of drugs, **liquid–liquid solvent extraction** is most widely used. In extraction procedures, the different classes of drugs are extracted from an aqueous solution at different pH values into a solvent such as methylene chloride or ether. Ether extracts should not be injected into the gas chromatograph if quantitative analysis is required because poor injection precision results from the high volatility of ether. The extract can be evaporated and taken up in methylene chloride or methanol for GC analysis.

Most drugs fall into one of three extractable groups based on acidity: organic bases, organic acids, and neutrals. The first two may exist in their "free" forms or as salts. The "free" forms are highly soluble in organic solvents, while the salts are soluble in water. Therefore, the acidic drugs will extract into organic solvents from acid solution, while the basic drugs will extract from alkaline solution, since under these conditions they will not be neutralized to salts. Neutral drugs generally will extract at any pH.

Acidic substances include barbiturates and salicylates, while basic substances include amines and alkaloids. Alkaloids extract at about pH 9, and amphetamines and phenothiazines above pH 10. A certain amount of crossover in extraction of the different groups occurs at each pH, and so for quantitative recovery of each group, it may be necessary to extract separate samples at each pH, keeping in mind that each extract may contain some of the other classes of drugs; or else the extracts from a single sample can be combined.

Marijuana can be extracted from dried and ground leaf with CH_2Cl_2. LSD can be extracted into CH_2Cl_2 after solubilizing with a carbonate buffer, and mescaline can be extracted with ethanol.

Drugs in powder or pill form are generally dissolved in aqueous KOH or HCl followed by extraction at the appropriate pH. To determine drugs in biological fluids, preconcentration and/or pretreatment of the extract or sample may be necessary. Free morphine, for example, can be extracted directly from urine at pH 8.6

into 4:1 CH_2Cl_2–isopropanol. However, only about 15% of excreted morphine is free, the rest of it being bound as the glucuronide salt. If morphine is believed to be present in high levels, that is, above the ppm level, the extract of free morphine simply needs to be concentrated by evaporation for analysis. But if trace levels are encountered, the urine must be hydrolyzed by acidifying with H_2SO_4 and heating at about 90°C for 15 minutes to free the bound morphine before the extraction is made. The hydrolysis step also frees other natural materials that are extracted from the urine. Hence, the extract must be cleaned up by back-extracting the morphine into dilute HCl, leaving many of the other materials behind. Then, the HCl solution is adjusted to pH 8.6 and the morphine again extracted with CH_2Cl_2–isopropanol.

Numerous extraction systems have been described for drugs in biological fluids. Alkaloids, antihistamines, barbiturates, and tranquilizers have been successfully extracted into a mixture of ether and acetone. Serum or urine is adjusted to pH 4 to 7.5 to extract many of the drugs. Additional extractions at pH 3 and 9 yields the remaining acidic or basic compounds. This procedure is not used for amphetamines.

For the analysis of barbiturates in micro samples or trace concentrations, they may be methylated with dimethyl sulfate before extraction into hexane. The derivatization time is only 4 to 5 minutes, and 0.01 to 0.05 μg of the drugs can be detected by gas chromatography. Blood or urine volumes of 100 μL can be handled.

Drugs in tissues are usually protein bound, making simple extraction very inefficient. Treatment of homogenized tissue with hot HCl solution will usually release most of the drugs. To determine trace levels of drugs, a cleanup procedure must be used. One successful procedure is to precipitate protein before extraction. This is accomplished by macerating (grinding) the tissue and treating it with sodium tungstate, followed by hydrolysis with hot acid (this forms tungstic acid, which is a protein precipitating agent—see Chapter 3).

If the lipid content of the particular tissue is high, additional cleanup steps may be necessary. For example, to determine alkaloids, they are ether-extracted at pH 10 and the ether extract is washed with 2.5% NaOH, followed by a water wash. The drugs in the ether phase are back-extracted with 0.05 M H_2SO_4 and finally reextracted from alkaline solution into ether.

Gas Chromatography Separations

The most common gas chromatography detector that is used in drug screening analysis is the flame ionization detector. This is due to its ease of operation, its high sensitivity, and its nonselectivity. Some specific applications, not screening procedures, may require a more selective and/or sensitive detector, for example, the electron capture detector, or the rubidium silicate bead detector for nitrogen and phosphorous. These are more difficult to use than the flame ionization detector.

Proper selection of the column is very important in the gas-chromatographic separation of drugs. A widely used column consists of a low-loaded methyl silicone rubber [SE-30 or OV-1 (nonpolar)] coated on silanized, acid-washed Chromosorb W. This is a nonpolar column that works well for most classes of drugs. Sample loss and poor peak shape may be obtained, however, for low concentrations of certain polar drugs.

The medium-polarity liquid-phase OV-17 column (phenylmethyl silicone fluid) has been found to be generally superior to the OV-1 column in both separation efficiency and peak shape, primarily because it more selectively retards the more polar compounds and it does a better job of coating the active sites on the solid support at low loading levels. Other useful columns are PPE-20 (medium polarity) and Carbowax 20M (high polarity).

Positive identification of unknown drugs is greatly enhanced by separating the sample on two or more different columns that give different retention times for the drugs. A successful combination that has been used is the nonpolar OV-1 column and the medium-polarity liquid OV-17 column. Alkaloids, barbiturates, amphetamines, and tranquilizers can be separated on these columns, although separation of amphetamine and methamphetamine is not too good. The last two compounds are easily separated on an auxiliary column of 10% Apiezon L on base-treated "Higher-Performance" Chromosorb W for confirmation, if desired.

In the gas-chromatographic determination of morphine, a glass column and a glass injection port liner must be used because large losses occur on metal columns. This problem can be avoided by converting the morphine to diacetylmorphine (heroin), at the expense of an extra step in the analysis. For drug analysis in general, it is good practice to use a glass column with direct column injection or glass injection port.

An example of the application of gas chromatography to drug screening is shown in Figure 20.2. The urine sample from a participant in a local methadone treatment program was worked up by the Dole technique (a combined ion exchange–solvent extraction isolation described in the next section) and subjected to gas chromatography. Several drugs are identifiable in the sample. Gas chromatography provides a relatively simple and rapid method for the separation and analysis

FIGURE 20.2
Gas chromatogram from a urine sample taken from a participant in a local methadone program. (Courtesy of Hewlett-Packard Company.)

of marijuana (THC compounds). The peaks are positively identified by comparing with those of known standards of the drugs.

To analyze for the three different classes of drugs, three separate chromatograms are frequently run using different required temperatures to separate the drugs. This requires 45 minutes or more. It is possible to analyze a combined extract with a single chromatogram by using temperature programming to reduce the measurement time to 10 to 15 minutes. This has been accomplished for the rapid general screening of pills and powders, for a mixture of amphetamines (basic), barbiturates (acidic), and alkaloids (basic).

A technique frequently used for drug screening is "head space" analysis, in which the sample is contained in a closed container and the volatile constituents are allowed to equilibrate with the atmosphere. An aliquot of the atmosphere is taken (through a rubber system with a syringe) and injected into the gas chromatograph.

Various instrument companies have developed multicolumn gas chromatographs for use in drug screening analysis. A number of companies are marketing combined **gas chromatography–mass spectrometry** instruments for drug analysis. Mass spectrometry identifies substances according to their mass. This combination allows positive identification of separated compounds without having to resort to a different column. In addition, nanogram quantities of drugs can be detected with the mass spectrometry system. By coupling the system to a small dedicated computer, identification of the drug can be automatically made by search and comparison with a stored library of compound retention time/mass spectra data.

20.3 THIN-LAYER CHROMATOGRAPHY SCREENING

Prior Drug Isolation

Both solvent extraction and ion exchange resins have been used for the isolation of drugs from urine and other fluids. Dole and co-workers in 1966 (Reference 3 at the end of the chapter) described a urine screening procedure combining the two for isolation of drugs prior to spotting on a thin-layer plate for thin-layer chromatographic identification. The drugs are removed from urine with cation exchange paper and are subsequently removed from the resin paper by a three-step extraction at different pH values. The extracts are spotted on the thin-layer plate. A modified procedure requires only a single extraction step. The pH of the urine is adjusted to 6 to 7 and a 25-mL sample is place in a screw cap test tube with a piece of SA-2 ion exchange paper (H. Reeve Angel & Co.). The capped tube is shaken on a reciprocating shaker for 1 or 2 hours or stored in a refrigerator overnight. The drugs that have been adsorbed to the ion-exchange-resin-loaded paper are eluted by shaking for 5 to 10 seconds with 5 mL saturated ammonium chloride. Then, 20 mL of 9:1 chloroform–isopropanol is added as extracting solvent and the tube is shaken for 30 to 60 seconds. The upper aqueous phase is drawn off and the organic phase, which contains the drugs, is evaporated on a water bath to concentrate the drugs. The bath temperature should not exceed 70°C, or loss of amphetamines can occur. Before evaporation to dryness is complete, 0.1 mL of 1% conc. HCl in methanol is added. The dried tube is cooled and the residue is dissolved in 50 μL absolute methanol for spotting on the thin-layer plate.

Davidow and co-workers, in 1968, described a simple one-step solvent extraction procedure for qualitative thin-layer chromatography screening purposes (Reference 2 at the end of the chapter). They showed that a single chloroform extraction at pH 9.5 will recover sufficient amounts of basic, neutral, and acidic drugs for detection.

Quam used a small disposable plastic column filled with Amberlite XAD-2 to collect for storage a large variety of drugs, including alkaloids, barbiturates, amphetamines, glutethimide, and phenothiazines. Commercially prepared kits are available based upon this procedure. In one system, the disposable column is wetted wtih 3 mL water and 20 mL urine is added, followed by 1 to 2 mL of pH 9.0 buffer. The buffered urine is allowed to pass slowly through the column (5 to 10 minutes). Finally, the column is washed with 10 mL distilled water and the column, containing the drugs, can be stored until the TLC analysis is performed. The drugs are eluted from the column with 15 mL of 4:6 1,2-dichloroethane:ethyl acetate solvent. After elution, one drop of 0.05 M H_2SO_4 in methanol is added, the solution is evaporated to dryness at 60 to 70°C, and the residue is dissolved in 25 μL methanol for TLC analysis.

The sample tubes can be mailed to a laboratory for processing. Under refrigeration, they can be kept for many weeks without danger of sample decomposition. The isolated drugs presumably contain fewer contaminants than occur in simple solvent extraction procedures. Sequential urine extractions from a single patient can be made using the same column over a period of one to two weeks, and then the combined samples are analyzed by a single TLC separation. Using this procedure, one cannot, of course, determine on what particular day a drug might have been misused or whether the patient was misusing it frequently or infrequently.

Using these columns, recovery of morphine at a pH of 8.5 to 9.5 is about 65 to 75%; of barbiturates, 50 to 60%; and of amphetamine, about 40%. If only certain groups or individual compounds are under investigation, recovery can be optimized by selecting a buffer with a pH equal to the isoelectric point of the compound to be isolated (pH at which the molecule has no net charge). For example, maximum recovery of barbiturates occurs at pH 1 to 3 and of amphetamines, at pH 11.

The above nonselective resin isolations can be used equally well for gas-chromatographic analysis if quantitative data are required. The efficiency of collection must be determined using standards.

Thin-Layer Chromatography Separations

Because of the large number of drugs now in use (or misuse), it is practically impossible to achieve a separation of all of them by thin-layer chromatography using one solvent system, let alone two solvent systems. Davidow and co-workers developed a screening procedure, however, that will resolve the most frequently abused acidic, neutral, and basic drugs.

The dissolved residue from the above resin isolation is spotted on a thin-layer plate using a capillary tube, and the chromatogram is developed (in a solvent-saturated chamber) using a 170:20:10 mixture of ethyl acetate: methanol: conc. ammonia as developing solvent. A standard solution containing a mixture of drugs is run on the same TLC plate for control purposes, because the R_f values of the various drugs may vary depending on the degree of vapor saturation achieved in the developing chamber. After predrying the developed plate, it is placed in an oven at 75°C for 10 minutes to remove the ammonia. The presence of drug is confirmed if its R_f value as well as its specific color are identical with those of one of the control standards; specific colors are obtained by consecutively applying five different spray reagents. This is illustrated in Figure 20.3.

Optional solvent systems and/or spray reagents may be used to confirm compounds not clearly identifiable in the first development.

20.4 SPECTROPHOTOMETRIC METHODS OF MEASUREMENT

Most drugs have moderate to strong absorption bands in the near-ultraviolet region, exhibiting absorption maxima in the 215- to 300-nm region. References 4, 5, 8 and 9 at the end of the chapter contain extensive compilation of the spectra for many drugs and narcotics. While UV methods are not as selective as chromatographic methods, they are simple and rapid, making them useful for screening purposes. The absence of significant absorption quickly eliminates many compounds, and the presence of absorption suggests further investigation or it can be used to confirm various compounds.

To obtain reliable identification and quantitative measurement, it is necessary to separate components using solvent extraction and/or chromatography. A widely used technique to improve specificity is to study absorption wavelength shifts on a change in pH. The band shifts usually occur due to changes in the chromophore groups. Tautomeric compounds exist as either the keto (in acid) or enol (in base) form:

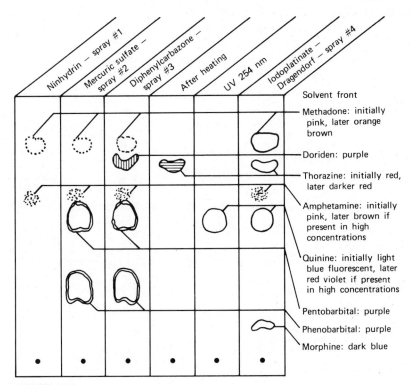

FIGURE 20.3
Spraying sequence and approximate position and color of drugs separated by thin-layer chromatography. (Courtesy of Brinkmann Instruments, Inc.)

$$-\overset{\displaystyle O}{\overset{\displaystyle \|}{C}}-CH_2-\overset{\displaystyle O}{\overset{\displaystyle \|}{C}}-\ \underset{H+}{\overset{OH-}{\rightleftharpoons}}\ -\overset{\displaystyle O^-}{\overset{\displaystyle |}{C}}=CH-\overset{\displaystyle O}{\overset{\displaystyle \|}{C}}-$$

keto form   enol form

The keto form contains two isolated $C{=}O$ chromophores, while the enol form contains a conjugated chromophore, $C{=}CH{-}C{=}O$, which causes a bathochromic shift to longer wavelengths. (See Chapter 13 for a detailed discussion of absorption by various chromophores.)

Infrared spectrophotometry is more specific than ultraviolet or visible spectrophotometry for the identification of drugs. The spectra should ideally be obtained on relatively pure components, but identification of mixtures can still frequently be made using specific absorption bands.

Spectrophotometric methods are generally not as sensitive as chromatographic methods, although micromethods do exist in which gas chromatography fractions or thin-layer chromatography spots can be collected and analyzed spectrophotometrically for positive identification.

QUESTIONS

1. What are the main types of drugs and narcotics?
2. What two steps are generally involved in drug analysis?
3. What are two commonly used drug screening procedures?
4. What types of drugs are better dermined in blood and what types are better determined in urine? Why?
5. How can drugs be classified based on their solvent extraction properties?
6. Suggest a procedure for determining traces of morphine in urine.
7. What is the Dole procedure for screening drugs?
8. What is the Davidow procedure for screening drugs?

RECOMMENDED REFERENCES

General

1. *The United States Pharmacopeia.* Rockville, Md.: United States Pharmacopeial Convention, Inc. Lists properties, identification, and assay procedures for drugs.

Thin-Layer Chromatography

2. B. Davidow, N. L. Petri, and B. Quame, "A Thin-Layer Chromatographic Screening Procedure for Detecting Drug Abuse," *Am. J. Clin. Pathol.*, **50** (1968) 714.
3. V. P. Dole, W. K. Kirn, and I. Eglistis, "Detection of Narcotic Drugs, Tranquilizers, Amphetamines, and Barbiturates in Urine," *J. Am. Med. Assoc.*, **198** (1966) 115.

Spectrophotometry/Fluorometry

4. C. N. Andres and S. W. Snow, "Ultraviolet Absorption Data of Pharmaceuticals," *Microgram*, **1** (1968) 145.
5. C. W. Bradford and J. W. Brackett, "Systematic Procedure for Identification and Dangerous Drugs, Poisons, and Narcotics by Ultraviolet Spectrophotometry," *Mikrochim. Acta*, **3** (1958) 353.

6. A. S. Curry, *Advances in Forensic and Clinical Toxicology*. CRS Press, Cleveland, Ohio, 1972.

7. "Drugs: A Bibliography of Fluorometric Methods of Analysis," in *Fluorometry Reviews*, G. K. Turner Associates, Palo Alto, Calif.

8. A. L. Hayden, O. R. Sammul, G. B. Selzer, and J. Carol, "Infrared, Ultraviolet, and Visible Absorption Spectra of Some USP and NF Reference Standards and Their Derivatives," *J. Assoc. Off. Agr. Chemists,* **45** (4) (1962) 798.

9. W. F. Ulrich, K. E. Steine, D. L. Blecha, and D. R. Harms, *Analytical Instrumentation in the Forensic Sciences*. Beckman Instruments, Inc., Fullerton, California, 1971. Presents infrared and Raman spectra of drugs.

chapter 21

ENVIRONMENTAL ANALYSIS

Environmental problems (air, water, and occupational health and safety) and their control have received a great deal of interest and publicity. We have come to realize that this is a very complex area; and while many advances have been made in recent years, much is yet to be learned concerning pollution. Analytical chemistry plays a very important role in both defining and controlling environmental pollution. In this chapter, we briefly describe some of the analytical techniques used to collect and determine air and water pollutants.

21.1 AIR ANALYSIS

It was once thought that when a particular pollutant was admitted into the atmosphere, its chemical composition did not change and that an analysis of that pollutant would give an indication of the degree of contamination. However, it is now recognized that many chemicals undergo photochemical decomposition and reaction in the atmosphere, forming different pollutants which may be even more toxic than their precursors. The familiar smog, for example, is generally considered to be related to the interaction of nitrogen oxide, hydrocarbons, and sunlight.

Air Sample Collection

One of the most important steps in air analysis is the collection of the sample, and so we will describe a number of techniques used. Once the sample has been collected, any number of standard measurement techniques may be employed.

1. GENERAL CONSIDERATIONS. In pollution analysis, samples are generally collected for one of three important reasons: to establish hazardous levels in the environment, to evaluate the efficiency of pollution control measures, or to determine the source of a pollutant. Hence, the mode of sampling can be important.

(a) Size of sample. The volume of air sample is governed by the minimum pollutant concentration that must be measured, the sensitivity of the measurement, and the information desired. The concentration range of a pollutant may be unknown, and the sample size will have to be determined by trial and error. Trial samples of more than 10 m^3 may be required to determine ambient concentrations.

(b) Rate of sampling. The useful sampling rate will vary with the sampling device and should be determined experimentally. Most sampling devices for gaseous constituents have permissible flow rates of 0.003 to 0.03 m^3/min. The collection efficiency need not be 100% as long as it is reproducible and can be calibrated with known standards. The efficiency should be at least 75%, however. All gaseous samplers have a threshold level below which their efficiency drops to near zero. This varies with the sampling device and must be determined under the conditions used. Some sampling devices are described below.

(c) Duration of sampling. The time of day and duration of sampling will be determined by the information that is desired. Remember that the sampling period will give an indication of only the *average* concentration during that period. It would be more meaningful to sample city air samples for lead content during the rush hour and between rush hours to obtain a realistic indication of the overall lead exposure of an individual in the city. Only a series of relatively short sampling times may reveal concentrations that are known to be deleterious. A sampling device capable of efficient operation at high flow rates will be required for short sampling times. It may be desirable in cases requiring many short interval samplings to employ instead automatic continuous monitoring using highly sensitive detection devices.

(d) Sample storage. Storage of air samples should be kept to a minimum. They should be protected from heat and light. Care should be taken that the desired test component does not react with other constituents or with the container. Gaseous samples are sometimes collected by adsorption onto a solid and the gases must not be lost by desorption prior to analysis.

2. THE SAMPLING TRAIN. The requirements for intermittent air sampling are a vacuum source, a means of measuring the amount of air sampled, and a collector or combination of collectors. An interval timer may be used to control the time and duration of sampling. The **sampling device** (collector) should be the first unit in the sampling train, followed by the **metering device** and then the **vacuum pump.** Some of the more commonly used devices are described below.

(a) Vacuum sources. A vacuum is used to draw the sample through the collection device. Motor- or hand-driven vacuum pumps, aspirators, and automobile vacuums are commonly used. When vacuum devices are being used to draw samples through a filter, in which pressure loss may build up during sampling (e.g., as the filter becomes partially clogged), it is recommended that some constant-flow device be attached. The device described in Reference 3 at the end of the chapter can be used with any vacuum pump. For a high-volume filter, the unit described in Reference 3 is suggested.

(b) Metering devices. Flow measurement devices are of two general types: those that measure rate and those that measure volume. The former are small and inexpensive but have the disadvantage of measuring only instantaneous rate of

FIGURE 21.1
Rotameter. (Courtesy of Fisher Scientific Co.)

FIGURE 21.2
Dry-test gas meter. (Courtesy of Arthur H. Thomas Co.)

flow; so they must be checked frequently during the sampling period. The latter record the total flow passing through them and are therefore more useful. They are, however, bulkier and generally more expensive. Some metering devices are illustrated in Figures 21.1, 21.2, and 21.3.

5648-B10

FIGURE 21.3
Wet-test gas meter. (Courtesy of Arthur H. Thomas Co.)

A **rotameter** is a rate-measuring device that consists of a spherical float within a tube that has a self-contained scale (see Figure 21.1). The bore of the tube is tapered, being larger at the top. The gas enters the bottom of the tube, and the stainless steel float rises and falls in direct proportion to the rate of gas flow. In effect, a variable orifice is supplied. The scale is calibrated for the particular gas (e.g., air) whose flow is to be measured. Graphs are frequently made available with the meters for applying temperature or pressure corrections, and the scale can be calibrated for measuring other gases.

The **dry-test gas meter** and the wet-test gas meter are volume-measuring devices. The first is shown in Figure 21.2. A set of plastic bellows is alternately filled and emptied, thereby driving the dial points via a system of bell cranks; very little pressure is required. A thermometer and manometer are provided with the meter for temperature and pressure corrections. Dry-test gas meters are useful for large-volume measurements. The **wet-test meter** (Figure 21.3) is generally more accurate that the dry-test meter for smaller volumes. The gas drives a rotor, which in turn drives the meter. The meter housing is partially filled with water through which the rotor turns. It is calibrated by the manufacture at a given level of water.

All gas-measuring devices should be calibrated before and after use. A usual method employed is to measure the volume of a liquid (such as water) displaced by the gas flowing through the meter. Figure 21.4 illustrates a simple calibration method. A saturator (to saturate the air with water) is placed before the meter to prevent evaporation of part of the water from the carboy. The displaced water is weighed or its volume is measured. By measuring the pressure with a manometer in series and the temperature, accurate calculation of the gas volume at standard conditions is obtained.

(c) Sampling devices. The third component of the sampling train is the collector, which may be of a variety of type, depending on the particular application. Included are filters, fritted-glass scrubbers, and impingers. These and others are described below for the sampling of aerosol contaminants and of gaseous contaminants.

Aerosol Contaminants. One of the most commonly used means of collecting aerosol contaminants is **filtration**. After collection on the filter, the aerosol content may be determined by weighing, by chemical analysis, or by particle sizing. Fiber

FIGURE 21.4
Calibration of gas meters.

FIGURE 21.5
Cascade impactor.

filters (wood fiber paper, glass fiber), granular filters (fritted glass or metal, porous ceramic, sand), and membrane filters (cellulose ester) are used. The last is best for particle sizing. Most filters cannot be used at high temperatures or under moist conditions, but some glass filters can be used at temperatures up to 800°C.

A second type of collection device for aerosol contaminants is the **impinger;** this collects both solid and liquid aerosols. In dry impingers, also called **impactors,** the aerosols impinge on a surface exposed to the air stream. The cascade impactor is shown in Figure 21.5. This consists of a series of progressively smaller size jets impinging at right angles on conventional microscope slides. Impactors are efficient for the collection of particles down to 2 μm in size. These impingers are well suited for collection of aerosols for microscopic examination.

In **wet impingers,** the aerosols impinge on a surface submerged in a liquid (Figure 21.6). A glass tube with an orifice at the end is directed toward a flat surface, which can be the bottom of the collection vessel or a glass platform connected to the tube. The particles are retained by the liquid in the collection tube. The liquid should not be a solvent for the particles to be collected. These impingers can collect particles as small as 0.1 μm.

More sophisticated collectors include **electrostatic precipitators** and **thermal precipitators** which will collect particles down to 0.001 to 0.01 μm.

Gaseous Contaminants.   Gases and vapors may be collected by absorption in a liquid, adsorption on a solid surface, freezing or condensation, or filling an evacuated container. The distinction between gases and vapors is that the latter exist as liquids at ordinary temperatures and can be easily condensed.

Air Sample Analysis

Described below are representative procedures for some important pollutants as indications of the general approaches taken.

A highly sensitive method for the determination of **nitrogen dioxide** in the atmosphere involves absorption of the nitrogen dioxide in a solution of sulfanilic acid that also contains an azo-dye-forming reagent. A stable pink color is produced within 15 minutes. Concentrations of 0.005 to 5 ppm nitrogen dioxide in the at-

(a)   *(b)*

FIGURE 21.6
Wet impingers. (a) Graduated impinger. (b) Midget impinger. (Courtesy of Arthur H. Thomas Co.)

mosphere may be measured. **Total oxides of nitrogen,** excluding nitrous oxide, may be determined in gaseous effluents from combustion by collecting the gas sample in an evacuated flask containing an oxidizing absorbent of hydrogen peroxide in dilute sulfuric acid. The oxides of nitrogen are converted to nitric acid by the absorbing solution, and the resulting nitrate ion is reacted with phenol disulfonic acid to produce a yellow product that can be measured colorimetrically. From five to several thousand parts per million oxides of nitrogen measured as nitrogen dioxide can be determined.

For the determination of **sulfur dioxide** in the atmosphere, the measured air sample is drawn through a solution of sodium tetrachloromercurate. The sulfur dioxide is absorbed by formation of the dichlorosulfitomercurate(II) complex ion, $HgCl_2SO_3^{2-}$. This complex resists oxidation by oxygen from the air and it reacts with formaldehyde and pararosaniline in acid solution to form a pararosaniline-methylsulfonic acid, which is highly colored and absorbs at 560 nm. This method is sensitive and relatively free from interferences. Metal traces such as iron and manganese interfere and, if they are not removed as particulates by prefiltration, their interference is eliminated by adding a chelating agent such as EDTA to the collection solution. As little as 0.003 ppm sulfur dioxide can be measured in the atmosphere. Automation of the procedure using an AutoAnalyzer (Chapter 17) has been widely adopted.

Total hydrocarbons in the air may be determined using infrared spectrophotometry. The hydrocarbons are collected in a condensation trap immersed in liquid oxygen. The hydrocarbons absorb in the 3- to 4-μm region of the infrared spectrum using a 20-m pathlength cell. They are expressed as parts per million hexane, and the instrument is calibrated using a hexane standard.

Needless to say, we have only touched on a few of the many analytical procedures involving air samples. Other analyses performed include the determination of acetylene, total aldehydes, ammonia, formaldehyde, formic acid, and total organic acids. Various aerosol fractions in the air are commonly analyzed.

21.2 WATER ANALYSIS

Some of the many sources of water contaminants include industries such as the petroleum industry, the iron and steel industry, the pulp and paper industry, the coal industry, the chemistry industry, and the food industry, in addition to the private sources in the home and public and private sewage disposal. Land runoff erosion and mine wastes are serious problems. This enumeration merely serves to point out the complexity involved in monitoring water purity. It is again apparent that water pollution analysis involves a multitude of substances with a correspondingly larger multitude of possible analytical methods.

Sampling of Water

In general, samples may be obtained from faucet outlets, at different points in pipe systems, from the surface of rivers and lake waters, and at different depths, and so forth. The most important consideration is that the frequency and duration of sampling be sufficient to obtain a representative and reproducible sample. In some cases, composite samples may be used, in which individual samples taken at frequent intervals are combined.

Depth samplers are used to collect samples from large bodies of water at a specific depth. These contain some mechanism for removing a stopper after the bottle has been lowered to the desired depth. Commercial samplers are available from laboratory supply houses. Samples to be shipped should have an air space of 10 to 25 mL to allow room for expansion.

Samples should be analyzed as soon after collection as possible for maximum accuracy. For certain constituents and physical values, immediate analysis in the field is necessary to obtain reliable results because the composition of the sample may change before it arrives at the laboratory. Included are pH and temperature and dissolved gases such as oxygen, hydrogen sulfide, and carbon dioxide. In certain cases, the gases may be fixed by reaction with a reagent and the analysis completed at the laboratory. Care must be taken to avoid contamination from oxygen or carbon dioxide in the air.

Analysis of Water Samples

The problem of analyzing water samples is essentially no different from that of analyzing aqueous solutions in general. You are referred to References 9 and 10 listed at the end of the chapter for a compilation of some of the many commonly analyzed substances in water and the procedures employed for their analysis. Measurements made include acidity or alkalinity, biochemical oxygen demand, carbon dioxide, chlorine, dissolved oxygen, electrical conductivity, fluoride, particulate and dissolved matter, ammonia, phosphate, nitrate, silica, sulfate, sulfite, sulfides, turbidity, various metal ions, bacteria, microorganisms, and so forth.

For a comparison of various techniques for the determination of heavy metals in water (atomic absorption and flame emission spectrometry, UV and visible spectrophotometry, anodic stripping voltammetry, X-ray fluorescence analysis, neutron activation analysis), see Reference 6.

QUESTIONS

1. What are the main components in a sampling train for air samples?
2. What are the principal uses of impingers? Scrubbers?
3. What precautions should be taken to protect air samples before analysis?
4. What are some of the commonly analyzed substances or parameters in water samples?
5. What are some parameters of water samples that are best measured in the field?

RECOMMENDED REFERENCES

Air Analysis

1. J. O. Ledbetter, *Air Pollution, Part* 1: *Analysis.* New York: Marcel Dekker, 1972.
2. G. Mamamtov and W. D. Schults, eds., *Determination of Air Quality.* New York: Plenum Press, 1971.
3. A. R. Meetham, D. W. Bottom, and S. Cayton, *Atmospheric Pollution,* 3rd ed. New York: The Macmillan Company, 1964.
4. *Quality Assurance Handbook for Air Pollution Measurement Systems.* U.S.E.P.A., Office of R&D, Environmental Monitoring Laboratory. Research Triangle, N.C. Vol. I, Principles; Vol. II, Ambient Air Specific Methods.

Water Analysis

5. W. S. Holden, ed., *Water Treatment and Examination.* London: The Society for Water Treatment and Examination, 1970.
6. K. H. Mancy, ed., *Instrumental Analysis for Water Pollution Control.* Ann Arbor, Mich.: Ann Arbor Science Publishers, 1971.
7. D. F. Martin, *Marine Chemistry, Vol. I: Analytical Methods,* 2nd ed. New York: Marcel Dekker, 1972.
8. M. S. Quinby-Hunt, "A Survey of Instrumentation Used for Monitoring Metals in Water," Am. Lab., **10 (13)** (1978) 17.
9. *Standard Methods for the Examination of Water and Wastewater,* 14th ed. New York: American Public Health Association, 1976.
10. "Water; Atmospheric Analysis," 1971 *Book of ASTM Standards,* Part 23. Philadelphia: American Society for Testing and Materials, 1971.

experiments

The following experiments follow the order of coverage in the text. Before beginning experiments, you should review the material in Chapter 2 on the basic tools of analytical chemistry and their use and the material in Appendix E on safety in the laboratory. It will also be helpful to review data handling in Chapter 3, particularly significant figures and propagation of errors, so you will know how accurately to make each required measurement.

USE OF APPARATUS

EXPERIMENT 1   Use of the Analytical Balance

Before beginning this experiment, familiarize yourself with the principles of the balance and the general rules for weighing, discussed in Chapter 2.

Principle

The zero point of the balance is determined and an unknown object is weighed, by difference and directly. The instructions are provided for use of a single-pan balance. Your instructor will guide you in the proper use of your particular balance. The experiment can be performed with double-pan balances, and your instructor will provide information on their use if necessary.

Solutions and Chemicals Required

None.

Procedure

1. *Determination of the zero point.* The zero point is to be adjusted to read zero by means of a zeroing knob before adding the object to be weighed. Dust the pan lightly with a small brush. Close the doors of the balance. The balance has a knob which will first partially release the beam and the pan and then completely release them (two positions). Release the beam and pan and adjust the zero reading. Arrest the pan and the beam and then repeat the operation. Throughout the remainder of this experiment, do all zero point and rest point determinations in *duplicate;* thereafter, single determinations should suffice.

The zero point of the balance will not remain constant. It may vary by several tenths of a scale division from day to day, or even from hour to hour. Hence, a new determination of the zero point must be made every time a series of weighings is to be made.

2. *Weighing of objects.* Weights are added (actually removed—see Chapter 2) to the nearest 0.1 g by means of knobs in the front of the balance, and the remainder of the weights are read to the nearest 0.1 mg on a vernier optical scale or a digital readout after the beam comes to a rest.

Determine the weights of three test objects supplied by the instructor. One of these may be the crucible cover to be weighed by difference in part 3 below. If more than an hour has elapsed or if the balance has been used by others since the zero point of the empty balance was adjusted, it should be rechecked. Place the test object on the pan using paper or tongs (never place or remove an object without the arrests engaged). Move the beam arrest knob to the first position. This partially releases the beam, but arrests it if it is so much off balance that it swings beyond a certain position. This prevents damage to the knife edges when heavy objects are on the pan and it is far off balance. Dial in weights from the weight knobs to the nearest 0.1 g until the beam is in motion (near balance). Start with the larger weights to find the nearest 10 g, then the unit weights (nearest 1 g), and finally the 0.1-unit weights (nearest 0.1 g). Then, completely release the beam and read the weight to the nearest 0.1 mg from the vernier or digital counter after it has come to rest. Record the total weight.

Some single-pan balances have a preweigh capability. Consult your instructor for the use of this feature.

3. *Weighing by difference.* Obtain a crucible and cover (or any two objects of about this weight) from your laboratory drawer. Weigh the cover of the crucible as above in part 2 and record this weight. Weigh the two objects combined and record this weight. Remove the crucible cover and obtain the weight of the crucible. The difference in the last two weighings represents the weight of the crucible cover. This should agree within 0.5 mg with the weight obtained by direct weighing.

EXPERIMENT 2   Use of the Pipet and Buret

Principle

Practice in the use of pipets is checked by weighing the amount of water delivered in successive pipettings. Precision of delivery of different volumes is determined. Similar experiments may be done with a buret. The experiments may also be used for calculation (calibration) of the volumes of the glassware if tables are supplied by the instructor and the temperature of the water is measured.

Solutions and Chemicals Required

Cleaning solution, distilled water.

Procedure

1. *Cleaning the glassware.* Check your buret and pipets for cleanliness by rinsing with distilled water and allowing to drain. Clean glassware will retain a continuous, unbroken film of water when emptied. If necessary, clean the buret with a detergent and a buret brush. For pipets, try warm water, then detergent plus warm water. If these do not work, then as a last resort, use chromic acid cleaning solution warmed to $60-70°C$. (**Caution!** Chromic acid cleaning solution is dangerously corrosive; do not spill any on your hands, clothing, or the desk top. Use it **only** for cleaning pipets, unless specifically instructed otherwise.) Fill the pipets with this solution and allow to stand for at least 15 minutes. (**Note:** When filling the pipet, **use a safety trap!**) Return the cleaning solution to the stock bottle. Rinse the buret or pipet several times with tap water, and finally with distilled water. Leave the buret filled with distilled water when not in use. Always check your volumetric glassware for cleanliness before use, and clean whenever necessary. Burets if thoroughly rinsed and filled with distilled water immediately after use should remain clean for weeks; pipets, however, become contaminated easily and must be cleaned frequently.

2. *Pipetting.* Practice filling a 25-mL pipet and adjusting the meniscus to the calibration mark until you are proficient. The forefinger used to adjust the level should be only slightly moistened. If it is too wet, it will be impossible to obtain an even flow.

Weigh a clean, dry 50-mL Erlenmeyer flask with a rubber stopper to the nearest milligram (0.001 g).

Transfer 25 mL water from the pipet to the flask; allow the water to run out with the tip of the pipet touching the side of the flask, being careful that no water splashes or is otherwise deposited on the neck of the flask. It is important that the neck of the flask and the stopper remain dry throughout the experiment. The pipet must be held vertically. Allow to drain for 10 seconds before removing the pipet. Do not blow out the last drop. Replace the stopper and weigh.

Perform the procedure at least two times. Only the outside and the neck of the flask need be carefully dried for subsequent weighings. Calculate the standard deviation and the range of delivery in parts per thousand as described in Chapter 3.

Repeat, using 1-, 5-, 10-, and 50-mL pipets. Compare the precision of delivery for the different pipets.

3. *Use of the buret.* Check the stopcock of your buret for proper lubrication, and be sure the bore and tip are clear. Fill with distilled water and place in the buret clamp; the water should be at room temperature. Displace any air bubbles from the tip of the buret, and adjust to near the zero mark. Allow to stand for a few minutes to check for leakage and read the initial volume to the nearest 0.01 mL. Use a meniscus reader, and be careful to avoid parallax.

Using a procedure similar to above for the pipets, draw off about 5 mL into the weighed flask. Touch the tip of the buret to the wall of the flask to remove the adhering drop, and read the volume to the nearest 0.01 mL. Insert the stopper and weigh to the nearest milligram. Repeat the operation, adding to the flask the next 5 mL water, and weigh again. Continue in this way until the entire 50 mL have been weighed.[1]

Subtract the weight of the empty flask from each of the succeeding weights to get the weight of 5, 10, 15, and so forth, mL water at the temperature observed.

Repeat the entire procedure. At each approximate volume, the weight, compared to the

---

[1] To minimize evaporation errors, your instructor may direct you to refill the buret for each volume delivery $(0-5, 0-10, \ldots 0-50$ mL$)$ and deliver it into an empty (dry) flask.

exact measured volume, should be within 0.03 g of that predicted from the first measurement. Recheck any points that are not.

4. *Calibration of glassware.* If your instructor wishes you to use the data from these experiments to calibrate the volumes of the glassware, you will be provided with appropriate tables to convert the weight in air to the volume. The temperature of the water delivered must be determined.

GRAVIMETRY

EXPERIMENT 3   Gravimetric Determination of Chloride

Principle

The chloride content of a soluble sample is determined by precipitating silver chloride with added silver nitrate, filtering, drying, and weighing the precipitate. The Cl content is calculated from the weight of AgCl.

Equation

$$Ag^+ + Cl^- \rightarrow \underline{AgCl}$$

Solutions Required

1. *Provided.*   Conc. HNO_3, conc. NH_3, dil. (3M) HCl.
2. *To prepare.*
(a) 0.1 M AgNO$_3$. Dissolve about 3 g solid AgNO$_3$ (need not be dried) in about 180 mL distilled water. Store in a brown bottle. Do not get AgNO$_3$ on your hands. If you do, wash immediately with a sodium thiosulfate solution. If left on the skin, brown-black spots (metallic silver) will appear within several hours and will take two to three weeks to wear off.
(b) Wash solution. Add approximately 2 mL chloride-free concentrated nitric acid to about 600 mL distilled water in a wash bottle.

Things to Do Before the Experiment

1. *Obtain your unknown from your instructor.*
(a) Solution. Give your instructor a clean, labeled, 250-mL volumetric flask with stopper. The flask should be cleaned and rinsed with three or four portions of distilled water. It need not be dry. After receiving the unknown solution in the flask, dilute to the mark with distilled water. Mix well (see Chapter 2 for proper procedure).
(b) Solid. Obtain a soluble chloride sample from your instructor and place it in a clean weighing bottle. The weighing bottle and cap are placed in a beaker marked with your name and covered with a watch glass during the drying. The weighing bottle should not be stoppered with the cap. Dry in the oven for one to two hours at 120–160°C (about 1 hour in a forced-air oven, or 2 hours in a gravity convection oven). Store in a desiccator after drying until ready for weighing.
2. *Prepare three filter crucibles as described below.*
3. *Prepare the 0.1 M AgNO$_3$ solution.*[1]

---

[1] Your laboratory will very likely have a collection bottle for leftover silver solutions and precipitates. Both economy and pollution abatement make it important not to dump these materials down the drain.

Preparation of Crucibles

1. *Sintered-glass filter crucible.* Prepare three crucibles. Use crucibles of fine or medium porosity and **not** coarse porosity. Clean each crucible from surface contamination using soap and water, rinse, then place it in a crucible holder in a suction flask. If chemical cleaning is required, follow the procedure at the end of the experiment. With gentle suction, draw several small portions of distilled water through the filter. Place three crucibles thus prepared in a beaker (marked with your initials), cover with a watch glass, and put the beaker and crucibles in the oven at 120°–130°C for one to two hours. With clean crucible tongs, transfer the hot crucibles to the desiccator, cool for 30 to 40 minutes, and weigh accurately. Repeat the heating for $\frac{1}{2}$ hour, cool, and weigh. Continue until weights constant within ±0.3–0.4 mg are obtained.

2. *Porcelain crucibles with glass fiber filter paper.* Place two glass fiber filter papers in the crucible to just cover the bottom of the porcelain crucible (Gooch crucible). Wash with water as above and heat to constant weight (±0.3–0.4 mg is obtained).

Procedure

1. *Preparation of the sample.*

(a) Solution. Pipet three 50-mL aliquots into three separate 500-mL beakers. Dilute each sample with 100–150 mL distilled water. Add about 0.5 mL concentrated nitric acid. Cover the beakers with clean watch glasses.

(b) Solid. After cooling the dried sample in a desiccator for at least 30–40 minutes, weigh accurately triplicate samples of about 0.2–0.3 g into numbered 400-mL beakers. (Samples may contain carbonate, which is hydroscopic. If so, your instructor will advise you to weigh the samples by difference.) Dissolve the sample in distilled water and dilute to about 150 mL. Add about 0.5 mL chloride-free concentrated nitric acid. (If carbonate-containing samples are used, add slowly until the foaming action stops.

2. *Precipitation.* Assume the solid sample to be pure sodium chloride, and calculate the millimoles of silver nitrate required to precipitate the chloride. The slow addition of silver nitrate to the solution is best accomplished by use of a buret, with stirring. Wash the buret well with tap water, rinse with three or four portions of distilled water, and finally fill it with approximately 0.1 M silver nitrate solution. To each sample, add silver nitrate solution slowly with good stirring until an excess of 10% over the calculated amount is added. Heat the suspensions nearly to boiling, with frequent or constant stirring, to coagulate the silver chloride; the stirring helps prevent bumping of the solution during heating and the danger of loss of precipitate. Let the precipitate settle, and test for complete preciptation by carefully adding a few drops of silver nitrate to the clear supernatant liquid; if more precipitate or a cloudiness appears, add a few more milliliters silver nitrate solution, stir well, heat, let the precipitate settle, and test again. Continue in this way until precipitation is complete. Let the covered beakers with their contents stand in the desk, **protected from light,**[2] for at least two hours before filtration; standing overnight, or from one laboratory period to the next, may be required, and this is all right as long as they are kept in the dark.

3. *Filtration and washing of the precipitate.* Decant the solution from the first sample through the first weighed crucible, pouring the solution down a stirring rod, and using gentle suction. The precipitate should be disturbed as little as possible. To the precipitate in the beaker add about 25 mL of the wash solution, stir well, let the precipitate settle, and decant

---

[2]When exposed to light, AgCl decomposes as follows: $AgCl \rightarrow Ag + \frac{1}{2}Cl_2 \uparrow$, resulting in a purple coloration on the surface of the precipitate due to metallic silver, and a loss in weight. Some coloration can not be avoided and is not serious as long as strong light (especially sunlight) is avoided as much as possible.

the solution through the filter crucible. The nitric acid wash solution replaces silver nitrate adsorbed on the surface of the precipitate. An electrolyte is required to prevent peptization of the sample. The nitric acid is volatilized during the drying step. Repeat the washing by decantation four times, and finally bring the precipitate onto the filter; use small portions of the wash solution for transfer. Remove with a rubber policeman any solid particles adhering to the beaker. Continue washing the precipitate in the crucible with the wash solution until the last portions of washings give a negative test for silver ion. Silver ion is tested for by adding a drop of hydrochloric acid. Ten or more washings may be required. Filter and wash the second and the third samples in the same way.

4. *Drying and weighing of the precipitate.* Place the crucibles containing the precipitates in a covered beaker in the oven for two hours at 120–130°C. Cool the crucibles in the desiccator, and weigh accurately. Reheat for 1-hour periods as necessary to obtain weights constant within ±0.3–0.4 mg.

Cleaning the Crucibles

After completing the analysis, clean the crucibles as follows. Remove the cake of silver chloride, then place the crucibles in a beaker, put about 5 mL concentrated ammonia solution in each crucible, and cover the beaker with a watch glass. After 10 to 15 minutes, transfer the crucible to the crucible holder in a suction flask and wash with several portions of distilled water. If the filter plate is dark, empty and rinse the ammonia from the suction flask and treat the crucible in the same way with a few milliliters concentrated nitric acid. Finally, wash well with distilled water.

Calculations

1. *Solution.* Calculate and report the grams of Cl contained in your unknown. Since one-fifth of the sample was taken for each determination (50 mL out of 250 mL), the weight determined in each aliquot must be multiplied by 5. Report each individual result, the mean, and the relative average deviation in ppm.

2. *Solid.* Calculate and report the percent Cl in your unknown for each portion analyzed. Report each individual result, the mean, and the relative standard deviation.

EXPERIMENT 4   Gravimetric Determination of SO₃ in a Soluble Sulfate

Principle

Sulfate is precipitated as barium sulfate with barium chloride. After the filtering with filter paper, the paper is charred off, and the precipitate is ignited to constant weight. The SO_3 content is calculated from the weight of $BaSO_4$.

Equation

$$SO_4^{2-} + Ba^{2+} \rightarrow \underline{BaSO_4}$$

Solutions and Chemicals Required

1. *Provided.* Dil. (0.1 M) AgNO₃, conc. HNO₃, conc. HCl.
2. *To prepare.* 0.25 M BaCl₂. Dissolve about 5.2 g BaCl₂ (need not be dried) in 100 mL distilled water.

Things to Do Before the Experiment

1. *Obtain your unknown and dry.* Check out a sample of a soluble sulfate from the

instructor and dry it in the oven at 110–120°C for at least 2 hours. Allow it to cool in a desiccator for at least $\frac{1}{2}$ hour.

2. *Prepare crucibles.*   Clean three porcelain crucibles and covers. Place them over Tirrill burners and heat at the maximum temperature of the burner for 10–15 minutes; then place in the desiccator to cool for at least 1 hour. Weigh accurately each crucible with its cover. Between heating and weighing, the crucibles and covers should be handled only with a pair of tongs.

3. Prepare the 0.25 M BaCl$_2$ solution.

Procedure

1. *Preparation of the sample.*   Weigh accurately to four significant figures, using the direct method, three samples of about 0.5–0.7 g each. Transfer to 400-mL beakers, dissolve in 200–250 mL distilled water, and add 0.5 mL concentrated hydrochloric acid to each.

2. *Precipitation.*   Assume the sample to be pure sodium sulfate and calculate the millimoles barium chloride required to precipitate the sulfate in the largest sample. Heat the solutions nearly to boiling on a wire gauze over a Tirrill burner and adjust the burner to keep the solution just below the boiling point. Add slowly from a buret, drop by drop, 0.25 M barium chloride solution until 10% more than the above calculated amount is added; stir vigorously throughout the addition. Let the precipitate settle, then test for complete precipitation by adding a few drops of barium chloride without stirring. If additional precipitate forms, add slowly, with stirring, 5 mL more barium chloride; let settle, and test again. Repeat this operation until precipitation is complete. Leave the stirring rods in the beakers, cover with watch glasses, and digest on the steam bath until the supernatant liquid is clear. (The initial precipitate is fine particles. During digestion, the particles grow to filterable size.) This will require 30–60 minutes or longer. Add more distilled water if the volume falls below 200 mL.

3. *Filtration and washing of the precipitate.*   Prepare three 11-cm No. 42 Whatman filter papers or equivalent for filtration; the paper should be well fitted to the funnel so that the stem of the funnel remains filled with liquid, or the filtration will be very slow. See the discussion in Chapter 2 for the proper preparation of filters. Filter the solutions while hot; be careful not to fill the paper too full, as the barium sulfate has a tendency to "creep" above the edge of the paper. Wash the precipitate into the filter with hot distilled water, clean the adhering precipitate from the stirring rod and beaker with the rubber policeman, and again rinse the contents of the beaker into the filter. Examine the beaker very carefully for particles of precipitate that may have escaped transfer. Wash the precipitate and the filter paper with hot distilled water until no turbidity appears when a few milliliters of the washings acidified with a few drops of nitric acid are tested for chloride with silver nitrate solution. During the washing, rinse the precipitate down into the cone of the filter as much as possible. Examine the filtrate for any precipitate that may have run through the filter.

4. *Ignition and weighing of the precipitate.*   Loosen the filter paper in the funnels and allow to drain for a few minutes. Fold each filter into a package enclosing the precipitate, with the triple thickness of paper on top. Place in the weighed porcelain crucibles and gently press down into the bottom. Inspect the funnels for traces of precipitate; if any precipitate is found, wipe it off with a small piece of moist ashless filter paper and add to the proper crucible. Place each crucible on a triangle on a tripod or the ring of a ring stand, in an inclined position with the cover displaced slightly. Heat gently with a small flame (Tirrill burner) until all the moisture has been driven off and the paper begins to smoke and char. Adjust the burner so that the paper continues to char without catching fire. If the paper inflames, cover the crucible to smother the fire, and lower the burner flame. When the paper has completely carbonized and no smoke is given off, gradually raise the temperature

enough to burn off the carbon completely. A red glowing of the carbon as it burns is normal, but there should be no flame. The precipitate should finally be white with no black particles. Allow to cool. Place the crucible in a vertical position in the triangle, and moisten the precipitate with three or four drops of dilute (1:4) sulfuric acid. Heat *very gently* until the acid has fumed off. (This treatment converts any precipitate that has been reduced to barium sulfide by the hot carbon black to barium sulfate.) Then, cover the crucibles and heat to dull redness in the full flame of the Tirrill burner for 15 minutes.

Allow the covered crucibles to cool in a desiccator for at least one hour and then weigh them. Heat again to redness for 10–15 minutes, cool in the desiccator, and weigh again. Repeat until two successive weighings agree within 0.3–0.4 mg.

Calculation

Calculate and report the percent SO_3 in your unknown for each portion analyzed. Report also the mean and the relative standard deviation.

EXPERIMENT 5 Gravimetric Determination of Nickel in a Nichrome Alloy

Principle

Nickel forms a red chelate with dimethylglyoxime (DMG) which is quite suitable for gravimetric analysis. Precipitation of the chelate is complete in an acetic acid–acetate buffer or in an ammoniacal solution. Acetate buffer is generally used when Zn, Fe, or Mn is present in the alloy. The sample given to you is a nichrome alloy which has Ni (approximately 60%), Cr, and Fe as the major constituents. Interference from Cr and Fe is removed by complexation with tartrate or citrate ions. Precipitation is then carried out in an ammoniacal solution. The Ni content is calculated from the weight of the precipitate (see Table 6.2 for the formula).

Equation

$$Ni^{2+} + 2HDMG \rightarrow Ni(DMG)_2 + 2H^+$$

Solutions and Chemicals Provided

HCl (1:1), HNO_3 (1:1), NH_3 solution (conc. and 1:1), 1% DMG in ethanol, 20% citric acid solution, 30% ethanol.

Things to Do Before the Experiment

1. *Obtain the alloy sample from your instructor.*
2. *Prepare three sintered-glass filter crucibles.* Dry to a constant weight of ±0.3–0.4 mg. (See Experiment 3).

Procedure

1. *Dissolving the sample.* Weigh accurately triplicate 0.10–0.12-g samples of the alloy. Place in separate numbered 400-mL beakers and add 25 mL each of HCl (1:1) and HNO_3 (1:1). Heat moderately (under the hood, please) until the brown fumes are driven off—40–50 minutes. Keep the beakers covered partially with watch glasses. (You may notice some undissolved carbon particles in the residue. If this occurs, then, using quantitative technique, filter the contents of each beaker through Whatman filter paper No. 40 into separate 600-mL beakers. Wash the 400-mL beaker several times with small amounts of

water and transfer the washings to the 600-mL beaker through the filter. Wash the filter several times with small amounts of water. Remove the filter paper and wash the funnel and stem with small amounts of water before removing from the receiving beaker.)

2. *Precipitation.* Add 4–6 mL citric acid solution and 100–150 mL water to each beaker. Add conc. NH_3 dropwise with stirring until the solution is slightly basic (smell of ammonia in the solution). This should take 2–3 mL. Test to see that the pH is about 8 by touching the wet stirring rod to pH paper. Be careful that no drops adhere to the rod in order to keep solution loss negligible. If a precipitate is formed, insufficient citric acid has been added. Dissolve the precipitate by adding dropwise HCl (1:1, about 3 mL) until the pH is about 3. Add 1–2 mL more citric acid solution. Check again by adding conc. NH_3. If no precipitate is formed add HCl (1:1) until the solution is slightly acidic (no ammonia odor in solution). Heat the beakers on the hot plate to about 70–80°C. Do not boil. Remove from the hot plate and add about 50 mL 1% DMG solution dropwise with stirring. If a red precipitate forms, add HCl (1:1) dropwise until it dissolves. The DMG should be homogeneous throughout the solution before the precipitation is begun. Add NH_3 (1:1) dropwise with stirring until the solution *strongly* smells of ammonia; this may take 3–4 mL. A red precipitate is formed. It is important that the solution be distinctly alkaline at this point. Since the nose may retain the odor of the ammonia reagent, take care that the odor comes from the beaker. Test to be sure the pH is greater than 8.5 by touching the wet stirring rod to pH paper to test the pH. This is best done *after* the bulk of the precipitate is formed and allowed to settle. Cover the beakers with watch glass and set aside for at least two hours (preferably overnight).

3. *Filtering and washing the precipitate.* Filter the precipitate through previously weighed glass crucibles using gentle suction. After transferring the bulk of the precipitate, wash the precipitate with lukewarm distilled water and transfer quantitatively into the crucible. If traces of the precipitate (which are difficult to transfer with a rubber policeman) stick to the sides of the beaker, dissolve in a few drops of hot HCl (1:1) and reprecipitate by adding a few drops of 1% DMG and NH_3 (1:1). The precipitate now obtained will not stick and can be transferred to the crucibles. Wash the precipitate in the crucible at least three to four times with warm water. Finally, wash the precipitate *once* with 30% alcohol, which dissolves any excess DMG from the precipitate.

4. *Drying and weighing the precipitate.* Place the crucibles in beakers (labeled with your name and covered with watch glasses). Dry in the oven for one to two hours at 110–130°C. Cool the crucibles in a desiccator (for 30–40 minutes) and weigh accurately. Reheat for 1-hour periods necessary to obtain weights constant to within ± 0.3–0.4 mg.

Calculation

Calculate and report the percent nickel in you unknown for each portion analyzed. Report each individual result, the mean, and the relative standard deviation.

ACID–BASE TITRATIONS

EXPERIMENT 6   Determination of Replaceable Hydrogen In Acid by Titration with Sodium Hydroxide

Principle

One-tenth molar sodium hydroxide is prepared and standardized against primary standard potassium acid phthalate (KHP). The unknown is titrated with the standardized sodium hydroxide.

Equations

$$HA^- + OH^- \rightarrow A^{2-} + H_2O$$

Solutions and Chemicals Required

1. *Provided.* KHP, 0.2% phenolphthalein solution in 90% ethanol, 50% NaOH solution.

2. *To prepare.* 0.1 *M* NaOH solution. This solution must be carbonate free. The Na_2CO_3 in NaOH pellets is insoluble in 50% NaOH solution. The precipitated Na_2CO_3 will settle to the bottom of the container after several days to a week, and the clear, syrupy supernatant (approximately 19 *M* in NaOH) can be carefully withdrawn or decanted from the container to obtain carbonate-free NaOH. The solution is prepared in a borosilicate (Kimax or Pyrex) beaker (it gets very warm) and after cooling is stored in a polyethylene bottle. Your instructor will provide the carbonate-free 50% solution to you.

Distilled water free from carbon dioxide will be needed in this experiment. To prepare this, fill a 1000-mL Florence flask nearly to the shoulder with distilled water, insert a boiling rod with the cupped end down (the cup must be empty), heat to boiling on a wire gauze over a Meker burner, and boil for 5 minutes. Cover the flask with an inverted beaker and allow to cool overnight, or cool under the cold water tap. Prepare an additional 800 mL the same way.

Fill a liter bottle, with rubber stopper, half full of CO_2-free water, add 5.0 mL clear 50% NaOH solution, stopper, and swirl to mix.[1] Avoid exposing the solutions to the atmosphere as much as possible. Finally, add more CO_2-free water to nearly fill the bottle, stopper, and shake thoroughly.

The sodium hydroxide thus prepared is approximately 0.1 *M*. It will be standardized against primary standard potassium acid phthalate.

Things to Do Before the Experiment

1. *Obtain KHP and dry.* Dry about 4 g primary standard potassium acid phthalate in a weighing bottle at 110–120°C for 1 to 2 hours. Cool in a desiccator for at least 30–40 minutes before weighing.

2. *Obtain your unknown and dry.* The unknown may be either a liquid or a solid. If it is a solid, obtain it in a weighing bottle before the day of the experiment and dry for two hours at 110–120°C. Store in the desiccator. If the unknown is a liquid, obtain it in a clean 250-mL volumetric flask and dilute to volume with CO_2-free water. Mix well.

Procedure

1. *Standardization of the 0.1 M NaOH solution against potassium acid phthalate.*

Weigh accurately three portions of the dried potassium acid phthalate of about 0.7–0.9 g each, and transfer to clean 250-mL wide-mouth Erlenmeyer flasks. (These quantities are designed for titrations using 50-mL burets. If you are supplied a 25-mL buret, your instructor will direct you to adjust the quantities of KHP and the unknown accordingly.) The direct method of weighing, using a tared weighing dish, may be used with this material. Dissolve

---

[1]If the NaOH were pure, only 4 mL (4 g NaOH) would be required for 1 L of 0.1 *M* NaOH. Five milliliters is required to compensate for the water and Na_2CO_3 content of the pellets.

each sample in about 50 mL CO_2-free distilled water. Rinse your buret with three small portions of the 0.1 M NaOH solution, fill, and adjust to near zero. Record the initial volume reading to the nearest 0.02 mL. Add 2–3 drops phenolphthalein indicator to each KHP sample and titrate with the 0.1 M NaOH to a faint pink end point. The color should persist at least 30 seconds. Split drops at the end of the titration. Estimate the buret reading to the nearest 0.02 mL.

Calculate the molarity of the NaOH to four significant figures from the weight of KHP used (three significant figures if the molarity is slightly less than 0.1 M). Use the average of the results.

Keep the NaOH bottle stoppered with a rubber stopper, and protect the solution from the air as much as possible to minimize absorption of CO_2. Proceed as soon as possible to the determination of the unknown acid; standard sodium hydroxide should be used within one week of standardization. After that time, changes in molarity may have occurred, and restandardization will be necessary.[2]

2. *Determination of replaceable hydrogen in an unknown acid.* If the unknown is a weak acid (for example, KHP, acetic acid, or vinegar) you will be instructed to use phenolphthalein indicator (color change same as above). If it is a strong acid, you may use another indicator, such as chlorophenol red (color change from yellow to violet).

If the unknown is a solid, dry it for 2 hours at 110°C. Cool for 30 minutes. Your instructor will inform you the approximate weight of sample to take so that about 30 – 40 mL NaOH will be used. If it is a KHP unknown, approximately 1 g may be taken. Weigh three samples to the nearest 0.1 mg and transfer them into numbered 250-mL wide-mouth Erlenmeyer flasks. Dissolve in 50 mL CO_2-free water, warming if necessary. Add two drops of indicator and titrate with 0.1 M NaOH until the color change persists 30 seconds.

If the unknown is a liquid, transfer with a pipet three 50-mL aliquots from the 250-mL volumetric flask and titrate as above.

Calculation

1. *Solids.* Calculate and report the percent of replaceable hydrogen or percent KHP in your unknown for each portion analyzed.

2. *Liquid.* Calculate and report the milligrams replaceable hydrogen in your unknown. Since one-fifth of the sample was taken for each determination (50 mL out of 250 mL), the weight determined in each aliquot must be multiplied by 5.

EXPERIMENT 7   Determination of Total Alkalinity of Soda Ash

Principle

One-tenth molar hydrochloric acid is standardized against primary standard sodium carbonate. Phenolphthalein is used to approximate the halfway point of the titration, and then either modified methyl orange indicator or bromcresol green indicator is used to detect the final end point; the last indicator is used with boiling of the solution near the end point to remove CO_2 (see discussion of Figure 8.11). Alternatively, the hydrochloric acid is standardized against standardized sodium hydroxide from Experiment 6 using phenophthalein indi-

---

[2]If you are to use this solution in Experiment 7 to standardize the HCl, your instructor will advise you to save the remaining solution after completing this experiment. In that event, you should prepare and standardize the HCl within one week of standardizing the NaOH.

cator. The soda ash sample is titrated with the hydrochloric acid solution, with the addition of two moles hydrogen per mole Na_2CO_3.

Equations

$$CO_3^{2-} + 2H^+ \rightarrow H_2CO_3$$

$$H_2CO_3 \rightleftharpoons H_2O + CO_2 \uparrow$$

Solutions and Chemicals Required

1. *Provided.* 0.2% Phenolphthalein in 90% ethanol, either modified methyl orange (mixture of methyl orange and xylene cylanol FF available from Eastman Kodak Co.) or 0.1% bromcresol green in 0.001 M NaOH, and either primary standard Na_2CO_3 or standardized 0.1 M NaOH.

2. *To prepare.* 0.1 M HCl. Concentrated hydrochloric acid has a density of 1.18 and contains 37% by weight HCl. Hence, about 8 mL concentrated acid should be diluted to 1 L to make 0.1 M acid. Measure about 1 mL more than this amount in a 10-mL graduated cylinder and pour into water in a 1000-mL glass-stoppered bottle that is filled to the shoulder with distilled water.[1] Shake until the solution is homogeneous.

Things to Do Before the Experiment

1. *Obtain and dry the Na_2CO_3.* If you are to standardize your HCl against Na_2CO_3, obtain about 2 g primary standard Na_2CO_3 in a weighing bottle. Place the weighing bottle and cover (removed) in a 150-mL beaker marked with your name and dry at 160°C for 2 hours or more. Cool at least 30 minutes in a desiccator before weighing.

2. *Obtain and dry your unknown.* Obtain the unknown in a weighing bottle and dry as above for the Na_2CO_3.

Procedure

Standardize the HCl by either method 1 or 2 below.

1. *Standardization of HCl against Na_2CO_3.* Weigh accurately (to 0.1 mg) the weighing bottle containing the dried sodium carbonate. Keep the stopper in place while weighing. Transfer quantitatively (for four-figure accuracy) about 0.2 g to a clean 250-mL Erlenmeyer flask. The difference is the weight of sodium carbonate taken. Anything between 0.16 and 0.24 g will be satisfactory. (This procedure is called "weighing by difference" and is necessary because Na_2CO_3 is hygroscopic.) Weigh out a second and third portion in the same way. Add about 50 mL distilled water and one to two drops phenolphthalein indicator solution to each flask.

The Na_2CO_3 will be titrated first to a phenolphthalein end point (pink to colorless; 1 H^+ added: $CO_3^{2-} + H^+ \rightarrow HCO_3^-$) to approximate where the final end point should be (modified methyl orange or bromcresol green indicator; 1 more H^+ added: $HCO_3^- + H^+ \rightarrow H_2CO_3$). Rinse the buret three times with small portions (about 5 mL each) of the approximately 0.1 M acid prepared above, then fill and adjust to near the zero mark. Record the volume to the nearest 0.02 mL. Titrate the first sample of the sodium carbonate, adding the acid no faster than 0.5 mL per second, swirling the flask constantly until the pink color disappears. At this point, about half the total volume of acid necessary has been added

---

[1] If you are to do Experiment 17, "pH Titration of Unknown Soda Ash," prepare twice this amount and use this standardization for that experiment.

(actually a slight excess). Use this number to estimate where the final end point will occur, keeping in mind that slightly less acid should be required than has been added. Continue the titration by either method (a) or (b) below.

(a) **Modified methyl orange indicator.** Before beginning, check with your instructor which indicator you are to use. Add two to three drops of modified methyl orange indicator and titrate until the indicator color changes from green to gray. The proper color can be more easily discerned by comparing with the color of two drops of the indicator in a solution prepared by adding 0.20 g potassium acid phthalate in 100 mL distilled water. The pH of this solution is 4.0, the same as the end point pH in the presence of CO_2. **Use the same technique for the standardization and the unknown titrations.**

(b) **Bromcresol green indicator (alternative method).** Add two to four drops bromcresol green. The color change for this indicator is from blue through pale green to yellow. Titrate to a blue-green color (just before the end point); interrupt the titration at this point and boil the solution carefully for 2 or 3 minutes to drive off the carbon dioxide. The color should revert to blue. Cool the solution to room temperature and continue the titration to the pale-green color. This marks the final end point.

Titrate the other two samples in the same manner as the first and calculate the molarity of the HCl from the weights of Na_2CO_3 taken, remembering that each carbonate reacted with two protons. Use the mean of the three determinations for calculations involving the unknown.

2. *Standardization of HCl against standardized 0.1 M NaOH (alternate procedure).*[2] Rinse your 25-mL pipet with 0.1 *M* HCl, and add with the pipet 25 mL to a 250-mL Erlenmeyer flask. Add two to three drops phenolphthalein indicator solution. Rinse your buret with three small portions of the 0.1 *M* NaOH solution prepared and standardized in Experiment 6. Fill and adjust to near zero. The NaOH is placed in the buret rather than in the flask in order to protect it more from atmospheric CO_2. Record the volume to 0.02 mL. Titrate the acid until a faint pink color is obtained that lasts for at least 30 seconds. Repeat the titration two more times, using your 50-mL pipet for the second and third titrations if less than 25 mL NaOH was required in the first. Calculate the molarity of the HCl.

3. *Determination of sodium carbonate in soda ash.* Accurately weigh out by difference three samples of about 0.25–0.35 g each, dissolve each in about 60 mL water, and titrate with 0.1 *M* HCl, following the same procedure with respect to indicators and end point used in the standardization procedure against Na_2CO_3 or, if not used, the one specified by your instructor.

Calculation

Calculate the percentage of Na_2CO_3 or Na_2O in your unknown for each portion analyzed. Your instructor will specify which one is to be reported.

EXPERIMENT 8 Determination of Bicarbonate in Blood Using Back-Titration

Principle

About 95% of the total carbon dioxide in human blood exists as HCO_3^-, the remainder existing as dissolved CO_2. The HCO_3^- concentration, for most clinical work, can be used as

---

[2]The sodium hydroxide solution is a **secondary standard,** and any errors in its standardization will be represented in the standardization of the hydrochloric acid. The sodium hydroxide should be used within one week of standardization. If this experiment is done before Experiment 6, this procedure can be used to standardize the sodium hydroxide solution for that experiment.

a diagnostic aid. It is determined by adding an excess of 0.01 M HCl, to volatilize the HCO_3^- as CO_2, swirling to allow the CO_2 to escape, and then back-titrating the excess HCl with 0.01 M NaOH. The 0.01 M HCl and NaOH solutions are prepared by diluting standardized 0.1 M solutions.

Equations

$$HCO_3^- + H^+ \rightarrow H_2O + CO_2 \uparrow$$

$$\text{excess } H^+ + OH^- \rightarrow H_2O$$

Solutions and Chemicals Required

1. *Provided.* 0.1% Phenol red (phenolsulfonphthalein) solution in 0.003 M NaOH, 1% saline (NaCl) solution in CO_2-free water, Antifoam A (Dow Corning Corp.)

2. *To prepare.* Standard 0.1 M and 0.01 M HCl and 0.1 M and 0.01 M NaOH solutions. The molarity of the standard HCl and standard NaOH needs to be known only to three significant figures. If your have prepared standard HCl or NaOH in Experiment 6 or 7, use these for the present experiment. If you have not, prepare standard 0.1 M NaOH as in Experiment 6. Since only three significant figures are required, you may use one-tenth the amount of KHP for titration, in which case the end point will occur at about $4.0 - 4.5$ mL. A 10-mL buret should be used in these titrations. Standardize a 0.1 M HCl solution by titrating 5.00 mL of it with the standard 0.1 M NaOH as in Experiment 7. The phenol red indicator may be used. If you have only a standard 0.1 M HCl solution, use this to standardize 0.1 M NaOH solution.

Prepare 500 mL of 0.01 M HCl and 0.01 M NaOH solutions by diluting 50 mL of the 0.1 M solutions to 500 mL with the saline solution. These should be prepared fresh on the day of use. The saline aids in the volatilization of the CO_2 from the acidified solution by decreasing its solubility.

Things to Do Before the Experiment

Prepare and standardize the 0.1 M HCl and 0.1 M NaOH. This will require drying primary standard KHP ahead of time if either standard HCl or NaOH is not available.

Procedure[1]

1. *Preparation of the sample.* Either serum or plasma (oxalated or heparinized) may be used for the determination. This may be freshly drawn blood from an animal or a volunteer student. (**Do not** do this yourself; your instructor will supply the sample.) See Chapter 1 for a discussion of the differences between serum, plasma, and whole blood. A 10- to 15-mL sample (20 – 30 mL whole blood) should be adequate for triplicate determinations by a class of 30 students. Fluoride should be added to prevent glycolysis, or breakdown of glucose, which can change the pH. The fluoride inhibits the enzyme catalysis causing glycolysis and stabilizes the pH for about 2 hours. The tube used for collecting the sample can be rinsed with a solution of 100 mg sodium heparin plus 4 g sodium fluoride per 100 mL. The sample should be kept anaerobically, that is, stoppered to keep out atmospheric CO_2. Since the analysis should be done on the day the blood is drawn, the solutions should be prepared ahead of time.

---

[1]This determination may be performed on a macro scale using 5.00 mL acid and 1.00 mL sample, or using 2.00 mL acid and 0.500 mL sample. In the former case, the back-titration will require about 2.4 mL of 0.01 M NaOH, and in the latter case, about 0.7 mL.

2. *Preparation of comparison solution.* Prepare a standard for color comparison at the end point as follows. Place 6 mL of 1% saline solution in a 25-mL Erlenmeyer flask and add 0.10 mL serum or plasma. Add two drops phenol red indicator, insert a stopper, and rotate gently to mix the contents. The transition range of this indicator is pH 8.4 – 6.7 (yellow to red). Because of the buffering capacity of the blood, the end point occurs in this range.

3. *Titration of the sample.* The pooled serum or plasma sample will have been prepared by touching the end of a stirring rod to some Antifoam A and rotating it in the pooled sample. This will prevent excess foaming when the sample is swirled. Place 0.100 mL of serum or plasma in a 25-mL Erlenmeyer flask and add 1.00 mL of 0.01 M HCl and 4 mL of 1% saline. Swirl the flask vigorously for at least 1 minute to allow the CO_2 to escape. Add two drops of indicator and then titrate with 0.01 M NaOH dropwise, but rapidly, until a pink color matching the standard persists for at least 15 seconds. The NaOH may be added carefully with a graduated 1-mL measuring pipet and read to the nearest 0.01 mL.

The normal value of blood bicarbonate is about 26 meq/L (25–32 meq/L), or 0.026 meq/mL. Meq HCO_3^- = mmol HCO_3^-. Since 0.1 mL blood was taken for analysis, it should consume about 0.0026 mmol HCl, or 0.26 mL of 0.01 M HCl. Hence, since 1 mL of 0.01 M HCl was taken, about 0.74 mL should remain unreacted, and the back-titration should take about 0.7 mL of 0.01 M NaOH.

Calculation

Calculate the bicarbonate content of the sample in meq/L. (See Chapter 1 for definition of milliequivalents of serum electrolytes.)

PRECIPITATION TITRATIONS

EXPERIMENT 9 Determination of Silver in an Alloy: Volhard's Method

Principle

Primary standard $AgNO_3$ is used to standardize a 0.1 M KSCN solution, using a ferric alum indicator. The unknown silver alloy is analyzed by titrating with the standardized KSCN solution.

Equations

$$SCN^- + Ag^+ \rightarrow \underline{AgSCN}$$

$$Fe^{3+} + SCN^- \rightarrow Fe(SCN)^{2+} \quad \text{(indicator reaction)}$$

$$\text{red}$$

Solutions and Chemicals Required

1. *Provided.* Primary standard $AgNO_3$, ferric alum indicator ($KFe(SO_4)_2 \cdot 12\ H_2O$, saturated solution), 6 M HNO_3 (free of oxides of nitrogen). Nitric acid free from lower oxides of nitrogen should be colorless and can be prepared, if necessary, by boiling 1:1 HNO_3 until NO_2 is expelled.

2. *To prepare.* 0.1 M KSCN. Weigh approximately 10.0 g KSCN, dissolve in water in a 1-L bottle and fill to the shoulder with distilled water. Shake well to ensure a homoge-

neous solution. This solution is approximately 0.1 M and will be standardized against primary standard $AgNO_3$.

Things to Do Before the Experiment

Prepare and standardize (below) the 0.1 M KSCN solution. This will require drying and cooling $AgNO_3$.

Procedure

1. *Standardization of KSCN solution.* Obtain about 3 g primary standard $AgNO_3$ in a weighing bottle from your instructor. Dry in the oven at 110°C for 1−2 hours, **but no longer.** Cool in a desiccator for 30−40 minutes. This will be weighed by the direct method.

Place a clean, dry weighing dish on the balance and determine its weight to the nearest 0.1 mg. Add 0.70−0.75 g $AgNO_3$ to the weighing dish and determine the increase in weight to the nearest 0.1 mg. Transfer quantitatively to a clean 250-mL wide-mouth Erlenmeyer flask. Keep away from strong light as much as possible until ready to titrate. Weigh two other 0.70- to 0.75-g portions of $AgNO_3$ and transfer to separate (numbered) 250-mL Erlenmeyer flasks.

Add 50 mL distilled water to the flask ready for titration, 50 mL of 6 M nitric acid free from lower oxides of nitrogen, and 2 mL ferric alum indicator. (Lower oxides of nitrogen form nitroso complexes with Fe^{3+} that are red in color and will interfere with the end point.) Fill your 50-mL buret with the KSCN solution, record the initial volume to the nearest 0.01 mL, and titrate with constant vigorous agitation until a faint reddish-brown color appears in the solution; this is more easily seen if the precipitate is allowed to settle after each addition near the end point. It will be helpful to compare the color with a solution made by adding 5 mL of 6 M nitric acid and 2 mL ferric alum to 75 mL water. The color must be permanent after strong shaking. Titrate the other two $AgNO_3$ samples in the same manner and calculate the molarity of the KSCN solution from each titration. Use the mean of the three determinations.

2. *Determination of silver in an alloy.*[1] Obtain a sample of a silver alloy from the instructor.[2] Place in a 250-mL beaker, add 20 mL dilute (1:1) HNO_3, cover with a watch glass, and warm on the steam bath in the hood until the alloy has completely dissolved. When solution is complete, remove the watch glass, rinse it off with a jet of water from your wash bottle, catching the rinsings in the beaker, and continue heating with the beaker uncovered until all brown fumes have disappeared and the solution is colorless. Cool to room temperature and transfer the solution quantitatively to a clean 250-mL volumetric flask with the aid of a funnel and stirring rod to pour the solution down. Rinse the beaker several times with distilled water, adding the rinsings to the flask. Dilute to the mark and shake thoroughly to insure a homogeneous solution.

Transfer with a pipet two samples, 50 mL each, into 250-mL Erlenmeyer flasks, add 2 mL ferric alum indicator, and titrate with standard 0.1 M KSCN solution to the appearance of a faint reddish-brown color, which is permanent even after strong agitation. The two titrations should agree within 0.05 mL. If they do not, pipet and titrate two more. When finished, put all silver-containing solutions in a jar for this purpose.

---

[1]The unknown may instead be impure silver nitrate, in which case dissolve it in 50 mL water and proceed as in the standardization of the KSCN above.

[2]You may be instructed to weigh out 0.3−0.4 g of the alloy and then report the percentage of silver in the sample.

Calculations

Calculate and report the number of grams silver in your alloy sample.

EXPERIMENT 10   Determination of Chloride in a Soluble Chloride: Fajans' Method

Principle

The sample is titrated with a standard $AgNO_3$ solution, using a dichlorofluorescein adsorption indicator end point.

Equations

$$Cl^- + Ag^+ \rightarrow AgCl$$

Solutions and Chemicals Required

1. *Provided.*   0.1% Dichlorofluorescein indicator solution, dextrin.
2. *To prepare.*   Standard 0.1 *M* $AgNO_3$. Obtain from the instructor about 4.5 g primary-standard-grade silver nitrate in a clean, dry weighing bottle; dry in the oven at 110°C for 1 – 2 hours, **but no longer.** This material will be used to prepare a standard solution by the direct method. Cool for 30 – 40 minutes in a desiccator.

Using a weighing dish, weigh accurately (nearest 0.1 mg) about 4.3 g $AgNO_3$. Transfer to a 250-mL beaker and dissolve in about 100 mL distilled water. Carefully pour this solution into a 250-mL volumetric flask and rinse the beaker several times with distilled water, adding the rinsings to the flask. Dilute to the mark, shake well, and pour into a clean, dry 500-mL glass-stoppered amber bottle. Shake again to ensure a homogeneous solution. Keep away from strong light as much as possible. Store away from light. Calculate the molarity of the solution.

Things to Do Before the Experiment

1. Prepare the 0.1 *M* $AgNO_3$ solution.
2. Obtain a sample of unknown chloride from your instructor and dry it in the oven at 120°C for 1 hour or longer. It is okay to dry it overnight. Cool in a desiccator at least 30 – 40 minutes before weighing.

Procedure

Weigh out in the weighing dish three samples of about 0.25–0.30 g each (vary the sample weights, but weigh each exactly), transfer to 250-mL Erlenmeyer flasks, and dissolve in about 50 mL distilled water. Add 10 mL 1% dextrin suspension (shake well before using) and 10 drops dichlorofluorescein indicator solution. The dextrin prevents excessive coagulation of the precipitate at the end point. This keeps a larger surface area for adsorption of the indicator, which enhances the sharpness of the end point. Instead of a suspension of dextrin, 0.1 g of the solid may be added.

The pH of the solution should be between 4 and 10. If it is too acid (e.g., due to hydrolysis of other chloride salts in the unknown), it may be neutralized by adding solid $CaCO_3$ until excess remains in suspension. The suspension does not interfere with the end point. Results are low at pH values above 4 because the indicator tends to displace chloride from the precipitate. (More accurate results are obtained by standardizing the $AgNO_3$ solution against dried reagent-grade NaCl, using the same conditions as for the unknown titration.)

Titrate with 0.1 M AgNO$_3$. Thorough mixing by vigorous swirling of the flask throughout the titration is essential to achieve a good end point. Do not let direct sunlight strike the flask; if you are working next to a window on a very bright day, it would be well to draw the shade. The end point is marked by a change from a pale yellow to a distinct pink color. If the titration is overrun, add a few grains of NaCl or KCl and practice the determination of the end point a few times before titrating the second and third samples. When finished, put all silver-containing solutions in a jar provided for this purpose.

Calculations

Calculate and report the percent Cl in your sample for each portion titrated.

COMPLEXOMETRIC TITRATIONS

EXPERIMENT 11   Mercurimetric Determination of Blood or Urine Chloride

Principle

After precipitation of the protein with sodium tungstate, the chloride ions in a serum or urine sample are titrated with a standardized mercuric nitrate solution. The appearance of a violet-blue color produced by a mercuric ion–diphenylcarbazone complex is used as the end point. Mercuric ions combine with chloride ions to form a soluble HgCl$_2$ complex[1]; mercuric ions also combine with diphenylcarbazone to form a violet-colored complex. The dissociation constant for HgCl$_2$ is smaller than the dissociation constant for the Hg–diphenylcarbazone complex; hence, essentially all the chloride ions must be complexed before the mercuric ions complex with diphenylcarbazone. The mercuric nitrate solution is standardized by titrating a standard solution of NaCl.

---

[1]The reaction is nowhere near as simple as the equation

$$Hg^{2+} + 2Cl^- = HgCl_2$$

would make it seem. Indeed, there are at least four mercury(II)-chloride equilibria occurring simultaneously:

(a) $Hg^{2+} + Cl^- = HgCl^+$ (I) $K_1 = 5.0\text{–}5.5 \times 10^6$
(b) $HgCl^+ + Cl^- = HgCl_2$ (II) $K_2 = 3.0\text{–}3.2 \times 10^6$
(c) $HgCl_2 + Cl^- = HgCl_3^-$ (III) $K_3 = 7\text{–}8$
(d) $HgCl_3^- + Cl^- = HgCl_4^{2-}$ (IV) $K_4 = 10$

Near the equivalence point, compounds (III) and (IV) have very low concentrations and can be disregarded in the stoichiometry of the titration. However, even though (II) is the major mercury(II)–chloride complex present at the end point, the concentration of (I) is significant, and suitable measures must be taken to account for this fact. There is also the added complication that the mercury(II)–diphenylcarbazone equilibrium is pH dependent. In order to minimize errors from these factors, it is necessary to either do a blank titration, or to use standards that give end points that are nearly like those with unknowns as possible (final pH and concentrations of Hg^{2+}, $HgCl^+$, $HgCl_2$, Cl^-, and diphenylcarbazone should be very similar). For this reason, conditions at the start of a titration in this determination are maintained as nearly similar as possible (volume of chloride standards and unknowns are 2.0 mL, etc.).

Equations

$$Hg^{2+} + 2Cl^- = HgCl_2 \quad \text{(see footnote 1)}$$

$$Hg^{2+} + C_6H_5—N{=}N—C{=}O \rightarrow \text{complex (violet colored)}$$
$$\underset{\displaystyle NH—NH—C_6H_5}{|}$$
diphenylcarbazone

Tungstic acid apparently precipitates protein by breaking the hydrogen bonds involved in the tertiary structure of a protein molecule, that is, it denatures the protein.

Solutions and Chemicals Required[2]

1. *Provided.* 0.009 M Mercuric nitrate solution in 0.04 M HNO_3, 100 mg/dL s-diphenylcarbazone solution in 95% ethanol (should be refrigerated and protected from light), reagent-grade NaCl, 10% (wt/vol) Na_2WO_4 solution, 0.33 M H_2SO_4.

2. *To prepare.* Standard 0.01 M NaCl solution. Weigh 0.125–0.150 g dried reagent-grade NaCl and dissolve in distilled water. The weighing need be made to only the nearest milligram (0.001 g). Dilute to 250 mL in a volumetric flask. Calculate the molarity of the solution.

Special Glassware Required

5-mL Serological pipet (graduated), 2-mL volumetric pipet, 0.5-mL volumetric pipet, two 25-mL Erlenmeyer flasks, 10-mL buret.

Things to Do Before the Experiment

1. *Dry about 0.5 g of reagent-grade NaCl at 120°C for at least 1–2 hours.* May be left in oven overnight. Cool in a desiccator at least 30–40 minutes before weighing.

2. *Clean and allow to dry three glass funnels and six 13 × 100 mm test tubes (may be disposable).*

Procedure

1. *Standardization of the 0.009 M Hg(NO₃)₂ solution.* Add with a pipet 2.00 mL of the chloride standard to a 25-mL Erlenmeyer flask, add four drops of the diphenylcarbazone indicator, and titrate with the mercuric nitrate solution using the 10-mL buret. The clear solution turns a faint purple on addition of the first drop of excess titrant. The tip of the buret should be very fine, so that very small drops can be added near the end point.

The mercuric nitrate solution is about 0.009 M. The reaction is $2Cl^- + Hg^{2+} \rightarrow HgCl_2$ (but see footnote 1.) So the millimoles of Hg^{2+} are equal to one-half the millimoles of Cl^-, and the end point should occur near 1.1 mL for 0.01 M NaCl. Read the volume to the nearest 0.01 mL.

The end point is somewhat difficult to observe, and it is advisable to maximize its visibility by:

(a) placing the titration vessel over a clean, white surface (e.g., a paper towel).

(b) having two standards nearby for comparison: one should be as the solution appears before the end point (e.g., a small beaker containing distilled water and added indicator),

---

[2]A Schales and Schales mercurimetric chloride kit is available from Sigma Chemical Company. It includes all the reagents listed.

and the other should be as the solution appears at or immediately after the endpoint (the first titration may be used in this way for all subsequent titrations).

Titrate at least two more aliquots of the NaCl solution. Calculate the molarity of the $Hg(NO_3)_2$ solution. Use the average of the determinations.

2. *Determination of chloride in serum.*[3]   Obtain a serum sample from your instructor in a 40 × 25 mm weighing bottle. Keep capped to avoid evaporation. It will probably have been refrigerated, so it should be allowed to warm up to room temperature before pipetting. All solution pipettings must be done to the indicated accuracy since an aliquot of the filtered solution will be taken for titration.

With a pipet, add three samples of 0.500 mL serum to separate 25-mL beakers or Erlenmeyer flasks containing 3.50 mL water. (The mixture may instead be centrifuged and an aliquot of the supernatant taken. In this case, the precipitation should be carried out in a dry 15-mL centrifuge tube.) The water may be added from a buret or a 5-mL measuring pipet. Add 0.50 mL 10% sodium tungstate solution, followed by 0.50 mL of the 0.33 M H_2SO_4 solution. Mix well and allow to stand for at least 10 minutes. Filter through dry filter paper and a dry funnel into a clean and dry 25-mL Erlenmeyer flask. The clear filtrate is the protein-free filtrate, and it is relatively neutral in pH. (Tungstic acid is precipitated when Na_2WO_4 and H_2WO_4 are added together: $Na_2WO_4 + H_2SO_4 \rightarrow \underline{H_2WO_4} + Na_2SO_4$. Since the Na_2WO_4 and H_2SO_4 are about equal in molar concentration, little excess acid should be present.)

With a pipet, transfer 2.00-mL aliquots of each of the protein-free filtrates into 25-mL Erlenmeyer flasks (the flasks need not be dry, but in order to reproduce the end point color exactly, it helps to have the flasks dry). Add four drops of indicator, and titrate as above for the standardization. This corresponds to 0.200 mL serum titrated. The normal range of chloride in serum is 100–110 meq/L, and so the end point should be in the neighborhood of 1.1–1.2 mL.

Gindler, *Clin. Chem.,* **14** (1968) 1172, states that "as always in this system, a violet color appeared for a short interval early in the serum titration. This was due to the pH initially exceeding 4. With the addition of more mercuric nitrate working solution, the pH decreased, and the color vanished, not to reappear until the equivalence point." This should explain serum samples that, at least at first, appear to have no chloride ions present.

3. *Determination of chloride in urine.*   The procedure is identical to that for serum, but 2.00 mL diluted urine (1:10) is titrated directly, or 2.00 mL of the Folin–Wu filtrate as described for serum is taken for analysis. If the chloride content is very low, repeat the analysis using a larger volume of diluted urine. If the urine is too alkaline, it will impart a pink color on addition of mercuric nitrate and the indicator, which will mask the end point. This can be circumvented by adding a few drops of 1 M HNO_3 after the pink color appears. Add only enough acid to remove the color.

The chloride content of urine varies with the diet and a normal adult excretes about 170–250 meq chloride per 24 hours in a volume of 1–2 L. Hence, the titration should require approximately 0.9–2.8 mL of 0.009 M $Hg(NO_3)_2$.

Calculations

1. *Serum.*   Calculate and report the chloride content for each portion of the serum sample analyzed, in meq/L. A 0.500-mL serum sample was diluted to 5.00 mL during precipi-

---

[3]Serum may be titrated without removal of the proteins, but the end point is then only a pale violet. This can still be recognized without difficulty, however. If the proteins are not removed, add a 0.200-mL serum sample to 1.8 mL water in a 25-mL Erlenmeyer flask and continue as above. The results obtained with a protein-free filtrate are about 1–2 meq/L lower than with untreated serum because of a small loss of chloride by adsorption on the protein precipitate.

tation of proteins and a 2.00-mL aliquot was taken for titration. Therefore, 0.200 mL serum was titrated; meq Cl^- = mmol Cl^-.

2. *Urine*.   Calculate and report the chloride content of each portion of the urine sample analyzed, in meq/L. As with the serum samples, 0.200 mL of the urine is titrated.

EXPERIMENT 12   Determination of Water Hardness with EDTA

Principle

Water hardness, due to Ca^{2+} and Mg^{2+}, is expressed as mg/L $CaCO_3$ (ppm). The total of Ca^{2+} and Mg^{2+} is titrated with standard EDTA using an Eriochrome Black T indicator.[1] A standard EDTA solution is prepared from dried (do not exceed 80°C) $Na_2H_2Y \cdot 2H_2O$ (purity 100.0 ± 0.5%). If the sample does not contain magnesium, Mg–EDTA is added to the titration flask to provide a sharp end point with the Eriochrome Black T, since calcium does not form a sufficiently strong chelate with the indicator to give a sharp end point. See the discussion in Chapter 9 for a more complete description.

Equations

Titration:

$$Ca^2 + H_2Y^{2-} \rightarrow CaY^{2-} + 2H^+$$
$$Ca^{2+} + MgY^{2-} \rightarrow CaY^{2-} + Mg^{2+}$$

End point:

$$Mg^{2+} + HIn^{2-} \rightarrow MgIn^- + H^+$$
$$MgIn^- + H_2Y^{2-} \rightarrow MgY^{2-} + HIn^{2-} + H^+$$

(red)   (colorless)   (colorless)   (blue)

The free acid parent of the indicator is H_3In, and that at the titrant EDTA is H_4Y.

Solutions and Chemicals Required

1. *Provided*.   0.5% (wt/vol) Eriochrome Black T indicator solution in ethanol, 0.005 M Mg–EDTA (prepared by adding stoichiometric amounts of 0.01 M EDTA and 0.01 M $MgCl_2$). The indicator solution should be prepared fresh every few days, as it is unstable. A portion of the Mg–EDTA solution, when treated with pH 10 buffer and Eriochrome Black T, should turn a dull violet color; one drop 0.01 M EDTA should change this to blue and one drop 0.01 M $MgCl_2$ should change it to red.

2. *To prepare*.

(a) NH_3–NH_4Cl buffer solution, pH 10. Dissolve 3.2 g NH_4Cl in water, add 29 mL conc. NH_3, and dilute to about 50 mL. The buffer solution is best stored for long periods of time in a polyethylene bottle to prevent leaching of metal ions from glass.

(b) Standard 0.01 M EDTA solution. Dry about 3 g reagent-grade $Na_2H_2Y \cdot 2H_2O$ in a weighing bottle at 80°C for 2 hours. Cool in a desiccator for 30 minutes and accurately weigh (to the nearest milligram) approximately 2.0 g. Transfer to a 500-mL volumetric flask. (This is the disodium salt of EDTA; the free acid is insoluble). Add about 400 mL distilled, **deionized** water and shake or swirl periodically until the EDTA has dissolved. EDTA dis-

---

[1]If it is desired to titrate only Ca^{2+}, this can be done at pH 12 (use NaOH), where $Mg(OH)_2$ is precipitated and does not titrate. Hydroxynaphthol blue indicator is used.

solves slowly and may take $\frac{1}{2}$ hour or longer. If possible, the solution should be allowed to stand overnight before using. If any undissolved particles remain, addition of three pellets of NaOH may aid dissolution, but there is danger of adding metallic impurities. After the EDTA is dissolved, dilute to 500.0 mL and shake thoroughly to prepare a homogeneous solution. Then, rinse a clean polyethylene bottle with three small portions of the EDTA solution and transfer the remainder of the solution to the bottle for storage. (Polyethylene is preferable to glass for storage because EDTA solutions gradually leach metal ions from glass containers, resulting in a change in the concentration of free EDTA.) Calculate the molarity of the EDTA solution.

Things to Do Before the Experiment

Dry the $Na_2H_2Y \cdot 2H_2O$ and prepare the standard EDTA solution.

Procedure

Obtain a water sample from your instructor. Add with a pipet or a buret a 50-mL aliquot of the sample to a 250-mL wide-mouth Erlenmeyer flask, add 2 mL of the buffer solution, 0.5 mL of the Mg–EDTA solution, and five drops of the indicator solution. (If the unknown contains magnesium, addition of Mg–EDTA is not necessary—consult your instructor.) Avoid adding too much indicator with dilute solutions or the end point change may be too gradual. The indicator will not become wine red until magnesium is added. **The buffer should be added before the indicator,** so that small amounts of iron present will not react with the indicator. (An end point color change from wine red to violet indicates a high level of iron in the water. This interference can be eliminated by adding a few crystals of potassium cyanide. **Caution must be used with this. Add it only after the alkaline buffer is added, since HCN is volatile and very toxic.** After the titration, add 1 g $FeSO_4 \cdot 7H_2O$ to convert CN^- to harmless $Fe(CN)_6^{4-}$, and then flush down the drain with a large amount of water.) If the water sample is likely to contain copper, add a few crystals of hydroxylamine hydro-chloride. This reduces copper(II) to copper(I), which does not interfere.

Titrate with 0.01 M EDTA until the color changes from wine red through purple to a pure blue. The reaction (color change) is slow at the end point, and titrant must be added slowly and the solution stirred thoroughly in the vicinity of the end point. A comparison solution for the proper color at the end point may be prepared by adding 2 mL of pH-10 buffer to 50 mL distilled water, five drops indicator, a few drops Mg–EDTA, and a few drops EDTA.

If the end point for the first titration is less than 10 mL, double the volume of sample for the remaining two titrations.

Calculation

Calculate and report the hardness of the water as ppm $CaCO_3$ for each portion analyzed.

REDUCTION–OXIDATION TITRATIONS

EXPERIMENT 13   Analysis of an Iron Alloy or Ore by Titration with Potassium Dichromate

Principle

An iron alloy or ore is dissolved in HCl and the iron is then reduced from Fe(III) to Fe(II) with stannous chloride ($SnCl_2$). The excess $SnCl_2$ is oxidized by addition of $HgCl_2$. The calomel formed (insoluble Hg_2Cl_2) does not react at an appreciable rate with the titrant. The Fe(II) is then titrated with a standard $K_2Cr_2O_7$ solution to a diphenylamine sulfonate end point.

Equations

Reduction:

$$2Fe^{3+} + Sn^{2+} \text{ (slight excess)} \rightarrow 2Fe^{2+} + Sn^{4+} + Sn^{2+} \text{ (due to excess)}$$
$$Sn^{2+} + 2Hg^{2+} + 2Cl^- \rightarrow Sn^{4+} + \underline{Hg_2Cl_2} \text{ (white precipitate)}$$

If too much Sn^{2+} is added, then

$$Sn^{2+} + \underline{Hg_2Cl_2} \rightarrow Sn^{4+} + 2Hg^0 + 2Cl^-$$

(Black Hg^0 precipitate makes end point determination impossible. Sample must be discarded because Hg^0 reacts with $Cr_2O_7^{2-}$.)
Titration:

$$6Fe^{2+} + Cr_2O_7^{2-} + 14H^+ \rightarrow 6Fe^{3+} + 2Cr^{3+} + 7H_2O$$

Solutions and Chemicals Required

1. *Provided.*   0.28% (wt/vol) of the sodium salt of p-diphenylamine sulfonate indicator, 0.5 M $SnCl_2$ in 3.5 M HCl (with mossy tin added to stabilize against air oxidation: Sn^{4+} + $Sn^0 \rightarrow 2Sn^{2+}$), saturated $HgCl_2$ solution, conc. HCl, 6 M HCl, conc. H_3PO_4–conc. H_2SO_4 mixture (15 mL each added to 600 mL water and cooled to room temperature), 0.1 M $FeCl_3$ in 6 M HCl.

2. *To prepare.*   Standard $\frac{1}{60}$ M $K_2Cr_2O_7$ solution (approx. 0.017 M). Dry about 3 g primary standard $K_2Cr_2O_7$ in a weighing bottle at 120°C for at least 2 hours. Drying in the oven for longer periods of time (i.e., until the next lab period) will do no harm. Place the $K_2Cr_2O_7$ in the desiccator to cool for 30 -- 40 minutes. Weigh accurately to the nearest milligram, about 2.5 g in a weighing dish, and transfer quantitatively to a 400-mL beaker. Dissolve in about 200 mL water and transfer quantitatively to a 500-mL volumetric flask. Dilute to volume and mix thoroughly. Rinse a clean 1-L bottle with three small portions of the solution and transfer the remainder of the solution to the bottle for storage. (If this solution is to be used for Experiment 14 also, save at least 150 mL of it after completing this experiment. It may also be used in Experiment 29, in which case 100 mL should be saved.) Calculate the molarity of the solution. $K_2Cr_2O_7$ may also be prepared approximately and standardized against electrolytic iron wire (primary standard) using the same procedure as given below for an alloy sample.

Things to Do Before the Experiment

1. *Dry the necessary amount of $K_2Cr_2O_7$.*
2. *Obtain and dry or dissolve your unknown as required.*
(a) Alloy sample. This may be in the form of a wire. Your instructor will provide you with three separate (weighed) pieces of the unknown, each to be placed in separate, labeled 500-mL Erlenmeyer flasks containing 10 mL conc. HCl. Cover with inverted 100-mL beakers and store these dissolving samples in your desk overnight or longer. Alternatively, the samples can be dissolved on the day of the experiment by heating on a hot plate or steam bath in 400-mL beakers (covered with ribbed watch glasses in the hood to hasten dissolution. (After dissolution, rinse the cover glass and the sides of the beaker, catching all the

rinsings in the beaker. Use as little water as possible. The final volume should not be more than 50 mL.)

(b) Ore sample. Check out a sample of an iron ore from the instructor. Dry in the oven at 110−120°C for at least 2 hours; longer drying will do no harm.

Procedure

1. *Reduction of the sample and trial titration.*   Before titrating your unknown, it is advisable to perform one or two trial titrations. This can be done while the (ore) samples are dissolving. Add approximately 10 mL 0.1 M $FeCl_3$ solution in 6 M HCl to a 600-mL beaker and add about 50 mL water. (This dilutes the sample sufficiently that when all the Fe^{3+} is reduced, the solution will be nearly colorless. If the volume is less, then the pale green of the Fe^{2+} makes detection of complete reduction more difficult). Place a ribbed watch glass on the beaker and heat nearly to boiling on a hot plate in the hood; the solutions must be very close to the boiling point, perhaps simmering gently, but not boiling violently since $FeCl_3$ can be lost due to volatilization. Add 0.5 M stannous chloride solution with a dropper through the lip of the beaker until the color beings to fade; then, continue the addition drop by drop, swirling the beaker and allowing each drop to reach before adding the next, until the solution is colorless. It will first become pale yellow and then will gradually turn more colorless. It may never get completely colorless but may instead go to a pale green due to the ferrous ion (this will depend on the amount of iron). Whichever you get (colorless or pale green) stop addition and allow the solution to heat for two more minutes. If the yellow color returns add a few more drops of $SnCl_2$ until it becomes colorless or pale green again. Repeat dropwise addition of $SnCl_2$ until the solution does not return to the yellow color. At this point, add two drops excess, no more. (If more than two drops are added, the stannous chloride can be oxidized with a few drops of potassium permanganate solution and the above reduction process repeated). Remove from the hot plate, rinse down the cover glass and sides of the beaker, and cool quickly to room temperature by immersing the bottom of the beaker in cold water. Two to three samples may be taken this far together; **the remainder of the procedure must be carried out with each sample individually without interruption through the titration.** If any sample turns yellow again while awaiting its turn, it must be reheated and sufficient stannous chloride added to discharge the color, with two drops excess. Fill your 50-mL buret with the standard $K_2Cr_2O_7$ and have it ready for titration.

To one sample, which must be at room temperature, add 100 mL water and then add rapidly 15 mL saturated mercuric chloride solution, previously measured out, while stirring and immediately mix thoroughly. A slight, white precipitate should form. If either a heavy gray or black precipitate (elemental mercury) or no precipitate forms, too much or not enough (to reduce all the Fe^{3+}) stannous chloride has been added; in either case, the sample must be discarded. Mix for 2 minutes, then add 100 mL of the H_3PO_4–H_2SO_4 mixture and six to eight drops diphenylamine sulfonate indicator. Titrate immediately with the $K_2Cr_2O_7$ solution, stirring constantly, until the green color changes to a purple or violet blue that remains for at least 1 minute. (The acid mixture provides the protons consumed in the titration reaction and forms a nearly colorless phosphate complex with the Fe^{3+} titration product, which sharpens the end point.)

2. *Alloy sample.*   The sample should by now be dissolved in the Erlenmeyer flasks (or in the heated 600-mL beakers). Adjust the volume to 40−60 mL with distilled water. Heat nearly to boiling and follow the same procedure as used in the trial titration, starting wtih addition of stannous chloride. All three samples can be taken up to the point to just before the addition of mercuric chloride and then must be treated one at a time up through the titration.

3. *Ore sample.* After cooling the dried sample in a desiccator for 30 – 40 minutes, weigh out by difference (the iron ore may be hygroscopic) three samples; consult the instructor as to the size of the samples. Transfer to 600-mL beakers. Add 10 mL water and swirl until the sample is completely moistened and in suspension; then cover with ribbed watch glasses and add 10 mL concentrated HCl, pouring it through the lip of the beaker. Heat on a hot plate in the hood until the iron ore has dissolved to give a clear, red-brown solution; with some samples there may be an insoluble sandy residue, which may be disregarded. [Silica or insoluble sulfides (black) or silicates may remain.] The hot plate should be adjusted to keep the solutions just barely at the boiling point; vigorous boiling should be avoided, since it may cause loss of material and excessive evaporation of acid. If necessary, add 6 M HCl to keep the volume about 20 mL. When all the iron has been dissolved, the insoluble residue (if any) will be gray or white, with no black or reddish particles, after adding stannous chloride to reduce the iron. When the solution appears clear, add distilled water to bring the volume to about 50 mL and follow the same procedure used in the trial titration starting with addition of stannous chloride to the hot solution. All three samples can be carried up to the point just before addition of mercuric chloride and then must be treated one at a time up through the titration.

Calculations

1. *Alloy sample.* Calculate and report the milligrams iron in each portion of the unknown analyzed, along with the mean and the precision.

2. *Ore sample.* Calculate and report the percent iron in each portion of the unknown analyzed, along with the mean and the precision.

EXPERIMENT 14   Analysis of Commercial Hypochlorite or Peroxide Solution by Iodometric Titration

Principle

The oxidizing power (percent NaOCl or H_2O_2) of the solution is determined iodometrically by reacting it with an excess of iodide in acetic acid solution and titrating the iodine produced (I_3^- in the presence of excess iodide) with standard sodium thiosulfate solution. The sodium thiosulfate is standardized against primary standard potassium iodate, and a starch indicator is used.

Equations

Standardization of $Na_2S_2O_3$:

$$IO_3^- + 8I^- + 6H^+ \rightarrow 3I_3^- + 3H_2O$$
$$I_3^- + 2S_2O_3^{2-} \rightarrow 3I^- + S_4O_6^{2-}$$

Determination of sample:

$$ClO^- + 3I^- + 2H^+ \rightarrow Cl^- + I_3^- + H_2O$$

or

$$H_2O_2 + 3I^- + 2H^+ \xrightarrow[\text{catalyst}]{\text{Mo(VI)}} 2H_2O + I_3^-$$

I_3^- is titrated with $S_2O_3^{2-}$ as above.

Solutions and Chemicals Required

1. *Provided.* $6M$ HCl, KI, primary standard KIO_3, glacial acetic acid, 3% ammonium molybdate solution (for peroxide samples), Na_2CO_3, dil. (1:4) H_2SO_4.

2. *To prepare.*

(a) Starch solution. Prepare a 1% solution by mixing 0.5 g soluble starch with 2–3 mL of distilled water; add a pinch of HgI_2. Pour this mixture into 50 mL boiling distilled water with stirring and continue heating for 2–3 minutes until the solution is clear or only faintly opalescent. Cool to room temperature. The HgI_2 stabilizes the starch indefinitely; otherwise, it should be prepared fresh on the day of use. Note: Approximately 0.4 g of the commercial indicator Thiodene may be used in place of the prepared starch solution.

(b) Standard 0.01 M KIO_3 solution. This will be used to standardize the $Na_2S_2O_3$ solu-tion. (If you prepared a standard $\frac{1}{60}$ M $K_2Cr_2O_7$ solution in Experiment 13, you may titrate 50-mL aliquots of this iodometrically for the standardization of the $Na_2S_2O_3$. If so, consult your instructor for directions.) A standard solution of KIO_3 is prepared and aliquots of this are titrated with the $Na_2S_2O_3$ solution. This procedure is used instead of titrating individually weighed portions of KIO_3. The reason is that KIO_3 has a low equivalent weight and only about 0.1-g portions can be titrated. Hence, it is more accurate to prepare a standard solu-tion. This requires special care in the accurate preparation of the solution since only one solution is prepared.

Dry about 1.5 g primary standard KIO_3 at 120°C for 1–2 hours and cool in a desiccator for 30–40 minutes. Accurately weigh (to the nearest 0.1 mg) 1.0–1.4 g and dissolve in a small amount of distilled water in a 200-mL beaker. Quantitatively transfer, with rinsing, to a 500-mL volumetric flask, using a glass funnel and stirring rod to direct the solution into the flask. Dilute to the calibration mark. Calculate the molarity of the solution.

(c) 0.1 M $Na_2S_2O_3$ solution. Boil about 1200 mL distilled water for 5–10 minutes to ensure sterility and to expel carbon dioxide. Cool to room temperature. (Sodium thiosulfate solutions are subject to bacterial attack, which may change the molarity after a time. So all water and glassware used to prepare and store the solution should be sterilized. If any turbidity or bacteria or mold growth appears, the solution should be discarded. Removal of carbon dioxide is also beneficial, because thiosulfate is more stable in neutral solution.) Sterilize a 1-L bottle with dichromate cleaning solution; rotate the bottle so the cleaning solution contacts the entire interior wall. Rinse very thoroughly with tap water, then with distilled water, and finally with the boiled distilled water. Weigh out on a watch glass, using a rough balance, 25 g sodium thiosulfate crystals, $Na_2S_2O_3 \cdot 5H_2O$. Transfer to the liter bottle, fill to the shoulder with the freshly boiled and cooled distilled water, add 0.1 g sodium carbonate, and shake thoroughly until the solution is homogeneous. (A small amount of sodium carbonate is added to keep the solution neutral or slightly alkaline and thereby stabilize it against decomposition to elemental sulfur.) Store in a refrigerator if pos-sible, but let warm up to room temperature before using.

Things to Do Before the Experiment

1. *Dry the required amount of primary standard KIO_3.*

2. *Prepare the 0.1 M $Na_2S_2O_3$ solution.*   Although this can be prepared on the day of the experiment, it is preferable to prepare it at least a day before it is to be standardized. The solution tends to lose some of its titer right after preparing.

Procedure

1. *Standardization of the $Na_2S_2O_3$ solution.*   The solution should be standardized on the day of the experiment. Consult your instructor if you are to use the standard $K_2Cr_2O_7$ solu-tion from Experiment 13 for standardization. Otherwise, proceed as follows.

Rinse the 50-mL buret several times with small portions of the thiosulfate solution and fill it with thiosulfate solution. Adjust to near the zero mark and record the volume reading to the nearest 0.02 mL. Add with a pipet a 50.00-mL aliquot of the potassium iodate solution to a clean 250-mL wide-mouth Erlenmeyer flask. Add about 2 g solid potassium iodide and swirl to dissolve. Add, with rapid mixing, 5 mL dilute H_2SO_4. Mix thoroughly.

Titrate **immediately** with thiosulfate solution. (In strongly acid solution, the excess iodide is rapidly air-oxidized to I_3^-, and so the titration must be performed quickly.) Thorough, continuous mixing throughout the titration is essential; the thiosulfate must not be allowed to accumulate in local excess in the acid solution or else some decomposition into H_2SO_3 and S may occur. Titrate until the yellow color (due to I_3^-) *almost* disappears. It will become a pale yellow. Then, add 2–3 mL starch solution and titrate until the blue color just disappears (properly done, this should occur within $\frac{1}{2}$ mL after adding the starch solution).

The standardization should be repeated until you are sure of the titration volume to within one part per thousand (± 0.03 mL at a titration volume of 30 mL, etc.) Calculate the molarity of the $Na_2S_2O_3$ solution.

2. *Determination of hypochlorite or H_2O_2 in unknown.*[1]

(a) Weighing and diluting the unknown. Roughly calibrate a weighing bottle by pouring into it about 12 mL water and noting the level to which it fills the bottle. Empty and thoroughly dry the weighing bottle and weigh it to the nearest milligram. The hypochlorite or peroxide solutions to be analyzed will be supplied in the commercial bottles, fitted with siphon delivery tubes. Clean off any solid crust on the tip and discard a few drops to flush out the tip. Withdraw about 12 mL into the calibrated and weighed weighing bottle; it is essential that the upper portion of the bottle, particularly the ground-glass part, remain dry. Replace the stopper and weigh to the nearest milligram. Empty the weighing bottle into a 250-mL volumetric flask containing about 100 mL water, using a funnel. Wash out the weighing bottle and the funnel with a jet of water from your wash bottle, catching the rinsings in the volumetric flask. Dilute to the mark and mix thoroughly. Transfer with a pipet three 50-mL aliquots of the solution into 250-mL Erlenmeyer flasks containing about 50 mL water; rinse down the walls of the flasks in such a way as to form a layer of water above the sample. From this point on, handle each sample individually through the remainder of the procedure.

(b) Titration. Fill your buret with standard 0.1 *M* sodium thiosulfate solution. Measure out and have ready 10 mL glacial acetic acid, 2 g potassium iodide, and, if the sample is a peroxide, a dropping bottle containing 3% ammonium molybdate catalyst solution. Hypochlorite requires no catalyst. When ready to titrate, add the acid to one of the samples, mix, add the potassium iodide, and, if the sample is a peroxide, add three drops of the catalyst solution. Titrate immediately, swirling the flask constantly. When the color has faded to a pale yellow, add about 2 mL starch solution and continue the titration drop by drop until the solution just becomes colorless. Complete the other samples in the same way.

Calculation

Calculate the percentage by weight of NaClO or H_2O_2 in the solution. (Note: Commercial bleach should contain at least 5.25% NaClO. If less than this, it cannot be called "bleach").

---

[1]This experiment should be completed in a single laboratory period by all students, including calculation and reporting of results. The hypochlorite and peroxide solutions are subject to decomposition, with a resultant change in concentration. The instructor may take the average of the class results as the correct value, or he or she may perform an analysis alone for comparison.

EXPERIMENT 15   Iodometric Determination of Copper

Principle

A copper metal sample is dissolved in nitric acid to produce Cu(II), and the oxides of nitrogen are removed by adding H_2SO_4 and boiling to SO_3 fumes. The solution is neutralized with NH_3 and then slightly acidified with H_3PO_4. (The H_3PO_4 also complexes any iron(III) that might be present and prevents its reaction with I^-.) Finally, the solution is treated with excess KI to produce <u>CuI</u> and an equivalent amount of I_3^-, which is titrated with standard $Na_2S_2O_3$ solution, using a starch indicator. KSCN is added near the end point to displace adsorbed I_2 on the CuI by forming a layer of CuSCN. For best accuracy, the $Na_2S_2O_3$ is standardized against high-purity copper wire, since some error occurs from reduction of copper(II) by thiocyanate.

Equations

$$2Cu^{2+} + 5I^- \rightarrow 2\underline{CuI} + I_3^-$$

$$I_3^- + 2S_2O_3^{2-} \rightarrow 3I^- + S_4O_6^{2-}$$

Solutions and Chemicals Required

1. *Provided.*   6 M HNO_3, conc. H_2SO_4, 3 M H_2SO_4, conc. H_3PO_4, 6 M NH_3, KSCN.
2. *To prepare.*
(a) Starch solution. Prepare the day of the experiment as described in Experiment 14, or use Thiodene indicator.
(b) 0.1 M $Na_2S_2O_3$ solution. Prepare as described in Experiment 14.

Things to Do Before the Experiment

Prepare the 0.1 M $Na_2S_2O_3$ solution. Although this can be prepared on the day of the experiment, it is preferable to prepare it at least a day before it is standardized. The solution tends to lose some of titer right after preparing.

Procedure

1. *Standardization of the 0.1 M $Na_2S_2O_3$ solution.*   The same procedure is used as will be used for analyzing the sample. Weigh out three 0.20- to 0.25-g samples of pure electrolytic copper foil and add to 250-mL Erlenmeyer flasks. In a hood, dissolve in 10 mL 6 M HNO_3, heating on a steam bath if necessary. Do in a fume hood. Add 10 mL conc. H_2SO_4 and evaporate to copious white SO_3 fumes. Cool and add carefully 20 mL water. Boil 1−2 minutes and cool. Add 6 M NH_3 dropwise with swirling of the sample solution until the first dark blue of the $Cu(NH_3)_4^{2+}$ complex appears. Then, add 3 M H_2SO_4 until the blue color just disappears, followed by 2.0 mL conc. H_3PO_4. Cool to room temperature.

From this point, each sample must be treated separately. Dissolve about 2 g KI in 10 mL water and add to one of the flasks. Titrate *immediately* with the $Na_2S_2O_3$ solution until the yellow color of I_3^- *almost* disappears. Add 2−3 mL of the starch solution or approximately 0.4 g Thiodene indicator, and titrate until the blue color begins to fade (should be less than $\frac{1}{2}$ mL). Finally, add about 2 g KSCN and continue the titration until the blue color just disappears.

Repeat with the other two samples. Calculate the molarity of the $Na_2S_2O_3$.

2. *Determination of copper in an unknown.*   Add 10 mL of 6 M HNO_3 to each of three clean 250-mL Erlenmeyer flasks and take these to your instructor, who will add an unknown

sample to each. (This unknown may be copper foil as used in the standardization.) Dissolve each and titrate as described for the standardization of $Na_2S_2O_3$. **Note:** In place of H_3PO_4, approximately 2 g ammonium bifluoride, NH_4HF_2 or $NH_4F \cdot HF$, may be added to complex any iron and at the same time adjust the solution to the proper acidity. This experiment is also suitable for determining copper in about 0.3 g brass.

Calculation

Calculate the grams copper in each unknown sample and report the values of each. (If a weighed brass sample is analyzed, report the percent copper.)

EXPERIMENT 16   Determination of Antimony by Titration with Iodine

Principle

Antimony(III) is titrated to antimony(V) in neutral or slightly alkaline solution with iodine to a blue starch end point. The iodine is standardized against primary standard arsenic(III) oxide. Tartaric acid is added to complex the antimony and prevent its hydrolysis to form insoluble basic salts such as SbOCl and SbO_2Cl (which form in slightly acid and neutral solution).

Equations

Standardization:

$$H_2AsO_3^- + I_3^- + H_2O \rightarrow HAsO_4^{2-} + 3I^- + 3H^+ \quad (pH\ 8)$$

Sample titration:

$$SbOC_4H_4O_6^- + I_3^- + H_2O \rightarrow SbO_2C_4H_4O_6^- + 3I^- + 2H^+$$

Solutions and Chemicals Required

1. *Provided.*   Primary standard As_2O_3, 1 M NaOH, Na_2CO_3, $NaHCO_3$, KI, tartaric acid, 1 M HCl. For stibnite ore: KCl, 0.1% (wt/vol) methyl red indicator in 60% ethanol, conc. HCl, 6 M HCl, 6 M NaOH.

2. *To prepare.*

(a) Starch solution. Prepare as described in Experiment 14, or use Thiodene indicator.

(b) 0.05 M Iodine solution. Weigh 6.5 g iodine crystals and 20 g potassium iodide. Grind the iodine in a mortar with repeated small portions of the weighed KI crystals and water, pouring off the solution frequently into a glass-stoppered bottle until the solids are completely dissolved. [I_2 is only slightly soluble in water, but forms soluble KI_3 (I_3^- complex) in the presence of excess KI.]

Avoid pouring undissolved iodine into the bottle. Dilute the solution to about 500 mL and mix thoroughly. Check for any undissolved iodine. Preferably let stand overnight before standardizing to insure complete dissolution of the iodine. Alternatively, before diluting the solution, add more KI until all iodine is dissolved.

Things to Do Before the Experiment

1. *Obtain and dry your unknown.*   Obtain a sample in a weighing bottle from your instructor and dry at 110–120°C for at least 1–2 hours. Cool for at least 30–40 minutes before weighing.

2. *Dry the As$_2$O$_3$.*   Obtain and dry about 1 g primary standard As$_2$O$_3$ at 110–120°C for 1–2 hours. Cool in a desiccator at least 30–40 minutes before weighing.

3. *Prepare the 0.05 M I$_2$ solution.*   If your unknown is a stibnite ore that will require some time to dissolve (as opposed to a water-soluble synthetic sample), it is advisable to also standardize the iodine solution before the day of the experiment to allow sufficient time to complete it (procedure below).

Procedure

1. *Standardization of iodine solution.*   Weigh directly and accurately three 0.15- to 0.20-g portions of dried primary standard As$_2$O$_3$. Transfer to 250-mL Erlenmeyer flasks and disssolve in 10–20 mL 1 *M* NaOH, heating if necessary to aid dissolution. No undissolved particles should remain. Cool and add 1 *M* HCl until the solution is just acidic to litmus paper. Add 3–4 g solid NaHCO$_3$. No further CO$_2$ evolution should occur on addition of the last portion of NaHCO$_3$. If it does, add more NaHCO$_3$. The pH of the solution should be 7–8. Wash the walls of the flask down and add 50 mL water and 3 mL starch solution or about 0.4 g Thiodene indicator. Titrate with the iodine solution to the appearance of the first tinge of blue that persists for at least 30 seconds. From the three titrations, calculate the molarity of the I$_2$ solution. Use the average of the results.

2. *Determination of antimony in unknown.*

(a) Water-soluble synthetic sample. Consult your instructor for the proper size sample to weigh so that it will contain about 2 mmol antimony. Weigh three portions of the dried unknown and dissolve in 50 mL water in 500-mL Erlenmeyer flasks. Dissolve 4 g NaHCO$_3$ and 2 g tartaric acid in 100 mL water and add this solution to the antimony solution. The solution should be clear at this point with no hydrolyzed antimony chloride. Add 3 mL starch solution and titrate to a blue color that persists at least 30 seconds.

(b) Insoluble stibnite ore. Consult your instructor for the proper size sample to weigh so that it will contain about 2 mmol antimony.

Into the dry 250-mL beakers weigh accurately triplicate samples of the dried ore. Add about 0.3 g finely powdered potassium chloride, nearly cover the beaker with a watch glass, and carefully add 10 mL conc. hydrochloric acid by pouring it down the side of the beaker. (A high concentration of chloride is necessary to prevent hydrolysis during the dissolution before tartaric acid is added. SbCl$_3$ is formed.) Warm (do not boil) in the hood until the ore is decomposed; the mixture should no longer give an odor of hydrogen sulfide, and any residue (silica) should be white or only slightly gray. A stibnite ore consists of antimony sulfide, Sb$_2$S$_3$, silica, and small amounts of other substances. When all the antimony is dissolved, no more hydrogen sulfide should be evolved. Do not allow the solution to evaporate to dryness, which might result in loss of antimony trichloride; add more HCl as necessary. When decomposition is complete, add 3 g finely powdered tartaric acid and continue the heating for 10–15 minutes. Add water in portions of about 5 mL with good stirring until the solution is diluted to about 100 mL. (The solution must be diluted slowly, since some of the antimony may be hydrolyzed by local excesses of water.) If, during dilution, a red-orange precipitate (Sb$_2$S$_3$) appears, heat gently until the precipitate has dissolved before continuing the dilution. If a white precipitate of basic salts forms, the determination should be discarded. When the dilution is complete, boil the solution for 1 minute.

Rinse off the watch glass into the solution, and carefully neutralize the solution with 6 *M* sodium hydroxide by using a few drops of methyl red indicator. Then, add 6 *M* HCl dropwise until the solution is just acidic, carefully avoiding an excess.

In 600-mL beakers or 500-mL Erlenmeyer flasks, prepare solutions containing 4 g sodium bicarbonate in 200 mL water. Pour the sample into the sodium bicarbonate solution, avoid loss by effervescence, and rinse several times with a stream of water from the wash bottle to obtain a complete transfer of the solution. Add 3 mL starch indicator or about 0.4 g Thiodene indicator and titrate with standard iodine solution to the appearance of the first permanent blue color. A fading or indistinct end point is due to insufficient sodium bicarbonate in the solution to consume the acid produced in the titration. Add 1 g additional $NaHCO_3$ and complete the titration to a permanent blue color.

Calculation

Calculate and report the percent Sb_2O_3 for each portion of your sample analyzed. Report also the mean of your values and the precision.

POTENTIOMETRIC MEASUREMENTS

EXPERIMENT 17   pH Titration of Unknown Soda Ash

Principle

The unknown soda ash is titrated with standard HCl using a potentiometric (pH) end point measured with a pH meter using a pH glass electrode–saturated calomel reference electrode combination. The end point breaks are compared with indicator color changes.

Equations

$$CO_3^{2-} + H^+ \rightarrow HCO_3^- \quad \text{(phenolphthalein end point)}$$
$$HCO_3^- + H^+ \rightarrow H_2O + CO_2 \quad \text{(methyl purple end point)}$$

Note that between the first and second end points, a gradual decrease in pH due to the HCO_3^-/CO_2 buffer system will occur. This will give a poor visual end point, unless the buffer couple is destroyed. In practice, the visual titration used for standardization is continued until the methyl purple end point is reached, at which time the solution is gently boiled to remove the CO_2, leaving only the remaining HCO_3^-, which is then titrated to completion (see Chapter 8 for a more detailed discussion).

Solutions and Chemicals Required

1. *Provided.*   0.2% Phenolphthalein in 95% ethanol, 0.1% methyl purple in water, primary standard Na_2CO_3, standard pH 7 buffer.

2. *To prepare.*   Standard 0.1 M HCl solution. Use the solution prepared in Experiment 7. If this solution is not available, prepare and standardize 1 L as described in Experiment 7. Alternatively, the acid may be standardized against the primary standard Na_2CO_3 by pH titration as described below for the unknown soda ash.

Things to Do Before the Experiment

Prepare and standardize the HCl solution.   This will require prior drying of primary standard Na_2CO_3.

Obtain the unknown soda ash from your instructor and dry for at least 2 hours at 160°C. Cool at least 30 minutes in a desiccator before weighing.

Procedure

The glass electrode to be used for pH measurements should have been soaked and stored in 0.1 M KCl for at least one day prior to its use. Always store the electrode in KCl solution when not in use. Calibrate the pH meter as described by your instructor, using the pH 7 standard buffer. This will consist essentially of adjusting the meter to read pH 7.00 with the electrodes immersed in the buffer solution. If only small quantities of buffer are used, it would be better to discard it rather than to chance contamination of the entire supply.

1. *Trial titration.* The purpose of this titration is to locate quickly and approximately the two end points. Weigh accurately by difference a dried sample of unknown soda ash (0.2– 0.3 g) and add it to a 400-mL beaker containing a magnetic stirring bar. Add approximately 50 mL water and a few drops phenolphthalein indicator. The indicators are for the purpose of making a comparison between the potentiometric end points and the indicator color changes. Place the beaker on a magnetic stirrer, immerse the electrodes, and start the stirrer, being careful not to touch the electrodes to the stirring bar. Titrate with standard HCl, taking readings about every 2 mL. After the phenolphthalein color disappears, add a few drops methyl purple indicator and titrate at 2-mL increments until the second end point is reached. Add a few increments beyond the end point. The correct color for the second end point can be determined by comparison with the color of a few drops of the indicator in a solution of 0.20 g potassium acid phthalate in 100 mL water. Plot a curve of pH versus volume of HCl and locate the approximate end points.

2. *Final titration.* Weigh accurately another sample of the unknown and titrate as before, but make pH readings every 5 mL to within 3 mL of each end point (both sides of end point). Then, make readings at 0.50- or 1-mL intervals within 1 mL of the end point. Near the end point, **take readings as quickly as possible because the pH will tend to drift as CO$_2$ escapes from the solution.** Note and record the points at which the indicators change color.

Plot a curve of pH (on the ordinate) versus HCl volume (on the abscissa) and indicate on this curve the range in which the indicators change color. Determine the end point from the second inflection point of the curve. Repeat the titration on two more portions of the unknown. Be sure to rinse the electrodes between titrations.

Calculations

Calculate and report the percent Na$_2$CO$_3$ and Na$_2$O in your unknown for each portion analyzed. Hand in the plots of the titration curves with your report. Report also the mean percent value and the precision.

EXPERIMENT 18   Potentiometric Titration of a Mixture of Chloride and Iodide

Principle

The mixture is titrated with a standard solution of silver nitrate, and the potentiometric end points are indicated with a silver wire electrode–glass electrode pair using a pH meter for potential measurements. Because the pH during the titration remains essentially constant, the glass electrode's potential remains constant, and this electrode serves as the reference electrode; this eliminates the necessity of preparing a chloride-free salt bridge for the reference electrode. AgI ($K_{sp} = 1 \times 10^{-16}$) precipitates first since it is less soluble than AgCl ($K_{sp} = 1 \times 10^{-10}$). The AgCl starts precipitating near the equivalence point of the iodide titration (when $[Ag^+][Cl^-] = 1 \times 10^{-10}$; $[Ag^+]$ at the iodide equivalence point is

$\sqrt{1 \times 10^{-16}} = 1 \times 10^{-8}$ M). The potential (i.e., pX) rise of the iodide titration curve will level off at the point when the chloride starts precipitating, that is, near the iodide equivalence point inflection. This will be followed by the typical S-shaped chloride potentiometric end point. The error in determining the iodide end point is small if it is taken at the point at which the potential levels off. (It should be noted that while mixtures of chloride and iodide can be titrated, mixtures of bromide with either chloride or iodide cannot generally be titrated because of mixed crystal formation—see isomorphous replacement in Chapter 6)

Equations

$$I^- + Ag^+ \rightarrow \underline{AgI}$$
$$Cl^- + Ag^+ \rightarrow \underline{AgCl}$$

Solutions and Chemicals Required

0.1 M Standard $AgNO_3$. Prepare as described in Experiment 10.

Things to Do Before the Experiment

Dry the primary standard $AgNO_3$ for 1–2 hours at 110–120°C **(no longer).** Store in a desiccator until ready for weighing.

Obtain and dry your unknown at 120°C for 1–2 hours. Store in a desiccator until ready for weighing.

Procedure

Weigh directly three 0.5- to 0.6-g samples of the dried unknown into 400-mL beakers. Dissolve in 150 mL distilled water, add a magnetic stirring bar and place the beaker on a magnetic stirrer. (Dissolve and titrate only one portion at a time to minimize air oxidation of the iodide.) Immerse the electrodes in the solution, taking care that they do not hit the magnetic stirrer. Connect the silver wire electrode to the reference terminal of the pH meter and the glass electrode to its usual terminal.[1] Stir the solution and titrate the sample with the standard $AgNO_3$. Take "pH" readings (actually pX) at 2-mL increments until the change is greater than 0.4 pH unit. Then add 0.2-mL increments. After the first end point is reached, add 2-mL increments until the second end point is approached and then 0.2-mL increments. Plot the potential versus volume of $AgNO_3$ and determine the end point for the iodide and the chloride (inflection point of second potential break). Use these values to estimate the end point for the other two samples and repeat the above procedure for these samples. Titrant may be added rapidly up to within 2 or 3 mL of the end point. Be sure to rinse the electrodes between titrations.

Calculations

Calculate and report the percent iodide (from the volume required to reach the first end point) and chloride (from the volume required to go from the first end point to the second

---

[1]The silver electrode is actually the indicating electrode and the glass electrode is the reference electrode. But most glass pH electrodes require a special plug and will not fit the reference terminal. The above arrangement is satisfactory and simply means the potential will be of the opposite sign from usual and change in the opposite direction.

end point) in your unknown for each portion analyzed. Hand in the plots of the titration curves with your report.

Report also the mean values and the precision.

EXPERIMENT 19 Determination of the pH of Hair Shampoos

Principle

A calibrated glass pH electrode–SCE pair is used to measure the pH of commercially available shampoos, hair conditions and rinses, and depilatories, both concentrated and diluted 1:10. An estimate is made whether the pH is controlled primarily by a buffered solution or a strong electrolyte. An evaluation is made of the relative importance of what is probably the best pH for a shampoo and why—does the pH affect the cleaning power of the shampoo or does it affect the hair itself?

Equations

Electrode response.

$$E = k - \frac{2.303RT}{F}\, pH$$

Solutions and Chemicals Required

Provided. Standard pH 3, 5, 7, 9, and 11 buffers (approximate—other values near these may be used). Each member of the class should bring one or more samples of commercially available hair shampoos, conditioners, rinses, and depilatories. These may include dandruff shampoos and hair removers such as "Neet" or "Nair." All the samples will be shared by the class. There should be enough of each to provide about 15–20 mL to each student.

Procedure

1. *Calibration of the electrodes.* The experiment will be performed by first calibrating the electrodes at pH 7 and making approximate pH measurements of the various samples. The samples are then separated into five groups: those with pH < 4, pH 4–5.9, pH 6–8, pH 8.1–10, and pH > 10. The gas electrode should have been soaked in 0.1 M KCl solution at least one day prior to its use; store the electrode in dilute KCl solution when not in use. Calibrate the pH meter for one group of samples at a time (pH 3 for the first group, pH 11 for the last group) using the procedure described by your instructor. (Your instructor may decide to use only three buffers for calibration, e.g., pH 4, 7, and 9.) This will consist essentially of adjusting the meter to read the pH of the standard buffer at room temperature with the electrodes immersed in the buffer solution. Rinse the electrodes with distilled water after calibration and blot with tissue paper; do not wipe them as this may impart a static charge to the glass membrane. Be careful to turn the pH meter to "standby" when removing electrodes from solution. If only a small quantity of buffer is used, it would be better to discard it rather than chance contamination of the entire supply.

2. *pH Measurements of samples.* Prepare a 1:10 dilution of each sample in a 100-mL beaker by pipetting 5.00 mL into 45.0 mL distilled water measured from a 50-mL buret. Measure the pH of first the diluted sample and then the concentrated sample. For the latter, if a small beaker is used (e.g., 25 mL) it may be possible to make the measurement with 5 mL of sample; just enough to cover the tips of the electrodes. Alternatively, a combination

pH–reference electrode may be used, in which case the sample can be placed in a test tube and the single probe dipped in this. (A combination electrode contains both a reference electrode and a glass electrode in one probe. A salt bridge wick for the reference electrode is above the pH glass bulb, and so the probe will have to be immersed a bit deeper to make contact with both electrodes. There will be two wire leads from the probe, one for each electrode.)

Immerse the electrode in the test solution and swish or agitate a few seconds. Then, allow the pH reading to equilibrate and record to the nearest 0.01 pH. Rinse the electrode well between measurements and blot off the water. It is best to make the measurements from low pH to high, or vice versa. If the pH reading of any of the samples falls outside the range of 4–10, the electrodes should be recalibrated with a buffer closer to the sample pH (if depilatories are measured, their pH will exceed 10 and a buffer at pH 11 should be used.)

Report. Arrange the samples into either shampoos or conditions and rinses (depilatories are separate). Arrange the samples in each group in order of increasing pH readings of the concentrated samples. Your instructor may compare your pH readings with the mean of the class. From the changes in readings on dilution, list whether you think the pH of each solution was determined by a buffer system or by a strong acid or base. Consult the reference J. J. Griffin, R. F. Corcoran, and K. K. Akana, *J. Chem. Educ.,* **54** (1977) 553 and include in your report a discussion of the relevance of pH in hair cleansing or conditioning and in hair damage. Propose mechanisms for the pH action. How do depilatories work? Within what pH range should children's shampoos be?

EXPERIMENT 20 Potentiometric Determination of Fluoride in Drinking Water Using a Fluoride Ion-Selective Electrode[1]

Principle

Fluoride in a water sample is determined by measurement with a fluoride ion-selective combination electrode (contains reference electrode built in). First, you will determine whether the electrode response is Nernstian over a wide range of concentrations. Then, you will determine fluoride in the unknown by comparing potential measurements with standards over a narrower range, bracketing the unknown; a calibration curve will be prepared.

Equations

$$E = k - \frac{2.303RT}{F} \log a_{F^-}$$

$$= k' - \frac{2.303RT}{F} \log [F^-] \quad \text{(if ionic strength is held constant)}$$

Solutions and Chemicals Required

1. *Provided.* TISAB (total ionic strength adjustment buffer) solution, which is prepared with 57 mL glacial acetic acid, 58 g sodium chloride and 4 g CDTA (cyclohexylene-dinitrilotetraacetic acid) in about 500 mL water, adjusted to pH 5.0–5.5 with 5*M* NaOH

---

[1] This experiment can also be used to determine % NaF or % SnF_2 in toothpaste by preparing a suspension of the paste in water and pipetting the supernatant, or to determine the fluoride content of children's fluoride tablets or drops. See T. Light and C. Cappucino, *J. Chem. Educ.,* **52** (1975) 247.

and diluted to a total volume of 1 L. A 1:1 dilution of **all** samples with this solution serves the following:

(a) It provides a high total ionic strength background, swamping out variations in ionic strength between samples.

(b) It buffers the samples at pH 5 to 6. (In acid media, HF forms; while in alkaline media, OH^- ion interferes in the electrode response.)

(c) The CDTA preferentially complexes with polyvalent cations present in water (e.g., Si^{4+}, Al^{3+}, Fe^{3+}, etc.) which otherwise would complex with F^- and change its concentration.

2. *To prepare.*

(a) Stock standard 0.1 M fluoride solution. Dry about 2 g NaF at 110°C for 1 hour, cool in a desiccator for 30 minutes. Weigh out 1.1–1.2 g of the dried NaF (to the nearest milligram), transfer to a 250-mL volumetric flask, dissolve, and dilute to volume with distilled deionized water. Shake thoroughly and transfer to a polyethylene bottle (rinse with a few small portions first). Fluoride tends to adsorb on glass and should be stored in plastic containers. **Caution:** Fluoride is poisonous. Handle with care. Commercially prepared fluoride solutions may be used.

(b) Linearity standards. By serial dilution of the stock solution with distilled deionized water, prepare a series of solutions of about 10^{-2}, 10^{-3}, 10^{-4}, 10^{-5}, and 10^{-6} M fluoride (calculate accurate concentrations). (Do **not** pipet by mouth!) For example, dilute the stock solution 10:100, 1:100, and 1:1000 mL to prepare the first three solutions. Then, dilute the 10^{-4} M solution 10:100 and 1:100 mL to prepare the last two. Transfer to polyethylene bottles. These solutions should be prepared on the day of use.

(c) Calibration standards. The unknown concentration will be within 1×10^{-3} and 1×10^{-2} M fluoride, and a calibration curve will be prepared using concentrations of fluoride to bracket the unknown. Using the above procedure, prepare additional standards of 2×10^{-3} and 4×10^{-3} M fluoride (calculate the accurate concentrations). Prepare on the day of use.

Things to Do Before the Experiment

Prepare the stock NaF solution. This will require drying of NaF.

Procedure

1. *Determination of range of response and range of linearity.* Connect the electrode leads to an expanded-scale pH meter. Add with a pipet 10 mL of the 10^{-6} M standard solution and 10 mL TISAB solution to the small plastic beaker provided. Place the electrode in the beaker. Stir the solution with a magnetic stirrer and small stirring bar during measurement. When a steady pH reading is obtained, record the value. Rinse and blot the electrode and repeat, going from dilute to concentrated standard solutions. Convert to millivolt readings (1 pH unit = 59.2 mV at 25°C; arbitrarily take pH 0 as 0 mV) and plot on the linear axis of semilog graph paper against concentration on the log axis. Determine the slope of the curve in mV/decade concentration and the range of linearity.

2. *Standardization for unknown.* Record the pH readings of the standard solutions from 1×10^{-3} M to $1 \times 10^{-2}M$ and plot a calibration curve of pH versus log concentration.

3. *Analysis of unknown.* After preparing the calibration curve, obtain an unknown fluoride sample. This may be a synthetic solution, in which case obtain the unknown in a 250-mL volumetric flask. Immediately dilute to volume with distilled deionized water and transfer to a polyethylene bottle. Add 10 mL of the unknown with a pipet to a small plastic beaker followed by 10 mL TISAB. Record the pH reading as above. Make at least three

separate runs (separate additions and potential readings). **Note:** The unknown should be measured at the same time the calibration curve is prepared. The pH scale should not be adjusted between calibration and sample measurement. If it is, take a new reading of one of the standards and readjust to the original reading.

Calculations

From the calibration curve, determine the concentration of fluoride in the unknown solution. Report the results in parts per million fluoride.

ELECTROCHEMICAL MEASUREMENTS

EXPERIMENT 21   Electrolytic Determination of Copper in an Ore

Principle

An unknown ore sample is dissolved and the copper is plated on a platinum gauze cathode using controlled direct current. The gain in weight of the cathode represents the copper in the unknown.[1]

Solutions and Chemicals Required

Conc. HNO_3, conc. H_2SO_4, urea, NH_4NO_3, acetone or ethanol.

Equations

$$Cu^{2+} + 2e^- \rightarrow Cu^0$$

Things to Do Before the Experiment

Obtain an unknown ore from the instructor and dry at 110–120°C for 1–2 hours if required.[2] Cool in a desiccator at least 30 minutes before weighing.

Procedure

1. *Preparation of the electrode.*   Clean the platinum electrodes by immersing in hot nitric acid (1:2) for 5 minutes. This removes any copper deposit and also grease, etc., from the platinum gauze cathode. Rinse with distilled water. From this point on, touch only the stems of the electrodes, particularly the cathode. Grease from the fingers will prevent a tight deposit of copper on the cathode gauze. Immerse the cathode in a beaker containing acetone or ethanol. Remove, let the bulk of the solvent drain off, then place the cathode in a clean, dry beaker. Dry the cathode in an oven at 110°C for 5 minutes. Cool in a desiccator and weigh accurately, being careful to touch only the stem of the electrode.

2. *Preparation of the sample.*   The analysis will be done in duplicate. Accurately weigh

---

[1] This procedure works satisfactorily only in the absence of more easily electrolyzed substances such as silver and iron. Silver may be removed by precipitation with chloride and filtration. One-half percent iron may be tolerated if only moderate amounts of nitric acid are present.

[2] The unknown may be a copper–nickel alloy, such as Monel metal, which need not be dried. Other possible unknowns include copper oxide plus inert matter, which is soluble in hot 1:2 sulfuric acid, or a mixture of $CuSO_4 \cdot 5H_2O$ and potassium sulfate, which must not be dried and is water soluble (should acidify with dilute sulfuric acid).

two dried samples. Consult your instructor for the size sample to weigh (about 1 g). Transfer the samples to 250-mL beakers and add 10 mL nitric acid. Cover with watch glasses and heat until the ore dissolves (in the hood!). A residue may remain. Cool, add 10 mL 1:1 sulfuric acid, and heat until dense fumes of SO_3 are evolved. This volatilizes the nitric acid, which would interfere. Cool and carefully dilute to 90 mL with hot distilled water. Filter the solutions through medium-porosity filter paper, collecting the filtrate in 200-mL tall-form electrolytic beakers. Wash the residue with three 5-mL portions of hot water.

3. *Electrolysis.* If there is apparatus for the electrolysis of only one sample at a time, proceed with only the sample to be electrolyzed. Otherwise, prepare and electrolyze both solutions simultaneously. Add 1 mL conc. nitric acid, 1 g ammonium nitrate, and 0.5 g urea. (A small amount of nitric acid is necessary because the nitrate ion acts as a depolarizer at the cathode and prevents the reduction of hydrogen ion to produce hydrogen gas. The urea is added to destroy any oxides of nitrogen or any nitrite ion produced by reduction of the nitrate ion, which inhibits the copper deposition.) Place the electrodes in the solution and turn on the stirring apparatus. The cathode should extend about 1 cm above the solution. Cover the beaker with a split watch glass, turn on the electrolysis current and adjust it to 1–3 amp. The voltage should be less than 3 V.[3] Continue the electrolysis until the blue color of the copper disappears. Then, remove the watch glass and the droplets collected on it into the beaker with distilled water (do not interrupt the electrolysis). Add sufficient water to cover the cathode and continue the electrolysis for another 15 minutes. If no copper has deposited on the top of the cathode, the electrolysis is complete. If copper has deposited, continue the electrolysis for another 15 minutes. At the completion of electrolysis, do not interrupt the current until the electrodes have been separated from the solution. Interruption of the current while the circuit is completed will cause some of the copper to redissolve spontaneously. Slowly lower the beaker from the electrodes, simultaneously washing the electrodes with distilled water. After thoroughly washing the electrodes, turn off the current and disconnect the cathode. Immerse it in a beaker of distilled water and then in a beaker of acetone or ethanol. Dry in the oven for 5 minutes (no longer) as before, cool, and weigh· longer drying periods will cause the surface copper to oxidize. Subtract the weight of the electrode determined above to obtain the weight of copper deposited. Clean the cathode by dissolving the copper in warm 1:1 nitric acid and dry as described before.

Calculations

Calculate and report the percent copper in your unknown for each portion analyzed.

EXPERIMENT 22   Polarographic Determination of a Trace Metal

Principle

An unknown containing lead, cadmium, or zinc is analyzed polarographically. The metal present is identified from the half-wave potential and is quantitatively determined by comparison of the wave heights with calibration curves. A 0.2 M KCl supporting electrolyte is used and gelatin is added as a maximum suppressor.

Solutions and Chemicals Required

1. *Provided.*   1% Gelatin solution, oxygen-free nitrogen.
2. *To prepare.*

---

[3] If nickel is present, some will be deposited if the voltage exceeds 4 V.

(a) 0.2 M KCl. Dissolve about 150 g KCl in about 1 L water.

(b) Stock 0.0100 M solutions of $Pb(NO_3)_2$, $ZnCl_2$, and $Cd(NO_3)_2$. Weigh 0.828 g $Pb(NO_3)_2$, 0.341 g $ZnCl_2$, and 0.591 g $Cd(NO_3)_2$ and transfer each to separate 250-mL volumetric flasks. Dissolve and dilute to volume, and shake thoroughly.

Things to Do Before the Experiment

Prepare the KCl supporting electrolyte and the stock 0.100 M solutions of lead, zinc, and cadmium ions.

Procedure

1. *Determination of half-wave potentials of known metals.* Prepare 10^{-4} M solutions of lead, cadmium, and zinc in 0.2 M KCl by dilution of the 0.01 M solutions. Run a polarogram on each of these as follows. Fill the polarographic cell with the solution. Add three drops gelatin solution. The cell will have a saturated calomel electrode as the anode. Deaerate the solution by bubbling with oxygen-free nitrogen for 10–15 minutes. Keep an atmosphere of nitrogen above the solution while the polarogram is being run by passing nitrogen over it. Lower the dropping mercury electrode into the solution with the mercury reservoir raised. (The capillary must never be placed in a solution unless the reservoir is raised and mercury is flowing from it. Otherwise, minute dust particles or other material may clog the capillary.) With the potential of the DME set at -0.2 V versus SCE, adjust the pen of the polarographic recorder to the bottom of the chart with the displacement knob.[1] Select a sensitivity (microamperes per division) on the polarograph and manually change the potential (increase the DME potential to more negative values) to see approximately at what potential the wave begins, and then increase the potential to see if the entire wave will remain on scale. If it goes off scale, decrease the sensitivity until the plateau of the wave remains on scale. The current should deflect the pen at least half-scale of the recorder. Set the starting potential at least 0.2 V more positive than the beginning of the wave and scan the potential automatically to record the polarogram. Determine the half-wave potential of the polarographic wave. Repeat for the other two solutions. Run a polarogram on the supporting electrolyte (0.2 M KCl), starting at -0.2 V and scanning until the current goes off scale. At the completion of a polarogram, lower the cell from the capillary and rinse the capillary thoroughly with distilled water before lowering the reservoir.

2. *Analysis of the unknown.* Take a labeled 250-mL volumetric flask to your instructor and have your unknown solution added to it. Dilute to volume with 0.2 M KCl solution and shake thoroughly to mix. The unknown will contain sufficient metal to give a resulting solution of 10^{-5} to 10^{-4} M when diluted to volume. Run a polarogram on the unknown as above, adding the same amount of gelatin.

From the half-wave potential, determine what metal your unknown contains. Prepare a calibration curve for this metal by running polarograms on 1, 2, 4, and 10×10^{-5} M solutions in 0.2 M KCl. (The above polarogram may be used for the 10×10^{-5} M solution.) Your instructor will give directions for properly doing this. Plot microamperes of diffusion current against concentration of the metal ion. From this plot, determine the concentration of the metal in the solution analyzed.

Calculations

Calculate and report the milligrams of metal in your unknown.

---

[1] You may have manual polarograph, in which case current readings are taken directly at different applied potentials.

EXPERIMENT 23   Amperometric Titration of Lead with Dichromate

Principle

Lead is precipitated as lead chromate by titration with standard potassium dichromate solution. The end point is detected amperometrically with a dropping mercury electrode as the indicating electrode. The titration is followed at both 0.0 V versus SCE and -1.0 V versus SCE. Dichromate is reduced at 0.0 applied volt, but lead is not. Therefore, the current will remain near zero until beyond the end point, where it will increase in direct proportion to the amount of excess dichromate added. At -1.0 applied volt, both dichromate and lead are reduced. Therefore, the current at this potential will decrease to a small value up to the end point as the lead is removed, and then will increase in proportion to the amount of excess to give a V-shaped titration curve.

Equations

$$2Pb^{2+} + Cr_2O_7^{2-} + 2OH^- \rightarrow \underline{2PbCrO_4} + H_2O$$

Solutions and Chemicals Required

1. *Provided.*   1% Gelatin solution, oxygen-free nitrogen, 2.5 M KNO_3–3% agar salt bridge, low-resistance saturated calomel electrode (prepared as described in Chapter 11 and Figure 11.3 with a saturated KCl–3% agar salt bridge). Prepare the KNO_3 salt bridge as follows. Bend an 8-mm glass tube into a U shape long enough so that both ends can be immersed in two separate beakers. Fire polish the ends. Prepare the agar suspension as follows. Heat 100 mL water in a 250-mL Erlenmeyer flask just to boiling and then place on a steam bath. Add 3.5 g agar powder, stir, and continue to heat until a uniform suspension is formed. Then, add 25 g KNO_3 and stir until it goes into solution. Cool under tap water to about 30°C (do not allow it to gel; if it does, reheat on the steam bath) and, with the aid of a rubber tube and suction, fill the glass tube with the solution, making sure no air bubbles are entrapped in the tube. Hold the tube upside down so that the solution does not run out the ends of the tube until it cools and gels. The gel process may be speeded up by running cold water over the tube. The ends of the tube should be immersed in potassium nitrate solution when not being used.

2. *To prepare.*

(a) Buffered supporting electrolyte solution. Dissolve 10 g KNO_3 and 8.2 g sodium acetate in water, add 10 mL glacial acetic acid and dilute to 1 L. This is 0.1 M in KNO_3 and is buffered at about pH 4.2.

(b) 2.5 M KNO_3 Solution. Dissolve 25 g KNO_3 in 100 mL water.

(c) Standard 0.00500 M $K_2Cr_2O_7$ solution. Add 0.735 g $K_2Cr_2O_7$ to a 500-mL volumetric flask, dissolve in water, and dilute to volume.

Procedure

Take a 100-mL volumetric flask to your instructor and have your unknown added to it. Dilute to volume with water and shake to mix thoroughly. Your unknown will contain sufficient lead to result in a final concentration of 0.002–0.004 M. Use a 100-mL beaker for the cell. Use a stopper to cover the beaker and to support the electrodes. It should contain holes for a dropping mercury electrode, salt bridge, a buret, and a nitrogen inlet tube. The holes for the buret, DME, and nitrogen inlet tube should not form tight seals. One end of the KNO_3 salt bridge is immersed in the test solution. The other end is immersed in a beaker containing 2.5 M KNO_3 into which is also immersed the KCl salt bridge from the SCE. A polarograph is used as the potential source and as the current-measuring device. The DME and SCE are connected to this. (If a polarograph is not available, simply short the SCE and the DME together and measure the current with a galvanometer of sensitivity of about 200 microamperes full-scale connected in series. In this way, the titration is performed only with zero applied volt between the SCE and the DME. The solution need not be deaerated in this case.)

Add with a pipet 25.0 mL of your unknown solution to the cell and add 25.0 mL of the supporting electrolyte and a few drops of 1% gelatin solution. Deaerate by bubbling with nitrogen for 10–15 minutes. Pass nitrogen over the surface of the solution when current readings are being taken. Determine zero current on the polarograph by shorting the two leads together and setting the applied potential at zero volt. The pen will move to the position of zero current. Displace it to near the bottom of the chart paper. Connect the DME and the SCE to the polarograph and adjust the sensitivity so that the current at an applied voltage to the DME of -1.0 V versus SCE remains on scale near the top of the chart. Record the current readings at -1.0 V and at 0.0 V versus SCE. Add 1.00 mL titrant from a 10-mL buret. Stir the solution with a magnetic stirrer. Bubble the solution with nitrogen for 2 minutes to remove oxygen added with the titrant. Stop the stirring and the bubbling and record the current at the above applied potentials. Continue the titration in this manner, adding 1.00-mL increments of titrant, until the current rises to off-scale. There are two chromium atoms per mole of dichromate available to form two chromate ions. Since lead and chromate react on a 1:1 basis, the end point should occur at 5–10 mL. If time permits, repeat the titration.

Correct the current readings for the volume change by multiplying by $(V + v)/V$, where V is the initial volume (50 mL) and v is the volume of added titrant. Plot on separate pieces of graph paper the corrected current readings at 0.0 V and at -1.0 V against volume of titrant. The two plots illustrate the effect of the choice of applied potential on the shape of the titration curve. Extrapolate the straight-line portions before and after the end point and take the intersections of these as the end point.

Calculations

Calculate and report the milligrams lead in your unknown. Since a 25.0-mL aliquot out of 100.0 mL was titrated, the amount titrated is multiplied by 4.

EXPERIMENT 24   Coulometric Titration of Arsenic with Iodine

Principle

Arsenic(III) is titrated in slightly alkaline solution (pH 8) with electrogenerated iodine. A biamperometric end point is employed with two platinum indicating electrodes. No current flows throughout the titration since generated I_3^- is consumed by the As(III). Beyond the equivalence point, excess I_3^- is generated, and the biamperometric current increases in proportion to the I_3^- concentration.

The iodine titrant is generated by oxidation of iodide (KI electrolyte) at a constant current at a platinum anode. The cathode is isolated from the test solution via a salt bridge or fritted tube to maintain 100% generating efficiency. The amount of arsenic titrated is calculated from Faraday's law.

Equations

$$3I^- \rightarrow I_3^- + 2e^-$$

$$H_2AsO_3^- + I_3^- + H_2O \rightarrow HAsO_4^{2-} + 3I^- + 3H^+$$

Solution and Chemicals Required

1. *0.2 M KI.*   Weigh about 3 g KI and dissolve in 100 mL water. Add 0.01 g Na_2CO_3 to prevent air oxidation.

2. *0.2 M Phosphate buffer, pH 8.*   Dissolve about 3 g K_2HPO_4 in 100 mL H_2O and adjust the pH to 8.0 $\pm$ 0.2 with 3 M HCl. Use a pH meter.

The Titration Equipment

If a commercial constant-current source is not available, prepare a circuit as follows.[1] Use a 45-V B-battery as the current source. In series with the cell, connect a 400-ohm, 5% resistor, a 5000-ohm potentiometer, a 100.0-ohm, 0.1% precision resistor, and a switch. Connect a student potentiometer across the ends of the precision resistor to measure the *iR* drop across it.

Use a 100-mL beaker for the cell and the setup as shown in Figure 12.3. Two platinum-foil generating electrodes will be used (ca. 1 cm^2). Isolate the cathode by placing it in a glass tube fitted with a fritted-glass plug (e.g., E. H. Sargent S-30417).

The indicating electrodes may be platinum wire or platinum foil electrodes. A polarograph may be used to impress the potential and measure the current. A fourth hole should be drilled in the stopper holding the electrodes for introduction of the samples.

Procedure

Before beginning, take a 100-mL volumetric flask to your instructor and have your unknown added to it. Dilute to volume and shake for thorough mixing. The unknown will contain sufficient arsenic(III) for the final concentration to be 0.01–0.025 M. Hence, 1 mL will contain 10–25 μmol, or 20–50 μeq arsenic. Most commercial coulometric current sources will read out directly the microequivalents of sample titrated. The number of microequivalents is equal to the number of micromoles times the *n* value used in Faraday's law.

1. *Pretitration and current standardization.*   Add 25 mL of 0.2 M KI and 25 mL of the phosphate buffer to the cell. Fill the tube for the isolated cathode with the buffer solution. Connect the isolated cathode to the negative terminal of the current source and the other generating electrode to the positive terminal. Apply about 0.15 V between the indicating electrodes with the polarograph or other potential source. Stir the solution at a constant rate with a magnetic stirrer. Turn the generating current on and note that the indicating current (polarograph or ammeter) immediately rises, due to the iodine being generated. If the generating current is unknown, measure the *iR* drop across the standard resistor with the student potentiometer. Adjust the 5000-ohm potentiometer until the *iR* drop reads about 1 V. The current will then be about 10 mA (amp = *iR* drop/100.0 ohms).[2] Shut the current off. Add just enough of the unknown arsenic solution for the iodine color to disappear and the indicating current to go back to zero. Then, add about 0.2 mL more. Titrate until the galvanometer or polarograph pen starts to deflect, and shut off the generating current just when the indicating current crosses a certain line (for example, 5 μA). Add another 0.2 mL (approximate) and titrate again until the indicating current deflects to the same value. Repeat until you can stop the generation at the same point each time. The last time you do this should be done accurately, because you will add your sample after this.

---

[1]If a Malmstadt–Enke Heathkit electronic station is available, the universal power supply may be used in place of the battery (EUW-15). The standard reference potential source may be used to apply the potential between the indicating electrodes (EUW-16A).

[2]Calculations are simplified if the current is adjusted to exactly 9.65 mA $\pm$ 0.1% (the *iR* drop is 0.965 V). This is a multiple of the Faraday constant, and at this value, seconds $\times$ 0.1 = microequivalents of sample titrated.

2. Titration of your unknown. Add 1.000 mL of your unknown with a pipet to the pretitrated cell. Any commercial current source available will have a timer built into it that is actuated simultaneously with the current. Otherwise, use a stopwatch. With the solution being stirred at the same rate as in the pretitration, start the generating current and, simultaneously, the timer. If the current is unknown, measure the *iR* drop across the standard resistor. With a battery, the current may tend to drift for the first few seconds. Monitor the current during the titration, and, if it drifts, adjust it with the 5000-ohm potentiometer to within 0.1% of the initial reading. When the indicating current starts to deflect, get ready to shut off the generating current. Stop the generation and the timer when the indicating current deflects to the same value as in the pretitration.[3] Record the time of the titration to the nearest 0.1 second. Add another 1-mL aliquot of your unknown and titrate in the same manner. Repeat for a total of three or four titrations.

Calculations

Calculate, using Faraday's law, and report the number of milligrams of arsenic in your unknown for each portion titrated. Since a 1.000-mL aliquot out of 100.0 mL was titrated, the amount titrated is multiplied by 100. **Note:** If the arsenic concentration is known, this experiment can be used to accurately calibrate micropipets [see G. D. Christian, *Microchem. J.*, **9** (1965) 16].

SPECTROMETRIC MEASUREMENTS

EXPERIMENT 25   Spectrophotometric Determination of Iron

Principle

A complex of iron(II) is formed with 1,10-phenanthroline, $Fe(C_{12}H_8N_2)_3^{2+}$, and the absorbance of this colored solution is measured with a spectrophotometer. The spectrum is plotted to determine the absorption maximum. Hydroxylamine (as the hydrochloride salt to increase solubility) is added to reduce any Fe^{3+} to Fe^{2+} and to maintain it in that state.

Equations

$$4Fe^{3+} + 2NH_2OH \rightarrow 4Fe^{2+} + N_2O + 4H^+ + H_2O$$

$\frac{1}{3}Fe^{2+}$ +   →

$Fe^{2+}/3$

1,10-phenanthroline   tris(1,10-phenanthroline) iron(II)

---

[3]The procedure described is a variation of the **dead-stop titration.** A pretitration is used and so a blank correction is not necessary. If a pretitration is not employed, perform three blank titrations using the same volume of reagents as used in the individual sample titrations (each sample would then be titrated in separate solutions).

Instead of stopping the titration at the same value of the indicating current each time, you may stop it arbitrarily and record the current after incremental additions of titrant. The end point would then be determined by plotting the indicating current against seconds and extrapolating the straight line back to zero current (no volume correction is required).

Solutions and Chemicals Required

1. *Standard iron(II) solution.* Prepare a standard iron solution by weighing 0.0702 g ferrous ammonium sulfate, $Fe(NH_4)_2(SO_4)_2 \cdot 6H_2O$. Quantitatively transfer the weighed sample to a 1-L volumetric flask and add sufficient water to dissolve the salt. Add 2.5 mL conc. sulfuric acid, dilute exactly to the mark with distilled water, and mix thoroughly. This solution contains 10.0 mg iron per liter (10 ppm); if the amount weighed is different than that above, calculate the concentration.

2. *1,10-Phenanthroline solution.* Dissolve 100 mg 1,10-phenanthroline monohydrate in 100 mL water. Store in a plastic bottle.

3. *Hydroxylammonium chloride solution.* Dissolve 10 g hydroxylammonium chloride in 100 mL water.

4. *Sodium acetate solution.* Dissolve 10 g sodium acetate in 100 mL water.

Procedure

Into a series of 100-mL volumetric flasks, add with pipets 1.00, 2.00, 5.00, 10.00, and 25.00 mL of the standard iron solution. Into another 100-mL volumetric flask, place 50 mL distilled water for a blank. The unknown sample will be furnished in another 100-mL volumetric flask. To each of the flasks (including the unknown) add 1.0 mL of the hydroxylammonium chloride solution and 5.0 mL of the 1,10-phenanthroline solution. Buffer each solution by the addition of 8.0 mL of the sodium acetate solution to produce the red color of ferrous 1,10-phenanthroline. [The iron(II)–phenanthroline complex forms at pH 2 to 9. The sodium acetate neutralizes the acid present and adjusts the pH to a value at which the complex forms.] Allow at least 15 minutes after adding the reagents before making absorbance measurements so that the color of the complex can fully develop. Once developed, the color is stable for hours. Dilute each solution to exactly 100 mL. The standards will correspond to 0.1, 0.2, 0.5, 1, and 2.5 ppm iron, respectively.

Obtain the absorption spectrum of the 2.5-ppm solution by measuring the absorbance from about 400 to 700 nm (or the range of your instrument). Take readings at 25-nm intervals except near the vicinity of the absorption maximum, where you should take readings at 5- or 10-nm intervals. Follow your instructor's directions for the operation of your spectrophotometer. The blank solution should be used as the reference solution. Plot the absorbance against the wavelength and select the wavelength of the absorption maximum. From the molar concentration of the iron solution and the cell path length, calculate the molar absorptivity of the iron(II)–phenanthroline complex at the absorption maximum.

Prepare a calibration curve by measuring the absorbance of each of the standard solutions at the wavelength of maximum absorbance. Measure the unknown in the same way. Plot the absorbance of the standards against concentration in ppm. From this plot and the absorbance of the unknown, determine the final concentration of iron in your unknown solution. Report the number of micrograms of iron in your unknown along with the molar absorptivity and the spectrum of the iron(II)–phenanthroline complex.

EXPERIMENT 26 Determination of Nitrate Nitrogen in Water[1]

Principle

Nitrate, NO_3^-, is reacted with phenoldisulfonic acid to give a yellow color with an absorption maximum at 410 nm. Chloride interference is removed by precipitating the chloride.

---

[1]Nitrate may also be determined fluorometrically by reacting with fluorescein in conc. H_2SO_4 and measuring the fluorescence quenching of the fluorescein. See *J. Chem. Educ.*, **51** (1974) 682.

Nitrite, NO_2^-, levels in excess of 0.2 mg/L cause positive interference, but these concentrations rarely occur in surface waters.

Solutions and Chemicals Required

1. *Provided.* 1 *M* NaOH solution; conc. NH_3, phenoldisulfonic acid reagent prepared by dissolving 25 g phenol in 150 mL conc. H_2SO_4, adding 75 mL fuming H_2SO_4 (15% free SO_3), stirring well, and heating for 2 hours on a hot water bath.

2. *To prepare.*

(a) Silver sulfate solution. (This will not be required if synthetic chloride-free unknowns are provided.) Dissolve 4.4 g Ag_2SO_4, free from nitrate, in 1 L distilled water. One milliliter is equivalent to 1 mg chloride.

(b) EDTA solution. Prepare a paste of 50 g $Na_2EDTA \cdot 2H_2O$ in 20 mL water, add 60 mL conc. NH_3, and mix well to dissolve the paste.

(c) Stock nitrate solution, 100 mg/L N. Dissolve 0.722 g anhydrous KNO_3 and dilute to 1 L.

(d) Standard nitrate solution, 10 µg/mL N (44 µg/mL NO_3^-). Dilute 50 mL of the stock solution to 500 mL with distilled water.

Procedure

1. *Removal of chloride interference.* (This step may be eliminated if synthetic chloride-free nitrate unknowns are prepared in distilled water.) Small amounts of chloride cause negative interferences. If the chloride content is above 10 mg/L the chloride should be removed. Determine the approximate chloride concentration in the water using the methods described in Experiment 10 or 11. (Your instructor may have already made this determination for you.) Treat a 100-mL sample with an equivalent amount of silver sulfate solution and remove the precipitated silver chloride by centrifiguation or filtration. If necessary, coagulate the precipitate by heating the solution or by allowing to stand overnight away from strong light (only if sample is free from nitryfing organisms—see below).

2. *Determination of nitrate.* The sample should not exhibit appreciable color. To prevent any change in the nitrogen balance through biologic activity, natural waters should be analyzed promptly after sampling. They may, however, be stored near freezing by adding 0.8 mL H_2SO_4/L or 40 mg $HgCl_2$/L as preservatives. If the sample is acidified, it should be neutralized just before the analysis is started.

Neutralize the chloride-free prepared sample from above or a 100-mL fresh sample if already chloride free to about pH 7 with dilute NaOH. Transfer to a casserole and evaporate to dryness. Mix the residue with 2.0 mL phenoldisulfonic acid reagent, using a glass rod to dissolve the solids; heat on a hot water bath if necessary to aid dissolution. Dilute with 20 mL distilled water and then add 6–7 mL ammonia until maximum color is developed. If a flocculent hydroxide forms, dissolve by adding the EDTA reagent dropwise with stirring. (Alternatively, the sample may be filtered.) Transfer the clear solution to a 50-mL volumetric flask and dilute to volume with distilled water.

Prepare standards in the same manner, using the same volumes of reagents, by evaporating 10, 25, and 50 mL of the standard nitrate solution, respectively; these represent 0.10, 0.25, and 0.50 mg N, respectively. Omit the chloride precipitation step. Prepare a blank using the same volumes of reagents.

Read the absorbance of the solution at 410 nm, correct for the blank, prepare a calibration curve, and calculate the concentration of nitrate nitrogen in your sample in mg/L.

As little as 1 µg nitrate nitrogen can be detected, representing 0.01 mg/L in a 100-mL sample. The nitrate concentration in drinking water usually falls below 10 mg/L. If concentrations are high, measurements can be extended sixfold by measuring at 480 nm, or twofold by diluting prepared samples to 100 mL instead of 50 mL.

EXPERIMENT 27 Spectrophotometric Determination of Lead on Leaves Using Solvent Extraction

Principle

Lead on the surfaces of leaves is dissolved by shaking with nitric acid solution. The lead is extracted as the dithizone complex into methylene chloride at pH above 9. The intensity of the color of the complex is measured spectrophotometrically and compared to a calibration curve prepared similarly from lead standards to calculate the amount of lead. Cyanide and sulfite are added as masking agents to eliminate most interference from other metals.

Equation

$$1/2Pb^{2+} + \quad \begin{matrix} \phi{-}NHNH \\ \phi{-}N{=}N \end{matrix} C{=}S \rightleftharpoons \begin{matrix} \phi{-}NHN \\ \phi{-}N{=}N \end{matrix} C{-}S \overset{Pb^{2+}/2}{|} + H^+$$

green   red

Solutions and Chemicals Required

1. *Provided.* 1 M HNO_3, 0.1 M HNO_3, thymol blue indicator solution (0.1% in water), 2 M NH_3 solution, ammonia–cyanide–sulfite solution (350 mL conc. NH_3 solution, 30 mL 10% NaCN, and 1.5 g Na_2SO_3 diluted to 1 L; the pH is about 11).

2. *To prepare.*

(a) Stock 1000-ppm standard lead solution. Dissolve 1.60 g $Pb(NO_3)_2$ and dilute to 1 L in a volumetric flask.

(b) Standard 10-ppm working solution. On the day of the experiment, dilute 1 mL of the stock solution to 100 mL in a volumetric flask.

(c) Dithizone solution. Dissolve 7.5 mg dithizone in 300 mL methylene chloride. This should be prepared fresh on the day of the experiment.

Caution: *Do not discard* any of the solutions used in this experiment down the drain. The aqueous solutions will contain cyanide and should be added to a reservoir provided for collection which contains $FeSO_4$ [to convert CN^- to $Fe(CN)_6^{4-}$]. **No acid must ever be added to this reservoir!** All glassware should be rinsed with an alkaline solution and the washings added to the reservoir. The methylene chloride solutions should also be added to a reservoir provided.

Things to Do Before the Experiment

Collect leaf samples. These can be from trees near a road or highway and from some that are more isolated for comparison. Collect at least two large leaves from each tree and place each in a clean plastic bag and seal. The leaves selected should be reasonably free from dirt or other visible contamination.

Procedure

1. *Preparation of calibration curve.* This should be done at the time the samples are to be analyzed. The chelate formation and solvent extraction are to be performed in clean 6-oz vials with caps. To each of six labeled vials add with pipets 0 (blank), 2, 4, 6, or 8 mL of the 10-ppm lead standard and sufficient water to bring the volume to about 20 mL. Then add about 60 mL of the ammonia–cyanide–sulfite solution using a graduated cylinder and 25 mL of the CH_2Cl_2–dithizone solution using a pipet (*not* by mouth). Stopper or cap the vial and shake for about a minute. Using a pipet, withdraw most of the heavier methylene chloride layer and filter through dry filter paper (Whatman No. 40) in a dry Bausch and Lomb Spectronic 20 measuring tube or the equivalent, or centrifuge. (The samples should be prepared for measurement now so they can be measured when the standards are.)

Using one of the standards, measure the absorbance from 400 to 600 nm in 20-nm increments to determine the wavelength of maximum absorption. Using this wavelength, measure the absorbance of each standard, using the blank to zero the instrument. Plot absorbance against micrograms of lead taken to prepare a calibration curve.

2. *Determination of lead on leaves.* For each plastic bag containing a leaf sample, heat 20 mL of 0.1 M HNO_3 to about 70°C. Add 20 mL to each bag, close, and shake for about 2 minutes. Pour into clean 100-mL beakers. Add one drop thymol blue indicator solution to each, followed by dropwise addition of 2 M NH_3 until the indicator color change is complete (to blue) and add a couple extra drops. Then, add with a pipet 25 mL of the CH_2Cl_2–dithizone solution and proceed with the extraction measurement as with the standards.

Calculation

Blot each leaf dry, place on a sheet of paper, and trace the outline of the leaf. Cut out the leaf outline and weigh on an analytical balance to three figures. Cut out a 10-cm × 10-cm (100 cm^2) square from the same paper and weigh. Calculate the area of the leaf in cm^2.

From the measured absorbance of each sample and the calibration curve, calculate the micrograms of lead on the leaf and report the amount of lead in μg Pb/100 cm^2 leaf. Is there any correlation of lead content with proximity of the tree to a roadway?

EXPERIMENT 28 Spectrophotometric Determination of Inorganic Phosphorus in Human Serum or Urine

Principle

The inorganic phosphorus in a protein-free filtrate is reacted with ammonium molybdate [Mo(VI)] to form ammonium phosphomolybdate. This is reduced with a mild reducing agent to produce "molybdenum blue," a heteropoly molybdenum(V) species. Molybdates are not reduced under these conditions. The blue color of the solution is measured spectrophotometrically.

Equations

$$7PO_4^{3-} + 12 (NH_4)_6Mo_7O_{24} + 36 H_2O \rightarrow 7 (NH_4)_3PO_4 \cdot 12MoO_3 + 51 NH_4^+ + 72 OH^-$$
$$(NH_4)_3PO_4 \cdot 12MoO_4 + \text{mild reducing agent} \rightarrow Mo(V) \text{ species (blue)}$$

[Although normal ammonium molybdate $(NH_4)_2MoO_4$ can be crystallized, the common crystalline form is $(NH_4)_6Mo_7O_{24} \cdot 4H_2O$ or $3(NH_4)_2O \cdot 7MoO_3 \cdot 4H_2O$.]

Solutions and Chemicals Required

1. *Provided.* 5% (wt/vol) Trichloroacetic acid solution; 5 *M* H_2SO_4; aminonaphtholsulfonic acid reducing solution prepared as follows: Add 0.50 g 1,2,4 -aminonaphtholsulfonic acid and 5.0 mL sodium sulfite solution (20 g anhydrous Na_2SO_3/100 mL) to 195 mL sodium bisulfite solution (15 g $NaHSO_3$/100 mL) in a brown glass-stoppered bottle. Stopper and shake until the powder is dissolved. If solution is not complete, add 1-mL increments of sodium sulfite solution with continued shaking until solution is complete. Avoid excess sodium sulfite. Store in a refrigerator. The solution is stable for about one month.

2. *To prepare.*

(a) Stock phosphorus standard solution (100 mg/dL P): dissolve 0.439 g KH_2PO_4 in water and dilute to 100 mL in a volumetric flask.

(b) Working phosphorus standards. Add with a pipet 1 mL of the stock solution to a 100-mL volumetric flask and dilute to volume with 5% trichloroacetic acid (TCA). (**CAUTION:** TCA is very corrosive. Avoid contact with the skin. It should never be pipetted by mouth.) This contains 1 mg/dL phosphorus and will be used to prepare the serial standards. Transfer with a pipet 2 and 5 mL of this solution into 10-mL volumetric flasks and dilute to volume with 10% TCA. You now have standards of 0.2, 0.5, and 1 mg/dL P. These correspond to serum concentrations of 2, 5, and 10 mg/dL, respectively, in the procedure below, since the sample is diluted 1:10.

(c) Ammonium molybdate solution. Dissolve 2.5 g ammonium molybdate, $(NH_4)_6Mo_7O_{24} \cdot 4H_2O$, in 80 mL water and add 30 mL of 5 *M* H_2SO_4. The solution should be stable indefinitely. Discard if blanks show a blue color.

Procedure

1. *Serum.* Perform the analysis in duplicate. Place 9.50 mL of 5% TCA in a 12-mL centrifuge tube. Add 0.500 mL serum, mix well, and let stand for 5 minutes. Centrifuge at 1500 rpm until the supernatant is clear (ca. 5 minutes). If a centrifuge is not available, the sample should be filtered through dry Whatman No. 42 filter paper into a dry beaker. Since the filter paper may contain reducing impurities, the blank should be prepared using filtered 5% TCA.

Transfer 5.00 mL of the clear supernatant to a 15 × 150 mm test tube. Prepare a blank and standards by pipetting 5.00 mL of 5% TCA and of the 0.2, 0.5, and 1.0 mg/dL P standard solutions into four separate test tubes. To all the test tubes, add 1.00 mL of the molybdate reagent and mix well. Finally, add 0.40 mL of the aminonaphtholsulfonic acid reagent and mix well. Allow to stand for 5–10 minutes or longer and measure the absorbance for each solution in a cuvet at 690 nm, setting the zero absorbance with distilled water. Subtract any blank reading from all standard and sample readings.

Plot the net absorbance of the standards against concentration. From this plot and the net absorbance of the sample, determine the concentration of phosphorus in the protein-free filtrate. Multiply by 20 to obtain the concentration in the original serum sample. The normal range of phosphorus in serum is about 3.0–4.5 mg/dL for adults and 4.5–6.5 mg/dL for children.

2. *Urine.* Use the same procedure as for serum, but dilute the urine 1:10 with water before beginning. Treat 0.500 mL of this as above for serum. The concentration determined in the protein-free filtrate will have to be multiplied by 100 to obtain the original urine concentration. The level of phosphorus varies greatly with the dietary intake, but widely accepted values fall in the range of 0.4 to 1.3 g/day.

EXPERIMENT 29   Spectrophotometric Determination of Manganese and Chromium in Mixture

Principle

Manganese and chromium concentrations may be determined simultaneously by measurement of the absorbance of light at two wavelengths, after having oxidized the metals to $Cr_2O_7^{2-}$ and MnO_4^-. Beer's law has been shown to apply closely if the solutions are at least 0.5 M in H_2SO_4. $Cr_2O_7^{2-}$ has an absorption maximum at 440 nm and MnO_4^- has one at 545 nm. (A somewhat more intense maximum is at 525 nm, but there is less interference from $Cr_2O_7^{2-}$ at 545 nm.) Equations similar to Equations 13.16 and 13.17 are solved for the unknown concentrations from the measured absorbances at the two wavelengths. The four constants ($\epsilon b = k$) are determined by measurements of absorbance at the two wavelengths using pure solutions of known concentration; a calibration curve is prepared at each wavelength for both $Cr_2O_7^{2-}$ and MnO_4^- and the slopes of the curves (A versus C) are used to obtain an average k value.

Equations

The unknown contains Cr^{3+} and Mn^{2+}. The former is oxidized to $Cr_2O_7^{2-}$ by heating with peroxydisulfate (persulfate) in the presence of a silver catalyst:

$$2Cr^{3+} + 3S_2O_8^{2-} + 7H_2O \xrightarrow[\Delta]{Ag+} Cr_2O_7^{2-} + 6SO_4^{2-} + 14H^+$$

Mn^{2+} is oxidized in part by peroxydilsulfate, but also by periodate:

$$2Mn^{2+} + 5S_2O_8^{2-} + 8H_2O \xrightarrow[\Delta]{Ag+} 2\,MnO_4^- + 10SO_4^{2-} + 16H^+$$

$$2Mn^{2+} + 5IO_4^- + 3H_2O \longrightarrow 2\,MnO_4^- + 5IO_3^- + 6H^+$$

For the mixture,

$$A_{440} = k_{Cr,440}C_{Cr} + k_{Mn,440}C_{Mn}$$
$$A_{545} = k_{Cr,545}C_{Cr} + k_{Mn,545}C_{Mn}$$

The k values are determined from the slopes of the calibration curves of the pure solutions:

$$k_{Cr,440} = A_{440}/C_{Cr} \quad k_{Cr,545} = A_{545}/C_{Cr}$$
$$k_{Mn,440} = A_{440}/C_{Mn} \quad k_{Mn,545} = A_{545}/C_{Mn}$$

Solutions and Chemicals Required

1. *Provided.*   18 M H_2SO_4, $K_2S_2O_8$, KIO_4, $AgNO_3$.
2. *To prepare.*

(a) Standard 0.002 M $MnSO_4$ solution. Dry about 1 g $MnSO_4$ at 110°C for 1 hour, cool for 30 minutes, and weigh out about 0.08 g (to the nearest milligram). Transfer to a 250-mL volumetric flask, dissolve, and dilute to volume. Calculate the molarity of the solution and the concentration of Mn in mg/L (ppm).

(b) Standard 0.017 M $K_2Cr_2O_7$ solution. Use the solution prepared in Experiment 13 or prepare 250 mL as directed there (weigh to the nearest milligram). Calculate the molarity of the solution and the concentration of Cr in mg/L (remember there are two Cr atoms per molecule $K_2Cr_2O_7$).

(c) 0.1 M $AgNO_3$. Dissolve about 0.2 g $AgNO_3$ in about 12 mL water.

Things to Do Before the Experiment

Prepare the standard $MnSO_4$ and $K_2Cr_2O_7$ solutions. This will require drying $MnSO_4$ and $K_2Cr_2O_7$.

Procedure

1. *Calibration (determination of k values).*   **Note:** The absorbance of the calibration solutions and the unknown should be read at the same time. Therefore, get all solutions prepared before making any readings. They are all sufficiently stable that they could be allowed to set until another laboratory period but it is best not to.

(a) Manganese. Add with pipets aliquots of 10, 15, and 25 mL of the standard $MnSO_4$ solution into three different 250-mL Erlenmeyer flasks. Add distilled water to bring the volume in each flask to about 50 mL. To each flask add 10 mL conc. H_2SO_4 (CAREFULLY, using a graduated cylinder) and 0.5 g solid KIO_4 (potassium periodate or metaperiodate, depending on the manufacturer). Heat each to boiling for about 10 minutes, cool, transfer quantitatively to 250-mL volumetric flasks, and dilute to the mark with distilled water. Determine the absorbance of each solution at 440 and 545 nm, using 0.5 M H_2SO_4 as a blank solution. Permanganate solutions containing periodate are stable. The absorbance at 440 nm will be less than 0.1 and, hence, the spectrophotometric error (precision) will be large (see Figure 13.21). But this is acceptable, in fact, desirable, because the correction for manganese absorption at this wavelength is small; that is, a relative large error in determining a small correction results in only a small error.

(b) Chromium. Add 10-, 15-, and 25-mL aliquots of the standard $K_2Cr_2O_7$ solution to 250-mL volumetric flasks, add about 100 mL distilled water and 10 mL conc. H_2SO_4, mix thoroughly, and dilute to 250 mL with distilled water. Determine the absorbance of each solution at 440 and 545 nm, using 0.5 M H_2SO_4 as the blank solution. The absorbance in this case will be small (<0.1) at 545 nm.

(c) Determination of k values. Plot absorbance versus concentration in units of mg/L for each solution at each wavelength, and draw the best straight line through each set of data points. The lines should intercept at zero absorbance and zero concentration. The slopes of these lines are the coefficients ($k = A/C$) to be used in determining the concentrations of chromium and manganese in the unknown. These slopes relate absorbance and concentration for the instrumental parameters used. Therefore, one should use the same instrument, cuvets, cuvet position, and volumes of solutions for all determinations in this experiment.

2. *Analysis of unknown.* Obtain a mixture of Mn^{2+} and Cr^{3+} or $Cr_2O_7^{2-}$ unknown in a 250-mL volumetric flask and dilute to volume. Transfer with a pipet three 50-mL aliquots into 250-mL Erlenmeyer flasks. The procedure may be stopped up to this point. But once the peroxydilsulfate is added, the oxidation should be completed.

To each flask add 5 mL conc. H_2SO_4 (CARE!), then mix well. Add about 1 or 2 mL of 0.1 M $AgNO_3$ solution and 1.0 g solid potassium peroxydilsulfate ($K_2S_2O_8$). CARE! PEROXYDISULFATE IS A STRONG OXIDIZING AGENT THAT CAN REACT VIOLENTLY WITH REDUCING AGENTS! USE ONLY AS DIRECTED. Do not spill. Dissolve peroxydisulfate and heat the solution to boiling and boil gently for about 5 minutes. Cool the solution and then add 0.5 g KIO_4. Again heat to boiling for 5 minutes.

Cool each solution to room temperature, quantitatively transfer to 250-mL volumetric flasks, and dilute to volume. The solutions at this point (or before dilution) are stable and can be saved until another laboratory period if necessary. Also save the serial standards to calibrate the instrument at the same time.

From the unknown absorbances at the two wavelengths, calculate the parts per million of Cr and Mn in your unknown using Beer's law for the mixture and the determined constants. The calculated results will have the same units as used in determining the constants. Keep in mind the dilutions made. Report the results for each portion analyzed.

EXPERIMENT 30 Ultraviolet Spectrophotometric Determination of Aspirin, Phenacetin, and Caffeine in APC Tablets Using Solvent Extraction

Principle

APC tablets are a mixture of aspirin, phenacetin, and caffeine. Each of these substances has characteristic absorption in the ultraviolet region, with the principle maxima lying at 277 nm for aspirin, 275 nm for caffeine, and 250 nm for phenacetin. In the procedure, a powdered tablet is dissolved in methylene chloride and the aspirin is separated from the phenacetin and caffeine by extracting it into aqueous sodium bicarbonate solution. The separated aspirin is back-extracted into methylene chloride by acidifying the aqueous layer and is then measured spectrophotometrically at 277 nm. The phenacetin and caffeine that remain in the original methylene chloride layer are determined in mixture as described in Chapter 13 (Equations 13.16 and 13.17).

Equations

aspirin (A)
(acetylsalicylic acid)

phenacetin (P)
(acetophenetidin)

caffeine (C)
(Theine; 1,3,7-
trimethylxanthine)

Solutions and Chemicals Required

1. *Provided.* CH_2Cl_2, 4% (wt/vol) $NaHCO_3$ solution (chilled), conc. HCl, 1 M H_2SO_4.
2. *To prepare.*[1]

Standard solutions. Prepare individual standard solutions of about 100 mg/L, 20 mg/L, and 10 mg/L each for aspirin, phenacetin, and caffeine in methylene chloride as follows. Weigh about 25 mg (to the nearest 0.1 mg) of each, transfer to 250-mL volumetric flasks,

---

[1]Aspirin and caffeine are available from Aldrich Chemical Co., and phenacitin from ICN Pharmaceuticals, Inc.

dissolve, and dilute to volume with methylene chloride. Dilute 10 and 5 mL of this solution to 50 mL in 50-mL volumetric flasks to prepare the 20 and 10 mg/L solutions, respectively.

Procedure[2]

Weigh accurately and record the weight of one tablet. This should be equivalent to about 220 mg aspirin, 160 mg phenacetin, and 30 mg caffeine. To minimize required dilutions and save on solvents, cut the tablet into quarters and weigh out a one-quarter portion to be analyzed. Crush to a fine powder in a beaker. Add, with stirring, 20 mL methylene chloride; then transfer the mixture quantitatively to a 60-mL separatory funnel, rinsing *all* particles in with a little more methylene chloride. Extract the aspirin from the methylene chloride solution with two 10-mL portions of chilled 4% sodium bicarbonate to which has been added two drops hydrochloric acid, and then with one 5-mL portion of water. Wash the combined aqueous extracts with three 10-mL portions of methylene chloride and add the methylene chloride wash solutions to the original methylene chloride solution. Leave the aqueous extract in the separatory funnel. Filter the methylene chloride solution through paper previously wetted with methylene chloride (to remove traces of water) into a 50-mL volumetric flask and dilute to the mark with methylene chloride. Then dilute further a 1-mL aliquot of this solution to 50 mL with methylene chloride in a volumetric flask.

Acidify the bicarbonate solution (aqueous extract), still in the separatory funnel, with 6 mL of 1 *M* sulfuric acid. This step should be performed without delay, to avoid hydrolysis of the aspirin. The acid must be added slowly in small portions. Mix well only after the most of the carbon dioxide evolution has ceased. The pH at this point should be 1 to 2 (pH test paper). Extract the acidified solution with eight separate 10-mL portions of methylene chloride and filter through a methylene chloride-wet paper into a 100-mL volumetric flask. Dilute to volume. Then, dilute further a 5-mL portion of this solution to 25 mL with methylene chloride in a volumetric flask.

Record absorbance versus wavelength curves for the standard solutions and unknown solutions between 200 and 300 nm. (This step may be deleted if you do not have a recording ultraviolet spectrophotometer.) Does the wavelength of 277 nm appear to be the most suitable wavelength for the determination of aspirin? Do the wavelengths of 250 and 275 nm appear to be the best wavelengths for the measurement of the absorbance for the mixture of phenacetin and caffeine? Explain.

Using the absorbances of the standard and the unknown aspirin solution at 277 nm, calculate the percent aspirin in the APC tablets and the number of milligrams of aspirin per tablet keeping in mind the dilutions.

To calculate the concentrations of phenacetin and caffeine, the absorbances of the phenacetin and caffeine standards and of the methylene chloride extract of the sample must all be read at both 250 and 275 nm. Using these absorbances, calculate the percent phenacetin and caffeine in the APC tablets and the milligrams of each per tablet. See Chapter 13 for the spectrophotometric determination of mixtures.

EXPERIMENT 31   Infrared Determination of a Mixture of Xylene Isomers

Principle

Meta- and *para*-xylene are determined in mixture using *ortho*-xylene as an internal standard, to compensate for variation in cell length between runs. The infrared spectrum of the un-

---

[2]Aspirin tends to decompose in solution, and analyses should be performed as soon as possible after preparing solutions.

known mixture is recorded and the relative height of peaks of the two compounds are compared with those of standard mixtures, using the baseline technique.

Solutions and Chemicals Required

1. *Provided.* Ortho-, meta-, and para-xylene.
2. *To prepare.* Meta-xylene/para-xylene standards. Prepare a series of standards (use available burets), all containing 30% (vol/vol) o-xylene as internal standard, by mixing the appropriate volumes of the three isomers to give 25, 35, and 45 vol-% of m-xylene. The corresponding concentrations of p-xylene will be 45, 35, and 25%, respectively.

Procedure

Consult your instructor on the proper operation of your instrument. Handle the infrared cell carefully, avoiding contact with water and the fingers. Fill the cell with pure m-xylene and obtain a spectrum on this from 2 to 15 μm, being sure to record the last peak just before 15 μm (692 cm^{-1}). Each time you run a sample, be sure to check 0% T by placing a card in the sample beam and adjust the pen to 0% T. Empty the cell, rinse and fill with p-xylene, and run a spectrum on this. Repeat for o-xylene. Run spectra on each of the standard mixtures. From the spectra of the pure substances, choose a peak of each isomer to measure. Using the baseline method (see Figure 13.11), measure P_0/P for the peak for each compound. Prepare a calibration curve of the ratio of log $(P_0/P)_{meta}$/log $(P_0/P)_{ortho}$ and of log $(P_0/P)_{para}$/log $(P_0/P)_{ortho}$ for the meta and para isomers, respectively.

Obtain an unknown mixture of meta and para isomers from your instructor. Prepare a mixture of this with o-xylene by adding 70 parts of the unknown to 30 parts o-xylene. Run the spectrum on this mixture and, using the baseline method and the same peaks as before, measure P_0/P for the three compounds and calculate log (P_0/P)/log $(P_0/P)_{ortho}$ for the two unknown isomers. Compare with the calibration curve to determine the percent concentrations of the meta and para isomers. Remember to divide by 0.7 to convert to initial concentrations.

EXPERIMENT 32 Fluorometric Determination of Riboflavin (Vitamin B$_2$)

Principle

Riboflavin is strongly fluorescent in 5% acetic acid solution. The excitation and fluorescence spectra are obtained to determine the wavelengths of excitation and emission to use, and an unknown is determined by comparison to standards.

Chemicals and Solutions Required

Riboflavin standards. Prepare a 100-ppm riboflavin stock solution by accurately weighing about 100 mg riboflavin, transferring to a 1-L volumetric flask, and diluting to volume with 5% (vol/vol) acetic acid. This should be stored in a cool, dark place. On the day of the experiment, dilute an aliquot of this 1:10 to obtain a 10-ppm working standard solution. Dilute aliquots of this with 5% acetic acid to prepare standards of 0 (blank), 0.2, 0.4, 0.6, 0.8, and 1.0 ppm riboflavin.

Procedure

Record the excitation and emission spectra of the 0.6-ppm solution to determine the best excitation wavelength and best detection wavelength. If a filter instrument rather than a recording spectrofluorometer is used, take readings with different arrangements of the filters to give the maximum reading. Using these wavelengths, set the instrument gain to give a reading of 100% with the 1-ppm solution. Read the fluorescence of the other standards and prepare a calibration curve. Obtain your unknown in a 50-mL volumetric flask and dilute to volume with 5% acetic acid. Read the fluorescence of this, and, from the calibration curve, calculate the micrograms riboflavin in your unknown.

EXPERIMENT 33   Determination of Calcium by Atomic Absorption Spectrophotometry

Principle

The effects of instrumental parameters and of phosphate and aluminum on calcium absorption are studied [see, for example, W. Hoskins et al., *J. Chem. Educ.*, **54** (1977) 128]. Calcium in an unknown synthetic or serum sample is determined by comparing the absorbance with that of standards.

Solutions and Chemicals Required

500 ppm Ca, 4% $SrCl_2$, 2000 ppm NaCl, 100 ppm phosphate, ethanol.

1. *Provided.*   Ethanol, 4%, $SrCl_2$ solution (wt/vol), stock solution of 140 meq/L Na and 4.1 meq/L K (see footnote 1).

2. *To prepare.*

(a) 500 ppm Ca stock solution. Dissolve 1.834 g (accurately weighed) $CaCl_2 \cdot 2H_2O$ in water and dilute to 1 L in a volumetric flask. Dilute this 1:10 to prepare a 50-ppm stock solution. Use this to prepare the solutions required below. (Commercial 1000-ppm Ca^{2+} solutions may be used.)

(b) 2000-ppm Na stock solution. Dissolve 0.51 g NaCl in 100 mL water.

(c) 100-ppm phosphate. Dissolve 0.15 g Na_2HPO_4 in 1 L water.

(d) 100-ppm Al stock solution. Dissolve 0.18 g $Al_2(SO_4)_3 \cdot K_2SO_4 \cdot 24H_2O$ in 100 mL water. ($AlCl_3$ may be used, but take care when adding to water.)

Study of Instrumental Parameters

Follow your instructor's directions for the operation of the instrument. If you have a single-beam instrument, the hollow-cathode lamp should be allowed to warm up for 30 minutes before the experiment. A few minutes should be adequate with a double-beam instrument. An air–acetylene flame should be used with a premix burner.

1. *Burner height.*   Adjust the fuel and support gas pressures until the flame is near stoichiometric (just a slight yellow color to the flame). Then, turn up the fuel pressure to impart a strong yellow glow to the flame (fuel rich). The yellow glow is due to unburned carbon particles in the rich flame. In a lean flame, an excess of oxidant is present and the flame appears blue. Prepare and aspirate a 5-ppm calcium solution and note its absorbance at 422.67 nm. Adjust the wavelength setting to obtain maximum absorbance. The monochromator is now set exactly at the calcium line. With the burner height adjusting knob, raise the burner so that the light beam just passes over the tip of it (base of the flame). Use distilled water to zero the instrument, and then measure the absorbance of the 5-ppm calcium solution. Lower the burner in increments (six to eight steps) and record the absorbance at each height.

Plot the absorbance against heights of observation in the flame and select the optimum height.

2. *Fuel/air ratio.*   Hold the air pressure constant and adjust the fuel pressure in increments from a very fuel-rich to a lean flame. Record the absorbance of 5 ppm Ca at each increment.

Select the optimum fuel pressure and vary the air pressure in a similar manner. Plot absorbance against gas pressure for both the fuel and the air, noting the pressure setting of the one held constant. Select the optimum fuel and air settings. Is this a rich, stoichiometric, or lean flame?

Interference Studies

1. *Effect of phosphate.* Prepare a solution containing 5 ppm Ca and 10 ppm phosphate. Record the absorbance of this solution, using the optimum conditions determined above, and compare to that of 5 ppm Ca. Explain the results.

Prepare a solution containing 5 ppm Ca, 10 ppm phosphate, and 1% $SrCl_2$. Prepare also a solution of 5 ppm Ca and 1% $SrCl_2$. Record the absorbance of the solutions. Compare the absorption of the first solution with that of the phosphate-containing solution above and with that of the 1% $SrCl_2$-containing solution. Explain.

2. *Effect of sodium.* Prepare a solution containing 5 ppm Ca and 1000 ppm Na. Record the absorbance and compare with that of 5 ppm Ca. Explain any difference.

3. *Effect of aluminum.* Prepare a solution containing 5 ppm Ca and 10 ppm Al. Record the absorbance and compare with that of 5 ppm Ca by itself. Suggest a possible reason for the results.

Organic Solvent Effect

Prepare a solution of 5 ppm Ca in 50% ethanol. Record the absorbance using 50% ethanol as the blank. (When an organic solvent is aspirated, a lean flame must be used in order to burn the solvent. With the blank solvent aspirating, increase the air pressure until a blue flame is achieved.) Compare with the absorbance of 5 ppm Ca in water.

Determination of Calcium in an Unknown

(The method of standard additions below may be used instead of the following procedures.)

1. *Synthetic unknown.* Obtain an unknown from your instructor and dilute to give a concentration of 5–15 ppm Ca. Prepare a series of calcium standards of 0, 2.5, 5, 7.5, 10, and 15 ppm from the 50-ppm stock solution. If the unknown contains phosphate, add $SrCl_2$ to standards and the unknown to give a final concentration of 1%. Record the absorbance (or % absorption and convert to absorbance) of these and prepare a calibration curve of absorbance versus concentration. Determine the concentration of the unknown in the usual manner.

2. *Serum.* Calcium in serum or an "artificial serum" as described in the footnote to Experiment 35 is determined by diluting 1:20 with 1% $SrCl_2$ solution. The normal calcium content of serum is about 100 ppm, and so that analyzed solution contains about 5 ppm Ca. Sodium and potassium equal to that in the sample are added to the standards.

Add 0.5 mL of the unknown serum or the "artificial serum" to a 10-mL volumetric flask and dilute to volume with 1% $SrCl_2$. (If the method of standard additions is to be used also for comparison, dilute 2.5 mL of unknown to 50 mL with 1% $SrCl_2$.) Prepare standards of 0, 3, 4, 5, 6, and 8 ppm Ca, each also containing 1% $SrCl_2$, 6.9 meq/L Na and 0.21 meq/L K.[1]

---

[1] A stock solution of 140 meq/L Na and 4.1 meq/L K (20 times the concentration in the standard solution) can be prepared by dissolving 8.1 g NaCl and 0.21 g KCl in 1 L water. This contains the normal levels of Na and K in serum and compensates for ionization interference due to these elements in the serum. Dilute this solution 1:20 in the standards.

Prepare a calibration curve from the absorbance of the standards and from this determine the concentration of calcium in the unknown.

Method of Standard Additions

This procedure may be used to analyze the unknowns, instead of the one above, and it illustrates the usefulness of the method of standard additions for compensating for matrix effects. Dilute your unknown as described above, using distilled water to give a concentration of about 5 ppm Ca (1:20 for serum, e.g., 2.5 mL diluted to 50 mL with 1% $SrCl_2$). Transfer with a pipet separate 10.0-mL aliquots of the diluted unknown to three separate clean and dry test tubes or flasks. Add to these 50.0, 100, and 150 μL, respectively, of the 500-ppm standard calcium solution (or 25.0, 50.0 and 75.0 μL of a 1000-ppm solution if available). This results in an increase in the calcium concentration in the diluted unknown of about 2.5, 5.0, 7.5 ppm, respectively, depending on the exact concentration of the standard, and brackets the unknown. The volume changes can be considered negligible. Use an appropriate syringe microliter pipet if available (e.g., a 50-μL Eppendorf pipet or Finn-pipette) or else a 0.1-mL graduated measuring pipet. For best accuracy, the pipet should be calibrated.

Zero the instrument with distilled water and aspirate the diluted unknown and the standard addition samples. The absorbance increases in the latter are due to the added calcium. Prepare a plot absorbance against added concentration of calcium (starting at zero added, that is, the sample). From the x-axis intercept of the plot, determine the concentration of calcium in the diluted unknown. Calculate the concentration in the original sample. How does this method account for phosphate interference? (See Reference 12 in Chapter 14 for a discussion of errors in calcium measurements in the presence of phosphate or aluminum using the method of standard additions.)

EXPERIMENT 34 Determination of Mercury in Laboratory Air or Water Using Flameless Atomic Absorption Spectrophotometry

Principle

Mercury in air is collected by bubbling through an acid potassium permanganate solution. This traps volatile elemental mercury (vapor from the air) by oxidizing it to Hg^{2+}. The excess permanganate is reduced with hydroxylamine, and the collected mercury (or mercury in a water sample) is then reduced to the element by stannous chloride. The finely divided elemental mercury has appreciable vapor pressure at room temperature, and by bubbling argon through the solution, mercury vapor is swept into a quartz-ended cell where its atomic absorption is measured using the 253.6-nm mercury line.

Equations

$$5Hg^0 + 2MnO_4^- + 16H^+ \rightarrow 5Hg^{2+} + 2\ Mn^{2+} + 8H_2O$$

$$6MnO_4^- + 5NH_2OH + 13H^+ \rightarrow 6Mn^{2+} + 5NO_3^- + 14H_2O$$

$$Hg^{2+} + SnCl_2 + 2Cl^- \rightarrow Hg^0 + SnCl_4$$

Solutions and Chemicals Required

1. *Provided.* 6 *M* HNO_3, 9 *M* H_2SO_4, argon.
2. *To prepare.*

(a) 10% $SnCl_2$ solution. Weigh out 10 g $SnCl_2$ and add 1 mL 9 M H_2SO_4 and sufficient distilled, deionized water to dissolve it. Add water to bring the volume to 100 mL. This should be prepared on the day of the experiment.

(b) 10% $KMnO_4$ solution. Weigh out 1 g $KMnO_4$ and dissolve in 10 mL water.

(c) 1.5 M hydroxylamine solution. Weigh out 10 g $NH_2OH \cdot HCl$ and dissolve in 100 mL water.

(d) Stock mercury solution. Prepare a 100-ppm mercury solution by dissolving 0.135 g $HgCl_2$ in water and diluting to 1 L.

(e) Mercury working standard. Dilute the stock mercury solution 1:100 on the day of use. This represents a 1-ppm mercury solution.

Collection of the Air Sample[1]

This should be started first, and then the reagents and calibration curve can be prepared while the sample is being taken. By then, you will be prepared to rapidly measure the unknown. Add 5 mL 6 M HNO_3, 5 mL 9 M H_2SO_4, two drops 10% $KMnO_4$ solution, and 90 mL deionized water to a fritted-glass scrubber. The solution should retain the permanganate color throughout the sampling. If it becomes colorless, add another drop of permanganate solution. Connect the inlet of the gas scrubber to a rotameter and the outlet to a laboratory aspirator mounted on a faucet. Turn the faucet on and adjust the air flow rate to 10–15 L/hour (about 0.2 L/minute).[2] Record the time the sampling is started. Continue the sampling for 3 hours or more, occasionally checking the flow rate. From the flow rate and the sampling time, calculate the volume of air sampled.

Determination of Mercury

1. *Preparation of atomic absorption cell.* Arrange the measuring apparatus on the atomic absorption spectrophotometer as directed by your instructor. The absorption cell will be wrapped with nichrome wire and asbestos to warm it. The wire is connected to a Variac. By bubbling argon through deionized water in the sampling vessel and into the cell, determine the Variac setting that will prevent condensation of appreciable water vapor in the absorption cell. This will be about 70°C (Variac setting approximately 15 depending on the amount of nichrome wire). The argon flow rate should be constant and just sufficient to give smooth aspiration. A fritted-glass bubbler should be used for most efficient operation. Adjust

---

[1]Choose a laboratory in which mercury is used, such as an electroanalytical chemistry laboratory. This will ensure sufficient mercury to measure. The recommended daily exposure to airborne mercury is 0.05 mg/m^3 or 0.05 μg/L; for organic mercury compounds, it is one-fifth this limit. A 50-L air sample may contain approximately 0.5 to 5 μg mercury, or it may be outside of this range. The limit of detection of mercury by this method is about 0.1 μg or, under favorable conditions, 0.01 μg. If insufficient mercury is collected in 4 hours to measure, then sampling overnight may be tried.

[2]If a rotameter is not available for measuring these rates of flow, a simply constructed eudiometer can be used to approximate the flow rate. This consists of an inverted 50-mL buret, the tip of which is connected to the inlet of the scrubber. The other end of the buret is immersed in water contained in a paper-chromatographic developing tube. The water level in the buret is adjusted to the zero-milliliter reading. The water aspirator is turned on and as the air contained in the buret is drawn into the scrubber, the buret is lowered in the chromatographic tube to keep the water level in the buret constant at zero-milliliter and prevent any backpressure. The time is noted for the 50-mL air sample to be drawn into the scrubber, and from this the flow rate is calculated. Disconnect the buret during the sampling process and then check the flow rate again at the end of the sampling period.

it to give an even bubbling rate and then keep the gas tank regulator setting constant. Turn the flow off only at the tank and not at the regulator diaphragm.

2. *Preparation of calibration curve.* This should be prepared shortly before the sample is to be run. Add 90 mL deionized water, 5 mL 6 M HNO_3 and two drops 10% $KMnO_4$ to the flask. Add successively in separate measurements 0.5-, 1-, 2-, and 5-mL portions of the 1-ppm mercury solution and swirl 15 seconds. The purpose of the permanganate is to oxidize and dissolve inorganic and organic mercury compounds. The solution should remain colored due to the permanganate. If it does not, add one more drop $KMnO_4$ solution. Then, add 5 mL 9 M H_2SO_4 and 5 mL 1.5 M $NH_2OH \cdot HCl$, swirl to mix, and wait 30 seconds for the solution to clear. If the permanganate color remains, add another portion of hydroxylamine hydrochloride. Add 5 mL 10% $SnCl_2$ and immediately stopper the flask with the aspiration stopper. Wait 10 seconds and then turn on the argon tank to aspirate the mercury into the cell. Record the absorbance on a strip chart recorder and note the maximum absorbance on the instrument meter. Run at least two blanks in the same manner, omitting the addition of mercury. The magnitude of the blank will generally govern the detection limit of the procedure. Extreme care must be taken to minimize reagent and glassware contamination. The cell should be flushed with argon between runs until the absorbance drops back to near the original baseline. Correct for the blank absorbance and plot a calibration curve of net maximum absorbance versus micrograms of mercury.

3. *Analysis of air sample.* Transfer the contents of the sampling scrubber to the sample flask. Add 5 mL 1.5 M $NH_2OH \cdot HCl$, swirl, and wait 30 seconds for the solution to clear. Add 5 mL 10% $SnCl_2$, stopper, wait 10 seconds, and then aspirate with argon. Record the absorbance as above. Correct for the reagent blank reading obtained above, and from the calibration curve determine the micrograms mercury in the analyzed sample. From the volume of sample collected, calculate the concentration of mercury in the laboratory atmosphere in μg/L.

4. *Analysis of water sample.* Tap water, river water, or other water sources may be analyzed. Alternatively, unknowns may be prepared by adding mercuric chloride to deionized water samples. Tap water should contain in the neighborhood of 1 ppb or less mercury, and so a 90-mL sample will contain 0.1 μg or less, near the detection limit. Analyze triplicate 90-mL water samples in the same manner that standards were run. Correct for the reagent blank absorbance, and from the calibration graph determine the quantity and concentration of mercury in the sample. Report the mean concentration from the three trials in parts per billion.

EXPERIMENT 35 Flame Emission Spectrometric Determination of Sodium

Principle

The intensity of sodium emission in a flame at 589.0 nm is compared with that of standards. If an internal standard instrument is available, the ratio of sodium to lithium emission is measured.

Solutions and Chemicals Required

1. *Stock standard NaCl solution (1000 ppm Na).* Dry about 1 g NaCl at 120°C for 1 hour and cool for 30 minutes. Weigh and dissolve 0.254 g NaCl in water and dilute to 1 L. Care must be taken to avoid sodium contamination, especially from the water and glassware. A blank must be run to correct for sodium in the water.

2. *LiNO₃ Internal standard solution (1000 ppm Li).*   (Lithium is not required if a direct-intensity instrument, rather than an internal standard instrument, is used.) Dissolve 9.94 g $LiNO_3$ in water and dilute to 1 L.

3. *Working standard solutions.*[1]

(a) Direct-intensity instrument. Prepare standards of 0, 10, 20, 30, and 40 ppm Na by diluting 0, 5, 10, 15, 20, and 25 mL of the stock NaCl solution to 50 mL. (It may be better with some instruments to prepare solutions five times more dilute than this by adding 0, 1, 2, 3, 4, and 5 mL of solution to the flasks. This may result in a more linear calibration curve. Follow your instructor's directions.)

(b) Internal standard instrument. Prepare the same solutions as above for the direct-reading instrument, but add 5 mL of the stock lithium solution to each flask. (This results in 100 ppm Li in each solution. The recommended concentration may vary from one manufacturer to another, and so your instructor may direct you to add a different amount.)

Things to Do Before the Experiment

Dry the NaCl at 120°C for 1 hour and cool in a desiccator.

Procedure

Have your instructor add your unknown to a 100-mL volumetric flask.

1. *Direct-reading instrument.*   Dilute your unknown to volume with water. Follow your instructor's directions for the operation of the instrument. Several atomic absorption instruments can be used for measuring emission. Set the zero reading while aspirating distilled water (the blank). Aspirate each standard and the unknown and record their emission intensity readings. With some instruments, the 100% reading is set with the most concentrated standard.

Plot the emission readings for the standards against concentration and determine the concentration in the unknown solution from the calibration curve. From this, calculate the micrograms of solution in your unknown if it is water, and ppm or meq/L if it is serum (see footnote 1).

2. *Internal standard instrument.*   Add the same amount of lithium solution to your unknown as was added to your standard, and dilute to volume with water. Follow your instructor's directions for the operation of the instrument. The lithium emission line is 670.8 nm. Prepare a calibration curve as above for the direct-reading instrument, but record the ratio of the Na/Li emission intensities. Determine the concentration of sodium in the unknown solution and report micrograms of sodium in the unknown if it is water, and ppm or meq/L if it is serum (see footnote 1).

---

[1]The unknown may be simply a sodium chloride solution, or it may be serum. If serum is analyzed, then the standards should be prepared over a narrower concentration range to better bracket the unknown. Sodium in serum (approximately 140 meq/L, or 3200 ppm Na) may be determined by simple 1:100 dilution (e.g., 0.1 mL diluted to 10 mL) or 1:500 if required by the instrument. An alternative unknown is an "artificial serum" prepared as follows:

NaCl	8.072 g
KCl	0.21 g
KH_2PO_4	0.18 g
$CaCl_2 \cdot 2H_2O$	0.37 g
$MgSO_4 \cdot 7H_2O$	0.25 g

This contains 138.1 meq/L Na, which can be varied from unknown to unknown.

CHROMATOGRAPHY

EXPERIMENT 36   Anion Ion Exchange Chromatography Separation of Cobalt and Nickel Followed by EDTA Titration of the Metals Using Back-Titration

Principle

Cobalt and nickel are separated on a strong-base anion exchange column (chloride form) by eluting with 9 M HCl and 3 M HCl, respectively. In 9 M HCl, the nickel does not form an anion chloro complex while the cobalt does; hence, the former will elute. In 3 M HCl, the anionic cobalt chloro complex is dissociated to form cobalt cation, which elutes. Following separation, the metals are titrated indirectly with standard EDTA for quantitation; excess EDTA is added and the unreacted EDTA is back-titrated with a standard zinc solution in slightly acid solution, using xylenol orange indicator.

Equations

Separation:

$$Ni^{2+} + Cl^- \rightleftharpoons NiCl^+$$
$$Co^{2+} + 4Cl^- \rightleftharpoons CoCl_4^{2-}$$

Titration:

$$Co^{2+} + H_2Y^{2-} \rightarrow CoY^{2-} + 2H^+$$
$$Ni^{2+} + H_2Y^{2-} \rightarrow NiY^{2-} + 2H^+$$
$$H_2Y^{2-} + Zn^{2+} \rightarrow ZnY^{2-} + 2H^+ \quad \text{(back-titration)}$$

End point:

$$H_4In + Zn^{2+} \rightleftharpoons ZnIn^{2-} + 4H^+$$
$$\text{yellow-green} \qquad \text{red-violet}$$

Solutions and Chemicals Required

1. *Provided.*   3 M NaOH, 9 M HCl, 3 M HCl, 0.5% (wt/vol) xylenol orange indicator in 10% ethanol (0.5 g dissolved in 10 mL ethanol and diluted to 100 mL with water-must be fresh), zinc granules, 0.2% phenolphthalein in 90% ethanol, hexamine (hexamethylenetetramine), Dowex 1-X8 anion exchange resin (chloride form).

2. *To prepare.*

(a) Standard 0.01 M EDTA solution. Prepare from $Na_2H_2Y \cdot 2H_2O$ dried at 80°C for 2 hours as described in Experiment 12 by accurately weighing about 1.9 g (to the nearest milligram), dissolving in distilled deionized water, and diluting to 500.0 mL in a volumetric flask. Calculate the molarity.

(b) Standard 0.01 M zinc solution. Weigh out accurately about 0.33 g pure zinc granules (to the nearest 0.1 mg) and transfer to a 400-mL beaker. Do not use zinc dust. For highest accuracy, the zinc granules should be treated with 2 M HCL to remove any zinc oxide coating. Decant the acid and wash the zinc repeatedly with water. Then, wash several times with ethanol and finally with ether in the fume hood. Dry the granules before weighing.

Dissolve with the minimum amount of HCl required. Heat on a steam bath to aid dissolution. Cover with a watch glass during dissolution. After the zinc is in solution, wash the droplets of water on the watch glass with distilled water into the beaker, wash down the sides of the beaker and quantitatively transfer the solution to a 500-mL volumetric flask. Dilute to volume with distilled water. Calculate the molarity of the solution.

Things to Do Before the Experiment

1. *Prepare the standard 0.01 M EDTA solution.* This will require drying and cooling the $Na_2H_2Y \cdot 2H_2O$.

2. *Prepare the standard 0.01 M zinc solution.*

Procedure

1. *Preparation of the ion exchange column.* Prepare a column using a 50-mL buret with glass wool in the bottom to hold the resin. Add a slurry of Dowex-1-X8 anion exchange resin (in the chloride form) in 9 *M* HCl to the buret until the height of the resin column is about 15–20 cm. Do not allow the liquid level to drop below the resin level. Keep about 2 mL liquid above the resin.

Wash the column with two 10-mL portions of 9 *M* HCl using a flow rate of 2 or 3 mL per minute. The column will darken somewhat when treated with HCl, but it will return to normal color if washed with water. Leave 2–3 cm of HCl above the resin level.

2. *Separation of the unknown mixture.* Obtain an unknown from your instructor in a 50-mL volumetric flask. Dilute to volume with 9 *M* HCl. The unknown will contain 5 mmol or less each of nickel(II) and cobalt(II) in 9 *M* HCl.

Add with a pipet 2 mL of the unknown solution onto the column. Elute the nickel with about 75 mL 9 *M* HCl added in 15-mL portions, using a flow rate of 1–2 mL per minute. Collect in a 250-mL Erlenmeyer flask. The pale yellow-green $NiCl^+$ complex will flow through the column and may darken the resin. The blue cobalt band (appears green due to yellow resin) will move partway down the column. While the sample is being eluted, it is advisable to perform a practice titration as described below.

After all the nickel is eluted and before the cobalt band reaches the end of the column, stop the flow and replace the collection flask with a clean one. Elute the cobalt with about five 10-mL portions of 3 *M* HCl, at a flow rate of 1 mL per minute. As the HCl is diluted on the column, the $CoCl_4^{2-}$ complex should eventually dissociate to a pink color. After the cobalt is seen to be eluted, stop the flow and proceed to titrate the separated metals. (Final elution with water may be used to assure complete removal of the cobalt.)

3. *Titration of the nickel and cobalt.* Evaporate the collected samples to near dryness on a hot plate. Cool and dilute with 50 mL distilled deionized water. Before titrating your unknown samples, and while they are evaporating, perform a practice titration or two using a prepared cobalt or nickel solution. Titrate the nickel and cobalt solutions indirectly as follows. Neutralize each solution to a phenolphthalein end point with 3 *M* NaOH, avoiding excess NaOH. Add 6 *M* HCl dropwise to remove red base color. Add 25.00 mL standard 0.01 *M* EDTA to the flasks, using a pipet, and add five drops 9 *M* HCl, 1 g hexamine, and four drops xylenol orange indicator solution. The hexamine buffers the solution at pH 5–6. If the solution is red violet, warm and add another 10.00 mL EDTA. Back-titrate with standard 0.01 *M* zinc solution until the indicator changes from yellow green to red violet.

Calculations

From the millimoles EDTA taken and the number of millimoles excess EDTA found in the back-titration, calculate and report the number of millimoles nickel and cobalt in your unknown. Since a 2.000-mL aliquot out of 50.00 mL was taken for analysis, the number of millimoles of each metal in the unknown is equal to 25 times the number titrated.

EXPERIMENT 37 Determination of Sulfate in Urine by Precipitation Titration Following Ion Exchange Separation[1]

Principle

Sulfate ion in urine is reacted with excess standard barium perchlorate to precipitate barium sulfate. The excess barium is back-titrated at about pH 2.5 with a standard sodium sulfate solution. Nitchromazo adsorption indicator is used. (The sulfate can be titrated directly with barium perchlorate, adding bromphenol blue to improve the color change, but the color change at the end point is slow.) To eliminate interfering cations, the sample is first passed through a cation exchange column in the acid form. The cations are replaced by an equivalent amount of hydrogen ions, while the sulfate ions are unretained. Titrations are carried out in water–propanol mixtures to give a more desirable precipitate for adsorption of the indicator and to prevent complexation of the indicator with barium ion.

Equations

$$SO_4^{2-} + Ba^{2+} \rightarrow BaSO_4 \quad \text{(sample reaction)}$$
$$Ba^{2+} + SO_4^{2-} \rightarrow BaSO_4 \quad \text{(titration reaction)}$$

Solutions and Chemicals Required

1. *Provided.* Nitchromazo (nitrosulfonazo III) indicator solution (0.1% in water), $Ba(ClO_4)_2$, Na_2SO_4, 0.1 M HCl, glacial acetic acid, 2-propanol, 50–100 mesh strong-acid cation exchange resin in acid form (Dowex 50 or 50w, or Amberlite IR-120).

2. *To prepare.*

(a) Standard 0.01 M $Ba(ClO_4)_2$ solution. Dry solid $Ba(ClO_4)_2$ overnight in a desiccator. Weigh to three significant figures about 0.34 g, dissolve in distilled water, and make up to 100 mL. Calculate the molarity.

(b) Standard 0.01 M Na_2SO_4 solution. Dry anhydrous Na_2SO_4 overnight at 120°C. Cool in a desiccator and weigh to three significant figures about 0.14 g, dissolve in distilled water, and make up to 100 mL. Calculate the molarity.

(c) Ion exchange column. Prepare a 10-cm column in a 1.5-cm-diameter tube (e.g., a 50-mL buret). Place glass wool in the bottom of the column to hold the resin. Add resin to water in a beaker and stir (this is to allow the resin to swell before adding to the column). Pour the slurry into the column and allow to settle to a height of about 10 cm. Wash the resin by passing distilled water through the column until the washings are nearly neutral (can be checked with a pH meter or with pH test paper). **Do not allow the liquid level to go below the top of the resin,** but allow the liquid to drain to just above the resin. The column will serve for three or four runs, and then the resins should be regenerated or placed in a used resin container for regeneration (can be accomplished in a large column by passing three or four volumes of 3 M hydrochloric acid through the column, followed by washing with water).

---

[1]This procedure measures inorganic sulfate in urine. Urine also contains acid-labile ester sulfates. Total (inorganic and organic) sulfate can be determined by hydrolyzing the esters before the ion exchange separations. This is accomplished by diluting 3 mL urine to 10 mL, adding 10 mL 2 M HCl, and evaporating to dryness in a boiling water bath. Add 10 mL water to the residue and again evaporate to dryness. Finally, add 10 mL water and treat as with inorganic sulfate (the ion exchange separation will effectively decolorize the mixture and give a clear filtrate). Total urine sulfate varies with the protein intake.

Things to Do Before the Experiment

Place the barium perchlorate (anhydrous) in a desiccator and the sodium sulfate in the oven at least a day before the experiment.

Procedure

1. *Ion exchange separation.* Urine samples should be fresh if possible. Add 1 mL glacial acetic acid per 50 mL urine to prevent precipitation of calcium salts with coprecipitation of sulfate on standing. This will stabilize samples for two or three days. Add slowly with a pipet 3 mL urine to the column. Place a 250-mL Erlenmeyer flask under the column and allow the solution to flow through column at a rate of about 2 mL/min. Stop the flow just before the liquid level reaches the top of the resin. Then, wash the sample through the column with four 5-mL portions of water. Check the completeness of elution by testing the pH of the last few drops. If the pH is less than 5, wash with another 5-mL portion of water. The pH of the eluted sample should be 2.0–3.0 (due to the replacement of urinary cations by protons). If it is not, add HCl or NaOH to bring it within this range.

Elute a second aliquot of the urine sample for a duplicate titration.

2. *Titration.* Add to the collection flask 20 mL 2-propanol and three drops nitchromazo indicator solution. Add with a pipet 5 mL standard 0.01 M barium perchlorate solution. Swirl for 1–2 minutes for complete precipitation. The indicator should be blue. If it is purple, insufficient barium has been added to react with all the sulfate or else the pH is less than 2. Check the pH. If it is properly adjusted, add another 5-mL aliquot of the barium perchlorate solution. Using a 10-mL buret, titrate the excess barium ion with the 0.01 M sodium sulfate solution. The indicator will change from blue to purple at the end point.[2]

Titrate the second sample aliquot and report the parts per million of sulfate in the urine.[3]

EXPERIMENT 38   Paper Chromatography Separation and Identification of Metal Ions

Principle

Zinc, cobalt, manganese, and nickel are separated by paper chromatography, by developing with a mixture of acetone, water, and hydrochloric acid. The spots are located by the color formed by reaction of the metals with a solution of a color-forming reagent. By comparison with the R_f values of knowns, the metals in an unknown mixture are identified.

Solutions and Chemicals Required

1. *Provided.* Metal salt solutions. Separate 0.3% solutions (wt/vol) in water of $Zn(NO_3)_2$, $Co(NO_3)_2$, $Ni(NO_3)_2$, $Mn(NO_3)_2$, and a mixture of all four of them (0.3% in each), with a few drops HCl added to prevent hydrolysis; conc. NH_3.

2. *To prepare.* (The locating reagents and wash solution can be prepared while the chromatogram is being developed.)

---

[2]It is helpful to prepare a comparison solution of the indicator at the end point by titrating an aqueous sodium sulfate solution (pH adjusted to 2.0–3.0).

[3]The amount of sulfate will vary widely from sample to sample depending on dietary intake. A typical range is 1000–10,000 ppm.

(a) Developing solvent. Mix 44 mL acetone, 3 mL water, and 4 mL conc. HCl.

(b) Locating reagents. Solution A: 50 mg sodium pentacyanoammine ferroate dissolved in 20 mL distilled water. Solution B: Saturated solution of rubeanic acid in ethanol (10 mg/10 mL ethanol).

(c) Wash solution. 2% acetic acid in water (5 mL glacial acetic acid plus 245 mL water).

Chromatographic Equipment

Glass tank, lid, filter paper (25 × 25 cm), plastic tongued clips, stainless steel clips, dip tray, stirring rod, large tray.

Procedure

Obtain an unknown mixture from your instructor. Place 50 mL of developing solvent in the bottom of the tank and replace the lid. Draw an "origin" line horizontally (lightly in pencil) approximately 3 cm from the edge of the paper. Locate where spots are to be applied by marking an X on the line for each spot and identify each with writing below the line (in pencil). Spots should be spaced at least 2 cm apart. Spot the 25 × 25-cm paper with each metal nitrate, the mixture of metal nitrates, and at least one unknown spot. Three drops of each solution are to be applied with the tip of glass stirring rod, allowing time for drying or drying with a heat lamp after each application. Never is the spot to exceed 8–10 mm in diameter. Where it is being spotted, the paper should not touch the table top. Extend the end of the paper beyond the edge of the desk.

Form the paper into a cylinder and secure with the tongued clips. Place the cylinder with the spotted end down in the tank, taking care not to let the paper touch the glass walls. The spots should not be immersed in the solvent. Close the tank with the lid.

Since no observations can be made while the chromatogram is running (the compounds are colorless in the small amounts employed), the time should be used to prepare the acetic acid solution and the locating reagents required to transform the separated metal ions into colored compounds when the chromatogram is finished. Analytical balances should be used to weigh the materials. Mix 20 mL of solution A with 1.5 mL of solution B. Shake a few minutes. If not then a dark purple, add more solution B and shake again. Filter. This reagent is unstable and must be made fresh daily and stored in cold, if it is not used within 2 hours.

After 2 hours, remove the chromatogram from the tank, mark the solvent front with a pencil, and open out the paper and dry it with a heat lamp. Be careful not to scorch the paper. Fume the dry chromatogram with vapors from concentrated ammonia solution until all smell of HCl is gone. Pour the locating reagent into the dip tray and pull the chromatogram—origin first—through the solution (catching all excess solution on the beaded edge of the tray). The fingers should not become wet. The whole paper should become dark purple. If this color change does not occur, expose the paper to NH_3 fumes again. Immerse the paper in the large, flat tray containing the 2% acetic acid and rock the tray slowly until the background color in the paper disappears completely. Deeply colored spots will remain, indicating the positions of the metal ions. (One change of acetic acid may be necessary to develop the colors fully.)

Dry the chromatogram by removing excess acetic acid by blotting with filter paper and then suspending it with stainless steel clips in front of the heat lamp (About 15 minutes—watch for scorching).

Compute and compare R_f values (make measurements from the centers of the spots) of each metal when run individually and in a mixture. Qualitatively identify the metal ions in your unknown.

EXPERIMENT 39   Thin-Layer Chromatography Separation of Amino Acids

Principle

The amino acids are separated on an Eastman Chromagram sheet, or the equivalent, using a choice of two developing solvents. The locating reagent is ninhydrin.

Solutions and Chemicals Required

Provided.

(a) Developing solvent: No. 1, butyl alcohol/acetic acid/water (80:20:20 vol/vol); No. 2, propyl alcohol/water (7:3 vol/vol).

(b) Locating reagent: Ninhydrin solution—0.3% ninhydrin (1,2,3-triketohydrindene, Eastman No. 2495) in butyl alcohol containing 3% glacial acetic acid.

Chromatographic Equipment

Eastman Chromagram sheet 6060 or 6061 (20 × 20 cm), sprayer (for application of locating reagent), developing apparatus (e.g., the Eastman Chromagram developing apparatus).

Procedure

Obtain an unknown mixture from your instructor. The mixtures to be separated will contain approximately 1 mg of each amino acid per milliliter 0.5 M alcoholic hydrochloric acid solution (see discussion below). Spot approximately 1 μL of the sample solution on the Chromagram sheet about 2 cm from the lower edge. (Activation of the sheet is not necessary for this separation.) The unknown should contain acids whose R_f values are sufficiently different with the solvent system used so that they will be readily separated. Your instructor will advise you which standard amino acid solutions to run with your unknown so that you can distinguish between two amino acids that might have an R_f value close to one in your mixture. Allow 15–20 minutes drying to ensure complete evaporation of the hydrochloric acid.

Develop the Chromagram sheet in the solvent of choice for a distance of 10 cm or for approximately 90 minutes. (See the list of R_f values in the table below and follow your instructor's directions for the solvent of choice for your unknown mixture.) Dry the developed Chromagram sheet and spray with the ninhydrin solution. Heat gently for several minutes until separated zones appear clearly visible.

Results

The table following lists the approximate R_f values obtained when the two solvent systems are applied to the separation of 13 different amino acids. From this table and the standard acids you run, determine what amino acids are present in your unknown.

Discussion

Due to the limited solubility of various amino acids, great care must be taken in the preparation of the sample before applying chromatography. The use of alcoholic 0.5 M HCl is suggested in the procedure. However, it may become necessary to add significant amounts of water to achieve solubility. When this is the case, it becomes extremely important to keep the spot area small and to allow enough time for complete evaporation of the spotting solvents before starting development.

When a greater degree of resolution is required, it is possible to improve results by using the two-dimensional technique of development. The spot containing the components to be separated is placed in the lower left-hand corner of a 20 × 20-cm sheet, 2 cm from each that of a vitamin C standard (*l*-ascorbic acid), the presence or absence of vitamin C is ascertained. Peak area measurements are used for quantitative determination of the concentration of vitamin C in those juices in which it is found.

Solutions and Reagents Required

1. *Provided.* Mobile-phase solution (1.36 g KH_2PO_4/L distilled deionized water), *l*-ascorbic acid.

2. *To prepare.* Vitamin C standard. Weigh 50 mg of *l*-ascorbic acid to the nearest 0.1 mg, dissolve in a 50-mL volumetric flask, and dilute to volume with distilled deionized water. Prepare the day of use. This is a 0.1% solution. Prepare serial dilution of this to obtain also 0.05% and 0.02% standards.

Procedure

1. *Calibration.* Consult your instructor for the operation of your instrument, including the proper attenuation. It will contain a column of a strong-base anion exchange resin, such as Zipax SAX (du Pont Instruments Co.). At a typical pressure of 1000 psi, the flow rate will be about 0.5 mL/min. Use a chart speed of about 0.5 in./min.

Using a 10- or 25-μL syringe (Hamilton 700 series, Unimetrics, or equivalent), inject a 10-μL aliquot of the 0.02% vitamin C standard and record the chromatogram. Repeat for the 0.05% and 0.1% standards. Measure the retention times and the areas. The peak areas may be obtained from the height times the width at half-height for each. Alternatively, the peaks may be cut out and weighed. Plot a calibration curve of the peak area against concentration.

2. *Unknown.* Your instructor will provide you three or more unknown juices. These may be such products as "Hi C" drinks, grape drinks, orange juice, and the like. Inject 10-μL portions of each and record the chromatograms. Several peaks may be obtained. If you have time, run at least two chromatograms on each. From a comparison of retention times, identify the juices that contain vitamin C. Measure the areas of the vitamin C peaks, and from the calibration curve determine and report the concentrations in the unknowns. If two chromatograms of the unknowns were run, report the average concentrations. The vitamin C peak may be partially overlapped by another. In this case, extrapolate to the baseline and measure the area from the baseline.

EXPERIMENT 42   Analysis of Analgesics Using High-Performance Liquid Chromatography

Principle

The compositions of the analgesics Bufferin, Anacin, Empirin, and several brands of aspirin tablets are qualitatively determined by comparing their chromatograms with those of standard aspirin, phenacitin, and caffeine samples. The aspirin content of each is quantitatively determined by comparison of peak areas with those of standards. Separations are performed on a strong anion exchange column, Zipax SAX, using a UV detector.

Solutions and Reagents Required

1. *Provided.*[1] Aspirin (acetylsalicyclic acid), phenacetin (acetophenetidin), caffeine

---

[1]Aspirin and caffeine are available from Aldrich Chemical Co. and phenacetin, from ICN Pharmaceuticals, Inc.

(theine; 1,3,7-trimethylxanthine), methanol, mobile-phase solution (0.01 M sodium borate + 0.005 M ammonium nitrate, or Fisher Grampac No. B-80). See Experiment 30 for the structures of aspirin, phenacetin, and caffeine.

2. *To prepare.*[2]

(a) Stock solutions of aspirin, phenacetin, and caffeine. Weigh 50 mg each of phenacetin and caffeine to the nearest milligram and transfer to labeled 50-mL volumetric flasks. Dissolve in methanol and dilute to volume. Weigh 300 mg aspirin to the nearest 0.1 mg, transfer to a labeled 100-mL volumetric flask, dissolve in methanol, and dilute to volume with methanol. Prepare on the day of use.

(b) Aspirin working standards. Prepare serial dilutions of the stock 300 mg/dL solution to obtain standards of about 200, 100, and 50 mg/dL in methanol.

Procedure

1. *Calibration.* Consult your instructor for the operation of your instrument, including the proper attenuation for each sample. It will contain a column of a strong anion exchange resin, such as Zipax SAX (du Pont Instrument Co.). At a typical operating pressure of 1000 psi, the flow rate will be about 0.5 mL/min. Use chart speed of about 0.5 in./min.

Using a 5- or 10-μL syringe (Hamilton 700 series, Unimetrics, or equivalent), inject a 5-μL aliquot of each 100 mg/dL standard solution to obtain chromatograms for aspirin, phenacetin, and caffeine. Similarly, obtain chromatograms for the serial aspirin standards (one chromatogram for each, unless you have time for more).

Measure the retention times. For the aspirin chromatograms, also measure the areas. The peak areas may be obtained from the height times the width at half-height for each. Alternatively, the peaks may be cut out and weighed. Plot a calibration curve of peak area against concentration.

2. *Unknowns.* Your instructor will provide you with five tablets each of Bufferin, Anacin, Empirin, Bayer aspirin, and one or two other brands of aspirin. Prepare solutions of each in methanol by grinding the five tablets with a mortar and pestle, weighing to the nearest 0.1 mg about 175 mg, and transferring to a 50-mL volumetric flask. Dissolve in methanol and dilute to the mark. Mix well and allow any residue to settle for about 10 minutes, then obtain chromatograms as above, taking 5-μL aliquots. If time permits, run two chromatograms on each. It will take about 10–15 minutes for each chromatogram.

From a comparison of retention times, identify the components of each analgesic. Measure the areas of the aspirin peaks, and from the calibration curve determine and report the concentrations in the various tablets. If there is any overlap of the aspirin peak with another, extrapolate to the baseline and measure the area from the baseline. If two chromatograms were run, report the average concentrations.

KINETIC MEASUREMENTS

EXPERIMENT 43   Catalytic Determination of Traces of Selenium

Principle

Trace amounts of selenium catalyze the reduction of methylene blue by sodium sulfide at pH 10.5–11. The time required for decolorization of the methylene blue is measured with

---

[2]Aspirin tends to decompose in solution, and measurements should be made as soon as possible after preparing solutions.

a stopwatch. This is inversely proportional to the amount of selenium present and is compared with the time required for decolorization by standards.

Solutions and Chemicals Required

1. *Provided.* Formaldehyde solution (assay 36–38%). Lachrymatory—take care!

2. *To prepare.*

(a) Standard selenium solutions. (**Caution:** Selenium is a toxic element.) Prepare a stock 100-ppm selenium solution by dissolving 0.10 g elemental selenium powder in a few drops (minimum required) concentrated nitric acid, heating if necessary to aid dissolution. After dissolution is complete, transfer to a 1-L flask and dilute to volume with water. The selenium is in the $+4$ valence state after dissolution (H_3SeO_3). Dilute this 1:100 to prepare a 1-ppm solution. By dilution of the 1-ppm solution, prepare a series of standards containing 0.01, 0.02, 0.04, 0.06, 0.08, and 0.10 ppm selenium. These should be used on the day of preparation.

(b) Alkaline sodium sulfide solution (ca. 0.1 *M*). Dissolve 2.40 g sodium sulfide, 2.40 g sodium sulfite, and 4 g sodium hydroxide in 100 mL distilled water. The sodium sulfite is added to overcome deleterious effects due to polysulfides present in the sodium sulfide by forming thiosulfates, which, like sodium sulfite, have no interfering actions on the reduction of methylene blue. This solution is stable for two days.

(c) Conditioner solution. Dissolve 25 g $Na_2EDTA \cdot 2H_2O$, 0.4 g ferric chloride, and 50 mL triethanolamine in distilled water and dilute to 1 L. The EDTA is added as a masking agent to prevent the interference of metal ions that react with sulfide. Copper (>10 μg) still interferes. Iron(III) has an enhancement effect on the selenium catalysis and enables the determination of smaller amounts of selenium.

3. *Procedure.* The reaction is temperature dependent, and samples and standards should be run on the same day. Take a 100-mL volumetric flask to your instructor who will add your unknown to it. Dilute to volume. This will contain 1 to 10 μg (gives 0.01–0.1 ppm) selenium. Into 30-mL test tubes with stoppers add with a pipet 10 mL of the standard and unknown solutions. Run the sample in duplicate. Add 5 mL conditioner solution, 5 mL formaldehyde solution, and 1 mL alkaline sulfide solution. Formaldehyde suppresses the reducing power of sodium sulfide and, therefore, stabilizes the blank. Mix the contents of the tube well after each addition of methylene blue solution (same dropper for all tubes), stopper, shake briefly and quickly, and start the stop watch. Measure the time required for complete decoloration of the methylene blue.

Construct a calibration graph by plotting $1/t$ (min^{-1}) versus micrograms selenium. The time of reaction will vary from about 4 minutes for 1 μg selenium to 16 minutes for 0.1 μg selenium. From the calibration graph, determine the number of micrograms selenium.

EXPERIMENT 44 Enzymatic Determination of Glucose in Blood[1]

Principles

A protein-free filtrate of a 0.5-mL sample of whole blood, serum, or plasma is prepared by precipitating proteins with zinc hydroxide. The glucose in an aliquot of this is reacted with a mixture of the enzymes glucose oxidase and horseradish preoxidase, and the hydrogen peroxide produced is coupled with the chromogenic hydrogen donor *o*-dianisidine to form a color with absorption maximum at 540 nm.

---

[1]See footnote 3, concerning aqueous unknowns.

Equations

$$\text{glucose} + O_2 + H_2O \xrightarrow{\text{glucose oxidase}} \text{gluconic acid} + H_2O_2$$

$$H_2O_2 + o\text{-dianisidine} \xrightarrow[\text{peroxide}]{\text{horse radish}} \text{chromogen} \quad (\text{absorption max 540 nm})$$

Solutions and Chemicals Required[2]

1. *Provided.*

(a) Zinc sulfate solution, 2.2% (from $ZnSO_4 \cdot 7H_2O$); 3 M H_2SO_4.

(b) Barium hydroxide solution, saturated. Add about 80 g $Ba(OH)_2 \cdot 8H_2O$ to 1 L boiling water. Remove heat, insert a stopper with a soda lime trap, mix thoroughly, and *set aside for several days to settle*. At 25°C, the solution will be about 0.22 M.

(c) Barium hydroxide solution, 0.06 M. Dilute 270 mL of the saturated barium hydroxide solution to 1 L with CO_2-free (boiled and cooled) water. A volume of 9.00 ± 0.10 mL of this should be required to neutralize 10.00 mL of the zinc sulfate solution. Test by adding 10.00 mL zinc sulfate and 25 mL water to a 100-mL Erlenmeyer flask, adding two drops 0.2% phenolphthalein solution (in alcohol), and titrating with the barium hydroxide solution to a faint but permanent pink color. Dilute one or the other solution to proper equivalence.

2. *To prepare.*

(a) Glycerol–buffer solution, pH 7.0. Dissolve 3.48 g Na_2HPO_4 and 2.12 g KH_2PO_4 in 600 mL water and add 400 mL reagent-grade glycerol.

(b) Enzyme reagent. Grind 500 mg glucose oxidase, 10 mg horseradish peroxidase, and 2 mL of the glycerol–buffer solution in a clean dry mortar. Wash into a 200-mL graduated cylinder using the glycerol–buffer solution, and bring to a volume of 200 mL with the same solution. Filter into a clean and dry flask. A Buchner funnel with light suction will facilitate the filtration.

Dissolve 20 mg o-dianisidine in 2.0 mL absolute methanol by intermittent shaking. Pour into an amber bottle and add the filtered enzyme solution. Pour the initial portion rapidly to prevent precipitation of the chromogen. This reagent is stable for three weeks when stored in a refrigerator.

(c) Glucose standards. Prepare a 1% stock solution by dissolving 1 g (accurately weighed) of oven-dried (110°C, 1 hour) reagent-grade glucose in water and diluting with 0.25% benzoic acid solution in a volumetric glask. The benzoic acid acts as a preservative.

Prepare working standards containing 100, 200, 300, and 400 mg glucose per 100 mL by dilution of the stock solution with 0.25% benzoic acid (e.g., 5.0, 10.0, 15.0, and 20.0 mL diluted to 50 mL).

Things to Do Before the Experiment

Dry the glucose at 110°C for 1 hour and cool in a desiccator.

Procedure

Add with a pipet a 0.5-mL blood, serum, or plasma sample[3] into a 50-mL Erlenmeyer flask and add 4.5 mL 0.06 M barium hydroxide solution. Swirl to mix and then slowly add 5.00

---

[2]Commercially packaged reagents are available, which eliminates much of the solution preparation. An example is Glucostat, produced by Worthington Biochemical Corp., Freehold, N.J.

[3]Standard commercial control serum preparations (for example, Versatol) may be used. Preparations with "abnormal" glucose concentrations are available, so the range of normal (ca. 90 mg/dL) to abnormal (300 mg/dL or greater) levels can be covered.

Alternatively, aqueous glucose unknowns may be used, in which case the protein precipitation step is eliminated and 9.5 mL water is added in place of the barium hydroxide and zinc sulfate solution.

mL 2.2% zinc sulfate solution. Mix thoroughly and allow to stand 5 minutes. Filter into a dry flask, or centrifuge in a dry centrifuge tube.

Prepare a blank filtrate and standard filtrates in the same manner by using 0.5 mL water for the blank and 0.5-mL aliquots of the working standards.

Add with a pipet a 0.2-mL aliquot of each of the filtrates into a clean, dry test tube, and at 30-second intervals add 1.00 mL enzyme reagent. Immediately place in a 37°C water bath. (Reactions may be run for 45 minutes at room temperature with some loss in sensitivity.) Run duplicate aliquots of the unknown filtrate. After exactly 30 minutes, remove the tubes one at a time (30-second intervals) and add 50 mL 3 M H_2SO_4.[4] The sulfuric acid effectively stops the enzyme reaction.

After 5 minutes, **but** before 1 hour, determine the absorbance of the solutions at 540 nm, using the blank as reference. Construct a calibration curve and report the mg/dL glucose in your sample.

---

[4]The reaction is pseudo first order at low concentrations of glucose, but a slower rate is observed at high concentrations of glucose. Thus, the absorbance may not be a linear function of incubation time at high concentrations.

appendix A

LITERATURE OF ANALYTICAL CHEMISTRY

When defining a problem, one of the first things the analyst does is to go to the scientific literature and see if the particular problem has already been solved in a manner that can be employed. There are many reference books in selected areas of analytical chemistry that describe the commonly employed analytical procedures in a particular discipline and also some of the not-so-common ones. These usually give reference to the original chemical journals. For many routine or specific analyses, prescribed standard procedures have been adopted by the various professional societies.

If you do not find a solution to the problem in reference books, then you must resort to the scientific journals. *Chemical Abstracts* is the logical place to begin a literature search. This journal contains abstracts of all papers appearing in the major chemical publications of the world. Yearly and cumulative indexes are available to aid in the literature search. The element or compound to be determined as well as the type of sample to be analyzed can be looked up to obtain a survey of the methods available. Author indexes are also available.

Following is a selected list of some references in analytical chemistry. References to the various methods of measurement are included at the ends of the chapters covering these methods throughout the text.

A.1 JOURNALS[1]

1. *American Laboratory*
2. *Analytical Biochemistry*

---

[1]The *Chemical Abstracts* abbreviation of each title is indicated by italics.

3. *Analytical Chimica Acta*
4. *Analytical Abstracts*
5. *Analytical Chemistry*[2]
6. *Analytical Instrumentation*
7. *Analytical Letters*
8. *Analyst, The*
9. *Applied Spectroscopy*
10. *Clinica Chimica Acta*
11. *Clinical Chemistry*
12. *Journal of the Association of Official Analytical Chemists*
13. *Journal of Chromatographic Science*
14. *Journal of Chromatography*
15. *Journal of Electroanalytical Chemistry and Interfacial Electrochemistry*
16. *Michrochemical Journal*
17. *Spectrochimica Acta*
18. *Talanta*
19. *Zeitschrift für analytische Chemie*

A.2 GENERAL REFERENCES

1. *ASTM Book of Standards, 7 vols.* Philadelphia: American Society for Testing and Materials, 1952.
2. H. H. Bauer, G. D. Christian, and J. E. O'Reilly, *Instrumental Analysis.* Boston: Allyn and Bacon, 1978.
3. R. Belcher and L. Gordon, eds., *International Series of Monographs on Analytical Chemistry.* New York: Pergamon Press. A multivolume series.
4. N. H. Furman and F. J. Welcher, eds., *Scott's Standard Methods of Chemical Analysis,* 6th ed., 5 vols. New York: D. Van Nostrand Co., 1962–1966.
5. I. M. Kolthoff and P. J. Elving, eds., *Treatise on Analytical Chemistry.* New York: Interscience Publishers. A multivolume series.
6. I. M. Kolthoff, E. B. Sandell, E. J. Meehan, and S. Bruckenstein, *Quantitative Chemical Analysis,* 4th ed. London: The Macmillan Co., 1969.
7. L. Meites, ed., *Handbook of Analytical Chemistry.* New York: McGraw-Hill, 1963.
8. C. N. Reilley, ed., *Advances in Analytical Chemistry and Instrumentation.* New York: Interscience Publishers. A multivolume series.
9. A. I. Vogel, *A Textbook of Quantitative Inorganic Analysis, Including Elementary Instrumental Analysis,* 3rd ed. London: Longman, 1972.
10. C. L. Wilson and D. W. Wilson, eds., *Comprehensive Analytical Chemistry.* New York: Elsevier Publishing Co. A multivolume series.

A.3 INORGANIC SUBSTANCES

1. *ASTM Methods for Chemical Analysis of Metals.* Philadelphia: American Society for Testing and Materials, 1956.

---

[2]This journal has a biannual volume every April that reviews in alternate years the literature of various analytical techniques and their applications in different areas of analysis.

2. F. E. Beamish and J. C. Van Loon, *Analysis of Noble Metals.* New York: Academic Press, 1977.

A.4 ORGANIC SUBSTANCES

1. J. S. Fritz and G. S. Hammond, *Quantitative Organic Analysis.* New York: John Wiley & Sons, 1957.
2. T. S. Ma and R. C. Rittner, *Modern Organic Elemental Analysis.* New York: Marcel Dekker, 1979.
3. W. McFadden, *Techniques of Combined Gas Chromatography/Mass Spectrometry.* New York: Wiley-Interscience, 1973.
4. J. Mitchell, Jr., I. M. Kolthoff, E. S. Proskauer, and A. W. Weissberger, eds., *Organic Analysis,* 4 vols. New York: Interscience Publishers, 1953–1960.
5. S. Siggia, Jr., and J. G. Hanna, *Quantitative Organic Analysis via Functional Group Analysis,* 4th ed. New York: John Wiley & Sons, 1979.
6. A Steyermarch, *Quantitative Organic Microanalysis,* 2nd ed. New York: Academic Press, 1961.

A.5 BIOLOGICAL AND CLINICAL SUBSTANCES

1. J. S. Annino, *Chemical Chemistry. Principles and Procedures,* 2nd ed. Boston: Little, Brown and Co., 1960.
2. G. D. Christian and F. J. Feldman, *Atomic Absorption Spectroscopy. Applications in Agriculture, Biology, and Medicine.* New York: Wiley-Interscience, 1970.
3. D. Glick, ed., *Methods of Biochemical Analysis.* New York: Interscience Publishers. A multivolume series.
4. R. J. Henry, D. C. Cannon, and J. W. Winkelman, eds., *Clinical Chemistry. Principles and Techniques,* 2nd ed. Hagerstown, Md.: Harper & Row, Publishers, 1974.
5. M. Reiner and D. Seligson, eds., *Standard Methods of Clinical Chemistry.* New York: Academic Press. A multivolume series since 1953.
6. N. W. Tietz, ed., *Fundamentals of Clinical Chemistry,* 2nd ed. Philadelphia: W. B. Saunders Co., 1976.

A.6 GASES

1. C. J. Cooper and A. J. DeRose, *The Analysis of Gases by Gas Chromatography.* New York: Pergamon Press, 1983.

A.7 WATER AND AIR POLLUTANTS

1. *Quality Assurance Handbook for Air Pollution Measurement Systems.* U.S.E.P.A., Office of Research and Development, Environmental Monitoring and Support Laboratory, Research Triangle, N.C. 27711. Vol. I, Principles. Vol. II, Ambient Air Specific Methods.

2. *Standard Methods for the Examination of Water and Wastewater.* New York: American Public Health Association.

A.8 OCCUPATIONAL HEALTH AND SAFETY

1. National Institute of Occupational Health and Safety (NIOSH), *Manual of Analytical Methods.* Washington, D.C.: U.S. Government Printing Office.
2. J. Peterson. *Industrial Health.* Englewood Cliffs, N.J.: Prentice-Hall, 1977.

appendix
B

REVIEW OF MATHEMATICAL OPERATIONS: EXPONENTS, LOGARITHMS, THE QUADRATIC FORMULA, AND CALCULATORS

B.1 EXPONENTS

It is convenient in mathematical operations, even when working with logarithms, to express numbers semiexponentially. Mathematical operations with exponents are summarized as follows:

$$N^a N^b = N^{a+b} \quad \text{e.g., } 10^2 \times 10^5 = 10^7$$

$$\frac{N^a}{N^b} = N^{a-b} \quad \text{e.g., } \frac{10^5}{10^2} = 10^3$$

$$(N^a)^b = N^{ab} \quad \text{e.g., } (10^2)^5 = 10^{10}$$

$$\sqrt[a]{N^b} = N^{b/a} \quad \text{e.g., } \sqrt{10^6} = 10^{6/2} = 10^3$$

$$\sqrt[3]{10^9} = 10^{9/3} = 10^3$$

The decimal point of a number is conveniently placed using the **semiexponential form.** The number is written with the decimal placed in the units position, and it is multiplied by 10 raised to an integral exponent equal to the number of spaces the decimal was moved to bring it to the units position. The exponent is negative if the decimal is moved to the right (the number is less than 1) and it is positive if the decimal is moved to the left (the number is 10 or greater). Some examples are:

Number	Semiexponential Form
0.00267	2.67×10^{-3}
0.48	4.8×10^{-1}
52	5.2×10^1
6027	6.027×10^3

Any number raised to the zero power is equal to unity. Thus, $10^0 = 1$, and 2.3 in semiexponential form is 2.3×10^0. Numbers between 1 and 10 do not require the semiexponential form to place the decimal.

B.2 TAKING LOGARITHMS OF NUMBERS

It is convenient to place numbers in the semiexponential form for taking logarithms or for finding a number from its logarithm. The following rules apply:

$$N = b^a$$
$$lob_b\, N = a$$

or

$$N = 10^a$$
$$\log_{10} N = a$$

e.g.,

$$\log 10^2 = 2$$
$$\log 10^{-3} = -3$$

Also,

$$\log ab = \log a + \log b$$

e.g.,

$$\log 2.3 \times 10^{-3} = \log 2.3 + \log 10^{-3}$$
$$= 0.36 - 3$$
$$= -2.64$$

$$\log 5.67 \times 10^7 = \log 5.67 + \log 10^7$$
$$= 0.754 + 7$$
$$= 7.754$$

The exponent is actually the **characteristic** of the logarithm of a number, and the logarithm of the number between 1 and 10 is the **mantissa.** Thus, in the example $\log 2.3 \times 10^{-3}$, -3 is the characteristic and 0.36 is the mantissa.

B.3 FINDING NUMBERS FROM THEIR LOGARITHMS

The following relationships hold:

$$\log_{10} N = a$$
$$N = 10^a = \text{antilog } a$$

For example,

$$\log N = 0.371$$
$$N = 10^{0.371} = \text{antilog } 0.371 = 2.35$$

In general, to find a number from its logarithm, write the number in exponent form, and then break the exponent into the mantissa (m) and the characteristic (c). Then, take the antilog of the mantissa and multiply by the exponential form of the characteristic:

TABLE B.5
Table of Four-Place Logarithms

No.	0	1	2	3	4	5	6	7	8	9
10	0000	0043	0086	0128	0170	0212	0253	0294	0334	0374
11	0414	0453	0492	0531	0569	0607	0645	0682	0719	0755
12	0792	0828	0864	0899	0934	0969	1004	1038	1072	1106
13	1139	1173	1206	1239	1271	1303	1335	1367	1399	1430
14	1461	1492	1523	1553	1584	1614	1644	1673	1703	1732
15	1761	1790	1818	1847	1875	1903	1931	1959	1987	2014
16	2041	2068	2095	2122	2148	2175	2201	2227	2253	2279
17	2304	2330	2355	2380	2405	2430	2455	2480	2504	2529
18	2553	2577	2601	2625	2648	2672	2695	2718	2742	2765
19	2788	2810	2833	2856	2878	2900	2923	2945	2967	2989
20	3010	3032	3054	3075	3096	3118	3139	3160	3181	3201
21	3222	3243	3263	3284	3304	3324	3345	3365	3385	3404
22	3424	3444	3464	3483	3502	3522	3541	3560	3579	3598
23	3617	3636	3655	3674	3692	3711	3729	3747	3766	3784
24	3802	3820	3838	3856	3874	3892	3909	3927	3945	3962
25	3979	3997	4014	4031	4048	4065	4082	4099	4116	4133
26	4150	4166	4183	4200	4216	4232	4249	4265	4281	4298
27	4314	4330	4346	4362	4378	4393	4409	4425	4440	4456
28	4472	4487	4502	4518	4533	4548	4564	4579	4594	4609
29	4624	4639	4654	4669	4683	4698	4713	4728	4742	4757
30	4771	4786	4800	4814	4829	4843	4857	4871	4886	4900
31	4914	4928	4942	4955	4969	4983	4997	5011	5024	5038
32	5051	5065	5079	5092	5105	5119	5132	5145	5159	5172
33	5185	5198	5211	5224	5237	5250	5263	5276	5289	5302
34	5315	5328	5340	5353	5366	5378	5391	5403	5416	5428
35	5441	5453	5465	5478	5490	5502	5514	5527	5539	5551
36	5563	5575	5587	5599	5611	5623	5635	5647	5658	5670
37	5682	5694	5705	5717	5729	5740	5752	5763	5775	5786
38	5798	5809	5821	5832	5843	5855	5866	5877	5888	5899
39	5911	5922	5933	5944	5955	5966	5977	5988	5999	6010
40	6021	6031	6042	6053	6064	6075	6085	6096	6107	6117
41	6128	6138	6149	6160	6170	6180	6191	6201	6212	6222
42	6232	6243	6253	6263	6274	6284	6294	6304	6314	6325
43	6335	6345	6355	6365	6375	6386	6395	6405	6415	6425
44	6435	6444	6454	6464	6474	6484	6493	6503	6513	6522
45	6532	6542	6551	6561	6571	6580	6590	6599	6609	6618
46	6628	6637	6646	6656	6665	6675	6684	6693	6702	6712
47	6721	6730	6739	6749	6758	6767	6776	6785	6794	6803
48	6812	6821	6830	6839	6848	6857	6866	6875	6884	6893
49	6902	6911	6920	6928	6937	6946	6955	6964	6972	6981
50	6990	6998	7007	7016	7024	7033	7042	7050	7059	7067
51	7076	7084	7093	7101	7110	7118	7126	7135	7143	7152
52	7160	7168	7177	7185	7193	7202	7210	7218	7226	7235
53	7243	7251	7259	7267	7275	7284	7292	7300	7308	7316
54	7324	7332	7340	7348	7356	7364	7372	7380	7388	7396
	0	1	2	3	4	5	6	7	8	9

No.	0	1	2	3	4	5	6	7	8	9
55	7404	7412	7419	7427	7435	7443	7451	7459	7466	7474
56	7482	7490	7497	7505	7513	7520	7528	7536	7543	7551
57	7559	7566	7574	7582	7589	7597	7604	7612	7619	7627
58	7634	7642	7649	7657	7664	7672	7679	7686	7694	7701
59	7709	7716	7723	7731	7738	7745	7752	7760	7767	7774
60	7782	7789	7796	7803	7810	7818	7825	7832	7839	7846
61	7853	7860	7868	7875	7882	7889	7896	7903	7910	7917
62	7924	7931	7938	7945	7952	7959	7966	7973	7980	7987
63	7992	8000	8007	8014	8021	8028	8035	8041	8048	8055
64	8062	8069	8075	8082	8089	8096	8102	8109	8116	8122
65	8129	8136	8142	8149	8156	8162	8169	8176	8182	8189
66	8195	8202	8209	8215	8222	8228	8235	8241	8248	8254
67	8261	8267	8274	8280	8287	8293	8299	8306	8312	8319
68	8325	8331	8338	8344	8351	8357	8363	8370	8376	8382
69	8388	8395	8401	8407	8414	8420	8426	8432	8439	8445
70	8451	8457	8463	8470	8476	8482	8488	8494	8500	8506
71	8513	8519	8525	8531	8537	8543	8549	8555	8561	8567
72	8573	8579	8585	8591	8597	8603	8609	8615	8621	8627
73	8633	8639	8645	8651	8657	8663	8669	8675	8681	8686
74	8692	8698	8704	8710	8716	8722	8727	8733	8739	8745
75	8751	8756	8762	8768	8774	8779	8785	8791	8797	8802
76	8808	8814	8820	8825	8831	8837	8842	8848	8854	8859
77	8865	8871	8876	8882	8887	8893	8899	8904	8910	8915
78	8921	8927	8932	8938	8943	8949	8954	8960	8965	8971
79	8976	8982	8987	8993	8998	9004	9009	9015	9020	9025
80	9031	9036	9042	9047	9053	9058	9063	9069	9074	9079
81	9085	9090	9096	9101	9106	9112	9117	9122	9128	9133
82	9138	9143	9149	9154	9159	9165	9170	9175	9180	9186
83	9191	9196	9201	9206	9212	9217	9222	9227	9232	9238
84	9243	9248	9253	9258	9263	9269	9274	9279	9284	9289
85	9294	9299	9304	9309	9315	9320	9325	9330	9335	9340
86	9345	9350	9355	9360	9365	9370	9375	9380	9385	9390
87	9395	9400	9405	9410	9415	9420	9425	9430	9435	9440
88	9445	9450	9455	9460	9465	9469	9474	9479	9484	9489
89	9494	9499	9504	9509	9513	9518	9523	9528	9533	9538
90	9542	9547	9552	9557	9562	9566	9571	9576	9581	9586
91	9590	9595	9600	9605	9609	9614	9619	9624	9628	9633
92	9638	9643	9647	9652	9657	9661	9666	9671	9675	9680
93	9685	9689	9694	9699	9703	9708	9713	9717	9722	9727
94	9731	9736	9741	9745	9750	9754	9759	9763	9768	9773
95	9777	9782	9786	9791	9795	9800	9805	9809	9814	9818
96	9823	9827	9832	9836	9841	9845	9850	9854	9859	9863
97	9868	9872	9877	9881	9886	9890	9894	9899	9903	9908
98	9912	9917	9921	9926	9930	9934	9939	9943	9948	9952
99	9956	9961	9965	9969	9974	9978	9983	9987	9991	9996
	0	1	2	3	4	5	6	7	8	9

$$\log_{10} N = mc$$
$$N = 10^{mc} = 10^m \times 10^c$$
$$N = (\text{antilog } m) \times 10^c$$

For example,

$$\log N = 2.671$$
$$N = 10^{2.671} = 10^{.671} \times 10^2$$
$$= 4.69 \times 10^2 = 469$$
$$\log N = 0.326$$
$$N = 10^{0.326} = 2.12$$
$$\log N = -0.326$$
$$N = 10^{-0.326} = 10^{0.674} \times 10^{-1}$$
$$= 4.72 \times 10^{-1} = 0.472$$

Whenever the logarithm is a negative number, the exponent is broken into a negative integer (the caracteristic) and a positive noninteger less than 1 (the mantissa), as in the last example. Note that in the example, the sum of the two exponents is equal to the original exponent (-0.326). Another example is

$$\log N = -4.723$$
$$N = 10^{-4.723} = 10^{0.277} \times 10^{-5}$$
$$= 1.89 \times 10^{-5} = 0.0000189$$

B.4   FINDING ROOTS WITH LOGARITHMS

It is a simple matter to find a given root of a number using logarithms. Suppose, for example, you wish to find the cube root of 325. Let N represent the cube root:

$$N = 325^{1/3}$$

Taking the logarithm of both sides,

$$\log N = \log 325^{1/3}$$

The $\frac{1}{3}$ can be brought out front:

$$\log N = \tfrac{1}{3} \log 325 = \tfrac{1}{3}(2.512) = 0.837$$
$$N = 10^{0.837} = \text{antilog } 0.837 = 6.87$$

B.6   THE QUADRATIC FORMULA

A quadratic equation of the general form

$$ax^2 + bx + c = 0$$

can be solved by use of the quadratic formula:

$$x = \frac{-b \pm \sqrt{b^2 - 4ac}}{2a}$$

Quadratic equations are frequently encountered in calculation of equilibrium concentrations of ionized species using equilibrium constant expressions. Hence, an equation of the following type might require solving:

$$\frac{x^2}{1.0 \times 10^{-3} - x} = 8.0 \times 10^{-4}$$

or

$$x^2 = 8.0 \times 10^{-7} - 8.0 \times 10^{-4}x$$

Arranging in the quadratic equation form above, we have:

$$x^2 + 8.0 \times 10^{-4}x - 8.0 \times 10^{-7} = 0$$

or $a = 1$, $b = 8.0 \times 10^{-4}$, $c = -8.0 \times 10^{-7}$. Therefore,

$$x = \frac{-8.0 \times 10^{-4} \pm \sqrt{(8.0 \times 10^{-4})^2 - 4(1)(-8.0 \times 10^{-7})}}{2(1)}$$

$$= \frac{-8.0 \times 10^{-4} \pm \sqrt{0.64 \times 10^{-6} + 3.2_0 \times 10^{-6}}}{2}$$

$$= \frac{-8.0 \times 10^{-4} \pm \sqrt{3.8_4 \times 10^{-6}}}{2}$$

$$= \frac{-8.0 \times 10^{-4} \pm 1.9_6 \times 10^{-3}}{2} = \frac{1.1_6 \times 10^{-3}}{2}$$

$$x = 5.8_0 \times 10^{-4}$$

A concentration can only be positive, and so the negative value of x is not a solution.

B.7 USE OF CALCULATORS

The wide availability of electronic calculators has simplified calculations required in analytical chemistry. If you have one of these calculators, the manual of operation will tell you how to perform calculations for which the calculator is designed. Two references that will aid in getting the most out of your calculator are

1. L. A. Currie, "On the Use of Small Calculators Having Stacked Register," *Anal. Lett.*, **6** (1973) 847.
2. J. M. Smith, *Scientific Analysis on the Pocket Calculator*, 2nd ed. New York: Wiley-Interscience, 1977.

If your calculator does not have a square-root function, the **square root of a number can be approximated** very closely by guessing at the value, dividing this into the number, and averaging the divisor and quotient. (If the guess is correct, the quotient will be equal to the divisor.) For example, suppose we wish to find the square root of 1236. This number is equal to 12.36×10^2, and a reasonable guess might be 3.5×10^1, or 35.

$$\frac{1236}{35} = 35.31$$

$$\frac{35.00 + 35.31}{2} = 35.16$$

To check our answer, $35.16 \times 35.16 = 1236$. So the answer is correct to four significant figures! Should the answer not be correct to the accuracy desired (as well may be the case if the original guess is relatively far off), a second calculation using the first answer as the assumed correct value will most certainly give the desired accuracy.

appendix C

BALANCING REDUCTION–OXIDATION REACTIONS

The calculations in volumetric analysis require that the balanced reaction be known. The balancing of redox reactions is, therefore, reviewed below.

Various methods are used to balance redox reactions, and we shall use the **half-reaction method.** In this technique, the reaction is broken down into two parts: the oxidizing part and the reducing part. In every redox reaction, an oxidizing agent reacts with a reducing agent. The oxidizing agent is reduced in the reaction while the reducing agent is oxidized. Each of these constitutes a *half-reaction,* and the overall reaction can be broken down into these two half-reactions. Thus, in the reaction

$$Fe^{2+} + Ce^{4+} \rightarrow Fe^{3+} + Ce^{3+}$$

Fe^{2+} is the reducing agent and Ce^{4+} is the oxidizing agent. The corresponding half-reactions are

$$Fe^{2+} \rightarrow Fe^{3+} + e^-$$

and

$$Ce^{4+} + e^- \rightarrow Ce^{3+}$$

To balance a reduction–oxidation reaction, each half-reaction is first balanced. There must be a net gain or loss of zero electrons in the overall reaction, and so the second step is multiplication of one or both of the half-reactions by an appropriate factor or factors so that, when they are added, the electrons cancel. The final step is addition of the half-reactions.

Half-reactions may be balanced in a number of ways. We will describe two ways, one involving determination of the change in the oxidation state or states in the species being oxidized or reduced, that is, the electron change. The other does not require a knowledge of the oxidation states involved. The same examples are worked by both methods for comparison. You may use the method with which you feel comfortable.

C.1 BALANCING USING THE ELECTRON CHANGE

The rules for balancing a half-reaction using this method are as follows:

1. Balance the number of atoms of the substance being reduced or oxidized. The half-reaction is written in the direction it occurs in the reaction.
2. Determine the electron change for the half-reaction by noting the change in oxidation state per atom and multiply this by the number of atoms reduced or oxidized. Add the electrons to the left side of the half-reaction if it is a reduction and to the right side if it is an oxidation.
3. Balance the charges including the electrons by adding protons if the reaction takes place in acid solution and hydroxyl ions if it takes place in alkaline solution.
4. Balance oxygens and hydrogens by adding water molecules on the appropriate side of the half-reaction.

If the acidity conditions are not specified, they can generally be deduced from the chemical species present in the reaction (e.g., NH_3 versus NH_4^+).

EXAMPLE C.1   Complete and balance the following reactions:
(a) $H_2O_2 + MnO_4^- \rightarrow O_2 + Mn^{2+}$ (acid solution)
(b) $MnO_4^- + Mn^{2+} \rightarrow MnO_2$ (slightly alkaline solution)
(c) $H_2O_2 \rightarrow H_2O + O_2$
(d) $PbS + H_2O_2 \rightarrow PbSO_4$

SOLUTIONS
(a) Following the above steps:
   1. $H_2O_2 \rightarrow O_2$ (oxygen is oxidized from -1 to 0 oxidation state
   2. $H_2O_2 \rightarrow O_2 + 2e^-$
   3. $H_2O_2 \rightarrow O_2 + 2H^+ + 2e^-$

   1. $MnO_4^- \rightarrow Mn^{2+}$
   2. $MnO_4^- + 5e^- \rightarrow Mn^{2+}$
   3. $MnO_4^- + 8H^+ + 5e^- \rightarrow Mn^{2+}$
   4. $MnO_4^- + 8H^+ + 5e^- \rightarrow Mn^{2+} + 4H_2O$

$$5(H_2O_2 \rightarrow O_2 + 2H^+ + 2e^-)$$
$$2(MnO_4^- + 8H^+ + 5e^- \rightarrow Mn^{2+} + 4H_2O)$$

$$\overline{5H_2O_2 + 2MnO_4^- + 6H^+ \rightarrow 5O_2 + 2Mn^{2+} + 8H_2O}$$

(b) 1. $MnO_4^- \rightarrow MnO_2$
   2. $MnO_4^- + 3e^- \rightarrow MnO_2$
   3. $MnO_4^- + 3e^- \rightarrow MnO_2 + 4OH^-$
   4. $MnO_4^- + 2H_2O + 3e^- \rightarrow MnO_2 + 4OH^-$

   1. $Mn^{2+} \rightarrow MnO_2$
   2. $Mn^{2+} \rightarrow MnO_2 + 2e^-$
   3. $Mn^{2+} + 4OH^- \rightarrow MnO_2 + 2e^-$

4. $Mn^{2+} + 4OH^- \rightarrow MnO_2 + 2H_2O + 2e^-$

$$2(MnO_4^- + 2H_2O + 3e^- \rightarrow MnO_2 + 4OH^-)$$
$$\underline{3(Mn^{2+} + 4OH^- \rightarrow MnO_2 + 2H_2O + 2e^-)}$$
$$2MnO_4^- + 3Mn^{2+} + 4OH^- \rightarrow 5MnO_2 + 2H_2O$$

(c) H_2O_2 is both oxidized (to O_2) and reduced (to H_2O), a disproportionation reaction:
 1. $H_2O_2 \rightarrow 2H_2O$ (each oxygen is reduced from -1 to -2)
 2. $H_2O_2 + 2e^- \rightarrow 2H_2O$
 3. $H_2O_2 + 2H^+ + 2e^- \rightarrow 2H_2O$ (can use either H^+ *or* OH^- to balance charge)

From (a) above,

$$H_2O_2 \rightarrow O_2 + 2H^+ + 2e^-$$
$$\underline{H_2O_2 + 2H^+ + 2e^- \rightarrow 2H_2O}$$
$$2H_2O_2 \rightarrow O_2 + 2H_2O$$

(d) S^{2-} is oxidized to SO_4^{2-} and H_2O_2 is reduced to H_2O:
 1. $PbS \rightarrow PbSO_4$
 2. $PbS \rightarrow PbSO_4 + 8e^-$
 3. $PbS \rightarrow PbSO_4 + 8H^+ + 8e^-$
 4. $PbS + 4H_2O \rightarrow PbSO_4 + 8H^+ + 8e^-$

From (c) above,

$$4(H_2O_2 + 2H^+ + 2e^- \rightarrow 2H_2O)$$
$$\underline{PbS + 4H_2O + PbSO_4 + 8H^+ + 8e^-}$$
$$PbS + 4H_2O_2 \rightarrow PbSO_4 + 4H_2O$$

C.2 BALANCING WITHOUT KNOWLEDGE OF OXIDATION STATES

The rules for balancing a half-reaction using this method are as follows:

 1. Balance the number of atoms of the substance being reduced or oxidized. Include other atoms that might be involved in the reaction but are not reduced or oxidized (other than hydrogen or oxygen). The half-reaction is written in the direction it occurs in the reaction.
 2. Balance the oxygens in the half-reaction by adding the correct number of H_2O molecules on the appropriate side of the half-reaction.
 3. Balance the hydrogens in the half-reaction by adding the correct number of protons (H^+) on the appropriate side of the half-reaction. *If the solution is alkaline,* then add the proper number of hydroxyl ions (OH^-) on each side of the half-reaction to neutralize the protons previously added (each H^+ will then be converted to a water molecule, H_2O; add to or cancel with previous H_2O molecules where possible).
 4. Balance the charges by adding the correct number of electrons (e^-) on the appropriate side of the half-reaction.

If the acidity conditions are not specified, they can generally be deduced from the chemical species present in the reaction (e.g., NH_3 versus NH_4^+).

After the half-reactions are balanced and multiplied by the appropriate factor(s), they are added. Any substances that occur on opposite sides of the reaction, such as protons, hydroxyl ions, water molecules, electrons, and so forth, are canceled.

EXAMPLE C.2   Complete and balance the following reactions:

(a) $H_2O_2 + MnO_4^- \rightarrow O_2 + Mn^{2+}$ (acid solution)
(b) $MnO_4^- + Mn^{2+} \rightarrow MnO_2$ (slightly alkaline solution)
(c) $H_2O_2 \rightarrow H_2O + O_2$
(d) $PbS + H_2O_2 \rightarrow PbSO_4$

SOLUTIONS
(a) Following the above steps,
    1. $H_2O_2 \rightarrow O_2$ (the oxygen is oxidized from -1 to 0 oxidation state)
    2. $H_2O_2 \rightarrow O_2 + 2H^+$
    3. $H_2O_2 \rightarrow O_2 + 2H^+ + 2e^-$

    1. $MnO_4^- \rightarrow Mn^{2+}$
    2. $MnO_4^- \rightarrow Mn^{2+} + 4H_2O$
    3. $MnO_4^- + 8H^+ \rightarrow Mn^{2+} + 4H_2O$
    4. $MnO_4^- + 8H^+ + 5e^- \rightarrow Mn^{2+} + 4H_2O$

$$5(H_2O_2 \rightarrow O_2 + 2H^+ + 2e^-)$$
$$2(MnO_4^- + 8H^+ + 5e^- \rightarrow Mn^{2+} + 4H_2O)$$
$$\overline{5H_2O_2 + 2MnO_4^- + 6H^+ \rightarrow 5O_2 + 2Mn^{2+} + 8H_2O}$$

(b) **1.** $MnO_4^- \rightarrow MnO_2$
    2. $MnO_4^- \rightarrow MnO_2 + 2H_2O$
    3. $MnO_4^- + 4H^+ \rightarrow MnO_2 + 2H_2O$
    4. $MnO_4^- + 4H_2O \rightarrow MnO_2 + 2H_2O + 4OH^-$
       $MnO_4^- + 2H_2O \rightarrow MnO_2 + 4OH^-$
    5. $MnO_4^- + 2H_2O + 3e^- \rightarrow MnO_2 + 4OH^-$

    1. $Mn^{2+} \rightarrow MnO_2$
    2. $Mn^{2+} + 2H_2O \rightarrow MnO_2$
    3. $Mn^{2+} + 2H_2O \rightarrow MnO_2 + 4H^+$
    4. $Mn^{2+} + 2H_2O + 4OH^- \rightarrow MnO_2 + 4H_2O$
       $Mn^{2+} + 4OH^- \rightarrow MnO_2 + 2H_2O$
    5. $Mn^{2+} + 4OH^- \rightarrow MnO_2 + 2H_2O + 2e^-$

$$2(MnO_4^- + 2H_2O + 3e^- \rightarrow MnO_2 + 4OH^-)$$
$$3(Mn^{2+} + 4OH^- \rightarrow MnO_2 + 2H_2O + 2e^-)$$
$$\overline{2MnO_4^- + 3Mn^{2+} + 4OH^- \rightarrow 5MnO_2 + 2H_2O}$$

(c) H_2O_2 is both oxidized (to O_2) and reduced (to H_2O), a disproportionation reaction.
    1. $H_2O_2 \rightarrow 2H_2O$
    2. $H_2O_2 + 2H^+ \rightarrow 2H_2O$
    3. $H_2O_2 + 2H^+ + 2e^- \rightarrow 2H_2O$

From (a) above,

$$H_2O_2 \rightarrow O_2 + 2H^+ + 2e^-$$
$$H_2O_2 + 2H^+ + 2e^- \rightarrow 2H_2O$$
$$\overline{2H_2O_2 \rightarrow O_2 + 2H_2O}$$

(d) S^{2-} is oxidized to SO_4^{2-}, and H_2O_2 is reduced to H_2O:

1. $PbS \rightarrow PbSO_4$
2. $PbS + 4H_2O \rightarrow PbSO_4$
3. $PbS + 4H_2O \rightarrow PbSO_4 + 8H^+$
4. $PbS + 4H_2O \rightarrow PbSO_4 + 8H^+ + 8e^-$

From (c) above,

$$\begin{array}{l} 4(H_2O_2 + 2H^+ + 2e^- \rightarrow 2H_2O) \\ \underline{PbS + 4H_2O \rightarrow PbSO_4 + 8H^+ + 8e^-} \\ PbS + 4H_2O_2 \rightarrow PbSO_4 + 4H_2O \end{array}$$

appendix
D

TABLES OF CONSTANTS

TABLE D.1
Dissociation Constants for Acids

Name	Formula	Dissociation Constant, at 25°C			
		K_{a1}	K_{a2}	K_{a3}	K_{a4}
Acetic	CH_3COOH	1.75×10^{-5}			
Alanine	$CH_3CH(NH_2)COOH^a$	4.5×10^{-3}	1.3×10^{-10}		
Arsenic	H_3AsO_4	6.0×10^{-3}	1.0×10^{-7}	3.0×10^{-12}	
Arsenious	H_3AsO_3	6.0×10^{-10}	3.0×10^{-14}		
Benzoic	C_6H_5COOH	6.3×10^{-5}			
Boric	H_3BO_3	6.4×10^{-10}			
Carbonic	H_2CO_3	4.3×10^{-7}	4.8×10^{-11}		
Chloroacetic	$ClCH_2COOH$	1.51×10^{-3}			
Citric	$HOOC(OH)C(CH_2COOH)_2$	7.4×10^{-4}	1.7×10^{-5}	4.0×10^{-7}	
Ethylenediamine-tetraacetic	$(CO_2H)_2NCH_2CH_2N(CO_2H)_2$	1.0×10^{-2}	2.2×10^{-3}	6.9×10^{-7}	5.5×10^{-11}
Formic	$HCOOH$	1.76×10^{-4}			
Glycine	$H_2NCH_2COOH^a$	4.5×10^{-3}	1.7×10^{-10}		
Hydrocyanic	HCN	7.2×10^{-10}			
Hydrofluoric	HF	6.7×10^{-4}			
Hydrogen sulfide	H_2S	9.1×10^{-8}	1.2×10^{-15}		
Hypochlorous	$HOCl$	1.1×10^{-8}			
Iodic	HIO_3	2×10^{-1}			
Lactic	$CH_3CHOHCOOH$	1.4×10^{-4}			
Leucine	$(CH_3)_2CHCH_2CH(NH_2)COOH^a$	4.7×10^{-3}	1.8×10^{-10}		
Maleic	$cis\text{-}HOOCCH{:}CHCOOH$	1.5×10^{-2}	2.6×10^{-7}		

TABLE D.1 (continued)
Dissociation Constants for Acids

Name	Formula	Dissociation Constant, at 25°C			
		K_{a1}	K_{a2}	K_{a3}	K_{a4}
Malic	$HOOCCHOHCH_2COOH$	4.0×10^{-4}	8.9×10^{-6}		
Nitrous	HNO_2	5.1×10^{-4}			
Oxalic	$HOOCCOOH$	6.5×10^{-2}	6.1×10^{-5}		
Phenol	C_6H_5OH	1.1×10^{-10}			
Phosphoric	H_3PO_4	1.1×10^{-2}	7.5×10^{-8}	4.8×10^{-13}	
Phosphorous	H_3PO_3	5×10^{-2}	2.6×10^{-7}		
o-Phthalic	$C_6H_4(COOH)_2$	1.2×10^{-3}	3.9×10^{-6}		
Picric	$(NO_2)_3C_6H_5OH$	4.2×10^{-1}			
Propanoic	CH_3CH_2COOH	1.3×10^{-5}			
Salicylic	$C_6H_4(OH)COOH$	1.0×10^{-3}			
Sulfamic	NH_2SO_3H	1.0×10^{-1}			
Sulfuric	H_2SO_4	$\gg 1$	1.2×10^{-2}		
Sulfurous	H_2SO_3	1.3×10^{-2}	5×10^{-6}		
Trichloroacetic	Cl_3CCOOH	1.29×10^{-1}			

[a] K_{a1} and K_{a2} for stepwise dissociation of the protonated form R—CH—CO$_2$H.

$$\underset{\underset{NH}{|}}{}$$

TABLE D.2
Dissociation Constants for Bases

Name	Formula	Dissociation Constant, at 25°C	
		K_{b1}	K_{b2}
Ammonia	NH_3	1.75×10^{-5}	
Aniline	$C_6H_5NH_2$	4.0×10^{-10}	
1-Butylamine	$CH_3(CH_2)_2CH_2NH_2$	4.1×10^{-4}	
Diethylamine	$(CH_3CH_2)_2NH$	8.5×10^{-4}	
Dimethylamine	$(CH_3)_2NH$	5.9×10^{-4}	
Ethanolamine	$HOC_2H_4NH_2$	3.2×10^{-5}	
Ethylamine	$CH_3CH_2NH_2$	4.3×10^{-4}	
Ethylenediamine	$NH_2C_2H_4NH_2$	8.5×10^{-5}	7.1×10^{-8}
Glycine	$HOOCCH_2NH_2$	2.3×10^{-12}	
Hydrazine	H_2NNH_2	1.3×10^{-6}	
Hydroxylamine	$HONH_2$	9.1×10^{-9}	
Methylamine	CH_3NH_2	4.8×10^{-4}	
Piperidine	$C_5H_{11}N$	1.3×10^{-3}	
Pyridine	C_5H_5N	1.7×10^{-9}	
Triethylamine	$(CH_3CH_3)_3N$	5.3×10^{-4}	
Trimethylamine	$(CH_3)_3N$	6.3×10^{-5}	
Tris(hydroxymethyl)aminomethane	$(HOCH_2)_3CNH_2$	1.2×10^{-6}	
Zinc hydroxide	$Zn(OH)_2$		4.4×10^{-5}

TABLE D.3
Solubility Product Constants

Substance	Formula	K_{sp}
Aluminum hydroxide	$Al(OH)_3$	2×10^{-32}
Barium carbonate	$BaCO_3$	8.1×10^{-9}
Barium chromate	$BaCrO_4$	2.4×10^{-10}
Barium fluoride	BaF_2	1.7×10^{-6}
Barium iodate	$Ba(IO_3)_2$	1.5×10^{-9}
Barium manganate	$BaMnO_4$	2.5×10^{-10}
Barium oxalate	BaC_2O_4	2.3×10^{-8}
Barium sulfate	$BaSO_4$	1.0×10^{-10}
Beryllium hydroxide	$Be(OH)_2$	7×10^{-22}
Bismuth oxide chloride	$BiOCl$	7×10^{-9}
Bismuth oxide hydroxide	$BiOOH$	4×10^{-10}
Bismuth sulfide	Bi_2S_3	1×10^{-97}
Cadmium carbonate	$CdCO_3$	2.5×10^{-14}
Cadmium oxalate	CdC_2O_4	1.5×10^{-8}
Cadmium sulfide	CdS	1×10^{-28}
Calcium carbonate	$CaCO_3$	8.7×10^{-9}
Calcium fluoride	CaF_2	4.0×10^{-11}
Calcium hydroxide	$Ca(OH)_2$	5.5×10^{-6}
Calcium oxalate	CaC_2O_4	2.6×10^{-9}
Calcium sulfate	$CaSO_4$	1.9×10^{-4}
Copper(I) bromide	$CuBr$	5.2×10^{-9}
Copper(I) chloride	$CuCl$	1.2×10^{-6}
Copper(I) iodide	CuI	5.1×10^{-12}
Copper(I) thiocyanate	$CuSCN$	4.8×10^{-15}
Copper(II) hydroxide	$Cu(OH)_2$	1.6×10^{-19}
Copper(II) sulfide	CuS	9×10^{-36}
Iron(II) hydroxide	$Fe(OH)_2$	8×10^{-16}
Iron(III) hydroxide	$Fe(OH)_3$	4×10^{-38}
Lanthanum iodate	$La(IO_3)_3$	6×10^{-10}
Lead chloride	$PbCl_2$	1.6×10^{-5}
Lead chromate	$PbCrO_4$	1.8×10^{-14}
Lead iodide	PbI_2	7.1×10^{-9}
Lead oxalate	PbC_2O_4	4.8×10^{-10}
Lead sulfate	$PbSO_4$	1.6×10^{-8}
Lead sulfide	PbS	8×10^{-28}
Magnesium ammonium phosphate	$MgNH_4PO_4$	2.5×10^{-13}
Magnesium carbonate	$MgCO_3$	1×10^{-5}
Magnesium hydroxide	$Mg(OH)_2$	1.2×10^{-11}
Magnesium oxalate	MgC_2O_4	9×10^{-5}
Manganese(II) hydroxide	$Mn(OH)_2$	4×10^{-14}
Manganese(II) sulfide	MnS	1.4×10^{-15}
Mercury(I) bromide	Hg_2Br_2	5.8×10^{-23}
Mercury(I) chloride	Hg_2Cl_2	1.3×10^{-18}
Mercury(I) iodide	Hg_2I_2	4.5×10^{-29}
Mercury(II) sulfide	HgS	4×10^{-53}
Silver arsenate	Ag_3AsO_4	1.0×10^{-22}
Silver bromide	$AgBr$	4×10^{-13}
Silver carbonate	Ag_2CO_3	8.2×10^{-12}
Silver chloride	$AgCl$	1.0×10^{-10}
Silver chromate	Ag_2CrO_4	1.1×10^{-12}

TABLE D.3 (continued)
Solubility Product Constants

Substance	Formula	K_{sp}
Silver cyanide	$Ag[Ag(CN)_2]$	5.0×10^{-12}
Silver iodate	$AgIO_3$	3.1×10^{-8}
Silver iodide	AgI	1×10^{-16}
Silver phosphate	Ag_3PO_4	1.3×10^{-20}
Silver sulfide	Ag_2S	2×10^{-49}
Silver thiocyanate	$AgSCN$	1.0×10^{-12}
Strontium oxalate	SrC_2O_4	1.6×10^{-7}
Strontium sulfate	$SrSO_4$	3.8×10^{-7}
Thallium(I) chloride	$TlCl$	2×10^{-4}
Thallium(I) sulfide	Tl_2S	5×10^{-22}
Zinc ferrocyanide	$Zn_2Fe(CN)_6$	4.1×10^{-16}
Zinc oxalate	ZnC_2O_4	2.8×10^{-8}
Zinc sulfide	ZnS	1×10^{-21}

TABLE D.4
Formation Constants for Some EDTA Metal Chelates
$(M^{n+} + Y^{4-} \rightleftharpoons MY^{n-4})$

Element	Formula	K_f
Aluminum	AlY^-	1.35×10^{16}
Bismuth	BiY^-	1×10^{23}
Barium	BaY^{2-}	5.75×10^7
Cadmium	CdY^{2-}	2.88×10^{16}
Calcium	CaY^{2-}	5.01×10^{10}
Cobalt (Co^{2+})	CoY^{2-}	2.04×10^{16}
(Co^{3+})	CoY^-	1×10^{36}
Copper	CuY^{2-}	6.30×10^{18}
Gallium	GaY^-	1.86×10^{20}
Indium	InY^-	8.91×10^{24}
Iron (Fe^{2+})	FeY^{2-}	2.14×10^{14}
(Fe^{3+})	FeY^-	1.3×10^{25}
Lead	PbY^{2-}	1.10×10^{18}
Magnesium	MgY^{2-}	4.90×10^8
Manganese	MnY^{2-}	1.10×10^{14}
Mercury	HgY^{2-}	6.30×10^{21}
Nickel	NiY^{2-}	4.16×10^{18}
Scandium	ScY^-	1.3×10^{23}
Silver	AgY^{3-}	2.09×10^7
Strontium	SrY^{2-}	4.26×10^8
Thorium	ThY	1.6×10^{23}
Titanium (Ti^{3+})	TiY^-	2.0×10^{21}
(TiO^{2+})	$TiOY^{2-}$	2.0×10^{17}
Vanadium (V^{2+})	VY^{2-}	5.01×10^{12}
(V^{3+})	VY^-	8.0×10^{25}
(VO^{2+})	VOY^{2-}	1.23×10^{18}
Yttrium	YY^-	1.23×10^{18}
Zinc	ZnY^{2-}	3.16×10^{16}

TABLE D.5
Some Standard and Formal Reduction Electrode Potentials

Half-Reaction	E^0, Volts	Formal Potential, Volts
$F_2 + 2H^+ + 2e^- = 2HF$	3.06	
$O_3 + 2H^+ + 2e^- = O_2 + H_2O$	2.07	
$S_2O_8^{2-} + 2e^- = 2SO_4^{2-}$	2.01	
$Co^{3+} + e^- = Co^{2+}$	1.842	
$H_2O_2 + 2H^+ + 2e^- = 2H_2O$	1.77	
$MnO_4^- + 4H^+ + 3e^- = MnO_2 + 2H_2O$	1.695	
$Ce^{4+} + e^- = Ce^{3+}$		1.70 (1 M HClO$_4$); 1.61 (1 M HNO$_3$); 1.44 (1 M H$_2$SO$_4$)
$HClO + H^+ + e^- = \frac{1}{2}Cl_2 + H_2O$	1.63	
$H_5IO_6 + H^+ + 2e^- = IO_3^- + 3H_2O$	1.6	
$BrO_3^- + 6H^+ + 5e^- = \frac{1}{2}Br_2 + 3H_2O$	1.52	
$MnO_4^- + 8H^+ + 5e^- = Mn^{2+} + 4H_2O$	1.51	
$Mn^{3+} + e^- = Mn^{2+}$		1.51 (8 M H$_2$SO$_4$)
$ClO_3^- + 6H^+ + 5e^- = \frac{1}{2}Cl_2 + 3H_2O$	1.47	
$PbO_2 + 4H^+ + 2e^- = Pb^{2+} + 2H_2O$	1.455	
$Cl_2 + 2e^- = 2Cl^-$	1.359	
$Cr_2O_7^{2-} + 14H^+ + 6e^- = 2Cr^{3+} + 7H_2O$	1.33	
$Tl^{3+} + 2e^- = Tl^+$	1.25	0.77 (1 M HCl)
$IO_3^- + 2Cl^- + 6H^+ + 4e^- = ICl_2^- + 3H_2O$	1.24	
$MnO_2 + 4H^+ + 2e^- = Mn^{2+} + 2H_2O$	1.23	
$O_2 + 4H^+ + 4e^- = 2H_2O$	1.229	
$2IO_3^- + 12H^+ + 10e^- = I_2 + 6H_2O$	1.20	
$SeO_4^{2-} + 4H^+ + 2e^- = H_2SeO_3 + H_2O$	1.15	
$Br_2(aq) + 2e^- = 2Br^-$	1.087[a]	
$Br_2(l) + 2e^- = 2Br^-$	1.065[a]	
$ICl_2^- + e^- = \frac{1}{2}I_2 + 2Cl^-$	1.06	
$VO_2^+ + 2H^+ + e^- = VO^{2+} + H_2O$	1.000	
$HNO_2 + H^+ + e^- = NO + H_2O$	1.00	
$Pd^{2+} + 2e^- = Pd$	0.987	
$NO_3^- + 3H^+ + 2e^- = HNO_2 + H_2O$	0.94	
$2Hg^{2+} + 2e^- = Hg_2^{2+}$	0.920	
$H_2O_2 + 2e^- = 2OH^-$	0.88	
$Cu^{2+} + I^- + e^- = CuI$	0.86	
$Hg^{2+} + 2e^- = Hg$	0.854	
$Ag^+ + e^- = Ag$	0.799	0.228 (1 M HCl); 0.792 (1 M HClO$_4$)
$Hg_2^{2+} + 2e^- = 2Hg$	0.789	0.274 (1 M HCl)
$Fe^{3+} + e^- = Fe^{2+}$	0.771	
$H_2SeO_3 + 4H^+ + 4e^- = Se + 3H_2O$	0.740	
$PtCl_4^{2-} + 2e^- = Pt + 4Cl^-$	0.73	
$C_6H_4O_2$ (quinone) $+ 2H^+ + 2e^- = C_6H_4(OH)_2$	0.699	0.696 (1 M HCl, H$_2$SO$_4$, HClO$_4$)
$O_2 + 2H^+ + 2e^- = H_2O_2$	0.682	
$PtCl_6^{2-} + 2e^- = PtCl_4^{2-} + 2Cl^-$	0.68	
$I_2(aq) + 2e^- = 2I^-$	0.6197[b]	
$Hg_2SO_4 + 2e^- = 2Hg + SO_4^{2-}$	0.615	
$Sb_2O_5 + 6H^+ + 4e^- = 2SbO^+ + 3H_2O$	0.581	

TABLE D.5 (continued)
Some Standard and Formal Reduction Electrode Potentials

Half-Reaction	E^0, Volts	Formal Potential, Volts
$MnO_4^- + e^- = MnO_4^{2-}$	0.564	
$H_3AsO_4 + 2H^+ + 2e^- = H_3AsO_3 + H_2O$	0.559	0.577 (1 M HCl, HClO$_4$)
$I_3^- + 2e^- = 3I^-$	0.5355	
$I_2(s) + 2e^- = 2I^-$	0.5345[b]	
$Mo^{6+} + e^- = Mo^{5+}$		0.53 (2 M HCl)
$Cu^+ + e^- = Cu$	0.521	
$H_2SO_3 + 4H^+ + 4e^- = S + 3H_2O$	0.45	
$Ag_2CrO_4 + 2e^- = 2Ag + CrO_4^{2-}$	0.446	
$VO^{2+} + 2H^+ + e^- = V^{3+} + H_2O$	0.361	
$Fe(CN)_6^{3-} + e^- = Fe(CN)_6^{4-}$	0.36	0.72 (1 M HClO$_4$, H$_2$SO$_4$)
$Cu^{2+} + 2e^- = Cu$	0.337	
$UO_2^{2+} + 4H^+ + 2e^- = U^{4+} + 2H_2O$	0.334	
$BiO^+ + 2H^+ + 3e^- = Bi + H_2O$	0.32	
$Hg_2Cl_2(s) + 2e^- = 2Hg + 2Cl^-$	0.268	0.242 (sat'd KCl–SCE); 0.282 (1 M KCl)
$AgCl + e^- = Ag + Cl^-$	0.222	0.228 (1 M KCl)
$SO_4^{2-} + 4H^+ + 2e^- = H_2SO_3 + H_2O$	0.17	
$BiCl_4^- + 3e^- = Bi + 4Cl^-$	0.16	
$Sn^{4+} + 2e^- = Sn^{2+}$	0.154	0.14 (1 M HCl)
$Cu^{2+} + e^- = Cu^+$	0.153	
$S + 2H^+ + 2e^- = H_2S$	0.141	
$TiO^{2+} + 2H^+ + e^- = Ti^{3+} + H_2O$	0.1	
$Mo^{4+} + e^- = Mo^{3+}$		0.1 (4 M H$_2$SO$_4$)
$AgBr + e^- = Ag + Br^-$	0.095	
$S_4O_6^{2-} + 2e^- = 2S_2O_3^{2-}$	0.08	
$Ag(S_2O_3)_2^{3-} + e^- = Ag + 2S_2O_3^{2-}$	0.01	
$2H^+ + 2e^- = H_2$	0.000	
$Pb^{2+} + 2e^- = Pb$	-0.126	
$CrO_4^{2-} + 4H_2O + 3e^- = Cr(OH)_3 + 5OH^-$	-0.13	
$Sn^{2+} + 2e^- = Sn$	-0.136	
$AgI + e^- = Ag + I^-$	-0.151	
$CuI + e^- = Cu + I^-$	-0.185	
$N_2 + 5H^+ + 4e^- = N_2H_5^+$	-0.23	
$Ni^{2+} + 2e^- = Ni$	-0.250	
$V^{3+} + e^- = V^{2+}$	-0.255	
$Co^{2+} + 2e^- = Co$	-0.277	
$Ag(CN)_2^- + e^- = Ag + 2CN^-$	-0.31	
$Tl^+ + e^- = Tl$	-0.336	-0.551 (1 M HCl)
$PbSO_4 + 2e^- = Pb + SO_4^{2-}$	-0.356	
$Ti^{3+} + e^- = Ti^{2+}$	-0.37	
$Cd^{2+} + 2e^- = Cd$	-0.403	
$Cr^{3+} + e^- = Cr^{2+}$	-0.41	
$Fe^{2+} + 2e^- = Fe$	-0.440	
$2CO_2(g) + 2H^+ + 2e^- = H_2C_2O_4$	-0.49	
$Cr^{3+} + 3e^- = Cr$	-0.74	
$Zn^{2+} + 2e^- = Zn$	-0.763	
$2H_2O + 2e^- = H_2 + 2OH^-$	-0.828	

TABLE D.5 (continued)
Some Standard and Formal Reduction Electrode Potentials

Half-Reaction	E^0, Volts	Formal Potential, Volts
$Mn^{2+} + 2e^- = Mn$	-1.18	
$Al^{3+} + 3e^- = Al$	-1.66	
$Mg^{2+} + 2e^- = Mg$	-2.37	
$Na^+ + e^- = Na$	-2.714	
$Ca^{2+} + 2e^- = Ca$	-2.87	
$Ba^{2+} + 2e^- = Ba$	-2.90	
$K^+ + e^- = K$	-2.925	
$Li^+ + e^- = Li$	-3.045	

[a]E^0 for $Br_2(l)$ is used for saturated solutions of Br_2 while E^0 for $Br_2(aq)$ is used for unsaturated solutions.
[b]E^0 for $I_2(s)$ is used for saturated solutions of I_2, while E^0 for $I_2(aq)$ is used for unsaturated solutions.

appendix
E

SAFETY IN THE LABORATORY

General Safety Rules

Good housekeeping practices will assure the safest working conditions in the laboratory. Always clean up spilled chemicals; do not leave broken or chipped glassware lying around; and put away all chemical bottles and apparatus when finished with them. Neutralize acid spills with sodium bicarbonate and alkali spills with boric acid. Mercury spills should be vacuumed up with a suction flask or dusted with sulfur powder. Clean the mercury thoroughly, because mercury vapors from fine droplets are highly toxic.

Many states by law require a person working in a chemical laboratory to wear protective eyeglasses. Even if your state does not require this, it is good practice to use them. You should locate fire extinguishers, exists, safety showers, eye fountains, and fire blankets in your laboratory. Any dangerous or potentially dangerous laboratory situation should be brought immediately to the attention of the laboratory supervisor.

Perform only the authorized experiments and never work alone. Never eat or drink in the laboratory. When working with volatile chemicals, as when heating acids or when using organic solvents, use the fume hood. Use a safety shield when working with potentially dangerous reactions. Special care should be taken when working with organic solvents. Many are flammable and many have been identified as acute or chronic toxic substances, frequently carcinogenic. Use rubber gloves when possible and avoid breathing fumes.

Waste Disposal[1]

Small quantities of acids or alkalis may be disposed of by slowly pouring into a stream of water and flushing down the drain with a large quantity of water.

Large quantities of any chemical should be returned to the plant for recovery or disposal. See your instructor for specific instructions.

Cyanide wastes must be placed in an appropriate waste bottle and the solution kept alkaline at all times.

Dry chemicals, in an appropriate container, should be placed in a drum labeled "Waste Chemicals for Disposal." Strong oxidizing and reducing agents (chlorates, bromates, peroxides, nitrates, iodides, metal dusts, hypochlorites, etc.) should not be placed in this drum. See your supervisor for instructions on the disposal of these reactive dry chemicals. They should never be placed with organic chemicals.

Small quantities of waste solvents should be placed in an approved container for disposal. Special arrangements should be made for disposing of large quantities of waste solvent.

Under no circumstances should amines, phosphorus compounds, acetic anhydride, acetyl chloride, or any other highly reactive substance be placed in the general disposal containers.

Incompatible Chemicals[2]

Many explosions, fires, and asphyxiations are caused by the accidental combination of potentially dangerous substances. The following is a partial list of such potentially dangerous combinations.

Do Not Contact:

ALKALI METALS, SUCH AS CALCIUM, POTASSIUM, AND SODIUM, with water, carbon dioxide, carbon tetrachloride and other chlorinated hydrocarbons.

ACETIC ACID with chromic acid, nitric acid, hydroxyl-containing compounds, ethylene glycol, perchloric acid, peroxides, permanganates.

ACETONE with concentrated sulfuric and nitric acid mixtures.

ACETYLENE with copper (tubing), fluorine, bromine, chlorine, iodine, silver, mercury, and their compounds.

AMMONIA, ANHYDROUS, with mercury, halogens, calcium hypochlorite, hydrogen fluoride.

AMMONIUM NITRATE with acids, metal powders, flammable fluids, chlorates, nitrates, sulfur, finely divided organics or combustibles.

ANILINE with nitric acid, hydrogen peroxide.

BROMINE with ammonia, acetylene, butadiene, butane, hydrogen, sodium carbide, turpentine, finely divided metals.

CHLORATES with ammonium salts, acids, metal powders, sulfur, finely divided organics or combustibles, carbon.

CHROMIC ACID with acetic acid, naphthalene, camphor, alcohol, glycerine, turpentine and other flammable liquids.

---

[1]Adapted from *Laboratory Safety Handbook,* Mallinckrodt Chemical Works, St. Louis, Mo., by permission.

[2]Reproduced from *Laboratory Safety Handbook,* Mallinckrodt Chemical Works, St. Louis, Mo., by permission.

CHLORINE with ammonia, acetylene, butadiene, benzene and other petroleum fractions, hydrogen, sodium carbides, turpentine, finely divided powdered metals.

CYANIDES with acids.

HYDROGEN PEROXIDE with copper, chromium, iron, most metals or their respective salts, flammable fluids and other combustible materials, aniline, nitromethane.

HYDROGEN SULFIDE with nitric acid, oxidizing gases.

HYDROCARBONS, GENERAL, with fluorine, chlorine, bromine, chromic acid, sodium peroxide.

IODINE with acetylene, ammonia, etc. (see BROMINE).

MERCURY with acetylene, fulminic acid, hydrogen.

NITRIC ACID with acetic, chromic, and hydrocyanic acids, aniline, carbon, hydrogen sulfide, flammable media, fluids or gases, substances that are readily nitrated.

OXYGEN with oils, grease, hydrogen-flammable liquids, solids, and gases.

OXALIC ACID with silver, mercury.

PERCHLORIC ACID with acetic anhydride, bismuth and its alloys, alcohol, paper, wood and other organic materials.

PHOSPHORUS PENTOXIDE with water.

POTASSIUM PERMANGANATE with glycerine, ethylene glycol, benzaldehyde, sulfuric acid.

SODIUM PEROXIDE with any oxidizable substances, for instance, methanol, glacial acetic acid, acetic anhydride, benzaldehyde, carbon disulfide, glycerine, ethylene glycol, ethyl acetate, furfural, etc.

SULFURIC ACID with chlorates, perchlorates, permanganates, water.

Laboratory Precautions with Some Potentially Dangerous Chemicals[3]

AROMATIC AMINES. Aromatic amines, such as aniline, *m*-nitroaniline, and benzidine, compose one of the few chemical groups that can readily be absorbed through the skin and cause rapid systematic poisoning. Toxic amounts of solid amines may be absorbed as readily as liquid amines. The amines react in the blood to convert hemoglobin to methemoglobin, a form that cannot carry oxygen. Poisoning can also occur by inhalation or ingestion.

Skin exposure requires prolonged washing with water, because most of the amines are only slightly soluble in water. Small laboratory spills can be removed with absorbent paper. Larger spills should be absorbed with sand or soil.

Work in the hood whenever possible and use rubber gloves when working with aromatic amines.

AROMATIC NITRO COMPOUNDS. Compounds such as nitrobenzene are similar to aromatic amines in that they are easily absorbed through the skin, and they convert hemoglobin to methemoglobin. Again, solid compounds are absorbed. The trinitro aromatic compounds explode at elevated temperatures. Trinitrotoluene (TNT) explodes at 240°C, and trinitrophenol (picric acid) explodes at 300°C. Nitrobenzene and nitric acid will form an explosive mixture in the absence of water.

Treatment of spills and skin exposure is similar to that for aromatic amines.

---

[3]Adapted from the *Laboratory Chemical Catalogue,* Safety Procedures section, Matheson Coleman & Bell, Norwood, Ohio, by permission.

CARBON DISULFIDE.   Carbon disulfide is toxic, and it is even more flammable than ethyl ether. Vapors can be ignited with static electricity.

Because of the toxicity of CS_2, small laboratory spills should be allowed to evaporate. If this is not possible, the CS_2 can be absorbed with a sponge, cloth, or paper and then allowed to evaporate in the hood. Small quantities can be flushed down the sink, using plenty of cold water.

CAUSTIC ALKALIS.   Sodium hydroxide and potassium hydroxide are commonly used in the laboratory and are highly caustic. Contact with the skin or eyes is the most common hazard of caustic alkali handling. Permanent eye damage may result if concentrated alkali (NaOH or KOH) is splashed in the eye. Wash skin or eyes continuously for at least 15 minutes following exposure.

Concentrated alkali solution spills should be flushed to a floor drain. They can be mopped using water, but the mop head may have to be discarded. Sand is also useful for absorbing spilled solutions.

CHROMIUM TRIOXIDE.   The toxic effects of chromium trioxide are due to its behavior as an acid and as an oxidizing agent. Contact with CrO_3 dust or concentrated solutions may cause skin inflammation or open sores. An ingestion of less than 6 g CrO_3 is fatal. Continued inhalation may cause damage to the respiratory tract.

Chromium trioxide should be thoroughly and immediately washed from the skin after contact. Spilled solutions can be converted to the less toxic trivalent chromium if reaction with a reducing agent such as sodium sulfite is brought about.

CYANIDES AND NITRILES.   Organic compounds combined with the —CN group are commonly called nitriles, although they may be called cyanides, as the inorganic compounds are. A commonly used organic cyanide is CH_3CN, called acetonitrile, or methyl cyanide. Cyanide compounds are powerful poisons that prevent utilization of oxygen by the body tissues, owing to the selective inhibition of respiratory enzymes: the transfer of oxygen from the blood to the tissues is prevented. Use amyl nitrite as an antidote.

Cyanide in the body is converted to the much less toxic thiocyanate, SCN^- Such conversion prevents any buildup of cyanide in the body. So poisoning from chronic, daily exposure to cyanide is less common than from acute exposures. Nitriles are less toxic than inorganic cyanides, but they do cause a greater irritation of the nose and eyes.

Wash cyanides and nitriles from the skin immediately. Spilled cyanides should be placed in a special container for disposal. Cyanide in (alkaline) solutions can be converted to thiocyanate by being boiled with sulfur, or to a polysulfide or ferrocyanide by addition of ferrous sulfate. Conversion should be done before solutions are washed down the sink (lot of water!). Nitriles can be converted to amides by reaction with hydrochloric acid.

ETHERS.   Ethers are not highly toxic, but they do cause dizziness, headaches, and other side effects. Dioxane and ethylene oxide are more toxic than ethyl and isopropyl ether; continual exposure may cause kidney and liver damage. Dioxane can be absorbed through the skin.

Ethers are highly flammable. Fumes are heavier than air and can accumulate in a low spot, flow to an ignition point, and flash back to the vapor source. Laboratory work with ethers should be confined to fume hoods. Static electricity can ignite the vapors.

Ethers react slowly with oxygen to form explosive peroxides. Frequent exposure to air and sunlight hasten the process. Evaporation can cause accumulation of peroxides around the cap or stopper of a container. Peroxides are less volatile than ethers and so distillation or evaporation of the ether will result in concentration of the peroxides. If a distillation goes too near to completion, an explosion may result.

Peroxides can be tested for by shaking a sample of the ether with an acidified solution of potassium iodide. Iodine is formed in the presence of the peroxides. The peroxides can be eliminated by shaking with a solution of ferrous sulfate or sodium sulfite. Addition of sodium metal to the container will prevent accumulation of peroxides.

Spilled ethers can be removed with a sponge or cloth and then allowed to evaporate in a hood.

HALOGENATED HYDROCARBONS. The most general toxic effect of halogenated hydrocarbons such as carbon tetrachloride is their anesthetic or narcotic action. Delayed effects are usually more dangerous than immediate effects. Damage to the kidney, liver, or nervous system may result, as well as cancer.

The saturated organic halides are generally more toxic than the halides of the unsaturated ethylene series. Bromo and iodo derivatives are usually somewhat more toxic than the chloro derivatives. Carbon tetrachloride is about 10 times more toxic than chloroform. Methylene chloride should be used when possible. The most common absorption route is by inhalation.

Carbon tetrachloride is sufficiently nonflammable to be used in fire extinguishers. However, at elevated temperature, many organic halides decompose to give carbonyl chloride (phosgene), $COCl_2$, which is far more toxic than the original compound. Hence, carbon tetrachloride should not be used as a fire extinguisher, nor should it be aspirated into flames.

MERCURY AND ITS DERIVATIVES. Mercury compounds are toxic by virtue of the fact that they interfere with or inhibit enzyme systems in the body.

Both metallic mercury and mercury compounds can be absorbed by inhalation, ingestion, or contact with the skin. Mercury readily amalgamates with gold, and you should never wear rings or other jewelry when working with it. Either continued chronic exposure or acute exposure can cause poisoning. In cases of chronic inhalation exposure, the poisoning symptoms usually disappear when the source of exposure is removed. Complete recovery may, however, require several years. Emotional disturbance, kidney damage, and other harmful effects are common.

Mercury oxide can react with ammonia, or ethyl alcohol and nitric acid, to form the highly explosive mercuric fulminate, $Hg(ONC)_2$.

Mercury spills are common in the laboratory. At room temperature, mercury-saturated air contains about 20 mg/m^3 of mercury. However, this point will rarely be reached, because of air movements and air exchanges. Nevertheless, because of the volatility of mercury, spills must be minimized and containers must be stoppered.

Complete recovery of mercury from spills is impossible. Pools and droplets can be pushed together and collected by suction into a filtering flask. Mercury sweeps are available for scooping up mercury. These contain a copper coil which can be amalgamated so it will collect the mercury and scoop it into a small shovel attachment. Small and unseen mercury droplets are always trapped in crevices. In reasonably ventilated laboratories, continued vaporization of the residual mercury presents no problem. Its volatility decreases rapidly as oxides, dust, and oils coat the surface. However, frequent spills combined with mercury from other sources may result in an airborne concentration in excess of the recommended limit.

Residual mercury can be fixed with sulfur dust, but a slurry of sulfur and calcium oxide is more effective as a fixing agent.

OXALIC ACID AND OXALATES.   Oxalic acid and its soluble salts, when absorbed into the blood or tissues, precipitate calcium oxalate. Ingestion is the most common form of absorption. Severe pain and vomiting rapidly occur and may be followed by convulsions and death.

The most common soluble compounds of oxalic acid used in the laboratory are sodium, potassium, and ammonium oxalate.

PERCHLORIC ACID AND PERCHLORATES.   The greatest hazard associated with these chemicals is the possibility of violent explosions when they come in contact with organic matter. Fumes from the evaporation of perchloric acid or of acid solutions of perchlorates should be collected or the evaporation should be carried out in fume hoods made for perchloric acid work. These are constructed with inert silica cements and are equipped with an exhaust scrubber or have a water spray in the exhaust duct. In the latter case, water collects in a trough at the rear of the hood and contains the perchlorates. In conventional hoods, perchloric fumes can combine with organic components, such as litharge-glycerine cement used for seals, and an explosion is likely to result if the hood is jarred.

Reagent bottles should be kept on stainless steel or plastic trays and away from wooden shelves and bench tops. A wooden surface soaked with perchlorate solutions may spontaneously ignite years later. Rinse perchloric acid bottles after use.

Remove spilled perchlorate immediately. If a mop is used (dilute the spill first), rinse thoroughly with water, allow it to dry, and dispose of it.

PEROXIDES.   In general, peroxides are irritating to the respiratory tract, skin, and eyes. A 30% hydrogen peroxide solution is commonly used in the laboratory, and this rarely presents an inhalation problem. It does not present an explosion hazard, but contaminants in the more concentrated 70% solution can initiate violent decomposition.

Peroxides in general should not be stored for long periods.

appendix F

ANSWERS TO EVEN NUMBERED PROBLEMS

CHAPTER 1

8.(a) 12.5 (b) 5.00 (c) 100 **10.**(a) 244.27 (b) 218.6 (c) 431.73 (d) 310.18 **12.**(a) 5.06 (b) 2.38 (c) 7.80 (d) 2.74 (e) 4.46 (f) 7.64 **14.**(a) 500 (b) 500 (c) 2.00×10^3 (d) 50.0 (e) 80.0 (f) 80.9 (g) 1.06×10^3 **16.** 0.00147 **18.**(a) 7.10 (b) 49.0 (c) 109 **20.** 8.06 **22.** 1.1, 2.3 **24.**(a) 5.88×10^{-6} (b) 2.9×10^{-6} (c) 2.27×10^{-5} (d) 1.58×10^{-6} (e) 2.73×10^{-5} (f) 1.00×10^{-5} **26.** 0.00702 g; 1.79×10^{-5} M **28.**(a) 0.254 (b) 0.165 **30.** 0.160 **32.**(a) 0.250 (b) 0.125 (c) 0.250 (d) 0.125 (e) 0.250 **34.** 10.0

CHAPTER 3

4.(a) 5 (b) 4 (c) 3 **6.** 68.9466 **8.** $162._2$ **10.** 0.01 g **12.**(a) 128.0 g (b) 128.1 g (c) 129.0−127.1 g **14.**(a) $s = 0.056\%$ (absolute); c.v. = 0.17% (relative) (b) $s = 0.0012\%$; c.v. = 8.8% **16.**(a) 0.052% (absolute) (b) 0.026% (absolute) (c) 0.028% (relative) **18.**(a) 0.141 ± 0.002 (b) 1.34 ± 0.03 (c) $(1.199 \pm 0.004) \times 10^4$ **20.** $(4.93 \pm 0.03) \times 10^4$ cm² **22.** $(2.0 \pm 0.1) \times 10^{-5}$ M **24.** $(2.2 \pm 0.3) \times 10^{-3}$ M **26.**(a) 139.6 ± 0.47 (b) 139.6 ± 0.64 (c) 139.6 ± 1.2 **28.** ± 0.9 meq/L **30.** Real at 90% conf. level **32.** 0.0258 **34.** yes **36.** yes **38.** yes **40.** 3.05 ppm **42.** 0.989

CHAPTER 4

2. $4.3 \times 10^{-7}M$; 0.30 M; 0.80 M **4.** 0.084 **6.** $H_2S = 2H^+ + S^{2-}$; 1.1×10^{-22} **8.**(a) 0.30 (b) 0.90 (c) 0.90 (d) 3.3 **10.** 0.96_5 **12.** 0.0019 M **14.**(a) $K_a^0 = K_a f_{H^+} f_{CN^-}$ (b) $K_b^0 = K_b f_{NH_4^+} f_{OH^-}$

CHAPTER 5

2.(a) 1.70; 12.30 (b) 3.89; 10.11 (c) -0.08; 14.08 (d) 7.00; 7.00 (e) 6.55; 7.45 **4.**(a) 3.8 $\times 10^{-10}$ M (b) 5.0×10^{-14} M (c) 1.0×10^{-7} M (d) 5.3×10^{-15} M **6.** 12.70; 1.30 **8.** 2.3×10^{-7} M; 6.64 **10.** $1.7_3 \times 10^{-2}$ M; 3.2% **12.** 182 **14.** 8.80 **16.** 0.014_5 M **18.** 11.54 **20.** 4 -fold **22.** 10.57 **24.** 2.92 **26.** 8.72 **28.** 7.00 **30.** 2.54 **32.** 4.16 **34.** 12.96 **36.** 10.98 **38.** 4.05 **40.** 5.12 **42.** 9.24 **44.** 7.3 g **46.** 2.62 **48.** 2.0×10^{-3} M, 9.6×10^{-4} M **50.** 0.24 mL; 5.4 g

CHAPTER 6

10.(a) $[Ag^+][SCN^-]$ (b) $[La^{3+}][IO_3^-]^3$ (c) $[Hg_2^{2+}][Br^-]^2$ (d) $[Ag^+][Ag(CN)_2^-]$ (e) $[Zn^{2+}]^2[Fe(CN)_6^{4-}]$ (f) $[Bi^{3+}]^2[S^{2-}]^3$ **12.** 1.3×10^{-4} M; 6.5×10^{-5} M **14.** 1.3×10^{-17} M **16.** 1.6×10^{-4} g **18.** AC_2 **20.** $[F^-] = 0.33$ M for precipitation. Feasible **22.** 2.0×10^{-5} M **24.** 16.2 g **26.** 0.2138; 1.902; 0.1314; 0.6474 **28.** 98.68% **30.** 1.071% **32.** 26 mL **34.** 24.74 g **36.** 42.5; 37.5 **38.** 0.846 g; 1.154 g

CHAPTER 7

6. 5.00 **8.** 100 **10.** 0.172 **12.** 390 mg **14.** 9.25 **16.** 15.3 **18.** 7.53 **20.** 22.7 mL **22.** 84.4% **24.** 47.1 **26.** 1.52 **28.** 20.0 **30.**(a) 128.1 (b) 64.05 (c) 64.05 **32.**(a) 151.91 (b) 17.04 (c) 17.01 (d) 17.01 **34.** 0.608 **36.** 0.180 **38.** 0.474 g $KHC_2O_4 \cdot H_2C_2O_4$/g $Na_2C_2O_4$

CHAPTER 8

14. 0.1025 **16.** 0.1174 **18.** 13.00; 12.70; 7.00; 1.90 **20.** 11.12; 9.84, 9.24; 5.27; 2.04 **22.** 9.72; 7.60; 7.12; 6.64; 4.54; 1.96 **24.** 35.5 mL **26.** 62.8 **28.** 0.0300 M; 0.0200 M **30.** 110; 404 **32.** 3.00 Na_2CO_3; 1.5_0 NaOH **34.** $25._2$

CHAPTER 9

6. 0.0017 M, 0.0033 M **8.** 2.0×10^{-5} M **10.**(a) $2.7_8 \times 10^7$ (b) $3.9_0 \times 10^{17}$ **12.** 1.2_6; yes **14.** 6.1×10^{-3} M; $8.0_1 \times 10^{-5}$ M; $1.2_0 \times 10^{-2}$ M **16.** 8.4×10^{-3} M; $5.9_6 \times 10^{-5}$ M; 8.4×10^{-3} M **18.** 0.1039_8 **20.** 10.01 **22.** 99.3, 4.95 **24.** 3.04 **26.** 5.434

CHAPTER 10

16. O_3; HClO; Hg^{2+}; H_2SeO_3; H_3AsO_4; Cu^{2+}; Co^{2+}; Zn^{2+}; K^+ **18.**(a) $Fe^{2+} - MnO_4^-$ (b) $Fe^{2+} - Ce^{4+}$ ($HClO_4$) (c) $H_3AsO_3 - MnO_4^-$ (d) $Fe^{3+} - Ti^{2+}$ **20.**(a) $2V^{2+} + PtCl_6^{2-} = 2V^{3+} + PtCl_4^{2-} + 2Cl^-$ (b) $Ag + Fe^{3+} + Cl^- = \underline{AgCl} + Fe^{2+}$ (c) $3Cd + ClO_3^- + 6H^+ = 3Cd^{2+} + Cl^- + 3H_2O$ (d) $2I^- + H_2O_2 + 2H^+ = \underline{I_2} + 2H_2O$ **22.** 0.419 V **24.** 0.65 V **26.** 0.735 V; 0.771 V; 1.218 V; 1.28V **28.** 0.360 V **30.** 1.35_7 V **32.**(a) 0.57 (b) 0.071 (c) 0.09 **34.** $2VO_2^+ + U^{4+} = 2VO^{2+} + UO_2^{2+}$; 0.67 **36.** 4.96 **38.** 1.47

CHAPTER 11

6. 0.799 V **8.**(a) 1.353 V (b) 0.034 V **10.** 0.446 V **14.**(a) ca. 0.02 pH unit, 1.2 mV, 4.8% (b) ca. 0.002 pH unit, 0.12 mV, 0.48% **16.**(a) -0.419 V (b) -0.472 V (c) -0.524 V **20.** 0.0079 M **22.**(a) 0.167 (b) 19.8 mV **24.** 3.8 $\times$ 10^{-3} M

CHAPTER 12

10. 4.9 $\times$ 10^{-6} **12.** 0.13 V vs. N.H.E. **14.** 0.1478_6 g **16.** 3.31 mA **18.** 7.7 $\times$ 10^{-5} M **20.** $[Fe^{3+}]/[Fe^{2+}]$ = 5:1

CHAPTER 13

26. 0.25, 250 **28.** 20,000–150,000, 5,000–670 **30.** 0.70, 0.10, 0.56, 0.10 **32.** 4.25 $\times$ 10^3 **34.** 5.15 $\times$ 10^4 **36.** 0.528 **38.**(a) 0.193 (b) 4.91 (c) 0.25 **40.** 0.405 **42.** 4.62, 2.61

CHAPTER 14

20. 13 **22.** 190 **24.** 84 ppm

CHAPTER 15

12. 48 **14.** 95 **16.** %MA_2/%MB_2 = 10^{10} **18.** 2.7 **20.** 5 **22.**(a) 4.6 $\times$ 10^{-10} (b) 1.9 $\times$ 10^{-2} (c) 1.2 $\times$ 10^{-2}

CHAPTER 16

20. 270 **22.** 6.7 g **24.** 0.041 **26.** 2810 **28.** 0.82, 0.631, 0.532, 0.13_5

CHAPTER 18

12. 48.5 s **14.** 12.7 h **16.** 0.016

index

A

Absolute error, 66
Absorbance, 372
Absorptivity, 372. *See also* Molar
    absorptivity
Accuracy, 64
  relative, 66
  ways of expressing, 66
Acid error, 299
Acidic solution, 120
Acidity constant, 116
Acids:
  dissociation constants of, 644
  see also Polyprotic acids; Weak acids
Acid in stomach, 122
Activation overpotential, 339
Activity, 107, 116
  enzyme, 509
Activity coefficient, 107
Activity standards, sodium and
    potassium, 312
A/D converter, 498
Adjusted retention time, 465
Air analysis, 548
  examples of, 552
Air sample collection, 548
Aliquot, 34

Alkaline error, 297
Alkaline solution, 120
Alkali reserve, 142
α-values, 136
Amino acids:
  separation of, 453
  thin-layer chromatography of, 620
  titration of, 212
Amperometric titrations, 350
Amperometry, 349
Amphiprotic solvents, 217
Amphoteric, 453
Amphoteric salts, 139
Analgesics, HPLC analysis of, 623
Analog-to-digital converter, 498
Analysis sample, 50
Analyte, 4
Analyte solution, preparation of, 274
Analytical balance, 25
  electronic, 29
  micro, 28
  semimicro, 28
  single pan, 27
  use of, 556
Analytical chemistry, literature of, 628
Analytical columns, 446
Anhydrone, 44
Anion exchange resins, 452

Anode, 249
Anodic-stripping voltammetry, 347
  sensitivity of, 347
Antibody, 529
  preparation of, 531
Antibody titer, 531
Antibonding orbitals, 365
Anticoagulants, 522
Antigen, 528
  radiolabeling of, 530
Antilogarithms, application of, 633
Antimony, iodimetric determination of,
  584
Antiserum, 528
APC tablets, UV analysis of, 606
Aprotic solvents, 217
Archimedes' principle, 31
Aromatic compounds, spectra of, 367
Arrhenius theory, 115
Arsenic, coulometric titration of, 596
Artificial serum electrolyte solution,
  614
Ascarite, 53, 216
Ashing, see Dry ashing; Low-temperature
  ashing
Aspirin, UV analysis of, 606
Asymmetry potential, 296, 297
Atomic absorption spectrophotometry,
  409
  background correction, 414
  detection limits in, 417, 418
  electrothermal atomizers, 418
  electrothermal detection limits,
    419
  ionization interference in, 415
  organic solvents in, 416
  physical interferences in, 416
  refractory compound formation in,
    415
  sensitivity, 417
  spectral interferences in, 414
Atomization, 411
AutoAnalyzer, 490
Autocatalytic decomposition, 272
Automated devices, 486
Automated instruments and process
  control, 486
Automatic devices, 485
Automatic instruments, 488
Autoprotolysis, 117, 217
Autoprotolysis constant, 117, 218
Auxiliary electrode, 330
Auxochrome, 365
Avogadro's number, 8

B

Back-emf, 327
Back-extraction, 433
Back-titration, 185
Balance, see Analytical balance
Balancing redox reactions, 639
Bandpass, 384
Baseline method, 377
Bases:
  dissociation constants of, 645
  see also Weak bases
Batch instruments, 488
Bathochromic shift, 365
Beer's law, 371, 372
  chemical deviations from, 389
  instrumental deviations from, 391
Biamperometry, 352
Bicarbonate determination in blood,
  568
Bio-Gel, 450
Biuret reagent, 368
Blank, 65
Blood:
  buffering capacity of, 142
  buffering systems in, 142
  composition of, 520
  measurement of pH in, 303
  normal concentrations in, 521
Blood gas analyzer, 533
Blood pH, and temperature, 122
  measurement of, 122
Bolometer, 383
Boltzmann constant, 408
Boundary potential, 297
Brønsted acids, 126
Brønsted base, 124
Brønsted theory, 115
Buffer(s):
  for biological measurements, 143
  constant ionic strength, 143
  definition of, 128
  effect of ionic strength on, 146
  optimum pH, 133
  physiological, 141
  selection of, 133
  see also Standard buffers
Buffering capacity, 130
  of blood, 142
  maximum, 130, 132, 203
  total, 204
Buffering mechanism, 129
Buffer region, 203
Buffer reserve, 131

Buret, 37
  use of, 557
β-values, 235

C

Caffeine, UV determination of, 606
Calcium:
  AAS determination of, 609
  titration of, with EDTA, 234
Calculators, use of, 637
Calibration curve, 5
Capacity factor, 465
Capillary columns, 458
Catalyst(s), 101, 505
Catalytic analyses, 505
  types of, 507
Cathode, 249
Cation exchange resins, 452
Cell, for potential measurements,
  292
Cell convention, 253
Cells without liquid junction, 285
Cellular elements, 520
Cell voltage, 252
  at equilibrium, 256
Centrifugal analyzer, 493
Cerium(IV), titrations with, 273
Characteristic, 633
Charge transfer transition, 368
Charging current, 343
Chelate(s), 169, 226
Chelating agent, 169, 176, 226
Chelometric titration, 226
Chemical equilibrium calculations, rules
  for, 106
Chloride:
  Fajans' determination of, 572
  gravimetric determination of, 559
  mercurimetric determination in
    blood, 573
  potentiometric titration of, 587
Chloridimeter, 334
Chromatography, 441
  classification of, 445
  resolution, 465
Chromatography detectors,
  amperometric, 354
Chromium, spectrophotometric
  determination of, 604
Chromium(II), titrations with, 274
Chromogen, 365
Chromophores, 365
  list of, 366

Cleaning solution, 39
Clinical analyses, 523
Closed-loop controls, 487
Closed-shell electrons, 364
Cobalt:
  anion exchange chromatography of,
    615
  EDTA titration of, 615
Coefficient of variation, 68
Coenzyme, 510, 512
Cofactor, 512
Colloids:
  hydrophilic, 160
  hydrophobic, 160
  sol, 160
Colors and wavelengths, 360
Common ion effect, 106, 151
Competitive inhibition, 510
Competitive protein binding, 529
Complementary color, 360
Computer interface, 499
Computers in analysis, 498
Concentration equilibrium constant,
    see Equilibrium constant
Concentration polarization, 328
Conditional formation constant, 229
Conditional solubility product, 165,
    237
Confidence interval, 77
Confidence level, 77
Confidence limit, 77
Conjugated bonds, 366
Conjugate pairs, 116
Constant weight, 162
Continuous analyzers, 486
Control chart, see Quality control chart
Controlled-current coulometry, 330,
    332
  advantage of, 334
  examples of, 335
Controlled-potential coulometry, 330
Control limits, 76
Control loop, 486
Copper:
  electrolytic determination of, 592
  iodometric determination of, 271,
    583
Coprecipitation, 160
Correlation coefficient, 88
Cotlove chloridimeter, 334
Coulometer, 331
  electronic, 331
  mechanical, 331
  titration, 331

Coulometric titration, *see* Controlled-current coulometry
Coulometry, *see* Controlled-current coulometry; Controlled-potential coulometry
Countercurrent distribution, 436
Crosslinkage, 453
Crucible:
  Gooch, 46
  porcelain filter, 46
  sintered glass, 46

D

D/A converter, 498
Dalton, 8
Dark current, 386
Davies modification of Debye-Hückel equation, 110
Dead-stop titration, 352, 598
Dead time, 487
Debye-Hückel equation, 109
  Davies modification of, 110
Decomposition potential, 337
Dedicated instruments, 489
Dehydrite, 44, 53
Dehydrogenase reactions, 513
Deionization of water, 455
Densitometer, 406
Density, 14
Depolarized electrode, 338
Depolarizer, 338
Depth samplers, 554
Derivative titrations, 315
Desiccants, 43, 44
Desiccator, 43
  vacuum, 43
Detection limit, 417
Determinate errors, 64
Dialysis, 56
Diffraction grating, 380
Diffusion coefficient, 345
Diffusion current, 341
Diffusion layer, 337
Diffusion potential, 297
Digestion, *see* Kjeldahl digestion; Wet digestion
Digestion of precipitates, 158
Digital-to-analog converter, 498
Dilution calculations, 180
Dimensional analysis, 10, 166
Dipole moment, 363
Discrete analyzers, 487

Discrete sampling instruments, 488
Dissociation constant(s), 224
  of acids, 644
  of bases, 645
Distribution coefficient, 425
Distribution ratio, 426
Dithizone, 430
Diverse ion effect, 111, 145
  on solubility, 154
  see also Thermodynamic equilibrium constant
DME, 340
Double-antibody technique, 531
Double layer, 344
Drierite, 44
Dropping mercury electrode, 340
Drug isolation:
  by ion exchange resins, 543
  by solvent extraction, 540, 543
Drugs, gas chromatography of, 541
Drug screening:
  by spectrophotometry, 545
  by thin-layer chromatography, 543
Dry ashing, 53. *See also* Low-temperature ashing
Drying agents, 44
Drying oven, 44
Drying of precipitates, 161
Dry-test gas meter, 551

E

$E_{1/2}$, table of values, 346
EDTA:
  α-values for, 228
  dissociation of, 226
  formation constants for, 647
  formula of, 226
  high purity, 233
  standard solution preparation, 576
  storage of, 577
  titration curve for calcium, 230
EGTA, 234
Electrical double layer, 343
Electrode, polarizable, 338
Electrode of first kind, 280
Electrode potential(s), 249
  limitations of, 267
Electrodes, hookup polarity of, 313
Electrode of second kind, 282
Electrogravimetry, 326
  determinations by, 329
Electrolytic cell, 249, 327

Electromagnetic radiation:
  absorption of, 360
  orders of, 380
Electromagnetic spectrum, 358
  regions of, 359
Electrometer, 291
Electronic calculators, use of, 637
Electronic spectra, and molecular
    structure, 364
Electronic transition, 360
Electrophoresis, 481
Electrostatic precipitators, 552
Electrothermal atomizers, 418
  detection limits with, 419
Eluent, 447
Emission spectrogram, 405
Emission spectrometry:
  flame, 406
  plasma, 407
Emission spectroscopy, 404
End point, 37, 175, 196, 198
Enol form, 546
Enthalpy, 99
Entropy, 99
Enzyme activators, 517
Enzyme activity, 509
Enzyme analyses, examples of, 513
Enzyme electrodes, 307, 353
Enzyme immunoassay, 532
Enzyme inhibitors, 517
Enzyme kinetics, 508
Enzyme nomenclature, 512
Enzyme properties, 509
Enzymes, 508
  determination of, 512
Enzyme specificity, 511
Enzyme substrates, determination of,
    513
Equilibria:
  concentration effects on, 101
  pressure effects on, 100
  see also Le Chatelier's principle
Equilibrium calculations, rules for,
    106
Equilibrium constant:
  calculations using, 102
  for chemical reactions, 103
  concentration, 111
  for dissociating species, 101, 105
  from free energy, 257
  molar, 98
  and Nernst equation, 256
  stepwise, 102

temperature effects on, 100
  see also Acidity constant;
    Autoprotolysis constant; Diverse
    ion effect; Hydrolysis constant;
    Thermodynamic equilibrium
    constant
Equilibrium potential, 255
Equivalence point, 175, 197
Equivalence point potential, 259
Equivalent(s), 11, 186
Equivalent weight, 11, 186. See also
    Normality
Eriochrome Black T, 231
Error:
  absolute, 66
  mean, 66
  relative, 66
  standard, 68
  see also Propagation of error
Eudiometer, 612
Exchange capacity, 452
Excited states, 361
Exclusion limit, 450
Exponents, use of, 632

F

Fajan's method, 243
Faraday constant, 254, 331
Faraday's law, 331
Far-infrared region, 359
Feedback mechanism, 487
Feedforward systems, 487
FIA, 495
Filtering of precipitates, 161
Filter paper, ashless, 47
First-order reactions, 502
First-order spectra, 380
Flame emission spectrometry, 406
  detection limits in, 418
  see also Atomic absorption
    spectrophotometry
Flame photometry, 406
Flames, list of, 413
Flow injection analysis, 495
Fluorescence, 393
  and chemical structure, 394
  dynamic range, 397
  quantitative relationship, 396
Fluorescence immunoassay, 532
Fluorescence instrumentation,
    396
Fluorescence quenching, 395

Fluoride:
    ion-selective electrode determination
        of, 590
    standard solution of, 591
Fluorometer, 397
    calibration of, 398
Fluorometry, 392
Fluorometry corrected spectra, 398
Formal, 14
Formality, 14
Formal potential(s), 259, 265
    table of, 648
Formation constant(s), 224
    table for EDTA, 647
    see also Conditional formation
        constant
Fraction collector, 448
Free energy, 99
    and equilibrium constant, 257
    for spontaneous reaction, 252
Frequency, 358
Freund's adjuvant, 531
F-test, 79
Fume hood, 45
Furnace, muffle, 44
Fusion, 52

G

Galvanic cell, 249
Gas chromatography:
    of drugs, 541
    principles of, 456
Gas chromatography detectors:
    argon ionization, 461
    β-ray, 461
    coulometric, 462
    electron capture, 461
    flame ionization, 459
    flame thermionic, 459
    hot wire, 459
    thermal conductivity, 459
Gas chromatography-mass spectrometry,
    458, 543
Gas constant, 254
Gaussian curve, 65
GC-MS, 458
GeMSAEC analyzer, 493
Gibbs free energy, 99
Gibbs-Stockholm convention, 251
Glass membrane electrodes, 304
Glass pH electrode, 295
    theory of, 297
Globar, 379

Glucose, enzymatic determination of,
    625
Glycolysis, 522
Golay columns, 458
Gooch crucible, 46
Grab sample, 50
Gram-atomic weight, 7
Gram-formula weight, 8
Gram-molecular weight, 8
Gran ruler, 321
Gran's plot, 318, 352
Grating:
    diffraction, 380
    holographic, 381
Gravimetric analysis:
    calculations in, 113
    conditions for precipitation in, 157,
        158
    examples of, 168
    preparing solutions for, 156
    steps of, 156
    see also Precipitates; Precipitation
Gravimetric factor, 166
Gross sample, 50
Ground state, 361

H

Hair shampoo, determination of pH in,
    589
Half-life:
    for first order reaction, 503
    for second order reaction, 504
Half-reaction, 250, 639
Half-wave potential, 338
    table of values, 346
Hanging mercury drop electrode, 347
Haptens, 531
Head space analysis, 543
Height equivalent to theoretical plate,
    444, 462
Hemolysis, 522
Henderson-Hasselbalch equation, 128
Hertz, 358
HETP, see Height equivalent to
    theoretical plate
Hierarchical computer system, 500
High performance liquid
    chromatography:
    columns for, 474
    equipment for, 471
    principles of, 469
    reversed-phase, 474
HMDE, 347

Hodgkin's disease, 535
Hollow cathode lamp, 410
Homogeneous solution, precipitation
    from, 163
Hood:
    fume, 45
    laminar flow, 45
Hydrocarbons, gas chromatography of,
    621
Hydrochloric acid:
    standardization of, 216, 567
    standard solution of, 216
Hydrogen electrode, 284
Hydrogen ion, titration of, 564
Hydrogen peroxide, iodometric analysis
    of, 580
Hydrolysis constant, 125
Hydrolysis of salts, 124
Hydronium ion, 116
Hydrophilic colloids, 160
Hydrophobic colloids, 160
Hyperchromism, 366
Hypochlorite, iodometric analysis of,
    580
Hypsochromic shift, 366
Hypsochromism, 366

I

ICP, 407
Igniting of precipitates, 161
Ilkovic equation, 344
Immunoassay, 526
    heterogeneous, 532
    homogeneous, 532
Impactors, 552
Impedance, 292
Impinger, 552
Indeterminate errors, 65
Indicator(s), 175
    acid-base, 198
    adsorption, 240, 242
    for chelometric titrations, 231
    pH transition range of, 200
    precipitation, 240
    redox, 263, 264
    selection of, 200
Indicator dyes, light absorption by,
    368
Indicator electrode, 260, 281, 285
Indirect potentiometry, see
    Potentiometric titrations
Inductively coupled plasma, 407
Inert solvents, 217, 219

Infrared absorption:
    and molecular structure, 369
    requirement for, 363
Infrared radiation, detectors for, 383
Infrared region, 359
    cells for, 381
    fingerprint, 369
    solvents for, 371
    sources for, 378
Infrared spectra, quantitative
    measurements from, 377
Inner-filter effect, 395
Input/output device, 499
Instability constant, 224
Internal conversion, 392
Internal standard, 420, 468
International units, 510
Intersystem crossing, 393
I/O device, 499
Iodide, potentiometric titration of, 587
Iodimetry, 267
    substances determined by, 269
Iodine, standardization of, 268, 585
Iodometry, 269
    substances determined by, 271
Ion association complexes, 429
Ion chromatography, 475
Ion exchange chromatography, 446
    applications of, 455
    mixed bed, 455
    removal of interferences by, 244
Ion exchange resins, 451
Ionic strength, 108
    effect on buffers, 146
Ionizable solvents, 217
Ion pair, 429
Ion-selective electrodes, 304
    glass membrane, 304
    liquid-liquid, 307
    mechanism of, 308
    in potentiometric titrations, 315
    precipitate, 305
    solid-state, 305
    use of, 311
Ion size parameter, 109
iR drop, 329
Iron:
    dichromate titration of, 577
    spectrophotometric determination of,
        598
Iron(II), titrations with, 274
Irreversible reaction, 267
Irreversible reduction or oxidation, 339
Isoelectric point, 453, 544

Isoenzymes, 516
Isomorphous replacement, 161
Isosbestic point, 390
Isozymes, 516
Iterative procedure, 236
I.U., 510

J

Johnson noise, 388
Jones reductor, 276

K

Karl Fischer titration, coulometric, 334
Keto form, 546
Key number, 60
KIMAX, 25, 26
Kinetics, 502
Kjeldahl analysis, 213
  micro, 215
Kjeldahl digestion, 55
Kjeldahl flask, 56
Known addition, 319
Kovats indices, 467

L

Laboratory safety, 651
Laminar flow burner, 411
Laminar flow hood, 45
Lead:
  amperometric titration of, 595
  spectrophotometric determination of, 601
Least squares, 87
Le Chatelier's principle, 100
Leveling effect, 218
Lewis acid, 223
Lewis base(s), 217, 223
Ligand, 223, 226
Limiting current, 337
Linear least squares, 87
Lineweaver-Burk equation, 510
Liquid junction, 286
  cells without, 285
Liquid junction potential, 286
  effect of pH on, 288
  residual, 294
Liquid-liquid electrodes, 307
Literature of analytical chemistry, 628
Logarithms:
  applications of, 633

for finding roots, 636
propagation of error in, 74
significant figures in, 63
table of, 634
Low-temperature ashing, 53
Lung, buffering in, 142

M

McReynolds constants, 467
Manganese, spectrophotometric
  determination of, 604
Mantissa, 633
Masking agents, 432
Mass, 26
Maxima, 344
  suppression of, 344
Maxwell-Boltzmann expression, 408
Mean error, 66
Mechanical slit width, 384
Median, 86
Meniscus, 33
Meniscus illuminator, 40
Mercury, flameless AAS determination
  of, 420, 611
Mercury dissolution current, 342
Metal chelates, extraction of, 429
Metal complexes, light absorption by,
  368
Metal ions, paper chromatography of,
  618
Metallic reductors, 275
Method of least squares, 87
Method of successive approximations,
  236
Michaelis constant, 510
Microcomputer, 496
Micropipets, coulometric calibration of,
  598
Microprocessor, 496
Microprocessor-controlled instruments,
  496
Migration current, 339
Milliequivalent(s), 18, 186
Millimole, 9
Mixed-bed ion exchange resin, 455
Mixed crystals, 161
Mixed potential, 294
Modified methyl orange, 567
Mohr method, 241
Molal, 14
Molality, 14
Molar, 10

Molar absorptivity, 365, 373
    typical values of, 365
    see also Absorptivity
Molar equilibrium constant, see
    Equilibrium constant
Molarity, 10
Mole, 8
Molecular activity, 510
Molecular gel chromatography, 449
Molecular sieve chromatography, 449
Molybdenum blue, 368
Monel metal, 592
Monochromator, 379
Monoclonal antibodies, 530
Muffle furnace, 44
Multichannel analyzers, 489
Multiple extractions, 434
Multiplicity, 392

N

NaCl-infrared region, 359
NAD$^+$, 512
    spectrum of, 514
NADH, 512
    spectrum of, 514
Near-infrared region, 359
Near-ultraviolet region, 359
Nebulization, 411
n electrons, 364
Nernst equation, 254
    for cell reaction, 255
    and equilibrium constant, 256
    for ion-selective electrodes, 310
Nernst glower, 378
Neutralization, 196
Neutral solution, 120
    at elevated temperatures, 122
NHE, see Normal hydrogen electrode
Nickel:
    anion exchange chromatography of,
        615
    EDTA titration of, 615
    gravimetric determination of, 563
Nicotinamide adenine dinucleotide,
    512
    spectrum of, 514
Nitrate, spectrophotometric
    determination of, 599
Nitrogen, Kjeldahl analysis of, 213
Nominal wavelength, 384
Nonaqueous solvents:
    classes of, 217

pH measurement in, 303
    reference electrode for, 314
    titration in, 217
    for titrations, 219
Noncompetitive inhibition, 510
Noncompetitive protein binding, 533
Nonionizable solvents, 217
Normal, 11, 186
Normal distribution, 65
Normal hydrogen electrode, 250
Normality, 11, 186
    acid-base, 189
    complexometric, 190
    precipitation, 190
    reduction-oxidation, 189
    see also Equivalent(s), Equivalent
        weight
Nuclear magnetic resonance
    spectroscopy, 360
Nucleation, 157

O

Occlusion, 160
"O-O" band, 393
"O-O" transition, 394
Open-loop controls, 487
Order of reaction, 502, 503
Ostwald ripening, 158
Oven, drying, 44
Overpotential, 328
    activation, 339
Overvoltage, 328
    of hydrogen on Pt, 328
    of oxygen on Pt, 328
Oxidation, 248
Oxidizing agents, titrations with, 272
Oxine, 432
Oxygen electrode, 353

P

Paper chromatography, 477
    two-dimensional, 479
Parallax error, 39
Parallel fast analyzer, 493
Partition chromatography, 446
Parts per billion, 16
Parts per million, 16
Parts per thousand, 16, 67
Parts per trillion, 16
Pearson correlation coefficient, 89
Peptization, 159, 161

PFF, see Protein-free filtrate
pH, 118
  acidic, 120
  alkaline, 120
  in blood, 122
  of blood, 303
  at elevated temperatures, 122
  measurement in blood, 122, 303
  measurement in nonaqueous solvents,
    303
  negative, 120
  neutral, 120
pH electrode, 295
  theory of, 297
Phenacetin, UV determination of, 606
pH glass electrode:
  calibration of, 589
  storage of, 587
pH of hair shampoos, determination of,
  589
pH measurement, accuracy of, 301
pH meter, 291
  measurements with, 301
Phosphate, spectrophotometric
  determination of, 602
Phosphorescence, 393
Phosphoric acid, dissociation of, 136
Photocell, 383
Photometer, 383
Photomultiplier tube, 383
  solar blind, 383
Photon, 358
Phototube, 383
pH scale, 118
pH-stat, 486
pH titrations, 314
Physiological buffers, 141
Pi orbitals, 364
Pipet(s), 34
  blowout, 35
  clinical, 34
  measuring, 34
  serological, 34
  syringe, 36
  transfer, 34
  use of, 557
  volumetric, 34
Planck's constant, 359
Plasma, 520
  blood, 3
  composition of, 527
  electrolyte composition of, 19
Plasma emission spectrometry, 407

Poising capacity, 294
Polarizable electrode, 338
Polarographic maxima, see Maxima
Polarography, 340
  inorganic, 346
  organic, 346
  sensitivity of, 344
  see also Pulse polarography
Polyprotic acids:
  α-values of, 136
  buffer calculations of, 134
  dissociation calculations of, 135
  equilibria of, 133
  fraction dissociated, 136
  salts of, 139
  titration of, 209
Pooled standard deviation, 82
Postprecipitation, 160
Potassium dichromate:
  standardization of, 578
  standard solution of, 578
  titrations with, 273
Potassium nitrate salt bridge,
    preparation of, 595
Potassium permanganate:
  standardization of, 272
  titrations with, 272
Potassium thiocyanate, standardization
    of, 571
Potential, 249, 252
  asymmetry, 296, 297
  for concentration determination,
    293
  decomposition, 337
  effect of complexation on, 266
  effect of pH on, 266
  equilibrium, 255
  at equivalence point, 259
  half-wave, 338
  measurement of, 291, 292
  mixed, 294
Potentiometer, 291
  hookup of electrode to, 313
Potentiometric measurements,
    accuracy of, 294
Potentiometric titrations, 313
  accuracy of, 315
  pH, 314
  precipitation, 314
  redox, 314
Potentiostat, 330
Ppb, see Parts per billion
Ppm, see Parts per million

Ppt, see Parts per thousand, Parts per trillion
Precipitate electrodes, 305
Precipitates:
  digestion of, 158
  drying of, 161
  filtering of, 161
  igniting of, 161
  washing of, 161
Precipitating agents, organic, 169
Precipitation:
  conditions for, 158
  from homogeneous solution, 163
  requirement for, 152
Precipitation process, 157
Precipitation titrations, 239
Precision, 64, 67
Premix chamber burner, 411
Preparative columns, 446
Preprocessing in computing, 500
Primary standard, 25, 175
  requirements of, 176
Prism, 379
Probability, 77
Process control, 486
Propagation of error:
  in addition and subtraction, 70
  in exponents, 74
  in logarithms, 74
  in multiplication and division, 71
  and significant figures, 75
Protein-free filtrate, 4, 56, 523
  $Ba(OH)_2$-$ZnSO_4$, 523
  Folin-Wu, 523
  trichloroacetic acid, 523
  tungstic acid, 523
Protein nitrogen, Kjeldahl analysis of, 213
Proteins, nitrogen in, 214
Pseudo first-order reaction, 504
Pseudo zero-order reaction, 509
Pulse polarography, 348
  differential, 348
  sensitivity of, 348
Pyrex, 25, 26

Q

Q test, 85
Quadratic formula, use of, 636
Qualitative analysis, 1
Quality control chart, 76
Quantitative analysis, 1

Quantized energy, 360
Quantum yield, 396
Quinhydrone electrode, 323

R

Radioimmunoassay, 528
  specificity of, 529
Radiolabeling of antigen, 530
Rate constant, 98
Rate expression, 503
Rate law, 503
  differential form of, 503
  integrated form of, 503, 504
Ratio method, 377
Reaction times, 504
Reagent blank, 65
Redox electrodes, 283
Redox reaction, see Reduction-oxidation reaction
Reducing agents, titrations with, 274
Reduction, 248
Reduction-oxidation, titration curve, 260
Reduction-oxidation reaction, 248
  balancing of, 639
  potential for complete, 253
  preparation of sample for, 274, 275
Reduction potential, 251
Reductors, metallic, 275
Reference electrode, 260, 281, 285, 288
  for Tris buffers, 145
Refractory elements, 414
Rejection of result, 84
Relative accuracy, 66
Relative error, 66
Residual current, 342
Residual liquid junction potential, 294
Resolution, 465
  and column length, 468
Resonance line, 410
Rest point, 27
Retention index, 467
Retention time, 457, 465
  adjusted, 465
Reversed-phase chromatography, 474, 480
R_f value, 478
Riboflavin, fluorometric determination of, 608
Rohrschneider constant, 467
Roots, from logarithms, 636
Rotameter, 551

Rotational transition, 360
Rounding off, 64
Rubber policeman, 48

S

Safety in laboratory, 651
Salt bridge, 249, 286
  agar-KCl, 288
  preparation of, 595
Salt gradient chromatography, 449
Salts, hydrolysis of, 124
Salts of weak acids, 124. *See also*
  Polyprotic acids
Salts of weak bases, 126
Sample, 50
  analysis, 50
  grab, 50
  gross, 50
  thief, 51
Saponification number, 220
Saturated calomel electrode, 286, 288
  sleeve-type, 314
SCE, *see* Saturated calomel electrode
SCOT, 458
Secondary standard, 216
Second-order reactions, 503
Selectivity ratio, 309, 310
Selenium, catalytic determination of,
  624
Self-ionization, 217
Self-ionization constant, 117
Semiautomatic instruments, 488, 489
Semiexponential form, 632
Sensitivity, 417
Separation factor, 432
Separatory funnel, 426
Sephadex, 449
Serum, 3, 520
Serum electrolytes, artificial solution of,
  614
Serum water, 312
SHE, *see* Standard hydrogen electrode
Shot noise, 388
Sigma electrons, 364
Significant figures, 59
  in addition and subtraction, 62
  key number in, 60
  in logarithms, 63
  in multiplication and division, 60
  and propagation of error, 75
Silver, Volhard determination of, 570
Single-channel analyzers, 488

Singlet state, 392
Size exclusion chromatography, 446,
  449
Slit width, 384
  mechanical, 384
  spectral, 384
"Smart" instruments, 496
SMDE, 349
Soda ash:
  pH titration of, 586
  titration of, 566
Sodium:
  activity coefficient of, in serum, 312
  flame emission determination of, 613
Sodium carbonate, titration of, 206
Sodium hydroxide:
  removal of sodium carbonate from,
    215
  standardization of, 216, 565
  standard solution of, 215
  storage of, 216
Sodium thiosulfate, standardization of,
  271, 581, 583
Sol, 160
Solar-blind photomultiplier, 383
Solid-state electrodes, 305
Solubility:
  effect of acidity on, 237
  effect of complexation on, 238
Solubility product, 151. *See also*
  Conditional solubility product
Solubility product constants, table of,
  646
Solvent extraction, 425
  of drugs, 540
  of metals, 429
  multiple, 434
Solvent gradient chromatography, 449
Sorption chromatography, 445
Specific activity, 510
Specific gravity, 14
Specific rate constant, 503
Spectral bandwidth, 384
Spectral slit width, 384
Spectrofluorometer, 398
Spectrograph, 404
Spectrometer, 377
  double beam, 387
  high absorbance, 387
  single beam, 385
Spectrometric error, 388
Spectrometry:
  instrumentation for, 377

quantitative calculations in, 371
solvents for, 369
Spectrophotometer, 383
Spectrophotometry, 383
Spectrum, see Electromagnetic
    spectrum; Electronic spectra, and
    molecular structure
Spontaneous reaction:
  cell voltage for, 257
  free energy for, 252, 257
Square root, estimation of, 637
Stability constant, 224
Standard acid solution, preparation of,
    216
Standard additions method, 319, 419,
    421, 468
Standard base solution, preparation of,
    215
Standard buffers, 299
  accuracy of, 299
Standard deviation, 66, 67
  of mean, 68
  pooled, 82
  relative, 68
Standard error, 68
Standard hydrogen electrode, 250
Standardization, 175
Standard potential(s), 250, 254
  table of, 648
Standard solution, 174
  preparation of, 175
Starch gel electrophoresis, 481
Starch indicator, 264
Starch solution, preparation of, 581
Stat analysis, 495
Static mercury drop electrode, 349
Statistical weights, 408
Stomach acid, 122
Strong electrolyte(s), 101, 116
Student's t test, 80
  with accepted value, 80
  with multiple samples, 83
  paired, 81
Substrate inhibition, 510
Substrates, 508
Sulfate:
  gravimetric determination of, 561
  ion exchange separation of, 617
  titration of, 617
  titration with barium, 244
Supersaturation, 157
Support-coated open tubular columns,
    458

Supporting electrolyte, 339
Suppressor column, 475
Systematic errors, see Determinate errors

T

Temperature gradient chromatography,
    449
Temperature programming, 468
THAM, 145
  primary standard, 216
THAM buffers, 145
Theoretical plates, 462
Thermal precipitators, 552
Thermistor, 383
Thermocouple, 383
Thermodynamic equilibrium constant,
    111, 145
Thief sampler, 51
Thin-layer chromatography, 479
  for drug screening, 543
  reversed-phase, 480
Thiodene, 581
TISAB, 306, 590
Titanium(III), titrations with, 274
Titer, 192
  antibody, 531
Titrant, 175
Titration(s), 37, 174
  classes of, 176
  dead-stop, 598
  K_a requirements for stepwise, 209
  K_b requirements for mixture, 212
  K_b requirements for stepwise, 206
  of mixtures of acids or bases, 210
  in nonaqueous solvents, 217
  of polyprotic acids, 209
  requirements of, 175
  strong acid, 196
  weak acid, 201
  weak base, 205
  see also Volumetric analysis
Titration coulometer, 331
Titration curve, 196
  calculation of, 197
TLC, see Thin-layer chromatography
Tolerances, of volumetric glassware,
    41
Total-consumption burner, 411
Trace elements in body, 533
Trace metals:
  paper chromatography of, 618
  polarographic determination of, 593

Transmittance, 372
  percent, 372
Triplet state, 392
Tris, 145
  primary standard, 216
Tris buffers, 145
t Test, see Student's t test
Turnover number, 508

U

Ultraviolet region, 359
  cells for, 381
  detectors for, 383
  solvents for, 370
  sources for, 378
Urine, composition of, 527

V

Vacutainer, 42
Vacuum-ultraviolet region, 359
van Deempter equation, 463
Variance, 70
  and F-test, 79
Venturi effect, 411
Vibrational relaxation, 393
Vibrational transition, 360
Visible region, 359
  cells for, 381
  detectors for, 383
  sources for, 378
Vitamin B_2, fluorometric determination
  of, 608
Vitamin C, HPLC determination of,
  622
Void volume, 465
Volhard titration, 242
Voltage, 252
Voltage drop, see iR drop
Voltaic cell, 249
Voltammetry, 335
  anodic stripping, see Anodic stripping
    voltammetry
  potential range of, 339
Voltammogram, 336
Volumetric analysis:
  calculations in, 177
  molarity calculations in, 177

normality calculations in, 186, 191
  see also Titration(s)
Volumetric flask, 33
Volumetric glassware, tolerances of, 41
von Wiemarn ratio, 157

W

Walden reductor, 276
Washing of precipitates, 161
Water:
  coulometric titration of, 334
  deionization of, 455
  sampling of, 554
Water activity error, 299
Water analysis, 554
  examples of, 554
Water hardness, determination of, 576
Wavelength, 358
  and colors, 360
  units of, 358
Wavenumber, 358
Weak acids:
  equilibria of, 122
  see also Titrations(s)
Weak bases:
  equilibria of, 124
  see also Titration(s)
Weak electrolytes, 101, 116
Weighing:
  by difference, 32
  direct, 32
Weighing bottle, 32
Weighing dish, 32
Weight, 26
Wet digestion, 53, 54
Wet-test gas meter, 551
Wilson's disease, 535
Working electrode, 330

X

Xylene, IR determination of, 607

Z

Zero point, 27
Zimmerman-Reinhart reagent, 273
Zone electrophoresis, 481